ADVANCED PHYSICS

Longman Group Limited

Longman House, Burnt Mill, Harlow, Essex CM20 2JE,
England and Associated Companies throughout the World.

First published 1995
© Longman Group 1995

Typeset in 10/12 pt Palatino, Aps 6
Produced by Longman Singapore Publishers Pte Ltd
Printed in Singapore

ISBN 0582 35596 6

All rights reserved. No part of this publication may be
reproduced, stored in a retrieval system, or transmitted in
any form or by any means, electronic, mechanical, photo-
copying, recording, or otherwise without either prior written
permission of the Publishers or a licence permitting
restricted copying in the United Kingdom issued by the
Copyright Licensing Agency Ltd, 90, Tottenham Court Road,
London W1P 9HE.

The publisher's policy is to use paper manufactured from
sustainable forests.

EDITOR
JONATHAN LING

AUTHORS
WENDY BROWN
TERRY EMERY
MARTIN GREGORY
ROGER HACKETT
COLIN YATES

ADVANCED PHYSICS

ABOUT THE AUTHORS

Jonathan Ling has taught physics in schools, colleges of FE, Teacher Education Establishments and in Higher Education. He is currently Director of Studies for Physical Sciences at the University of Hertfordshire. He is a past Director of the Engineering Education Scheme which provides engineering experience for 16+ students in schools and colleges, and has wide experience of dealing with the relationships between Physics and industry.

Wendy Brown attended Bishop Auckland Grammar School and Stevenage Girls' School. She went to Birmingham University where she obtained a BSc in Physics. She started teaching at the George Dixon School in Birmingham, and is currently teaching at Loreto College, a comprehensive girls' school in St. Albans, where she is Head of Physics. She is an active committee member of the Hertfordshire Science Teaching Association.

Terry Emery graduated from the University of Hull with a BSc in Physics and a Postgraduate Certificate in Education. After a short spell of teaching in Welwyn Garden City in the late 1960s, he moved to Canada, where he taught Science at a High School in Hamilton, Ontario. In 1972, he took up a Physics teaching post at Richard Hale School in Hertford, where he subsequently became Head of Physics. He is presently an AS Awarder for the Oxford and Cambridge Schools Examination Board. In 1990, he was appointed to the permanent staff of what is now the University of Hertfordshire and is currently involved in various aspects of Initial Teacher Training.

Martin Gregory has been on the staff of Winchester College for over thirty years, much of that time as Head of the Physics Department. He was involved with the Nuffield A-level Physical Science project and took up Nuffield A-level Physics twenty years ago. He has been a chief examiner for A-level Physics for many years. He was a member of the SEC Physics committee throughout its existence and has been involved with several A-level scrutinies for SEC and SEAC. His previous publications include books on Alternating Currents and Systems Electronics.

Roger Hackett is Head of Physics at Christ's Hospital Horsham. He has been a chief examiner for A-level Physics since 1981. He played a major role in the revision of the Nuffield A-level Physics syllabus and examination from 1979 to 1985 and is still involved in the setting of the examination. He also assists in the setting of both the Cambridge STEP paper and the Oxford November entry paper in Physics. He helped to set up the new style of modular syllabus for one of the examining boards. He has recently completed four years on the editorial board of the IOPP magazine, Physics Education.

Colin Yates has taught Science in comprehensive schools since 1973. In his second year he was put in charge of Physics and later promoted to Head of Physics at John Bunyan School and Community College, Bedford. He was appointed Head of Science at The Barclay School, Stevenage in 1982. Colin chaired the East Anglian Northern CSE Physics panel for five years. He has been ASE Bedfordshire Section Secretary, Home Counties Region Chairman and is currently Treasurer of the Hertfordshire Science Teaching Association.

PREFACE

This book has been arranged in such a way that anyone reading it with a basic level of mathematical ability should be able to follow the main text, picking up the physical concepts and ideas along the way. Areas of particular interest can be studied in greater depth by referring to the boxed sections. These contain details of the applications of the ideas, mathematical explanations, questions, ideas for experiments and practical activities, and historical anecdotes.

The book has been written in such a way that it may also be used as a reference book. You can access any topic through the contents list, the index or via the comprehensive system of references given throughout the text. The authors believe that this combination of a sequential structure within a random access framework will prove particularly helpful to both student and teacher alike.

The editor and authors are indebted to many people who have given help and advice during the preparation of this book. In particular we would like to thank the editorial and publishing staff of Longman who have taken great pains to ensure that the book has all the attributes that we would wish to see. Our colleagues in the schools, colleges and universities in which we work have provided tremendous help and encouragement. However, most of all we would like to thank our spouses who have been hugely patient as the manuscript has undergone a number of metamorphoses over several years.

We are grateful to the following for permission to reproduce copyright material:

Oxford and Cambridge Schools Examination Board for questions from past *Nuffield Physics Papers* 1976, 1978, 1979, 1980, 1981, 1985, 1986 & 1988 and Longman Group Ltd/Nuffield Chelsea Curriculum Trust for questions from *Nuffield Advanced Physics Student's Guide 1, 1985*.

ACKNOWLEDGEMENTS

We are grateful to the following for permission to reproduce photographs; Ace Photo Agency, figures 12.9 and page 1 (photo Mauritius Bildagentur), 5.28d and page 17 (Journalism Services), 5.12 (Pantechnicon), 6.1b (Kidulich), 7.17 (Roger Howard), 7.18 (Berry Bingel), 9.13 right (Paul Thompson) 12.9 (Mauritius) and 13.14 (Vibert Stokes); Action Plus, 4.10 (Tony Henshaw); AEA Technology, 38.25, 38.26, 38.28, 39.11 and 39.14; Allsport, 6.13 (Vandystadt), 8.19 (David Leah) and 14.5c (Chris Cole); Aviation Picture Library, 6.1a and 16.1; John Birdsall, 4.1; Birkbeck College Department of Crystallography, 25.14b; Professor P.J. Black, 25.14a; British Petroleum, 4.8 and 5.2; British Steel, Port Talbot Welsh Laboratory, 25.17c; Bubbles, 27.14a (Ian West); Civil Aviation Authority, 34.17; CERN, Geneva, 40.19; Michael Cole Camerawork, 14.5b; EEV, 32.17a; E.G. Photography, Manchester, 20.26; Environmental picture Library, 28.5b (Vanessa Miles); Leslie Garland Picture Library, 4.9 and 5.1; Geoscience Features, 17.2 (Dr. B. Booth); Peter Gould, West Lothian, 4.19, 18.9, 18.10, 19.3, 20.1, 20.17, 22.14, 25.2, 25.21, 25.26b, 27.15, 27.36, 29.5, 30.6, 30.9, 30.11, 30.15, 30.21, 30.22, 30.23, 31.3, 31.4a, 31.16, 32.3, 32.16, 32.31b, 34.15, 35.7a, 35.8, 35.12a, 35.13a, 36.21a, 36.23, 37.5, 37.7b, 37.21, 38.27 and 38.29b; G.R. Graham, CCAT, 31.2a–d and 33.28b; J.M. Gregory, 28.13; Hewlett Packard, 16.12; Holt Studios International, 16.2 (Nigel Cattlin); JET Joint Undertaking, 40.29a; Kew Bridge Steam Museum, 28.3b; Andrew Lambert, Ilkley, 5.15, 6.9, 6.14, 6.18, 8.1, 8.16, 9.19, 12.7, 12.10, 12.22, 12.39, 12.48, 14.3, 17.6a, 17.7a, 17.18, 17.22c, 18.12b, 21.1, 25.26a, 25.30a, 27.5a, 27.11, 27.37b, 28.8, 29.18, 31.2e, 31.5, 31.17, 33.4, 33.21, 34.9a, 35.9b and 35.10a; Frank Lane Agency, 16.13 above and page 165 (H. Binz), 16.13 below (D. Kinzler) and 16.22 (J. Tinning); Life Science Images/Ron Boardman, 16.3, 17.1a, 27.5b and 32.22; Longman Photographic Unit, 18.12a, 25.30b, 28.4 a and b; J.W. Martin, *The Elementary Science of Metals*, 25.3 and 25.4; J.W. Martin, Department of Metallurgy, University of Oxford, 25.19; Microscopix, 37.24b (Andrew Syred); Milepost $92\frac{1}{2}$, 30.3b; Dr. H. Judith Milledge, Department of Geology, University College London, 32.24; NASA, Washington, 7.9; National Power, 18.12c; Nuffield Chelsea Curriculum Trust, 16.6, 25.25, 25.32 and 36.12; Oxford Scientific Films, 27.3a (London Scientific Films), 27.3b (G.I. Bernard); Planet Earth, 8.10 (Neville Coleman); Quadrant Picture Library, 7.3, 7.4, 9.13 left and 18.5; Rolls Royce, 8.7; Ann Ronan at Image Select, 35.1; Science Photo Library, 4.2 (Professor Howard Edgerton, 4.20 (Heini Schneebeli), 6.12 (European Space Agency), 6.19, 7.10 (Edgerton), 8.13 and page 75 (NASA), 8.17 (Lawrence Berkeley Laboratory), 9.1 (J.L. Charmet), 14.1 (NASA), 14.2 (Tom van Sant/Geosphere Project, Santa Monica); 14.4a (Alex Bartel); 14.4b (University of Birmingham Consortium on High TC Superconductors); 14.10 (NASA), 15.11 (NASA), 15.23 (NASA), 17.1b, 17.3 (Peter Menzel), 18.1, 19.20 (James Stevenson), 27.24a and page 371 (Roger Ressmeyer, Starlight), 25.5 (G. Muller Struers), 25.12 (Dr. G.K.L. Cranstound) 25.15 (Professor H. Hasimoto, Osaka University), 27.37a (Dr. Ray Clark), page 433 (David Parker), 29.15 (Martin Bond), 30.1 and page 687 (CNRI), 32.1 (NASA), 32.18 (NOAA), 32.33 (Peter Aprahamian/Sharples Stress Engineers), 33.25 (Gordon Garradd), 34.14b (Dr. K.F.R. Schiller), 34.42 (Philippe Plailly), 34.43 (Ray Ellis), 36.22a and page 555 (NASA), 35.15b (Jack Finch), 35.15c (Pekka Parviainen), 35.17b (Peter Menzel), 35.19 (Erwin Mueller), 36.1 (Library of Congress), 36.22b (Jeremy Burgess), 38.10 (N. Feather) , 38.18, 40.13 (Lawrence Berkeley Laboratory), 40.14 and 40.29a (Los Alamos National Laboratory); C.A. Taylor, *The Physics of Musical Sounds*, 29.23; Telegraph Colour Laboratory, page 247 (F.P.G/J. Divine Scenics); Topham Picture Source, 9.11 (Associated Press); Transport Research Laboratory, 14.5a; A.W. Trotter, 20.10b and c; University of California, 31.4b; University of Cambridge Cavendish Laboratory, 35.11a; UPI/Bettmann, 19.1.

Picture research by Marilyn Rawlings

CONTENTS

	About the authors	iv
	Preface	v
	Acknowledgements	vi
Unit A	INTRODUCING PHYSICS	1
1	What is Physics?	3
2	Obtaining data and measurements	7
3	Investigations and problem solving	13
Unit B	MATTER AND ITS INTERACTION: FORCES	17
4	Elastic materials	19
5	Structures in equilibrium	31
6	Forces and motion	45
7	Circular motion	63
Unit C	CONSERVATION LAWS	75
8	Conservation of momentum	77
9	Energy transfer in mechanical systems	86
10	Model for an ideal gas	100
11	Energy conservation in mechanical, thermal and electrical systems	112
12	Conserving charge and energy: Simple electrical circuits	123
13	Storing charge and energy: capacitors	150
Unit D	FORCES AND FIELDS	165
14	Forces of nature	167
15	Gravitational fields	176
16	Electric fields	192
17	Magnetic fields	213
18	Electromagnetic induction and magnetic flux	230
Unit E	OSCILLATIONS	247
19	Simple mechanical oscillations	249
20	Damped and forced oscillations: resonance	271
21	Electrical oscillations	289
22	Energy and power in ac circuits	312
23	Analogue electronic systems	333
24	Digital electronic systems	352

Unit F	MATERIALS: THEIR PROPERTIES AND STRUCTURE	371
25	Nature of solids	373
26	Inter-particle forces and energies	390
27	Fluids and transport	398
28	Thermodynamics	421

Unit G	WAVES: TRANSFERRING ENERGY THROUGH SPACE	433
29	Nature of waves	435
30	Wave behaviour	453
31	Wave nature of light	474
32	Electromagnetic waves	488
33	Superposition of electromagnetic waves	513
34	Optical systems	535

Unit H	INSIDE THE ATOM	555
35	Quantisation of charge: the electron	557
36	Quantisation of radiation	579
37	Quantisation of matter: atoms	598
38	Radioactivity	620
39	Radioactive decay	645
40	Nuclear model of the atom	664

Unit I	INFORMATION AND DATA NEEDED IN PHYSICS	687
41	Tables of constants	689
	Mathematical processes	690
	Table of mathematical formulae	699
	The Greek alphabet	702
	Table of nuclides	702
	Index	707

UNIT A
INTRODUCING PHYSICS

INTRODUCING PHYSICS

Physicists seek to understand the physical universe in order to predict its behaviour. Physics provides a picture of **how** the world behaves and **how** the laws of nature operate. It cannot answer the question: **why** do they operate in this way?

Although physics is a fundamental science, its principles underlie much of technology, which is humankind's means of solving its problems. It is, therefore, important for technologists and for engineers, who actually design solutions to problems, to have a thorough understanding of the principles of physics.

1 WHAT IS PHYSICS?

1.1 ABOUT THIS BOOK

Throughout this book run several themes, such as energy, waves and fields. These reflect physicists' attempts to understand the world in terms of a relatively small number of concepts, laws and models. The main aim for you, the reader, is to understand them at this level. However, you can most easily assess your understanding of physics by applying your knowledge of it to other situations.

You will find questions in boxes throughout the text. Think about and work through the questions as you read the text. They are there as prompts to direct your thoughts, and as practice to give you confidence in handling ideas as they are discussed.

The questions themselves often contain numerical data in order to enable you to calculate the value of a certain quantity. Sometimes, the data you will need to use consists of standard physical constants. These have well known values which have been measured very accurately by physicists and others over a long period. The symbols of any constants needed for a question will be listed at the end of that question, and their values are given in Section 41.2. Take care to use an appropriate degree of accuracy in your answer. It is only rarely that you will need to use the most accurate value available.

Question

An electron is accelerated from rest in a vacuum using a voltage of 200 V. What speed will it have? (e, m_e)

Answer

From Section 41.1: $e = 1.6 \times 10^{-19}$ C, $m_e = 9.1 \times 10^{-31}$ kg

Kinetic energy at end = Potential energy at start

$$\frac{1}{2} m_e v^2 = eV.$$

So,
$$v^2 = \frac{2eV}{m_e}$$

or
$$v = \sqrt{\frac{2eV}{m_e}} = \sqrt{\frac{2 \times 1.6 \times 10^{-19} \times 200}{9.1 \times 10^{-31}}}$$

$$= 8.4 \times 10^6 \, \text{m s}^{-1}.$$

A INTRODUCING PHYSICS

Until you have made observations on the real world, you do not have the factual data on which to base a concept or model. Scientific theories develop from evidence and predict new evidence, and investigating this requires experiments which are generally practical in nature. Important aspects of suggested experiments are provided in boxes marked with **P** and full details of these are given in the separate book, *Advanced Physics – A Practical Guide*.

You should always, if at all possible, do experiments and investigations yourself. Do not feel you always have to take somebody else's word for it. Unfortunately, your experiments are unlikely to provide a large enough database of information, so extra data is given in some of the experiment boxes.

There are also boxes containing historical notes, applications and theory. The theory boxes, often include mathematical models. This is because a major use of physics is to quantify and generate numerical information, which can be used for further investigations or applications. The mathematical models are there to help you do just that. For example, gravity is not just an abstraction to talk about, but a concept that enables detailed calculations to be made, such as those providing the orbit of the satellite Giotto which passed within a few hundred kilometres of Halley's comet. It was not possible to calculate precisely the whole path – there are too many unknown perturbations for that. However, it did enable a mathematical model to be used with known data, and numerical methods to be introduced to simulate the expected trajectory.

To solve quantitative problems, you will need a facility in handling formulae. To make numerical predictions from data, you will need a calculator and, for a lot of data, a computer. The necessary formulae are given in the text and in an appendix. Although you can always look up a formula, it does help you to understand the underlying theory if you can follow through its derivation – particularly the scientific model on which the theory is based.

Much of the mathematical development of physics depends on mathematical techniques such as calculus. In this book, we use the minimum of mathematics in the main text, often giving the mathematical development in theory boxes. If you have not yet covered the relevant mathematics, you can bypass a theory box as the results are given in the text.

At the end of each chapter is a summary of the ideas introduced and some further questions for revision.

1.2 THE PHYSICIST AT WORK

A physicist starts by observing the world and recording her or his observations as data. Faced with this raw data she or he looks for patterns. When the data obey simple mathematical rules, the patterns are called **laws**. For example, when you stretch a steel spring elastically, so that it is not permanently deformed, you find that a 2 N force extends it by twice the extension produced by a 1 N force. When you find that the data continue to follow this pattern, you are able to suggest a **hypothesis**: a spring's extension is proportional to the force applied to it.

A **hypothesis** is based on the initial set of data obtained. Once this data has been confirmed (in many different ways) by other scientists, the experimental data enables the physicist to develop the **empirical** law:

applied force $\propto$ extension.

This leads to a **mathematical model**:

$F \propto x$
or $F = kx$

where k is called the **stiffness of the spring**, and is the applied force per unit extension. This model has been used again and again, and found to be true for all situations as long as the spring is not stretched beyond its elastic limit. It is therefore referred to as a **law**. Because it was first described by the physicist Robert Hooke, it is often referred to as **Hooke's law**.

Hooke's law helps you to answer the question 'how does a spring behave?' and enables you to predict the effect of stretching a spring with a particular force. It does not help you to answer such questions as 'why does a spring behave elastically only over a limited range?' or 'why do some metals make good springs and others not?'

After looking for simple patterns, the next step is to imagine a framework in which these laws operate on a wider scale. This involves defining **concepts – abstract ideas** which physicists use to organise mentally patterns of data and interpret them. You will be familiar with one of the most important of them: the concept of energy. The concepts of energy transfer and energy conservation help us to understand a wide variety of phenomena. By drawing up rules for calculating energy in various parts of a system, you can answer a large number of 'how?' questions and make predictions.

Concepts are linked to the structure of **models**, which are used a lot in physics. When, you played with dolls' houses or model trains, you recognised them as houses or trains because they had many properties in common with full-sized houses or trains. They were, of course, scaled down by a common factor. Further, with a bit of imagination they could be made to behave like real houses or trains. Engineers frequently make use of scaled models of systems to understand the behaviour of the full-sized system, and answer questions such as 'what?' and 'how?'. For instance, models of future aircraft are built for wind tunnel studies. They enable the aeronautical engineer to decide what should be designed and how it should be structured. The models used in physics are rather different. They represent the physicist's guess at an answer to the question 'why?'. Physicists frequently use ideas from other areas of knowledge to explain new phenomena. For example, to help understand why light produces interference patterns, a physicist uses a wave model of light. This model is based on previous knowledge of the properties of other waves, such as water waves. Although the links between water and light are imagined, the model helps explain some of the properties of light.

In the seventeenth century Huygens imagined light as waves. Fifty years later, Newton imagined light as a stream of particles, which he called corpuscles.

The successors to these models are with us today as the modern wave model and the photon model. Neither can answer the question 'what is the nature of light?' Using the wave model will enable you to understand some of the properties of light. Using the photon model, you will be able to understand some of its other properties. One of the skills of a physicist is to select a model that seems to work, and then adapt it and build on it as understanding grows.

1.3 CHANGE AND RATE OF CHANGE

The world around you is not static, but changes as time passes. A lot of physics is concerned with changes and the rate at which they take place.

CHANGE

If you drop a ball from a height of 1 m above the floor and record the time for it to reach the floor you will, within the likely accuracy of such an experiment,

INTRODUCING PHYSICS

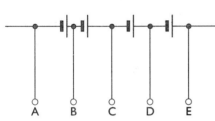

Figure 1.1

record the same time whether you perform the measurement on the ground floor, in the cellar or on an upper floor. The time depends on the change in height not on the actual height. In this book, Δ is used to mean 'the change in' a quantity. In the measurement above, the time equals $\sqrt{2h/g}$ where h is the distance fallen and g is the acceleration due to gravity.

It is the change in a quantity that is important in many measurements:

i Figure 1.1 shows four 1 V cells connected in series. A 2.5 V lamp would light normally when connected between A and C, or B and D, or C and E. The change in potential or pd is the same in each case.

ii When the internal energy of 1 kg of water is increased by 4 200 J then, assuming no change of state occurs, its temperature rises by 1 °C, say from 15 °C to 16 °C. An energy change ΔE produces a temperature change ΔT.

RATE OF CHANGE

The physicist tries to predict not only what change will happen, but also how fast the change will take place. In some cases different phenomena are observed when the rate of change is varied. For example, in electromagnetism, different effects are observed in each of the following cases:

i charges at rest relative to each other produce electric forces,
ii charges moving with constant relative velocity produce magnetic fields,
iii charges accelerated at a constant rate cause electromagnetic induction,
iv charges with a changing acceleration radiate electromagnetic waves.

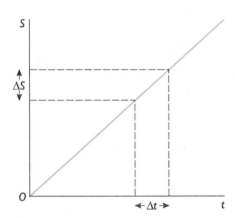

Figure 1.2

Distance-time graphs provide another example. Speed equals distance divided by time. The speed measured on a car speedometer is the rate of change of position with time. On a graph of distance from the starting point s against time taken t, the change in s in time t is the slope or gradient of the graph as shown in Figure 1.2. Similarly, the acceleration is the rate of change of speed. It was to solve such problems that Newton and Leibniz invented the mathematics of calculus. A more complicated distance-time graph, as in Figure 1.3, shows regions in which the rate of change of position is not constant:

- from O to P the car accelerates and the graph gets steeper, implying that the rate of change of speed is positive;
- from P to Q the speed is constant;
- from Q to R the car decelerates and the graph is getting less steep;
- from R the speed is zero – the car is at rest.

There are situations that may appear unchanging only because two or more rates of change cancel each other out; population will appear constant when the birth rate equals the death rate. This is known as **dynamic equilibrium** ($\rightarrow$ 27.2).

When you are working on a topic, make sure you understand whether the important quantity is the absolute value, the change or the rate of change.

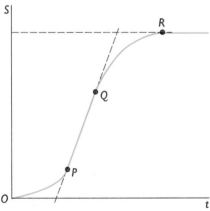

Figure 1.3

Physics is a challenge and to enjoy it you will need:

- curiosity to observe and investigate phenomena,
- perseverance to look for patterns and link them with empirical laws,
- imagination to understand the concepts and models used in physics, to communicate ideas and to work out the applications of physics to the solution of real problems.

You may find some concepts difficult to grasp first time around. If so, do not be dispirited. Only by thinking through ideas over a period of time do most people gain the understanding to handle physics with confidence.

2 OBTAINING DATA AND MEASUREMENTS

In this chapter, we discuss the art of experimentation. Note that SI units, or Système International d'Unités, are always used unless a clearly stated reason is given.

2.1 GUIDELINES FOR EXPERIMENTS

Physics starts with observing the world, but just sitting in your chair and watching is rarely sufficient. Great discoveries usually come from patient experimentation. Becoming a good experimenter is an art you will have to aspire to. Here are a few guidelines to bear in mind when you design an experiment:

i identify a single problem or phenomenon you wish to observe;
ii design the experiment so that you can observe the phenomenon in the presence of other effects;
iii make quantitative measurements whenever you can, and always repeat measurements to check that they are reproducible;
iv when you vary the conditions in an experiment, never change more than one variable at a time;
v always keep an open mind. Record the unexpected as well as the expected, and be on the look out for readings that do not fit in with your preconceived ideas.

2.2 UNITS AND DIMENSIONS

As soon as you start making quantitative measurements, you need a system of units to enable you to communicate with others and compare your results. The system used throughout this book is SI, which has seven **base units**. The base units are defined arbitrarily. For instance, one second is the duration of a large number of oscillations, or periods, of the electromagnetic waves emitted by a particular atom – to be precise, 9 192 631 770 periods of the radiation corresponding to the transition between the two hyperfine levels of the ground state of the caesium 133 atom.

Six of the seven SI base units are used in this book. They are the **metre** (m), **kilogram** (kg), **second** (s), **ampere** (A), **kelvin** (K) and **mole** (mol). SI is said to be a unitary system because each derived unit is defined from combinations of base units of magnitude one. For example, the newton is defined as the force which accelerates 1 kg at 1 m s^{-2}, or 1 N = 1 kg m s^{-2}.

INTRODUCING PHYSICS

Nearly all physical quantities have units. Therefore, whenever you write down or calculate a value for a quantity, write down the unit as well as the number giving the magnitude of the quantity. 'One' is not the same as 'one metre' which is not the same as 'one second'. . . .

Systems of units

All units are arbitrary. The units originally used in each country were based on simple measurements of readily available objects. In Great Britain, these were embodied in the Imperial system of units. In France, the metric system was developed after the revolution of 1789. The metre was defined as 1/10 000 000 of the distance between the north pole and the equator on the meridian through Paris.

The many systems meant that a large number of different sets of units, defined in quite different ways, were used for the same quantities. For example, there were several different units of mass: the kilogram (kg), the pound (lb) and the slug (sl). The kg was originally based on the mass of 1 dm^3 of water at 20°C and normal atmospheric pressure. The lb was based on a standard mass kept in London. Mechanical engineers defined the slug as the mass that is accelerated at 1 foot s^{-2} by a force equal to the force of gravity on a mass of 1 lb. The latter force was normally referred to as 1 pound force (lbf).

Force had an even greater number of basic units than mass, including:

1 kilogram-force (kgf) : the force of gravity on 1 kg mass
1 gram-force (gf)   : the force of gravity on 1 g mass
1 pound-force (lbf)   : the force of gravity on 1 lb mass.

All of these had previously been referred to as 'weights' rather than 'forces'. For example, 1 kgf had been called 1 kilogram weight (1 kg wt). In addition, there were fundamental units of force, based on Newton's second law of motion:

1 newton (N) : the force necessary to accelerate 1 kg by 1 m s^{-2}
1 dyne (dyn) : the force necessary to accelerate 1 g by 1 cm s^{-2}
1 poundal (pd) : the force necessary to accelerate 1 lb by 1 ft s^{-2}.

In the SI system, the unit of mass is the kilogram and the unit of force is the newton. The kilogram is the mass of a standard mass kept at Sèvres, Paris, while the N is defined by 1 N = 1 kg m s^{-2}.

Today, scientists and engineers use only the SI system. However, if you refer to books and papers published before 1970, you will need to know what the older units mean. Be careful when you use older or mixed units, because under them the numerical values of quantities and constants change. For instance, g has the value 9.8 N kg^{-1}, in SI and 32 pd lb^{-1} in the Imperial system. In a horrible system of mixed units it could be 142 N slug^{-1}.

DIMENSIONS

The base units of SI define arbitrary values for the physical quantities length (L), mass (M), time (T), electric current (I), temperature (θ) and amount of substance (n). These base quantities are often called the **base dimensions**. The dimensions of all other quantities can be expressed in terms of these base dimensions. For example, speed is distance divided by the time taken, and so its dimensions are LT^{-1}, using the base dimensions length and time. Similarly, the value of a force equals the acceleration of an object multiplied by its mass, and has the dimensions MLT^{-2}.

A knowledge of the dimensions of a quantity is useful in two ways. First, all equations must be dimensionally homogeneous – that the dimensions of the terms on the left-hand side of the equation must be the same as the dimensions of the terms on the right-hand side. This can provide a check on whether you have written down the correct formula. Second, it enables you to work out the appropriate SI unit in terms of the base units.

2 OBTAINING DATA AND MEASUREMENTS

Question

1 In your studies, you will probably have come across the ideal gas law:

$$\frac{\text{pressure} \times \text{volume}}{\text{absolute temperature}} = \text{a constant} = R.$$

Satisfy yourself that this equation is dimensionally homogeneous, given that the dimensions of each are:

pressure : $ML^{-1}T^{-2}$
volume : L^3
R : $ML^2T^{-2}\theta^{-1}$
temperature : θ.

UNIT-DEPENDENT CONSTANTS

In any quantitative system, there are equations containing constants that change value when the system of units changes. For example, the price P, of a piece of cheese of mass M may be written as

$$P = kM$$

where k is the price per kg. Other such constants include R in Question 1 and G in Newton's law of gravitation. This law may be written

$$\text{force} = G \times \frac{\text{mass} \times \text{mass}}{\text{distance}^2}.$$

Force has the dimensions MLT^{-2}. The terms on the righthand side are $G \times M^2 L^{-2}$. For the equation to be homogeneous, G must have dimensions $M^{-1}L^3T^{-2}$. From this, we can see that the SI unit for G is $kg^{-1} m^3 s^{-2}$. Its numerical value will be appropriate to this unit.

In SI units, G will always have the same value. In any other system, it would have the same dimensions, but its value and its units would be different.

As you work through this book you will come across several other constants whose value depends on the system of units used. So if you look up values in a data book, make sure you use SI values.

2.3 ERRORS AND UNCERTAINTIES

Uncertainties are present in every measurement you make. Some uncertainties arise from the design of the experiment, some from the calibration of the measuring instruments and some from you, the person carrying out the experiment. For instance, suppose you try to measure g, the acceleration due to gravity, by timing the swing of a simple pendulum. The uncertainty in your value for g at the end of the experiment will include:

design uncertainties : the rigidity of the pendulum support and damping
calibration uncertainties : the clock used to measure the period and the rule used to measure the length
personal uncertainties : your reaction time starting and stopping the clock
'parallax' uncertainties : using the rule at a slight distance.

This is not a complete list but shows some possible sources of uncertainty.

INTRODUCING PHYSICS

Whenever you make a measurement, you should try to identify as many sources of uncertainty as you can.

> **Questions**
>
> **2** A metre rule has 1 mm divisions. Give a reasonable estimate of the uncertainty in using this metre rule to measure the length of this page. Express the uncertainty as a percentage.
>
> **3** Suppose you measure the time for 20 swings of the pendulum of a 'grandfather' clock. You record times, in s, of 40.0, 40.1, 39.8, 39.8 and 39.9. Calculate the mean period of the pendulum and estimate the uncertainty in this result.
>
> **4** Suppose you measure the current in an electrolysis experiment and record the current at 1 minute intervals. Your readings, in A, are 1.5, 1.6, 1.5, 1.6, 2.6, 1.6, 1.7, 1.7, 1.8, 1.6, 1.7, 1.8. The ammeter could be read to ± 0.1 A. Write a few lines commenting on these readings. Do you think the current was constant during the 12 minute period? What should you do about the reading of 2.6 A?

THE SIGNIFICANCE OF UNCERTAINTIES

To be able to use the value of a particular measurement, we need to know how much confidence we can place on it. What limits does it lie within? No measurement can be 100% accurate unless it is a simple counting technique: 1, 2, 3, 4, etc. For example, 95 and 100 differ by 5%. If your experimental data have an uncertainty of 10%, you could not distinguish between 95 and 100. A measurement of 95 ±10% implies a value between 85.5 and 104.5, and 100 ±10%, a value between 90 and 110. These ranges overlap. To distinguish between 95 and 100, you would need to make each measurement to an uncertainty of 2% or less.

SIGNIFICANT FIGURES

You can learn something of the uncertainty in a quantity from the number of significant figures to which it is quoted. For example, a 5 Ω resistor implies a value of 5 Ω rather than 4 Ω or 6 Ω. A value given as 5.0 Ω tells you that its value lies between 4.9 Ω and 5.1 Ω.

A problem can arise with values such as 2 000. This could imply a value halfway between

$$\begin{aligned} &\qquad\qquad 1\ 000 \text{ and } 3\ 000 \\ &\text{or between } \quad 1\ 999 \text{ and } 2\ 001. \end{aligned}$$

Such an ambiguity can be resolved by using index notation or by using prefixes such as kilo- and Mega-, as follows:

2 000 meaning between 1 000 to 3 000, can be written 2×10^3 or 2 k,
2 000 meaning between 1 999 to 2 001, can be written 2.000×10^3 or 2.000 k.

When evaluating mathematical expressions, you should not give the answer to more significant figures than that of any component quantity in the expression. When using a calculator, always remember to round off your answer to the correct number of significant figures. For example, 237 896.53 ±12.01 is meaningless, and needs to be written: 237 900 ± 10.

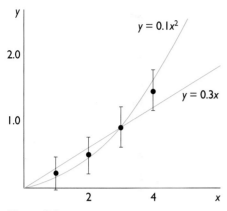

Figure 2.1

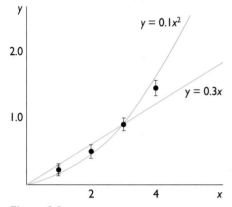

Figure 2.2

The former value means

between 237 884.52 and 237 908.56,

while the latter means

between 237 890 and 237 910.

ERROR BARS AND GRAPHS

When you plot a graph, it is important to indicate the uncertainty in the data you use. You can do this by putting error bars on the points as you plot them. Figure 2.1 shows a plot of two quantities x and y, where x is assumed to be known to high accuracy and y is only known to 0.25. Error bars corresponding to 0.25 are therefore added to the y coordinates of the points. Figure 2.1 shows that the points, with their error bars, fit both:

$y = 0.3x$ and $y = 0.1x^2$.

The data does not enable you to distinguish between the two solutions.

Figure 2.2 shows the same points plotted with an uncertainty in y of 0.1. With this improvement in the accuracy of y, you can now distinguish between the two solutions.

It is very important to make a reasoned estimate of the uncertainties in your measurements, so that you can make an estimate of the uncertainty in your experimental result. It can then be used with confidence by physicists, who can compare it with other results, and by engineers who wish to use the value of the result with a known tolerance.

Uncertainties and orders of accuracy

Suppose you make a number of measurements and then use them to calculate a result. How can you know the order of accuracy of the result? You need to find the **relative uncertainty** in each measurement and add them together. The relative uncertainty is the uncertainty divided by the value, and can be expressed as a vulgar fraction, a decimal fraction or a percentage. For example, a measurement of 40 mm with an uncertainty of 1 mm has a relative uncertainty of 1/40, 0.025 or 2.5%. In the following example, we use percentages.

Suppose you wish to know how much water will be needed to fill a constant depth swimming pool and the order of accuracy of this result. You measure its length, breadth and depth with a rule and obtain the following results:

measurement	length	breadth	depth
value and uncertainty	100.0 ± 0.1 m	60.00 ± 0.05 m	3.00 ± 0.01 m
relative uncertainty	0.1/100 = 0.10 %	0.05/60 = 0.08 %	0.01/3 = 0.33 %

The relative uncertainty in the volume is the sum of these:

relative uncertainty = 0.10 % + 0.08 % + 0.33 %
= 0.51 %
= 0.5 %, approximately.

Then you need to find the order of accuracy, or uncertainty, in the volume, so that it can be expressed as a value. This is done by calculating the result and finding the value represented by the relative uncertainty as follows:

volume = length × breadth × depth
= 100 × 60 × 3
= 1.8×10^4 m^3.

A INTRODUCING PHYSICS

The order of accuracy of a result = relative uncertainty × result
$$= 0.5\% \times 1.8 \times 10^4$$
$$= 90 \text{ m}^3 = 0.009 \times 10^4 \text{ m}^3.$$

So the uncertainty in the volume of water is 90 m³. Normally, such a result is written:

volume = $(1.800 \pm 0.009) \times 10^4$ m³.

When a quantity needed to calculate a result is raised to a power then the uncertainty is multiplied by that power. For example, suppose the swimming pool is circular with a radius of 40 ± 0.05 m. Then its volume is its circular area, πr^2, multiplied by its depth, 3 ± 0.01 m.

Volume = $\pi r^2 \times$ depth = $\pi \times r \times r \times$ depth.

The radius occurs twice and so we need to add the uncertainty twice. The relative uncertainty in r is $0.05/40 = 0.125\%$.

So, the relative uncertainty in the volume = uncertainty in r + uncertainty in r
+ uncertainty in depth

$$= 2 \times 0.125\% + 0.33\%$$
$$= 0.25\% + 0.33\%$$
$$= 0.58\%$$
$$= 0.6\%, \text{ approximately.}$$

Calculating gives the volume as 15 080 m³ with an order of accuracy of ± 90 m³. Once again, this is normally written:

volume = $(1.508 \pm 0.009) \times 10^4$ m³.

If a quantity is cubed, you should multiply the uncertainty by 3 because it occurs three times. The same principle applies to any other power.

Question

5 You want to know how the pd across and the current in a resistor vary under various conditions. So you connect a voltmeter across the ends of the resistor and an ammeter in series with it, and obtain the following results:

ammeter reading ± 0.1 A	0.3	0.5	0.7	1.0	1.3
voltmeter reading ± 0.2 V	1.2	1.9	2.8	4.1	5.2

a Plot each point on a graph and put in the uncertainty or 'error bars' on each point.
b Draw the steepest and least steep graph through these points with the vertical and horizontal bars just touching the graphs. The graphs, of course, do not necessarily go through the origin.
c Measure the gradients of these two straight-line graphs to obtain maximum and minimum values.
d Express these as a ± uncertainty in the gradient of the best straight line through the points.

3 INVESTIGATIONS AND PROBLEM SOLVING

An investigation is a systematic practical inquiry into some effect or phenomenon. For example, suppose it is suggested to you that hot water freezes more quickly than cold. How could you find out whether this statement is true or not? You might begin by assembling equipment that allows you to cool water in a container, and to monitor its temperature and the time as the water cools and freezes. You might then do a set of trials using identical quantities of water, each with a different starting temperature.

Alternatively, you might wish to investigate the behaviour of 'cling film' which is, after all, rather unusual in its properties. Why does it behave as it does? How does it work? Finding out would be an example of investigating technology rather than a natural phenomenon, since cling film is a humanmade product.

In either case, the purpose of the investigation is to seek to gain a better understanding of the situation. The outcome will be essentially unknown until completion of the investigation. This contrasts with other practical activities in *Advanced Physics – A Practical Guide*, which have been designed to illustrate particular concepts. The apparatus for these has been chosen so that the illustrations are clear and unambiguous, and you avoid 'wrong paths'.

You do not have to do all, or indeed any, of these investigations. However, they will help you to work as a physicist does and to find out first hand what it is like. You may, if you wish, think up your own investigations, just as a professional physicist does.

The need for physicists to investigate phenomena and to solve problems is very important. It is, after all, essential to find out more about the universe by investigating unexplained observations and events. Mankind's technological achievement has led to a plethora of devices and systems, many of which depend on the principles of physics. So there is a need for the engineer, who has to be a physicist amongst many other attributes, to understand how things work. This is important for further development as well as maintaining existing equipment. A car engine, for example, depends on many physics principles: Newton's laws, gas laws, the principles of electricity, and many more.

In addition both physicists and engineers need to solve problems in order to make something happen which otherwise wouldn't! Design is just as important here as physics. For example, a physicist often needs to design pieces of equipment to enable him or her to carry out an investigation.

3.1 CARRYING OUT AN INVESTIGATION

There are three stages involved in carrying out an investigation – planning, operating and evaluating.

A INTRODUCING PHYSICS

PLANNING STAGE

During the planning stage, you need to:

- understand the nature of the problem to be investigated, and the appropriate and relevant scientific models
- identify the variables in the system under investigation, and ways of controlling and possibly measuring them
- identify apparatus and equipment that will allow control and measurement of the variables. Such apparatus may be readily available in the laboratory, but you will need to choose measuring instruments of suitable range and resolution. Alternatively, you may need to design apparatus specifically for the purpose
- identify the plan of action, at least for the initial stages. What measurements are to be taken? How are they to be recorded and presented?

OPERATING STAGE

This is the stage at which you actually carry out the investigation. During this, you need to:

- carry out rough trial experiments to ensure that the proposed plan is feasible, and that the appropriate measurements are possible and within the range of the measuring instruments
- maintain awareness of peculiar or interesting results or observations that could lead to other paths of investigation
- maintain awareness of the limitations of the measuring equipment
- be thorough over recording observations and results, and make use of appropriate methods for processing and presenting them
- maintain awareness of statistical variation in measurements taken.

EVALUATING STAGE

At this stage, you need to:

- identify how well the investigation has been tackled
- be aware of the validity of any conclusions drawn
- reflect on the way the investigation was performed – would any changes in the design give a more conclusive outcome?
- think about how the investigation might be developed further, given more time.

3.2 PLANNING TO INVESTIGATE AN EXAMPLE

While blowing up a balloon, a student notices that the balloon initially seems quite hard to inflate but, as it gets bigger, it seems to become easier to blow more air into it. Then, when the balloon is very large, it again becomes difficult to inflate further. This fairly simple situation lends itself quite nicely to a practical investigation, once we have identified the relevant aspects of the physics involved.

For air to flow into the balloon, there must be an excess of air pressure on the inside compared with the outside of the balloon. We therefore need to consider the pressure of the air inside the balloon. The size of the balloon is clearly another important variable. Because we are dealing with the bulk properties of a mass of air, it would be sensible to try to relate the volume of the air in the balloon to the pressure it exerts. Other factors such as ambient temperature and external pressure changes may also affect the balloon size and we would need to ensure that these variables were adequately controlled.

To help us provide some explanation of any pressure-volume relationship we found, we might usefully investigate the stress-strain characteristic of the balloon rubber. We could take a sample of the rubber and place it under load. When drawing a link between the pressure-volume relationship and the behaviour of the rubber under stress, we should bear in mind that the radial force causing the balloon to expand outwards is found by the product of internal air pressure and the surface area of the balloon. We would also need to decide whether to consider the force-extension or the stress-strain characteristics of the rubber.

Such thoughts would need to be considered fully before embarking on any practical work. We could then start planning what apparatus we might need for such an investigation.

We would certainly need a way of monitoring the balloon's internal pressure while inflating or deflating it. Figure 3.1 shows a T-connector with flexible tubing to connect to the balloon, a pressure gauge and a clean plastic tube to blow through.

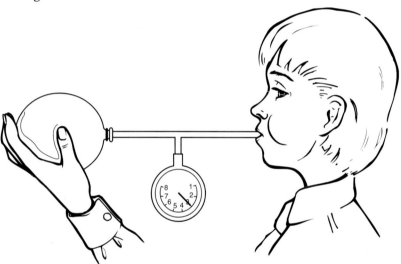

Figure 3.1 A T-connector with flexible tubing connected to a balloon, a pressure gauge and a clean plastic tube for inflation and deflation.

A rough experiment with a laboratory bourdon gauge would probably lead us to the conclusion that it was not sensitive enough for the relatively small changes of pressure in this context. A manometer is likely to be much more suitable. Again, a rough trial experiment with a length of clear plastic tubing would give an indication of the maximum height of manometer required.

3.3 SUGGESTED INVESTIGATIONS

Overleaf are listed two sets of investigations. Those in the first set relate directly to observed physical phenomena and tend to be open ended. Those in the second set concern a particular human-made artifact or system and beg the question 'how does it work?'

In both cases, areas of scientific understanding beyond the realms of physics may well become involved. Certainly, it is unlikely that investigations of either type can be confined to a particular topic in physics. So your approach will have to be inter-disciplinary. This book provides starting points, leads and aids to understanding across physics. You should also consult other books, papers, videos and so on, when the investigation is in progress, to gain new perspectives and understanding.

A INTRODUCING PHYSICS

INVESTIGATIONS OF PHYSICAL PHENOMENA

1 What is the effect of air resistance on falling objects? We all know that a feather falls more slowly than a pin, but what about lead shot and pieces of paper? Try dropping similar objects, such as a cricket ball and a tennis ball. A leaning tower is not essential for this investigation!

2 Do springs have the same stiffness when compressed as when expanded? Try using compression springs of various dimensions. Extend them and compress them, applying particular values of force.

3 Is water at its most dense at 4 °C? There is a special piece of apparatus for this – Hope's. You could actually measure the density of water at temperatures around and at 4 °C.

4 Does rubber display plastic properties – do rubber bands stretch permanently? Stretch rubber bands with known forces and measure their extension. Do they return to the same length every time you remove the force? How about time effects? If it has deformed, does the rubber revert to its original size after a time?

INVESTIGATIONS OF TECHNOLOGY

Here we are concerned with human-made objects or systems which depend on physics for their operation. Some examples are:

1 How is it that electric motors can be a.c. or d.c. but can only be reversed by changing the supply polarity if they are d.c. in type?

2 Do heavy duty motors run more smoothly with three-phase than single phase electricity?

3 Do 6-cylinder automobile engines provide greater torque and run more smoothly than 4-cylinder ones?

4 How does two-way household switching work?

3.4 PROBLEM SOLVING

Problem solving is an inter-disciplinary activity. You need to make use of all your background knowledge, very often not just in physics but in all the sciences. You may also need skills which you have gained from technology and other subjects, for example the humanities. Some examples are given here of the sort of problems in which the physics plays a major part.

1 Operate a living room light with more than two switches.

2 Find the quickest way to heat 1 litre of water from 20 °C to boiling point.

3 Produce an air speed indicator which will provide relative wind speed and direction information for winds of between 0.1 and 10 m s^{-1}.

4 Design aircraft made from A4 sheets of plain paper which will deliver a message over a distance of 50 m in light winds or still air.

5 Develop a system which will indicate the thickness of a pile of paper between 0.1 mm and 10 mm thick to an accuracy of ± 1 sheet.

6 Design an accelerometer which will indicate accelerations of between 0.1 and 2 m s^{-1} in a car.

UNIT B
MATTER AND ITS INTERACTION: FORCES

MATTER AND ITS INTERACTION: FORCES

Mechanical, structural and civil engineers need to know how matter and materials behave so that the structures they design are able to bear the forces acting on them. Knowing the nature of forces and their effects on the behaviour of materials and structures is as important as understanding the meaning of equilibrium and the effects of forces on the motion of bodies. It is the interaction between objects through physical contact and fields, such as those due to gravity and electricity, that give rise to forces.

4 ELASTIC MATERIALS

If you drop a glass, it will shatter into pieces which fit back together exactly – just like the pieces of a jigsaw puzzle. But a tennis ball struck with a tennis racket becomes considerably deformed and then returns to its original shape, while cars involved in a collision remain permanently deformed. Figures 4.1 and 4.2 show these effects.

The understanding of the behaviour of materials under stress ($\rightarrow$ 26) has led to improvements in the mechanical properties of existing materials. It has also led to the development of new materials.

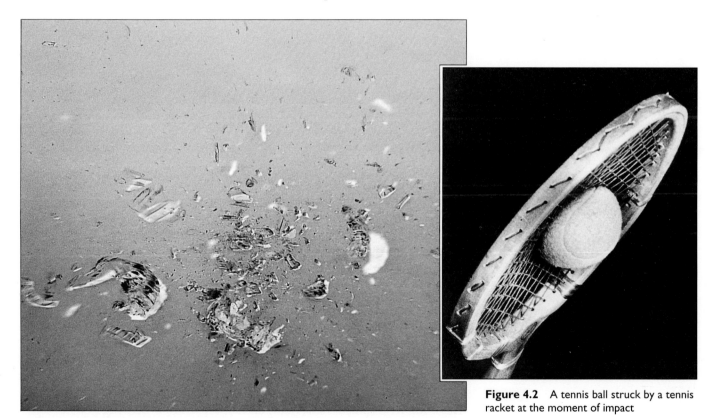

Figure 4.1 A glass tumbler shattering on impact

Figure 4.2 A tennis ball struck by a tennis racket at the moment of impact

4.1 ELASTICITY AND SPRINGS

When you increase the load on a helical steel spring, as in Figure 4.3, the length of the spring increases. A graph of the total increase in length or extension against the applied force can then be drawn.

MATTER AND ITS INTERACTION: FORCES

EXPERIMENT 4.1 Stretching a helical steel spring

Add loads to a helical steel spring, as in Figure 4.3. You can then plot a graph of the applied force in newtons against the extension of the spring in millimetres, as in Figure 4.4.

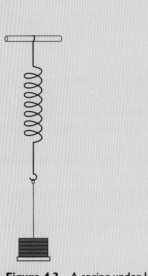

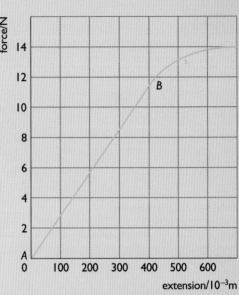

Figure 4.3 A spring under load   **Figure 4.4** Force-extension graph for a spring

Robert Hooke

Born on the Isle of Wight in 1635, Robert Hooke established the Royal Society and acquired a reputation for claiming that many discoveries made at the time originated from his own ideas. He was a contemporary of Isaac Newton with whom he had many clashes.

While developing clocks using a balance wheel and spiral springs in around 1680, he suggested that we can gain an understanding of the behaviour of solids under stress from a study of the behaviour of springs.

Figure 4.4 shows a graph obtained from experiment 4.1. The section labelled AB is linear. There the extension, x, of the spring is proportional to the force, F, applied. This relationship is known as **Hooke's law**.

So, $F \propto x$

giving $F = kx$

where the constant k is the **stiffness** or **spring constant** of the spring.

The constant k represents the force necessary to produce unit extension. The force is measured in newtons (N). A stiff spring will have a small extension for a given force, so k will be large. A weak or flexible spring will have a low stiffness. The unit of k is the newton per metre, $N\,m^{-1}$. The spring constant k can be found from the gradient of the force-extension graph of the spring.

For reasonable extensions, Hooke's law is obeyed. Beyond this, as long as the spring returns to its original length when the force is removed, the spring is said to possess **elastic behaviour**. The **elastic limit** marks the end of the elastic behaviour of the spring. If the spring is stretched beyond this limit, it will not return to its original length, but will remain permanently stretched or deformed. This is known as **plastic behaviour**, the name 'plasticine' deriving from it. Materials that undergo considerable plastic deformation are said to be **ductile**.

SPRINGS IN SERIES AND IN PARALLEL

If two similar springs joined in series have a load added, as in Figure 4.5, each will be stretched by the load. Together, they will have twice the extension of a single spring for the same applied force, or tension. Therefore the stiffness of this system, $k = F/x$, will be half that for a single spring. Two springs in parallel, as

4 ELASTIC MATERIALS

in Figure 4.6, will share the load. So they will have half the extension of each spring on its own for the same applied force. So, the stiffness overall will be twice that of each spring.

The rules for calculating the effective stiffness k_n for n springs each having a stiffness k will be

$k_s = k/n$ for springs connected in series, and

$k_p = nk$ for springs connected in parallel.

These rules apply equally to elastic solid materials that are being stretched, such as metals.

EXPERIMENT 4.2 Springs in series and in parallel

You can demonstrate the behaviour of springs in series and in parallel using two similar helical springs, as shown in Figures 4.5 and 4.6.

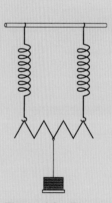

Figure 4.5 Two springs in series

Figure 4.6 Two springs in parallel

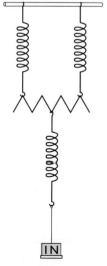

Figure 4.7

Questions

1 a What is the stiffness of a spring stretched 10 mm by a force of 1 N?
b A second identical spring is connected to the end of the spring in **a** and the same load is hung from the pair. Calculate the total extension and the stiffness of the two springs in series.
c The two springs in **b** are disconnected and hung side by side from the same support. The ends are connected together to the same load. Calculate the extension and the stiffness of the two springs in parallel.
d Calculate the extension when three similar springs are arranged as in Figure 4.7.

2 The equations that describe the behaviour of series and parallel spring systems bear a close resemblance to those that describe the behaviour of series and parallel combinations of resistors and capacitors ($\rightarrow$ 12.6 and 13.1). What are the similarities and differences? Why is it helpful to draw such analogies?

MATTER AND ITS INTERACTION: FORCES

> **3 a** Estimate the stiffness of a car suspension coil spring by sitting on the wing above the wheel – how much is it compressed by your weight?
> **b** What factors would need to be considered in choosing the value of the stiffness of the spring to be used in a forcemeter, or spring balance?

4.2 STRESS, STRAIN AND THE YOUNG MODULUS

There are different ways of putting a material under stress:

The steel cables used to tow an oil platform out to sea are under tension …

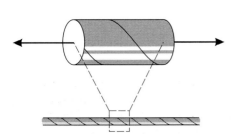

Figure 4.8 Tensile stress

The vertical pillars supporting a roof structure are under compression …

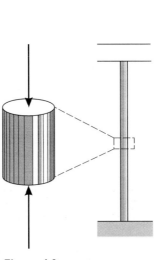

Figure 4.9
Compressional stress

4 ELASTIC MATERIALS

The horizontal bar supporting a gymnast suffers both tensile and compressional stress – the outside edge of the bent bar is under tension and the inside edge is under compression …

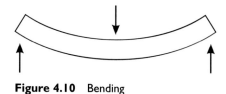

Figure 4.10 Bending

and torsional or twisting stresses are set up in the blade of a screwdriver as it tightens a screw in a block of wood.

TENSILE STRESS AND STRAIN

As we saw in experiment 4.2, two springs in series undergo twice the extension of one spring when stretched by a given force. We can apply the same idea to a rod of material. This can be thought of as consisting of particles linked by springy bonds. The long rod in Figure 4.11b contains twice as many bonds in a line, as the shorter rod in Figure 4.11a. Each rod is stretched by a force F and the shorter rod has an extension x. The total extension of the long rod is $2x$. So the extension of a rod of material, for a given force, depends on its length. But the fractional change in length is the same for a given force, irrespective of the rod's length whenever a material is placed under a tensile stress, so it is convenient to think in terms of the extension per unit length. This is called **strain**. It is usually expressed as a decimal fraction or as a percentage, and has no unit.

$$\text{Tensile strain} = \frac{\text{extension}}{\text{original length}}.$$

Metals undergo strains of less than 1% before they stop being elastic.

Two identical rods in parallel, shown in Figure 4.11c, act like a single rod of twice the cross-sectional area. The two rods together can support twice the load of one rod for the same extension. This is because the load is supported by twice as many bonds, side by side, in the two rods. So it seems reasonable to suppose that the force needed to break a rod is proportional to its cross-sectional area and is a measure of the **strength** of the rod. For this reason, when comparing the behaviour of different materials under tension, we use the force acting per unit area of cross-section of the material. This is defined as **stress**.

The cross-sectional area of a material decreases as the material stretches. For convenience, the stress is calculated using the original cross-sectional area. This is often called the **nominal tensile stress**.

$$\text{Tensile stress} = \frac{\text{force}}{\text{original cross-sectional area}}.$$

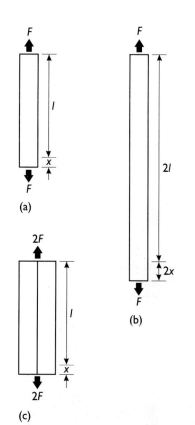

Figure 4.11 Rods under tension

MATTER AND ITS INTERACTION: FORCES

The unit of stress is the newton per square metre, or **pascal (Pa)**.

1 Pa = 1 N m^{-2}.

The **ultimate tensile stress (UTS)** is the breaking force divided by the original cross-sectional area, and is the maximum stress that a material can undergo before breaking. Clearly, materials used in permanent structures such as bridges and buildings must not remain deformed when any applied forces are removed, but must behave elastically. So it is important for structural engineers to know the stress at which plastic deformation of the material begins. This is known as the **yield stress**.

THE YOUNG MODULUS

For materials that are elastic, stress is proportional to strain. The ratio of stress to strain is a constant for a particular material, if all other variables, such as temperature and pressure, are not changed. This constant is known as the **Young modulus, E**, and is a measure of the stiffness of the material. The larger the Young modulus, the greater the force needed to stretch or compress it, and so the stiffer the material.

$$E = \frac{\text{tensile stress}}{\text{tensile strain}}.$$

The unit of the Young modulus is the same as the unit of stress – the pascal. **Stiff** materials have a high Young modulus and **flexible** materials have a low Young modulus. The Young modulus of metals in the form of wires can be measured using simple apparatus, as in Figure 4.14.

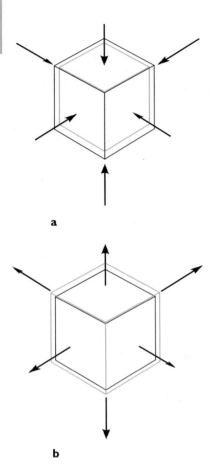

Figure 4.12 Change in volume under
a uniform compression,
b uniform tension

Other Elastic Moduli

The Young modulus applies to measurements of tensile stress and strain. When materials are subjected to a change in pressure, they undergo a change in volume (see Figure 4.12). Then we use the volume or **bulk modulus, K**, given by

$$K = \frac{\text{bulk stress}}{\text{bulk strain}}$$

$$= \frac{\text{change in force per unit area}}{\text{fractional change in volume}}$$

or, $$K = \frac{\text{change in pressure}}{\text{change in volume/original volume}}.$$

Forces operating in opposing directions, as in Figure 4.13, cause a distortion or shear. The **shear stress** is defined as the force per unit area tangential to the surface of the material. It is necessary to provide a net compensating couple of forces, or torque, in the opposite sense to maintain rotational equilibrium ($\rightarrow$ 5.6). The angle of shear, γ, is defined as the strain. The **shear modulus, G**, is given by:

$$G = \frac{\text{shear stress}}{\gamma}.$$

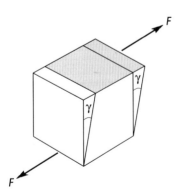

Figure 4.13 Shear

MATERIALS UNDER STRESS

Experiment 4.3, enables you to look in detail at the way some materials behave under stress. Three very different materials are suggested: copper, rubber and glass. Figure 4.17 shows the stress/strain graphs for each material.

4 ELASTIC MATERIALS

EXPERIMENT 4.3 Measuring the mechanical properties of different materials

Warning: Eye protection must be worn for this experiment.

a Thin copper wire
Using the apparatus set up as in Figure 4.14, you can plot force against extension. This enables you to find the ultimate tensile stress of the material, the yield stress and the Young modulus.

b Rubber
Using a rubber band as in Figure 4.15, you can plot a force-extension graph to see whether rubber obeys Hooke's law and to obtain the Young modulus of rubber.

c Glass
Support a thin soda glass rod or a fibre of about 0.5 mm in diameter as in Figure 4.16, and then add masses until the glass breaks. This enables you to find the ultimate tensile stress and the Young modulus of glass.

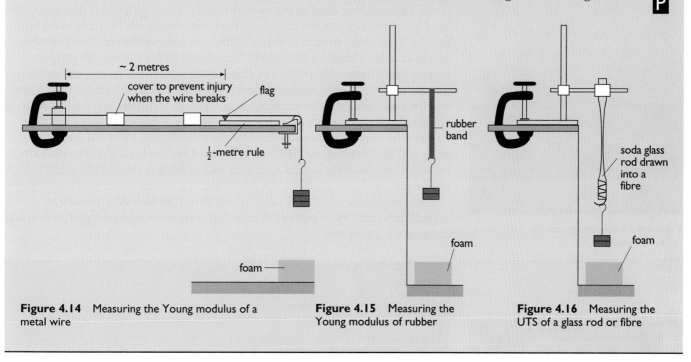

Figure 4.14 Measuring the Young modulus of a metal wire

Figure 4.15 Measuring the Young modulus of rubber

Figure 4.16 Measuring the UTS of a glass rod or fibre

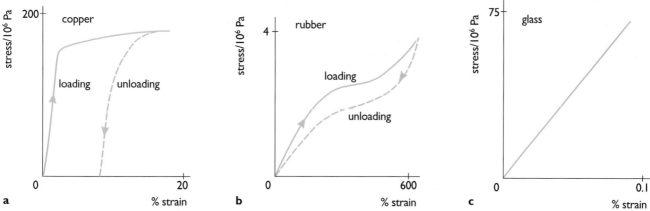

Figure 4.17 Load-extension graphs for copper, rubber and glass

MATTER AND ITS INTERACTION: FORCES

material	E / 10^{10} Pa	UTS / 10^7 Pa
iron	21.1	21
mild steel	20	25
high tensile steel	20	35
copper	13	22 to 43
rubber		3
sheet glass	7	0.4 to 15
aluminium	7	10
earthenware pottery	5	6
soda glass	7	4 to 9
perspex	0.3	6
oak (hardwood along grain)	0.9	2

Figure 4.18 Values of the Young modulus and the UTS for some materials

Figure 4.18 gives values for some of the mechanical properties of a number of materials, including steel. In the motor industry, many different types of steel are used for components in the motor car. For example, the steel used to manufacture the discs in the front brakes is of a different composition from that used to construct a drive shaft for the engine. Steel designed for a particular purpose may acquire chemical impurities or structural irregularities during its manufacture. So samples of manufactured steel are tested, often to destruction, to ensure that their mechanical properties (stiffness, strength, hardness and so on) are up to the standards required for a particular application. Force-extension graphs for the samples may be obtained automatically by clamping a test rod into a machine and gradually elongating it as in Figure 4.19. The load and extension are recorded continuously and an automatic plot of the force-extension graph is produced.

What properties would you need to look for in the materials to make an artificial hip joint? Here part of the bone has been replaced by a metal section and the socket is a polymer.

Figure 4.19 An extensometer being used to test a sample

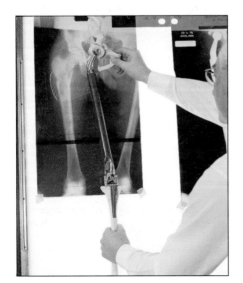

Figure 4.20 An artificial hip joint

Questions

4 a Calculate the extension resulting from a 1% strain on a 2 m length of wire.
b Explain why long metal wires are often used for measuring the Young modulus.
c A steel wire of cross-sectional area 1 mm² supports a load of 10 N. Find the stress in the wire.
d The ultimate tensile stress of steel is 4×10^8 Pa. Calculate the maximum load that the wire in **c** can support.
e If the wire in **d** has a 1% strain, what is the Young modulus of the steel?

5 A tensile testing machine has vice-like 'friction grips' for clamping a rod of solid material and breaking it in tension. These grips, however, damage the metal and cause it to fail prematurely at the ends. Explain how a metal test piece that is thinner in the middle overcomes this problem.

6 When designing cables for the electrical supply industry, engineers require materials that conduct electricity and also have high mechanical strength. Find out how they build such properties into power cables used in the national grid.

4 ELASTIC MATERIALS

4.3 STRESS-STRAIN GRAPHS

Solid materials can be roughly classified into four groups:

i   ceramics, which include china, brick and glass;
ii metals;
iii polymers, which include polythene, melamine and rubber;
iv composite materials, such as glass-fibre reinforced plastic or carbon fibres.

GLASS RODS AND FIBRES

The stress-strain graph for glass, shown in Figure 4.17c is a straight line so glass obeys Hooke's law and is elastic. The graph stops abruptly because glass breaks suddenly. If you look at the two ends of a broken glass rod you will notice that the clean breaks fit together perfectly. Materials that break in this way are described as **brittle**. Brittle materials are susceptible to the propagation of cracks.

ELASTIC POTENTIAL ENERGY

A force moves through a distance when a material is deformed, so work is done. In an elastic material this work is stored as potential energy and released when the force is released. This elastic potential energy is equal to the area under the force-extension graph for the material. For a linear elastic material, such as a spring, which has an extension, x, for a force, F, the elastic potential energy, E_p is given by

$$E_p = \frac{1}{2} Fx.$$

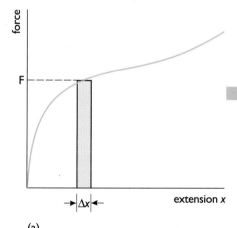

Figure 4.21 Force-extension graph for evaluating elastic potential energy

Energy stored in an elastic material

When a constant force, F, moves through a distance, x the work done equals Fx which is equal to the shaded area under the force-distance graph shown in Figure 4.21. The force required to stretch rubber however, changes as the extension increases, as in Figure 4.22a. Here let us assume that when the extension is x, a constant force F is required to cause a further small extension. In mathematical symbols this small extension is written as Δx, pronounced "delta x". Δx means a small change or increment in x. The work done by F in stretching the rubber by Δx causes potential energy $F\Delta x$ to be stored. This is equal to the area of the shaded strip.

In Figure 4.22b the whole area under the graph has been divided into strips. Each strip corresponds to the different force F required to stretch the band through the same extra distance Δx.

So, as the work is done, the energy E_p transferred to the stretched material is given by:

E_p = sum of each of the values of $F \cdot \Delta x$ for each strip

$\quad = F_1 \Delta x + F_2 \Delta x + F_3 \Delta x + \cdots$

or $E_p = \Sigma F \Delta x.$

The smaller Δx, the more accurate this equation becomes. In the limit where x becomes vanishingly small, the summation becomes an integral ($\rightarrow$ 41.3) and we can say that the total energy transferred is:

$$E_p = \int F\,dx.$$

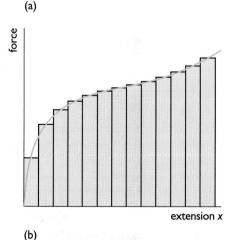

Figure 4.22 Graphs of force against extension

So, the total work done transfers energy equal to the area under the graph. For an elastic material, the shape of this area is a triangle and the total increase in potential energy is:

$$E_p = \tfrac{1}{2} \text{height} \times \text{base} = \tfrac{1}{2} Fx = \tfrac{1}{2} kx^2.$$

Questions

7 What happens to the energy stored in a steel wire when the stretching force is halved?

8 You have two tow-ropes of identical material. They have equal areas of cross-section but one is twice as long as the other. They obey Hooke's law up to their breaking point.
a How will the tensile force needed to break each of the ropes compare?
b How will the extension of each rope compare at the breaking point?
c How will the energy stored in each rope compare at the breaking point?
d Which is the better rope to use when towing a car?

ENERGY STORED IN RUBBER

If a force-extension curve for the loading and unloading of rubber is plotted, the unloading curve will be seen not to coincide with the loading curve. This is known as **elastic hysteresis**. Hysteresis means 'lagging behind'. The strain on unloading lags behind the loading strain.

EXPERIMENT 4.4   Elastic hysteresis of rubber

Rapidly stretching a rubber band as far as possible without breaking it and releasing it several times causes a change in its temperature. You can detect this against your lips.

Figure 4.23 shows the elastic hysteresis of rubber. The work done in stretching the rubber band is equal to the area under the loading curve. However, the elastic potential energy is equal to the area under the unloading curve: less than the work done in stretching. The difference – the area between the curves – is the increase in internal energy, which explains the rise in temperature of stretched rubber. A similar phenomenon, called **magnetic hysteresis**, occurs in electromagnetism and contributes to the inefficiency of transformers.

Car tyres

Elastic hysteresis has to be taken into account by the manufacturers of car tyres. Bumps in the road and changes in speed and direction of the car cause the tyres to be deformed and relaxed continually.

The major energy loss in tyres comes from their rolling resistance. It amounts to nearly 6 % of the energy derived from the car's fuel. The materials used in building a tyre have different elasticities: the cords that encircle the tyre and give it its strength; the breakers, or steel sheets, which help to keep the tread firmly on the road; and the rubber and other flexible materials used in the side walls which enable it to give under the strains of carrying a car along an uneven surface. Some of the energy absorbed by the tyre raises its temperature, and the rest is transferred by vibration to the car and to the air, as sound waves.

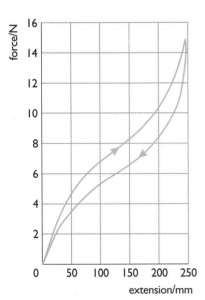

Figure 4.23 Graph showing elastic hysteresis in rubber

4 ELASTIC MATERIALS

> The amount that a tyre's component parts flex under load depends on the air pressure within. A drop of about 40 kPa (6 lbf per in^2) below the recommended pressure can completely negate any improvements in design to save energy. However, exceeding the recommended pressure reduces the tyre's grip on the road, causes it to wear quickly in the centre and reduces passenger comfort.

SUMMARY

An elastic material is one that returns to its original shape after having been deformed, and has an extension proportional to the applied force. The ratio of force to extension measures the stiffness of the particular wire or spring being stretched.

Most materials behave elastically to some extent. Many materials reach an elastic limit and then show plastic behaviour, remaining deformed after removal of the force applied.

The extension of a material for a given force is proportional to its length. In order to compare materials of different lengths, we use strain – the extension per unit length.

The strength of a material depends on its cross-sectional area. To take this into account, we use stress – the force per unit area. The ratio of stress to strain is the Young modulus and measures the stiffness of the material.

The ultimate tensile stress is the maximum stress a material can withstand before failure. The yield stress gives the maximum stress a material can withstand before undergoing plastic flow.

Materials, such as ceramics, that do not undergo plastic behaviour but fail by cracking are brittle materials.

When a material is deformed elastic potential energy is stored. The energy stored is equal to the area under the force-extension graph.

A tough material requires a large amount of energy to break it.

The energy stored in an elastic material is equal to $\frac{1}{2}Fx$ or $\frac{1}{2}kx^2$.

Rubber undergoes elastic hysteresis: the force-extension curve obtained from loading a rubber sample does not coincide with the unloading curve for the same sample. When rubber is stretched and released, some of the work done increases the internal energy of the material, so increasing its temperature.

SUMMARY QUESTIONS

9 A car has a suspension consisting of four vertical helical springs, and sinks by 35 mm when four adults get inside.
a Assuming that the suspension obeys Hooke's law, estimate the stiffness k (in N m^{-1}) of the suspension.
b Calculate the stiffness of each spring.
c The stiffness of car springs often varies with depression. Why might this be so?

10 a While measuring the Young modulus of copper, a 2 m length of copper wire of diameter 0.30 mm is stretched elastically by 7.5 mm when a load of 30 N is added. Calculate the stress, strain and the Young modulus of copper.

b The length was measured using a metre rule and the weights are accurate to 5 %. Estimate the uncertainties in the answers to **a**.

11 Use the force-extension graph for a spring in Figure 4.4.
a Calculate the stiffness of the spring.
b Calculate the maximum recoverable energy that could be stored in the spring.
c Estimate the extra energy needed to break the spring.

12 A spring, which has a spring constant of 20 N m^{-1}, is stretched by 10 cm and used to propel a trolley of mass 1 kg along a friction-compensated runway. Calculate the velocity that the trolley will acquire. Explain what assumptions you have made.

13 A model of a motorway bridge is built using the same materials but to $\frac{1}{32}$ scale. Work out the ratios of:
a the mass of the bridge to that of the model
b the area of the pillars of the bridge to the area of the pillars of the model
c the stress in the pillars of the bridge to the stress in the pillars of the model.

14 A man of mass 70 kg stands still with both feet on the ground. The cross-sectional area of the lower leg bone is 4×10^{-4} m^2. Calculate the stress in his leg bones.
Explain why this has to be much less than the ultimate compressive stress of bone, which is 1.5×10^8 Pa. (*g*)

5 STRUCTURES IN EQUILIBRIUM

The Humber bridge (see Figure 5.1) was designed to contend not only with the stresses imposed upon it by traffic but also factors such as wind, temperature variation and other weather conditions. Its design also utilised a minimum amount of material to ensure economic viability in its construction and provided an aesthetically pleasing appearance. Compare its structure with that of a typical oil platform, such as the one shown in Figure 5.2. The design of this has to be even more exacting because the stresses due to high seas and winds can be enormous.

The bridge and oil platform are examples of structures that are in equilibrium, where forces acting on and within the structures balance each other. An understanding of the effect of these forces and a knowledge of the behaviour of materials under stress are needed to achieve efficiency in design and construction of such structures.

Figure 5.1 The Humber suspension road bridge

Figure 5.2 The Miller 16/8B oil platform

5.1 VECTORS AND SCALARS

For some quantities, direction is important as well as size. These are called **vector** quantities. Because a force is one side of an interaction between two bodies, it

MATTER AND ITS INTERACTION: FORCES

operates in a particular direction – it is a vector quantity. Suppose two people both push on a crate with a force of 500 N so that they both push the crate to the right (see Figure 5.3 a). There will be a total force of 1000 N pushing the crate to the right. However if they push in opposite directions, as in Figure 5.3b, the resulting force on the crate will be zero.

A vector can be represented on a diagram by an arrow that points in the direction of the vector. The length of the arrow represents the size of the vector, as shown in Figure 5.3 c.

 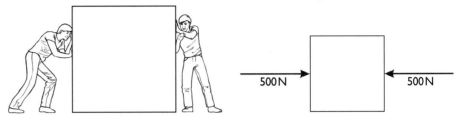

Figure 5.3 a Pushing a crate **b** Forces in balance **c** Arrows representing vectors

Displacement is the name given to the distance travelled in a certain direction and so is also a vector quantity. Imagine walking 2 km east then 4 km west. You will have travelled a total distance of 6 km, but your displacement will be 2 km west of your starting place.

Velocity, another vector quantity, is a speed in a certain direction. A car travelling north at 30 m s^{-1} (about 70 mph) along a motorway has the same speed as a car travelling south at 30 m s^{-1}. But their velocities are different because they are travelling in different directions. Momentum and acceleration are examples of other vector quantities.

Scalar quantities, on the other hand, are specified by their size only. Mass, energy and volume are common examples of scalars. Unlike vectors, scalars always add arithmetically. A mass of 1 kg added to another of 1 kg always gives 2 kg. They cannot add together algebraically to make zero, as vectors can.

5.2 FORCES IN EQUILIBRIUM

When a body is stationary or moving at constant velocity, it follows from Newton's laws of motion ($\rightarrow$ 6.4) that the algebraic sum of the forces acting on it must be zero. However, such forces can still cause the body to rotate ($\rightarrow$ 5.6).

A simple example is a book resting on a table. The interaction between the book and the Earth, caused by gravity, attracts the book towards the Earth and pulls down on it with the force of gravity – the book has weight. The weight of the book causes a slight compression of the surface of the table which pushes up on the book with an equal and opposite force. Whenever an object rests on or against a surface, the compression of the surface results in a force that acts at right angles to the surface. This is called the **normal force**, or **reaction**. Under these two forces, the book is in equilibrium and remains at rest.

Where the distinction between vectors and scalars is important, a letter in bold type, **F**, represents a vector and the corresponding letter in normal type, F, represents a scalar.

So, in Figure 5.4

$$\mathbf{N} + \mathbf{W} = \mathbf{0},$$

or $\mathbf{N} = -\mathbf{W}$.

5 STRUCTURES IN EQUILIBRIUM

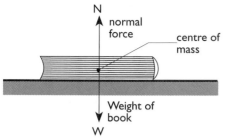

Figure 5.4 Forces on a book

The forces have the same magnitude but the minus sign means that **W** acts in the opposite direction to **N**.

Actually the book in Figure 5.4 consists of many particles each of which is attracted towards the centre of the Earth by the force of gravity. The weight of the book is the sum of all of these forces, and the point through which the weight effectively acts is called the **centre of mass** of the book.

Question

1 A student of weight 500 N sits on a laboratory stool which weighs 30 N. Calculate: **a** the force that the stool exerts on the student, **b** the force that the floor exerts on the stool.

EXPERIMENT 5.1 Equilibrium of three forces acting through a point

Using two forcemeters, as shown in Figure 5.5, you can draw the vector polygon described in the text. Results obtained from this experiment are:

$\theta = 48°$,
$\mathbf{F}_1 = 7.4$ N
and $\mathbf{F}_2 = 5.3$ N.

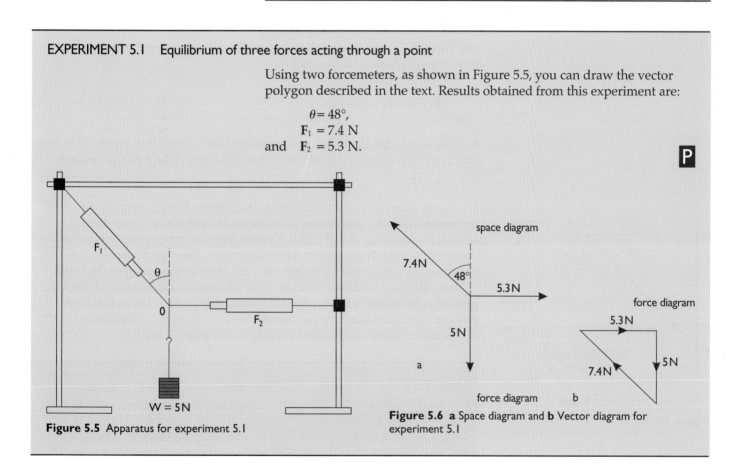

Figure 5.5 Apparatus for experiment 5.1

Figure 5.6 **a** Space diagram and **b** Vector diagram for experiment 5.1

THE VECTOR POLYGON

In experiment 5.1, three forces act through a point, O, and the system is in equilibrium. Figure 5.6a shows the point with the three forces acting on it, using the results obtained from the experiment. This is called a **space diagram**. The three forces can be removed from their positions and drawn in a separate **vector diagram**, as in Figure 5.6b. They are placed pointing head to tail with the length of each representing its magnitude. They join up to form a closed triangle – the **triangle of forces**.

The triangle of forces is just an example of the more general **vector polygon**. In this example, the three forces act in the same plane and so are coplanar forces. For any point in equilibrium under the action of any number of coplanar forces

MATTER AND ITS INTERACTION: FORCES

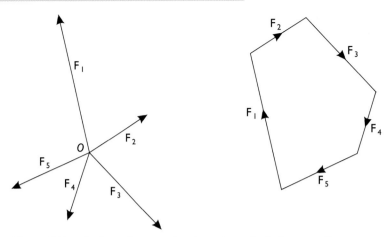

Figure 5.7 **a** Coplanar forces acting through a point **b** A vector diagram of the same forces

(see Figure 5.7a), the forces will form a closed polygon when drawn head to tail in a vector diagram, as in Figure 5.7b.

The polygon can be summarised

$$\mathbf{F}_1 + \mathbf{F}_2 + \mathbf{F}_3 + \mathbf{F}_4 + \mathbf{F}_5 = 0$$
or $$\Sigma \mathbf{F} = 0.$$

In other words, the algebraic or **vector sum** of the forces acting at point O is zero. This is not necessarily the case when the magnitudes of the forces are added:

$$F_1 + F_2 + F_3 + F_4 + F_5 <> 0.$$

The vector sum takes into account both the magnitude and direction of each of the forces. When the vector sum of the forces at a point is **not** zero, there is a resultant force acting at that point. A vector diagram can be drawn representing these forces, but it will not form a closed polygon. The magnitude and direction of the resultant is given by the algebraic sum of the forces acting on the body.

These ideas are useful for describing the addition and subtraction of vectors generally, including velocity, momentum and displacement as well as force. Phasors (quantities that vary periodically with time), such as displacements of waves and oscillations, can also be added in this way (→ 19.3).

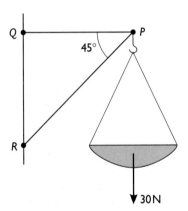

Figure 5.8

Question

2 A hanging basket of weight 30 N hangs from a metal framework as shown in Figure 5.8.
a Draw a space diagram of the forces acting at point P.
b Say whether PQ and PR are in tension or compression.
c Draw a vector diagram to find the forces in PQ and PR.
d The force exerted by PQ on the wall acts horizontally. What must be the magnitude and direction of the reaction of the wall at Q?
e Find the magnitude and direction of the reaction of the wall at R. Note that the force exerted by PR on the wall acts at 45° to the wall.

5.3 ADDING FORCES

Suppose the point O is acted upon by the two forces, $\mathbf{F}_1$ and $\mathbf{F}_2$, as in Figure 5.9a. Clearly, O is then not in equilibrium. Creating a vector diagram representing the

5 STRUCTURES IN EQUILIBRIUM

two forces, as in Figure 5.9b, enables you to see that a third force F_3, the equilibriant, is needed to balance them and make a closed triangle.

So $\quad F_1 + F_2 + F_3 = 0$

which means that $\quad F_1 + F_2 = -F_3,$

or the sum of F_1 and F_2 equals a force that is equal in size to F_3 but acts in the opposite direction.

In Figure 5.9c, $\quad F_1 + F_2 = F$

where F is the sum or **resultant** of the two forces and the direction of F opposes the direction of the other two forces around the triangle.

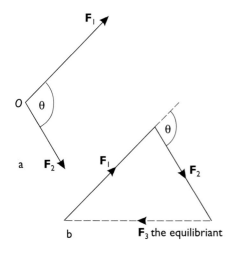

Figure 5.9 a Space diagram
b Vector diagram c Adding vectors

> **Question**
>
> **3** A girl of weight 300 N sits on a swing of weight 40 N. The girl's father holds the swing, so that the chains make an angle of 30° to the vertical, by exerting a force at right angles to the chains.
> **a** Draw a diagram showing the three forces that act on the swing, assuming that the weight of the chains is so small that it can be ignored.
> **b** Draw a vector triangle of the three forces, and calculate the force exerted by the girl's father and the tension in the chains of the swing.
> **c** When her father lets go of the swing there is a resultant force at right angles to the chains. Work out the value of the resultant force and the new tension in the chains.

5.4 RESOLVING VECTORS

In Figure 5.9c, the two forces F_1 and F_2 could be replaced by the single force F. Sometimes it is useful in solving problems to replace a single force by two forces acting at right angles to each other.

For example, in Figure 5.10a a motor boat is towing a dinghy with a force F at a direction θ above the horizontal. This towing force can be considered to be made up of two perpendicular components, one acting vertically upwards and the other horizontally. Figure 5.10b shows how the horizontal component F_x pulls the boat along while the vertical component F_y lifts the bows of the dinghy out of the water.

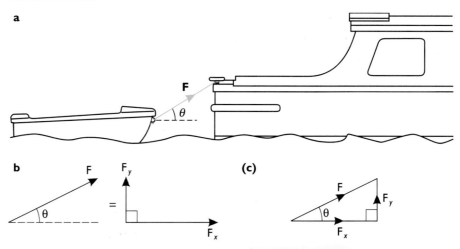

Figure 5.10 a Towing a dinghy
b Components of the force
c Resolution of the force

MATTER AND ITS INTERACTION: FORCES

Using $\mathbf{F}_x + \mathbf{F}_y = \mathbf{F}$

then, from the vector triangle of Figure 5.10c:

$\mathbf{F}_x = \mathbf{F} \cos \theta$

and $\mathbf{F}_y = \mathbf{F} \sin \theta$.

This leads to the rather more general conclusion that the component of a force at an angle θ to the force can be found by multiplying the magnitude of the force by the cosine of that angle.

You might wonder how this statement can apply to $\mathbf{F}_y$. Since $(90 - \theta)$ is the angle between the force F and its vertical component, then

$\mathbf{F}_y = \mathbf{F} \cos (90 - \theta) = \mathbf{F} \sin \theta$.

Any force can be resolved into components in this way. When a number of forces act on an object, and the object is in equilibrium, the sum of all the components in any given direction is zero. All vectors, not just forces, can be added and resolved in the same way.

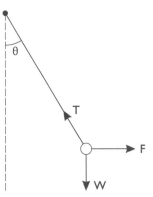

Figure 5.11 Force on a pendulum bob

Questions

4 Calculate the component of a force of 100 N in a direction
a 30° **b** 45° **c** 90° to the direction of the force.

5 Look back at question **2** and then
a using your space diagram for point P, resolve the forces vertically and then horizontally to check your answers to part **2c**.
b resolve the forces in a direction parallel to PR and check that the components balance.

6 A pendulum is held at an angle θ to the vertical as shown in Figure 5.11.
a Resolve the forces horizontally and vertically and show that

$F = W \tan \theta$.

b For small values of θ, we can say that F is proportional to the horizontal displacement of the pendulum bob. Why is this? Is 5° a small enough angle? Is 10°? Is 20°?
c Demolition contractors sometimes use a very massive ball on the end of a chain attached to the jib of a crane as shown in Figure 5.12. Estimate the force needed to displace a 1000 kg ball, 5 m horizontally from its undisplaced position. The length of the supporting chain is 25 m. (g)

Figure 5.12 A demolition crane

5.5 FRICTION

An important force contributing to the equilibrium of bodies is that due to friction. **Static friction** tends to resist motion when a force is applied to a stationary object and **dynamic friction** tends to resist the motion of a moving object.

STATIC FRICTION

Look back at Figure 5.4 and imagine that the table is tilted slightly, as in Figure 5.13. The weight, **W**, of the book acts vertically downwards and there is a normal force, **N**, at right angles to the table as before. A third force, the frictional force, **F**, acts along and between the surfaces of the book and the table (→ 9.5). This force

5 STRUCTURES IN EQUILIBRIUM

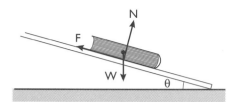

Figure 5.13 Forces on a book on a sloping surface

acts up the slope because it always opposes a tendency towards motion. These three forces on the book form a closed vector triangle, and there is no resultant force.

When the slope of the table is increased, the static frictional force increases until it reaches a maximum value, which depends on the surfaces in contact. Any further increase in the slope will cause the book to accelerate down the table.

A force is required to slide one material over another. A frictional force acts in a direction to oppose such motion. Observed at a microscopic level, any two surfaces are very uneven. When placed in contact, they deform, and pressure welds occur where they touch. For sliding to occur, these points of contact must be sheared off giving rise to forces parallel to the surfaces, where the solids rub on each other. The study of friction, lubrication and wear is called tribology (from the Greek word for strife, τριβοσ).

The ratio of the maximum frictional force, **F** to the normal force, **N**, is called the **coefficient of static friction**, μ, as shown in Figure 5.14. This depends on the type of materials in contact and the state of their surfaces.

$$\mu = \frac{\text{maximum force before motion starts}}{\text{normal force}}.$$

When the book in Figure 5.13 is on the point of slipping, the coefficient of static friction μ is given by

$$\mu = \tan \theta,$$

where θ is the angle that the slope makes with the horizontal.

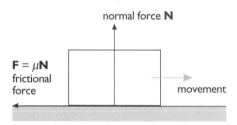

Figure 5.14 A block sliding on a horizontal table

Question

7 Support a metre rule horizontally by placing your two index fingers at the 10 cm and 90 cm marks, as in Figure 5.15. Slowly bring your hands together. What do you notice? Now put one finger at the 10 cm mark and the other at the 70 cm mark and repeat the experiment. What do you notice now? What happens when one finger is at the 10 cm mark and the other under the middle of the rule?
How do you explain all of these observations?

Figure 5.15 Fingers supporting a metre rule

MATTER AND ITS INTERACTION: FORCES

DYNAMIC FRICTION

Suppose the table in Figure 5.13 is tilted at such an angle that the book, when given a push, slides down at constant velocity. Because it is not accelerating, the book must be in some form of equilibrium. This is known as **dynamic equilibrium**. The three forces acting on it will form a closed vector triangle as before. The frictional force between the book and the table when the book travels with a constant velocity is called the dynamic frictional force. The **coefficient of dynamic friction**, μ' is the ratio of the dynamic frictional force when an object moves at constant velocity to the normal force, **F**.

$$\mu' = \frac{\text{frictional force at constant velocity}}{\text{normal force}}.$$

For two given surfaces, the coefficient of dynamic friction is less than the coefficient of static friction.

Friction on a slope

Suppose that the book in Figure 5.13 is just on the point of slipping down the slope. The ratio **F/N** will be equal to μ, the coefficient of static friction between the book and the surface. However the book is in equilibrium, so there is no net force acting on it. The frictional force **F** is balanced by the component of the weight of the book parallel to the slope:

$$F = W \sin \theta.$$

The normal force **N** is balanced by the component of the weight at right angles to the slope:

$$N = W \cos \theta.$$

Combining these two equations by dividing **F** by **N**:

$$\frac{F}{N} = \frac{W \sin \theta}{W \cos \theta}$$

or, $\mu = \tan \theta.$

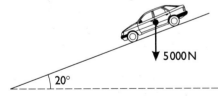

Figure 5.16

Questions

8 A block of wood is placed on a board which can be tilted. Estimate the angle of tilt for the block to start sliding, and also its acceleration once moving. Use the following data: coefficient of static friction = 0.4, coefficient of dynamic friction = 0.3.

9 A dynamics trolley can be made to travel down a runway at constant speed, when given a push, by tilting the runway to compensate for friction in the trolley's bearings.
a Draw a diagram showing the forces acting on the trolley.
b Explain how the tilting has compensated for friction.
c If the trolley has a weight of 5.0 N and the angle of the slope is 5°, work out the frictional and normal forces on the trolley.

10 A car of weight 5000 N is parked on a hill as shown in Figure 5.16.
a Calculate the frictional force, **F**, between the locked wheels and the road surface, which stops the car from moving down the slope.
b What minimum coefficient of static friction is required between the rubber tyres and the road surface for the car to remain stationary on the slope with its handbrake on?

Friction and Skis

Skiers sometimes hydroplane on a thin layer of water because the friction between the ski and the snow generates enough heat for some of the snow to melt. As the temperature falls, more of this heat is conducted away through the ski. Below a certain temperature (for the early skis made of waxed wood this temperature is about –10°C) so much heat is conducted through the ski that the snow no longer melts and the coefficient of friction between the ski and the surface increases. Skis with metal rims are slower at low temperatures than those without. Today ski under-surfaces are coated with polymers such as PTFE ("Teflon") which have a low coefficient of friction even when the temperature is so low that the snow does not melt.

5 STRUCTURES IN EQUILIBRIUM

EXPERIMENT 5.2 Measuring the coefficient of static friction between two wooden surfaces

Using the apparatus shown in Figure 5.17, you can find the average value of the coefficient of static friction between the two surfaces.

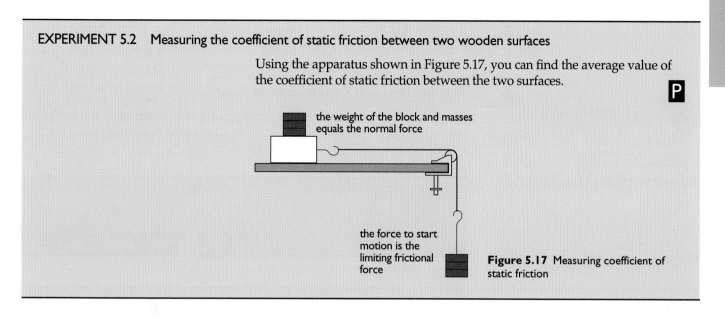

Figure 5.17 Measuring coefficient of static friction

5.6 TURNING FORCES

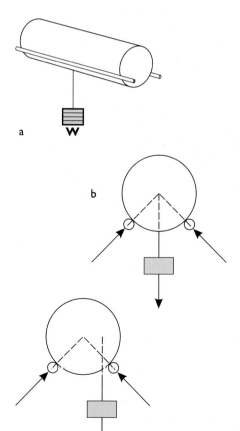

Figure 5.18 a and b Cylinder in equilibrium c Unbalanced forces

FORCES NOT ACTING THROUGH A POINT

When the resultant of several forces acting on an object is zero, the object does not accelerate because it is in equilibrium (← 5.2). Suppose you rest a cylinder, such as an empty aluminium soft drinks can, on two thin rollers or rods so that it can rotate about its own axis. Then you attach a weight so that it hangs freely from a point halfway along the lowest part of the cylinder, as in Figure 5.18a. Forces between the rollers and the cylinder act perpendicularly through the surfaces in contact. So they act through the cylinder's centre. The force due to the weight hanging from the lowest point of the cylinder will also act through the centre of the cylinder. So all three forces act through the centre of the cylinder with no resultant force and the cylinder is in equilibrium and remains at rest (see Figure 5.18b).

Rotating the cylinder about its axis, as in Figure 5.18c, means that the force due to the weight no longer acts through the centre, and the equilibrium is upset. By imagining that the cylinder is released, you can see that the combined result of the forces is to rotate the cylinder back to its equilibrium position. In the displaced position, the sum of the resolved components of the forces on the cylinder vertically and horizontally is still zero, causing no acceleration of its centre of mass. However, the three forces are no longer passing through one point. So, they cause the cylinder to rotate.

THE TURNING EFFECT OF FORCES

A turning effect is produced when a nut is tightened with a spanner. A spanner is an example of a **lever** – a bar that turns about a hinge or pivot to exert a force on a load. As you might expect, the turning effect increases when you push harder on the spanner. It also depends on the length of the spanner: the longer the handle of the spanner, the greater the turning effect of a certain force. Experiment 5.3 investigates the relationship between the turning effect of a force and its distance from the pivot.

The turning effect of a force, often called its **moment** or **torque**, is defined as the product of the force **F** and the perpendicular distance, d, from its line of action to the pivot as shown in Figure 5.19.

Moment or torque = **F**d.

The unit of **torque** is the **newton metre (N m)**.

An electric drill may produce a torque of around 10 N m whilst a 1600 cm³ petrol engine may give a torque of up to 150 N m. The torque required to undo the wheel bolt on a car is typically around 80 N m.

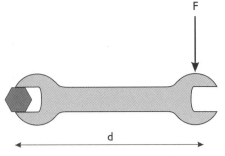

Figure 5.19 Moment of a turning force

Question

11 When working on a car engine, why is it important not to over- or under-tighten nuts and bolts? How does a torque-wrench help with this?

EXPERIMENT 5.3 Investigating the turning effect of a force

You can investigate how the turning effect of a force is influenced by

- the length of a lever, such as a spanner, using the apparatus shown in Figure 5.20,
- the direction of the force by using a pulley, as in Figure 5.21.

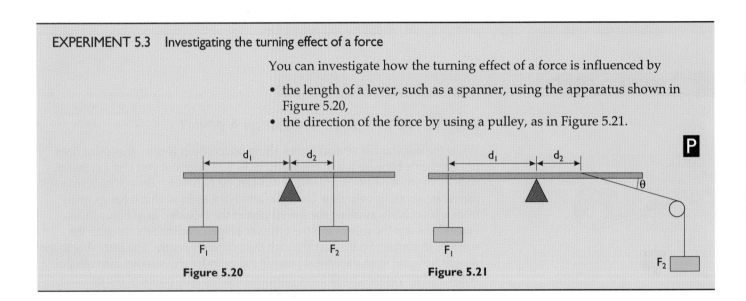

Figure 5.20   **Figure 5.21**

ROTATIONAL EQUILIBRIUM

If you calculate the moment of F_1 and the moment of F_2 in experiment 5.3, you will see that, at equilibrium, the sizes of the moments are equal. The experiment shows that when a lever is balanced under the action of any number of forces, the sum of the moments having an anticlockwise turning effect about the pivot will equal the sum of the moments having a clockwise turning effect, or the algebraic sum of the moments about the pivot is zero. This is known as the **law of moments**. For any object in rotational equilibrium, the **net torque** about any axis will be **zero**.

In Figure 5.22, if you take moments about an axis through A, the force N_1 will have no moment as it acts through the axis. The clockwise turning moment T_c about A is given by:

$$T_c = W_1 l_1 + W_2(l_1 + l_2).$$

Similarly, the anticlockwise turning moment T_A is given by:

$$T_A = N_2 (l_1 + l_2 + l_3).$$

Figure 5.22

5 STRUCTURES IN EQUILIBRIUM

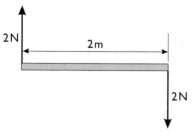

Figure 5.23 A couple

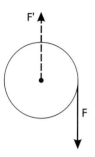

Figure 5.24

But, for equilibrium, the total torque, **T**, must be zero, so

$$\mathbf{T} = \mathbf{T}_c - \mathbf{T}_A = 0,$$

which means that $\mathbf{T}_c = \mathbf{T}_A$

or $\mathbf{W}_1 l_1 + \mathbf{W}_2 (l_1 + l_2) = \mathbf{N}_2 (l_1 + l_2 + l_3).$

TORQUE PROVIDED BY A COUPLE

In Figure 5.23, there is translational equilibrium because the resultant force is zero. However taking moments about any axis shows that there is a turning moment, or torque, of 4 N m. A pair of equal but opposite parallel forces like this is called a **couple** and its turning effect, or torque, is calculated by finding the product of one of the forces and the perpendicular distance between them. The state of rotation of an object is changed when a couple is applied. A single couple will make a stationary object start to spin or a spinning object spin faster or slower. When an object spins at a constant rate, there must either be no couple acting, or the resultant torque must be zero.

A single turning force always gives rise to a couple when it acts on an object that is pivoted. In Figure 5.24, a force **F** applied to the rim of a pivoted wheel gives rise to an equal but opposite force, **F'** at the pivot. The net result is a couple which will change the rotation of the wheel. If the wheel was not pivoted, **F'** would not exist and the force **F** would cause both a change of rotational and translational motion, as Frisbee throwers will know!

5.7 CONDITIONS FOR THE EQUILIBRIUM OF COPLANAR FORCES

An object under the action of a number of forces acting in the same plane will be in:

- translational equilibrium when the sum of their resolved components in any two perpendicular directions is zero, and
- rotational equilibrium when the sum of the clockwise moments is equal to the sum of the anticlockwise moments about any point in the system.

For any system where all the forces act through the same point, the second statement must be true and it just requires that the first is satisfied for equilibrium. Using these rules, we can predict the forces that exist in structures ranging from simple scaffolding systems to suspension bridges.

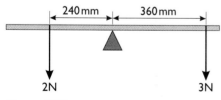

Figure 5.25

Questions

12 a The rule shown in Figure 5.25 weighs 1 N. Explain why the weight of the rule has no moment about the pivot when its centre of mass rests over the knife edge.
b Show that the rule is not balanced.
c Work out where a 4 N weight would have to be placed in order to balance the rule.

13 Using two stands and clamps, some string and some known masses, explain how you could find the mass of one of the stands.

MATTER AND ITS INTERACTION: FORCES

14 One of the arms of a cantilever bridge is shown in Figure 5.26. The arm, of weight 15 MN, is pivoted on a pier at P. One side of the arm rests on an abutment, A, and the other side supports one end of the central span.

a Show that the cantilever arm exerted a force of 4.8 MN on the abutment before the centre span was placed in position.

b When the centre span was in position, the force on the abutment became 4 MN. Calculate the force that the centre span exerts on the cantilever.

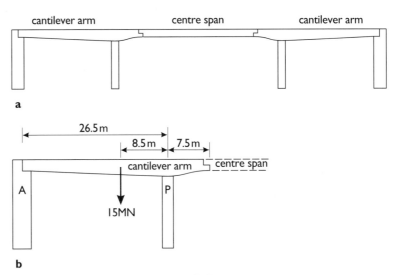

Figure 5.26 **a** A cantilever bridge **b** Dimensions

SUMMARY

A scalar quantity, such as mass, has magnitude only. A vector quantity, such as force, has both magnitude and direction. In a diagram, a vector quantity can be represented in both size and direction by an arrow.

The weight of an object acts through a single point called its centre of mass. When an object rests on a surface, its weight gives rise to a normal force which acts at right angles to the surface.

A point with coplanar forces acting on it will be in equilibrium if the arrows representing the forces form a closed polygon when drawn pointing head to tail in a vector diagram.

To add vectors, draw the arrows representing the vectors pointing head to tail in a vector diagram. The arrow needed to close the polygon, but pointing in the opposite direction around the polygon, will be the sum or resultant.

The part of a vector **F** that has an effect in a direction θ to the vector is given by $F \cos \theta$.

Two surfaces sliding over each other produce a frictional force which opposes their relative motion.

The coefficient of friction, μ is the ratio of the frictional force to the normal force. When the surfaces are just stationary before sliding, μ is the coefficient of static

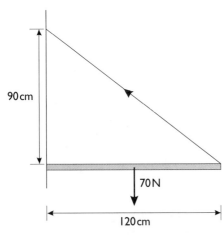

Figure 5.27

5 STRUCTURES IN EQUILIBRIUM

friction. When the surfaces are sliding, μ is the coefficient of dynamic friction. When the forces acting on an object do not pass through a single point, they have a turning effect. The turning effect of a force about an axis is known as its moment or torque. It is given by the product of the force and the perpendicular distance from its line of action to the pivot.

When an object is not accelerating linearly in a given direction, the resultant force on it in that direction is zero and it is in translational equilibrium. When an object is not accelerating about an axis, the resultant moment about that axis is zero and it is in rotational equilibrium.

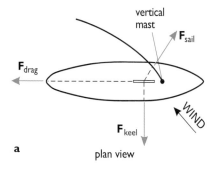

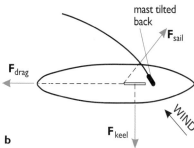

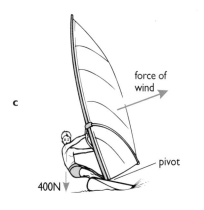

SUMMARY QUESTIONS

15 a A bridge 90 m long is made of a uniform beam of mass 7000 tonnes which rests on a pier at each end. What is the force on each pier?
b Work out the force on each pier when a locomotive of mass 100 tonnes is one third of the way across the bridge. (g)

16 A gardener pulls a garden roller along a level lawn at a steady speed. The mass of the roller is 100 kg and its diameter is 0.6 m. The gardener exerts a force of 200 N on the handle, which is inclined at 45° to the horizontal.
a Work out the horizontal component of the force at the handle.
b What must be the value of the frictional force at the ground?
c Calculate the torque causing the roller to rotate.
d What must the frictional torque be at the axle of the roller?
e Calculate the total reaction at the ground. (g)

17 A trapdoor is supported as shown in Figure 5.27. Calculate the tension in the string, and the magnitudes and directions of the forces at the hinge.

18 Figure 5.28a shows the horizontal forces acting on a moving sailboard. These are: the force of the wind on the sail, which acts about one third of the way up the sail; the force on the centreboard or keel, which is almost perpendicular to the direction of motion; and the frictional force.
a Suppose the sailboarder leans the sail towards the back of the board so that the forces no longer act at a point, as in Figure 5.28b. Which way will the board turn?
b Draw a diagram to show the forces when the sail is moved forwards. What will happen to the board?
c Unlike that in a conventional sailing boat, this sail is normally made to lean into the wind (towards the direction from which the wind is coming). To do this, sailboarders use their weight to produce a moment to counteract the moment due to the wind on the sail, as in Figure 5.28c. If the wind increases, how should the sailboarder react?
d A sailboarder leans so that his weight, 600 N, acts along a line 40 cm away from the base of the mast. Calculate the turning moment of the sailboarder. When the sailboarder is balanced, what is the turning moment of the force of the wind on the sail?
e Advanced sailboards may not be big enough to support the sailboarder while stationary. In Figure 5.28c the force due to the wind is not horizontal and can be resolved into a vertical and a horizontal component. What will be the effect of the vertical component? How is it possible to sail these advanced sailboards? (Look at Figure 5.28d.)

Figure 5.28 a Board in equilibrium
b Tilting mast back to turn
c An advanced sailboard
d A sailboard in action

43

19 Figure 5.29 shows a skier on a drag or 'button' lift. The button is attached to a cable on a spring-loaded drum, which is fixed to an overhead cable. The cable is driven by electric motors via pulley wheels at the top and bottom of the slope.
a On a copy of the diagram, put in the forces acting on a skier as she is pulled up the slope at a steady speed.
b What factors determine the angle, θ that the towing cable makes with the slope?
c The speed of the overhead cable increases slightly and then stays steady at the new value. What effect will this have on the skier and the towing cable?

Figure 5.29

6 FORCES AND MOTION

Forces in equilibrium cause no changes in motion (← 5). An aircraft will remain in level flight with a steady forward motion if the lift on the wings balances the weight of the aircraft and the forward thrust of the engines balances the drag exerted by the air on the aircraft, as in Figure 6.1a. However, for an aircraft to take off the forces cannot be in equilibrium, the thrust of the engines must be greater than the combined weight and drag forces.

Establishing laws on the behaviour of objects under the action of unbalanced forces enables us to predict their behaviour.

Figure 6.1 **a** Concorde in level flight **b** Concorde taking-off

speeds/ms^{-1}	
electromagnetic waves	3×10^8
Earth around the Sun	3×10^3
Concorde	6×10^2
sound in air (STP)	3×10^2
motorway speed limit	30
human walking	1
snail	10^{-3}
growth of human hair	3×10^{-9}

Figure 6.2 Some typical speeds

6.1 SPEED AND VELOCITY

Average speed is defined as the total distance travelled divided by the time taken, and in SI units is measured in metres per second (m s^{-1}). Figure 6.2 shows some values of speeds for different objects, given to one significant figure.

MATTER AND ITS INTERACTION: FORCES

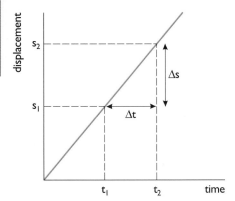

Figure 6.3 Displacement-time graph for car at constant velocity

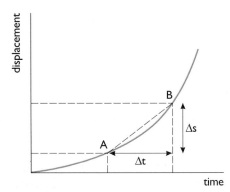

Figure 6.4 Displacement-time graph for accelerating car

distance/m	time/s
100	9.86s
200	19.72s
400	43.29s
800	1m 41.73s
1500	3m 29.46s
5000	12m 58.39s
10000	27m 13.81s
Marathon	2h 10m 0.00s

Figure 6.5 Times for athletics races

Velocity is speed in a given direction, so it is a vector quantity (← 5.1). Displacement is distance moved in a given direction, so it is also a vector quantity. Average velocity is defined as change in displacement divided by time taken.

Figure 6.3 shows a displacement-time graph for a car travelling in a straight line from s_1 to s_2. If we consider the average velocity of the car from s_1 to s_2 then

$$\text{average } v = \frac{\text{change in displacement}}{\text{time taken}}$$

$$= \frac{s_2 - s_1}{t_2 - t_1} \text{ or } \frac{\Delta s}{\Delta t},$$

which is the same as the slope or gradient of the graph.

This graph has a constant slope, so the car moves with constant velocity. In practice, a car will seldom travel very far with a constant velocity, even on a straight motorway. Figure 6.4 shows a displacement-time graph for a car accelerating from rest. The slope of this graph, and therefore the car's velocity, is always changing. The velocity of the car at any moment is called its **instantaneous velocity**.

Questions

1 Figure 6.5 shows typical times to complete short, middle and long distance races in athletics. Calculate the average speed of the athletes in each case. How do you think the actual, or instantaneous, speed might vary during the course of each of the races?

2 In a game of tennis, the ball might leave the racquet on the serve at up to 50 m s^{-1}. Estimate how long it will take to travel the length of the court at this speed. What maximum reaction time will the opponent need to have a chance of returning the serve?

3 The distance between London and Edinburgh is about 550 km. Sketch a graph showing the variation of time taken to travel that distance with the speed of travel. Use speeds ranging from walking pace, which is about 1 m s^{-1}, to that of a passenger aircraft, about 250 m s^{-1}.

In order to calculate average velocity, we measure distance travelled and time taken. A stopwatch can be used to measure a time interval that is sufficiently long to make human reaction time a small error in comparison. For shorter time intervals, a millisecond timer controlled by an optical sensor, such as a phototransistor, can be used. Such a sensor is often called a 'light gate'. The object whose speed is to be measured carries a card of known length. As the card starts to interrupt the light beam to the sensor, the timer is activated. As the card leaves the light beam, the timer is stopped and records the period during which the card was passing. Experiment 6.1 uses this method.

The shorter the card, the closer the average speed recorded will be to an instantaneous speed. However it will also mean that the response time of the sensor will have to be fast and there will be a greater uncertainty in the measurement of the shorter time.

RELATIVE VELOCITY

Imagine sitting in a train travelling north at 50 m s^{-1}. Your velocity relative to the train is clearly zero because you are not moving within the train. Relative to the track and the Earth, your velocity is 50 m s^{-1} in a northerly direction. Suppose

6 FORCES AND MOTION

Instantaneous velocity

On the graph in Figure 6.4, the average velocity of the car from point A to point B is given by:

$$\text{average velocity} = \frac{\Delta s}{\Delta t}.$$

This is the slope of the straight line from A to B. When the time interval is made smaller and smaller, the straight line from A to B becomes a closer approximation to the actual curve. When it is small enough to be considered infinitesimal, the average velocity approaches the instantaneous velocity at A. This can be written

$$v = \lim_{\Delta t \to 0} \frac{\Delta s}{\Delta t}$$

or $v = \frac{ds}{dt}$,

which is called the derivative of s with respect to t.

The instantaneous velocity is, therefore, the derivative of displacement with respect to time. It can be found for a point on a displacement-time graph by working out the slope of the curve at that point. This will be the same as the slope of the tangent to the curve at that point. Alternatively, if the displacement-time equation is known, the velocity-time equation can be found by using the mathematical technique of calculus and differentiating the displacement with respect to time.

you now get up and walk towards the back of the train at 1 m s^{-1}. This speed is relative to the train and in a southerly direction. So, your velocity relative to the Earth is now 49 m s^{-1} in a northerly direction.

All velocities are relative to each other and it is **not** possible to state an absolute value of velocity.

When you are moving, it is useful to be able to work out velocities relative to your own. For example, if you travel along a motorway at 20 m s^{-1} and a car travels in the opposite direction at 30 m s^{-1}, it passes by very quickly. A car overtaking at 30 m s^{-1} takes much longer to pass because its velocity relative to yours is very small. A speed of 20 m s^{-1} is about 45 mph and one of 30 m s^{-1} is about 70 mph.

Suppose a car A travels north at 30 m s^{-1} and another car B travels south at 20 m s^{-1} towards car A. From car A's point of view, the Earth is moving south at 30 m s^{-1} ($-v_A$) carrying a car B which is also moving south at 20 m s^{-1} ($+v_B$) relative to the Earth. So the velocity of B relative to A is $v_B + (-v_A)$ or 50 m s^{-1} south. Similarly, the velocity of A relative to B is 50 m s^{-1} north.

Therefore, **to work out the velocity of B relative to A, subtract the velocity of A from the velocity of B**. Subtracting a vector is the same as reversing its direction and adding it.

$$v_B - v_A = v_B + (-v_A).$$

When vectors do not act along the same line, they must be added according to the vector diagram method (← 5.2).

Questions

4 A car A travels north at 30 m s^{-1} and overtakes a car B also travelling north at 20 m s^{-1}. Calculate the velocity of B relative to A and the velocity of A relative to B.

5 Two Intercity 225 electric high speed trains, one travelling north and the other south, pass each other on adjacent tracks. A passenger in one of them observes that the other train took about two seconds to pass by the carriage window.

Estimate the length of such a train. Then calculate the speed of each train relative to the track, assuming that both were travelling at the same speed.

6.2 ACCELERATION

Whenever an object changes its velocity, it is said to accelerate. This can involve a change in speed or direction or both. Average acceleration is defined as the change in velocity divided by time taken. Its unit is the metre per second squared (m s^{-2}).

	m s^{-2}	g
acceleration due to gravity on the Sun	~270	27
maximum acceleration a human can withstand	~60	6
acceleration due to gravity on Jupiter	~30	3
acceleration due to gravity on the Earth	9.8	1
acceleration of a car starting from rest	2.5	0.25
acceleration due to gravity on the Moon	1.6	0.16

Figure 6.6 Some typical accelerations

MATTER AND ITS INTERACTION: FORCES

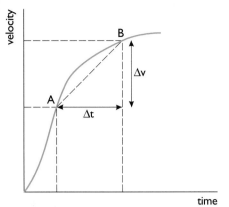

Figure 6.7 Velocity-time graph for a car

Figure 6.6 shows some typical values of accelerations. On the Earth's surface the acceleration due to gravity, g is 9.81 m s^{-2}. This is often used as a reference acceleration, other accelerations being expressed as so many 'g's.

When something has a constant acceleration of 1 m s^{-2} its velocity will increase by 1 m s^{-1} every second, and a graph of its velocity against time will be a straight line. In many real situations, though, accelerations are rarely constant. The acceleration of a car from rest to a cruising speed of 30 m s^{-1} will be far from uniform. In Figure 6.7, the acceleration is not constant. The average acceleration between A and B will be the slope of the straight line from A to B:

average acceleration = $\Delta v / \Delta t$.

As with velocity, the acceleration at any moment is the **instantaneous acceleration**.

Question

6 Figure 6.8 contains data from the *Highway Code* showing stopping distances of a car from various speeds. Using the facts that 70 mph is about 30 m s^{-1} and 3.3 ft is about 1 m, answer the following questions.
a What do 'thinking distance' and 'reaction time' mean?
b Find the average retardation, often called deceleration, of the car in each case.
c Why is the braking distance not proportional to the initial speed of the car? What, do you think, is the relationship between these quantities?

velocity/ mph	thinking distance/ m	braking distance/ m	stopping distance/ m
20	6	6	12
30	9	14	23
40	12	24	36
50	15	38	53
60	18	55	73
70	21	75	96

Figure 6.8 Stopping distances for a car

Instantaneous acceleration

The acceleration of an object at a particular instant of time is its rate of change of velocity with time. So, we can write:

$$a = \frac{dv}{dt}.$$

Now, $v = \frac{ds}{dt}$

so its acceleration is given by $a = \frac{dv}{dt} = \frac{d}{dt}\frac{(ds)}{(dt)}$

$$= \frac{d^2s}{dt^2}.$$

Velocity is the rate of change of displacement with respect to time. So acceleration is the 'rate of change of the rate of change' of displacement with respect to time – the second derivative of displacement s with respect to time t.

MEASURING ACCELERATION

In order to measure the acceleration of, say, a car pulling away from traffic lights, we need to know the change in velocity and the time taken for that change. Calculating change in velocity/time will give us an average acceleration over that time interval.

To measure instantaneous accelerations, some form of accelerometer must be used. Aeronautical engineers use tiny piezo-electric sensors as accelerometers to monitor vibrations in vital parts of an aircraft. The bigger the amplitude or frequency of a vibration, the larger the acceleration and the larger the stresses on the component involved.

6 FORCES AND MOTION

EXPERIMENT 6.1 Measuring accelerations

a Using a ticker-tape timer

This simple device displays the distance travelled by an object in 1/50th of a second intervals (see Figure 6.9). In the experiment, the tape of the timer is attached to a trolley, which is on a slope so that it will accelerate (see Figure 6.10). We can see how the velocity of the trolley changes with time, and work out average velocities over different lengths from

$$\text{average velocity} = \frac{\text{change in displacement}}{\text{time taken}}.$$

Then we can calculate average acceleration from

$$\text{average acceleration} = \frac{\text{change in velocity}}{\text{time taken}}.$$

b Using a computer and light gates

The very high-frequency oscillator in a microcomputer enables it to be used as an electronic timer. A computer also has the advantage that it can be programmed to calculate and display the acceleration. Using a series of accelerating weights or increases in the tilt of the runway, you can find different accelerations.

Figure 6.9 Apparatus for experiment 6.1a

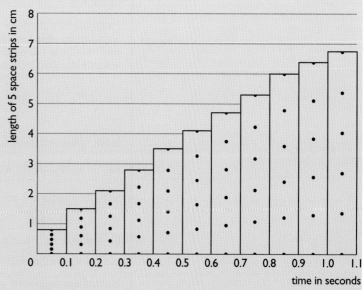

Figure 6.10 A ticker-tape chart

MATTER AND ITS INTERACTION: FORCES

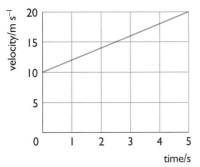

Figure 6.11

Questions

7 Work out the average velocity over each 5-space length shown in Figure 6.10. Use these values to calculate the acceleration
a between the first two strips of tape
b between the last two strips of tape
c between the first and last strips of tape.

8 A car accelerates steadily from 10 m s^{-1} to 20 m s^{-1} in a time of 5 s. The velocity-time graph of its motion is shown in Figure 6.11.
a What is the acceleration of the car in m s^{-2}?
b Explain why the car travels 11 m during the first second of its motion.
c This distance is the area under the velocity-time graph between $t = 0$ and $t = 1$ second. Why?
d How far in total does the car travel during the 5 s it is accelerating?
e Check your answer to **d** using one or more of the equations of motion.
f Suppose that the acceleration was not constant but you had an accurate velocity-time graph of its motion. How would you estimate the distance travelled by the car during the 5 s interval of time? Why could you not use the equations of motion?

9 A fast car is advertised as accelerating from 0 to 60 mph in 8.4 s. Calculate its average acceleration in mph s^{-1} and m s^{-2}. (70 mph ≈ 30 m s^{-1})

10 Using the same time axes, sketch **a** a velocity-time graph and **b** a displacement-time graph of the following motions.
i A trolley accelerates uniformly from rest down a slope. Displacement is measured from the point at which the trolley is released.
ii An air track vehicle moves back and forth at a steady speed along an air track. Displacement is measured from the middle of the track. Consider two complete cycles of motion.

Equations of motion at constant acceleration

Some useful equations can be derived for objects travelling at constant acceleration. Suppose an object starts off with a speed u, and accelerates at a constant rate a for time t to reach a final speed v. Its acceleration is given by change in velocity divided by time taken, so

$$a = \frac{v - u}{t}$$

or, $\quad v = u + at. \qquad (1)$

Now, displacement = average velocity × time

so, $\quad s = \frac{(u + v)}{2} \cdot t. \qquad (2)$

Substituting for v from equation 1 into equation 2 gives,

$\quad s = ut + \frac{1}{2}at^2. \qquad (3)$

By eliminating t from equations 1 and 3, we get,

$\quad v^2 = u^2 + 2as. \qquad (4)$

Note: these equations are useful only for linear motion with constant acceleration. In real situations, accelerations invariably are changing. Also, the quantities u, v, a and s are all vectors, so once a particular direction has been chosen as positive, negative values of these quantities indicate the opposite direction.

6 FORCES AND MOTION

MOTION UNDER GRAVITY

Without air resistance, all objects in free fall near the surface of the Earth move towards the Earth with a constant acceleration. This acceleration, known as the acceleration due to gravity, is denoted by the letter g.

Its value at sea level varies slightly from about 9.78 m s^{-2} at the equator to 9.82 m s^{-2} at the poles. The variation occurs for two reasons:

- the Earth is not a sphere – at the poles, the surface is about 10 km closer to the Earth's centre than at the equator ($\rightarrow$ 15.1)
- the Earth is rotating with a point on the equator moving at a speed of around 320 m s^{-1} (1700 km h^{-1}) relative to the poles. ($\leftarrow$ 6.1)

For most purposes we may take g as 10 m s^{-2}.

The acceleration due to gravity is the same for any object, irrespective of its size, shape or mass. This can be demonstrated in the laboratory by releasing different objects at the same instant in a vertical glass tube. If the objects are, for example, a feather with a small steel hook and a ball bearing, they can be released simultaneously using an electromagnet. When the tube is evacuated, the feather and the ball bearing reach the bottom of the tube at the same instant, whereas in air, the feather takes longer to fall.

The equations for motion at constant acceleration, derived above, can be applied to objects falling freely near the Earth provided that air resistance is negligible. This will be so when the object is made from dense material and its speed is low. For example, when someone dives from the top board in a swimming pool, their motion is little influenced by air resistance. Even as they accelerate towards the water, they are simply not moving fast enough to experience significant air resistance.

Air resistance opposing the motion of a moving object increases with speed, being roughly proportional to the square of the speed of the object. If the velocity of a falling object becomes so great that the frictional force balances the force of gravity on it, it will reach a constant velocity called the **terminal velocity**. A free-fall parachutist reaches a terminal velocity of around 60 m s^{-1} after many seconds. However, this reduces to a velocity of less than 9 m s^{-1} after the parachute has opened.

Figure 6.12 The Earth

Figure 6.13 Free-fall parachutists

MATTER AND ITS INTERACTION: FORCES

To see an object reach its terminal velocity, drop a small ball bearing into a measuring cylinder containing a viscous liquid, such as glycerol or washing up liquid. The resistive forces on the bearing are sufficient to balance its weight at a much lower speed than when travelling in air, so it stops accelerating within a few centimetres.

EXPERIMENT 6.2 Measuring the acceleration due to gravity

a Using a computer

Using the apparatus described in experiment 6.1b, you can find the value of the acceleration due to gravity.

b Using free fall apparatus

By setting up the apparatus shown in Figure 6.14, you can find the time t it takes for an object, such as a steel ball, to fall a known distance s. This enables you to find a, the acceleration, from:

$$s = ut + \tfrac{1}{2}at^2$$

where u, the initial velocity, is usually zero. Figure 6.15 gives some typical values.

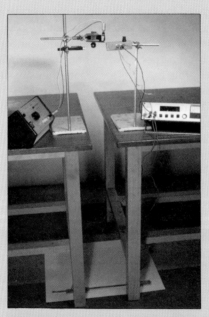

distance fallen/m	average time/s
0.84	0.421
0.78	0.410
0.73	0.397
0.68	0.381
0.63	0.366
0.58	0.352

Figure 6.14 Apparatus for experiment 6.2b

Figure 6.15 Some typical results

Questions

11 A stone, dropped from the top of a well, takes 2.50 s to reach the water below. How deep is the well? How fast is the stone travelling as it hits the water, assuming that the effect of the air is negligible? (g)

12 A ball, thrown vertically upwards, takes 2.0 s to reach its highest point.
a At what speed was the ball projected?
b What maximum height did it reach?

6 FORCES AND MOTION

c What height would the ball have reached had it been projected at the same velocity on the Moon, where g is only 1.6 m s^{-2}.(g)

A NUMERICAL METHOD FOR STUDYING MOTION

The displacement-time graph for an object can be calculated numerically, if its acceleration and initial motion are known.

In Figure 6.3, the displacement, s_2 at the end of a time interval, Δt, equals the displacement at the beginning, s_1, plus a small increase, Δs.

$$s_2 = s_1 + \Delta s.$$

Since $v = \Delta s / \Delta t$,

then $s_2 = s_1 + v\Delta t$.

Suppose an object is travelling at a constant speed of 5 m s^{-1}. By choosing a value for the time interval, say 1 s, successive calculations show how the displacement from the start changes with time.

Initially $s_0 = 0 \text{ m}$

and $t_0 = 0 \text{ s}$.

Now $v = 5 \text{ m s}^{-1}$

and $\Delta t = 1 \text{ s}$

so $v\Delta t = 5 \text{ m}$.

After each second, the displacement will have increased by 5 m. A table can be drawn up as in Figure 6.16.

This is a simple example but the method can be used for more complex types of motion. Under constant acceleration, the velocity is not constant. In the same time interval Δt, the change in velocity Δv will be $a\Delta t$,

so $v_1 = v_0 + a\Delta t$.

displacement/m	time/s
0	0
5	1
10	2
15	3

Figure 6.16 Numerical analysis of motion at constant speed

Suppose an object accelerates from rest with a constant acceleration of 5 m s^{-2}. Initially, $s_0 = 0$, $v_0 = 0$ and $t_0 = 0$.

Since $a = 5 \text{ m s}^{-2}$

and $\Delta t = 1 \text{ s}$,

then $\Delta v = a\Delta t$

$= 5 \text{ m s}^{-1}$.

So, after the first second, the velocity will have increased by 5 m s^{-1}.

Because the object is accelerating, the average velocity, 2.5 m s^{-1}, must be used to calculate the displacement. After 1 s, the displacement s_1 is given by

$s_1 = v\Delta t$

$= 2.5 \text{ m}$.

So after the first second, the displacement will have increased by 2.5 m. By the same process, in the next second, the velocity will change from 5 to 10 m s^{-1} with an average of 7.5 m s^{-1}, so

$s_2 = 2.5 + 7.5$

$= 10 \text{ m}$.

MATTER AND ITS INTERACTION: FORCES

Figure 6.17 shows some results for this second example. Computers can perform such repetitive calculations quickly and accurately. Spreadsheets and modelling programs that program a computer to carry out numerical analysis are available.

displacement/m	time/s	velocity/ms^{-1}	average velocity/ms^{-1}
0	0	0	
2.5	1	5	2.5
10	2	10	7.5
22.5	3	15	12.5

Figure 6.17 Numerical analysis of motion at constant acceleration

Question

13 For both examples in 'A numerical method for studying motion', continue working out the values in the tables to at least 6 s from the start. Then sketch graphs to show how the displacement, velocity and acceleration vary with time. Does it matter what value is chosen for Δt?

6.3 PROJECTILES

Figure 6.18 A stroboscopic photograph of balls falling, (taken at 0.02s intervals)

Figure 6.18 shows a stroboscopic photograph of two balls which were released at the same time. Such photographs are taken by illuminating an object using a flashing light, and with the shutter of the camera left open for the duration of the motion. In this case, the ball on the left was dropped while that on the right was given a horizontal velocity. We can see that, at each instant shown, the vertical displacements of the two balls are identical. The only difference is that the ball on the right acquires an increasing horizontal displacement and falls along the path of a parabola. Its horizontal and vertical motions can be treated separately. Ignoring air resistance, the only force acting on the ball on the right is gravity, which accelerates it vertically downwards. So, its vertical velocity changes in the same way as the ball on the left.

Questions

14 A golf ball is struck so that it leaves the head of the club with a speed of 50 m s^{-1} at an angle of 30° to the ground. Assume air resistance effects are negligible.
a What is the vertical component of the velocity?
b What maximum height is reached by the golf ball?
c For how long is the ball in the air?
d What is the horizontal component of the velocity?
e How far away from the golfer does the ball land?
f The golf ball is not smooth but has 'dimples' over its surface. How might these affect its motion? (g)

Shape of projectile trajectory

The vertical displacement of the ball on the right in Figure 6.19 is given by

$$s_y = \tfrac{1}{2}gt^2.$$

Horizontally no forces act, and the projectile proceeds with a constant horizontal velocity, v_x. The horizontal displacement is given by

$$s_x = v_x t.$$

Eliminating the time of flight from these two equations gives

$$s_y = \frac{g}{2v_x^2} s_x^2.$$

So, s_y is proportional to s_x^2, a relationship that gives the observed parabolic trajectory.

---

15 A projectile is fired into the air with a speed of u m s^{-1} at an angle of $\theta°$ to the ground.
a What is the vertical component of the velocity?
b Show that the maximum height s_y and the time of flight t are given by:

$$s_y = u^2 \sin^2 \theta / 2g$$
and $t = 2u \sin \theta / g.$

c What is the horizontal component of the velocity?
d Show that the range, or horizontal distance travelled, s_x, is given by:

$$s_x = u^2 \sin 2\theta / g.$$

e At what angle to the ground must the projectile be fired to attain the maximum range for a given velocity of projection?
f A long jumper reaches a speed of about 10 m s^{-1} at the end of her 'run-up' and takes off at an angle of 25° to the horizontal. What value would this give for the length of the jump? Are the values for her speed and angle of take-off reasonable? What other factors might be involved in making a more accurate prediction of the jump? (*g*)

6.4 NEWTON'S LAWS OF MOTION

Isaac Newton realised that unbalanced forces cause acceleration. An unbalanced force arises when the forces acting on an object are not in equilibrium, leaving a resultant force. He formulated a theory for the motion of objects due to the forces acting on them. The theory is based on three laws known as Newton's laws of motion. Newton's laws are adequate for speeds that are low compared with the speed of light. For fast-moving objects, such as atomic particles in an accelerator, **relativistic mechanics** must be used ($\rightarrow$ 9.6).

Figure 6.19 Isaac Newton

Isaac Newton

Born in 1642, Isaac Newton has probably contributed more to the fields of mathematics and physics than any other person, living or dead. He is particularly noted for his work on light and colour, and for his formulation of the laws of motion and gravitation. In his studies on velocity and acceleration, he invented a form of differential and integral calculus, which he called the method of 'fluxions', to enable motions to be calculated more precisely. He did his best creative work in the two years 1665–66 when he was staying at Woolsthorpe, near Grantham in Lincolnshire, because of The Plague. Among the offices he held were President of the Royal Society and Master of the Royal Mint.

DEVELOPMENT OF THE FIRST LAW OF MOTION

Newton's first law was not entirely his own creation. Galileo conducted a thought experiment in which he imagined the motion of a ball rolling down a

V-shaped slope as in Figure 6.20a. Because of friction, the ball does not quite reach the same height on the other side. However, without frictional forces, the ball would reach exactly the same height on the other side and could slide there without rolling. Suppose the other side did not have such a steep slope, as in Figure 6.20b. The ball would travel further along the slope to reach the same height. If the other side had no slope at all but was horizontal, as in Figure 6.20c, the ball would travel forever!

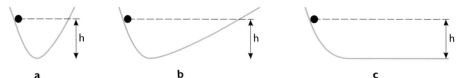

Figure 6.20 a, **b** and **c** Galileo's "thought" experiment

Newton's first law: Unless acted upon by an unbalanced force, an object remains at rest, or continues moving with a constant speed in a straight line.

This law is difficult to verify, and may also seem to contradict everyday experience. You will have found that a force is needed to keep an object moving at a constant speed in a straight line. For example, if you are cycling along a level road and then stop pedalling, you eventually stop moving. But this is because frictional forces are always present. So when you stop pedalling there is an unbalanced force acting which opposes the motion of the bicycle. You can show the effect of reducing friction by placing a small block of material on different surfaces. For a similar push, the block will travel the greatest distance before stopping on the smoothest surface ($\rightarrow$ 9.5).

Newton's first law implies that objects tend to resist changes to their state of motion. This property is called **inertia**, and is demonstrated whenever a moving vehicle stops suddenly. The passengers continue moving until unbalanced forces, such as the tension in a seatbelt, stop them. Similarly, when a stationary vehicle starts to move the passengers tend to remain at rest, as shown by the motion of passengers standing in an accelerating bus or an underground train.

Questions

16 Modern car safety belts are called 'inertia-reel' belts. They allow freedom of movement in the car seat until the car speeds up, slows down or changes direction. How do you explain their behaviour in terms of inertia and Newton's first law? Find out how the belts work.

17 It is a common misconception that a force is required to keep a projectile moving. The impression gained by many people is that as the force gets 'used up', the projectile slows down. What really happens?

18 Spin **a** a raw egg and **b** a hardboiled egg on a smooth tray. What differences do you notice in their behaviour? Momentarily stop them by placing a finger on the top of each egg as they spin. What difference do you notice in their behaviour now? What has this to do with inertia and Newton's first law of motion?

Large masses need a larger force than small masses to undergo the same acceleration. For this reason, the mass of an object is often referred to as its inertia. It is measured in kilograms (kg) in the SI system of units. The standard kilogram is the mass of a cylinder of platinum which is kept in a laboratory at Sèvres, Paris. There are replicas in other countries which are compared with the standard from time to time.

MOMENTUM

For his second law, Newton needed to quantify the 'amount of motion' possessed by an object. This is now called **momentum**. It is defined as the product of the mass of the object and its velocity, and is a vector quantity.

Momentum = mass × velocity
(in kg m s^{-1}) (in kg) (in m s^{-1})

So a 10 tonne truck travelling at 30 m s^{-1} has more momentum than a 1 tonne car with the same velocity. Momentum is a particularly useful quantity when dealing with motions involving collisions or explosions (→ 8), situations in which all the forces are between the interacting objects.

Newton's second law: When an unbalanced force acts on an object, the object gains momentum in the direction in which the force acts. The change of momentum is proportional to the size of the force and the time for which it acts.

From this statement,

$$\text{force} \propto \frac{\text{change of momentum}}{\text{time taken}}.$$

Since momentum = mass × velocity, if the mass is constant

$$\text{force} \propto \text{mass} \times \frac{\text{change of velocity}}{\text{time taken}}$$

or $\qquad \propto \text{mass} \times \text{acceleration}$

$$F \propto ma.$$

So the acceleration of an object is proportional to the net force acting on it as long as its mass is not changing. Similarly, when objects of differing masses are accelerated equally each requires a force proportional to its mass. A 2 kg mass requires a force twice that for a 1 kg mass and four times that for a 0.5 kg mass, and so on.

We use the relationship between force, mass and acceleration to define the newton, the unit of force in the SI system. The **newton, N**, is defined as the force that gives a mass of 1 kilogram an acceleration of 1 metre per second each second.

This means, for example, that if an object has mass m of 1 kg and a force of 1 N is applied, then its acceleration a is 1 m s^{-2}. Alternatively, if m = 3 kg and a = 2 m s^{-2} then the force required F = 6 N.

$\qquad F \ = \ m \ \times \ a$
$\qquad$ (in N) (in kg) (in m s^{-2})

or $\qquad F = ma.$

In general $\quad F = \dfrac{d(mv)}{dt}$

which enables mass changes to be taken into account.

MATTER AND ITS INTERACTION: FORCES

IMPULSE

A tennis player hitting a ball follows the stroke through with the racket. This means that the ball is acted on by the force for a longer time and acquires a greater velocity. Multiplying the value of a particular force acting on an object by the period of time for which it acts produces a quantity called the **impulse** of the force.

Impulse = force × time.
(in N s) (in N) (in s)

From Newton's second law, force is proportional to the rate of change of momentum. If we measure the force in N then:

$$\text{force} = \frac{\text{change in momentum}}{\text{time}}.$$

So the impulse of a force on an object equals the change in its momentum, or

impulse = change in momentum.

In fact, we can express Newton's second law as: the change in momentum of an object is equal to the impulse it experiences.

Relating impulse and momentum

Force is equal to the rate of change in momentum, so

$$F = \frac{\text{change in momentum}}{\text{time}}$$

or $F = \frac{d(mv)}{dt}$.

For a mass m, which undergoes a change in velocity from v_1 to v_2,

$$F = \frac{mv_2 - mv_1}{\Delta t}.$$

So if a force F acts for a time Δt,

$F\Delta t = mv_2 - mv_1$

or impulse = change in momentum.

Questions

19 Imagine jumping off a wall, about a metre in height, so that it takes you about 0.5 s to reach the ground.
a Calculate your velocity on reaching the ground.
b Calculate your momentum on reaching the ground.
c What is your change in momentum as you come to rest?
d What impulse do you receive from the ground?
e Suppose you bend your knees as you come to rest and that the decelerating force acts on you for 0.5 s. Calculate the average value of this force.
f What might be the consequences of not bending your knees so that the decelerating force acts for a much shorter time? (*g*)

20 In 1972, a Czechoslovakian air stewardess fell, without a parachute, from an aircraft flying at an altitude of 10 000 m. She landed in a bank of snow and survived. What nature of individual 'impulses' must she have experienced to have survived?

EXPERIMENT 6.3 Verifying F = ma

Using ticker-tape timers or a suitably programmed computer and light gates, as in experiment 6.1b, you can the measure the acceleration of dynamics trolleys on a runway or gliders on an air track. Tilting either of these enables you to compensate for friction.

There are three variables in the equation $F = ma$, so the experiment needs to be done in two parts.

a With mass constant, see how acceleration depends on force

Weights suspended by a string over a pulley enable you to apply a set of forces of different values to a vehicle of a certain mass. Figure 6.21 shows some results obtained.

force/ N	acceleration/ cm s^{-2}
0.1	18.1
0.2	38.4
0.3	58.2
0.4	76.3
0.5	94.5

Figure 6.21 Results obtained from experiment 6.3a

mass/kg	acceleration/ cm s^{-2}
1	95.1
2	48.7
3	31.0
4	24.5
5	17.2

Figure 6.22 Results obtained from experiment 6.3b

Plotting a graph of force against acceleration enables you to find the relationship between acceleration and force.

b With force constant, see how acceleration depends on mass

Applying just one force to trolleys or gliders of different mass enables you to obtain results similar to those in Figure 6.22.

You can then plot a graph of acceleration against mass. The shape of this graph indicates that acceleration and mass may be inversely proportional. Plotting acceleration against 1/mass should show you that this is indeed likely.

MASS AND WEIGHT

The **weight** of an object is the force of gravity on it, so weight is measured in newtons. Since the gravitational field ($\rightarrow$ 15.1) varies from place to place, the weight of an object will depend on where it is. Its mass does not depend on position.

Using $F = ma$, the weight W of a mass m is given by $W = mg$. Where the acceleration due to gravity is 10 m s^{-2}, the weight of a 1 kg mass is about 10 N.

Because $F = ma$, a mass m can be measured by applying a force and comparing the acceleration it acquires with the acceleration of a standard mass m_s under the same force. Then

$$ma = m_s a_s$$

so, $m = \dfrac{m_s a_s}{a}$.

In practice, mass is conveniently found by using a spring balance, which is a forcemeter calibrated in mass units. This works because mass is proportional to weight, as long as g remains constant. A spring balance or an electronic balance has to be calibrated using a standard mass. A beam balance compares the weight of a mass with that of a standard mass and, therefore, works anywhere without the need for calibration.

Questions

21 A simplification has been made in experiment 6.3: actually the pull of gravity on the falling mass accelerates not only the trolley's mass but also its own mass. The total mass of the accelerated object is the sum of these two masses. Can this simplification be justified?

22 a A forcemeter with a 1 kg mass on it reads 9.8 N on Earth. What value would it display on the Moon where the acceleration due to gravity is 1.6 m s^{-2}?

b At the equator, g is 9.78 m s^{-2} so that the Earth pulls on each kilogram with a force of 9.78 N. In Britain, g is 9.81 m s^{-2}. How much do you weigh? What is your mass? Suppose you measure your weight on some bathroom scales in Britain, and then travel with them to Kenya, on the equator. How many kg will the scales indicate you have lost?

MATTER AND ITS INTERACTION: FORCES

23 Astronauts on extended space flights need to carefully monitor their body mass. However, methods employed on Earth would be unsuitable. Why is this? How, in practice, is body mass measured?

NEWTON'S THIRD LAW

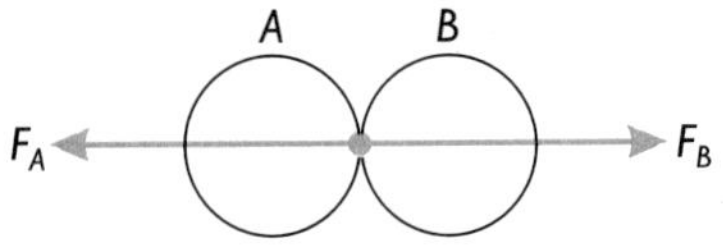

Newton realised that a force is just one side of an **interaction**. For example, when an air-gun is fired, the gun exerts a force on the pellet, accelerating it in one direction, and the pellet exerts an equal but opposite force on the gun, causing it to recoil. When you walk, your feet interact with the surface of the Earth pushing back along the surface. The Earth's surface provides an equal forward force on your feet. Of course, the effect of the force on the Earth is negligible because of its large mass.

Newton's third law: Whenever an interaction occurs between two objects, each object exerts the same force on the other, but in the opposite direction and for the same length of time.

Put another way, both objects receive the same impulse. The impulse received by one body is identical but in the opposite direction to that received by a body with which it interacts.

Figure 6.23 shows an instant during the collision of two objects, B exerts a force F_A on A, which changes A's momentum, while A exerts an equal but opposite force $-F_B$ on B, which changes B's momentum. The forces must act for the same length of time – that of the interaction. The product of force and time gives the impulse, or change in momentum. So the change in A's momentum must be equal and opposite to the change in B's momentum. In other words, the **total momentum** of both A and B **remains constant** before and after the impact.

Additionally, we can speak of **paired forces**. For example, there are two pairs of forces involved when a book sits on a table – gravitational forces between the book and the Earth and the forces between the book and the table, as shown in Figure 6.24.

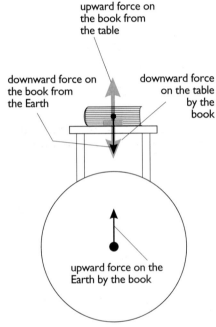

Figure 6.23 Newton's third law

Figure 6.24 Paired forces

Question

24 Draw the two pairs of forces acting in each of the following cases.
a A rocket leaves the Earth's surface against gravitational pull because of the thrust produced by the high velocity particles emerging from its base.
b Standing in a lift which descends from one floor to another causes your apparent weight to:
i   decrease
ii return to normal
iii increase
iv return to normal when the lift stops.

As shown above, the law of conservation of momentum between two objects ($\rightarrow$ 8) is another way of stating Newton's third law. Broadening this, imagine a system in which many masses are moving and interacting with each other through collisions and fields. The momentum of the whole system is conserved when the resultant force on its centre of mass is zero. All three of Newton's laws need to be considered here, of course, not just the third one.

Question

25 Newton wrote the following example to explain the third law of motion. 'If a horse draws a stone tied to a rope, the horse (if I may so say) will be equally drawn back towards the rope: for the distended rope, by the same endeavour to relax or unbend itself, will draw the horse as much towards the stone as it does the stone towards the horse, and will obstruct the progress of the one as much as it advances that of the other.'

Explain how a horse is able to pull a stone along.

SUMMARY

Velocity is the rate of change of displacement with time and acceleration is the rate of change of velocity with time. Objects moving under Earth's gravity at its surface have a constant acceleration of approximately 10 m s^{-2}, where the effect of air resistance can be ignored.

Provided that their speed is small compared with the speed of light c, the motion of all objects is described by Newton's laws of motion. These are:

Law 1 The velocity of an object will be constant if the unbalanced or resultant force on it is zero.

Law 2 When an unbalanced force acts on an object, the object gains momentum in the direction in which the force acts, and the rate of change of momentum is proportional to the size of the force.

Alternatively, we can say:

The change in momentum of an object is proportional to the impulse acting on it.

The momentum of an object is the product of its mass and its velocity.

The impulse provided by a force is the product of the force and the time for which it acts.

Law 3 When two objects interact, they exert equal and opposite forces on each other for the same length of time, and so receive equal and opposite impulses.

This can also be expressed as conservation of momentum: when objects exert forces on each other, the total momentum remains unchanged.

SUMMARY QUESTIONS

Assume air resistance is low enough to be neglected.

26 Use the measurements below to work out the average speed of the object and the uncertainty in the result.
a A bullet, which travels a distance of 5 m, measured with a metre rule, in a time of 10 ms, taken by a millisecond timer.
b A dynamics trolley, which travels a distance of 2 m, measured with a metre rule, in a time of 2.00 s, taken by a person with a reaction time of 0.2 s using a stopwatch.

27 The take-off speed of a jet is 360 km h^{-1}. What is this in m s^{-1}? The jet takes off from a runway 2100 m long. Assuming that the jet's acceleration is constant, calculate its value.

28 In a motorway accident, a car left skid marks 60 m long. Friction between the tyres and the road surface cause a maximum deceleration of approximately 10 m s^{-2}. Calculate the initial speed of the car before braking.

29 A racing car travelled a quarter of a mile (402 m) from rest in 6 s.
a Calculate the average acceleration of the car.
b The car reached a final velocity of 100 m s^{-1}. Show that its acceleration was not uniform.

30 A ball is thrown upwards with an initial velocity of 30 m s^{-1}.
a Work out the greatest height reached by the ball and how long it takes to get there.
b Work out the velocity and height of the ball after 4 s. ($g = 10$ m s^{-2})

31 A stunt rider wants to clear 11 parked cars by driving off a ramp as shown in Figure 6.25. Calculate the minimum speed at which his motorcycle must leave the ramp. (g)

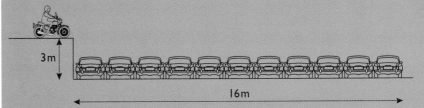

Figure 6.25 A stunt rider wants to clear 11 parked cars

32 A bullet is fired into the air with a speed of 200 m s^{-1} at an angle of 30° to the ground.
a What is the vertical component of the velocity?
b Work out how long it will take to reach the maximum height and, therefore, the time of flight.
c What is the horizontal component of the velocity?
d Work out the horizontal distance travelled by the bullet before it hits the ground.

33 At lift-off, the mass of the Space Shuttle Columbia and its fuel was 2×10^6 kg. It was used to send satellites into orbit around the Earth and carried a payload of 29 500 kg. The Shuttle was launched by two solid rocket boosters, each exerting a thrust of 11.8×10^6 N, and three liquid fuel rocket engines, each exerting a thrust of 1.6×10^6 N.
a What was the total downward force on the Shuttle at launch?
b What was the total upward force?
c What was the resultant force on the Shuttle?
d What was the initial acceleration of the Shuttle?

34 In a car accident simulation, a car was driven at 13 m s^{-1} (about 30 mph) into the back of a stationary car and the crash was filmed using high-speed filming techniques.
a The car took 0.2 s to stop. What was its average acceleration?
b How far did the main body of the car travel before coming to rest?
c The mass of the car was 850 kg. Calculate its change in momentum.
d What was the average force on the car during the crash?
e The mass of the driver was 65 kg. What was the driver's change in momentum?
f The driver came to rest in 0.1 s. What was the average force exerted by the seatbelt?

7 CIRCULAR MOTION

Satellite television and worldwide telephone communications depend on the ability of scientists to place satellites into a circular orbit around the Earth at about 43 000 km from its centre. Such satellites move at a steady speed around their orbits, appearing to hover above the same fixed point on the equator. There are many other situations involving circular or near circular motion, including planets orbiting the Sun, a car negotiating a roundabout, particles in a synchrotron and a passenger on the 'big wheel' at a fairground.

Newton's laws of motion (← 6.4) can be applied to motion in a circle, and used to predict and explain the behaviour of objects moving in circular paths or rotating.

7.1 OBJECTS MOVING IN A CIRCULAR PATH

From Newton's first law of motion (← 6.4), we know that every object remains at rest or travels in a straight line at constant speed unless a resultant force acts on it. When an object travels in a circle, its direction is constantly changing, it is accelerating, even though it is moving at constant speed. A resultant force must be acting on it to change its direction from the straight line path.

Figure 7.1a shows a gun firing a shell from the top of a tower. The force due to gravity acts vertically downwards on the shell, which follows a nearly parabolic path towards the sea. If it were not for air resistance, the path would be a perfect parabola.

Suppose we now take the gun to a much higher position, anywhere above the atmosphere, and fire the shell at a higher speed. The shell will, as before, fall in a curved path towards the Earth. However, at a particular speed of projection the shell falls towards the Earth but gets no closer to it because the Earth 'falls away' from beneath the path of the shell and the shell becomes a satellite of the Earth, as in Figure 7.1b. It does not change its distance from the surface, neither does it gain nor lose gravitational potential energy. The only force on the shell is that due to gravity, which always acts at right angles to the shell's motion at any instant. Since there is no component of the gravitational force along the direction of motion, the force only changes the direction of motion of the shell. Its kinetic energy stays constant and it orbits at a steady speed. But because it moves in a curved path, its velocity is changing. The shell is accelerating in the same direction as the force – towards the centre of the Earth.

A rocket used to launch a satellite from the surface of the Earth has two purposes:

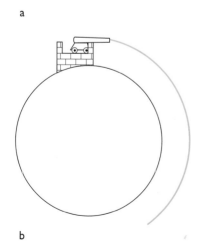

Figure 7.1   **a** Firing horizontally from a tower **b** Firing horizontally above the atmosphere

i   to increase the gravitational potential energy of the satellite and
ii to give it the correct tangential velocity to insert it into orbit.

MATTER AND ITS INTERACTION: FORCES

Once the satellite is high enough and travelling at the appropriate speed parallel to the surface of the Earth, it freely orbits the Earth. Objects in the satellite appear 'weightless' because they, as well as the satellite, are falling towards the Earth. The satellite orbits the Earth **because** of gravity, not through lack of it. Without gravity, the craft would continue in a straight line at a steady speed.

Any object that moves in a circular path must have a constant force on it at right angles to its direction of motion at all times. It is always accelerating towards the centre of the circle, and is said to have a **centripetal**, or centre-seeking, acceleration. If an object is travelling with speed v in a circle of radius r, the magnitude of the centripetal acceleration is v^2/r.

Centripetal acceleration

Figure 7.2a shows a particle travelling at constant speed v in a circle centre O of radius r. At A, the velocity of the particle is represented by $\mathbf{v}_1$. After a time interval Δt, the particle is at B where its velocity is $\mathbf{v}_2$. The velocity at B must equal the velocity at A plus the change in velocity. The change in velocity, $\Delta \mathbf{v}$, is shown in the vector diagram of Figure 7.2b,

where $\mathbf{v}_1 + \Delta \mathbf{v} = \mathbf{v}_2$.

using the rule for adding vectors (← 5.3).

The angle AOB has a value θ. The direction of the velocity changes by angle θ. Triangle AOB is similar to the triangle XYZ, as they are both isosceles triangles, and so angle AOB equals angle XYZ. This means that:

$$\frac{XZ}{XY} = \frac{AB}{AO}$$

or $$XZ = \frac{AB.XY}{AO}$$

so $$\Delta v = AB . \frac{v}{r}.$$

Now acceleration $= \frac{\Delta v}{\Delta t}$

$= \frac{AB}{\Delta t} . \frac{v}{r}.$

As we take a smaller and smaller time interval Δt, B moves closer to A and θ becomes very small. As θ becomes closer to zero, line AB becomes equal to the arc of the circle Δs.

So acceleration $= \frac{\Delta s}{\Delta t} . \frac{v}{r}.$

Because speed equals rate of change of distance:

$$v = \frac{\Delta s}{\Delta t}$$

and acceleration $= \frac{v^2}{r}.$

Since θ is very small, points A and B almost coincide. The change in velocity must be at right angles to v_1 – towards the centre of the circle. So for the particle to travel in a circular path it must have an acceleration of v^2/r towards the centre of the circle.

Question

1. A satellite orbits the Earth at a constant distance of 200 km above its surface. It accelerates towards the centre of orbit at 9.8 m s^{-2}.
   a Using the equation for centripetal acceleration, find how fast the satellite is travelling around its orbit.
   b How long will the satellite take to complete one orbit? (E$_R$)

ANGULAR VELOCITY

When a stone is whirled around on a string it travels at constant speed v in a circle of radius r. The string sweeps out the same angle in equal time intervals. Angular velocity is the angle swept out in radians per second and is usually denoted by ω. The linear speed of the stone and its angular velocity are related by the equation, $v = \omega r$.

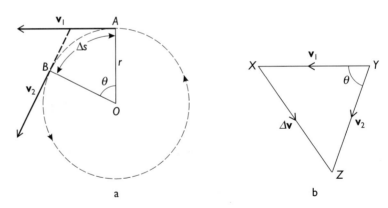

Figure 7.2 a Showing centripetal acceleration is v^2/r
b Δv is the difference between the vectors v_1 and v_2

Questions

2. The drum of a spin drier rotates at 600 revolutions per minute. Calculate:
   a the angular velocity of the drum,
   b the speed of a point on the edge of the drum, if the drum is 400 mm in diameter,
   c the acceleration of this point towards the central axis of the drum.

3. A toy train moves around a circular track of diameter 0.7 m at 6 revolutions per minute.

7 CIRCULAR MOTION

Angular velocity

Look back at Figure 7.2a. In t seconds the particle has travelled a distance s along the circumference, while the radius has swept through an angle of θ radians. Angular velocity is the rate of change of angle, so:

$$\omega = \frac{\theta}{t}.$$

An angle in radians is defined as the ratio of arc to radius, so

$$\theta = \frac{s}{r},$$

giving $\omega = \dfrac{s}{rt}.$

Using $v = \dfrac{s}{t},$

then $\omega = \dfrac{v}{r},$

or $v = \omega r.$

a How long does it take for each revolution?
b What is the speed of the train?
c What is its angular velocity?
d What is its centripetal acceleration?

CENTRIPETAL FORCE

A particle of mass m travelling at constant speed in a circle has an acceleration towards the centre given by:

$$a = \frac{v^2}{r}.$$

From Newton's second law of motion, $F = ma$. So a resultant force of:

$$F = m \cdot \frac{v^2}{r} = m \cdot \omega^2 r$$

must act on the object. The force must act in the same direction as the acceleration it causes – towards the centre of the circle. This is the **centripetal force**. In the case of a stone on the end of a piece of string, it is provided by the tension in the string. If the string breaks causing the centripetal force to disappear, the stone will move off in a straight line at a tangent to the circle.

For a car travelling in a curved path, the centripetal force is usually produced by the frictional force between the tyres and the road surface, as shown in Figure 7.3. If this frictional force is not large enough – for example, if the car is trying to corner on ice – the car cannot turn. Instead, it tends to carry on moving in a straight line and skids.

Sometimes cornering does not rely only on friction. The road can be 'banked' as in some car and cycle race tracks (see Figure 7.4). Figure 7.5a shows the relevant forces acting on the car. The vector diagram of Figure 7.5b shows the resultant of the normal force and the weight of the car, which provides the centripetal force.

Figure 7.3 Frictional force causing a car to turn on a bend using its wheels

65

MATTER AND ITS INTERACTION: FORCES

Figure 7.4 A car on a banked track

An aircraft has a lift force caused by the different rates of air flow above and below the wings. The lift force balances its weight when it is flying on a straight and level path. In order to bank, the rudder is used to point the aircraft in a different direction, and the ailerons cause it to tilt. The speed is also increased, which increases the lift. The aircraft then has a horizontal resultant force at right angles to its motion, which causes it to turn (see Figures 7.6a and b).

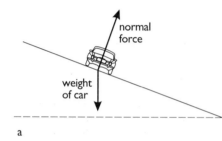

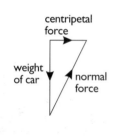

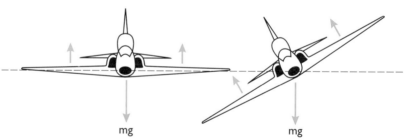

Figure 7.5 a The forces acting on a car on a banked track **b** The vector diagram of these forces

Figure 7.6 a Forces on an aeroplane in level flight **b** Forces on an aeroplane banking

Questions

4 An aircraft from the Red Arrows display team makes a steeply banked horizontal turn of radius 200 m. It is travelling at a speed of 100 m s^{-1} (230 mph). Calculate the centripetal acceleration of the aircraft and its pilot. How many times gravitational acceleration is this? Supposing the mass of the pilot is 80 kg, estimate the force exerted on him by the seat of the aircraft in order that he maintains this path. (g)

5 The Moon orbits the Earth at an average radius of 3.82×10^8 m. It takes 27.3 days to complete an orbit.
a What is the centripetal acceleration of the Moon towards its centre of orbit?
b Show that your answer to **a** is consistent with the inverse square law for the Earth's gravity ($\rightarrow$ 15.1). (g, E_R)

7 CIRCULAR MOTION

EXPERIMENT 7.1 Testing $F = mv^2/r$

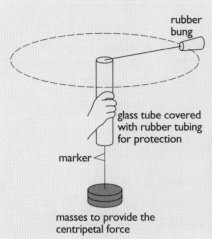

Figure 7.7 Apparatus for testing $F = mv^2/r$

Using the apparatus shown in Figure 7.7, you can whirl a rubber bung in a horizontal circle. When you increase the rate of rotation, you will find that the centripetal force has to be increased in order to keep the rubber bung moving in a circle.

Figure 7.8 shows the results of such an experiment, where the mass, m is 0.015 kg. By calculating the values of mv^2/r, you can find how close these are to the corresponding values of the centripetal force in the table.

radius (m)	force (N)	time for 50 revs (s)
0.60	0.98	30
0.60	0.78	35
0.60	0.54	39
0.30	0.98	21
0.30	0.78	23
0.30	0.54	27

Figure 7.8 Typical results from experiment 7.1

CENTRIFUGAL FORCE: REAL OR FICTITIOUS?

Imagine you are on a rotating roundabout and you place a ball on the surface. It will roll in a curve towards the edge. From your point of view in a rotating frame of reference, it will seem that there is a force pushing the ball **away from** the centre of the roundabout. This is called a centrifugal force.

The concept of a centrifugal force is unnecessary and, indeed, can be positively misleading. We can quite adequately explain the circular motion of an object by

Centrifuges

Particles of solid suspended in a liquid will eventually settle under the force of gravity. This and other similar processes are dependent upon the relatively weak force of gravity. A centrifuge is a device that accentuates gravity by rotating the sample and its container at high speed. As a result, the whole sample is subjected to a sustained centripetal force. More dense particles or liquids experience greater relative forces than less dense ones and migrate away from the axis of rotation, leaving the less dense ones closer to the axis. This effectively provides considerably increased gravity.

Astronauts suffer large accelerations on launching and re-entry. Space research centres, such as NASA, use centrifuges large enough to spin a person, as shown in Figure 7.9, to find the effects of large accelerations on the human body.

Figure 7.9 A centrifuge at the NASA manned spacecraft centre

stating that it is caused by a force at right angles to the direction of motion – the **centripetal force**. The object is accelerating in this direction, towards the centre of the circle.

> Questions
>
> **6** A centrifuge rotates at 3 500 rpm. What is the centripetal acceleration at a radius of 20 cm from the axis of rotation? How does this value compare with the gravitational acceleration? (g)
>
> **7 a** What provides the centripetal force that causes each of the following to travel in a circular path:
> **i** a train on a curved track
> **ii** a passenger in a car which is going round a roundabout
> **iii** a child on a playground roundabout
> **iv** a satellite orbiting the Earth?
> **b** Explain how a spin drier removes water from wet clothes.
>
> **8** A space station in Earth orbit is freely falling towards the Earth. An astronaut inside the station is also undergoing free fall and, because her acceleration is the same as that of the station, she is weightless in relation to the station. Artificial gravity can be produced by rotating the station about an axis. Design a structure that might, in principle, give an acceleration of 10 m s^{-2}, the same as that on Earth. How large would the structure be? How quickly would it rotate? Where would the 'floor' be in such a space station?

7.2 ROTATING OBJECTS

All objects are made up of particles. All particles move around the axis of rotation with the same angular speed.

Figure 7.10 shows a stroboscopic photograph of the motion of a diver through the air. The motion of the diver's centre of mass through the air is seen to be a smooth parabolic path, such as a particle would follow. However, the motion of the diver is quite complex, consisting of a combination of rotational motion about the centre of mass, and translational motion of the centre of mass through the air.

The model we use to explain the behaviour of rotating objects is based on them being made up of many interconnected particles. The particles are at various distances away from the axes of rotation, and so travel in circles of different radii.

ROTATIONAL INERTIA

Just as the mass of an object is its inertia or resistance to change of linear motion (← 6.4), so there is a resistance to change in the rotation of an object because of its **rotational inertia**, I.

To find out what determines rotational inertia try spinning a swivel chair, first when it is empty and then when someone is sitting on it. You will find it easier to get the chair to spin at a particular rate when it is empty. This suggests that the rotational inertia of an object depends upon its mass. Now, suppose the person on the chair holds a heavy mass in each hand. It is easier to get the chair spinning at a particular rate when the hands are close to the body than when they are

Figure 7.10 A stroboscopic photograph of a diver in free fall

outstretched. So the rotational inertia of an object depends on its mass and on the distribution of the mass about the axis of rotation.

> Question
>
> 9 Which flywheel will have the greatest rotational kinetic energy when spinning with a particular angular velocity:
> - one with its mass concentrated around its rim as for a spoked wheel, or
> - one with the same mass and radius but in the shape of a disc?

In defining rotational inertia, I, for a particular shaped object, the masses of all the particles that make it up and their distances to the axis of rotation have to be taken into account. Using the particle model, the rotational inertia of symmetrical shapes can be obtained mathematically relatively simply. For complex shapes, it is easier to find it by experiment.

Moment of inertia is the most frequently used name for rotational inertia. It is defined by $I = \Sigma mr^2$ about a particular axis of rotation.

Because the mass is distributed differently for each axis of rotation, the moment of inertia of an object will depend on which axis of rotation is used. Some expressions for it and values are given in Figure 7.11.

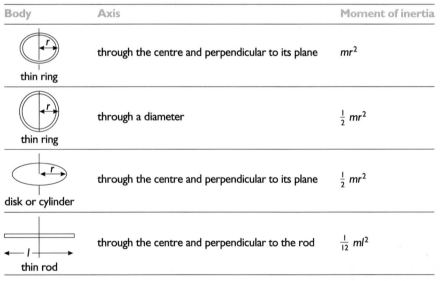

Body	Axis	Moment of inertia
thin ring	through the centre and perpendicular to its plane	mr^2
thin ring	through a diameter	$\frac{1}{2}mr^2$
disk or cylinder	through the centre and perpendicular to its plane	$\frac{1}{2}mr^2$
thin rod	through the centre and perpendicular to the rod	$\frac{1}{12}ml^2$

Figure 7.11 Expressions for the moment of inertia of some simple shapes about a main axis

EXPERIMENT 7.2 Energy of a rolling cylinder

Tilting a flat runway so that a heavy cylinder will roll down it without slipping, enables you to measure the time it takes to reach the bottom. By measuring the vertical height of the cylinder's starting position, you can calculate the time that it should take for an object to fall this distance, assuming that frictional forces are negligible.

You should find that the time you measure is longer than the one you calculate. This is because gravitational potential energy is transferred into both rotational and translational kinetic energy as the cylinder rolls.

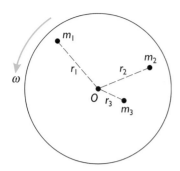

Figure 7.12 All parts of a disc rotate at the same angular velocity ω

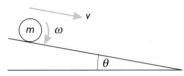

Figure 7.13 A cylinder rolling down a slope

Rotational kinetic energy of an object

The disc in Figure 7.12 can be considered to consist of many particles having masses m_1, m_2, m_3, etc and lying at different distances r_1, r_2, r_3 from O. The particles will all have the same angular velocity, if the flywheel is rigid. But, since $v = \omega r$, they will travel at different speeds. If the particle of mass m_1 travels with velocity v_1 then its kinetic energy E_{K1} will be given by

$$E_{K1} = \tfrac{1}{2} m_1 v_1^2$$
$$= \tfrac{1}{2} m_1 r_1^2 \omega^2.$$

The kinetic energy E_K of the whole object is found by adding together the kinetic energies of all of its particles:

$$E_K = E_{K1} + E_{K2} + E_{K3} + \cdots$$
$$= \tfrac{1}{2} m_1 r_1^2 \omega^2 + \tfrac{1}{2} m_2 r_2^2 \omega^2 + \tfrac{1}{2} m_3 r_3^2 \omega^2 + \cdots$$
$$= \Sigma \tfrac{1}{2} m r^2 \omega^2.$$

Because ω is constant for all parts of the object, assuming it is rigid,

$$E_K = \tfrac{1}{2} \omega^2 (\Sigma m r^2).$$

Here, $\Sigma m r^2$ is the sum of all values of $m r^2$ for the whole object. This is defined as the rotational inertia I of the object and:

$$E_K = \tfrac{1}{2} I \omega^2.$$

Because of its definition, I has the unit kg m².

Question

10 A cylinder of mass 0.17 kg is rolled down a slope so that its vertical drop in height is 0.08 m. (g)
 a Calculate the loss in potential energy of the cylinder.
 b It took 2.0 s to travel down the slope, which was 1.0 m long. What was its average velocity?
 c Assuming that its acceleration down the slope is uniform, calculate the cylinder's velocity at the end of the slope.
 d What is the translational kinetic energy of the cylinder at the bottom of the slope ($\rightarrow$ 9.2)?
 e Explain why the gain in translational kinetic energy is much less than the loss in potential energy. (g)

ENERGY OF A ROTATING OBJECT

Figure 7.12 shows a disc spinning at a constant angular velocity ω about an axis through its centre and perpendicular to its plane. There is no linear motion of the disc's centre of mass and, therefore, no translational kinetic energy. However, the particles that make up the disc are travelling in circles and each possesses some kinetic energy. The kinetic energy of a particle of mass m travelling with velocity v is given by $\tfrac{1}{2} m v^2$. The rotational kinetic energy of the whole disc is equal to the sum of the kinetic energies of all these particles. This is equal to $\tfrac{1}{2} I \omega^2$ where I is the rotational inertia of the disc about the axis concerned.

Notice that these two equations are similar in form, or analogous. The translational kinetic energy of a mass m is given by $\tfrac{1}{2} m v^2$. If mass is replaced by rotational inertia and linear velocity replaced by angular velocity, it becomes the equation for rotational kinetic energy, $\tfrac{1}{2} I \omega^2$.

Imagine an object spinning through space or the cylinder rolling down the slope in experiment 7.2. Figure 7.13 shows that the cylinder has both translational kinetic energy of the centre of mass down the slope, and rotational kinetic energy about its axis. So

total kinetic energy = $\tfrac{1}{2} m v^2 + \tfrac{1}{2} I \omega^2$.

Question

11 The flywheel inside the engine of a motor car consists essentially of a flat disc of steel of radius 0.2 m and with a mass of about 10 kg. It rotates about an axis through its centre and perpendicular to its plane. Figure 7.11 gives the moment of inertia for such a disc.
 a Satisfy yourself that, when its rate of rotation is 3 000 rpm, the rotational kinetic energy stored in the flywheel is around 10 000 J.
 b What linear speed would the flywheel have to be given in order to possess 10 000 J of translational kinetic energy?

TORQUE AND ANGULAR ACCELERATION

The analogy between rotational and translational motion can be extended further. For example, in translational motion, a force F acting on a mass m causing it to accelerate (see Figure 7.14a) is given by:

$$F = ma.$$

7 CIRCULAR MOTION

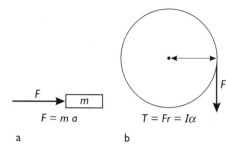

Figure 7.14 **a** A force F accelerating a body of mass m **b** A torque T rotationally accelerating a body with rotational inertia I

In rotational motion, a turning force is needed to cause a change in rotation. In Figure 7.14b, the size of the torque T (← 5.6) is given by:

$T = Fr$.

This causes an angular acceleration α, and

$T = I\alpha$.

Just as linear acceleration is the rate of change of linear velocity

$a = dv/dt$

so angular acceleration is the rate of change of angular velocity

$\alpha = d\omega/dt$.

ANGULAR MOMENTUM

Linear momentum is an important quantity when dealing with collisions and explosions. It is defined as the product of the mass of an object and its velocity (← 6.4). The analogous quantity in rotational motion is **angular momentum**. The angular momentum of a particle about an axis is defined as the **moment of its momentum** about the axis. In Figure 7.15, the angular momentum of the particle A about the axis through O, or the moment of its momentum, is given by the product of its momentum and the perpendicular distance from the particle to the axis.

Angular momentum of $A = m_1 v_1 \times r_1$.

The angular momentum of a solid object is equal to the sum of the angular momenta of all the particles that make up the object, and is equal to $I\omega$. This formula is equivalent to that for linear momentum, but with mass replaced by rotational inertia and linear velocity replaced by angular velocity. Figure 7.16 lists the formulae and equations in linear motion and their equivalents in rotational motion.

In a similar way to linear motion (← 6.4), Newton's laws of motion applied to rotational motion lead to the conclusion that **angular momentum is conserved in an isolated system**.

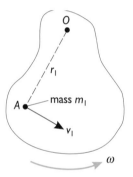

Figure 7.15 Rotational momentum as 'moment of momentum'

Angular momentum

In Figure 7.15, the angular momentum of particle A about O is given by

$m_1 v_1 \times r_1$.

In general, $v = \omega r$, so this becomes:

$m_1 r_1^2 \omega$.

The angular momentum of the whole object will be the sum of the angular momenta of all of its particles, or

angular momentum $= m_1 r_1^2 \omega$
$+ m_2 r_2^2 \omega$
$+ m_3 r_3^2 \omega + \cdots$
$= \Sigma (mr^2)\omega$.

But Σmr^2 equals the rotational inertia I about axis O, so

angular momentum $= I\omega$.

Linear motion	Rotational motion
mass = m	moment of inertia = I
linear velocity = v	angular velocity = ω
acceleration $a = \dfrac{dv}{dt}$	angular acceleration $\alpha = \dfrac{d\omega}{dt}$
force = F	torque = T (= Fr)
$F = ma = m\dfrac{dv}{dt}$	$T = I\alpha = I\dfrac{d\omega}{dt}$
momemtum = mv	angular momentum = $I\omega$

Figure 7.16 Equivalent quantities in translational and rotational motion

SUMMARY

When an object moves in a circular path, it is accelerating, even though it may be travelling at a constant speed, because its direction is constantly changing.

To make an object move in a circular path, it must have a force pulling it always towards the centre of the circle – the centripetal force.

This force is equal to mv^2/r.

The angular velocity of a rotating object ω is its rate of change of angle in radians where $\omega = v/r$.

To study the rotation of an object, we must consider the motion of the particles of which it is made. Each particle is moving in a circle around the axis of rotation.

The rotational inertia of an object is its resistance to changes in its rotation. The equations of rotational mechanics are analogous to those for linear motion, but with mass replaced by rotational inertia, linear velocity replaced by angular velocity, force by torque and so on. The sets of equations are listed in Figure 7.16.

Figure 7.17 A fairground rotating drum

SUMMARY QUESTIONS

12 The Earth has a mass of 6.0×10^{24} kg and moves around the Sun in an orbit of radius 1.5×10^{11} m.
a Calculate the average speed of the Earth in its orbit in m s⁻¹, using 3×10^7 for the number of seconds in a year.
b Calculate the Earth's angular velocity.
c Calculate the magnitude of the centripetal force acting on the Earth.

13 The coefficient of static friction between a car's tyres and the road surface is 0.8. Work out the maximum speed at which a car can go round a roundabout of radius 50 m without skidding. (*g*)

14 One type of fairground ride consists of a large cylindrically shaped enclosure which can rotate about its vertical axis as shown in Figure 7.17. Typically, the diameter of the enclosure is about 5 m. A person is supported by the floor when the enclosure is stationary. When the enclosure rotates at a sufficiently high speed, the floor is gradually lowered. The person is then supported by the wall alone.
a Describe what the person would feel as the enclosure gradually speeds up from rest.
b What force keeps the person 'fixed' to the wall as the enclosure rotates?
c What provides the centripetal force on the person?
d Suppose the coefficient of friction between the person and the wall is 0.5 and the person's mass is 60 kg. What is the normal force when the person is on the point of slipping down the wall?
e At what speed does the enclosure rotate when the person is just supported?
f To be supported, does a heavier person need to rotate at the same, a higher or a lower speed? Explain your answer. (*g*)

15 A curve, of radius 500 m, on a race track is banked at an angle of 20°.
a Find the maximum speed at which a racing car can travel round the curve without having to rely on the frictional force between the road and the tyres.
b What would happen to a car that went **i** faster **ii** slower than this speed? (*g*)

16 A centrifuge at the NASA manned spacecraft centre is used to test the effects of large accelerations on the human body. Work out the acceleration of an astronaut who is rotated 20 times a minute in a circle of radius 15 m by the machine.

17 Describe the changing forces a person will feel as they loop the loop in the "corkscrew" ride in Figure 7.18 at constant speed. What is the minimum speed needed to keep someone in their seat in a vertical circle of radius 6 m? (*g*)

18 A record turntable is a thin disc of mass 0.5 kg and radius 157 mm, and rotates at 33 revolutions per minute.
a Calculate **i** the angular velocity, **ii** the rotational inertia and **iii** the rotational kinetic energy of the turntable.
b It takes 12 s for the turntable to come to rest when switched off. What is its angular deceleration?
c Calculate the frictional torque acting on the turntable.

19 A car is travelling at 30 m s^{-1} and each wheel has a mass of 27 kg.
a What is the translational kinetic energy of each wheel?
b The wheel diameter is 0.76 m. How many revolutions does it make in one second, and what is its angular velocity?
c The rotational inertia of the wheel is 1.55 kg m^2. What is the rotational kinetic energy of the wheel?
d Work out the total kinetic energy of the wheel.

Figure 7.18 A corkscrew ride

UNIT C

CONSERVATION LAWS

CONSERVATION LAWS

The laws of conservation occupy a prime position in science. The idea of a quantity that is conserved – one whose value remains constant even when other quantities change is of considerable importance in the development of models in physics.

Conservation applies right across the Universe it seems. If you were to list the quantities and phenomena that you know do not change overall with time, your list might well include mass, energy, momentum and charge. All of these are conserved, but the conditions under which each is conserved are different.

Mass m has energy mc^2 associated with it, or we can say energy E has mass E/c^2 ($\rightarrow$ 9.6). Either way, mass and energy can be seen to be part of the same phenomenon. However, potential and kinetic energy are the two fundamental ways in which energy exists. Any interaction, whether a collision or just action at a distance, involves interchange between them. Kinetic energy is only conserved in elastic collisions, but energy overall is always conserved. Momentum is conserved in any collision, whether elastic or not. Charge seems to be indestructible – like energy, it can neither be created nor destroyed.

8 CONSERVATION OF MOMENTUM

Drop an egg on the ground and it stays there. What has happened to its momentum? Has it been destroyed? If so how can momentum be conserved?

When two or more bodies interact within a system, momentum possessed by one of them (← 6.4) may be transferred to another. The **total** momentum possessed by the bodies in the system does not change because each experiences the same force, oppositely directed, for the same length of time. This principle enables us to predict or explain the behaviour of any system involving the motion of bodies, whether it be alpha particles colliding with nuclei in a cloud chamber, vehicles colliding at a road junction or the motion of a comet around the Sun.

8.1 ONE-DIMENSIONAL COLLISIONS

Momentum is a vector quantity (← 5.1) and so has direction associated with it. An object of mass m moving at a velocity v has a momentum mv, whereas another of the same mass travelling in the opposite direction at the same speed has a velocity $-v$ and a momentum of $-mv$. The momentum of an object is in the same direction as its velocity.

EXPERIMENT 8.1 Investigating collisions on a linear air track

Using a typical linear track as shown in Figure 8.1, you can investigate interactions, such as collisions and repulsions between gliders. Affixing small magnets to the gliders so that they repel each other can be used to

Figure 8.1 A linear air track

produce 'springy' collisions without contact. Sticky material on the gliders enables them to become a combined mass on collision and to proceed as one body – a 'sticky' collision. Placing on each glider a flag which interrupts a light beam controlling a clock, enables you to measure the time for the flag to pass the light beam and, hence, calculate the glider's velocity. For example, travelling at $1.0\,\mathrm{m\,s^{-1}}$, a flag 0.10 m long would interrupt the beam for 0.10 s.

before collision			after collision		
glider 1		glider 2	glider 1		glider 2
time/s	velocity/m s^{-1}			time/s	velocity/m s^{-1}
0.21	0.48	at rest	at rest	0.23	0.43
0.35	0.29	at rest	at rest	0.35	0.29
0.26	0.38	at rest	at rest	0.28	0.36
0.30	0.33	at rest	at rest	0.31	0.32

Figure 8.2 Data on 'springy' collisions between gliders of equal mass (each 0.40 kg)

before collision			after collision				
glider 1		glider 2	glider 1		glider 2		
time/s	velocity/m s^{-1}		time/s	velocity/m s^{-1}	time/s	velocity/m s^{-1}	
0.24	0.41	at rest	0.41	−0.24	0.32	0.31	
0.36	0.28	at rest	0.60	−0.17	0.46	0.22	
0.20	0.50	at rest	0.34	−0.29	0.25	0.40	
0.28	0.36	at rest	0.48	−0.21	0.36	0.28	

Figure 8.3 Data on 'springy' collisions between gliders with masses in the ratio 2:1 (glider 1 mass 0.40 kg, glider 2 mass 0.20 kg)

before collision			after collision	
glider 1		glider 2	glider 1 + 2	
time/s	velocity/m s^{-1}		time/s	velocity/m s^{-1}
0.22	0.45	at rest	0.45	0.22
0.27	0.37	at rest	0.56	0.18
0.35	0.29	at rest	0.70	0.14
0.16	0.63	at rest	0.33	0.30

Figure 8.4 Data on 'sticky' collisions between gliders of equal mass (each 0.40 kg)

Figures 8.2–8.4 show some typical data obtained using a flag 0.10 m long. In each case, glider 1 is given an initial velocity and hits glider 2, which is at rest.

The three sets of data for the collisions in experiment 8.1 (Figures 8.2–8.4), show that there are many obvious differences between them. But, in each case, the total momentum of the gliders before the collision is approximately equal to the total momentum after the collision. Momentum has been conserved within the range of uncertainties in the readings.

Motion experiments are very difficult to carry out with high accuracy. Compared with trolleys on runways, the linear air track appears to have almost no frictional losses. Yet data from air track experiments do not show accurately and conclusively that momentum is conserved. The gliders on the air track in

8 CONSERVATION OF MOMENTUM

experiment 8.1 do not form an isolated system but are subject to external forces of friction. We could include these frictional forces as part of our isolated system but we would then need to consider the motion of the air track and the Earth to which it is attached – clearly an impossible task. To show conclusively the principle of conservation of momentum, we need to apply Newton's second and third laws (← 6.4) to a truly isolated system and consider only the forces exerted by objects on each other within the system.

Conservation of momentum

Suppose two objects collide with one another. Each will exert on the other a force F for the same time interval Δt. The momentum of A will change by an amount $\Delta(mv_A)$ and that of B by an amount $\Delta(mv_B)$.

The impulse received by an object is equal to its change in momentum, according to Newtons' second law (← 6.4). So, for the effect of the interaction on object B, we can say:

$F\Delta t = \Delta(mv_B)$.

From Newton's third law, the interaction causes an identical but opposite force $-F$ to act on A. Since both A and B are in contact for the same time Δt, we can say that the momentum of A will change by an amount $\Delta(mv_A)$. This means that:

$-F\Delta t = \Delta(mv_A)$.

The total momentum change of A and B together is:

$\Delta(mv_A) + \Delta(mv_B) = F\Delta t + (-F\Delta t) = 0$.

So, the total momentum change of A and B is zero and **momentum is conserved**. Note that no assumption has been made about the initial velocities of A and B or about whether the collisions are 'springy' or 'sticky'.

A collision is only one type of interaction between two bodies. Other investigations show that this **law of conservation of momentum** applies to all types of interaction.

Questions

1 Consider the following situations:
a A glider runs into the end of the air track and stops.
b A mass is released and accelerates under gravity.
c A train accelerates steadily from rest to a velocity of 30 m s⁻¹.
In each case, the momentum of the object concerned changes. Why is the principle of conservation of momentum not violated?

2 A rifle fires a bullet into a box filled with sand, leaving the bullet embedded in the sand. The mass of the bullet is 0.035 kg and that of the box is 20 kg. The box recoils with an initial velocity of 0.5 m s⁻¹. Calculate the velocity of the bullet.

3 Two trolleys are tied together with a compressed spring between them, as shown in Figure 8.5. When the string is cut, the trolleys fly apart. Ignoring frictional effects:
a under what conditions will one trolley move at twice the speed of the other?
b in which direction does the centre of mass of the pair of trolleys move?
c what is the net external force acting on the trolley system?

We can illustrate the connection between the conservation of momentum and Newton's laws by a simple numerical example.

A car of mass 800 kg travelling at 30 m s⁻¹ runs into the back of a truck of mass 3 000 kg travelling at 10 m s⁻¹ as shown in Figure 8.6. The interaction, or collision, is between the car and the truck. The force that each exerts on the other is one side of this interaction, having the same value but oppositely directed. Clearly, this force varies during the collision and its value at any instant will depend upon how the two vehicles have been designed.

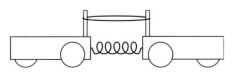

Figure 8.5 Exploding trolleys

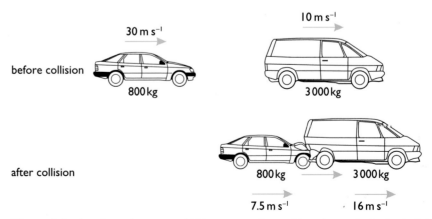

Figure 8.6 A collision between a 800 kg car travelling at 30 m s⁻¹ into the back of a 3 000 kg truck travelling at 10 m s⁻¹

Suppose that the interaction lasts for 2 s, and each force has an **average** value of 9 000 N. The impulse delivered to the truck in the direction of its motion will be 2 × 9 000 = 18 000 Ns. So, its momentum will change by the same amount, 18 000 Ns. This will increase the truck's velocity:

change in momentum = mass × velocity change

so, velocity change = momentum change / mass
= 18 000 / 3 000 m s^{-1}
= 6 m s^{-1}.

The truck will now be travelling at 10 + 6 = 16 m s^{-1}.

The **average** force on the car involved in the collision will be the same as on the truck but in the opposite direction, −9 000 N. Lasting, as it must, for the same length of time, the impulse experienced by the car will be −18 000 Ns. So the change in the car's momentum will be 18 000 Ns, and:

change in car's velocity = −18 000 / 800
= −22.5 m s^{-1}.

After the collision the car will be travelling at a velocity v given by:

v = 30 − 22.5
= 7.5 m s^{-1}.

Question

4 Modern cars are designed so their front ends crush more easily, lengthening the duration of the impact. Suppose the impulse received by the vehicles is the same as described in the main text. What effect will this have on the forces experienced by the vehicles and their occupants, and on their final velocities after the collisions?

THE JET ENGINE

An aircraft gas turbine or 'jet' engine is shown in Figure 8.7. Air taken in at atmospheric pressure is compressed and heated before being allowed to escape at high velocity through the tailpipe. Much of the energy of the hot gases is extracted by a turbine to drive a fan which accelerates the stream of air flowing past the engine as shown in Figure 8.8.

Figure 8.7 An aircraft gas turbine (jet) engine

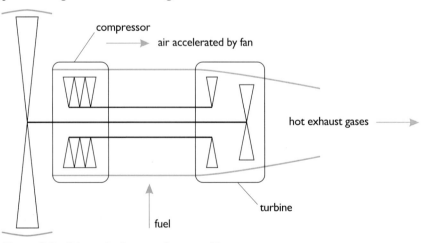

Figure 8.8 Schematic diagram of a gas turbine

8 CONSERVATION OF MOMENTUM

The momentum of the air and burnt fuel leaving the engine plus the momentum of that air accelerated past the engine by the fan is greater than the momentum of the air entering the engine. So the engine can develop a thrust. For example, under normal cruising conditions, the engines of a Boeing 747 aircraft enable it to fly at 250 m s^{-1} and at an altitude of 11 km.

> ### Questions
>
> **5** Use the data in Figure 8.9 to calculate:
> **a** the momentum of the air entering the engine each second,
> **b** the momentum of the air plus combustion products leaving the engine each second,
> **c** the change in momentum per second – the thrust developed by the engine.
>
> **6** Early jet engines did not have a fan and produced only a hot, high velocity gas stream. Why do present-day subsonic jets have fans?

Data for RB211 type engine at cruising altitude of 10 500 m	
mass of fuel burnt in engine per second	0.67 kg s^{-1}
mass of air passed through engine per second	220 kg s^{-1}
airspeed of aircraft	250 m s^{-1}
average speed of exhaust gases	420 m s^{-1}
average specific energy of fuel	43 MJ kg^{-1}
net thrust (after correction)	38 kN

Figure 8.9 Data on the Boeing 747 and its jet engines

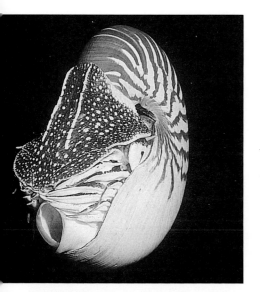

Figure 8.10 The pearly nautilus – a jet-propelled mollusc

A jet-propelled animal

The 'pearly nautilus', a type of mollusc which grows to about 200 mm across, propels itself with water jets (see Figure 8.10). Many molluscs, such as mussels and clams, tend to be rather immobile creatures. They feed by taking in water through one end of a mantle cavity and ejecting it, after filtration, through a tube at the other end. This process has evolved in the pearly nautilus and, to a lesser extent, the squid and the octopus, such that they can use the change in momentum given to the ejected water for propulsion.

ROCKETS

A rocket motor produces thrust to propel the rocket by giving momentum to a stream of material, such as combustion products, ejected from the motor (see Figure 8.11). Suppose R is the mass of material ejected per second with a velocity v relative to the rocket. Then the change in momentum per second of the combustion products is vR. This equals the thrust produced by the motor on the body of the rocket. So,

the acceleration of the rocket = vR/M

where M is the mass of the rocket. When the fuel in the rocket is burnt and ejected, the mass M decreases and so the acceleration increases. Most of the lift-off mass of a satellite launcher, such as the Space Shuttle, is fuel. Once that fuel is burnt, the boosters and fuel tank are jettisoned.

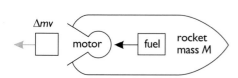

Figure 8.11 Schematic diagram of a rocket

CONSERVATION LAWS

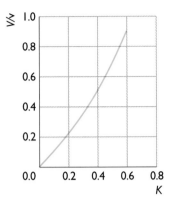

Figure 8.12 Graph of velocity of rocket against fuel burnt

Figure 8.13 The American Space Shuttle

take off mass of space shuttle and boosters, fuel and fuel tank	2×10^6 kg
take off thrust from main engines	5×10^6 N
take off thrust from solid fuel boosters	2.4×10^7 N

Figure 8.14 Data on mass and thrust of Space Shuttle

How rockets are propelled

The mass M of a rocket at time t will be given by:

$$M = M_0 - Rt$$

where M_0 is the initial mass of the rocket when $t = 0$, at lift-off. At any instant of time after this it will have a velocity V.

Because the interaction between ejected material and rocket causes forces that are equal but in opposite directions, the equation of motion of the rocket is given by:

rate of change of momentum
of ejected material
r = accelerating force on rocket

$$Rv = M \, dV/dt$$
$$= (M_0 - Rt) \, dV/dt.$$

Separating the variables and integrating this expression gives

$$\int dV = Rv \int dt / (M_0 - Rt)$$
$$V = -Rv/R \, [\ln(M_0 - Rt)]$$
$$= -v \ln(1 - Rt/M_0).$$

Enough fuel has been burnt to halve the initial mass of the rocket when

$$Rt = \tfrac{1}{2} M_0.$$

At this time the speed of the rocket V will be $v \ln 2$. Ln 2 has a value of 0.693, so V will be nearly 70% of the speed of the exhaust gases relative to the rocket. Figure 8.12 shows how the speed of the rocket varies as more fuel is burnt. In this graph, V is the velocity of the rocket expressed as a fraction of the velocity v of the exhaust gases relative to the rocket, and K is the fuel burnt expressed as a fraction of the initial mass of the rocket.

Questions

7 From the data in Figure 8.12, and the lift-off mass of the Space Shuttle and its boosters in kg and the available thrust from the main engines and boosters in N (see Figure 8.14), estimate the initial acceleration of the Shuttle as it leaves the launch pad.

8 Communications satellites carry several small rocket motors to perform small corrections to their orbits after launch. What sequence of rocket motor burns is needed to increase the radius of orbit of a satellite?

9 A rocket motor ejects material at 3 km s^{-1} relative to the rocket.
a Use the formula for V to calculate the proportion of the lift-off mass that must be ejected for the motor to power a rocket to escape from the Earth (escape velocity = 11 km s^{-1}).
b Why are multi-stage rockets generally used to launch satellites?

8.2 CONSERVATION OF MOMENTUM IN MORE THAN ONE DIMENSION

Because momentum is a vector quantity, the momentum of a moving body can be resolved into three components perpendicular to each other – the x, y and z components. When two bodies collide, we can apply the law of conservation of momentum separately to each set of components in turn. This procedure simplifies the problem to three one-dimensional problems.

8 CONSERVATION OF MOMENTUM

EXPERIMENT 8.2 Collisions in two dimensions

Using dry ice or air pucks you can produce two-dimensional collisions. You can record puck positions by making multiple image photographs using a stroboscope or by using a video recording with frame by frame replay.

Pucks of the same and different masses can be used and they may be magnetised to repel each other, producing elastic, or 'springy', collisions. Differing angles of approach are worth investigating.

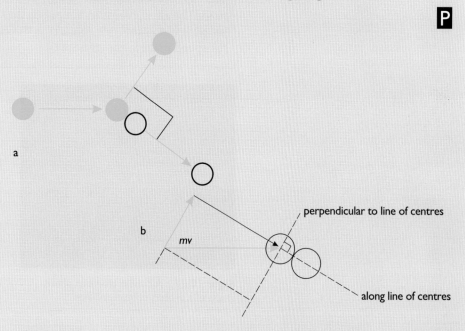

Figure 8.15 a The collision between a moving puck and a stationary puck of equal mass
b Resolution of momenta for the collision

One particular type of collision has a very characteristic shape (see above). When a moving puck makes a 'springy' collision with a stationary puck of the same mass, the two pucks move at right angles to each other after the collision (see Figure 8.15a). We can analyse the collision by resolving the momentum of the incident puck into components along and perpendicular to the line of centres of the two pucks as in Figure 8.15b. We find that:

- along the line of centres, the components behave as in the one-dimensional case. The moving puck loses this component of its momentum to the second puck, which moves away with the same velocity
- perpendicular to the line of centres, there is no force between the pucks. So, this component of the incident puck's momentum is unchanged and the second puck does not acquire any momentum in this direction.

Questions

10 Figure 8.16 is a scaled photograph of two pucks colliding. The images are recorded at 0.1 second intervals. By resolving the momenta of the pucks into components parallel to the sides of the photograph, find out if momentum is conserved in the collision.

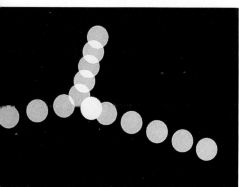

Figure 8.16 The collision of two pucks of equal mass

CONSERVATION LAWS

11 Figure 8.17 shows a beam of high energy protons crossing a bubble chamber filled with liquid hydrogen. What changes occur in the momentum of the proton involved in the collision with a hydrogen atom at A?

12 A van of mass 1 500 kg, proceeding along a main road at 20 m s^{-1}, collided with a car of mass 650 kg coming out of a side road at right angles to the main road, as in Figure 8.18. The two vehicles locked together and skid marks showed them to move initially at an angle of 30° to the main road. The car driver told the police that he was only travelling at 5 m s^{-1} as he came out of the side road. Was the car driver telling the truth? Estimate his speed prior to the impact.

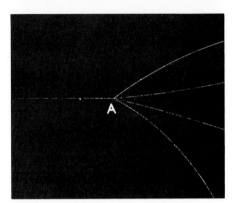

Figure 8.17 A bubble chamber photograph of a proton-proton collision

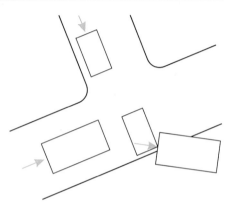

Figure 8.18 A van collides with a car

8.3 CONSERVATION OF ANGULAR MOMENTUM

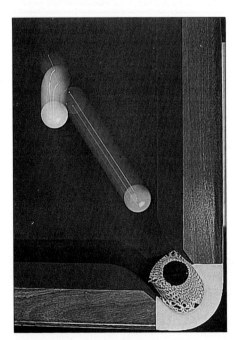

Figure 8.19 Collisions on the snooker table, cue ball hitting stationary ball

A set of relationships for rotational motion can be developed which are analogous to those for linear motion (← Figure 7.16). The analogy can be extended to include conservation of momentum: **for an isolated system of bodies, the total angular momentum remains constant**.

You can easily demonstrate the conservation of angular momentum. Sit on a swivel chair with a large mass in each hand and set the chair spinning. You can alter the rate of rotation of the chair by extending your arms or pulling them in closer to the body. With arms extended, the moment of inertia I of your body increases. This is because the mass is now at a greater distance from the axis of rotation. The angular momentum $I\omega$ cannot increase, and so the angular velocity ω decreases as I increases.

Divers make use of this principle. On leaving the board, they curl up their body so that their rate of rotation increases. This allows them to complete more twists and turns before entering the water. Ice skaters and gymnasts also make use of the conservation of angular momentum.

Question

13 Figure 8.19 shows a strobe photograph of a collision involving snooker balls. This picture shows very different motion from those taken with pucks. Why is this?

8 CONSERVATION OF MOMENTUM

SUMMARY

The momentum of a body is defined as its mass multiplied by its velocity. Like velocity, it is a vector quantity.

For any isolated system, the total momentum remains constant. The momenta of all bodies in a system add up to the same total momentum at all times.

For collisions in more than one dimension, momentum may be resolved into component momenta at right angles and the conservation of momentum applied to each component in turn.

When a body rotates, it possesses angular momentum and this is also conserved.

The change in momentum, or impulse, for each body in any interaction is the same but in the opposite direction. It equals the average force between them multiplied by the time for which they interact.

SUMMARY QUESTIONS

Some of these questions also use the ideas of conservation of energy in mechanical systems ($\rightarrow$ 9).

14 A rocket of mass 10^5 kg leaves the ground with an initial acceleration of 5 m s^{-2}. Calculate the thrust of its engines.

15 A car of mass 900 kg travelling at 15 m s^{-1} runs into the rear of a stationary car of mass 750 kg. The stationary car is shunted forward with an initial speed of 11 m s^{-1}. Calculate:
a the speed of the 900 kg car after the collision, **b** the 'loss' in kinetic energy after the collision. Explain where the energy 'loss' is dissipated.

16 An α-particle, travelling at 10^6 m s^{-1}, collides head on with a stationary gold nucleus. Calculate the change in velocity and change in kinetic energy of the α-particle. (Mass of α-particle = 6.6×10^{-27} kg, mass of gold nucleus = 3.3×10^{-25} kg.)

17 An astronaut repairing a satellite becomes separated from her spacecraft and her rocket backpack fails. She is carrying a hammer and a spanner. What should she do in an attempt to return to the spacecraft?

18 A hammer used to drive home a tempered nail rebounds in an elastic collision. The same hammer deforms a soft nail and does not rebound. Compare the impulse given to the nail in each case. Which type of nail would be easier to drive into, say, brick?

19 Two skaters, each of mass 60 kg, move with speeds of 5 m s^{-1} and 3 m s^{-1} at an angle of 30° to each other. They then collide and embrace. Neglecting friction, estimate their speed and direction after the collision, if they do not fall over.

20 A skater of mass 70 kg carrying a 10 kg mass is moving with a velocity of 4 m s^{-1}. Calculate his new velocity and change in direction when he throws the 10 kg mass:
a straight ahead, **b** sideways with a velocity of 1 m s^{-1} relative to himself.

21 A proton moving at 10^6 m s^{-1} collides elastically with a stationary proton, and is deflected through an angle of 50°. Estimate the speed and direction of the second proton after the collision.

9 ENERGY TRANSFER IN MECHANICAL SYSTEMS

To make a journey, whether by bus, car, train, plane or any other form of transport, a source of energy is needed. We can consider the energy transfers taking place during such a journey. For example, on a train energy is transferred from the overhead power line to the motors that drive the wheels. Work done by the motors transfers energy to the train so that it increases its speed or climbs an incline. Kinetic energy is transferred from the movement of the train to the brakes whenever the train slows. Additionally, energy is transferred away from the power lines and train by the emission of sound waves and electromagnetic waves. Friction may transfer energy and increase the internal energy of the power lines, the surrounding air, the track and the train.

If we were to account for **every** transfer taking place, we should find that the total energy present in the whole system remains constant at all times – the **total energy is conserved**. However, to show this experimentally is quite a challenge.

As energy in a system is transferred between its various parts, it becomes increasingly spread out amongst the components of the system. So it becomes very difficult to account for each and every transfer that may take place.

Whenever energy is transferred, there is an **interaction** between one part of a system and another, which transfers potential or kinetic energy. In the interaction, oppositely directed forces act on each part (← 6.4).

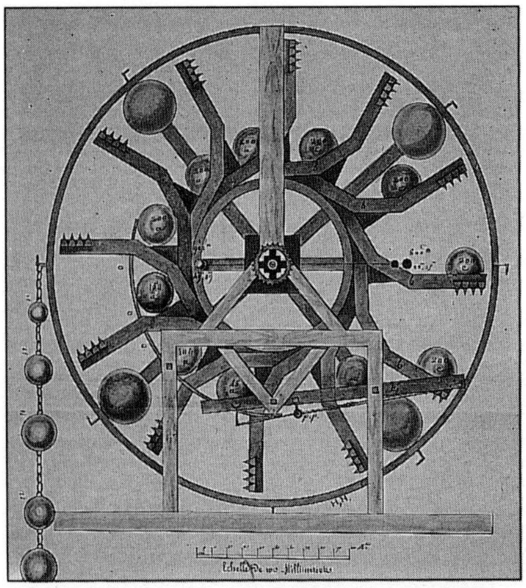

Momentum and energy

The idea of conservation of momentum (← 8) emerged from considering Newton's laws of motion (← 6.4). It was accepted by the mid-eighteenth century as a helpful model for mechanical systems. However, the conservation of energy in mechanical systems was not generally accepted for another century. Even then, scientists continued to produce designs for perpetual motion machines – systems where the energy output is greater than the input such as that in Figure 9.1.

People discussed and debated the relative meanings of what we now refer to as impulse and work:

$$\text{Impulse} = \text{transfer of momentum} = F.\Delta t \rightarrow \int F.dt.$$

$$\text{Work} = \text{transfer of energy} = F.\Delta x \rightarrow \int F.dx.$$

The key concept in understanding the conservation of energy in mechanical systems was the transfer of energy by doing work.

Figure 9.1 A 'perpetual motion' machine

9 ENERGY TRANSFER IN MECHANICAL SYSTEMS

9.1 WORK

Work is done by a force whenever it moves in its own direction of operation. It transfers energy from one system, or part of a system, to another.

Energy transferred = force F × distance moved in the
when work is done   direction of the force x.

This defines the unit of energy, the joule (J):

1 joule of energy is transferred when a force of 1 newton moves 1 metre in the direction of the force.

Note that:

- for work to be done on a body, the force, or a component of it, must act along the direction of movement of that body. Any movement at right angles to the line of action of the force does not involve a transfer of energy
- both the force and the distance moved are vectors but the energy transferred is a scalar. There is no direction associated with energy.

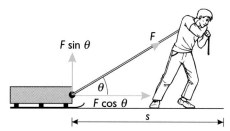

Figure 9.2

Figure 9.2 shows an example of work being done. The boy towing the sledge exerts a force F, which may be resolved into two components:

horizontally, $F \cos \theta$ draws the sledge along

vertically, $F \sin \theta$ tends to lift up the front of the sledge.

When moving the sledge a distance s horizontally, the boy transfers energy $Fs \cos \theta$. On the other hand, once the front of the sledge has been lifted, the force $F \sin \theta$ does no work because there is no movement vertically.

The relative direction of force and distance is very important. The girl in Figure 9.3 does work when she lifts up the case in the Earth's gravitational field, but does no work when carrying the case horizontally. This is because the weight of the case acts perpendicularly to her direction of motion. The muscles of her hand and arm will eventually tire because they are contracted to grip the case. While contracted, a muscle continues to transfer energy, even if it does not move, because of the processes within the muscle keeping it taut. The energy transferred from the arm muscle to the case when lifting it to a higher position can be calculated from the shaded area underneath the graph of force against distance (← 4.3).

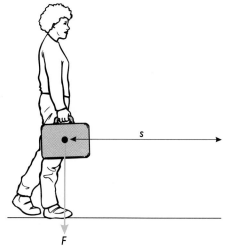

Figure 9.3

Questions

1 Estimate the number of joules of energy transferred in each of the following situations:
a climbing upstairs at home,
b stretching a rubber band to twice its length,
c accelerating a car from rest to 20 m s^{-1}.

> **2** Two barrels, each of weight 500 N, are loaded on to a lorry whose load platform is 1.5 m above the ground. One barrel is lifted directly on to the platform, while the other is slid up a plank of length 5.0 m. Neglecting frictional forces, find in each case:
> **a** the force applied to the barrel to move it,
> **b** the energy transferred in loading it.

9.2 POTENTIAL AND KINETIC ENERGY

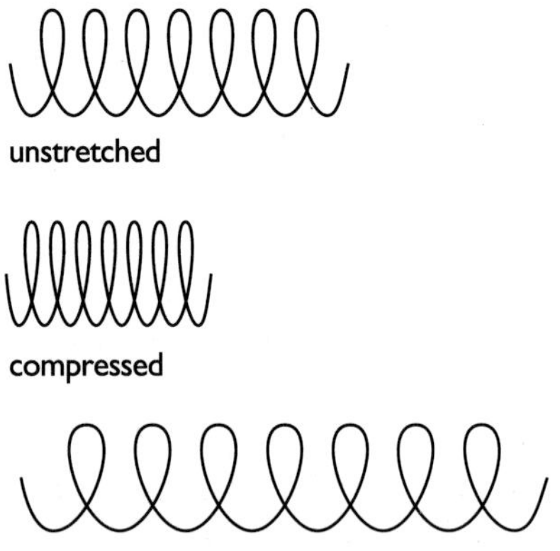

Figure 9.4 Coil springs in various states

Potential energy is stored energy. A simple example occurs when work is done on a spring to stretch or compress it (see Figure 9.4). Much of the energy transferred is stored as elastic potential energy in the spring. This potential energy enables the spring to do work, transferring energy to another system. For a spring or any other material that obeys Hooke's law,

$$\text{stored energy} = \tfrac{1}{2}kx^2,$$

where k is the stiffness of the spring (the force needed to produce unit extension) and x is the extension (← 4.3).

Potential energy also changes when you lift a suitcase off the floor and place it on a table. Its gravitational potential energy is increased as you raise it. The gravitational force equals the weight mg of the case. The distance moved in the direction of the force is the height h by which you raise the case. So, the energy transferred is the increase in potential energy of the case:

increase in potential energy = force × distance moved in direction of force

$$= mg \times h$$
$$= mgh.$$

Suppose you push the case off the table, allowing it to fall back to the floor. It will lose potential energy and gain kinetic energy as it accelerates towards the floor. Finally, energy will be transferred from the falling case to the floor as work is done on the floor, possibly making a dent.

Kinetic energy is energy associated with motion. When work is done to accelerate an object, the energy transferred increases its kinetic energy. Often its internal energy increases because of frictional forces and air resistance. If we assume that the object is completely free to move and there are no resistive forces, all the energy is transferred to the movement of the object. In this case, we can derive an expression for its increase in kinetic energy:

energy transferred to object = change in its kinetic energy.

The acceleration a is given by:

$$a = F/m$$

(← 6.4). The speed v of an object accelerated from rest over a distance s with acceleration a is given by:

$$v^2 = 2as.$$

Combining these two expressions gives:

$$v^2 = \frac{2Fs}{m}$$

or, $Fs = \frac{1}{2}mv^2$.

So the work done by the force F in accelerating an object from rest through a distance s in the direction of the force provides its kinetic energy E_k of $\frac{1}{2}mv^2$.

EXPERIMENT 9.1 The transfer of gravitational potential energy to kinetic energy

Raising one end of an air track height h above the other end and releasing a glider at the top enables you to measure the kinetic energy of the glider after falling h metres (see Figure 9.5). Alternatively, the glider can be accelerated by attaching it with a piece of string to a falling weight over a pulley. After calculating the loss in gravitational potential energy, you can compare this with the kinetic energy of the glider at the end of the track.

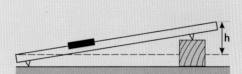

h/m	time/s	velocity/ms^{-1}
0.010	0.22	0.45
0.015	0.18	0.55
0.020	0.16	0.63
0.025	0.14	0.71
0.030	0.13	0.77
0.035	0.12	0.83

Figure 9.5 A tilted air track

Figure 9.6 Data obtained in experiment 9.1

Figure 9.6 shows some typical results obtained when a glider, of mass 0.40 kg, was released from rest at the raised end of the track. It carried a flag 0.10 m long which interrupted a light beam placed 1.00 m along the track.

Questions

3 Two runways, A and B, are set up, each having the same total length. They both start at the same height above the floor level and both end at floor level. However, they have different hills, valleys and curves, with no hill being higher than the starting point, and A having twice as many hills valleys as B.
  Starting at the same instant, two identical toy cars are rolled from rest down each runway. Ignoring frictional effects, compare their:
a potential energies at the top of the runway,
b kinetic energies when they reach the bottom of the runway,
c speeds at the bottom of the runway,
d times taken to reach the bottom of the runway.
Suppose a car of twice the mass is used on runway A. How will this affect your answers?

4 In experiment 9.1, the loss in potential energy of a falling weight is plotted against the gain in kinetic energy of a glider and a falling weight. From the data in Figure 9.6, calculate:
a the average impulse $F.\Delta t$,
b the average change in potential energy $F.\Delta x$.

CONSERVATION LAWS

EXPERIMENT 9.2 The transfer of elastic potential energy to kinetic energy

This is similar to experiment 9.1 but uses a horizontal air track, and a stretched rubber band to 'catapult' the glider along the track. By attaching a newtonmeter, or spring balance, to the rubber band you can find the force necessary to pull back the catapult over a range of distances. Plotting a graph of force against distance pulled back enables you to calculate:

i the energy stored in the band for various pull back distances,
ii the energy transferred to the glider when it is pulled back through each of these distances and released.

Questions

5 Calculate the kinetic energies of
a a car of mass 650 kg moving at 30 m s^{-1},
b a bullet of mass 0.025 kg moving at 300 m s^{-1},
c an oxygen molecule of mass 5.0×10^{-26} kg moving at 400 m s^{-1}.

6 Starting from rest, a car rolls down a hill marked with a gradient of 8 %. Neglecting friction, how fast will the car be travelling after rolling 50 m down the hill?

7 Rotating flywheels are often used as temporary energy stores in machinery. The rotational kinetic energy of a body is $\frac{1}{2}I\omega^2$ ($\leftarrow$ 7.2). Calculate the energy stored in a flywheel of rotational inertia 5.0 kg m^2 rotating at 3 000 rpm.

8 When you make measurements similar to those in experiment 9.1 on a ball rolling down a hill, the speed of the ball after falling a given distance is much less than that of the glider, even when compensated for friction. Explain this observation.

9.3 ELASTIC AND INELASTIC COLLISIONS

Investigation of the transfer of energy between objects involved in collisions enables us to identify two types of collision: those which are elastic, or 'springy', and those which are inelastic, or 'sticky'. Momentum is conserved in either situation ($\leftarrow$ 8.1), but very different results occur.

ELASTIC COLLISIONS

Elastic or 'springy' collisions are those in which **both momentum and kinetic energy are conserved**. The collision on an air track, between two gliders with magnets that repel each other is an example of an approximately elastic collision. In experiment 8.1, when a moving glider collided with a stationary glider of equal mass, the moving glider stopped and the second glider moved off with the same velocity as the original glider. This may have come as a surprise. However, you can now predict it from the two conservation rules.

A mass M travelling with velocity u collides with an equal mass at rest (see Figure 9.7). Suppose after the collision, the original mass has a velocity v_1 and the second mass a velocity v_2.

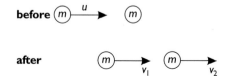

Figure 9.7 Elastic collision between two equal masses

Conservation of momentum gives: momentum before = momentum after

$$mu = mv_1 + mv_2$$

and so $u = v_1 + v_2.$

Conservation of kinetic energy gives: E_K before = E_K after

$$\tfrac{1}{2}mu^2 = \tfrac{1}{2}mv_1^2 + \tfrac{1}{2}mv_2^2$$

and so $u^2 = v_1^2 + v_2^2.$

Solving these equations for u and u^2 gives

$v_1 = 0$, and

$v_2 = u$

in agreement with the experimental observations.

Analysing two-body elastic collisions in one dimension

A mass m travelling with velocity u collides with a stationary mass M. After the collision, m is moving with velocity v and M with velocity V (see Figure 9.8). During the interaction or collision, they exert the same but oppositely directed force on each other for the same length of time. So the net change in momentum must be zero and momentum is conserved:

momentum before = momentum after
$$mu = mv + MV.$$

Assuming the collision is perfectly elastic, the total kinetic energy is unchanged:

E_K before = E_K after
$$\tfrac{1}{2}mu^2 = \tfrac{1}{2}mv^2 + \tfrac{1}{2}MV^2.$$

From the first equation $\qquad v = u - (M/m)V$

and from the second equation $\quad v^2 = u^2 - (M/m)V^2.$

Eliminating v gives $\qquad V = 2u/(1 + M/m)$

$\qquad\qquad\qquad\qquad\qquad = 2u/(1 + R),$

where R is the ratio of the masses, M/m.

Eliminating V gives $\quad v = u \cdot \dfrac{(1-R)}{(1+R)}.$

From this we can work out the final velocities of both masses for any values of the masses.

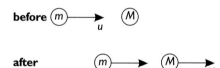

Figure 9.8 Elastic collision between two unequal masses

Many collisions between sub-atomic particles are elastic, for example, collisions between fast moving electrons and atoms in a gas discharge tube, the scattering of α-particles by gold nuclei, and the collisions between incident neutrons and the protons in paraffin wax (→ 40.2 and 40.3).

The following three situations may occur in elastic collisions:

  i the masses are equal so that $M = m$. The incident mass stops, and the second mass acquires the velocity of the incident mass. This means that
$V = u$ and $v = 0$

 ii a very small mass hits a stationary large mass so that $M \gg m$. The small mass rebounds with its velocity reversed in direction, while the large mass hardly moves. So, V is negligible and, approximately, $v = -u$

iii a large mass hits a very small stationary mass so that $M \ll m$. The large mass carries on with almost unchanged velocity. The small mass, on the other hand, is projected forward with velocity $2u$. So, approximately,
$V = 2u$ and $v = u.$

One important point to arise from a detailed analysis is that, in an elastic collision, the kinetic energy transferred between the colliding bodies is very dependent on their relative masses. If their masses are very different, small amounts of kinetic energy are transferred. In contrast, a mass colliding elastically head-on with a stationary object of equal mass transfers all of its kinetic energy to the one with which it collided.

Matching

In many physical situations, it is necessary to transfer energy. For example, the output signal energy from a tape deck or radio tuner has to be passed to an amplifier and the output from that to a loudspeaker. To transfer the maximum energy from one part of the system to the next we must **match** the output of the first part to the input of the next. Unless this happens, the signal energy will be reflected rather than transmitted. In electrical systems, transformers are often used to match the different parts of the system. The collision of particles of equal mass is a mechanical example of matching. Other examples include the gearbox of a car, the horn of a musical instrument and the blooming on the lenses of cameras and binoculars.

Questions

9 Using the three cases, i, ii and iii noted in the main text, predict what will happen in each of the following collisions.
a A shuttlecock is struck firmly by the head of a badminton racquet. The collision is elastic and the mass of the racquet is much larger than the mass of the shuttle.
b A fast moving electron collides head-on with a gas atom. The atom is much more massive than the electron and the collision is elastic.
c An α-particle collides head-on with the nucleus of a nitrogen atom in a cloud chamber. The collision is elastic and the mass of the α-particle is less than the mass of the nitrogen nucleus in the ratio 4:14.

10a Use the formula for V and v in the Theory box to show that the final kinetic energy of the initially stationary particle is $4Mm/(M+m)^2$ of the kinetic energy of the incident particle.
b From this, show that $4m/M$ of the kinetic energy is transferred between the particles when the masses are very different.
c Estimate this value of the transferred kinetic energy for the situation in question **9b**.

INELASTIC COLLISIONS

A collision is described as **inelastic** when **kinetic energy is not conserved**. After such a collision, the kinetic energy of the colliding bodies is always less than that before the collision. Usually, the drop in kinetic energy is transferred away from the systems in collision by increases in internal energy and, therefore, heat and the emission of electromagnetic or sound waves.

Generally, real collisions are inelastic. When two cars collide in a motoring accident, much of the kinetic energy of the cars is dissipated in crushing the structure of the cars. Inelastic collisions also occur at an atomic level. Low-energy electrons bombarding neutral atoms can be scattered, losing some of their kinetic energy to the atom. This excites an electron into a higher energy level in the atom. The Franck-Hertz experiments use this phenomenon (→ 37.1).

9 ENERGY TRANSFER IN MECHANICAL SYSTEMS

There are no simple formulae to predict the outcome of inelastic collisions, since kinetic energy is transferred out of the system. Each case must be worked out from the conditions prevailing at the time.

Questions

11 Why **must** the interactions of gas molecules with each other and with the walls of their container on average behave as elastic collisions?

12 Describe what would happen if a car could collide elastically with a concrete wall. What would be the effects of such a collision on the driver of the car?

13 A rubber ball is dropped on to a concrete floor from a height of 1.3 m and rebounds to a height of 1.0 m. Calculate the fraction of the energy transferred in the collision with the floor. Assuming that the fraction transferred is the same in each subsequent bounce, estimate the number of bounces the ball makes before it fails to rise 0.05 m above the floor.

14 The hammer used by a sculptor on his chisel is very different from the one used by a carpenter to drive nails into wood or that used by a panel beater to knock a dent out of a car door. What have elastic and inelastic collisions to do with these processes?

9.4 POWER

Power is defined as the **rate of transfer of energy**. The unit of power, the watt (W), equals one joule per second ($J\,s^{-1}$). Power, like energy, is a scalar quantity.

In many practical systems, the rate of transfer of energy, or power, is limited by the system itself. For example, a diesel engine powering a railway locomotive develops constant power by driving a generator at constant speed. The electrical output from the generator supplies the traction motors, which provide the force to propel the train.

The power delivered is the energy transferred per second, or the force multiplied by velocity. So, at constant power, there must be a reduction in the force available to propel the train as its speed increases, assuming it is on level track. This is shown in Figure 9.9. As the tractive force reduces, the acceleration of the train is reduced until the tractive force just balances the air resistance and frictional forces. Then the train maintains a constant speed.

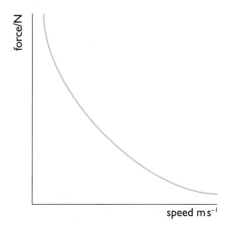

Figure 9.9 Graph of force against speed for constant power

C CONSERVATION LAWS

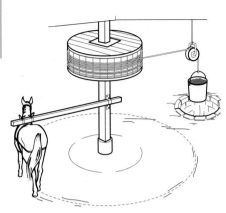

Figure 9.10 The 'horse power'

Figure 9.11 An attempt at 'human-powered' flight

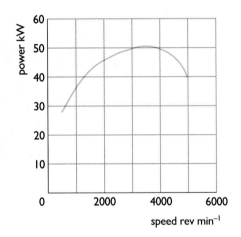

Figure 9.12 Graph of power against angular speed for a petrol engine

Questions

15 James Watt defined the first unit of power, the horse power, from the work done by a horse walking round and round in circles raising water from a mine (see Figure 9.10). One horse power is equivalent to 746 W. Estimate the mass of water raised per minute by a horse from a mine 50 m deep. What assumptions do you have to make?

16 Estimate
a the power you develop when running upstairs
b the power necessary to sustain human-powered flight, if a speed of 2 m s^{-1} is necessary to produce the required 'lift'. Assume this produces air resistance of 400 N.

17 Figure 9.12 shows how power varies with angular speed for a small petrol engine. The rotational analogue of 'power = force × speed' is

power = torque × angular speed.

Use this to sketch a graph of torque against angular speed for the same engine.

9.5 DISSIPATIVE FORCES AND EFFICIENCY

Forces often arise from deforming materials, such as springs or rubber (← 4), or from fields, such as those due to gravity and electricity (→ 15 and 16). When forces do work they transfer potential energy into kinetic energy or into potential energy in another system. **Dissipative forces**, like friction, air resistance and viscosity, also transfer energy. This energy appears as increased kinetic energy of the particles of the bodies concerned. So their internal energy increases and heating takes place.

FRICTION

Friction between solids is not a constant force. Static friction, occurs just before there is relative movement of the surfaces (← 5.5). The coefficient of static friction is usually greater than the coefficient of kinetic friction when the surfaces are moving.

9 ENERGY TRANSFER IN MECHANICAL SYSTEMS

AIR RESISTANCE

You will have noticed that you have to exert a larger force when walking into a strong wind than when walking on a calm day. The extra force necessary to keep you moving on a windy day is a dissipative or **non-conservative** force, because energy is lost by you. It is difficult to develop equations to model a force like this, because it depends on your shape and speed and any air turbulence that occurs. However, this force is increasingly important at high speeds, becoming more or less proportional to the square of your speed relative to the air (← 6).

Modern trains and cars tend to be designed so that their shapes produce the least 'drag'. Computer-aided design techniques have optimised the shape of cars, in particular, so that air resistance is at a minimum (see Figure 9.13).

VISCOSITY

Viscosity (→ 27.4) is sometimes called 'fluid friction'. It is associated with low-speed streamline flow in liquids and gases. As with the other dissipative forces, work done against viscous forces is dissipated causing increased internal energy, and the temperature rises.

Figure 9.13 Photograph of 1930s Austin 7, 1990s Metro

> **Question**
>
> **18** The resistance to motion of a car may be represented by a constant term plus a term dependent on the speed:
>
> resistive force = $250 + 1.1\,v^2$,
>
> where the force is in N and the velocity in m s^{-1}.
> Suppose the car is powered by an engine developing a constant power of 40 kW at all speeds.
> Using these data, on one set of axes plot graphs of:
>
> **a** tractive force against speed,
> **b** resistive force against speed,
>
> and so estimate the maximum speed of the car.

CONSERVATION LAWS

Conservative and non-conservative fields

The gravitational field around the Earth and other bodies has quite a remarkable property. The work that needs to be done in moving an object from one position to another in the field is totally independent of the path taken between the two positions (→ 14.3). We can illustrate this by supposing that a skier wishes to reach the top of the mountain from a village in the valley, and has a choice of routes:

i she can take a cable car, which will lift her along the shortest and steepest route to the top. So, a large force will act on her over a relatively short distance. The work done by this force on the skier will transfer energy from the supply feeding the cable car to increase her gravitational potential energy.

ii she can take a series of 'drag' lifts. These often follow less steep routes, requiring a smaller force to be exerted on the skier but over a longer distance.

Ignoring dissipative forces, such as friction, the total work done is the same in each case, as is the total potential energy gained. If the skier were to take a precipitous route down the mountain, the gravitational potential energy transferred to kinetic energy would be just the same as that exchanged on the way up – but would occur a little faster!

So, whichever route you take between two points, the same energy is needed, and the sum of the potential and kinetic energies is constant at all times. The gravitational field around the Earth or other bodies is described as a **conservative field** and the gravitational force as a **conservative force**.

In reality, the longer slower routes on a ski run will cause the skier to transfer most energy through friction. The fast routes transfer energy largely due to air resistance. This energy cannot be reclaimed by the skier. So, the frictional field and the air resistance field are **non-conservative**.

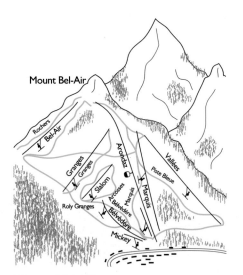

Figure 9.14 Piste map

Question

19 Figure 9.14 shows a piste map. Try to identify the least steep routes for a skier to take.

DISSIPATIVE FORCES AND CONSERVATION

Dissipative or non-conservative forces, such as friction, cause energy transfer through heating. The effect of heating an object is to increase the kinetic energy of the molecules within the object. So, in one sense, we can say that the sum of the potential energy and kinetic energy of an object is still constant. If we continue to analyse energy transfer at this more fundamental level, we find that we can explain all energy transfer in terms of changes of just potential and kinetic energy:

- chemical changes when fuel is burnt involve changes in the electrical potential energy of the electrons in the atoms of the fuel
- nuclear fuels make use of the changes in potential energy of particles within the nucleus (→ 40.6)
- light and other forms of electromagnetic radiation transfer kinetic and potential energy by means of photons (→ 36.2).

It seems that, even at the sub-atomic level, the sum total of the kinetic and potential energy within any system is constant.

EFFICIENCY

When work is done, energy is transferred into potential or kinetic energy. However, in a real situation, some work is done against dissipative forces and the temperature increases due to changes in internal energy. The **efficiency** of an energy transfer is defined as the **proportion of the energy input** to a system that is **transferred into useful output energy**.

$$\text{Efficiency} = \frac{\text{useful energy output}}{\text{total energy input}}$$

$$= \frac{\text{useful power output}}{\text{total power input}}.$$

This may be expressed as a fraction or as a percentage.

For a machine to have an efficiency of 1, it would have to transfer the total energy input into useful energy output. It would have to be an idealised machine with no friction. The efficiency of many machines varies with the conditions under which they are operated ($\rightarrow$ 12.3). The only practical ways in which energy transfers occur with an efficiency of 1 are those in which all the energy is transferred to internal energy – all the work input is done against dissipative forces.

Question

20 The food intake of a healthy human male is about 10 MJ of energy per day. He is capable of doing useful work at such a rate that he can deliver an average power of 60 W for 7 hours of the day. Estimate his overall efficiency. For what other purpose is the energy content of the food we eat used?

9.6 CONSERVATION OF MASS AND ENERGY

Many simple experiments can be done in chemistry to show that there is no detectable change in mass during a chemical change. For example, the reactants can be placed in a stoppered flask and weighed. Then, when the reaction is complete, reweighing the flask and its contents shows no detectable change in mass. It was concluded from such observations that mass is conserved independently of energy conservation.

However, in the early years of the twentieth century, Albert Einstein developed a new model of mechanics and electromagnetism called the **theory of special relativity**. This predicts that **the mass of a body is not a constant** but depends on the velocity of the body relative to the observer.

According to the theory of special relativity, the mass m of a body moving with velocity v relative to the observer is given by:

$$m = \frac{m_0}{(1 - (v^2/c^2))^{\frac{1}{2}}}$$

where m_0 is the rest mass, or the value of the mass when stationary relative to the observer, and c is the speed of electromagnetic waves (see Figure 9.15).

All our common experience is of bodies moving with velocities very much less than c. In such cases $m = m_0$ and the mass does not appear to vary. Only when dealing with sub atomic particles moving at speeds approaching c will the change be detectable, and it is indeed detected in such cases.

The theory of special relativity also predicts the **equivalence of mass and energy**. Energy change ΔE has an associated mass Δm given by:

$$\Delta m = \Delta E / c^2$$

or $\quad \Delta E = \Delta m c^2.$

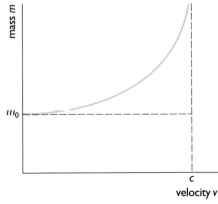

Figure 9.15 Relativistic variation of mass with velocity

CONSERVATION LAWS

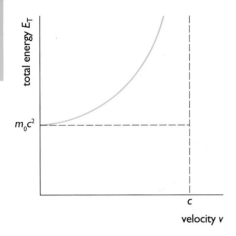

Figure 9.16 Relativistic variation of total energy with velocity

Evidence to support the equivalence of mass and energy comes from nuclear physics. Detailed measurements show that the mass of a nucleus is always less than the total mass of its component protons and neutrons. This **mass defect** represents the energy needed to separate the component parts of the nucleus – the **binding energy**.

The equation $\Delta E = \Delta m c^2$ relates the change in mass Δm in forming the nucleus to the energy ΔE released on its formation from individual particles. This relationship is found, experimentally, to hold consistently. It is extremely useful in nuclear physics and nuclear engineering ($\rightarrow 40$).

Question

21 Use the data below to work out the total nuclear binding energy of a helium atom made from two hydrogen atoms and two neutrons.

$$2\,^1_1\text{H} + 2\,^1_0 n \rightarrow\,^4_2\text{He}$$

mass of hydrogen atom = 1.6730×10^{-27} kg

mass of neutron = 1.6744×10^{-27} kg

mass of helium atom = 6.6443×10^{-27} kg

Total energy and mass

The theory of special relativity predicts expressions for the momentum and total energy of a body. The momentum p is given by

$$p = mv$$
$$= \frac{m_0 v}{(1 - (v^2/c^2))^{\frac{1}{2}}}.$$

At low speeds ($v \ll c$), the expression gives

$$p = m_0 v,$$

which agrees with Newton's laws ($\leftarrow 6.4$).

Using the relativistic expression for momentum enables the conservation of momentum to be extended to collisions between high-energy sub-atomic particles. The relativistic expression for the total energy E_T is:

$$E_T = \frac{m_0 c^2}{(1 - (v^2/c^2))^{\frac{1}{2}}}.$$

Figure 9.16 shows the total energy as a function of velocity. The expression for E_T can be expanded using the binomial series to give:

$$E_T = m_0 c^2 + \tfrac{1}{2} m_0 v^2$$

at low velocities where $v \ll c$. In this equation, the total energy is the sum of the rest energy $m_0 c^2$ and the kinetic energy of the mass.

SUMMARY

Work transfers energy from one system to another. The energy transferred is equal to the force multiplied by the distance moved in the direction of the force. Its unit is the joule (J).

Potential energy is stored energy and depends on position. In a stretched spring it is $\tfrac{1}{2} k x^2$, and in a gravitational field it is mgh.

Kinetic energy is energy due to movement and depends on velocity. For a moving body with velocity v, it is $\tfrac{1}{2} m v^2$.

Elastic collisions conserve both momentum and kinetic energy. In inelastic interactions, some of the energy is transferred by heating and dissipative forces, such as friction, air resistance and viscosity, so increasing the internal energy of nearby bodies.

Power is the rate of transfer of energy. The unit of power is the watt (W).

The efficiency of a system is the ratio of the useful energy output to the total energy input.

The conservation of energy includes mass, using the Einstein mass/energy equation $E = mc^2$.

SUMMARY QUESTIONS

Some of these questions also involve the conservation of momentum ($\leftarrow 8$).

22 Two masses, of 5.0 kg and 8.0 kg, are held together by an explosive bolt. When detonated, a small explosive charge liberates 25 J as the kinetic energy of the masses moving apart. Calculate the velocity of each mass after the explosion.

9 ENERGY TRANSFER IN MECHANICAL SYSTEMS

The rest mass term m_0c^2 leads to the idea of Einstein's famous equation:

$$E = mc^2$$

or $\quad \Delta E = c^2 \Delta m$

for the total energy, so linking energy to mass. This broadens the idea of the conservation of energy to include the conservation of mass.

A quantity of energy E has a mass of E/c^2.

23 A bullet of mass 25 g moving at 400 m s^{-1} embeds itself in a box of sand of mass 20 kg suspended as shown in Figure 9.17. Calculate the height h to which the box will rise after the bullet hits it.

24 A jet engine, powering an aircraft travelling at 250 m s^{-1}, takes in 70 kg s^{-1} of air and burns fuel at a rate of 5 kg s^{-1}. The combustion products leave the engine at 350 m s^{-1}. Estimate the thrust and power of the engine.

25 A trolley of mass 25 kg moving at 2 m s^{-1} collides with a spring buffer and is brought to rest. The spring constant of the buffer is 1600 N m^{-1}. Estimate the compression of the spring when 85 % of the kinetic energy of the trolley is stored in the spring.

26 The pulley system shown in Figure 9.18 raises a mass of 7 kg using a force of 25 N. Calculate **a** the work done in raising the load through 0.5 m, **b** the distance moved by the 25 N force when the load moves 0.5 m, **c** the efficiency of the machine.

27 When an electron and a positron meet, they annihilate each other and two γ-ray quanta are produced (→ 40.3). The momentum of a γ-ray quantum is equal to its energy/c^2.
a Use $E = mc^2$ to find the energy carried by the two γ-ray quanta, given that both electron and positron have a mass of 9.1×10^{-31} kg. (c)
b Why are two γ-ray quanta produced?

28 The patterns of motion of the suspended spheres in Newton's cradle (see Figure 9.19) can be explained using the elastic collision ideas above. When one or more spheres are pulled aside and released each ball in sequence makes a collision with its neighbour, even when they appear to be touching. Use these ideas to explain the complete motion when **a** one ball is, **b** two balls are pulled aside and released.

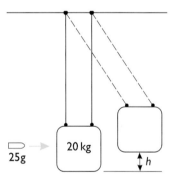

Figure 9.17 Ballistic balance

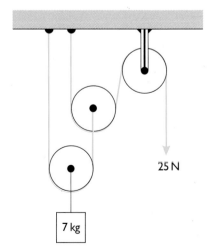

Figure 9.18 Force of 25 N lifting 7 kg using three pulleys

Figure 9.19 Newton's cradle

10 MODEL FOR AN IDEAL GAS

You can recognise gases by their well-defined properties: they expand to fill the container in which they are placed, they are easy to compress, and they have densities one thousand or more times smaller than liquids or solids. When you increase the pressure on most gas samples at normal room temperature, their densities increase proportionally to the pressure increase. This leads to Boyle's law. Similarly, when heating a gas sample, we find that its volume increases proportionally to its absolute temperature, leading to Charles's law.

These laws apply to an 'ideal' gas, under all conditions. Actual gases, such as hydrogen, helium and oxygen, behave like ideal gases at normal temperatures and pressures. However, at high pressures, when its molecules are very close together, the attractive forces between the molecules tend to alter the properties of a gas, so that it does not obey these gas laws.

To predict the behaviour of gases, you need a model for the molecules of a gas which provides these properties. It must also contain the important ideas associated with the **conservation of energy** and **momentum**.

10.1 THE KINETIC MODEL

Imagine a gas to consist of a very large number of small particles with large random velocities, rattling around in a container. This is the basis for the kinetic model of gases, one of the most successful models in physics. Starting from some simple assumptions, you can predict correctly the behaviour of gases in a wide range of situations. To build the model, you can apply Newton's laws of motion to these particles. You will need to be familiar with these laws and with the principle of the conservation of momentum, and of energy in mechanical systems ($\leftarrow$ 6.4 $\leftarrow$ 8 $\leftarrow$ 9).

Evidence to support the kinetic model comes from observations, such as Brownian motion. When smoke or dust particles are viewed under a microscope they are seen to be in a jittery random motion. This was first observed by the botanist Robert Brown in 1827, using pollen grains suspended in water. The motion can be interpreted as the bombardment of the particles by hordes of small, fast moving air molecules moving at random, a model put forward by W. Ramsay in 1876.

10.2 THE PROPERTIES OF AN IDEAL GAS

Experiments carried out by Robert Boyle in the seventeenth century led him to formulate the well-known **Boyle's law** in 1660. He found that when a sample of gas is compressed, the pressure and volume of the gas are related by the equation:

pressure × volume = a constant

or $PV = $ a constant

for a particular mass of gas at constant temperature (see experiment 10.1 and, particularly, Figure 10.3).

EXPERIMENT 10.1 Compressing a gas at constant temperature

You can compress a sample of gas at constant temperature using the apparatus shown in Figure 10.1 a or b.

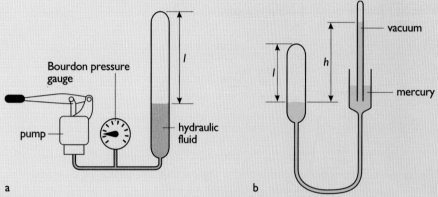

Figure 10.1 Apparatus for compressing a gas at constant temperature

P/kPa	l/cm
101	15.8
150	10.5
200	8.0
250	6.4
300	5.3
350	4.5
400	4.0

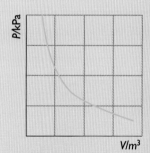

Figure 10.2 Data on pressure and volume for a fixed mass of gas at constant temperature

Figure 10.3 Graph of pressure against volume for a fixed mass of gas at constant temperature

Figure 10.2 tabulates data taken using apparatus of type a. A mass of air is trapped in a glass tube of uniform cross-section, so the volume of air is proportional to the length of the air column. Measurements taken of the pressure P in the system and length l of the air column are used to plot Figure 10.3. Whatever method you use, satisfy yourself that it gives similar characteristics.

When the temperature of the gas in a container is changed, the pressure changes. If the container is able to change in volume, the change in temperature can produce a volume change at constant pressure. So, we can investigate the expansion of gases by changing one variable at a time. We can observe the change in volume with temperature at constant pressure and the change in pressure with temperature at constant volume, as in experiments 10.2 and 10.3.

EXPERIMENT 10.2 Heating a gas at constant pressure

Figure 10.4 shows how you can investigate the expansion of a sample of a gas heated at constant pressure in a capillary tube. You can trap the sample in the tube with a drop of mercury or concentrated sulphuric acid.

The volume of the gas is proportional to the length l of the gas and, because the top of the capillary tube is open, the pressure is atmospheric.

A typical set of data from this experiment is plotted in Figure 10.5.

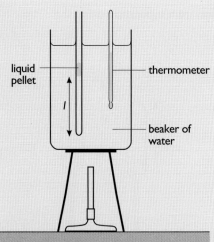

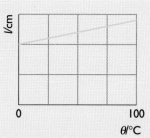

Figure 10.4 Apparatus for measuring the expansion of a gas at constant pressure

Figure 10.5 Graph of volume against temperature for a fixed mass of gas at constant pressure

EXPERIMENT 10.3 Heating a gas at constant volume

You can investigate the change in pressure for a heated gas sample kept at constant volume using the apparatus shown in Figure 10.6 a or b.

A typical set of data from this experiment is plotted in Figure 10.7.

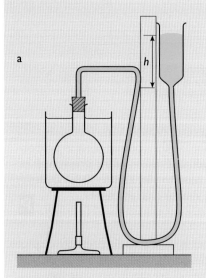

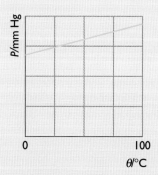

Figure 10.6 a Constant volume gas thermometer immersed in a large beaker of water b Bourdon gauge attached to a flask

Figure 10.7 Graph of pressure against temperature for a fixed mass of gas at constant volume

10 MODEL FOR AN IDEAL GAS

Real gases and liquefaction

At temperatures greater than 100 °C, you might expect a real gas to continue to expand, or the pressure to continue to rise. This is the case for most gases. However, as the temperature is reduced well below 0 °C, gases liquefy causing the volume to fall dramatically. For most real gases this happens at atmospheric pressure or above. But below a particular temperature, known as the **critical temperature**, liquefaction does not occur, however high the pressure. This is because the gas molecules remain too far apart for them to attract one another to form the liquid state and are so energetic that they are not, on average, ever close enough to do so. Figure 10.8 shows critical temperatures and the associated critical pressures for some gases.

The data from experiments 10.2 and 10.3, plotted in Figures 10.5 and 10.7, show that both **volume and pressure vary linearly with temperature**, if the other is kept constant during the process.

So, if pressure is kept constant, the volume's variation with temperature may be written:

volume = constant × temperature (in °C) + volume at 0 °C.

This is known as Charles's law. Similarly, if volume is kept constant, the pressure's variation with temperature may be written:

pressure = constant × temperature (in °C) + pressure at 0 °C.

When each graph in Figures 10.5 and 10.7 is extrapolated, to reach the temperature axis, it cuts it at about −273 °C. More accurately, this occurs at **−273.15 °C**, which is normally referred to as **absolute zero**. From this temperature, the volume and the pressure are proportional to temperature. Figure 10.9 shows a graph of pressure against temperature with the origin moved to zero pressure and −273 °C.

You can now define an absolute or thermodynamic temperature scale. The Kelvin scale takes absolute zero as 0 K, with the melting point of water at 273.15 K.

The ideal gas temperature scale

You can use the properties of an ideal gas to define a temperature scale in which the pressure exerted by the ideal gas at constant volume, or the volume occupied by the ideal gas at constant pressure, is **proportional** to the temperature. This is called the **ideal gas temperature scale**.

A look at Figure 10.9 shows that the ideal gas temperature may be made equal to the celsius temperature plus 273.15 °C. The ideal gas scale is identical to the Kelvin thermodynamic scale, where the unit of temperature is measured in kelvin (K) and the absolute zero of temperature, 0 K, is defined as −273.15 °C.

So, keeping pressure the same, the volume's variation with temperature may be written:

$$\text{a constant} = \frac{\text{volume}}{\text{temperature in °C} + 273.15}$$

$$= \frac{\text{volume}}{\text{temperature in °C above absolute zero}}$$

$$= \frac{\text{volume}}{\text{absolute temperature in kelvin (K)}}.$$

The value of the constant, in $m^3 K^{-1}$, depends on the amount of gas present and the value of the constant pressure.

An otherwise identical relationship for the variation of pressure with temperature was first deduced from measurements by Amontons in 1702. In this case, the constant is in units of pascals per kelvin ($Pa\ K^{-1}$).

At a precise temperature and pressure, solid, liquid and vapour can exist together in equilibrium for a given substance. This is known as its **triple point**, and is always exactly reproducible for a pure liquid. The pressure at the triple point of water is 610.5 Pa, which is much less than atmospheric pressure. The temperature at the triple point is just 0.01 K above its freezing point under normal atmospheric pressure – that is, 0.01 °C or 273.16 K. This precise temperature of the triple point of water is used to define the kelvin, one of the base units of SI:

the **kelvin (K)** is 1/273.16 of the thermodynamic temperature of the triple point of water.

gas	critical temperature /K	critical pressure /MPaA
He	5	0.23
H_2	33	1.29
O_2	155	5.06
H_2O	647	21.8
CO_2	304	7.36

Figure 10.8 Critical pressures and temperatures for various gases

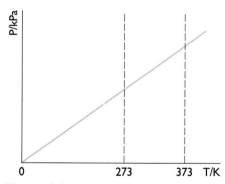

Figure 10.9 Graph of pressure against temperature for a fixed amount of ideal gas at constant volume

CONSERVATION LAWS

EQUATION OF STATE

Boyle's and Charles's laws, derived from experiments 10.1–10.3, can be used to define the properties of an ideal gas in an equation of state:

$$\frac{\text{pressure} \times \text{volume}}{\text{temperature in K}} = \text{a constant}.$$

The constant depends on the unit used and the amount of gas present. For unit mass, the constant is the gas constant per kg of that particular gas. Often, the amount of gas considered is 1 mole. Then the constant is the **universal gas constant** R. When n moles are considered, the constant is written in the form nR.

In SI units, the pressure is in Pa, the volume in m³ and n is the quantity of gas in moles. Then, for n moles:

$$PV = nRT$$

where R has the value 8.3 J mol⁻¹K⁻¹.

Questions

1 A motor car tyre is inflated to a total pressure of 250 kPa at 10 °C. At the end of a journey, the tyre has warmed up to 30 °C. Estimate the pressure in the tyre at 30 °C, stating any assumptions you make.

2 A high-altitude balloon is only partially filled at ground level so that, even at altitude, the skin is not taut. It contains 2 mol of an ideal gas. Calculate the volume of the gas bubble in the balloon at:
a ground level, where the pressure is 100 kPa and the temperature is 15 °C,
b its operating altitude, where the pressure is 6.0 kPa and the temperature is –30 °C.

3 a When using a gas thermometer, such as that in Figure 10.6a, why is it important that the volume of the tube connecting the bulb to the manometer is negligible compared with the volume of the bulb?
b Mercury is commonly used in gas thermometers and other gas handling apparatus to seal the gas space and measure the pressure. Apart from the dangers associated with its vapour, why is it considered to be such a suitable material?

10.3 THE KINETIC MODEL AT WORK

When using the kinetic model, you can assume the following properties for the particles of an ideal gas:

　i the particles occupy a negligible volume compared with that of the container,
　ii the collisions between particles are elastic and so kinetic energy is conserved,
　iii the particles exert negligible forces upon each other except when colliding,
　iv the collisions take a negligible time compared with the time between collisions.

You also have to make assumptions about the motions of the particles – that these are random in direction and have a distribution of speeds such that, in a given time, as many gain as lose a given speed through collision. This results in a particular distribution of speeds known as the **Maxwell–Boltzmann distribution** (see Figure 10.10). The distribution is not symmetrical: there are a few particles moving very slowly, a large number moving at speeds near to the most probable speed at the peak and some particles moving much faster than this. Because the distribution is not symmetrical, the average speed is greater than the most probable speed (→ 28.4).

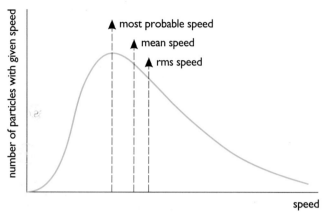

Figure 10.10 Maxwell-Boltzmann distribution of speeds

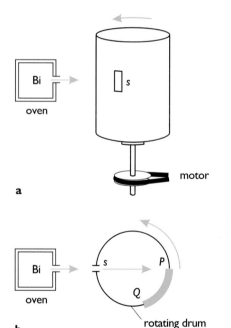

Figure 10.11 Zartmann and Ko experiment a general view b plan view

Questions

4 Suggest one prediction that would result from each of the following changes in the assumed properties of the particles:
a the volume occupied by the particles is not negligible compared with that of the container,
b kinetic energy is lost, on average, in collisions,
c the particles exert attractive forces on each other at all distances,
d the particles stick together on collision for a significant portion of the total time.

5 The Maxwell-Boltzmann distribution of speeds was demonstrated experimentally by Zartmann and Ko in 1930. Their apparatus is shown diagramatically in Figure 10.11. Groups of bismuth atoms from the oven enter the rotating drum through the slit s, when the slit and the oven opening coincide. They drift across the drum and are deposited between P and Q on the far side.
a Assuming the drum rotates at a constant rate, why do all the bismuth atoms not arrive at the same point on the far side of the drum?
b Do the fastest atoms arrive nearer to P or Q?
c Given that the atoms have a Maxwell-Boltzmann distribution of speeds, describe how you would expect the thickness of deposit to vary between P and Q.

10.4 INTERPRETING PRESSURE USING THE KINETIC MODEL

Particles rattling around in a box collide with the walls of the box, as shown in Figure 10.12. Although an individual particle's kinetic energy is conserved in

CONSERVATION LAWS

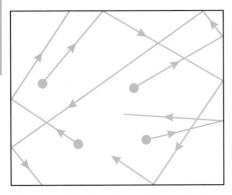

Figure 10.12 Particles colliding with the walls of a box

each collision, its momentum changes as it is reflected during its interaction with one of the walls. Over many collisions, the average force exerted on the wall is dependent on the values of:

i the average change in momentum per collision,
ii the average number of collisions with the wall per second.

The average number of collisions per second depends on the density of particles and on their average speed.

The pressure is the average value of this force per unit area. It is given by the expression:

$$P = \tfrac{1}{3}\rho \overline{c^2}$$

or $\quad P = \tfrac{1}{3}\dfrac{mN\overline{c^2}}{V}$

where P is the pressure, ρ the density of the gas, $\overline{c^2}$ the mean of the square of the speed of the particles, m the mass of each particle and N the number of particles in volume V.

THE AVERAGE SPEED OF PARTICLES

There are several ways of calculating the average speed of the particles. One is the mean speed c. For N particles with speeds $v_1, v_2, v_3 \ldots v_N$ the mean speed is given by:

$$\overline{c} = \frac{v_1 + v_2 + v_3 + \cdots v_N}{N}.$$

However, the pressure from the collisions of the particles with the walls depends on the momentum change in each collision and the number of collisions per second. Both of these are proportional to the speed of the particle, so pressure is proportional to v^2 for each particle. Therefore, the average must be calculated from the mean value of the speed squared:

$$\overline{c^2} = \frac{v_1^2 + v_2^2 + v_3^2 + \cdots v_N^2}{N}.$$

This is the mean square speed, which is directly related to the average kinetic energy of the gas molecules and, therefore, the temperature of the gas (→ 10.5).

A third average speed often used is the root mean square speed, c_rms, or $\sqrt{(\overline{c^2})}$.

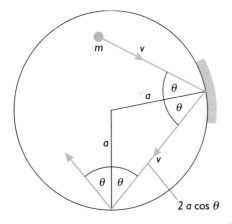

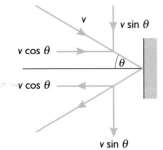

Figure 10.13 Collisions of one particle in a spherical container

The pressure equation

Consider a sphere of radius a, containing only one particle of mass m moving with speed v, as shown in Figure 10.13. Every time the particle collides elastically with the wall, its component velocity parallel to the wall is unchanged and its component velocity perpendicular to the wall is reversed in direction. The momentum of the particle, measured perpendicular to the wall, changes from $+mv\cos\theta$ towards the wall to $-mv\cos\theta$ away from the wall. So the change in momentum is $2mv\cos\theta$.

The force F perpendicular to the wall is the change in momentum per second. This equals the change in momentum per collision multiplied by the number n of collisions per second:

$$F = 2mv\cos\theta \times n.$$

Now, $\quad n = \dfrac{\text{distance travelled per second}}{\text{average distance between collisions}}$

10 MODEL FOR AN IDEAL GAS

$$= \frac{v}{2a\cos\theta},$$

so $F = mv^2/a$.

This force is distributed over the whole of the surface of the container. So the average pressure due to this one particle is given by:

$$\text{average pressure} = \frac{\text{force}}{\text{surface area of spherical container}}$$

$$= \frac{mv^2/a}{4\pi a^2}$$

$$= \frac{mv^2}{4\pi a^3}.$$

Since the volume of the container V is $4\pi a^3/3$,

$$\text{average pressure} = \frac{mv^2}{3V}.$$

This is the pressure due to only one particle. With N particles, the pressure will be N times as great and given by:

$$\text{pressure} = mv_1^2/3V + mv_2^2/3V + mv_3^2/3V + \cdots$$

$$= \frac{m}{3V}\sum_1^n v_n^2.$$

Since the particles do not all move at the same speed, the average value of v^2, the mean square speed $\overline{c^2}$, is used.

$$\overline{c^2} = \frac{1}{N}\sum_1^n v_n^2.$$

So the pressure P exerted on the walls is given by:

$$P = \frac{1}{3}\frac{mN\overline{c^2}}{V}.$$

The total mass of the particles is mN, and mN/V is the density ρ. Substituting this gives:

$$P = \frac{1}{3}\rho\overline{c^2}.$$

The root mean square speed

In deriving the expression for the pressure exerted by the particles of the gas, the mean square speed $\overline{c^2}$ was used (see box above). In most discussions of the kinetic model in which an *average* speed is required the square root of the mean square speed c_{rms} is used in place of the mean speed $\overline{c}$. c_{rms} is called the *root mean square* speed.

Questions

6 Ten particles have speeds, in m s^{-1}, of 50, 100, 120, 130, 135, 130, 140, 150, 170, 200. Calculate
a the mean speed $\overline{c}$,
b the mean of the speeds squared $\overline{c^2}$, and
c the square root of the mean of the speeds squared $\sqrt{(\overline{c^2})}$, c_{rms}.

7 For the Maxwell-Boltzmann distribution, would you expect the rms speed to be different from the mean speed? Explain your answer.

10.5 TEMPERATURE AND INTERNAL ENERGY

INTERPRETING TEMPERATURE USING THE KINETIC MODEL

In the kinetic model, the temperature of a gas is imagined to be related to the mean kinetic energy of the particles making up that gas. The kinetic energy of the particles is small at low temperatures and large at high temperatures.

CONSERVATION LAWS

The total kinetic energy of the gas particles $= \Sigma \tfrac{1}{2} mv^2$
$= \tfrac{1}{2} mN\overline{c^2}$,

where m is the mass of each particle, N the number of particles and $\overline{c^2}$ the mean square speed of the particles.

Comparing this with the relation for the pressure exerted by the gas

$$P = \frac{mN\overline{c^2}}{3V}$$

gives $PV = \tfrac{1}{3} mN\overline{c^2}$.

From the equation of state (← 10.2), for n moles of an ideal gas:

$PV = nRT$.

Combining this equation with that from the kinetic model relates the particles' energy to temperature:

$$PV = \tfrac{1}{3} mN\overline{c^2}$$
$$= \tfrac{2}{3} (\tfrac{1}{2} mN\overline{c^2}).$$

PV is two-thirds of the total kinetic energy of the particles and equals nRT. So the total kinetic energy of the particles of an ideal gas is directly proportional to the temperature on the kelvin scale. This is a very important link joining the concept of temperature to that of energy.

Now, $N = nN_A$, where N_A is the Avogadro constant – the number of particles in a mole (→ 35.2). Also, mN_A is the molar mass M of the gas. So $nM = mN$ and the expression becomes:

$$PV = \tfrac{2}{3} (\tfrac{1}{2} nM\overline{c^2})$$
$$= nRT.$$

The total kinetic energy of n moles is $\tfrac{1}{2} nM\overline{c^2} = \tfrac{3}{2} nRT$.

Each mole contains N_A particles. So, the kinetic energy of n moles of any gas, regarded as an ideal gas, depends only on the temperature of the gas and not on the identity of the gas. We can take this argument a little further and consider what the average kinetic energy of one particle must be. Since there are N_A particles in a mole, then:

$$\overline{E_k} = \tfrac{3}{2} \frac{nRT}{nN_A}$$

$$= \tfrac{3}{2} \frac{RT}{N_A}$$

$$\overline{E_k} = \tfrac{3}{2} kT.$$

The constant k is called the **Boltzmann constant** and is the gas constant per particle.

Questions

8 Use the link between temperature and kinetic energy to make the following calculations for gases, assuming that they behave as the ideal gas.
a Calculate the total energy and root mean square speed of oxygen molecules at a warm room temperature of 300 K. The mass of a mole of oxygen gas is 32×10^{-3} kg.

10 MODEL FOR AN IDEAL GAS

b The molar mass of hydrogen gas is 2.0×10^{-3} kg. Do hydrogen molecules move faster or slower than oxygen molecules at a given temperature? Calculate the ratio between their speeds. (R)

9 Increasing the temperature of a gas generally increases the rms speed of its particles. Predict what will happen to the distribution of speeds of the particles as the temperature rises. Will there still be some very slow particles? Will the most probable speed rise?

INTERNAL ENERGY

The energy content of a gas is made up of many components. There are the kinetic energies of translation, rotation and vibration of the particles. There are also many potential energies such as that arising from the attraction of the particles for each other.

In setting up the ideal gas model we have assumed that the particles are very small and without any structure, such as molecular bonding. So, the kinetic energies of rotation and vibration are assumed to be negligible. The forces of attraction between particles are also neglected. So the internal energy of an ideal gas consists only of the kinetic energy of translation of its particles.

On this simple model, temperature and internal energy are proportional at all temperatures ($\rightarrow$ 28.1). At the absolute zero of temperature, the internal energy of the ideal gas is zero.

10.6 HOW CLOSE ARE REAL GASES TO THE 'IDEAL'?

The ideal gas is defined to obey $PV = nRT$ for all values of P, V, n and T. The ideal gas temperature scale is defined so that $P \propto T$ at all temperatures. How useful is this ideal gas model for real gases?

Experiments such as 10.2 and 10.3 show that air behaves nearly like an ideal gas. At temperatures near to room temperature, gases such as oxygen, nitrogen, hydrogen and helium behave as ideal gases within a precision of ±1% for pressures of between $\frac{1}{2}$ and 5 atmospheres (50–500 kPa). So, normally, your experiments will show good agreement with ideal gas behaviour.

Figure 10.14 shows PV/RT plotted against P for 1 mol samples of gas at 300 K. For the ideal gas, PV/RT equals 1 at all pressures. The deviations from this of real gases increase at high pressures. Some gases such as carbon dioxide can be liquefied by increasing the pressure at room temperature. They are not behaving as ideal gases under these conditions.

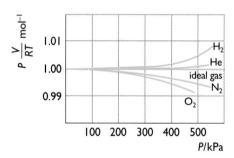

Figure 10.14 Comparison between real gases and the ideal gas

10.7 DIFFUSION AND EFFUSION

When you enter a room containing some hyacinths in flower, you are immediately aware of their scent. Molecules evaporated from the flowers diffuse through the air and can be detected throughout the room. A familiar laboratory demonstration of diffusion uses a drop of liquid bromine in a stoppered flask. As the bromine evaporates, the gas in the flask becomes uniformly brown. Although the bromine molecules are heavier than oxygen or nitrogen molecules, the random collisions between the large numbers of oxygen and nitrogen molecules and the small number of bromine molecules results in the bromine being randomly and uniformly distributed throughout the flask.

It can be difficult to make quantitative measurements of diffusion. However, it is easy to measure **effusion**, which is the rate at which a gas effuses or leaks out of a small hole in a container (→ experiment 10.4).

EXPERIMENT 10.4 Comparison of rates of effusion

Using the apparatus shown in Figure 10.15, you can compare the rates of effusion of gases by recording the time for the same volume of each gas to effuse through a pinhole in a piece of aluminium foil. Typical data from the experiment are plotted in Figure 10.16.

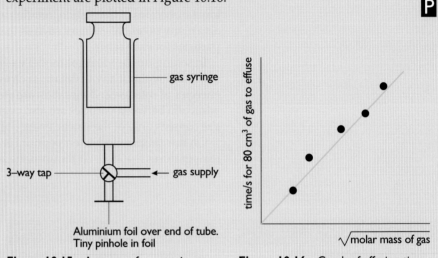

Figure 10.15 Apparatus for experiment 10.4

Figure 10.16 Graph of effusion time against $\sqrt{\text{molar mass}}$

Figure 10.16 shows the time of effusion of 50 cm³ plotted against the square root of the molar mass of various gases. The straight line graph suggests that time is proportional to molar mass. The time per 50 cm³ is proportional to 1/(rate of effusion). Combining these gives

$$\text{rate of effusion} \propto \frac{1}{\sqrt{\text{molar mass of gas}}}.$$

This empirical relationship is called **Graham's law of effusion**, and can be predicted from the kinetic model.

The kinetic model suggests that faster particles should effuse more rapidly because they have more collisions with the wall and, therefore, more chance of 'hitting' the pinhole. The rate should be proportional to the average speed, and also to the rms speed. At a given temperature, the faster particles should be those with smaller masses, suggesting that the rate must be higher for gases with small molar mass M. All this is borne out in practice, showing that the kinetic model is an extremely effective one. For example, according to the model:

$$c_{\text{rms}} \propto 1/\sqrt{M}$$

giving rate $\propto 1/\sqrt{M}$ and predicting Graham's law.

10 MODEL FOR AN IDEAL GAS

SUMMARY

The kinetic theory of gases models the behaviour of gases in terms of particles in rapid random motion. It can be used to predict the properties of the ideal gas, which obeys the relation $PV = nRT$ for all values of P, V and T. Direct evidence for the kinetic model comes from Brownian motion.

The kelvin temperature of a gas is directly proportional to the internal energy of the gas. The internal energy of the ideal gas equals the kinetic energy of its particles.

SUMMARY QUESTIONS

10 Explain in terms of the kinetic model
a Brownian motion,
b diffusion,
c Dalton's law of partial pressures which states: the pressure exerted by a mixture of gases equals the sum of the pressures which would be exerted by each component occupying the volume alone.

11 A toy balloon is inflated at a pressure of 10^3 kPa to a diameter of 0.30 m on a day when the temperature is 10 °C. Estimate its diameter when the temperature rises to 23 °C, assuming the pressure inside remains constant.

12 a A flask of volume 1.0 dm^3 is connected by a narrow tube to another of volume 3.0 dm^3. Both are at 300 K and the pressure is 120 kPa throughout. Calculate the amount of gas in the flasks.
b The larger flask is immersed in boiling water at 373 K. Calculate the new pressure throughout, and the amount of gas transferred along the connecting tube.

13 Estimate the rms speed and internal energy per mole of argon gas at 400 K. The molar mass of argon = 40 g mol^{-1}.

14 Explain the origin of the force causing the lift in **a** a hot air balloon, **b** a balloon filled with helium gas.

11 ENERGY CONSERVATION IN MECHANICAL, THERMAL AND ELECTRICAL SYSTEMS

In a power station, there seems to be no efficient way of transferring the chemical potential energy in coal, oil or gas directly into electricity. A similar situation applies to nuclear fuels. It is necessary to transfer the energy between thermal, mechanical and electrical systems before it can be used to do useful work.

Energy transfer frequently occurs between mechanical, thermal and electrical systems. We can link them using the model of an ideal gas ($\leftarrow$ 10) and energy transfer in electrical circuits ($\rightarrow$ 12.2). In any model for a combined system, the internal energy must be related to its temperature, and energy must be transferred by heating and working. Such a model must also be consistent with the law of conservation of energy.

11.1 HEATING AND WORKING

The particles of a gas are moving in random directions. So the average velocity for the whole volume occupied by the gas is zero. However, the average kinetic energy of each gas particle is not zero, but depends directly on the absolute, or kelvin, temperature ($\leftarrow$ 10.5). The internal energy of the gas is the sum of the kinetic energies of each particle.

The average linear, or translational, kinetic energy of a gas molecule is $\frac{3}{2}kT$, where k is the Boltzmann constant ($\leftarrow$ 10.5). For a mole of an ideal gas, the internal energy is $\frac{3}{2}RT$, where $R = N_A k$ and N_A is the Avogadro constant, ($\rightarrow$ 35.2). So, when we increase the temperature of a mole of an ideal gas by 1 K its internal energy rises by $\frac{3}{2}R$. This is known as the **molar heat capacity**. It is similar to **specific heat capacity** but refers to 1 mol rather than a unit mass such as 1 kg.

From this, you might expect that real gases that behave like the ideal gas model would all have the same value for the molar heat capacity:

$C = \frac{3}{2}R = 12.5 \text{ J mol}^{-1}\text{K}^{-1}$.

A glance at the table in Figure 11.1 will show you that this is not so. If

gas	C_p/J mol^{-1} K^{-1}	C_v/J mol^{-1} K^{-1}
helium (He)	21	12.5
argon (Ar)	21	12.5
oxygen (O_2)	29	21
hydrogen (H_2)	28	20
nitrogen (N_2)	29	20.5
carbon dioxide (CO_2)	37	30
steam (H_2O)	36	25.5
sulphur dioxide (SO_2)	41	–
methane (CH_4)	35	–

Figure 11.1 Molar heat capacities of gases at 273 K

11 ENERGY CONSERVATION IN MECHANICAL, THERMAL AND ELECTRICAL SYSTEMS

you carried out experiments 10.2 and 10.3, you made measurements at constant volume and at constant pressure. These specialised conditions are also used when measuring the heat capacities of gases. So, two molar heat capacities are quoted for each gas: C_v at constant volume and C_p at constant pressure.

To understand the apparent poor agreement with the kinetic model, it is necessary to look in more detail at the energy transfers in heating and cooling gases.

Questions

1 Using the data in Figure 11.1, compare the specific heat capacities of hydrogen and oxygen at constant pressure. Why do you think the values are so different?

2 Explain why most experimental data are measured at constant pressure rather than at constant volume.

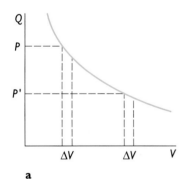

Figure 11.2 Applying a force to an ideal gas in an ideal pump cylinder

THE ENERGY TRANSFERRED IN COMPRESSING AN IDEAL GAS

When you pump up a bicycle tyre, the pump cylinder becomes warm. Some of this heating comes from friction between the piston and the cylinder but much of it is due to the work done in compressing the air in the pump.

Imagine 1 mole of ideal gas trapped in an ideal pump cylinder with a frictionless piston at one end, as in Figure 11.2. A force F applied to the piston compresses the gas by moving the piston in a distance Δx. The energy W transferred as work is done by the force is $F\Delta x$. The pressure P on the gas is the force F per unit area. The area over which the force acts is A, providing a pressure of F/A. So:

$$W = F\Delta x = PA\Delta x = P\Delta V,$$

where ΔV is the change in the volume of the gas. To use this expression, the temperature under which the compression takes place must be specified.

Consider three sets of conditions for the compression of 1 mole of ideal gas.

i **Constant volume** For the volume to remain constant, ΔV must be zero, so that the energy transferred is zero.

ii **Constant pressure** Suppose you cool the gas in the cylinder at such a rate that, as the piston is pushed in, the pressure remains constant. The graph of pressure against volume you would obtain is shown in Figure 11.3. The energy transferred as work is done in compressing the gas is:

$$P\Delta V = P(V_2 - V_1)$$

which is the shaded area on the graph. Such a change is called an **isobaric change**.

iii **Constant temperature** Suppose you compress the gas in a cylinder made of a good thermal conductor so that the gas temperature remains constant during the compression. The gas would obey Boyle's law and P and V would be related as shown in Figure 11.4. The energy transferred as work is done in compressing the gas by ΔV changes as P changes. So each ΔV has to be multiplied by a different P. The total energy transferred in compressing the gas from V_1 to V_2 is still the shaded area under the graph. In algebraic terms:

energy transferred to gas = $RT \ln(V_2/V_1)$ per mole.

Such a change is called an **isothermal change**.

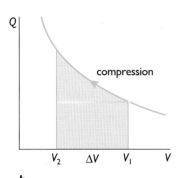

Figure 11.4 Shaded area shows work done in an isothermal change

113

CONSERVATION LAWS

Having considered these three conditions, it is possible to make a better prediction of the molar heat capacity. When 1 mole of ideal gas is heated at constant volume, the work done on or by the gas is zero. So the heating causes all the energy to be transferred to the internal energy of the gas. The value of $\frac{3}{2}R$ for the molar heat capacity of the ideal gas is the prediction for C_v.

When 1 mole of ideal gas is heated by 1 K at constant pressure, the internal energy of the gas is increased **and** the gas must do work against its surroundings by expanding. The work done is equal to $P(V_2 - V_1)$ where V_1 and V_2 are the initial and final volumes. At constant pressure, an ideal gas expands in proportion to the temperature, so we can write:

$$V \propto T$$

giving

energy transferred by work = $R(T_2 - T_1)$ per mole,

where $(T_2 - T_1)$ is the rise in temperature (see Figure 11.5). For a rise of 1 K, the work done by the gas in expanding at constant pressure is the molar gas constant R. Therefore C_p, the molar heat capacity at constant pressure, is predicted to be 8.3 J mol⁻¹ K⁻¹ greater than C_v.

The data in Figure 11.1 show that the inert gases helium and argon, fit these predictions well. However, for other gases the **difference** between C_p and C_v agrees with the prediction, but the measured values are all larger than predicted.

When working out problems on gases, it turns out that the ratio (C_p/C_v) is more useful than the difference between C_p and C_v. This **ratio of the principal molar heat capacities** of the gas is denoted by γ. γ_p is found to be important for explaining the speed of sound in gases ($\rightarrow$ 29.4) and explaining adiabatic changes where no energy transfer, due to a heating process, occurs as a gas is compressed or decompressed ($\rightarrow$ 11.3). The value of γ is 1.67 for monatomic gases, such as helium and argon, and 1.4 for diatomic gases, such as oxygen and nitrogen at room temperature.

Compressing a mole of gas

The energy W transferred when a force F moves a distance x is $F \Delta x$. So,

$W = F \Delta x$

$\quad = PA \Delta x$

$\quad = P \Delta V,$

where V is the volume.

For a gas at constant pressure P, the energy transferred is:

$P \Delta V = P(V_2 - V_1) \rightarrow \int P \, dV$.

Now, $PV = RT$, so: $P = \dfrac{RT}{V}$.

This means that at constant temperature T, the energy transferred is:

$\int P \, dV = RT \int \dfrac{dV}{V}$

$\qquad\quad = RT \ln(V_2/V_1)$

where R is the molar gas constant.

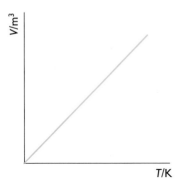

Figure 11.5 Graph of volume against temperature in K

Question

3 Helium and argon are monatomic gases. The other gases in Figure 11.1 exist at room temperature as polyatomic molecules. Consider the properties of ideal gas particles ($\leftarrow$ 10.3) and then give reasons why you might expect helium and argon to give the best agreement with the ideal gas model.

11.2 ENERGY TRANSFER

Energy can be transferred in many ways. James Prescott Joule was one of the first people to investigate these transfers. The following examples of energy transfers use modern models and ideas, many of which were unknown to Joule.

A A falling weight driving a dynamo whose output is dissipated in a heating coil:

1 potential energy of weight in Earth's gravitational field
↓
2 kinetic energy of weight due to falling through gravitational field
↓

11 ENERGY CONSERVATION IN MECHANICAL, THERMAL AND ELECTRICAL SYSTEMS

    3 potential energy of electrons in electromagnetic fields produced in dynamo
↓
    4 kinetic energy of electrons in wires
↓
    5 transfer of energy to atoms of coil, so increasing its internal energy and therefore its temperature
↓
    6 transfer of energy from coil by conduction, convection and radiation, so heating surroundings.

B A falling weight churning a liquid such as water, paraffin or mercury:

    1 potential energy of weight in Earth's gravitational field
↓
    2 kinetic energy of weight due to falling through gravitational field
↓
    3 transfer of energy to molecules of fluid, so increasing its internal energy and therefore its temperature
↓
    4 transfer of energy from fluid by conduction, convection and radiation, so heating surroundings.

C Battery and heating coil:

    1 potential energy of charged particles, such as ions and electrons, in electric field in battery
↓
    2 kinetic energy of charged particles
↓
After this the process is the same as in **A**.

D Compressed gas escaping from a cylinder and cooling:

    1 decrease in potential energy of gas molecules as they separate from one another
↓
    2 decrease in kinetic energy of gas molecules
↓
    3 decrease in internal energy and therefore temperature of gas
↓
    4 transfer of energy to gas by conduction, convection and radiation, so heating it.

James Prescott Joule

In the period 1838–78, Joule, a Manchester brewer, carried out a long series of experiments transferring energy from one situation to another. Some of his experiments involved several energy transfers and, in most of them, the energy was finally transferred by heating.

Joule's experiments were not very accurate, his numerical values showing variations of up to ±25%. Their importance lies in Joule's realisation that the exchange of energy always gave the same conversion coefficient. His contemporary Mayer called this the 'mechanical equivalent of heat'.

TRANSFERRING ENERGY BY WORKING AND BY HEATING

Rubbing causes work to be done against friction. This transfers energy to the rubbing surfaces, so raising the temperature and therefore increasing the internal energy. Experiment 11.1 investigates this process.

The change in internal energy of a material when the temperature changes by 1 K is called the heat capacity. For 1 kg, it is called the specific heat capacity, and for 1 mole, the molar heat capacity. The **energy transferred by heating**, Q is given by:

Q = mass × specific heat capacity × change in temperature.

EXPERIMENT 11.1 Energy transferred by working and heating using friction

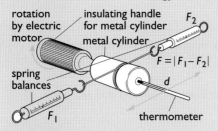

Figure 11.6 Apparatus for experiment 11.1

Figure 11.6 shows apparatus you can use to investigate this. By wrapping the string round the metal cylinder and moving the cylinder a distance d as shown, you transfer energy Fd, because of the frictional force F between the string and the cylinder. This causes an increase in internal energy and a rise in temperature of ΔT kelvin.

Energy transferred, $Fd = m \times \Delta T \times$ specific heat capacity of the metal.

From your data, you should be able to calculate a value for the specific heat capacity of the metal.

EXPERIMENT 11.2 Dissipating energy from electrical systems

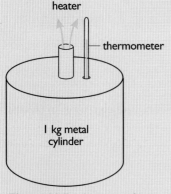

Figure 11.7 Apparatus for experiment 11.2

Figure 11.7 shows how you can measure the energy transferred from a heater into an aluminium block by using a joulemeter or by measuring the current I, voltage V and time t. Then,

energy dissipated from the charge flowing in heater = increase in internal energy of aluminium,

or $\Delta V I t = m \times \Delta T \times$ specific heat capacity of aluminium.

You can do the experiment using other metals, such as iron or copper. In each case, you can calculate values for the specific and molar heat capacities of the metal.

EXTERNAL AND INTERNAL WORK

For the ideal gas (← 10) a distinction is made between measurements made at constant volume and at constant pressure. The change in internal energy is proportional to the heat capacity at constant volume. However, when we heat solids and liquids, we cannot do so at constant volume because we cannot easily supply the enormous pressures necessary to stop them expanding. So the energy supplied both increases the internal energy and does **external work** through expansion. In most cases the change in volume ΔV due to expansion is so small that, unless very high pressures are used, the external work $P\Delta V$ can be neglected.

When a substance expands, its particles move further apart so **internal work** is done against the inter-particle forces. Like the external work, this requires a negligible correction to be made for solids and liquids. It is zero for the ideal gas, since its particles are assumed not to exert forces on each other. However, for real gases it cannot be neglected because considerable inter-molecular forces exist, especially at high pressure.

INTERNAL ENERGY AND ENTHALPY

The total energy transferred into a system by heating is called the **enthalpy change** ΔH. The enthalpy change is the increase in the internal energy plus the energy necessary to cause any expansion. If ΔU is the internal energy change and $P\Delta V$ is the energy that expands the system by ΔV at the prevailing pressure P, then ΔH is defined by:

$$\Delta H = \Delta U + P\Delta V.$$

ΔH is the minimum value of Q necessary to cause a particular chemical reaction. So, we usually use enthalpy when dealing with energy transfers in chemical reactions.

> ### Questions
>
> **4** Explain why simple measurements, such as those in experiments 11.1 and 11.2 usually give values for the heat capacity greater than the values obtained from more accurate experiments.
>
> **5** Experiments 11.1 and 11.2 suggest you measure the heat capacities of metals. Discuss the extra practical problems in trying to measure the heat capacity of a material such as polyethene or glass.
>
> **6** Given the following data for copper:
>
> $$\text{molar heat capacity} = 25 \text{ J mol}^{-1}\text{K}^{-1}$$
> $$\text{coefficient of volume expansion} = 5.0 \times 10^{-5}\text{K}^{-1}$$
> $$\text{molar volume} = 7.1 \text{ cm}^3,$$
>
> calculate the value of $P\Delta V$ for 1 mol at 100 kPa pressure and express it as a percentage uncertainty in the value for the molar heat capacity.

HEAT CAPACITIES OF SOLIDS

The heat capacities of solids vary with temperature. The variation of the molar heat capacity C_p of copper with temperature, shown in Figure 11.8, is typical. The molar heat capacity is very small near absolute zero and increases to approach a steady value of approximately 25 J mol^{-1}K^{-1}. For copper, it has almost reached this steady value by 300 K. The table in Figure 11.9 shows the specific heat capacities and molar heat capacities for a range of materials at 300 K. From it you can see that, although values for the specific heat capacity vary widely, the molar heat capacity is almost independent of the material.

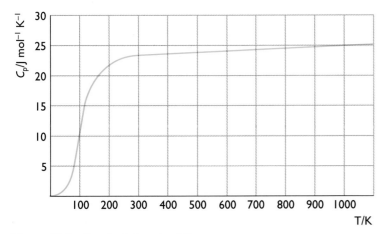

Figure 11.8 Graph of C_p against T for copper

material	specific heat capacity J kg^{-1} K^{-1}	molar heat capacity J mol^{-1} K^{-1}
ice	2100	38
copper	390	24
aluminium	900	24
iron	450	25
sulphur	720	23
nylon	600	
limestone	820	
glass	640	

Figure 11.9 Specific and molar heat capacities of solids

The molar heat capacity is the energy required to change the internal energy of 6×10^{23} particles so that the temperature rises by 1 K. Why do you think it is nearly independent of the type of material?

HEAT CAPACITIES OF LIQUIDS

The heat capacities of liquids can be measured in similar ways to those for solids – frictional heating by churning the liquid (used by Joule) or electrical heating. As with solids, the measurements are made at constant pressure and the external and internal work terms are negligible because liquids do not expand much. A typical apparatus for heat capacity measurements on liquids, called a **calorimeter**, is shown in Figure 11.10. When taking measurements, it is important to ensure that all the liquid is at the same temperature and to make a correction for the heat capacity of the container.

The table in Figure 11.11 shows the heat capacities of some liquids. As you can see, no simple pattern is apparent between values for either the specific or molar heat capacities.

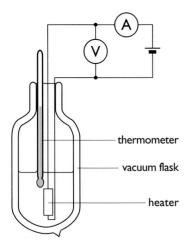

Figure 11.10 Simple calorimeter for measuring the heat capacities of liquids

liquid	specific heat capacity $J\ kg^{-1}\ K^{-1}$	molar heat capacity $J\ mol^{-1}\ K^{-1}$
water	4200	75
ethanol	2410	110
octane	2200	250
glycerol (propan–1,2,3 triol)	2420	223
mercury	140	28

Figure 11.11 Specific and molar heat capacities of liquids

Questions

7 What are some of the consequences of water having a very large specific heat capacity compared with most liquids and solids?

8 A car of mass 800 kg travelling at 30 m s^{-1} is stopped by a disc braking system whose heat capacity is 12 kJ K^{-1}. Estimate the rise in temperature of the brakes. Is your estimate a valid one?

9 Estimate for how long an electric immersion heater dissipating 3 kW must be switched on to heat a domestic hot water supply containing 0.1 m^3 from 10 °C to 65 °C.

CHANGES OF STATE AND LATENT HEAT

When you heat a substance at a constant rate, its temperature does not go up uniformly. The heat capacity varies with temperature and, when the substance melts or vaporises, large quantities of energy are absorbed to increase the internal energy of the substance with no change in temperature.

Figure 11.12 shows how the temperature of 1 kg of ice/water/steam changes with time when heated at a rate of 1 kW. The horizontal sections of the graph occur when the ice melts to form water and the water vaporises to form steam. On these sections, energy is absorbed by the ice/water with no change in temperature. The energies are known as the **latent heats** of **fusion or melting** and **vaporisation or condensation**, the **specific latent heat** being the energy per kg. They result in changes in internal energy because the particles are rearranged, and external transfers of energy by work because there is a volume change.

11 ENERGY CONSERVATION IN MECHANICAL, THERMAL AND ELECTRICAL SYSTEMS

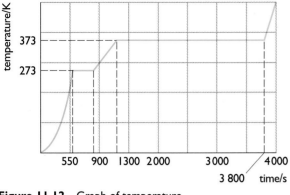

Figure 11.12 Graph of temperature against time for heating 1 kg of water at a constant rate of 1 kW

material	specific latent heat capacity kJ kg^{-1}	molar latent heat capacity kJ mol^{-1}
fusion melting		
ice/water	330	5.9
ethanol	108	5.0
copper	205	13
sodium chloride	480	28
vaporisation (condensation)		
water/steam	2260	41
ethanol	840	39
copper	4840	305
sodium chloride	2900	170

Figure 11.13 Specific and molar latent heats

In all cases, the latent heat of vaporisation is much greater than the latent heat of fusion. In many cases, it is greater than the energy required to heat the sample from its melting point to its boiling point temperature at one atmosphere pressure. Figure 11.13 gives some data on latent heats for common substances.

THE UNUSUAL THERMAL PROPERTIES OF WATER

It is well known that liquid water has a lower density than its solid state, ice. In addition, water is at its most dense at 4 °C and it has very high values for its specific heat capacity and latent heat. These properties, it has been found, are due to **hydrogen bonding** – the bonds between the electrons of the oxygen of one water molecule with nearby hydrogen atoms in other water molecules.

EXPERIMENT 11.3 The latent heat of vaporisation

You can measure the specific latent heat of vaporisation of liquids using apparatus similar to that shown in Figure 11.14. Liquid in the boiler is heated and evaporated by transferring energy to it electrically. The vapour is condensed and, once a steady state is reached, the rate of condensation of vapour is related to the power input by:

power × time = latent heat + energy transferred out
$$VI \times t = mL + Q$$

where t is the time during which mass m of liquid was collected, L is the specific latent heat of vaporisation of the liquid, and Q the energy supplied but not absorbed by the liquid.

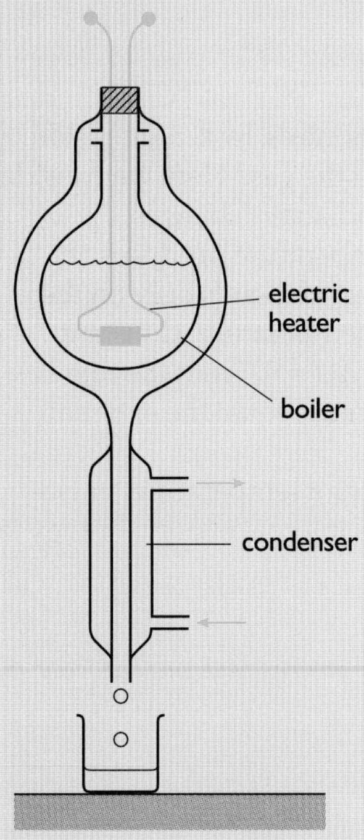

Figure 11.14 Apparatus for measuring the latent heat of vaporisation

CONSERVATION LAWS

> **Questions**
>
> **10** An immersion heater element in a copper tank is rated at 2 500 W. The tank has a heat capacity of 4000 J K^{-1} and contains 100 l of water, initially at 15 °C. Estimate how long after switching on the water will take to reach its maximum safe temperature of 70 °C.
>
> **11** Why are burns caused by steam at 100 °C more serious than those from boiling water at 100 °C?

11.3 FIRST LAW OF THERMODYNAMICS

In working out the results of heat capacity measurements, it has been assumed that the conservation of energy includes transferring energy as internal energy. In fact, the equations have been set up to ensure that this is so. The realisation that the conservation of energy extended beyond mechanical systems came from the experiments of Joule and others, and led to thermodynamics.

The first law of thermodynamics unites the conservation of energy in mechanical systems and energy transferred by heating in a formal way. It can be written in the form:

$$Q = \Delta U + W,$$

where Q is the energy supplied to the system by heating, ΔU is the change produced in the internal energy of the system, and W is the energy transferred out and represents the external work done by the system.

Note that the distinction between energy transferred by heating and that transferred by working is an important one in theremodynamics ($\rightarrow$ 28).

ISOTHERMAL AND ADIABATIC CHANGES

Imagine 1 mole of gas trapped in a cylinder by a frictionless, leak-tight piston. What happens when changes are made to the state of the gas? Here we consider two extreme cases, which are called isothermal and adiabatic changes.

Isothermal change Heating the gas at constant temperature produces an isothermal change. For an ideal gas, the internal energy depends only on the kinetic energy of the particles and is proportional to temperature. Constant temperature implies constant internal energy and so $\Delta U = 0$. From the first law of thermodynamics, the energy supplied by heating Q must equal the external energy output W, which does work.

Similarly, when an ideal gas is compressed at constant temperature, or isothermally, work done on the gas does not increase its internal energy. Energy is rejected to the external world, so

$$-W = -Q.$$

The sign convention used here is that Q is:

positive when heat is supplied **to** the gas
negative when heat is supplied **by** the gas

and W is:

positive when work is done **by** the gas
negative when work is done **to** the gas.

The changes must be reversible (→ 28.2).

Adiabatic change Imagine the gas to be confined to a cylinder with walls that are perfect thermal insulators. In this case, no energy can get in or out by heating, so $Q = 0$. Such a change is called an adiabatic change.

If $Q = 0$, then

$$\Delta U + W = 0.$$

When the gas expands and work is done in transferring energy W **out**, ΔU must be negative and the gas must cool. Conversely, transferring energy $-W$ **in** by doing work compressing the gas leads to a rise in its temperature.

The work done in moving a piston to compress or expand a gas is the sum of $P\Delta V$ over the change in volume V (← 11.1). For the isothermal change, PV is a constant, and the change traces a path AB along the isothermal curve shown in Figure 11.15. The energy transferred by the work done is the area under that curve.

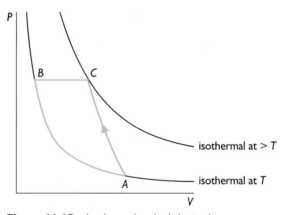

Figure 11.15 Isothermal and adiabatic changes

For the adiabatic change, the temperature varies so that another path AC is followed. This adiabatic path is steeper than the isothermal one because the pressure **and** the temperature rise as the gas is compressed adiabatically.

Questions

12 Explain why compressing air in a bicycle pump is more likely to be adiabatic if done quickly, and isothermal if done very slowly.

13 Sound waves in air involve compressions and rarefactions which are found to be adiabatic. Suggest one reason why isothermal compression waves are not found.

SUMMARY

The work done on a system results in the transfer of energy to increase its potential, kinetic or internal energy. The work done in compressing a gas is given by $P\Delta V$.

The energy needed to increase the temperature of a system by a given amount can be found by multiplying the appropriate heat capacity by the change in temperature. The heat capacity of a body is the energy required to raise its temperature by 1 K. The specific heat capacity of a substance is the energy required to heat 1 kg of the substance through 1 K. The molar heat capacity of a substance is the energy required to heat 1 mol of it through 1 K.

Energy, known as latent heat, is required to change the state of a substance. The specific latent heat of fusion or vaporisation is the energy per kg needed to melt a solid into a liquid or vaporise a liquid into a gas without a change in temperature.

Most energy changes in thermal systems involve both heating and working. The first law of thermodynamics links heating, working and changes in internal energy.

Isothermal changes involve no change in internal energy, so the temperature remains constant. Adiabatic changes involve no transfer of energy by temperature differences.

SUMMARY QUESTIONS

14 An air compressor cylinder of volume 1 000 cm^3 takes in air at 100 kPa and 290 K, and compresses it to $\frac{1}{8}$th of its volume isothermally. Calculate the work done on each stroke of the pump.

15 Estimate the difference in temperature between the bottom and the top of a waterfall of height 80 m. Do you think it would be possible to measure this change?

16 An electric kettle has an element rated at 2 500 W. From the data below, estimate:
a the time taken for the kettle to boil, starting from 20 °C,
b the extra time after boiling to boil dry.

$$\text{initial mass of water} = 1.0 \text{ kg}$$
$$\text{mass of aluminium kettle} = 0.45 \text{ kg}$$
$$\text{specific heat capacity of water} = 4.2 \text{ kJ kg}^{-1}\text{K}^{-1}$$
$$\text{specific heat capacity of aluminium} = 900 \text{ J kg}^{-1}\text{K}^{-1}$$

specific latent heat of vaporisation of water = 2.2 MJ kg^{-1}.

17 A 20 g piece of ice at 0 °C is added to 300 g of water at 15 °C. Estimate the temperature just after the ice has melted. Take the specific latent heat of fusion of ice to be 334 kJ kg^{-1}.

18 When water boils, the steam produced has to displace the air in the atmosphere above the water. 1 mol of water requires 40 kJ to convert it into steam, which occupies 0.03 m^3 at 373 K. Estimate the work done by the steam displacing the air, and express this as a percentage of the latent heat of vaporisation.

12 CONSERVING CHARGE AND ENERGY: SIMPLE ELECTRICAL CIRCUITS

When you switch on a table lamp, the lamp filament glows almost immediately. What causes it to radiate energy, and how is this energy transferred through the conducting wires? It is important to understand electrical circuits and analyse their performance if you wish to use them or design your own.

Charge flowing round a circuit is confined to the components and the connecting wires. The total charge and the energy associated with the circuit are always conserved. These laws of conservation were formalised by Gustav Kirchhoff (1824–87) and are often used to solve circuit problems.

12.1 FLOW OF CHARGE

An electric current is a net flow of electric charges (→ 16.1). By convention, charge moves **from the positive to the negative** of an electrical supply. We use this convention in this book. Positive currents travel in the direction of positive charges. Electrons, carrying a negative charge, move in the opposite direction. The unit of electric current is the **ampere (A)**, which is one of the base units of SI. A current of one ampere corresponds to a **flow of one coulomb (C) of electric charge per second**.

A simple method to detect a flow of charge is illustrated in Figure 12.1. When charge flows through a fine wire in a lamp, energy is transferred to the wire and its internal energy is increased, so its temperature rises. This makes the wire glow, indicating a flow of charge. Another common method of detection makes use of the magnetic field set up by a moving charge. If you set up the circuit in Figure 12.2, using a coil of wire and a small plotting compass, you will observe a deflection of the compass needle when there is a current in the coil. A moving coil meter works on a similar principle.

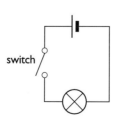

Figure 12.1 When charge flows, the lamp glows

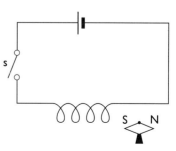

Figure 12.2 Detection of current using the deflection of a compass needle in the magnetic field produced in a coil

C CONSERVATION LAWS

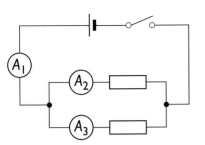

Figure 12.3 Current in parallel resistors

In the circuit in Figure 12.3, charge can flow round the circuit via either of the two resistors in parallel. Conservation of charge suggests that the current recorded by ammeter A_1 should be the sum of the currents recorded by ammeters A_2 and A_3, and we find that this is indeed the case. The circuit, and experiment 12.1, illustrate

Kirchhoff's first law, which states:

The sum of the currents leaving a junction is equal to the sum of the currents entering it.

This is also known as **Kirchhoff's current law**.

EXPERIMENT 12.1 Conservation of charge

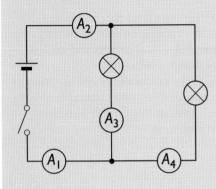

Figure 12.4 shows a circuit with two lamps, and ammeters in each of four positions A_1 to A_4. By constructing the circuit with either four ammeters or with one ammeter used in each position in turn, you should note that the ammeters change in unison when the circuit is completed or broken. Similarly, you should note that ammeters A_1 and A_2 both register the same current - the sum of the currents registered by A_3 and A_4.

Figure 12.4 Conservation of charge in a circuit

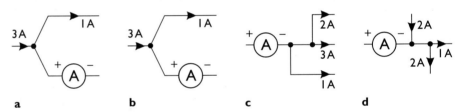

a b c d

Figure 12.5

Questions

1 Use Kirchhoff's current law to calculate the current registered by the ammeter A in each of the circuit elements in Figure 12.5.

2 A car lighting system consists of four side/rear lights, each normally drawing a current of 0.5 A, and two headlamps, each drawing 4.0 A. All the lamps are connected in parallel.
a Calculate the current drawn with only the side/rear lights on, and then with all the lights on.
b Why do cars have all lights connected in parallel rather than in series?

3 An analogy can be drawn between electric current and the traffic flow on a motorway. A section of the M1 motorway carries a steady flow of 4 000 vehicles per hour north out of London. The average speed of the vehicles is 80 km h^{-1}.
a Which figure corresponds to the 'current' along the M1?
b Calculate the average number of vehicles on each km of the M1.
c Does the fact that all vehicles do not move at the same speed affect your answers to **a** and **b**?

12 CONSERVING CHARGE AND ENERGY: SIMPLE ELECTRICAL CIRCUITS

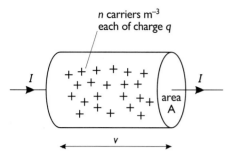

Figure 12.6 The current in a wire

Repairs to the motorway reduce the speed on one section to 40 km h^{-1}.
d Has the current along the M1 changed?
e Has the number of vehicles on each km of motorway changed?
f Is there any phenomenon in an electric circuit corresponding to a vehicle breaking down on the motorway and not completing its journey?

CONDUCTION IN METALS

In all metals, the conduction of electricity takes place by the movement of electrons (→ 27.6). The current depends on the density n of charge carriers, such as electrons, their charge q, the velocity v with which they drift through the metal and the area of its cross-section A. The current I can be predicted using the equation:

$$I = nqAv.$$

The charge on the electron e is only 1.6×10^{-19} C, so 1 A corresponds to a very large number of about 6×10^{18} electrons moving through the wire each second. However, there are as many as 10^{29} m^{-3} electrons free to move in a metal. So, the drift velocity of electrons along a wire is small, about 0.1 mm s^{-1} for a 1mm^2 wire. Compare this with an electron's average random thermal speed, at room temperature, of about 1×10^5 m s^{-1}.

The current in a wire

Consider a wire of cross-sectional area A containing n charge carriers per cubic metre, each of charge q, as shown in Figure 12.6. The current I equals the charge crossing a plane across the wire in 1 second. The charges move very fast at random velocities but there is a resultant, or **drift velocity**, v in the direction of the current along the wire. On average, all charges within a length v will cross a particular plane in 1 second. The length v contains nAv charge carriers whose total charge is $nqAv$. So, the current I, is given by:

$$I = nqAv.$$

Question

4 Copper has a density of 8 900 kg m^{-3} and a relative atomic mass of 63.5. Use these data, and the value of the Avogadro constant, N_A to calculate:
a the number of copper atoms in 1 cubic metre of copper,
b the number of conduction electrons per cubic metre in copper, assuming that each copper atom contributes one electron to the conduction electrons,
c the drift velocity of these conduction electrons along a copper wire of diameter 1.0 mm carrying a current of 5.0 A,
d the number of electrons per second passing through an ammeter reading 1.5 mA.

12.2 ENERGY TRANSFER IN ELECTRICAL CIRCUITS

Since the charge flowing round a circuit is conserved, what is transferred or 'used up' in an electrical circuit? The answer is energy – potential energy from the cell. In Figure 12.7a, energy is transferred from the circulating charge to the lamp. The lamp filament rises in temperature and radiates energy, as shown in Figure 12.7b.

The energy transferred to each unit charge is called the **electromotive force**, or **emf** of the cell or supply. It is not a force, but the electrical potential energy transferred by each coulomb of charge when moving between the 'poles' of the supply. When 1 joule of potential energy is possessed by each coulomb of charge, a source is said to have an emf of 1 joule per coulomb, or 1 **volt (V)**.

1 volt = 1 joule per coulomb.

A circuit requires an emf for a current to flow in it. The emf can be thought of as the potential energy a 'source' of electricity provides per coulomb.

CONSERVATION LAWS

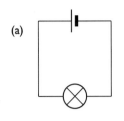

Figure 12.7 **a** Battery transferring energy to a lamp **b** Energy transfers in an electrical circuit

When 1 coulomb of charge passes through any part of a circuit and the potential energy of the charge decreases by 1 joule, the charge has moved through a **potential difference**, or **pd** of 1 volt. The pd in volts across any component is the energy transferred to the component by each coulomb passing through it. Both **emf** and **pd** are therefore measured in volts and the term **voltage** is often used for both quantities. In this book, we use the term most appropriate for the situation under consideration.

KIRCHHOFF'S SECOND LAW

In Figure 12.8 the 2 V cell has negligible internal resistance (→ 12.3). The voltmeters in the circuit display the voltage between the points across which they are connected. When the switch is open and then closed, the following are observed:

- with **switch open**, the ammeter reads zero and the voltmeter V_1 reads 2 V because, although the current is zero, the cell has an emf of 2 V. V_2 and V_3 read zero because no energy is being dissipated in the lamps,
- with **switch closed**, the lamps light. V_1 still reads 2 V, but now V_2 and V_3 both register voltages which add up to 2 V.

When the switch is closed, the emf of the cell is the same as the sum of the voltages across the two lamps because there are no other components. The energy provided by the cell to each coulomb is transferred to the two lamps as it passes through them. This is consistent with the law of conservation of energy, and is formalised in **Kirchhoff's second law**:

The total emf in a closed circuit equals the sum of the potential differences, or voltages, across all components in the circuit.

Questions

5 What readings would you expect a voltmeter to register when connected across the switch shown in Figure 12.8, with the switch: **a** open, **b** closed?

6 In a real circuit like that in Figure 12.8 both the ammeter and the cell would dissipate energy. Suggest how you could detect these energy 'losses'.

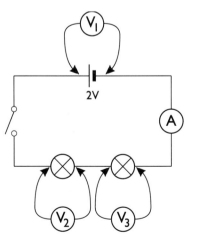

Figure 12.8 Voltages across two lamps in a circuit

ELECTRICAL POTENTIAL ENERGY – AN ANALOGY

Consider a 'big-dipper' ride at a fairground, as in Figure 12.9. At the start of the ride, the car and its occupants are lifted by a chain drive to the highest point in the ride. Each kilogram of their mass is given an extra amount of gravitational potential energy. As the car continues its journey, energy is steadily dissipated until the car reaches its lowest position before being lifted ready to make another journey. The loss in potential energy of each kilogram of the mass and its occupants, in travelling from the highest point through the circuit to the bottom, is equal to the extra potential energy given to each kilogram at the start.

A 'big dipper' car can be compared to a coulomb of charge provided by a cell. Each coulomb is given extra electrical potential energy by the chemical processes within the cell. However, it is as though there is a continuous chain of cars, or coulombs – while some are losing potential energy in the circuit, others are gaining it inside the cell.

12 CONSERVING CHARGE AND ENERGY: SIMPLE ELECTRICAL CIRCUITS

Figure 12.9 A fairground 'big dipper'

Figure 12.10 Joulemeter

Question

7 In what ways do you think the 'big dipper' analogy is:
a helpful,
b misleading,

when trying to gain a better understanding of Kirchhoff's second law?

POWER IN ELECTRICAL CIRCUITS

The commercial supply of electricity centres on the transfer of energy rather than charge. An electricity meter, shown in Figure 12.10, is a joulemeter calibrated in **kilowatt hours (kW h)**. One kW h is the total energy transferred when 1 000 joules per second are transferred for 1 hour. As there are 3 600 s in an hour, 1 kW h is equivalent to 3.6 MJ.

Recording the voltage across and the current through a component enables us to calculate the power rather than the energy dissipated in a component. **Power** is the rate of energy transfer and is measured in **watts** (W), where 1 watt = 1 joule per second. When V joules per coulomb is the pd across a component and I coulombs flow through the component per second, the power dissipated by the component is VI joules per second. We can therefore find the power dissipated by multiplying the pd by the current. The units agree with this:

C CONSERVATION LAWS

$$1 \text{ volt} \times 1 \text{ ampere} = 1 \text{ joule/coulomb} \times 1 \text{ coulomb/second}$$
$$= 1 \text{ joule/second}.$$

To calculate the energy transferred, we need to know the power and the time for the transfer.

$$\text{Energy transferred} = \text{power} \times \text{time}$$
$$= VIt.$$

So, we need to measure voltage, current and time. We need a voltmeter, an ammeter and a clock or, alternatively, we can just use an electrical joulemeter.
Figure 12.11 shows the power rating of some common electrical systems.

system	power	voltage	current
large alternator in power station	500 MW	33 kV	15 kA
electric kettle	2.7 kW	240 V	11.3 A
colour television set	200 W	240 V	0.8 A
reading lamp	60 W	240 V	0.25 A
two car headlamps	80 W	12 V	6.7 A
pocket torch	0.9 W	3 V	0.3 A
pocket calculator with LCD display	0.3 mW	3 V	0.1 mA

Figure 12.11 Power ratings of some systems

EXPERIMENT 12.2 Energy in electrical circuits

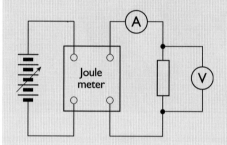

The circuit in Figure 12.12 allows you to use both a joulemeter and a voltmeter plus ammeter to measure the energy supplied by a battery to a load of resistance R. By using a 10 Ω 100 W resistor and then a 12 V, 24 W lamp for R, and setting the supply to an appropriate voltage in each case, you can see how the current and the energy transferred vary with time.

Figure 12.12 Circuit for experiment 12.2

Question

8 Use the data in Figure 12.11 to
a compare the energy needed to boil water in an electric kettle in 3 minutes with that used by a television in 3 hours,
b calculate the useful energy available from a calculator battery, given that the battery will operate the calculator for 1 500 hours.

12.3 RESISTANCE

When studying individual components, we often wish to know the relationship between the current through and the voltage across the component. The **voltage per amp**, or unit current, is called the **resistance** of the component. It has the unit **ohm, or Ω**, where 1 ohm is the resistance of a conductor that carries 1 ampere when a pd of 1 volt is applied across it. So,

$$1 \, \Omega = 1 \text{ volt/ampere} = 1 \text{ VA}^{-1}.$$

Figure 12.13 gives the resistances of some common components. Under any particular condition, the electrical resistance of a component is the voltage per unit current. For some materials and components the resistance is constant over a large range of applied voltages and currents. Such a conductor is said to be an **ohmic** conductor. Georg Simon Ohm, a German physicist, demonstrated that:

component	resistance Ω
carbon film resistors used in electronics	10 to 10^7
electric kettle element	21
filament of 60 W 240 V lamp bulb hot	1 000
5 m of '6 A rating' mains cable	0.1
insulation resistance of 5 m of cable	$>10^7$
filament of 150 W 24 V projector bulb	3.8
hi-fi loudspeaker	8
crystal microphone	10^5

Figure 12.13 Resistances of some common components

the steady current flowing in a metallic conductor is directly proportional to the pd, or voltage, across the conductor, provided that the temperature and other physical conditions remain unchanged.

This statement is known as Ohm's Law and can be written:

$$V \propto I$$

or, using the definition of resistance,

$$V = IR.$$

The only really **ohmic** resistors, where resistance is constant, are metals used at constant temperature and pressure so it might be thought that Ohm's law is of very limited application. However, it enabled the concept of resistance measured in **ohms** or Ω to be defined. Also, the relationship that resistance $R = V/I$ holds whether a component is ohmic or not. For a non-ohmic component, R will change at different voltages. Only for an ohmic component does R stay the same at constant temperature as the voltage is changed.

EXPERIMENT 12.3 Voltage–current relationship for a metal resistor

The circuit in Figure 12.14 enables you to apply different voltages across a 10 Ω 10 W wire-wound resistor by taking readings of the voltage across and current passing through it. You can then plot a graph of voltage against current, as in Figure 12.15. This passes through the origin because the current is zero when there is zero voltage across R.

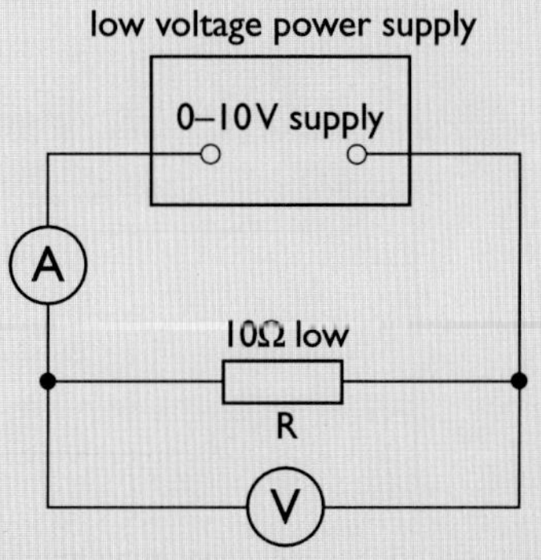

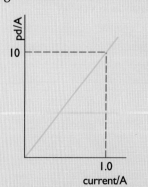

Figure 12.14 Circuit for experiment 12.3

Figure 12.15 Graph of voltage against current for a 10 Ω wire-wound resistor

CONSERVATION LAWS

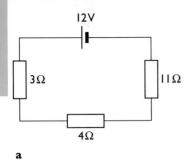

a

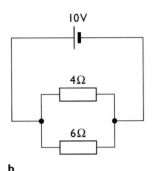

b

Figure 12.16

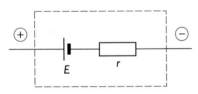

Figure 12.17 The circuit equivalent to a real cell

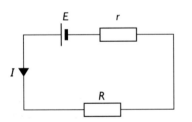

Figure 12.18 Power dissipated in internal and external resistances in a circuit

> ## Questions
>
> 9 Use Ohm's law to calculate the voltage across and the power delivered in each of the circuits in Figure 12.16. ($\rightarrow$ 12.6)
>
> 10 A resistor marked 330 Ω, 0.5 W is connected across a 10 V supply.
> a Calculate the current and power delivered in the resistor. Is the power dissipated less than the maximum quoted by the manufacturer?
> b Calculate the maximum current the resistor can safely carry.

INTERNAL RESISTANCE

In all the circuits looked at so far, the instruments and cells are assumed to be ideal – voltmeters draw no current, ammeters have zero resistance, and all the energy provided by a cell is dissipated in the circuit external to the cell. In fact, cells dissipate energy because they have an internal resistance. A real cell can be represented by an ideal cell of emf E in series with a resistance r (see Figure 12.17). The energy dissipated in the internal resistance represents energy that is not transferred to the rest of the circuit.

The current in the circuit of Figure 12.18 is given by:

$$I = \frac{E}{(R+r)}.$$

Across the load, the voltage is given by $V = IR$

$$= \frac{ER}{(R+r)}.$$

Because there is always some internal resistance, the voltage across the load is always less than the emf of the cell.

To transfer a large proportion of the energy to the external load, resistance R must be much greater than r. Although this provides maximum efficiency, the current is small and so results in low power delivery. This is used in devices such as portable radios, in which power requirements are low and only a small current is needed. In other devices, such as powerful torches and model cars, the

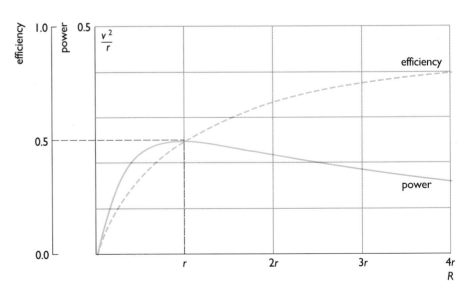

Figure 12.19 Variation of power and efficiency with load resistance

12 CONSERVING CHARGE AND ENERGY: SIMPLE ELECTRICAL CIRCUITS

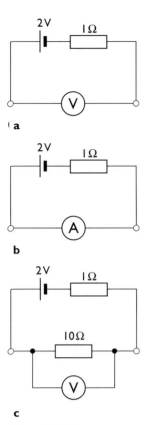

Figure 12.20

power delivered is more important than the efficiency. In these cases, a considerable proportion of the energy is dissipated in the electrical supply itself.

Figure 12.19 shows how the power and efficiency of energy transfer vary with the resistance of the load. Note that maximum power is delivered by the cell when the load resistance equals the internal resistance. Under this condition only half the energy supplied is dissipated in the load, and the voltage across the load is half the voltage of the cell. The importance of 'matching' the source with the load is discussed further in Chapter 22.

> **Question**
>
> **11** For the circuits in Figure 12.20 calculate:
> **a** the voltage measured by the voltmeter in circuit a,
> **b** the current measured by the ammeter in circuit b,
> **c** the voltage measured by the voltmeter in circuit c.

RESISTIVITY

When comparing materials rather than components, it is useful to remove geometrical factors, such as the size of the component, from a discussion of resistance. Consider a metal wire cut in half: each of the new pieces has half the length and half the resistance of the original. When two wires of the same material whose cross-sectional areas are in the ratio 1:2 are compared, the thinner wire has twice the resistance per metre of the thicker wire. Combining these observations we may write:

$$\text{resistance} \propto \text{length/area of cross-section}$$
$$= \text{constant} \times \text{length/area of cross-section}.$$

This constant is called the resistivity ρ, of the material and is a characteristic property of that material. It is defined from

$$\text{resistivity } \rho = \frac{\text{resistance} \times \text{area of cross-section}}{\text{length of conductor}}$$

and has the unit ohm metre (Ωm).

For a given material, ρ may vary with temperature and other physical conditions. Figure 12.21 shows the resistivities of some materials at room temperature. There is an enormous range of values, with the best insulators having resistivities more than 10^{20} times those of the best conductors.

material	resistivity/Ωm at 300 K
silver	1.50×10^{-8}
copper	1.55×10^{-8}
aluminium	2.45×10^{-8}
tungsten	4.82×10^{-8}
iron	8.70×10^{-8}
constantan (copper/nickel alloy)	5.0×10^{-7}
nichrome (nickel/chromium alloy)	1.1×10^{-6}
pure silicon	$\sim 10^{-3}$
pure water	$\sim 10^{4}$
pyrex glass	$\sim 10^{8}$
vitreous silica	$\sim 10^{12}$
most plastics (e.g. polyethene)	$\sim 10^{14}$

Figure 12.21 Resistivities of various materials

CONSERVATION LAWS

CONDUCTANCE AND CONDUCTIVITY

In the previous text, we could instead have considered the **reciprocal of resistance – the conductance G**, the current per volt. A good conductor has a low resistance and a large conductance. The SI unit of conductance is the siemens (S), where 1 siemens is 1 ampere/volt. Similarly, the conductivity σ, of a material is the reciprocal of its resistivity. The unit of conductivity is siemens/metre (S m^{-1}).

Questions

For these questions, refer to the data in Figure 12.21.

12 Calculate the resistance of a coil wound from 25 m of copper wire of diameter 0.60 mm.

13 Calculate the length of a tungsten lamp filament of diameter 0.020 mm and having a resistance of 8.0 Ω at 300 K.

14 An engineer wishes to design a heater to dissipate 1.0 kW from a 240 V supply, using a 5.0 m length of wire.
a Calculate the resistance of and the current in the wire.
b Advise her which is the most suitable – iron, nichrome or constantan – for this purpose.

15 Two square plates, each of side 0.10 m, are immersed in water 0.15 m apart. Calculate the resistance between the plates. Give two reasons why your calculated value would be much higher than that obtained by experiment.

16 A metal conductor of surface area 0.025 m^2 is insulated by a layer of polyethene of thickness 0.50 mm. Estimate the leakage current through the insulation when there is 500 V across it.

Preventing electric shocks

An electric shock can be fatal if enough current passes for long enough through the body. Clearly, death will be more likely if the current passes through a particularly vulnerable area of the body, such as the heart or the head. Shocks often occur through the hand touching a live conductor, perhaps the metal casing of a faulty electrical appliance such as an electric kettle. If the feet make good electrical contact with the ground, or **earth**, a current may flow through the heart to produce a fatal shock. For safety, effective 'earthing' of electrical appliances is essential. The external metal casing must be electrically connected securely to earth via the earth pin in the plug attached to the appliance. Should the casing become live for whatever reason, a large current will then flow to earth and cause the fuse in the plug to blow. Shocks from mains electricity occur because of lack of appropriate earthing or because of failure of earthing cables. Good earthing enables the metal parts of appliances to be maintained at earth voltage.

Unlike direct current electricity, the mains alternates at 50 Hz. There are essentially two conductors involved in the distribution of mains electricity. One of them, the **neutral**, is connected to the earth at places such as the power station, each transformer and the local sub-station. The voltage of the other, the **live** conductor, oscillates sinusoidally around zero at 50 cycles per second. In the home, a third conductor, the **earth line**, is available via mains sockets. This is connected to an earthing rod or plate, near the point of entry to the house of the live and neutral conductors. Anyone touching these 'live' parts will complete the circuit to earth and a current will pass to earth through that person's body. The earth line provides a low resistance path, so ensuring that any metal

12 CONSERVING CHARGE AND ENERGY: SIMPLE ELECTRICAL CIRCUITS

parts are at earth voltage. Only when there is a possibility of current between the live supply and earth, passing through the body, does danger exist.

Fuses or miniature circuit breakers are used as current limiting devices to protect wiring and appliances. Unfortunately, these are not able to prevent electric shocks. Improved electrical safety can be produced by using an **earth leakage circuit trip** or a **residual current circuit breaker**. The residual current is the difference between the live and the neutral currents and is equal to the earth current. A break of circuit is triggered when the residual current rises above a preset level. Figure 12.22 shows such a power breaker.

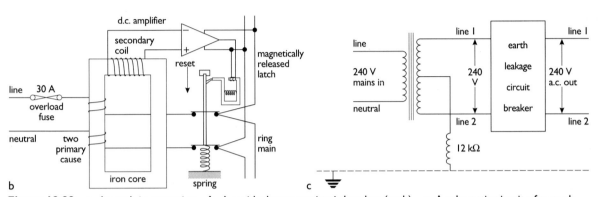

Figure 12.22 **a** An rccb in operation **b** A residual current circuit breaker (rccb) **c** A schematic circuit of an rccb

12.4 VOLTAGE–CURRENT RELATIONSHIPS

The resistance of a component is very often considered to be constant. However, resistance can vary with temperature, incident light intensity, applied voltage, time and several other quantities.

To select the right component to be incorporated in a particular circuit design, we need to know the component's resistance – how the voltage across and the current in that component are related. Figure 12.23 shows a suitable circuit for investigating the characteristics of components connected first between P and Q, and then between Q and P. Typical graphs obtained are shown in Figures 12.24 and 12.26 – 12.28, and we discuss these in the following text.

Experiment 12.3 gave part of the graph for a metal wire resistor at constant temperature. The full graph is drawn in Figure 12.24. Such resistors obey Ohm's law, so the ratio V/I is a constant. These resistors are very common in low

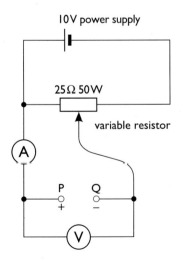

Figure 12.23 Circuit for experiment 12.4

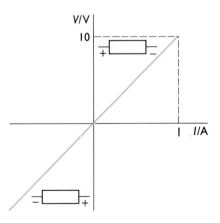

Figure 12.24 Graph of voltage against current for a resistor

133

CONSERVATION LAWS

power circuitry. Generally, carbon resistors are identified by coded coloured bands or a number code (see Figure 12.25). Wire-wound ones are usually uncoded.

international code for resistors

1st letter (shows position of decimal point):   R ohms   K kilohms   M megohms
2nd letter (shows tolerance):   F ± 1%   G ± 2%   J ± 5%   K ± 10%   M ± 20%
thus: 1R0M denotes 1.0 Ω ± 20%   100KK denotes 100 KΩ ± 10%   6K8G denotes 6.8 kΩ ± 2%
       4R7J denotes 4.7 Ω ± 5%   4M7F denotes 4.7 MΩ ± 1%

preferred values for resistors

bold type: available in ± 5%, ± 10% and ± 20% tolerance ranges
normal type: available in ± 5%, and ± 10% tolerance ranges only
italic type: available in ± 5% tolerance range only
figures are repeated over each decade from 0.22 Ω to 22 MΩ. Values outside this range are not always available.

10   11   12   13   **15**   16   18   20   **22**   24   27   30   **33**   36   39   43   **47**   51   56   62   **68**   75   82   91   **100**

old colour code for resistors

colour	black	brown	red	orange	yellow	green	blue	purple	grey	white	silver	gold
band A 1st sig. fig.	0	1	2	3	4	5	6	7	8	9	–	–
band B 2nd sig. fig.	0	1	2	3	4	5	6	7	8	9	–	–
band C multiplier	1	10	100	1000	10^4	10^5	10^6	–	–	–	0.01	0.1
band D tolerance	–	–	–	–	–	–	–	–	–	–	± 10%	± 5%

composition resistors: all bands equal width. Wire wound resistors: band A double width. No band D implies ± 20%.

Figure 12.25   Colour code for resistors

EXPERIMENT 12.4   Voltage-current relationships

By using the circuit of experiment 12.3 (Figure 12.14) and inserting a lamp and various other devices in place of a resistor, you can plot a graph of V against I for each device. You can then compare its shape with the graph for a lamp, shown in Figure 12.26.

The graph of V against I for the lamp in experiment 12.4 looks like Figure 12.26. The resistance of any specimen of any material is the ratio V/I at any point. Generally, it increases for larger values of V and I (→ 12.5).

Question

17 a Both the wire resistor and the lamp filament in experiment 12.4 are metallic. Explain why the lamp filament glows white hot whereas a wire-wound resistor of equivalent power rating is only warm. **b** Why do the graphs in Figures 12.24 and 12.26 differ in shape?

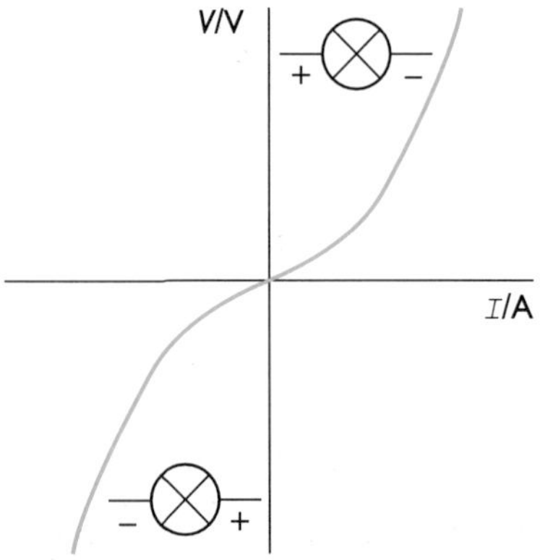

Figure 12.26   Graph of voltage against current for a lamp

Next, consider a resistor made from a semiconductor, the **thermistor**. The thermistor should be coated with an insulating varnish, if not already encapsulated, and mounted in a water bath. A voltage-current graph of data taken at a constant temperature is a straight line, as shown in Figure 12.27. A second set of

12 CONSERVING CHARGE AND ENERGY: SIMPLE ELECTRICAL CIRCUITS

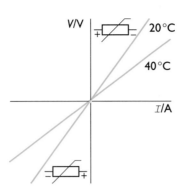

Figure 12.27 Graphs of voltage against current for a thermistor

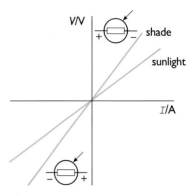

Figure 12.28 Graphs of voltage against current for an LDR

data taken at a different temperature also gives a straight line graph but with a different gradient (→ 12.5).

Figure 12.28 shows the voltage-current graphs for another resistor made from a semiconductor, the **light-dependent resistor (LDR)**. The graphs are straight lines so long as the light intensity falling on the LDR is constant. The higher the light intensity, the lower the resistance of the LDR. This change in resistance with illumination is used in light-operated switches and camera exposure meters.

DIODES

Other devices have characteristics that vary with the applied voltage. One such is the silicon diode, whose characteristic voltage-current graph is shown in Figure 12.29. When the arrow head points in the direction of the conventional current, the diode will conduct when the voltage across it is greater than about 0.6 V. There is only a tiny leakage current when the voltage has the opposite polarity. So the diode behaves as a 'one-way valve', requiring a forward voltage of about 0.6 V to 'turn on'. For a germanium diode, the forward voltage is about 0.3 V, while for a red LED, or light-emitting diode, it is about 1.4 V. For most LEDs, a forward voltage of 2 V is necessary for them to conduct and glow.

One major use of diodes is to **rectify** ac into dc (→ 23.1). Figure 12.30a shows a circuit for a half-wave rectifier to convert ac into a pulsed dc. Figure 12.30b shows the time variation of the voltages at points A and B.

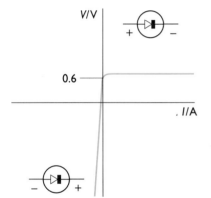

Figure 12.29 Graph of voltage against current for a silicon diode

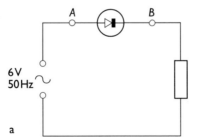

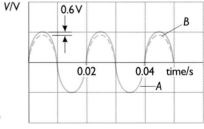

Figure 12.30 **a** Half-wave rectifier circuit **b** Time variation of voltages on either side of diode

Questions

18 An LDR is connected in the circuit in Figure 12.31. Calculate the change in the voltage across the LDR when sunlight falling on the LDR is shaded. The resistance of the LDR is 150 Ω in sunlight and 10 kΩ in the shade.

Figure 12.31 A light-dependent resistor in a circuit

CONSERVATION LAWS

19 Explain why silicon diodes are not suitable for rectifying a.c. voltages less than 0.5 V.

20 Silicon diodes are used to protect moving coil meters, as in the circuit in Figure 12.32. Explain how this circuit would prevent a voltage greater than about 0.6 V appearing across the ammeter, if a large voltage was accidentally connected across it.

12.5 THE VARIATION OF RESISTANCE WITH TEMPERATURE

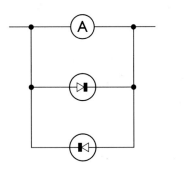

Figure 12.32 Meter protection circuit

The variation of electrical resistance with temperature is of interest for several reasons: it provides a means of measuring temperature; it provides an electrical signal directly related to temperature in 'control circuits'; and it provides an understanding of the mechanism of electrical conduction.

METALS

Figure 12.33 contains data for a 240 V, 60 W lamp. Note that, as more power is dissipated in the lamp and the filament temperature rises, the resistance of the filament increases. This is true of nearly all metals.

Figure 12.34 shows how the resistivities of two metals vary with temperature. Over a considerable range around room temperature there is a linear relationship between resistivity and temperature, which is useful when designing **resistance thermometers**. The resistance thermometer scale is defined using platinum as the conducting metal.

voltage/V	current/A
5	0.05
25	0.10
50	0.13
75	0.16
100	0.19
125	0.21
150	0.23
175	0.25
200	0.27
225	0.29

Figure 12.33 Voltage and current data for a 240 V, 60 W lamp

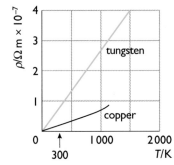

Figure 12.34 Variation of resistivity with temperature for tungsten and copper

SUPERCONDUCTIVITY

At very low temperatures, the resistances of metals become very small. The resistivity of copper at 20 K, for example, is only 1/2000 of its value at 300 K. Small as that is, it is still finite. In 1911, Kamerlingh Onnes discovered that when some metals are cooled below a **critical temperature** (7.2 K for lead) their resistances fall abruptly to zero. This is known as **superconductivity** and means that a current in a lead ring, cooled below 7.2 K, continues indefinitely after the supply is removed. No energy is dissipated as heat from the ring. However, any other interaction with the current will transfer energy and result in the current in the ring decreasing.

In the late 1980s, composite ceramic superconductors with critical temperatures of over 100 K were discovered. Imagine how useful a superconductor that could operate at room temperature would be!

12 CONSERVING CHARGE AND ENERGY: SIMPLE ELECTRICAL CIRCUITS

Questions

21 Plot the data in Figure 12.33 as a graph of resistance against current and then:
a estimate the 'cold' resistance as the current tends to zero,
b estimate the maximum current on switching on the lamp,
c suggest a suitable value for the fuse required for the lamp.

22 A commercial platinum resistance thermometer uses a foil element of resistance 100 Ω at 273 K. Calculate the resistance of the element at the extremes of its range, 220 K and 500 K, given that the resistance increases by 0.385 Ω K⁻¹. Explain why it is important not to have a large current in the element of a resistance thermometer.

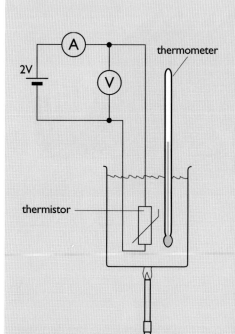

Figure 12.35 Variation of the resistance of a thermistor with temperature

SEMICONDUCTORS

Investigating the variation of resistance with temperature for a thermistor, as in experiment 12.5, shows that the resistance decreases with increase of temperature. This variation is shown in Figure 12.35 and is approximately exponential, obeying the relationship:

$$R_T = R_0 e^{A/T}$$

where R_T and R_0 are the resistances at T K and 273 K, and A is a constant dependent on the material.

Thermistors make sensitive thermometers, since a small temperature change results in a large resistance change. However, the non-linear relationship between resistance and temperature means that they require careful calibration.

EXPERIMENT 12.5 Calibration of a thermistor

Using the apparatus in Figure 12.36, you can investigate the variation of resistance with temperature for a thermistor. Figure 12.37 gives some typical results for a thermistor, which are plotted in Figure 12.35.

Figure 12.36 Apparatus to measure the variation of resistance of a thermistor with temperature

temperature/°C	temperature/K	resistance/Ω
0	273	945
10	283	670
23	296	425
30	303	350
38	311	268
49	322	194
63	336	131
74	347	99
82	355	80
94	367	62
100	373	53

Figure 12.37 Data for thermistor type TH3

CONSERVATION LAWS

EXPERIMENT 12.6 A diode thermometer

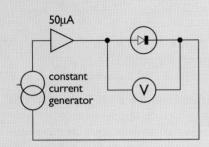

A silicon diode conducting in the forward direction can be used as a semiconductor resistance thermometer. By setting up the circuit in Figure 12.38 to maintain a constant current of 50 μA through such a diode, you can plot a graph of forward voltage, typically around 0.6 V, over a range of temperatures.

P

Figure 12.38   Circuit for a diode thermometer

THERMAL RUNAWAY

The decrease in resistance of semiconductors with increase in temperature has important consequences for circuit design. When the current through a component increases, the internal energy increases and a rise in temperature of the component occurs. This rise in temperature reduces the component's resistance and the current increases. This larger current leads to more heating and so on, in a process known as **thermal runaway**. To prevent catastrophic failure, susceptible semiconductor components must be adequately cooled by heat sinks, such as those in Figure 12.39.

Figure 12.39   Transistor heat sinks

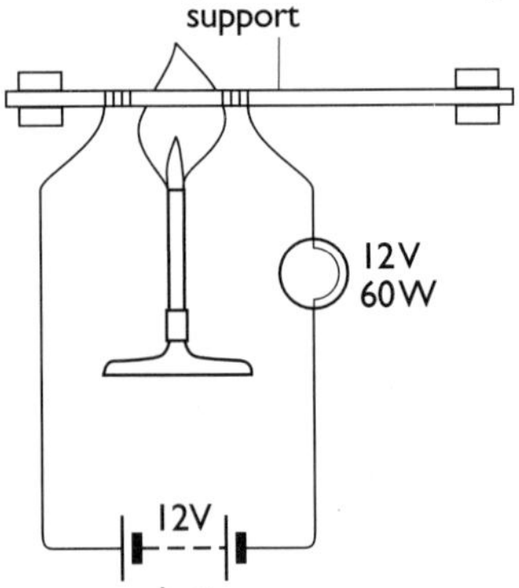

Figure 12.40   Conduction of electricity by a glass rod

INSULATORS

No insulator is perfect. The resistivity of an insulator is very large, but decreases as the temperature rises. This is shown in a dramatic way in the demonstration 'Glass from insulator to conductor'.

12 CONSERVING CHARGE AND ENERGY: SIMPLE ELECTRICAL CIRCUITS

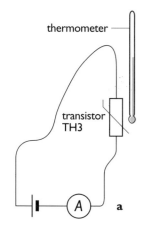

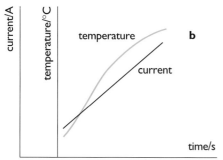

Figure 12.41 **a** Circuit containing thermistor **b** Graphs of current in and temperature against time for thermistor in **a**

Demonstration: Glass from insulator to conductor

Figure 12.40 shows how to set up the demonstration. The burner used to heat the rod must be insulated as its flame contains ionised gases which conduct electricity. Two coils, each of a few tens of turns of bare copper wire spaced a few mm apart, are wrapped round the centre of a glass rod held between insulating supports. The assembly is connected in series with a 12 V, 60 W lamp, to a 12 V car battery.

When the glass rod is cold, the lamp is unlit. When the glass rod is heated to a yellow colour, the resistivity of the glass begins to fall and the lamp begins to glow. As the temperature of the glass is increased, its resistance decreases and the lamp glows more brightly.

Questions

23 When used as a thermometer, a thermistor has a small current to minimise self heating. When used in free air in the circuit of Figure 12.41a, the thermistor gets hot. Explain, in terms of thermal runaway, the shape of the graphs in Figure 12.41b.

24 In a 'Glass from insulator to conductor' demonstration, there was a current of 0.20 A in the rod with a voltage of 50 V across it when hot. The rod was 20 mm long and 6.0 mm in diameter. Calculate the resistivity of glass when hot and compare it with that when cold (← Figure 12.21).

12.6 RESISTORS IN SERIES AND IN PARALLEL

We often need to calculate the **equivalent resistance** R_E of combinations of resistors in series and parallel. In series, the equivalent resistance equals the sum of the individual resistances, as shown in Figure 12.42. So:

$$R_E = R_1 + R_2 + \cdots.$$

In parallel, as in Figure 12.43, the conductances add so that the equivalent conductance is the sum of the individual conductances. Conductance is the reciprocal of resistance, so:

$$1/R_E = 1/R_1 + 1/R_2 + \cdots.$$

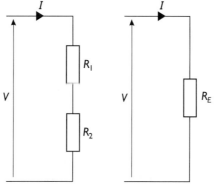

Figure 12.42 Resistors in series

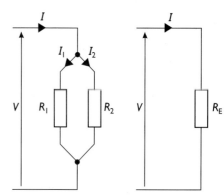

Figure 12.43 Resistors in parallel

CONSERVATION LAWS

Resistors in series

Consider the energy transfers taking place in circuits containing resistors in series and parallel. In Figure 12.42, the resistors R_1 and R_2 can be replaced by an equivalent resistance R_E. The currents in both circuits are the same and the sum of the energies dissipated per coulomb in R_1 and R_2 must equal the energy dissipated in R_E. The energy transferred per coulomb in a resistance is the voltage across that resistance, so:

$$V_E = V_1 + V_2.$$

Since $V = IR$, we can write:

$$IR_E = IR_1 + IR_2$$

or $\quad R_E = R_1 + R_2.$

To find the equivalent resistance of a number of resistors in series, simply add the individual resistances.

Resistors in parallel

In Figure 12.43, the individual resistors R_1 and R_2 in parallel can be replaced by an equivalent resistance R_E. The energy dissipated by each resistor for every coulomb that passes is the same. Therefore, the pd V across each resistor is the same as that across the equivalent resistance.

The sum of the currents through R_1 and R_2 is equal to the current through R_E:

$$I_E = I_1 + I_2.$$

Since $\quad I = V/R$

$$V/R_E = V/R_1 + V/R_2$$

or $\quad 1/R_E = 1/R_1 + 1/R_2.$

Questions

25 Explain the observation that the equivalent resistance of two resistors in parallel is always less than the resistance of either resistor alone.

26 Write down the values of the equivalent resistances of all the combinations you can make up using four identical 10 Ω resistors.

12.7 VARIABLE VOLTAGE

THE VOLTAGE DIVIDER

One important application of two resistors in series is the **voltage divider**, often called a **potential divider** (see Figure 12.44). The voltage V across one resistor is a fraction of the supply voltage V_s. The current I in both resistors is the same and is given by:

$$I = V_s/(R_1 + R_2).$$

The voltage V across R_2 is equal to IR_2, so

$$V = V_s[R_2/(R_1 + R_2)].$$

By a suitable choice of the two resistors, the voltage divider can produce an output voltage that is any desired fraction of the supply voltage.

12 CONSERVING CHARGE AND ENERGY: SIMPLE ELECTRICAL CIRCUITS

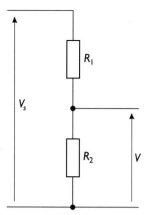

Figure 12.44 Voltage divider

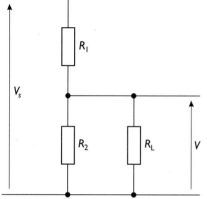

Figure 12.45 A load resistance and the voltage divider

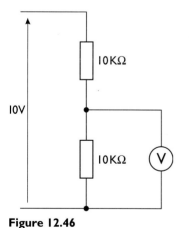

Figure 12.46

Consider the effect of a load resistance R_L in parallel with R_2, as in Figure 12.45. The equivalent resistance of R_L and R_2 is less than either individually. So V becomes a smaller fraction of V_s than with R_2 alone.

Question

27 The voltage divider in Figure 12.46 uses two equal resistors. A student measures the voltage across one of the 10 kΩ resistors, first using a digital meter with a resistance of 10 MΩ and then using a moving coil meter of resistance 10 kΩ.
a Explain why the digital meter would read 5.0 V, whereas the moving coil meter would read 3.3 V.
b Suppose both meters were connected in parallel across the 10 kΩ resistor. What would each read?
c Another student says that the digital meter is a much better meter because it is more accurate. What comments do you have on this statement?

THE POTENTIOMETER

When a variable voltage is required from a supply of output voltage V_s, a **potentiometer** is used, as shown in Figure 12.47. The supply is connected across the complete resistor and the output is taken between one end and a sliding contact which moves along the resistor. Potentiometers come in a range of sizes from those handling many kilowatts of power, controlling lights and motors, down to those in the milliwatt range in electronics (see Figure 12.48). Potentiometers get very hot in high power applications. So for high power ac, **variable voltage transformers** or 'variacs' are used which have much lower energy dissipation.

Figure 12.47 Potentiometer

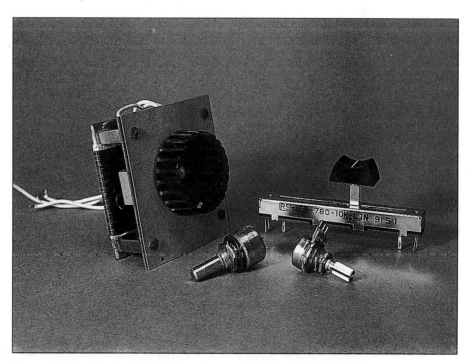

Figure 12.48 Commercial potentiometers

141

CONSERVATION LAWS

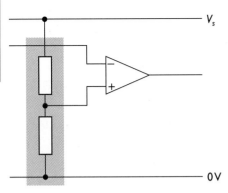

Figure 12.49 Voltage divider on the input of an op-amp

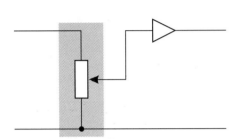

Figure 12.50 Potentiometer used for volume control

Using variable voltage

Voltage dividers and potentiometers occur in many circuits. Figure 12.49 shows two resistors defining the input voltage of an op-amp. Another use in amplifier circuits is in volume and tone control circuits. Figure 12.50 shows a potentiometer as a volume control on the input of an amplifier circuit. (→ 21)

A more complex circuit, containing both fixed and variable resistors, is used to provide a range of voltages to control the brightness and focus the electron beam in the cathode ray tube of a laboratory oscilloscope (see Figure 12.51).

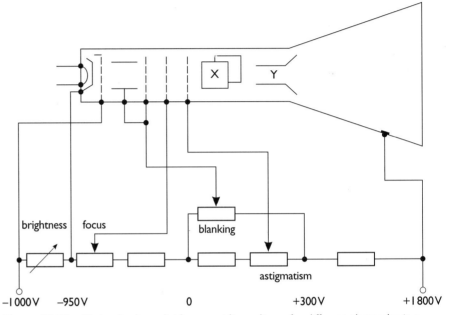

Figure 12.51 Chain of voltage dividers providing voltages for different electrodes in a cathode ray tube

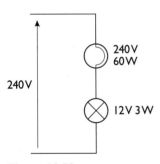

Figure 12.52

Questions

28 Calculate the current in each of the lamps in Figure 12.52 when operating at their rated voltage. What would you expect to observe when the lamps are connected, in series, to a 240 V supply?

29 A 15 Ω load resistance is connected across a potentiometer of total resistance 50 Ω to a 12 V supply (see Figure 12.53). The resistance of the section in parallel with the load varies linearly with the position of the sliding contact. Calculate the voltage across the load for various positions of the sliding contact. Then plot a graph of the voltage across the load against the position of the sliding contact.

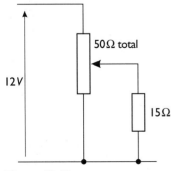

Figure 12.53

POTENTIOMETER EXPERIMENTS

Experiment 12.7 suggests a way of using a potentiometer to find the ratio between two different voltages. Before doing such an experiment, note that:

i reference voltages can be the emf of a cell, such as a Weston Standard Cell with an emf of 1.0183 V at 20 °C, or the reverse voltage of a Zener or avalanche diode, which has been previously ascertained,

12 CONSERVING CHARGE AND ENERGY: SIMPLE ELECTRICAL CIRCUITS

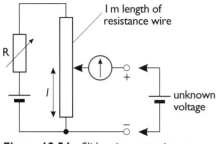

Figure 12.54 Slide-wire potentiometer

ii if the potentiometer is a rotary type, it is necessary to measure the angle through which the sliding contact is turned,

iii one simple form of measuring potentiometer uses a length of resistance wire stretched along a metre rule (see Figure 12.54). The resistor R enables the voltage across the wire to be adjusted to cover the range of measurement required.

With different circuits and methods, you can use a potentiometer to measure a wide range of voltages.

WARNING if you use voltages of over 50 V, you must take special precautions. Do not attempt to use high-tension or extra high-tension supplies except in a suitable laboratory with qualified technical help at hand.

EXPERIMENT 12.7 The potentiometer as a voltmeter

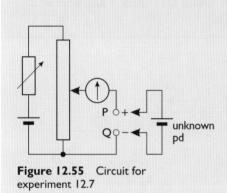

Figure 12.55 Circuit for experiment 12.7

Using the potentiometer circuit shown in Figure 12.55, you can find the ratio between two different voltages V_2. The circuit is driven by a supply of voltage V_1, which must be greater than V_2 and capable of maintaining a constant current during the period of use.

When the position of the sliding contact on the potentiometer is adjusted so that the galvanometer reads zero, the voltage across section x of the potentiometer balances voltage V_2. So when voltage V_2 gives a balance length x_1, and another voltage V_2', one of x_2, you can calculate their ratio from:

$$\frac{V_2'}{V_2} = \frac{x_2}{x_1}.$$

12.8 SOME USEFUL COMPONENTS AND CIRCUITS

VOLTMETERS AND AMMETERS

Analogue meters are those with a pointer moving over a continuous scale. They are usually moving coil meters. Typically, they have a resistance of 1 kΩ and need a current of 100 μA for **full scale deflection** (fsd). To use the meter as an ammeter, a low resistance shunt is connected in parallel with the meter. When used as a voltmeter, we assume it obeys Ohm's law with a potential drop of 0.1 V fsd across it and connect a high resistance R in series. The resistance R and the meter resistance act as a voltage divider across the unknown voltage.

A typical electronic **digital meter** has a range of 200 mV fsd and a resistance of 10 MΩ.

When using meters in circuits, remember that the meter itself has a resistance. The higher the resistance of a voltmeter and the lower the resistance of an ammeter, the less the instrument changes the operation of the circuit.

By adding an internal battery of known voltage, a meter can be used to measure resistance. The current scale is recalibrated to read resistance, and then the current through the unknown resistor is measured.

CONSERVATION LAWS

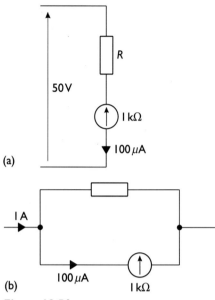

Figure 12.56

Questions

30 A moving coil meter has a resistance of 1 kΩ and an fsd of 100 mA. Calculate the values of the resistors required to convert it into
a a 0-50 V voltmeter using the circuit in Figure 12.56 a
b a 0-1 A ammeter using the circuit in Figure 12.56 b.

31 Figure 12.57 shows circuits for using a 0-200 mV digital voltmeter of very high resistance to measure
a voltages up to 20 V,
b currents up to 2 A.
Calculate the values of the appropriate resistors.

32 Figure 12.58 shows two possible arrangements of meters to measure the resistance of a 1 kΩ resistor. The basic meter movement of the ammeter requires 1 mA for fsd and has a resistance of 100 Ω. The basic meter movement of the voltmeter requires 0-200 mV for fsd and has a very high resistance. Which is the best way of inserting the 0-20 mA ammeter and 0-20 V voltmeter?

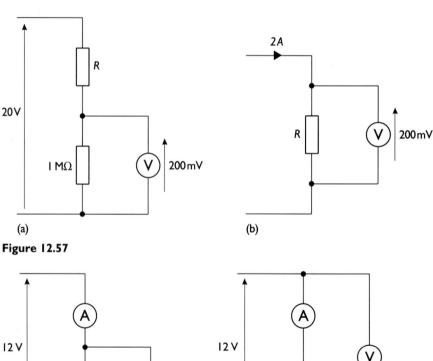

Figure 12.57

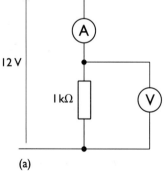 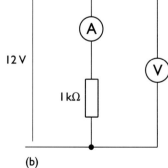

Figure 12.58

EQUIVALENT CIRCUITS

We have looked at the properties of some individual components, but the increasing complexity of modern electrical systems means that we cannot

12 CONSERVING CHARGE AND ENERGY: SIMPLE ELECTRICAL CIRCUITS

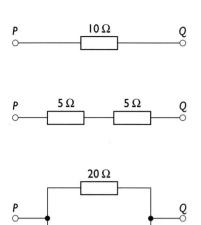

Figure 12.59 Three simple equivalent circuits

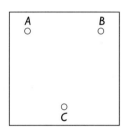

Figure 12.60 A three-terminal circuit

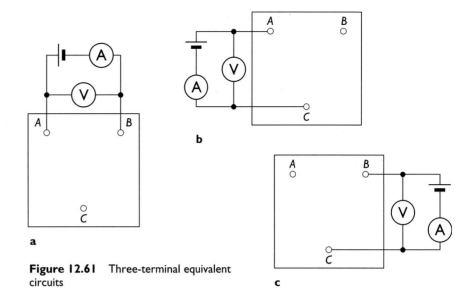

Figure 12.61 Three-terminal equivalent circuits

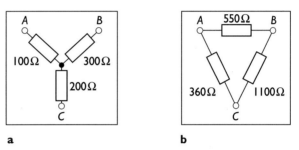

Figure 12.62 **a** Star and **b** delta configurations for three-terminal circuit

dismantle a circuit to investigate these components. Instead, we can infer a possible set of components from measurements made on the accessible connections. This is called the **equivalent circuit** approach. For example, the circuits in Figure 12.59 all behave in exactly the same way when a cell is connected across PQ. If all three were sealed in separate boxes with only the terminals P and Q accessible, we could not work out, from measurements between P and Q, which arrangement was present in a given box.

A more difficult example, shown in Figure 12.60, has three terminals A, B and C. There are many circuits we could set up to make measurements on this box. Figure 12.61 shows three circuits that would enable us to measure the equivalent resistances between AB, BC and CA. Figure 12.62 shows two arrangements of three resistors connected between A, B and C. These are the basic **star** and **delta configurations** for a three-terminal circuit. When we work out the equivalent resistances between AB, BC and CA, we find that both configurations give the same values. No measurements made using the circuits shown in Figure 12.61 can distinguish between star and delta arrangements.

Questions

33 Show that the two circuits in Figure 12.62 are indeed equivalent. Do this by finding the equivalent resistance between A and B with C unconnected, for each circuit. Then repeat the calculation with A and C, and B and C.

34 Figure 12.63 shows three further circuits for making measurements on the three-terminal box. Assuming the box contains the resistors of Figure 12.62, deduce the readings of the ammeter and voltmeter in each circuit.

35 A box has two terminals. Measurements on it give the voltage–current graph shown in Figure 12.64. Deduce what you can about the contents of the box.

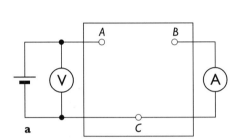
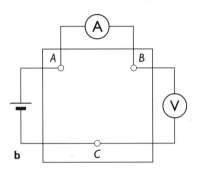
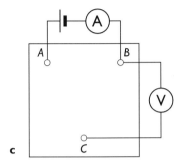

Figure 12.63

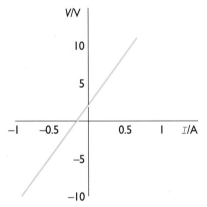

Figure 12.64 Graph of voltage against current

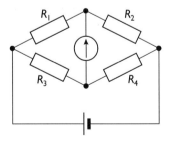

Figure 12.65 Wheatstone bridge circuit

RESISTANCE BRIDGE CIRCUITS

Charles Wheatstone (1802–75) developed **bridge circuits** (see Figure 12.65) to measure resistance accurately. When one or more of the four resistors are varied, a set of values for the four resistances can be found at which there is zero current in the galvanometer. The bridge circuit is then said to be balanced, the **balance condition** being that:

$$R_1 \times R_4 = R_2 \times R_3.$$

When the resistances of three of the resistors are known, that of the fourth may be found. Note that the balance condition does not depend on the supply voltage or on the calibration of the galvanometer.

Questions

36 When the bridge circuit in Figure 12.65, is balanced, the current through the galvanometer is zero, the current through R_1 and R_2 is I and the current through R_3 and R_4 is I'.
 a What can be said about the pd across the galvanometer?
 b Why is the current through R_1 the same as that through R_2?
 c Why is the current through R_3 the same as that through R_4?
 d Explain why, for R_1 and R_3, we can write
  $IR_1 = I'R_3$.
 e Similarly, explain why we can write
  $IR_2 = I'R_4$.
 f Derive the balance equation eliminating I and I' from the equations in d and e.

37 The bridge circuit in Figure 12.66 has a 6.0 V battery, and $R_1 = 20\,\Omega$, $R_2 = 22\,\Omega$, and $(R_3 + R_4)$ make up a 10 Ω potentiometer.
 a Calculate the value of R_3 at balance.
 b The slider is moved to increase R_3 by 0.01 Ω above the value at balance. Calculate the voltage across the galvanometer. Would you expect to be able to detect this?

12 CONSERVING CHARGE AND ENERGY: SIMPLE ELECTRICAL CIRCUITS

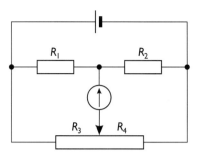

Figure 12.66 Bridge circuit

Bridge circuits

When one or more of the resistances in a bridge circuit is changed, the change can be measured as an out of balance voltage across the galvanometer. Alternatively, the galvanometer current can be returned to zero by making a known change in a calibrated resistor. Nowadays the galvanometer is often replaced by an op-amp comparator.

Figure 12.67 shows a bridge circuit for a resistance thermometer. The sensor is R_2, which has a large temperature coefficient of resistance. Changes in the resistance of R_2 unbalance the bridge and generate an out of balance voltage across the detector D.

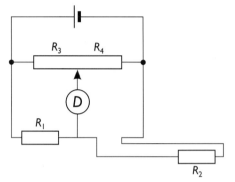

Figure 12.67 Bridge circuit for a resistance thermometer

Question

38 A platinum resistance thermometer element has a resistance of 100.0 Ω at 273 K and its resistance increases by 0.4 Ω K^{-1}. When it is used in the circuit of Figure 12.67 with R_1 equal to 100.0 Ω and $V = 5$ V, calculate the temperature rise to unbalance the bridge by 0.1 mV.

THE STRAIN GAUGE

When a wire is stretched, it gets longer and thinner. The resistivity does not normally change so the resistance increases. Although the effect is small, it is widely used to monitor strain in structures. Highly sensitive methods of detecting the small changes in resistance are used, often employing bridge circuits.

Figure 12.68 shows a typical **strain gauge**. It consists of a very thin etched metal foil element on a plastic backing sheet which is firmly bonded to the structure being monitored. When the structure is deformed, the strain gauge deforms with it. The resulting change in the resistance of the strain gauge is detected and monitored. Strain gauges are usually used in pairs to compensate for variations in temperature which can produce larger changes in resistance than the strain being monitored.

Figure 12.69 shows four strain gauges mounted on a beam. When the beam is loaded, R_1 and R_4 are compressed and their resistance drops, whilst R_2 and R_3 are stretched, increasing their resistance. This arrangement gives a larger output at the detector. If the relationship between the load applied and the strain produced is known for the beam, the output of the strain gauge bridge can be calibrated in terms of the load applied. Such devices are called **load cells** and are used in heavy duty weighing machines.

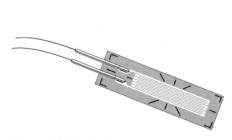

Figure 12.68 Strain gauge

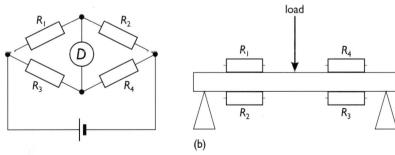

Figure 12.69 Strain gauge bridge **a** circuit **b** arrangement

CONSERVATION LAWS

SUMMARY

The behaviour of simple circuits can be analysed using two laws.
- **Kirchhoff's first law** or current law: the total current out of a junction equals the total current into it: **the conservation of charge**.
- **Kirchhoff's second law**: the total emf of the sources in a closed circuit equals the sum of the voltages, or pds, across all components in the circuit: **the conservation of energy**.

Ohm's law states that the voltage across and the current in a component are proportional at constant temperature. It applies to most metals and carbon resistors. However, it does not apply to many semiconductor materials, insulators, gases, liquids and electronic devices. Such components are often described as non-linear or non-Ohmic.

The resistivity ρ of a material is the resistance of unit cross-sectional area of the material per unit length. It generally varies with temperature, incident light, pressure, applied voltage or other physical conditions. In general:
- the resistivity of metals increases with increase in temperature,
- the resistivity of semiconductor materials decreases with increase in temperature over their active range and increases with tempertuire either side of this,
- the resistivity of insulators decreases with increase in temperature.

Conductance G is the ability of a conductor to pass current and is measured in siemens, or Ω^{-1}. It is equal to the reciprocal of resistance:

$$G = 1/R.$$

Conductivity σ is the conductance of unit length of a material, per unit cross-sectional area. It is equal to the reciprocal of resistivity:

$$\sigma = 1/\rho.$$

The equivalent resistance of resistors in series is the sum of the individual resistances. The equivalent conductance of resistors in parallel is the sum of the individual conductances.

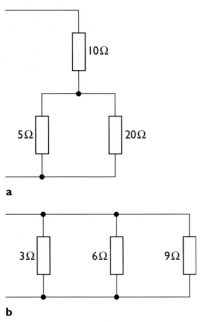

Figure 12.70 Resistor combinations

SUMMARY QUESTIONS

39 Calculate the resistances of the combinations of resistors in Figure 12.70.

40 Calculate the power dissipated in each resistor, and the total power, when 10 Ω and 22 Ω resistors are connected
a in series,
b in parallel
to a 12 V supply.

41 An ideal silicon diode is a perfect insulator in the reverse direction and a 25 Ω resistor in the forward direction for voltages greater than 0.6 V. Draw a voltage-current graph for such a diode in series with a 100 Ω resistor.

42 An LED is connected in series with a resistor to a 5.0 V supply. Calculate the value of the series resistor, given that the current in the LED is 12 mA and the voltage across it is 2.0 V when lit. Estimate the maximum voltage that can be applied across the combination if the current in the LED is not to exceed 25 mA.

12 CONSERVING CHARGE AND ENERGY: SIMPLE ELECTRICAL CIRCUITS

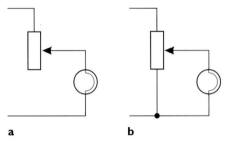

Figure 12.71

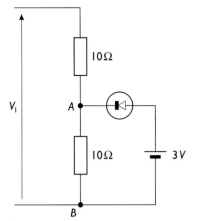

Figure 12.72 Circuit with voltage divider

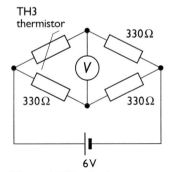

Figure 12.73 Bridge circuit

43 An electric toaster element is wound from a length of nichrome ribbon of length 3.0 m and width 3.0 mm. It dissipates 500 W when connected to a 240 V supply. Calculate the resistance and thickness of the ribbon, taking the resistivity of nichrome to be $1.1 \times 10^{-6}\,\Omega\,\text{m}$.

44 What is the importance for power delivery of matching the load resistance in a circuit to the characteristic of the supply? (← Figure 12.19)

45 What are the effects of increasing the supply voltage applied to a circuit containing components with
a positive,
b negative
temperature coefficients of resistance.

46 A lamp is connected in series with a variable resistor, as in Figure 12.71a. Explain why adjusting the variable resistor dims the lamp but will not reduce the current to zero. When the variable resistor is replaced by a potentiometer, as in Figure 12.71b, explain why the lamp can now be dimmed to zero current.

47 Figure 12.72 shows a circuit incorporating a voltage divider. Deduce the voltage between A and B when V_1 is
a 0 V,
b 2 V,
c 4 V,
d 6 V.
Assume the cells and diode are ideal components.

48 A bridge circuit, shown in Figure 12.73, contains a TH3 thermistor whose data are given in Figure 12.37. In the other arms are three 330 Ω resistors connected to a 6.0 V supply. Estimate the out of balance voltage V at
a 20 °C,
b 50 °C.

13 STORING CHARGE AND ENERGY: CAPACITORS

Electrical charge can be stored. But you certainly cannot store electrons in a box or fill up a can with protons! What you can do is to cause charge to flow on to a conductor until the flow stops. The voltage of the conductor will rise until it repels any further charges. Since the charges flowing on to the conductor experience repulsive forces and work has to be done, energy is transferred and the electrical potential energy of the conductor is raised. The higher the charge for a given voltage, the greater the capacitance of the conductor.

Because electric charge is conserved, the total current remains the same, as it flows through a circuit. So, when circuits containing a capacitor are switched on or off, the resulting movements of charge, called transient currents ($\rightarrow$ 21), transfer energy into and out of the capacitor.

13.1 CHARGE STORAGE IN A CAPACITOR

EXPERIMENT 13.1 Charging a capacitor

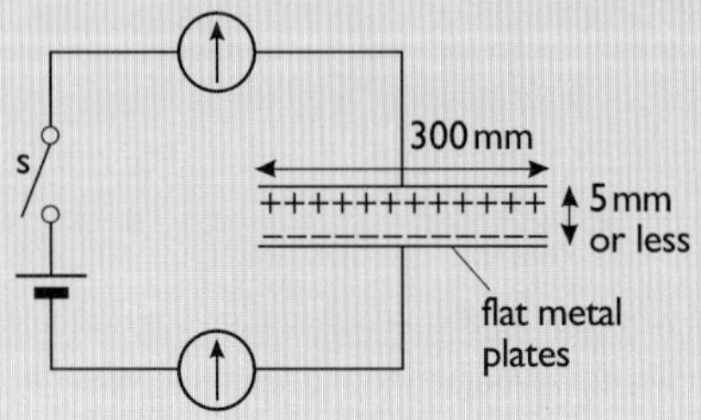

Using the circuit in Figure 13.1 you can sketch a graph of the variation of current in each part of the circuit with time after the switch is closed.

Try using a 1000 μF capacitor instead of the two plates.

Figure 13.1 Charging a capacitor made from two plates of conductor

Experiment 13.1 enables you to charge a capacitor made from two separated metal plates (see Figure 13.1). The current registered is zero when the switch has been open or closed for a long time. This is not surprising since the circuit is not complete. However, when the switch is closed, both meters give a momentary deflection in the same direction and then return to zero. This momentary deflection indicates that, on closing the switching, there is a small current for a short time even though the circuit is incomplete.

Capacitors are often said to **store** charge. However, the quantity of charge in a circuit is constant. The cell in Figure 13.1 drives a momentary current through the circuit and **redistributes** the charge, but the net charge on the capacitor remains unchanged. The 'pulse' of current might appear as in Figure 13.2. The area under this graph represents the charge transferred.

13 STORING CHARGE AND ENERGY: CAPACITORS

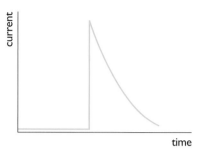

Figure 13.2 The pulse of current on charging a capacitor

After the pulse, there is an excess of positive charge on the plate connected to the positive terminal of the cell and an excess negative charge of the same magnitude on the plate connected to the negative terminal. This redistribution of charge sets up a voltage across the capacitor. If you connect another similar cell in the circuit in Figure 13.1, you would obtain a larger pulse of current, and would store twice the charge on the capacitor at the new voltage. The ratio of the charge stored, or redistributed, on each plate to the voltage set up is called the **capacitance** of the capacitor.

$$\text{Capacitance} = \frac{\text{charge}}{\text{voltage}}$$

or in symbols,
$$C = \frac{Q}{V}.$$

The unit of capacitance is the coulomb per volt and is called the **farad (F)**. Most practical capacitors with a voltage of 1 V across them only store a tiny fraction of a coulomb. So, most of the capacitors in common use have values expressed in picofarads (pF), nanofarads (nF), microfarads (μF) and millifarads (mF).

EXPERIMENT 13.2 Charging and discharging a capacitor

By setting up the circuit in Figure 13.3, you should observe opposite deflections of the galvanometer as the switch is moved from position 1 to position 2 and back. Connecting an oscilloscope across the resistor R enables you to monitor the voltage across it (see Figure 13.4). You can observe the pulses when different supply voltages are used.

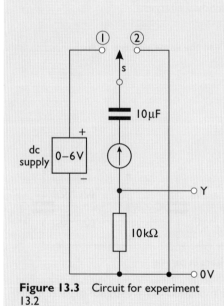

Figure 13.3 Circuit for experiment 13.2

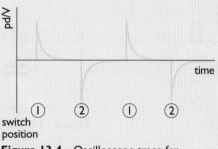

Figure 13.4 Oscilloscope trace for experiment 13.2

CAPACITORS IN SERIES AND PARALLEL

The equivalent capacitance C_E of capacitors in parallel is the sum of their capacitances:

$$C_E = C_1 + C_2 + \cdots.$$

For capacitors in series, the equivalent capacitance is given by:

$$1/C_E = 1/C_1 + 1/C_2 + \cdots.$$

The relationships are the reverse of those for resistances. We increase the capacitance of a system by adding extra capacitors in parallel, and reduce the effective capacitance by connecting them in series.

CONSERVATION LAWS

Capacitors in series

Figure 13.5 shows two capacitors C_1 and C_2 in series. When connected to a supply, the current charges both with charge Q. The total voltage V across both capacitors is the sum of the voltages across each individual capacitor:

$$V = V_1 + V_2.$$

Since the equivalent single capacitor C_E would store the same charge Q at the same voltage V, then:

$$\frac{Q}{C_E} = \frac{Q}{C_1} + \frac{Q}{C_2}$$

so $\dfrac{1}{C_E} = \dfrac{1}{C_1} + \dfrac{1}{C_2}.$

Capacitors in parallel

For the two capacitors in parallel in Figure 13.6, the current charges each capacitor to the same voltage. They will carry charges Q_1 and Q_2 proportional to their individual capacitances. So:

$$Q_1 = C_1 V$$

and $Q_2 = C_2 V.$

The total charge Q is made up of these individual charges:

$$Q = Q_1 + Q_2.$$

Q would charge the equivalent capacitor C_E to the same voltage V, so:

$$C_E V = C_1 V + C_2 V.$$

Dividing by V gives:

$$C_E = C_1 + C_2.$$

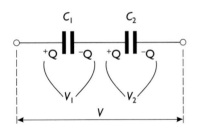
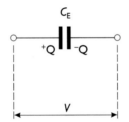

Figure 13.5 Capacitors in series

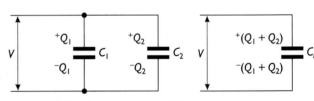

Figure 13.6 Capacitors in parallel

The circuit used in experiment 13.2 can be modified to investigate capacitors in series and parallel (see Figure 13.7). Alternatively, you can check the predicted equivalent capacitances for assemblies of capacitors using a direct reading electronic capacitance meter.

Questions

1 Calculate the capacitances of all arrangements that can be made up from three 1.0 μF capacitors.

2 a Figure 13.8 a shows two capacitors of 1.0 μF and 3.3 μF connected in series to a 12 V supply. Calculate the voltage across each capacitor.
b A 10 kΩ resistor is connected across the 1.0 μF capacitor (see Figure 13.8 b). Deduce the voltage across each capacitor once equilibrium has been established and there is no net movement of charge.

13 STORING CHARGE AND ENERGY: CAPACITORS

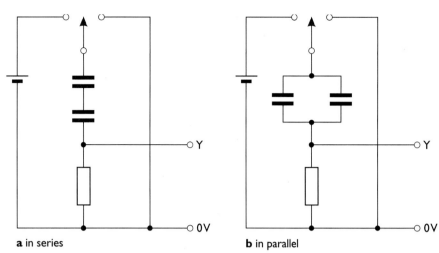

a in series **b** in parallel
Figure 13.7 Investigating capacitors in **a** series **b** parallel

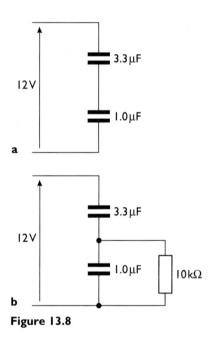

Figure 13.8

13.2 ENERGY STORAGE IN CAPACITORS

When a capacitor is charged, work is done in removing charge from one plate and adding the same charge to the other (see Figure 13.9). Energy is transferred from the charging supply and appears as electrical potential energy in the capacitor. This potential energy stored E_P is given by:

$$E_P = \tfrac{1}{2}QV$$
$$= \tfrac{1}{2}CV^2$$
$$= \tfrac{1}{2}\frac{Q^2}{C}.$$

where V is the voltage across the capacitor, Q the displaced charge and C the capacitance.

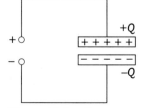

Figure 13.9 Charging a capacitor

CONSERVATION LAWS

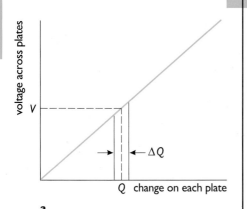

a

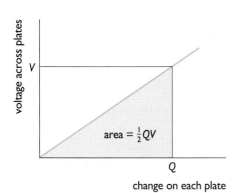

b

Figure 13.10 a Graph of voltage against charge for a capacitor b Area under graph gives energy transferred

The energy stored in a capacitor

The voltage V across a capacitor varies linearly with the charge Q on the capacitor, as shown in Figure 13.10a. Suppose we wish to displace the charge on the capacitor by an amount ΔQ from a little under Q to a little over Q. The supply has to do work to achieve this. The energy transferred will be $V\Delta Q$ assuming that ΔQ is so small that V does not change appreciably whilst ΔQ is being displaced. The quantity $V\Delta Q$ is the area of the strip of width ΔQ under the graph.

The area under the graph, and therefore the work done, increases as the charge on the capacitor increases. More and more energy is transferred as each extra amount of charge ΔQ is redistributed from one plate to the other. Each increase in charge ΔQ causes an increase in potential energy per coulomb of ΔV volts, and

total energy transferred = $\Sigma\, V\Delta Q$.

So as the charge on a capacitor is increased from 0 to Q, the energy transferred from the supply into the capacitor can be found by summing the areas of the strips of width ΔQ under the graph. In other words, the energy transferred is the total area under the graph. This area is equal to $\frac{1}{2}QV$ (see Figure 13.10b).

We know that $Q = CV$, so the potential energy stored E_P is given by:

$$E_P = \tfrac{1}{2}QV$$
$$= \tfrac{1}{2}CV^2$$
$$= \tfrac{1}{2}\frac{Q^2}{C}.$$

Connecting a circuit to a supply of voltage V causes each coulomb that passes between the poles of the supply to give up V joules of potential energy. When a capacitor is connected across a cell, as in Figure 13.11, charge Q is displaced and so the energy supplied is QV. Yet the capacitor only stores energy $\tfrac{1}{2}QV$. What happens to the other $\tfrac{1}{2}QV$?

The 'lost' half is represented by the area of the upper triangle of the graph in Figure 13.10b. It is lost by the charge as it passes through the resistance of the circuit – the connecting leads, the capacitor materials and the internal resistance of the cell. The voltage across this resistance decreases from V to 0 as it increases from 0 to V across the capacitor. The energy supplied to a capacitor is always half 'dissipated' and half 'stored', whatever the resistance of the circuit.

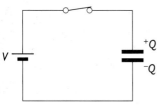

Figure 13.11 Charging a capacitor

Questions

3 a Compare the charge and energy stored in a 1.0 μF capacitor charged to 300 V with that stored in a 1 000 μF capacitor charged to 12 V.
b Would capacitors provide a viable way of storing electricity on a large scale?

4 Describe what you would expect to observe when charging a capacitor to the same final charge through **a** a high resistance, and **b** a low resistance to the same final voltage.
If the charging circuit consisted of a superconductor, and therefore had zero resistance, what would happen?

13 STORING CHARGE AND ENERGY: CAPACITORS

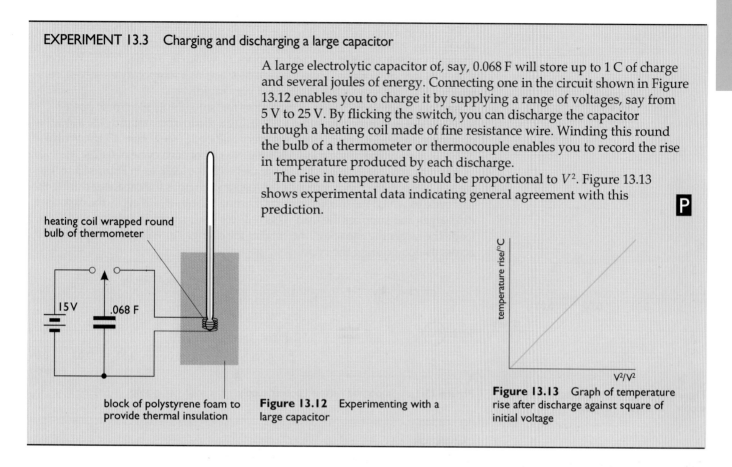

EXPERIMENT 13.3 Charging and discharging a large capacitor

A large electrolytic capacitor of, say, 0.068 F will store up to 1 C of charge and several joules of energy. Connecting one in the circuit shown in Figure 13.12 enables you to charge it by supplying a range of voltages, say from 5 V to 25 V. By flicking the switch, you can discharge the capacitor through a heating coil made of fine resistance wire. Winding this round the bulb of a thermometer or thermocouple enables you to record the rise in temperature produced by each discharge.

The rise in temperature should be proportional to V^2. Figure 13.13 shows experimental data indicating general agreement with this prediction.

Figure 13.12 Experimenting with a large capacitor

Figure 13.13 Graph of temperature rise after discharge against square of initial voltage

USING CAPACITORS TO STORE ENERGY

Capacitors store only small quantities of energy. The maximum energy stored in the large capacitor in experiment 13.3 is about 10 J. Capacitors are used as energy stores because they can be **loaded and unloaded very quickly**.

An electronic flash gun used in photography (see Figure 13.14) is powered by a small battery, yet is required to provide energy very rapidly during the flash.

Figure 13.14 Camera flash gun

CONSERVATION LAWS

Typically, the light flash emits 1 J of energy in about 0.1 ms. This is a power of 10 kW, well beyond the power rating of the battery. However, the battery can charge a capacitor in, say, 10 s and supplies 2 J, which is, a power of 0.2 W, well within its rate of working. Here the capacitor acts as a temporary energy store which is charged slowly but discharged very rapidly.

Another example is the use of capacitors for 'smoothing' electronic power supplies. An ac supply converted to dc using a diode does not provide a steady dc supply. Instead it provides pulses every half cycle (see Figure 13.15). With a 50 Hz ac supply, the diode conducts for 0.01 s out of each cycle of 0.02 s.

When a capacitor is connected in parallel with the load, as in Figure 13.16, the current through the diode supplies the load and charges the capacitor during the half cycle in which the diode conducts. During the other half cycle, the capacitor discharges and supplies the load. This reduces the changes in voltage across the load. The larger the capacitor, the smaller the changes in voltage, or 'ripples', across the load during the cycle.

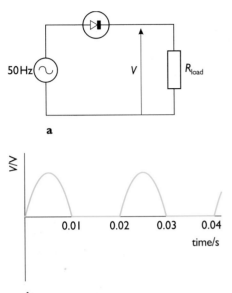

Figure 13.15 Half–wave rectified power supply **a** circuit **b** graph

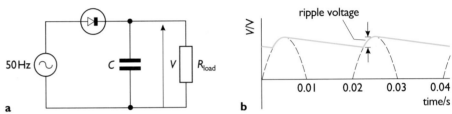

Figure 13.16 Smoothed rectified power supply **a** circuit **b** graph

Question

5 A power supply with the circuit in Figure 13.15 provides d.c. from a 50 Hz a.c. source. The average voltage across the load of resistance 24 Ω is 12 V. The capacitor has a capacitance of 5 000 μF. Calculate
a the average current in the load,
b the charge flowing through the load in half a cycle,
c the drop in voltage across the capacitor over the half cycle from supplying the charge calculated in **b** to the load.
Has the designer provided an adequately large capacitor for the power supply to appear 'smooth'?

13.3 GROWTH AND DECAY OF THE CHARGE ON A CAPACITOR

In experiment 13.2, a capacitor is charged through a resistor and a moving coil meter and the variation in the charging current with time is observed by monitoring the voltage across the resistor on an oscilloscope. The discharge current is observed by discharging the capacitor through the same resistor and meter. In this experiment, the current instantly rises to a maximum value when the supply is connected, and then decays, at an ever decreasing rate until it is effectively zero. How do we explain this?

When the current is initially at its peak, charge is building up on the capacitor at a maximum rate: the voltage across the capacitor is rising quickly and so the voltage across the resistance and therefore the current in the circuit are decreasing rapidly. As the current decays, the capacitor charges less rapidly which, in turn, causes its voltage to rise less rapidly. The voltage across the resistance and

13 STORING CHARGE AND ENERGY: CAPACITORS

therefore the current through it fall more slowly, and so on. During discharge, a similar process occurs. The voltage across the charged capacitor initially drives a peak current through the resistance. Charge leaving the capacitor at a rapid rate causes the voltage across the capacitor to fall rapidly at the beginning. This falling voltage drives less and less current through the resistance. Charge therefore leaves the capacitor at a decreasing rate which, in turn, causes its voltage to fall at a decreasing rate.

Clearly, the time taken to charge or discharge the capacitor depends upon the total amount of charge transferred to or from the capacitor at a given charging voltage and the rate at which it is transferred. We can investigate the nature of the charging and discharging processes more effectively if we slow down the changes that take place. We can do this by increasing the resistance or the capacitance or both (see experiment 13.4).

EXPERIMENT 13.4 Charging and discharging a capacitor through a resistor

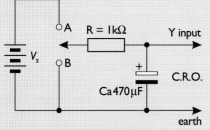

Figure 13.17 Circuit for experiment 13.4

Connecting up the circuit shown in Figure 13.17, and moving the switch from B to A, should produce a trace on the screen similar to Figure 13.18a. Returning the switch to B you should produce the trace shown in Figure 13.18b. If you interchange the capacitor and resistor, as shown in Figure 13.19, and move the switch from B to A, you will notice that the trace on the screen appears as in Figure 13.20a immediately afterwards. When the switch is taken back to B, the trace looks like that in Figure 13.20b.

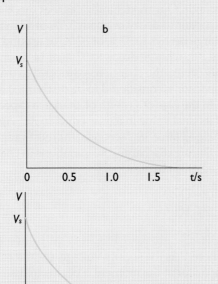

Figure 13.18 **a** The charge and **b** discharge current

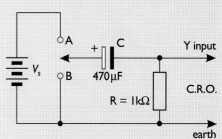

Figure 13.19 Behaviour of an RC circuit

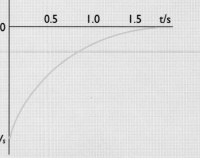

Figure 13.20 Variation of voltage across the capacitor with time

CONSERVATION LAWS

> **Questions**
>
> **6** Explain why the oscilloscope displays the current variation with time in the circuit in Figure 13.19. Why does the trace go negative?
>
> **7** What is the voltage across **i** C and **ii** R
> **a** several seconds after moving the switch to B?
> **b** several seconds after returning the switch to A?
> **c** What is the current in the circuit **i** several seconds after moving the switch to B, and **ii** several seconds after returning it to A?

The growth or decay of charge on a capacitor with time shares similarities with many other natural processes. For example, the growth of a stable daughter product and the decay of a parent radionuclide with time ($\rightarrow$ 39.4) are also **exponential changes**. ($\rightarrow$ 41)

In radioactivity, we use the idea of half-life as an indication of the time taken to decay. For example, a radionuclide has a half-life of 1 hour when half of the atoms within a sample decay in 1 hour. When dealing with the charging or discharging of a capacitor C through a resistance R, we can use a concept similar to that of half-life – the **time constant** of the circuit. This is measured in seconds and is **equal to the product RC** when R is measured in ohms and C in farads. The larger the value of C or R, the larger the time constant and the longer it takes for a steady charged or discharged state to be reached.

The time constant of the circuit in Figure 13.17 is $1\,000 \times 470 \times 10^{-6}$ s or 0.47 seconds. Our model indicates that this is the time taken for the charge on the capacitor to decay to about 37 % of its initial value. After a time equal to 2RC, the charge decays to 37 % **of** 37 %, or around 14 % of its initial value. After a time of 3RC, the charge drops to 37 % of 14 %, or 5 %. After 5RC, or about 2.4 seconds in this circuit, the charge has dropped to 0.67 %. So, for most practical purposes, a capacitor is considered to be discharged after 5RC seconds.

In the case of growth of charge, a capacitor will charge to 100 % – 37 %, or 63 % of its final value in a time equal to RC. Again, it can be considered to be fully charged after 5RC seconds, for most practical purposes.

During discharge, the variations in charge Q, voltage V and current I with time t are given by:

$$Q = Q_0 e^{-t/RC}$$
$$V = V_s e^{-t/RC}$$
$$I = I_0 e^{-t/RC}$$

where V_s is the supply voltage, Q_0 and I_0 are the charge on and current from the capacitor at the moment it begins to discharge, and e is the exponential function. ($\rightarrow$ p. 160, 'Equations for the discharge of a capacitor'.)

> **Question**
>
> **8 a** By using the basic equations $Q = CV$ and $V = IR$, show that the product RC is measured in 'seconds' if R is in Ω and C in farads.
> **b** Why do we need the concept of half-life in radioactivity? Why not the 'whole-life'?
> **c** How is it that we can say a capacitor is effectively discharged after 5RC seconds?

13 STORING CHARGE AND ENERGY: CAPACITORS

Explanation of RC curves

We can explain more quantitatively the shape of the growth and decay curves in experiment 13.4.

Suppose a supply voltage V_s is connected to resistance R and an uncharged capacitor C in series, as in Figure 13.21. At an instant t_0, there is a voltage V_s across R and C according to Kirchhoff's second law ($\leftarrow$ 12.2). The voltage across C is zero at t_0 since C is uncharged. So the voltage across R is V_s and the current in R is of value V_s/R.

Charge is moved round the circuit by the supply charging the capacitor. At time t_1, there is a charge Q stored on the capacitor and a voltage V_c across it given by:

$V_c = Q/C.$

The voltage across the resistor is now $V_s - V_c$ because the sum of the voltages across R and C must always equal V_s. The current in the circuit is reduced to

$I = (V_s - V_c)/R.$

Charge is therefore flowing round the circuit more slowly, so the capacitor is charging more slowly. The process continues more and more slowly until the voltage across the capacitor reaches V_s when a charge:

$Q = CV_s$

has been moved round the circuit from one plate of the capacitor to the other. The current is zero so the voltage across the resistor is zero from $V = IR$. The voltage across the capacitor exactly balances the supply voltage, and a steady state has been reached.

When the supply voltage is reduced to zero at time t_2, the capacitor discharges with the charge flowing the opposite way round the circuit, giving a negative voltage across R. The voltage across the capacitor drops, causing the current to fall so the discharge becomes slower. By Kirchhoff's second law, at all times the sum of the voltages across R and C is zero because the supply voltage is zero. After a sufficiently long time, both V_R and V_c fall to zero and the capacitor is discharged.

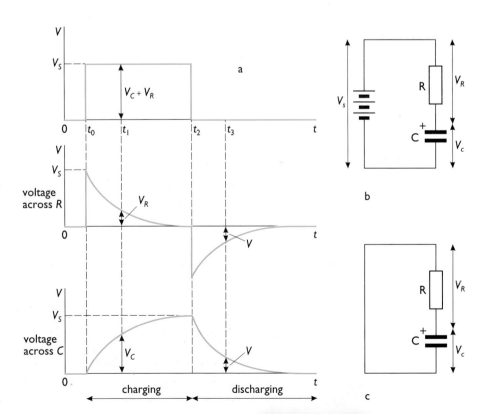

Figure 13.21 **a** Graphs of V_s, V_R and V_c against time **b** charging circuit **c** discharging circuit

CONSERVATION LAWS

Equations for the discharge of a capacitor

Consider some time t_3 after t_2 when the capacitor is discharging. The current in the circuit is I and charge Q remains on the capacitor, which is charged to voltage V (see Figure 13.21a). The charge ΔQ leaving the capacitor in the following small interval of time Δt is given by:

$$\Delta Q = - I\Delta t.$$

The current I is equal to V/R at this time, so

$$\Delta Q = - (V/R)\, \Delta t.$$

The voltage V across C is given by Q/C, so we can write:

$$\Delta Q = - (Q/RC)\, \Delta t$$

or $\Delta Q/\Delta t = - Q/RC$.

Therefore, the rate of flow of charge from the capacitor is directly proportional to the charge on the capacitor at the time. This relationship is fundamental to all exponential changes.

The equation can be solved numerically by a repeated or **iterative** process. Many computers use this type of process both in their processors and in applications programs. It can also be written as a differential equation by taking very small changes of Q and t so that $\Delta Q/\Delta t$ tends to dQ/dt as Δt tends more and more towards zero (→ 41.3):

$$dQ/dt = - Q/RC.$$

This can be solved analytically by integration (→ 41.3), to give:

$$Q = Q_0\, e^{-t/RC}$$

where Q_0 is the initial charge on the capacitor at $t = 0$, the moment it begins to discharge, and e is the exponential function. A similar relationship holds for radio-active decay (→ 39.4).

Because the voltage across the capacitor is proportional to Q, we can write:

$$V = V_s\, e^{-t/RC}.$$

Since the discharge current is proportional to the voltage across the capacitor, we can also write:

$$I = I_0\, e^{-t/RC}$$

where I_0 is the initial current at the moment the capacitor begins to discharge.

The equations relating the voltage and current variations with time are consistent with the oscilloscope traces of Figures 13.18 and 13.20. This can easily be confirmed by substituting values for C and R and plotting values of V and I against t.

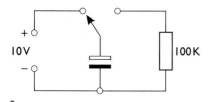

a

t/s	Q/C		ΔQ/C
0	5 × 10⁻³ ⟶		5 × 10⁻⁴
5	4.5 × 10⁻³ ⟶		4.5 × 10⁻⁴
10	4.05 × 10⁻³ ⟶		
15			
20			
25			

b

Figure 13.22

Questions

9 This question asks you to use a numerical method to find the variation of charge with time for a capacitor discharging through a resistor and then compare the solution with that obtained analytically by integration. A 500 μF capacitor is charged to 10 V, and then discharged through a 100 kΩ resistor (see Figure 13.22a).
a What is the initial charge on the capacitor?
b What is the current as the capacitor begins to discharge, at $t = 0$?
c Suppose this current continues for 5 s without appreciably reducing. How much charge leaves the capacitor during this time?
d How much charge is left on the capacitor after 5 s? What proportion is this of the initial charge?

13 STORING CHARGE AND ENERGY: CAPACITORS

You could have obtained the answer to **d** using the equation $\Delta Q/\Delta t = -Q/RC$.

e Show that $\Delta Q = -1/10$ in a time $\Delta t = 5$ s. What fraction of the charge remains at the end of the interval?

f What charge is lost in the next 5 s? What is left on the capacitor at $t = 10$ s?

g Repeat the process, copying and completing the table in Figure 13.22 b, to find the charge at $t = 15$ s, 20 s, up to 100 s. Plot a graph of Q against t. How should you join up the points, bearing in mind your answer in **c**?

h Using the same axes as in **g**, draw the graph of the exact relationship $Q = Q_0 e^{-t/RC}$, obtained analytically. Compare the two graphs.

i Explain why your graph in **g** would have been closer to the graph of the exact expression in **h**, if you had taken 1 s intervals rather than 5 s intervals.

A computer spreadsheet program provides an ideal method for the numerical, or iterative, solution of differential equations because the values can easily be changed and the results recalculated ($\rightarrow$ 43.1).

10 Look at the V_R and V_C curves in Figure 13.21a. The V_R charge curve is the same as the V_C discharge curve. During charging, $V_C = V_S - V_R$.

a Use this equation to show that the voltage V across a capacitor during charging increases according to the relationship $V = V_s(1 - e^{-t/RC})$.

b How does the charge Q on a capacitor vary with time as the capacitor charges up under a constant supply voltage?

c Write down equations for I_R and V_R during the charging period.

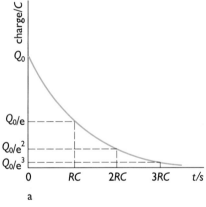

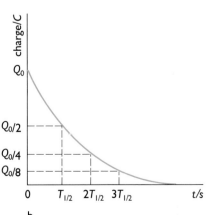

Figure 13.23 Decay of charge with time showing **a** time constant and **b** 'half–life'

TIME CONSTANT

Charge, current and voltage all decay with time in the same way. The curves of the variation of each quantity with time have the same shape. In the capacitor discharge, the current and charge both change by a constant fraction of themselves in equal intervals of time.

The curves of their variation with time are the same shape, whatever point we start our measurements of t. Experiment 13.5 suggests a way of plotting the discharge current against time.

A capacitor would completely discharge in a time RC if it continued to discharge at the rate of the initial current. However, as we have seen, charge, voltage and current all change exponentially in the capacitor discharge circuit. This is because the rate of discharge is proportional to the charge on the capacitor at any time.

Figure 13.23a shows the decay of charge with time. From the decay equations we can see that after a time interval RC, the time constant of the circuit, the charge remaining on the capacitor is Q_0/e. This is 0.37 ($= 1/e$) of its initial value. After a time equal to $2RC$, it is 0.14 ($= 1/e^2$) of its initial value, and so on.

We can choose any constant interval of time we like. In Figure 13.23b, we take the time for the charge stored to halve, known as the **half life**, $T_{1/2}$. After two half–lives, the charge remaining is 0.25 of its initial value, after three, 0.125, and so on. The half–life is the quantity usually referred to in radioactive decay experiments ($\rightarrow$ 39.3). It is more convenient to use the time constant in resistance–capacitance circuits because this is easily calculated by multiplying R and C.

CONSERVATION LAWS

Questions

11 Use the equation in question **10 a** to show that, in Figure 13.18a, the voltage grows to 0.63 of its final value V_S in a time RC and to 0.86 V_S after $2RC$.

12 Show that the time taken for the charge on a capacitor to decay to one half of its initial value $T_{1/2}$ is equal to $0.69\,RC$.

EXPERIMENT 13.5 Plotting the decay curve for a capacitor discharging through a resistance

The circuit of Figure 13.24 has a time constant of about 50 s, giving adequate time to take voltmeter and ammeter readings. With a fully charged capacitor, starting a timer and disconnecting the supply to the positive terminal of the capacitor enables you to take readings of the current I at 10 s intervals as the capacitor discharges. You can then plot a graph of the discharge current against time. Figure 13.25 shows a typical set of results.

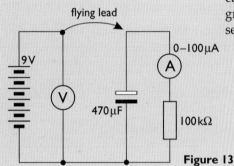

Figure 13.24 Circuit

t/s	0	10	20	30	40	50	60	70	80	90	100
I/μA	90	73	59	48	39	31	25	21	17	14	11
V/V	9.1										
Q/c											

Figure 13.25 Typical results

Question

13 a Copy and complete the table in Figure 13.25, calculating values of the voltage V across the capacitor and the charge Q remaining for each of the times shown.
b Plot a graph of charge remaining on the capacitor against time. What does a graph of ln (charge) against time look like?
c Does the current halve in 33 s ($= 0.69RC$)? Is it at 0.37 of its initial value after 48 s ($= RC$)?

SUMMARY

Capacitors store energy by the displacement of charge. Capacitance C is the ratio of charge displaced Q to voltage V across the capacitor. The energy stored E_P equals one half the final voltage multiplied by the charge displaced:

$$E_P = \tfrac{1}{2}QV = \tfrac{1}{2}CV^2 = \tfrac{1}{2}Q^2/C.$$

When capacitors are connected in series and in parallel, the equivalent capacitance C_E is given by:

in series $1/C_E = 1/C_1 + 1/C_2 + \cdots$,

in parallel $C_E = C_1 + C_2 + \cdots$.

The growth and decay of charge on a capacitor is exponential in nature.

13 STORING CHARGE AND ENERGY: CAPACITORS

The time constant of a circuit containing capacitance and resistance is equal to RC seconds when R is in ohms and C in farads. It is the time taken for the charge on a capacitor to decay to 0.37 of its initial value, or to discharge completely at its initial rate. A capacitor is effectively fully charged or discharged after a time of $5RC$.

The charge Q on a capacitor at time t as it discharges through a resistance R is given by:

$$Q = Q_0 e^{-t/RC}$$

where Q_0 is the charge on the capacitor at time $t = 0$.

The charge Q on a capacitor at time t as it charges through a resistance R is given by:

$$Q = Q_0(1 - e^{-t/RC})$$

where Q_0 is the final charge acquired by the capacitor.

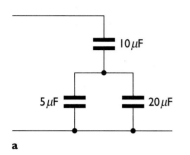
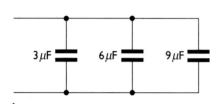

Figure 13.26

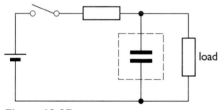

Figure 13.27

SUMMARY QUESTIONS

14 Calculate the equivalent capacitance of the combinations of capacitors in Figure 13.26.

15 A 1.0 µF capacitor, initially uncharged, is connected to a 12 V supply through a 4.7 kΩ resistor. What is the current in the resistor
a just after the circuit is connected?
b a long time after?

16 Two 3.3 µF capacitors are connected in series to a 6.0 V supply. Calculate
a the pd across each capacitor,
b the charge on each capacitor,
c the total energy stored.
d The capacitors are isolated and reconnected in parallel + to + and – to –. Discuss any changes that take place in the pd's, charges and energies stored in the capacitors.

17 Discuss the effect of placing a capacitor in the circuit in Figure 13.27. Sketch how the pd across the load resistor changes with time after the circuit is switched on.

18 A 5.00 V supply, derived from a 50 Hz ac supply, is required to have a ripple pd of <0.01 V. The load resistance is 10 Ω. Estimate a value for a capacitance to achieve this low ripple. Would you expect such a value to be available?

19 In an experiment, a 1 000 µF capacitor is charged to 10 volts, as recorded by a digital voltmeter connected across its terminals. The digital voltmeter has a resistance of the order of 10 MΩ so that an insignificant amount of charge leaves the capacitor during the course of the experiment.
a What is the charge stored on the capacitor?
b How much energy is stored by the capacitor?
An uncharged 2 000 µF capacitor is then connected in parallel with the first, leaving the voltmeter connected so that it records the pd across the combination.
c Explain why the voltmeter reads 3.3 volts.

CONSERVATION LAWS

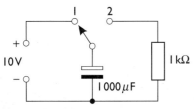

Figure 13.28

d What is the charge on each capacitor?
e What is the energy stored by each capacitor?
f What is the total energy stored by the two capacitors? Why is this value less than that calculated in **b** ?

20 Figure 13.28 shows a circuit containing a 1 000 μF capacitor, a 1 kΩ resistor and a 10 V d.c. supply. The capacitor is charged by moving the switch to position 1. It is then discharged by switching to position 2 at $t = 0$.
 a Sketch three graphs showing the variation with time after $t = 0$ of:
  i charge on the capacitor,
  ii voltage across the capacitor,
  iii current through the resistor.
Label the axes carefully, including suitable scales.
 b Another 1 000 μF capacitor is added in parallel with the first. Repeat part **a** of the question using the same axes, showing clearly any differences you would expect to see.
 c The second capacitor is then removed and another resistor of 1 kΩ is added in parallel to the original resistance. Repeat part **a**, again showing clearly any differences.

UNIT D
FORCES AND FIELDS

FORCES AND FIELDS

Vast forces shape our Universe, from galaxies to nuclei, acting both within and between systems. Models have been developed to explain the forces that control all natural phenomena. Each is an approximation and relates to a limited range of observations.

In recent years, physicists have been moving towards a single unified model that will predict accurately our observations of effects involving gravitation, electromagnetism and the nuclei of atoms. Fields, particles and their interactions form the basis of this model. Fields are regions of space, or matter, in which particular forces act in particular directions. They cause the interactions which are the basis for Newton's laws (← 6.4) and all the forces in our Universe.

14 FORCES OF NATURE

During the last three centuries, physicists have extended our knowledge of the forces that control all natural activity. We now have understanding of:

- gravitational forces, which control the motion of the planets and stars,
- electromagnetic forces, which determine the nature and behaviour of materials, and
- nuclear forces, which are responsible for sunlight and nuclear power.

Exploring the nature of these forces, and investigating the suggestion that there may be a single model that could bring together these ideas, are major quests of physicists.

Figure 14.1

a Jupiter and its moons form a system held together by gravity

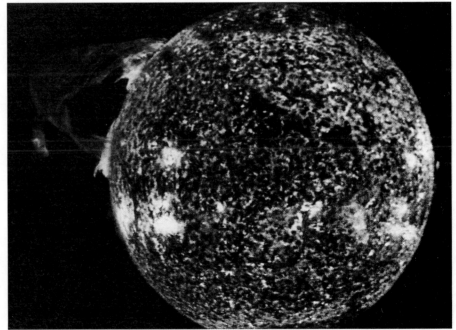

b The Sun's nuclear power sustains life on Earth

14.1 ACTION AT A DISTANCE

If you hold a ball about a metre above the floor and let it go, it will fall to the floor, travelling faster all the time it is falling. Also, it seems to get faster at a steady rate. This suggests that a constant force acts upon it (← 6.4). A spring

D FORCES AND FIELDS

balance attached to the ball will record the same value – its weight – no matter where the reading is taken in the room. The weight of the ball is, of course, the net force of gravity acting upon it.

Satellites in circular orbits are actually falling towards the Earth in the vacuum of space, but do not get any closer to the surface. They are able to maintain their orbits (← Chapter 7 and Figure 14.2). This is due to the force of gravity, which **acts at a distance** from the Earth through all media, including a vacuum.

Similarly, when a plastic comb is rubbed on a sleeve or a jumper, small pieces of paper placed near the comb jump up as they are attracted by the comb, as in Figure 14.3. A sheet of paper is quite noticeably attracted when placed near to the screen of a television, again showing action at a distance. Figure 14.4 shows other examples of forces acting at a distance.

The falling mass and the attraction and repulsion between charges and magnets are clear examples of forces acting at a distance. Even at an atomic level, the forces that bind particles together to make the strong solids with which we are all familiar act at a distance. This is a very much smaller distance, of course.

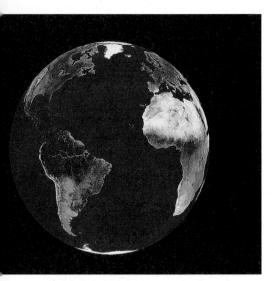

Figure 14.2 This photograph of Earth was taken from a satellite, which can orbit the Earth because of gravity

Figure 14.4

Figure 14.3 Electrical forces acting at a distance

a Magnetic levitation used in a hover-train in Australia. Friction between train and track is eliminated as the train floats a few milimetres above the track.

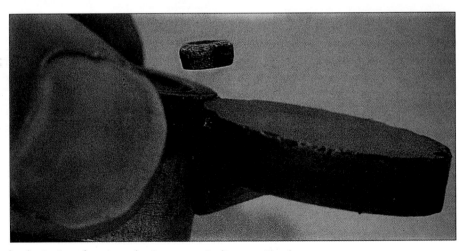

b Levitation of superconducting a ceramic disc carrying a large current

14 FORCES OF NATURE

Figure 14.5

a Are the cars 'touching'?

> Question
>
> 1 The following are some examples of collisions and impacts:
> - two cars collide head-on
> - a squash ball is struck by the head of a racquet
> - a boxer strikes the chin of his opponent.
>
> In each of these examples, we would normally consider that two objects have come 'into contact' with each other.
>
> All materials consist of atoms which, in turn, are made up of positively charged nuclei surrounded by negatively charged electrons. Contact between objects arises when electrons in one object come close to electrons in another, giving rise to repulsion. This is action at a distance, so do two objects ever actually touch each other? Consider all three of the above interactions. Can the boxer above claim that he never touched his opponent? Do any objects **touch** each other when they come into contact?

b Does the ball 'touch' the racket?

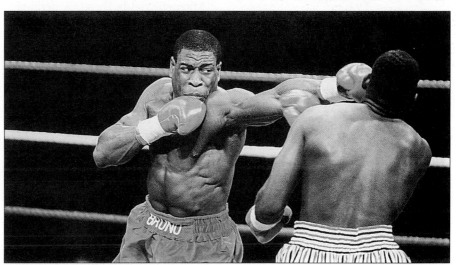

c Does the glove 'touch' the chin?

14.2 THE CONCEPT OF A FIELD

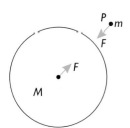

Figure 14.6 Gravitational forces

In order to explain such observations, physicists have developed the concept of a **field**, such as those due to gravity, electricity and magnetism.

Gravitational fields Because of its mass, the Earth interacts with any mass near to, or out in space. The mass experiences a force towards the centre of the Earth. Figure 14.6 shows the Earth and a mass just above it. The field created by the Earth's mass causes a force F to be experienced by any mass m in the field. The interaction causes the same force F to be experienced by the Earth, but towards the mass m. Putting these two ideas together, we can say that around mass m there is a tiny field which interacts with the large field due to the much bigger mass of the Earth.

There are, of course, many other masses in the Universe. The **total** gravitational field surrounding both the Earth and m is due to both of them **and** all the other masses in the Universe. This complete field is the gravitational field we need to investigate. The **field's strength** at any point in space, is g N kg^{-1}: the

FORCES AND FIELDS

force experienced there by a mass of 1 kg. Its direction is the same as that of the force. So at point P in Figure 14.6:

$$g = \frac{F}{m}.$$

Electric fields Similarly, there is an electric field around any charged object, such as a piece of plastic that has been rubbed with another material. When another charge is placed in the field it experiences a force. The force on 1 coulomb at a point in space is the strength of the electric field E at that point.

Suppose a positive charge of Q coulombs is introduced. The direction of the field E is the same as the direction in which the force F acts on it and is measured in NC^{-1}:

$$E = \frac{F}{Q}.$$

When a negative charge is introduced, both the field and the force will act in the opposite direction. Note that the charge Q must be very small, much less than 1 C, not to upset the existing field significantly.

Magnetic fields These are defined in terms of the force on a moving charge ($\rightarrow$ 17).

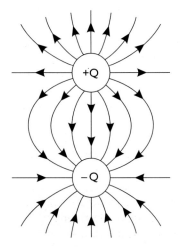

Figure 14.7 Field around 2 oppositely charged objects

> ### Questions
>
> **2** Why do you think that physicists developed the concept of the field? Is it necessary or is there an alternative way of dealing with forces between masses or charges?
>
> **3** The gravitational field strength g at the surface of the Earth is 9.81 N kg^{-1}. Why will a mass placed freely in that field accelerate at 9.81 m s^{-2}?

VARIATION OF FIELD WITH DISTANCE

The further we travel from the Earth, the smaller in value is the Earth's gravitational field, and so the smaller the force on any mass in the field. A similar lessening occurs when a charged conductor is gradually moved further away from the dome of a Van de Graaff generator or a compass needle moved further from a magnet.

Fields diminish with distance from their source, or origin. The gravitational field around the Earth or the electric field around a point or uniform sphere of charge varies according to an **inverse square law**: doubling the distance from the source reduces the field to one quarter, trebling the distance reduces it to one-ninth, and so on.

FLUX

To explain some of the properties of fields, it is convenient to use the idea of **flux**, which is the Latin word for 'flow'. When water flows from a 'source' to a 'sink', water is transferred at a certain rate or flux, measured in kg s^{-1}. The associated **flux density** is the mass of water per second crossing unit area perpendicular to the flow.

You can think about energy flux in a similar way. It is measured in joules per second, or watts. Energy flux density, usually called **intensity**, is measured in W m^{-2}. Field strength and energy flux density are related. The strength of a field

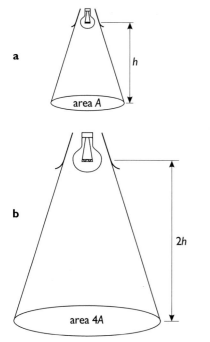

Figure 14.8 Inverse square law: lamp **a** and lamp **b**

falls off proportionally to the square of the distance from its source, by the inverse square law, and so does the intensity of electromagnetic waves.

For example, lamp **b** in Figure 14.8 is twice as far from the desk as lamp **a**. The area illuminated by lamp **b** is four times that illuminated by lamp **a**. Suppose you fit identical bulbs to the two desk lamps and assume that light is not being significantly absorbed between the lamp and the desk. Then the level of illumination, the flux density, on the desk from **a** will be four times greater than that from **b**. Similarly, raising **b** to three times the height will cut the level of illumination on the desk to one-ninth.

Other electromagnetic waves that are not absorbed very much when passing through air behave very much like light. The intensity of radiation from a gamma-ray or X-ray source, for example, varies according to an inverse square law and diminishes rapidly with distance from the source.

In these examples, we associate flux with energy or the flow of an incompressible fluid. The idea of flux can also be usefully applied to fields in which there is no evidence that anything is actually transferred, such as static electric fields, gravitational fields and magnetic fields. The mathematics that model flux are the same whatever the field.

In SI units, electric, magnetic and gravitational flux densities are defined to be identical in value to the corresponding field strengths. For example, electric flux from an electric charge spreads out radially. The flux density or electric field strength E diminishes as the inverse square of the distance from the source ($\rightarrow$ 16.2).

14.3 ENERGY AND FIELDS

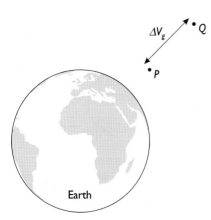

Figure 14.9 Energy changes in a gravitational field

Suppose you lift a box from the floor on to a table. You need to do work to raise the box because of the downward gravitational force on it. The box, once raised, has increased potential energy. Letting the box fall off the edge of the table causes this extra potential energy to be transferred to kinetic energy which reaches its maximum just as the box hits the floor. The maximum kinetic energy is exactly equal to the increase in potential energy given to the box in raising it.

This demonstrates that gravitational fields are **conservative** – the change in potential energy between two points has a unique value whatever the path taken between those two points. A **non-conservative** field, such as one in which friction acts, causes energy to be transferred away from an object in the field. Energy is transferred to the internal energy of the object and its surroundings ($\leftarrow$ 9.5).

Whenever a mass is moved through a gravitational field other than at right angles to it, energy is transferred causing a change in potential energy. The change in gravitational potential energy per kilogram of mass as it is moved from one point P to another point Q is called the **gravitational potential difference** ΔV_g between points P and Q (see Figure 14.9). It is measured in J kg^{-1}.

Question

4 The surface area of a sphere of radius r is $4\pi r^2$.
a Calculate the surface area of a sphere of radius **i** 1 m, **ii** 2 m, **iii** 3 m. Use these results to justify the inverse square law relationship for gravitational fields.
b Energy is only conserved in a conservative field. Why do you need to assume that a gravitational field is conservative?

D FORCES AND FIELDS

Again, bringing two similarly charged polyethene strips closer together requires that work is done in increasing the potential energy of the system. Separating them further causes a fall in the potential energy. Oppositely charged strips produce an attractive force and so potential energy increases with increasing separation, as for gravity.

The potential energy change of a charge of 1 coulomb as it moves through a field is called the **electrical potential difference** ΔV_E and is measured in $J\,C^{-1}$ or **volts**, since

$$1\,V = 1\,J\,C^{-1}.$$

The concepts of gravitational potential difference V_g and electrical potential difference V_E are analogous:

V_g = potential energy difference per unit mass

V_E = potential energy difference per unit charge.

> ### Question
>
> **5** At the Earth's surface, the gravitational field strength is about $10\,N\,kg^{-1}$. Calculate:
> **a** the potential energy change of a high diver of mass 60 kg when diving off a 5 m board,
> **b** the potential energy change per kilogram of mass that this represents,
> **c** the gravitational potential difference between the diving board and the water surface.

14.4 THE FOUR FUNDAMENTAL FORCES

There are many different forces on Earth:

- the attractive forces that maintain the Earth's motion around the Sun,
- the devastatingly large forces generated during an earthquake,
- the upward forces exerted on the wings of an aircraft as it moves through the air,
- the forces within the cables supporting the road deck of a suspension bridge,
- the frictional forces exerted by the brakes of a motor car,
- the forces controlling the movement of an electron beam in a television,
- the attractive forces binding electrons in atoms and protons within the nucleus.

Despite the enormous variety of forces, physicists have identified just four fundamental forces of nature from which all known forces originate.

GRAVITATIONAL FORCES

Figure 14.10 shows our planet viewed from the Moon. The Moon orbits Earth due to gravity, a force that was given much attention by Galileo and Newton in the sixteenth and seventeenth centuries. It is an attractive force that exists between all particles of matter. Only when there are exceedingly large numbers of particles involved are significant gravitational forces observed. Two people standing next to each other, on the Earth's surface, show no noticeable gravitational attraction towards one another. However, both are very much aware of their own weight – the force that the Earth's gravity exerts on their

14 FORCES OF NATURE

Figure 14.10 Earth rising over the lunar horizon

mass. The effect of gravity becomes more significant on an astronomical scale as the number of particles involved increases.

ELECTROMAGNETIC FORCES

Two electric charges will attract or repel each other depending upon whether they are similarly or oppositely charged. Charges in motion – electric currents – produce magnetic fields which can also interact to produce forces. In the mid-nineteenth century, James Clerk Maxwell, brought together ideas concerning electric and magnetic forces to form a **unified electromagnetic field** theory or model.

Electromagnetic forces, originating from charged particles at rest or in motion, are very much larger than the gravitational forces that arise between them.

Question

6 Gravitational forces are much more significant on the human scale on the Earth's surface than at the atomic, or sub-atomic, level. Even so, they are less significant in the laboratory than electrical forces. However, on an astronomical scale, gravitational forces dominate electrical forces. What does this suggest about the net charges possessed by the Earth, the Moon and other heavenly bodies?

STRONG NUCLEAR FORCE

The nucleus of an atom contains positively charged protons confined within a space of about 10^{-14} m across ($\rightarrow$ 40.2). Assuming no other forces are involved, these will electrically repel each other according to Coulomb's law ($\rightarrow$ 16.2):

$$F = \frac{Q_1 \cdot Q_2}{4\pi\epsilon r^2}.$$

where Q_1 and Q_2 are the charges on protons a distance r apart, and ϵ is the permittivity of the medium between them.

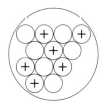

Figure 14.11 Nucleus of carbon-12

FORCES AND FIELDS

This force turns out to be over 2 N, which is very large indeed – approximately the force of the Earth's gravity on a couple of apples! So a very large force, non-electrical in origin, must be acting within the nucleus to bind its protons together and to maintain its stability. This **attractive** force is the **strong nuclear force** (→ 40.5).

Recent research leads physicists to believe that neutrons and protons are not fundamental particles but seem to be made up of simpler building blocks, called **quarks**. There are six different types of quark, each carrying a charge equivalent to either one-third or two-thirds that of the electron. The smallest unit of charge that has been measured in isolation is that possessed by a single electron, so it appears that quarks cannot exist in isolation.

Combinations of quarks lead to the charge of neutrons, protons and the other known particles. However, these electrical charges do not, of themselves, produce the strong nuclear force. It is the interaction between quarks and **messenger particles** called **gluons** that is responsible for this.

WEAK NUCLEAR FORCE

The fourth force in nature, known as the weak force, is about 1/1000th as strong as the strong force, but is repulsive. Usually it is masked by the effects of the strong force inside nuclei. However, the decay of nuclear particles, such as neutrons, indicates that another force is in operation. When a neutron decays into a proton and an electron, a third particle, a neutrino, is emitted. Electrons interact electromagnetically with each other and with protons, but they also interact with each other and with neutrinos via the weak force. Neutrons have no charge and can only interact with each other and electrons via the weak force.

The so-called weak force also has its origins deep in the nucleus. It has a range of around 10^{-17} m, and plays an important part in the radioactive decay of unstable nuclei by β^- and β^+ emission of electrons and neutrinos (→ 40). Only in recent years have physicists gained an understanding of the weak force and established models that unified this force with the electromagnetic force. The **weak force** is not really weak but acts as a **short range repulsive force**.

UNIFYING THE FUNDAMENTAL FORCES

It has long been a dream of physicists to understand the origin of the forces of nature and to establish a single model that explains and describes their behaviour. Maxwell moved towards this goal about a century ago when he brought together the forces of electricity and magnetism into a single model of electromagnetism. Gravity as proposed by Newton has been further explained and extended by Einstein's General Theory of Relativity in the early part of the twentieth century. Since then, the strong and weak forces existing between subatomic particles has been described.

A **unified model** would draw together our ideas of nuclear physics and radioactivity, of electricity and magnetism, and of the behaviour of the planets, stars and galaxies in the cosmos.

It is now generally accepted that around 15 thousand million years ago at the time of the 'Big-Bang' the Universe was so small that quantum effects dominated its behaviour. The Universe has evolved and expanded since the 'Big-Bang'. Just after the 'Big Bang', it is now thought that all the forces of nature were unified. There was only one force. As particles separated from each other in the first few fractions of a second, the four forces of nature became established. The question physicists want the answer to is 'why?'

Estimating the force between protons

The charge on each proton is the same as that of an electron although positive. Let us assume that the permittivity is that of free space, so that ϵ is equal to ϵ_0. Suppose, also that we take r to be 1×10^{-14} m, its maximum value. Using approximate values, the lowest value of the force F is given by:

$$F = \frac{Q_1 . Q_2}{4\pi r^2 \epsilon}.$$

Using $Q_1 = Q_2 = 1.6 \times 10^{-19}$, $r = 10^{-14}$ m, $\epsilon_0 = 8.9 \times 10^{-12}$,

$$F = \frac{2.56 \times 10^{-38}}{4 \times 3.14 \times 10^{-28} \times 8.9 \times 10^{-12}}$$

$$= 2 \text{ N}.$$

Therefore, the force between two protons in the nucleus is at least 2 N.

14 FORCES OF NATURE

SUMMARY

The concept of a field is useful when describing the nature of these forces. For example, a gravitational field exists around a mass. The magnitude and direction of the field is the same as the magnitude and direction of the force acting on unit mass within it.

Similarly, an electric field exists around a charge. The magnitude and direction of the field is the same as the magnitude and direction of the force acting on unit charge within it.

Flux density has the same value as field strength when defined in SI units.

There are just four fundamental forces that exist in nature: the gravitational, electromagnetic, and the strong and weak nuclear forces.

SUMMARY QUESTIONS

7 Identify four different situations that you meet in everyday life in which forces are involved. For each situation, classify the forces in terms of one or more of the four fundamental forces of nature.

8 Explain in terms of the fundamental forces acting why you need to do work to stretch a rubber band or compress a steel spring.

9 Progression towards a unified model of forces entails research requiring vast sums of money being spent on accelerators, storage rings and other high energy physics systems. These are used to discover and investigate the properties of the most elusive of fundamental particles. What are the pros and cons of such research?

15 GRAVITATIONAL FIELDS

During the last two thousand years or so, our knowledge about the Earth and its place in the Universe has increased dramatically. This has enabled our understanding of force, motion and energy to develop beyond all previous bounds.

Greek models of the Universe, which put the Earth at its centre, dominated thinking until around the fifteenth century. Then Copernicus proposed a new model, which had the Earth and other planets in circular orbits around the Sun. Later, Tycho Brahe collected very precise data on planetary positions and, in 1609, Kepler used this data to define much more clearly the nature of planetary orbits. Kepler and his predecessors were essentially concerned with producing physical models of the Universe. Such models were developed for the purpose of calculating planetary positions very accurately (see Figure 15.1).

In the sixteenth and seventeenth centuries, Galileo, and subsequently Isaac Newton, formalised laws relating to force and motion, and established the concept of gravity. We still make wide use of Newton's ideas concerning gravitation and apply his laws of motion (← 6.4) to satellite and planetary motion. However, there are situations where Newton's Laws cannot explain all our observations. Einstein, at the beginning of this century, developed theories of 'Relativity' to deal with such situations.

Figure 15.1 Our changing view of the Universe

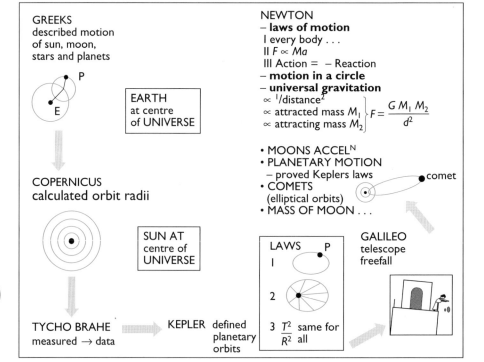

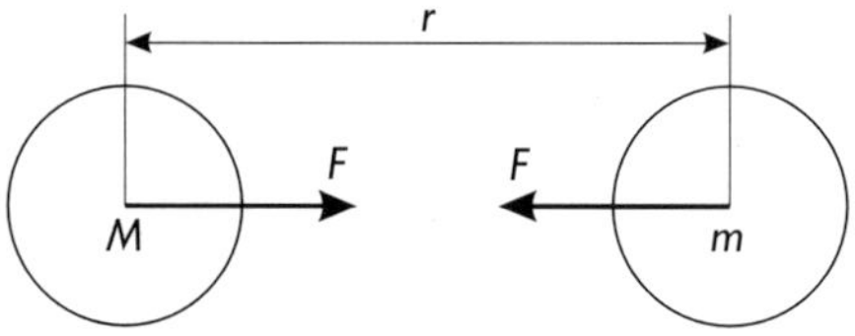

Figure 15.2 Newton's law of gravitation: $F \propto Mm$ and $F \propto 1/r^2$

15 GRAVITATIONAL FIELDS

15.1 GRAVITATIONAL FORCES

In the seventeenth century, after extensive consideration of the available evidence, Isaac Newton put forward the following:

All objects attract one another with a force that depends upon their masses and their separation.

Consider two masses M and m, with their centres of mass separated by a distance r, as shown in Figure 15.2. Newton suggested that the force F between them is proportional to each of the masses:

$$F \propto M$$
and $F \propto m,$

Newton and the inverse square law

Newton was aware that gravity did not remain at a constant strength as it extended through space. He arrived at the conclusion that gravity must 'thin out' with distance from the Earth. Even this 'diluted' gravitational field was able to control the motion of the Moon in its orbit around the Earth.

The apple that fell on Newton's head in the orchard accelerated towards the Earth at $9.81\,\mathrm{m\,s^{-2}}$. Assuming that an inverse square law applies to gravity, then the Moon, which is about 60 Earth radii away, would fall freely with an acceleration of $(9.81/60^2)\,\mathrm{m\,s^{-2}}$ or $0.00272\,\mathrm{m\,s^{-2}}$ towards the Earth (see Figure 15.3). Because the Moon is moving in an essentially circular path, this acceleration is the centripetal acceleration (← 7.1) of the Moon in its orbit. It takes the value v^2/r where v is the speed of the Moon along its orbit and r the orbital radius.

The average orbital speed is the total distance travelled in one orbit divided by the orbit period:

$$\text{orbital speed} = \frac{\text{circumference of orbit}}{\text{time for one orbit}}$$

$$= \frac{2\pi R}{T}.$$

The Moon takes 27.3 days to complete an orbit of radius $3.82 \times 10^8\,\mathrm{m}$, so:

$$\text{orbital speed} = \frac{2\pi \times 3.82 \times 10^8}{27.3 \times 24 \times 3600}$$

$$= 1\,020\,\mathrm{m\,s^{-1}}.$$

This gives a value for v^2/r of $0.00272\,\mathrm{m\,s^{-2}}$. The predicted value of g using the inverse square law is clearly consistent with the actual value we observe for the Moon.

Newton performed this test on the Moon's motion and confirmed beyond reasonable doubt his idea that gravitational forces between masses vary according to an inverse square law.

but what does this mean? Suppose we consider the effect of gravity on a mass on the surface of the Earth. You know that a 1 kg mass is pulled towards the Earth with a force of 9.81 N. It will accelerate towards the Earth at $9.81\,\mathrm{m\,s^{-2}}$. A 2 kg mass must have twice the gravitational force acting on it to cause it to accelerate at the same rate. A 10 kg mass has 10 times the force. Similarly, if you could double the mass of the Earth but keep its size the same, you would also double the force of attraction on any mass at its surface.

Newton used different arguments to deduce the relationship between force and the separation of masses. He arrived at the **inverse square law** of gravity (← 14.2) by working back from Kepler's laws of planetary motion.

$$F \propto \frac{1}{r^2}.$$

We can write the full relationship as:

$$F \propto \frac{Mm}{r^2}.$$

This means that:

$$\frac{F}{Mm/r^2} = \text{a constant}.$$

The constant, usually denoted by G, is called the **universal constant of gravitation**, and

$$F = G.\frac{Mm}{r^2}.$$

You may sometimes see this equation with a minus sign. It is there to indicate the attractive nature of the gravitational force F towards a mass, whereas distance r increases away from the mass.

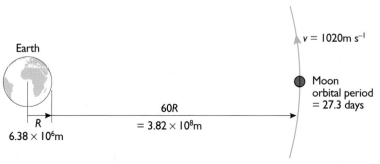

Figure 15.3 Motion of the Moon

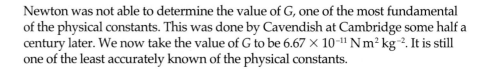

Newton was not able to determine the value of G, one of the most fundamental of the physical constants. This was done by Cavendish at Cambridge some half a century later. We now take the value of G to be $6.67 \times 10^{-11}\,\text{N}\,\text{m}^2\,\text{kg}^{-2}$. It is still one of the least accurately known of the physical constants.

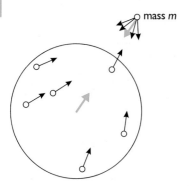

Figure 15.4 Imagine the Earth to be made up of a very large number of smaller masses, all of which are attracting and are attracted by the mass on the surface

Question

1 When applying Newton's law of gravitation, you often assume that:
- the mass of an object is concentrated at its centre of mass, and
- the distance between two masses is the distance between their centres of mass.

Suppose you want to calculate the gravitational force between the Earth and a mass just above its surface. It seems odd to consider the Earth's mass to be concentrated at a point. Nevertheless, you can still use the equation. Using Figure 15.4 in a qualitative way, why are you able to do this? (Imagine the Earth to be made up of a very large number of smaller masses, all of which are attracting and are attracted by the mass on the surface).

Cavendish experiment

Various attempts were made after Newton's time to determine the gravitational constant G. Since, the force F between two masses M and m of separation r is given by:

$$F = GMm/r^2,$$

then $G = F r^2/Mm.$

So, to find G, you must measure the force between two known masses a known distance apart.

One method compares the Earth's vertical pull on a pendulum bob with the sideways pull of a natural mass, such as a mountain. Early experiments of this type were carried out by Pierre Bouguer in the mid-eighteenth century but they did not yield very good results.

Towards the end of the eighteenth century, Henry Cavendish, devised a method which essentially depended upon measurement of the force between two lead spheres, one 30 cm and the other 5 cm in diameter. The force measurement was made with a torsion balance, using the arrangement shown in Figure 15.5. Two small spheres were attached to the ends of a light rod supported by a fine torsion wire. Two larger spheres were placed close to the small spheres so that the gravitational attraction between the spheres caused the torsion wire to twist one way very slightly. Placing the large spheres on the other side of the small ones made the torsion wire twist in the opposite direction. Measurement of the angle of twist between the two extreme equilibrium positions gave a measure of the force between the two unequal masses. This was achieved using a light beam pointer reflecting from a mirror attached to the torsion wire suspension. The natural period of oscillation of the torsion balance was measured without the masses in position to find the torsion constant of the wire suspension.

From his experiment, Cavendish obtained a value for G within 1.2% of our currently accepted value, $6.6720 \times 10^{-11}\,\text{N}\,\text{m}^2\,\text{kg}^{-2}$. This was a remarkable achievement for such an early and difficult measurement, and was not improved upon for another hundred years. More modern methods, such as that used by Heyl and Chrzanowski in 1942, are based on the same principle as that used by Cavendish but employ more sophisticated measuring techniques.

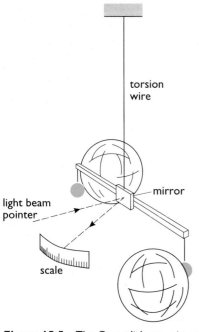

Figure 15.5 The Cavendish experiment

15 GRAVITATIONAL FIELDS

Question

2 a Use Newton's law of gravitation to find the gravitational force between two 1 kg masses, 1 m apart. If their separation is reduced to 10 cm, one-tenth of its initial value, what value will the force have? (G)
b Calculate the force between two lead spheres like those Cavendish used to measure G, given that the density of lead is 11 000 kg m^{-3} and the spheres were on the point of touching each other. (G)
c Suppose the diameters of the spheres used in the Cavendish experiment were doubled.
  i How would this affect the masses of the spheres?
  ii How would it affect the minimum separation of the spheres?
  iii How would the force between them be affected?
d It might seem a good idea to go on increasing the sphere diameters. What factors would put a limit on their maximum size?

15.2 GRAVITATIONAL FIELD g

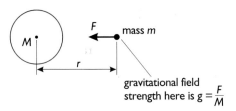

Figure 15.6 Gravitational field strength

Figure 15.6 shows a mass m placed a distance r from another mass M. The force F between them is given by Newton's law of gravitation as GMm/r^2.

Gravitational field strength g is a measure of the **force per unit mass** (F/m) placed in the field (← 14.2). A field of g N kg^{-1} will cause a mass to accelerate at g m s^{-2}.

Rearranging the equations gives:

$F/m = g = GM/r^2$.

Question

3 A mass of 1 kg at the Earth's surface, which is 6.38×10^6 m from the centre, experiences a force of gravity of 9.81 N.
a Calculate the mass M of the Earth, using the relationship $F/m = g = GM/r^2$.
b Show that the value of GM for the Earth is about 4×10^{14} N m^2 kg^{-1}.

VARIATION OF g WITH DISTANCE FROM THE EARTH'S CENTRE

At the Earth's surface, 6.38×10^6 m from its centre, the gravitational field strength is typically 9.81 N kg^{-1}. Newton's law of gravitation predicts that g varies according to the inverse square of distance. So at twice this distance from the centre, g will be one-quarter of its value at the surface, 2.45 N kg^{-1}. At three Earth radii (E_R) from the centre, g will have dropped to 1.09 N kg^{-1} or one-ninth of its value at the surface.

Figure 15.7 illustrates the variation of g with distance from the centre of the Earth. Notice that g changes more rapidly with distance closer to the Earth.

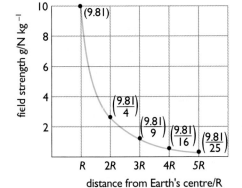

Figure 15.7 Variation of g with distance from the centre of the Earth

Questions

4 On a manned space journey to the Moon a spacecraft left Earth's orbit and followed a path directly away from the Earth's centre with its rocket motors shut down. It was coasting freely and slowing down under the influence of gravity alone, much as a ball would do when thrown directly upwards but, of course, with no air resistance.

Figure 15.8 shows the velocity of the craft at four pairs of distances from the Earth's centre. Each pair of points is separated fairly closely in distance and time.

Time as on launch in h:m:s	Distance from Earth's centre $r/10^3$ km	Velocity v/ms^{-1}	Mean distance $\bar{r}/10^6$ m	Mean acceleration g/ms^{-2}
03:58:00	26.3	5374		
04:08:00	29.0	5102		
05:58:00	54.4	3633		
06:08:00	56.4	3560		
09:58:00	95.7	2619		
10:08:00	97.2	2594		
19:58:00	169.9	1796		
20:08:00	170.9	1788		

Figure 15.8 Data on an Apollo spacecraft

a Complete a fourth column containing the estimated mean distance r of each pair of points from the Earth's centre.
b Calculate the mean acceleration of the craft between each pair of points.
c Complete a fifth column containing the mean value of gravitational field strength g for each pair of points.
d Calculate the value of $1/r^2$ using your answers to **a**. Plot a graph of g against $1/r^2$. What conclusions can you draw from your graph?
e Do the corresponding values of g and r in parts **a** and **c** lie on the curve plotted in Figure 15.7?

5 a What do you think the value of g is at the Earth's centre?
b How do you think g might vary between the surface of the Earth and its centre?

Discovery of the outermost planets

For a model to be viable, it must account for all the observed phenomena to which it relates. However, the real test of a model is its ability to predict, and Newton's law of gravitation enabled people to predict the existence of the outermost planets of our solar system.

In 1781, the planet Uranus was discovered by William Herschel, who made a careful plot of its orbit in relation to the pattern of fixed stars. Before this date it was recorded as a star on the star maps of the day. However, the observed motion of Uranus was not in accordance with the elliptical orbit predicted by Newton's law of gravitation: sometimes it seemed to be ahead of and at other times behind its expected position. Figure 15.9 shows the extent of the 'error' in its position.

Adams in England and Leverrier in France independently suggested that these small perturbations, or variations, were due to an unknown planet further out. Before 1822, it seemed, this unknown planet pulled Uranus faster along its orbit so that it was ahead of its predicted position. After that date it had a slowing

15 GRAVITATIONAL FIELDS

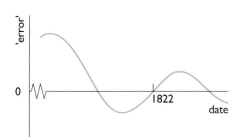

Figure 15.9 The 'error' in the motion of Uranus

effect on Uranus. This is shown in Figure 15.10. Adams and Leverrier, again independently, used Newton's model to predict the mass and the orbit of the unknown planet. The two astronomers identified the location of the planet in relation to the stars and, in 1843, the planet Neptune was seen. Then, they only saw a featureless small disk through a telescope. Today, we can send TV cameras on spacecraft and use electronic image enhancement to produce pictures, such as that in Figure 15.11.

Subsequent observation of the perturbations in the orbit of Neptune led to the discovery, in 1930, of the planet Pluto. This smaller planet is in a rather elliptical orbit. In the late twentieth and early twenty-first centuries it is closer to the Sun than Neptune. However, when discovered it was further from the Sun than Neptune, as it is during most of its orbital period.

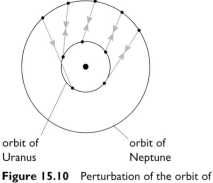

Figure 15.10 Perturbation of the orbit of Uranus

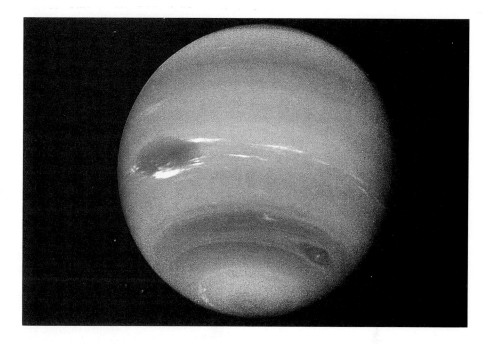

Figure 15.11 Neptune taken from Voyager 2

15.3 ORBITAL MOTION

The laws of motion can be applied to a body moving in a circular path (← 7.1). A centripetal force of mv^2/r is required for a body of mass m to continue to move at a speed v along a circular path of radius r. The centripetal acceleration of such a body is v^2/r.

We can use the concepts of centripetal force and acceleration in the context of gravitation to develop a better understanding of the motion of planets and satellites. Where, for example, would we place a satellite to ensure that it orbited the Earth and at all times appeared to be above the same point on the Earth's surface? Figure 15.12 shows an object of mass m in orbit around a central mass M, such as an artificial satellite orbiting the Earth. The centripetal force required to maintain the orbit is equal to mv^2/r, where v is the speed of the satellite along its orbit and r the radius of the path.

The centripetal force is clearly provided by the gravitational force GMm/r^2 between the Earth and the satellite, separated by a distance r.

Centripetal force = gravitational force
$$\frac{mv^2}{r} = G.\frac{Mm}{r^2}.$$

FORCES AND FIELDS

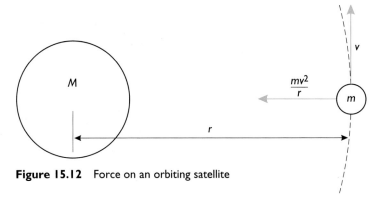

Figure 15.12 Force on an orbiting satellite

This equation may be simplified by dividing both sides of the equation by m:

$$\frac{v^2}{r} = \frac{G.M}{r^2}.$$

So, m does not play a part in determining the orbital speed for a given orbital radius. Multiplying both sides by r gives:

$$v^2 = \frac{G.M}{r}.$$

This equation shows that there is a unique orbital speed for a given radius of orbit around a central mass M.

Geostationary orbits

Communication satellites move in orbit at the same rotational speed as the Earth. They stay in position above the Atlantic, Pacific and Indian Oceans and provide worldwide communications (see Figure 15.13). Using one of these satellites, a person in one part of the world can have a telephone conversation or conduct a television interview with a person in any other part (→ 32.2).

To place such satellites in orbit, you need to work out how far away they should be and how fast they should be travelling. You must take into account that:

- the satellite needs to be orbiting in the same direction as the Earth is spinning with the same period of rotation as that of the Earth,
- the only possible orbit that will allow a satellite to remain above the same point on the Earth's surface is one directly above the Earth's equator,
- the centre of any stable orbit around the Earth is at the centre of the Earth since that is where the gravitational force on the satellite is directed.

The satellite will have to make one complete orbit in 24 hours and in that time will cover a distance of $2\pi r$ metres, where r is the orbital radius in metres. The speed v of the satellite will therefore be $2\pi r/(24 \times 3600)$ m s^{-1}, and the exact orbit radius and speed is found from the equation $v^2 = GM/r$:

$$(2\pi r/24 \times 3600)^2 = GM/r$$

giving $\quad 4\pi^2 r^3 = GM(24 \times 3600)^2.$

Because GM for the Earth is 4×10^{14} N m^2 kg^{-1}, we can write:

$$r^3 = 75.6 \times 10^{21} \text{ m}^3$$

or $\quad r = 4.23 \times 10^7$ m

$\quad\quad = 42\,300$ km.

So the satellite will be in a geostationary orbit at a radius of just over 42 000 km. Since the Earth's radius is approximately 6 000 km, the satellite will be about 36 000 km or 6 Earth radii above the surface of the Earth.

15 GRAVITATIONAL FIELDS

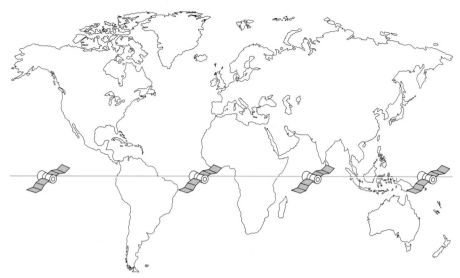

Figure 15.13 Locations of communication satellites

Questions

6 a When a person is being interviewed in, say, Australia by an interviewer in England, a slight delay occurs before there is a response to a question. Assuming a geostationary satellite is relaying the broadcast, calculate the minimum value of this delay.
b Estimate the speed of a geostationary satellite along its orbit.
c Gas jets on the satellite are used to finely adjust the orbit by giving an impulse to the satellite. What are the implications of giving too large an impulse **i** in the direction of its motion and **ii** against the direction of its motion along the orbit?

7 The NOAA-7 and NOAA-8 weather satellites are in circular polar orbits (→ 32.2), passing over the north and south poles with the Earth spinning underneath them. They orbit about 900 km above the Earth.
a Estimate the radius of their orbit.
b Calculate their speed along their orbit.
c How long does it take them to complete an orbit?

ESTIMATING THE MASS OF THE SUN

To find the mass of the Earth, you need to know the strength of the Earth's field at a given distance from its centre. Since you know the Earth's radius and the value of g at its surface, it is simple to calculate the Earth's mass (← 15.1).

You can also estimate the mass of the Sun by finding the value of the gravitational field strength at a known distance from its centre. To do this, you can use the motion of the Earth in its orbit around the Sun. The Earth is 'falling' towards the Sun with a centripetal acceleration of v^2/r, where v is the orbital speed and r the radius of orbit.

As shown in Figure 15.14, the Earth travels along a path of radius approximately 150 million kilometres. It completes an orbit in 365 days, or about 3.2×10^7 s. It is therefore travelling along its orbit at a speed v of 30 km s^{-1}. From these data, you can calculate that its centripetal acceleration (v^2/r) is 0.006 m s^{-2} or, from more exact data, 0.0058 m s^{-2}.

FORCES AND FIELDS

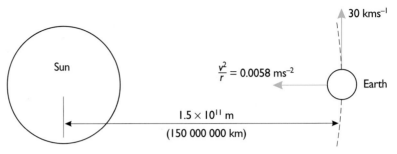

Figure 15.14 The Earth as a satellite of the Sun

This means that the gravitational field of the Sun at the distance of the Earth must be 0.0058 N kg^{-1}, and

$g = GM/r^2$.

So, for the Sun with mass M_s:

$$M_s = \frac{r^2 g}{G}$$

or $M_s = \dfrac{(1.5 \times 10^{11})^2 \times 0.0058}{6.67 \times 10^{-11}}$

$= 2 \times 10^{30}$ kg.

Similarly, to obtain the mass of the Moon, you need to know orbital details about a satellite of the Moon or the value of g on its surface. Newton realised that the tides on Earth were caused by the gravitational pull exerted on oceans by the Moon. By treating them as 'satellites' of the Moon, he was able to estimate the mass of the Moon.

We now know from manned missions to the Moon that g on its surface is equal to about 1.7 N kg^{-1}. Given that the radius of the Moon is about 1 700 km, we can calculate its mass.

> **Question**
>
> 8 Use the data given in the main text to show that the Moon has a mass of about 7×10^{22} kg, about one-hundredth that of the Earth.

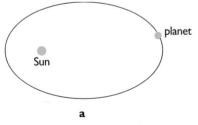

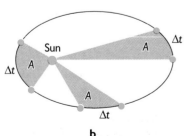

Figure 15.15 Kepler's laws of motion

KEPLER'S LAWS

One of the achievements of Newton's work was to enable Kepler's laws of planetary motion to be deduced. These laws were based on careful observation of planetary movement made by Tycho Brahe. They are a natural consequence of Newton's laws of motion and gravitation. Kepler's laws state that:

1. **All planets move in elliptical orbits, with the Sun at one focus (see Figure 15.15a).**
2. **The imaginary line joining the planet with the Sun sweeps out equal areas in equal times (see Figure 15.15b).**
3. **Denoting R as the average orbital radius and T as the period of the orbit, then R^3/T^2 is the same for all planets.**

Note that although these laws specify elliptical orbits, they also apply to a circular orbit. A circle is a special form of ellipse with the two foci coinciding with each other.

15 GRAVITATIONAL FIELDS

> **Question**
>
> **9 a** Use the equation $v^2 = GM/r$ to deduce Kepler's third law:
>
> $$\frac{R^3}{T^2} = \text{constant}$$
>
> where the planet travels a distance equal to the circumference of the orbit in time T.
>
> **b** What is the value of the constant for:
>   **i** any planetary orbits around the Sun?
>   **ii** all satellites around the Earth, including the Moon?
>   **iii** the twelve moons around the planet Jupiter?

DYNAMIC MODELLING OF ORBITAL MOTION

There are a number of ways of analysing and predicting the motion of a body under the influence of known forces. One way, ideally suited to the use of a computer, is a numerical method which analyses the behaviour of the body over successive, short intervals of time.

Suppose there is a body at a particular moment, $t = 0$, travelling with a known velocity at a given distance from the centre of the Earth. How will it move under the influence of the Earth's field? There will be a force of gravity on it towards the Earth which will cause an acceleration. After a short interval of time, say 5 s, the craft will have changed its velocity to a new value. Let us assume that the interval is so short that the craft does not move far enough to experience a significant change in the gravitational field during this time. You can calculate its average velocity and the distance it has moved during the interval. You can also find its new position and direction of travel. You can then consider the next interval of time, and calculate the new acceleration and from that the new position and velocity. Repeating this process many times over allows you to plot out the path of the body.

There are many computer packages that will, amongst other things, model dynamically the motion of a spacecraft under gravity ($\rightarrow$ 41.3).

15.3 GRAVITATIONAL POTENTIAL V_g

When launching a satellite into an orbit around the Earth, energy needs to be transferred from the fuel in the rocket motors to the rocket ($\leftarrow$ 8.1) and the satellite. The energy needs to be transferred in **two** ways:

- into the potential energy ($\leftarrow$ 9.2) to lift the satellite to its orbit
- into the kinetic energy ($\leftarrow$ 9.2) to enable the satellite, once released, to move along its orbit.

We can calculate the speed and the kinetic energy for any particular orbit radius. But how can we calculate the potential energy gained as it moves from one place to another in the Earth's gravitational field?

Suppose we wish to lift a 10 kg mass 3 metres from one floor to the next in a house. The **increase** in **potential energy** necessary is **mgh**, which is 300 J, assuming that $g = 10$ N kg^{-1}. This is the energy transferred when work is done to move the mass upstairs. The mass's weight is the force due to its interaction with the

FORCES AND FIELDS

Gravitational field and potential difference

Suppose a 1 kg mass is taken from a point at distance r to another at a small distance Δr further from the centre of a large mass, as shown in Figure 15.16. The gravitational potential difference ΔV_g between these two points is given by:

$$\Delta V_g = g\,\Delta r$$

where g is the gravitational field over that region, assumed constant. This equation also gives another interpretation of **field** – that it is a measure of the rate of change of potential with distance, the **gravitational potential gradient**.

In the limit as Δr tends to zero,

$$\frac{\Delta V_g}{\Delta r} \rightarrow \frac{dV_g}{dr} = g.$$

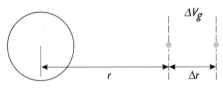

Figure 15.16 Gravitational potential difference in a constant field

Earth's gravitational field, and has the value:

weight = 10 N kg^{-1} × 10 kg = 100 N.

Because the vertical distance in the uniform gravitational field of the Earth is 3 m, the energy required is:

100 N × 3 m = 300 J

which is the increase in the potential energy of the mass.

The **gravitational potential difference** ΔV_g (← 14.3) between two places in a gravitational field is the potential energy change that a mass, 1 kg, experiences when moved between them. Therefore, in the above example, the gravitational pd is equal to 30 J kg^{-1}.

Question

10 a For a 500 kg satellite, calculate the increase in potential energy that has occurred since its launch from the Earth's surface, when it is at **i** 10 m, **ii** 1 km, **iii** 10 km, above the surface. (g)
b Why can you not use the same method to find its gain in potential energy when raising it through a large distance of, say, 10 000 km?
c Calculate the gravitational pd ΔV_g between a point on the surface of the Earth and one 10 km above its surface.

Clearly, the change in potential energy depends very much upon the value of the gravitational field g. Moving a mass against a smaller field will result in a smaller gain in potential energy. So, to calculate the gain in potential energy as a mass is gradually moved away from the Earth's surface, you must take into account the changing value of g.

Suppose you take a 1 kg mass from the Earth's surface, 6 400 km from its centre, to twice that distance R. The energy required – the gravitational pd between these two positions – can be found by estimating an average value of g over this distance. At $2R$, the gravitational field has dropped to 2.5 N kg^{-1}, as shown in Figure 15.17. Since g decreases at a smaller rate the further we are from

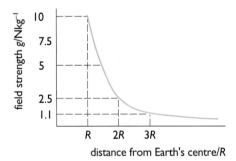

Figure 15.17 Gravitational field strength against distance

the centre, we might estimate an average value of g of about 5 N kg^{-1}.

The potential energy change is given by:

$$\Delta V_g = 5\text{ N kg}^{-1} \times 6\,400\text{ km}$$
$$= 3.2 \times 10^7 \text{ J kg}^{-1}.$$

Taking 2 kg would transfer twice this energy, taking 10 kg, ten times as much, and so on.

15 GRAVITATIONAL FIELDS

Energy needed to move a mass in a changing field

The main text suggests a way of calculating the energy needed to take a 1 kg mass from a distance r_1 to a distance r_2 in a changing field. Another way is to break the journey down into small steps Δr. Over each of these intervals, assume that g is reasonably constant then calculate the value of $g\Delta r$ for each interval and sum these values over the distance from r_1 to r_2.

The value of $g\Delta r$ is the area of a strip of width Δr under the graph of g against r (see Figure 15.18). So the energy needed to take 1 kg from one distance to another can be found by measuring the area under the graph between the required values of r. The calculated energy transfer is independent of the path taken between the distances (← 9.5).

Question

11 Use Figure 15.7 to estimate:
a the energy required to take 1 kg from the surface of the Earth to three times the Earth's radius, R_E,
b the energy required to place a satellite of mass 800 kg into a geostationary orbit around the Earth, about 42 000 km from its centre.

ZERO GRAVITATIONAL POTENTIAL

Energy is required to lift a mass against the gravitational field, and as the mass moves further and further away from the Earth, its potential energy increases. These ideas raise two questions:

- where is the zero level of gravitational potential energy? (at the Earth's surface? at the centre of the Earth?)
- in what manner does the potential energy increase with distance? (at a steady rate? at an increasing or decreasing rate?)

The zero level can be chosen to be at any position. Normally, near the Earth's surface, we take it to be the floor, a table top, the bottom of a cliff or any other convenient level. Below this chosen level, potential energies will be negative, and above that level, they will be positive. However, when considering gravitational forces on a large scale, such as those between the Earth and satellites or between stars, it is convenient to take the position of zero potential energy to be as far away as possible. Since gravitational forces are attractive, this makes all potential energies negative.

So, the **zero of potential** is defined as being **at infinity**. The gravitational potential of any point is therefore the energy needed to take 1 kg from that point to infinity. The gravitational potential V_g at a distance r from the centre of a mass M is given by:

$$V_g = -GM/r$$

where G is the constant of gravitation.

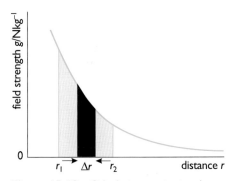

Figure 15.18 Calculating gravitational potential difference

Question

12 The field g is a direct measure of the gravitational potential gradient dV/dr. Use the graph in Figure 15.19 to show that its gradients at $r = 6380$ km and $r = 12760$ km are about 10 N kg^{-1} and 2.5 N kg^{-1} respectively.

THE VARIATION OF V_g WITH DISTANCE

Figure 15.19 shows the variation of V_g with distance r. At the surface of the Earth, where r is $R = 6380$ km, the gravitational potential is $-GM/r$ or $-4 \times 10^{14}/6.38 \times 10^6$ giving:

$$V_g = -6.26 \times 10^7 \text{ J kg}^{-1}.$$

In other words, you need to supply 62.6 MJ to take a mass of 1 kilogram from the surface of the Earth to infinity. This leads to the idea that **gravitational potential at a point is the energy required to take unit mass** from that point **to infinity**.

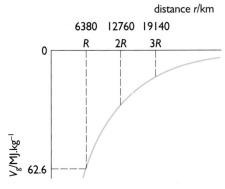

Figure 15.19 Variation of gravitational potential with distance

FORCES AND FIELDS

Gravitational potential difference

We have established that the change in gravitational potential ΔV_g in moving a distance Δr against the field is given by:

$$\Delta V_g = g\,\Delta r.$$

As Δr approaches zero, the gravitational potential V_g is given by:

$$V_g = \Sigma\,g\,\Delta r$$
$$= \int g\,dr$$
$$= \int \frac{GM}{r^2}\,dr$$

or $\quad V_g = \dfrac{-GM}{r} + \text{constant}.$

We **choose** the value of gravitational potential V_g to be zero at an infinite distance r from the Earth. This makes the value of the constant zero. The value of the gravitational potential at distance r from a mass M is therefore given by:

$$V_g = -GM/r.$$

The gravitational potential difference between any two points at distance r_1 and r_2 is given by the difference between the two potentials:

$$\Delta V_g = -GM/r_2 - (-GM/r_1)$$
$$= GM(1/r_1 - 1/r_2).$$

Since $V_g \propto \dfrac{1}{r}$

the gravitational potential at twice the radius ($2R$) is about -31.3 MJ kg^{-1}. At three times the radius ($3R$) it is about -27 MJ kg^{-1} (see Figure 15.20).

To calculate the gravitational potential difference ΔV_g between a point at distance r_2 and another point at a closer distance r_1, we simply find the difference in potentials at the two points. For example, to take 1 kg from the Earth's surface to twice the distance from the centre requires:

$$\text{energy} = -31.3 - (-62.6)\text{ MJ}$$
$$= 31.3\text{ MJ}.$$

You can obtain this from the graph or by calculation using the formula:

$$\Delta V_g = GM\,(1/r_1 - 1/r_2).$$

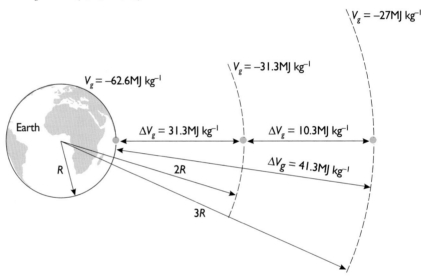

Figure 15.20 Three 'equipotential lines' in the Earth's field and the gravitational potential differences between them

Questions

13 a Calculate the gravitational pd between the Earth's surface and
  i an orbit of radius 8 000 km, like that of a polar orbiting satellite
  ii an orbit of radius 42 000 km, like that of a geostationary satellite.
  b How much potential energy is gained by putting into each orbit a satellite of mass 1 000 kg?
  c Calculate the sum of the potential and kinetic energies of such a satellite in a geostationary orbit.

14 Calculate the gravitational potential of a point on the Moon's surface, using the following data for the Moon: mass = 7×10^{22} kg, radius = 1 700 km.

THE POTENTIAL WELL

The concept of the potential well helps you to understand the nature of field and potential. Figure 15.21 illustrates a mechanical model of a potential well. The bottom of the well represents, for example, the surface of the Earth. The sides of

15 GRAVITATIONAL FIELDS

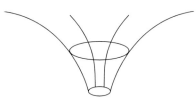

Figure 15.21 A $1/r$ potential well

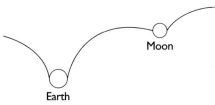

Figure 15.22 Variation of gravitational potential between two masses

the well near the bottom are very steep, giving a high rate of change of potential. So a large force pulls an object placed in this region towards the bottom of the well.

Climbing further out of the well, the sides become less and less steep until, a large distance from the centre, they become almost horizontal. Many people refer to this region as 'outside the gravitational field' but, of course, we never are! A ball placed near the edge of the well will accelerate downwards into the centre of the well. Given a small sideways push, it can be made to 'orbit' the centre of the well, providing a model of satellite motion.

Questions

15 What does the term 'model' mean in the context of Figure 15.21? Why do we construct such models? In what ways is this model an unsatisfactory one? In what ways is it useful?

16 Figure 15.22 shows the changes in potential that might occur on a journey from the Earth to the Moon. Describe the various stages that might be involved in a journey leading to a soft landing on the surface of the Moon. Pay particular attention to the changes in potential and kinetic energies of the spacecraft and to the forces acting on it.

ESCAPING FROM THE EARTH

As we have seen, 62.6 MJ needs to be supplied to take a 1 kg mass from the Earth's surface to as far away as possible. Near the surface of the Earth but above its atmosphere the kinetic energy E_k is given by:

$$E_k = \tfrac{1}{2}mv^2.$$

So $v^2 = \dfrac{2E_k}{m} = \dfrac{2 \times (62.6 \times 10^6)}{1}$

or $v = 1.13 \times 10^4 \text{ m s}^{-1}$.

Therefore, we need to project our mass at a speed of at least 11.3 km s^{-1} to cause it to leave completely the influence of the Earth's field (assuming that the Earth is isolated and the field is not influenced by other bodies in space). This speed is called the **escape velocity**.

Air resistance makes it virtually impossible for spacecraft to be launched on Earth with speeds anywhere near the escape velocity. Yet very high launch speeds are reached. Thinking about multi-stage rockets, satellites above the atmosphere and the values of g locally in space should help you workout how this can happen.

Questions

17 Suppose you project a 100 kg mass into space.
a Will it require a velocity greater than, the same as or smaller than the escape speed of 11.3 kms^{-1} calculated for a 1 kg mass? It helps here to imagine each kilogram as separate.
b Will it matter what direction this speed is in? In thinking about this, consider what might happen if you were to roll a ball up a shallow slope and then, at the same initial speed, up a steeper slope.

c How will the vertical heights reached with different initial directions compare, assuming air and other frictional effects are absent?

18 Figure 15.23 shows a photograph of a manned spacecraft taking off from the surface of the Moon. Why is much less fuel required to escape from the Moon's field than that required to escape from the Earth? Calculate the escape velocity. (mass of Moon = 7×10^{22} kg, radius of Moon = 1.7×10^6 m)

Figure 15.23 Apollo 16 after its lauch from the Moon

SUMMARY

The force between two masses M and m, distance r apart, is given by:

$F = GMm/r^2$.

The gravitational field g at distance r from a mass M is given by:

$g = GM/r^2$.

The gravitational field provides the centripetal acceleration for a satellite travelling in a circular orbit at speed v and radius r around a mass M, so:

$v^2/r = GM/r^2$.

Gravitational potential V_g at a point is the energy required to take unit mass from that point to infinity. Gravitational potential V_g at distance r from a mass M is $-GM/r$.

A mass is defined to have zero potential energy at infinity and, therefore, a negative potential energy at closer distances.

The energy needed to take 1 kg from one point to another in a gravitational field can be obtained from the area under the graph of field g against distance r between the two distances concerned.

15 GRAVITATIONAL FIELDS

The gravitational field g is equal to the gravitational potential gradient dV_g/dr and so, for small distances Δr:

$\Delta V_g = g \Delta r$.

Over large distances r_1 to r_2, the gravitational potential difference:

$\Delta V_g = GM(1/r_1 - 1/r_2)$.

SUMMARY QUESTIONS

19 a The NASA Space Shuttle, when on the launch pad and about to take off, consists mainly of fuel. Most of the fuel is burnt in the first few minutes after launch. Why is this?
b The energy available from burning 1 kg of kerosene, or paraffin, is about 35 MJ. It is used for stage 1 of some rockets. Why is it not usually used for the last stages of multi-stage rockets (assuming that 63 MJ is needed for 1 kg to escape from the Earth's field)?

20 A satellite is in a circular orbit of radius r around a mass M.
a Calculate the gravitational potential energy per kilogram of the satellite.
b Use the relationship $mv^2/r = GMm/r^2$ to obtain an expression for the kinetic energy per kilogram of the satellite.
c Show that the total energy of the satellite is equal to $-GM/2r$.

21 The Earth's centre is approximately 6 380 km from the surface. By considering this to be the mean radius R, find:
a the gravitational potential at the Earth's surface,
b the gravitational potential in space at $2R$, $3R$, and $4R$.
Why does the potential change less rapidly as the distance from the Earth's centre increases?

22 The temperature at the surface of the Moon is about 400 K when illuminated by radiation from the Sun. An oxygen molecule has an average speed of about 500 m s^{-1} at this temperature.

Find the escape speed for an object at the Moon's surface and argue whether this is consistent with the fact that there is no atmosphere on the Moon. (mass of moon = 7×10^{22} kg, radius of Moon = 1.7×10^6 m)

16 ELECTRIC FIELDS

We are very aware of the effects of gravitation in our environment but less so of the effects of electric fields. For example:

- tall buildings and trees are vulnerable under the intense electric fields created during a thunderstorm,
- to avoid explosions ignited by sparks from charged clouds of particles, industries handling large quantities of powdered materials have to take special precautions when moving the particles through pipes at high speed,
- similarly, fuel pumped into an aircraft's fuel tank, as shown in Figure 16.1, can become highly charged during the pumping process, with perhaps even more disastrous consequences should sparks arise.

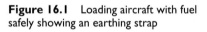

Figure 16.1 Loading aircraft with fuel safely showing an earthing strap

The effects of electric fields can also be useful:

- electric fields are used in photocopiers, in a process known as xerography,
- the presence of electric fields helps to increase the efficiency of paint and crop spraying (see Figure 16.2),
- electric fields can prevent the release into the atmosphere of dust particles from industrial processes,
- ink-jet printers make use of electric fields to control the movement of the stream of ink particles towards the paper.

All of the above can be explained in terms of the forces between charges, and the fields and potential around them. This chapter looks at the nature of these forces and fields, and establishes the mathematical models that describe them.

Figure 16.2 Electrostatic crop spraying

16 ELECTRIC FIELDS

16.1 INVESTIGATING ELECTRIC FIELDS AND FORCES

Suppose you rub a polythene strip with a soft duster and support it on a watch glass resting on a smooth surface, as shown in Figure 16.3. Another charged polythene strip brought close will experience a force of repulsion. A charged acetate strip will attract the polythene rod. This is because friction during rubbing causes the electrically insulated rods to become charged. Polythene gains electrons and so becomes negatively charged, while acetate loses electrons and becomes positively charged. Two similarly charged rods repel each other and oppositely charged rods attract one another.

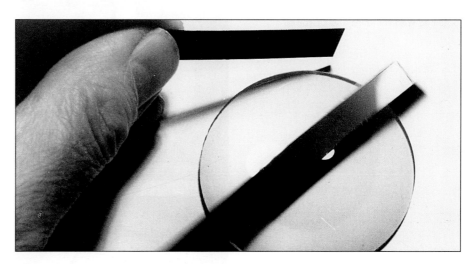

Figure 16.3 Electrical repulsion and attraction

You can interpret these observations in terms of an **electric field** (← 14.2) which is produced around the charge on the rod. When placed in this field, the other charged rod experiences a force which is attractive or repulsive according to the sign of its charge.

You can extend the idea to the situation in Figure 16.4, where oppositely charged plates produce a strong electric field. A small strip of metallised plastic film is attached to an insulating rod and placed in between the plates. The uncharged strip experiences no electric force, even though it is in the field. However, if you touch the strip to the negative plate, it will pick up negative charges, electrons, and be repelled by that plate and attracted by the other. So it will show a visible deflection. The deflection will be reasonably uniform over the whole of the region between the plates, indicating a fairly constant electric field. Should the strip touch the positive plate, it will lose electrons, become positively charged and deflect in the other direction.

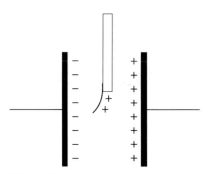

Figure 16.4 Force on a charge between charged plates

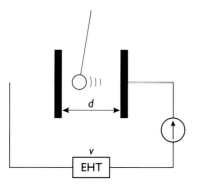

Figure 16.5 The shuttling ball

Question

1 Figure 16.5 shows a pair of plates connected to a high voltage supply. A ping-pong ball, coated with conducting paint, is suspended by a nylon thread between the plates.
a Initially, the ball is uncharged. What would you expect to observe?
b The ball is gently pushed towards one of the plates which it touches. The ball then bounces to and fro between the plates. Using the idea of the electric field, explain why this happens.
c Sketch a graph showing how you would expect the velocity of the ball to change with time.

FORCES AND FIELDS

ELECTRIC FIELD PATTERNS

Just as iron filings will display magnetic field patterns, so electrically polarisable particles can be used to investigate the electric field patterns around charged conductors. For example, we can use electrically conducting paper or semolina floating on a non-conducting oil.

Figure 16.6 shows photographs of some field patterns that might be obtained with variously shaped charged conductors. Notice the radial nature of the field pattern from point or circular arrangements of charge and the parallel or uniform nature of the field associated with a flat distribution of charge.

Figure 16.6 Some typical field patterns obtained with semolina and castor oil

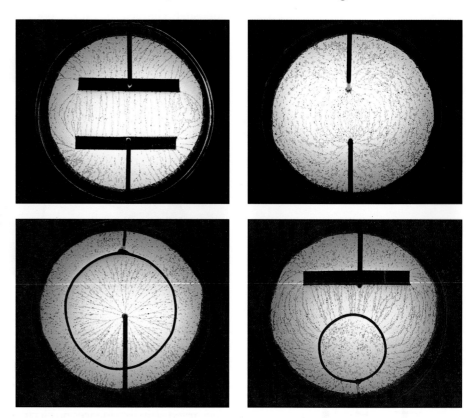

EXPERIMENT 16.1 Investigating electric field patterns

You can use an arrangement, such as that shown in Figure 16.7, to investigate field patterns around charged conductors of various shapes. Connecting a high voltage supply and sprinkling semolina lightly over the surface of the oil enables you to produce electric field patterns, shown in Figure 16.6.

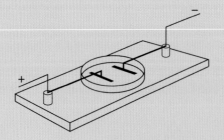

Figure 16.7 Observing field patterns

16 ELECTRIC FIELDS

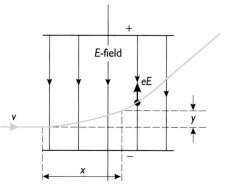

Figure 16.8 E-field deflection of an electron beam

Questions

2 The electric field patterns shown in Figure 16.6 are 'models' of real situations. Which model most closely resembles each of the following?
a the field between the electron beam deflecting plates in a cathode ray oscilloscope,
b the field between the inner and outer conductor in a co-axial cable used to convey television signals between the aerial and the television set,
c the field between a charged thunder cloud and the tip of a lightning conductor above the spire of a church.

3 Figure 16.8 shows an electron gun which accelerates electrons into the space between two horizontal plates inside the tube. The plates are 10 cm long and a distance $d = 5$ cm apart. The accelerating voltage of the gun is 3.0 kV. The potential difference between the plates is 1.5 kV.
a Use energy transferred and $E = V/d$, to show that the beam will just touch the edge of the positive plate.
b Suggest two different adjustments you can make so that the beam just misses the plate's edge.
c What happens to the beam once it is beyond the plates?
d What two quantities of the electron's motion at the plate's edge do you need to know to calculate the direction in which it is travelling?

Paint and crop spraying

Spraying paint on to a metal or pesticides on to crops can be a wasteful process. Much of the paint will miss the object to be coated and, equally, parts of the object will not be exposed directly to the spray and so will not be painted. Similarly, parts of the underside of leaves will be missed by a spray directed from above.

However, if we charge the droplets of paint or pesticide positively to a high potential as they emerge from the sprayer, the charged particles will follow the lines of the electric field pattern. So they will move towards an earthed object, such as the metal to be painted or the plants to be treated (← Figure 16.2).

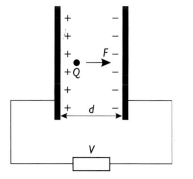

Figure 16.9 Calculating field strength

CALCULATING ELECTRIC FIELD STRENGTH

Figure 16.9 shows a pair of parallel plates charged to a voltage V and separated by a distance d. Imagine that you place a small positive charge Q near the negative plate and steadily move the charge towards the positive plate. You would have to do work against the electric force F_E on the charge. The **energy transferred** as work is done against the field over a distance d will be $F_E \times d$. This is equal to the **increase in electrical potential energy** of the system, $Q \times V$, since V is the energy transferred by each coulomb of charge. So,

$$F_E \times d = Q \times V$$

or $F_E/Q = V/d$.

This shows that the electric field E is equal to the electrical potential gradient V/d in volts per metre between the plates. However, E is actually defined as the electrical force per unit charge F_E/Q in newtons per coulomb (← 14.2), so

$$1\,\text{N C}^{-1} = 1\,\text{V m}^{-1}.$$

FORCES AND FIELDS

FIELD AND POTENTIAL GRADIENT

Taking the negatively charged plate, the reference plate, to be at zero potential, we can consider how the electrical potential changes with distance between the plates. In the case of a **uniform** electric field, the potential gradient is constant, as shown in Figure 16.10. You can imagine equipotential surfaces drawn at regular intervals between the plates, in much the same way that contour lines drawn on a map at regular intervals represent a uniform gradient (see Figure 16.11). Around a small source of charge, the electric field decreases with distance from the charge, giving a **non-uniform** potential gradient (→ 16.3).

For both uniform and non-uniform fields, you can define the electric field E at a point as a function of the potential gradient at that point:

$$E = -dV/dx.$$

The negative sign indicates that, in Figure 16.10, the direction of E is the direction of the force on a positive charge in that field. This direction is opposite to that in which x and V are increasing.

Ink-jet printers

In an oscilloscope, an electron gun fires a beam of electrons between a pair of parallel plates. When the plates are charged, the beam is deflected to a particular part of the screen. An ink-jet printer (see Figure 16.12) operates in much the same way. Ink particles are charged as they leave an ink gun. The ink jet is deflected by an electric field produced by an applied voltage between a pair of parallel plates. This voltage is under computer software control, which enables the ink stream to be deflected to produce characters and graphics at high speed. Printing can be in colour and on almost any surface.

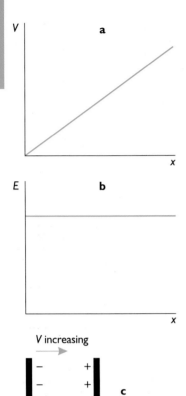

Figure 16.10 Variation of potential between two plates

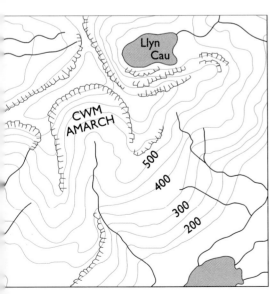

Figure 16.11 Contour lines on a map, representing points of equal gravitational potential

Figure 16.12 Ink-jet printer

16 ELECTRIC FIELDS

Figure 16.13 'Forked' and 'sheet' lightning

Questions

4 In Figure 16.9, the pd across the plates is 1 kV, the plate separation is 10 mm and the charge Q is 10^{-12} C. Calculate:
a i the electric field between the plates,
  ii the electric force on charge Q,
  iii the energy transferred by work done in moving the charge Q against the field from one plate to another.
b Suppose that the plate separation is doubled, other factors remaining unchanged. What will your answers to **a i, ii** and **iii** now be? Why must you assume that the charge Q is small compared with the charge stored on each of the plates at a voltage of 1 kV?

5 An electron beam passes between a pair of parallel plates, separated by a distance of 40 mm and connected to a 5 000 V supply. Calculate the force on a single electron in the beam. (*e*)

Lightning

Certain types of weather, notably hot and humid conditions, cause water vapour in the atmosphere to rise due to convection. As it rises, the vapour cools and condenses into water droplets which, in turn, freeze to produce hailstones. These fall under gravity and interact with the rising water droplets to become charged. It is thought that the mechanism depends on the different mobilities of H^+ and OH^- ions. When the centres of the water droplets freeze, the outer layers crack off because of the volume change on freezing. As the hailstones descend, their temperature increases again and they turn to rain.

The result of this process is an uneven distribution of charge within the cloud. The top of the cloud becomes positively charged and the bottom negatively charged. If the field within the cloud becomes too large, ionisation will occur and, the air will break down and become electrically conducting, giving rise to 'sheet' lightning. However, if the field between the base of the cloud and the Earth's surface exceeds the 'breakdown value', a forked lightning flash will occur as negative charge leaves the base of the cloud and travels to the ground (see Figure 16.13).

In dry air at normal atmospheric pressure, fields of 3 000 V mm^{-1} are sufficient to cause air to break down due to ionisation and become conducting. Wet air requires a rather smaller field of about 1 000 V mm^{-1} to cause breakdown. Because the base of a thundercloud is typically 2 000 m above the Earth, there is momentarily, a pd of around 2 000 MV between the cloud and the Earth. However, the voltage quickly reduces to less than 100 MV because of the ionisation of 'step leaders' before the main flash. At the moment a lightning flash is produced, currents of up to 10 000 A occur for a very short time.

ELECTRIC FLUX AND GAUSS'S LAW

A charge Q has electric flux Ψ associated with it ($\leftarrow$ 14.2). Gauss, a German mathematician born a little over 200 years ago, established a link between the electric flux Ψ and charge Q. **Gauss's law** can be expressed by stating that the total flux associated with charge Q is:

$$\Psi = \frac{Q}{\epsilon}$$

where ϵ is the **permittivity** of the medium in which the flux exists. In a vacuum, ϵ is the permittivity of free space, ϵ_0 ($\rightarrow$ 16.2).

FORCES AND FIELDS

The **flux density** Ψ/A in a given region of space at distance r from the charge Q is defined as the electric field E. Consider Figure 16.14, which shows a sphere of radius r with Q at centre. The flux Ψ is spread out over an area $4\pi r^2$ giving an electric field:

$$E = \frac{\Psi}{4\pi r^2}$$

or, using Gauss's law: $E = \dfrac{Q}{4\pi\epsilon r^2}.$

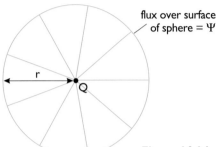

Figure 16.14 Electric flux

Question

6 Assuming that the total flux remains constant and that it is not absorbed or diminished in any way as it spreads out from a source of charge, how would the electric field vary with distance from the charge, according to Gauss's law? Is it reasonable to suppose that flux does stay constant and is not absorbed?

EXPERIMENT 16.2   Equipotentials around charged conductors

Fixing a 150 mm by 150 mm sheet of 'resistance paper' to a drawing board enables you to plot equipotentials, or lines of equal voltage, around any conductor. The conductor is pressed on to the paper and held at a particular voltage with respect to another also placed on the paper. Connecting the electrodes as shown in Figure 16.15, you can produce a pd of, say, 9 volts between them. Placing the probe in the centre, you can move it to obtain a reading of, say, 4.5 volts and move the probe again to identify a number of other positions giving similar readings. You can then repeat the process for other readings to obtain other equipotentials. Figure 16.16 illustrates the equipotentials for three different arrangements.

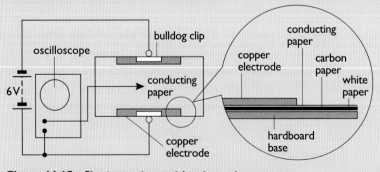

Figure 16.15 Plotting equipotentials using resistance paper

16 ELECTRIC FIELDS

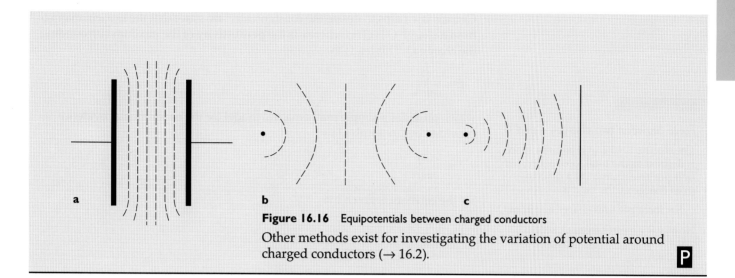

Figure 16.16 Equipotentials between charged conductors

Other methods exist for investigating the variation of potential around charged conductors (→ 16.2).

Question

7 Look at the equipotentials in Figure 16.16. The electric field E provides the potential gradient.
a Do the largest fields occur near flat or near pointed parts of a charged object?
b Why are pointed conductors used for protection against lightning strikes?
c Why is it important that the dome of a Van de Graaff generator is as smooth as possible?
d What is the link between equipotential diagrams and the field patterns shown in Figure 16.16? How would we deduce the field pattern given the equipotentials?

16.2 THE ELECTRICAL INVERSE SQUARE LAW

Much of our present understanding of the nature and stability of atoms comes from an appreciation of the interaction between point and spherical charges. From Newton's work, two neighbouring masses experience an attractive gravitational force, the size of which depends upon their masses and their separation (← 15.1). Following on from this, Charles Coulomb, a French physicist, established some two hundred years ago the nature of the electrical force between charges.

He found that the force F between two point sources of charge Q and q, separated by a distance r, is proportional to the size of the charges and to the inverse square of their separation, as shown in Figure 16.17. We can write **Coulomb's law**:

$$F \propto \frac{Qq}{r^2}$$

or $\quad F = \frac{kQq}{r^2}.$

The value of the constant is dependent upon the nature of the medium between the two charges.

Figure 16.17 Coulomb's law

FORCES AND FIELDS

Force between two charges q and Q

The electric field E at a point in space is defined as the total force acting per unit positive charge. At a distance r from charge Q, the field is:

E = force per unit charge at r from Q.

Now, q is r metres from Q, so:

$$E = \frac{F}{q}$$

or $F = Eq$.

From Gauss's law, we can write:

$$E = \frac{Q}{4\pi\epsilon r^2}.$$

Combining these last two equations gives the force between q and Q as:

$$F = \frac{Qq}{4\pi\epsilon r^2}.$$

Gauss's law helps to define the constant. His law states that the flux emanating from a charge Q is Q/ϵ, where ϵ is the **permittivity** of the medium surrounding the charge Q. This leads to

$$E = \frac{Q}{4\pi r^2 \epsilon},$$

which provides a simple mathematical model for the experimental evidence provided by Coulomb's law:

$$F = \frac{Qq}{4\pi\epsilon r^2}.$$

So, the constant is given by:

$$k = \frac{1}{4\pi\epsilon}.$$

PERMITTIVITY

The permittivity ϵ varies from one medium to another. Its value in a vacuum, **the permittivity of free space, ϵ_0** is an important physical constant, which is found to be 8.854×10^{-12} F m^{-1}. This constant relates specifically to the case where there is a vacuum between the charges. When air or another medium is present, you can find ϵ by multiplying the value of ϵ_0 by a number ϵ_r, the **relative permittivity**.

The permittivity ϵ of a medium is related to the permittivity of free space ϵ_0 by the equation

$$\epsilon = \epsilon_r \epsilon_0.$$

In air, ϵ_r is 1.000 6, so it is usually unnecessary to distinguish between ϵ_0 and ϵ in air. Figure 16.18 shows other values of ϵ_r.

Note that the higher the value of ϵ, the smaller the forces between charges in the medium. So, perhaps a better term than permittivity would be preventivity!

material	ϵ_r
air	1.000 6
polyethene	2.4
glass	4.0 to 8 depending on type
paper	2.7
mica	6.0
water (pure)	80
TiO$_2$	400

Figure 16.18 Relative permittivity of some materials

Questions

8 This question concerns a possible method for testing Coulomb's law.
a A polystyrene sphere coated with a conducting paint and about 10 mm in radius will, when charged to around 1 000 V, store a charge of about 1 nC. Calculate the force you would expect between two spheres of this type placed 20 mm apart.
b Describe how you might use a top-pan balance with a resolution of 0.01 g or 0.001 g to measure such a force.
c How would you use the apparatus to test for an inverse square law of force between the spheres?
d In a test like this, the inverse square law seems to break down at small separations. Why do you think this might be?
e How would you show that doubling the charge on each sphere quadruples the force between them.
   Indicate any problems you might expect to encounter in each of your experiments, explaining how you would minimise or overcome them.

9 Two lead spheres, each 100 mm in diameter, are placed on a table top with their centres 200 mm apart. Suppose we remove an electron from 1%

16 ELECTRIC FIELDS

of the atoms in each sphere. The molar mass of lead is 0.207 kg mol^{-1} and its density is 11 300 kg m^{-3}.
a Calculate what the electrical force between the spheres would be. (N_A, e)
b Does your answer suggest anything about the maximum size of the charge we can have on 'ordinary' lumps of matter?

16.3 ELECTRIC FIELD AND POTENTIAL AROUND A CHARGE

Imagine an electrically isolated conducting sphere charged to a potential of V above its surroundings, as shown in Figure 16.19a. It will store a positive charge Q on its surface which will generate an electric field around it. The field outside the hollow sphere is radial in shape as shown in Figure 16.20. The field inside the sphere is zero.

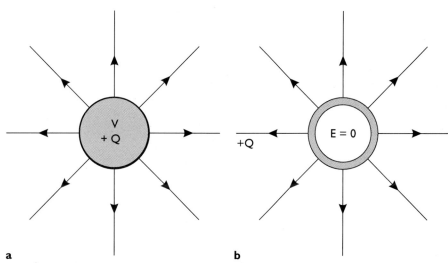

a **b**
Figure 16.19 a A charged spherical conductor **b** A hollow charged conductor

A small positive charge q placed in that field will be repelled by the charge Q. The repulsion will depend on the distance of q from the sphere because **electrical potential** V_E varies with distance from the sphere. Experiment shows that, as with gravitational potential, V_E **varies inversely with distance** from the centre of charge. A mathematical analysis based on Coulomb's law also demonstrates this, as in the box 'Potential around a point charge'.

The potential V_E at a distance r from the centre of the sphere of charge Q is given by:

$$V_E = \frac{Q}{4\pi\epsilon r}.$$

At twice the distance from the centre of the sphere, the potential drops to one half its value, at three times the distance it drops to one-third and so on.

Note that, as with gravitation, we choose the **zero level** of potential to occur at an **infinite distance** from the source of charge. Unlike gravity, the potential is positive and increases when approaching the source of charge. This is due to the repulsive nature of the force between two like charges. If the central charge was negative, you would have attractive fields and negative potential energies, just as you do for gravity.

FORCES AND FIELDS

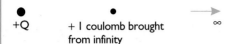

Figure 16.20 Electrical potential

The potential V_E at any point is the energy transferred when a charge of one coulomb is brought from infinity to that point, as shown in Figure 16.20.

The **electrical potential energy** E_p stored in a system consisting of two charges Q and q, separated by a distance r, is therefore given by:

$$E_p = \frac{Qq}{4\pi\epsilon r}.$$

Question

10 Why is the electric field inside a charged, hollow conductor zero, do you think? What are the implications of this for passengers in motor cars or aeroplanes during thunderstorms?

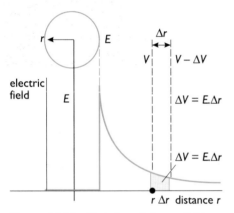

Figure 16.21 Variation of electric field with distance

Potential around a point charge

The change in electrical potential when moving from r to $r + \Delta r$ from a charge Q is equal to $E\Delta r$ where E is the electric field at distance r from the charge (see Figure 16.21). Since

$$E = \frac{Q}{4\pi\epsilon r^2}$$

and $\quad -\Delta V_E = E\Delta r$

then $\quad -\Delta V_E = \dfrac{Q}{4\pi\epsilon r^2}\Delta r.$

The electrical potential at a point is found by summing the ΔV's from distance r to infinity. This amounts to finding the area under the graph of E against r between these limits. We can therefore write:

$$V_E = \Sigma \frac{-Q}{4\pi\epsilon} \frac{\Delta r}{r^2}$$

$$V_E = \int_{\infty}^{r} \frac{Q}{4\pi\epsilon} \frac{dr}{r^2},$$

giving $\quad V_E = \dfrac{Q}{4\pi\epsilon r}.$

Lightning conductors

The lightning conductor is a device designed to reduce the damage caused to buildings as a result of lightning strikes. The main reason for them is to prevent lightning striking, rather than giving it a target to hit. The Earth is negatively charged, generating a field at its surface of a little over 100 V m^{-1}. Negative charge on the base of a cloud will repel electrons in the Earth's surface, leaving the surface less negatively charged.

A pointed conductor above the surface has a heavy concentration of charge at its tip, producing a sufficiently large electric field around it to ionise air particles, as shown in Figure 16.22. The lightning conductor therefore allows transfer of charge to take place, preventing a large build-up of charge on the cloud.

16 ELECTRIC FIELDS

Figure 16.22 Ionisation around a charged conductor

EXPERIMENT 16.3 Measuring electrical potential around a charged sphere

Setting up the apparatus shown in Figure 16.23, with the EHT supply set to 1 000 or 3 000 V and connected to a metal sphere, enables you to mount the end of a probe so that it is inside a small gas flame. By moving the probe, you can measure the potential at various distances from the charged sphere.

Plotting a graph of potential V against distance r from the centre of the spherical charge should indicate to you that they are likely to be inversely proportional. You could confirm this by plotting a graph of V against $1/r$ to give a straight line graph.

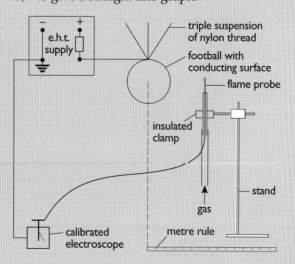

Figure 16.23 Apparatus for experiment 16.3

203

FORCES AND FIELDS

distance r	0.085	0.1	0.3	0.4	0.5
potential V	1000	430	290	210	165

Figure 16.24 Typical results from experiment 16.3

Question

11 Figure 16.24 shows some typical results obtained in experiment 16.3.
a Plot a suitable graph to show that V_E is inversely proportional to distance r.
b In experiment 16.3, why is it important to electrically isolate the charged sphere from its surroundings? Why is the flame probe mounted in an insulated stand?
c The flame is used to produce ions in the air. What other method might be used?

EXPERIMENT 16.4 Measuring the constant k in Coulomb's law

By measuring the potential V_E of a sphere, its radius r and the charge stored on its surface, you can obtain a value for ϵ. Using the conducting ball in experiment 16.3, you can momentarily charge the ball to a potential of 2 000 V and measure the charge deposited on the ball, as shown in Figure 16.25. This will give you a value for k from $k = 1/4\pi\epsilon$ and you can then find ϵ.

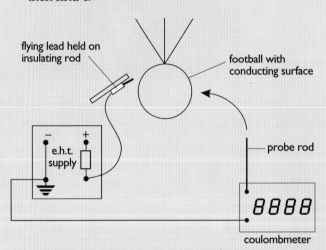

Figure 16.25 Finding the constant in Coulomb's law

Question

12 This question concerns experiment 16.4.
a A ball has a diameter of 170 mm. When charged to 2 000 V, it stores 10 nC of electric charge. What value of ϵ do these results lead to? How close is this to the accepted value for free space ϵ_o?
b What advantages or disadvantages would a larger charging voltage have in this experiment?
c What precautions would you need to take in carrying out the experiment to ensure the best accuracy in your final value for ϵ?

CAPACITANCE OF A SPHERE

The **capacitance C** of a body is the ratio of the electric charge residing on the body to its potential. It is measured in coulombs per volt ($\leftarrow$ 13.1).

The sphere in experiment 16.3 is charged to a potential V above its surroundings, stores a charge Q on its surface and has a radius r. So:

$$V = \frac{Q}{4\pi\epsilon r}$$

from which $C = \frac{Q}{V}$

and $C = 4\pi\epsilon r$.

> **Question**
>
> 13 **a** Calculate the capacitance of a sphere of diameter 170 mm.
> **b** Estimate your own capacitance. Calculate the charge you would store, if your potential was raised to 4 kV. (This voltage can easily be achieved by just walking over some types of floor carpeting.)
> **c** Why do you think you might need to take great care when handling some types of electronic integrated circuits?

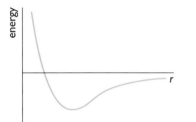

Figure 16.26 Electrical potential well

ELECTRICAL POTENTIAL WELLS AND HILLS

As in gravitation (← 15.3), it is convenient to use the concept of the **potential well** when considering electrical potential. The potential energy of a hydrogen atom or a singly ionised helium atom (or any single electron atom) depends on the distance between the electron and the nucleus. Figure 16.26 shows the variation of potential energy with distance for an electron near a nucleus.

The model for electrical potential due to a charge Q leads to a potential energy E_p given by:

$$E_p = \frac{Qq}{4\pi\epsilon r}.$$

So for the atom: $E_p = \frac{-Ze.e}{4\pi\epsilon r} = \frac{-Ze^2}{4\pi\epsilon r},$

where e is the electronic charge and Ze is the nuclear charge. The negative sign is present because the electronic charge e has a negative value and the nuclear charge Ze has a positive value. The field is one of attraction and so work has to be done to separate the charges. This means that energy must be supplied and the potential energy of the system is negative.

In its lowest energy state, the electron is close to the proton and deep down in the well. The electron can be made to climb to higher levels in the well when the atom becomes 'excited'. It rises out of the well completely when ionisation occurs.

Geiger and Marsden performed an experiment involving the firing of α-particles at gold foil (→ 40.2). An α-particle has a charge of +2e and a gold nucleus, +79e. When an α-particle makes a head-on collision with a gold nucleus, it is scattered back in the direction from which it came. You can explain this by imagining that the kinetic energy of the incoming alpha is sufficient to allow it to climb some way up an electrical **potential hill** created by the gold nucleus. As it reaches the highest point, its kinetic energy has been transferred into electrical potential energy. Figure 16.27 shows the 1/r variation in a potential hill.

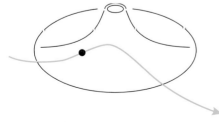

Figure 16.27 A 1/r potential hill

FORCES AND FIELDS

	Gravitation	Electric
Potential gradient or field strength	$-\dfrac{dV_G}{dr}$	$-\dfrac{dV_E}{dr}$
Potential at distance r	$-G\dfrac{M}{r}$	$\dfrac{Q}{4\pi\varepsilon_0 r}$
Potential energy at distance r	$-G\dfrac{Mm}{r}$	$\dfrac{Qq}{4\pi\varepsilon_0 r}$
Field strength at distance r	$G\dfrac{M}{r^2}$	$\dfrac{Q}{4\pi\varepsilon_0 r^2}$
Force at distance r	$G\dfrac{Mm}{r^2}$	$\dfrac{Qq}{4\pi\varepsilon_0 r^2}$

Figure 16.28 Formulae for gravitational and electric fields

Question

14 a Calculate the kinetic energy, in joules, of a 5 MeV α-particle.
b Find an expression for the electrical potential at distance r from the centre of the nucleus of an atom of gold.
c Write down the potential energy of an α-particle and a gold nucleus separated by a distance r.
d Estimate the closest distance of approach when a 5 MeV α-particle 'collides' head-on with a stationary gold nucleus in a sheet of gold foil. What assumption have you made?
e Estimate the maximum radius of a gold nucleus.

ELECTRIC AND GRAVITATIONAL FIELDS

Many of the concepts involved in a discussion of electric fields around a point or a charged sphere are essentially the same as those for gravitational fields around a spherical mass (← 15). The mathematical models that describe these two types of field are also very similar. Figure 16.28 compares some formulae for gravitational and electric fields. The differences arise because mass is the important property in gravitation and charge is the important property when dealing with electric fields. Also, gravitational forces are always attractive, while charges can be either negative or positive giving rise to repulsive as well as attractive forces.

Question

15 Two electrons are 1 metre apart in a vacuum. Calculate the ratio of the electrical force to the gravitational force between them. (m_e, e, ε_o, G)

UNIFORM ELECTRIC FIELD

Electric fields vary considerably in shape. For some purposes, e.g. in a mass spectrometer, we need to produce one that is uniform and occupies a reasonably large region of space.

Close to a flat conducting plate carrying a 'carpet' of positive charge, E is fairly constant over the central area of the plate. This is also true over distances above the plate that are small compared with the dimensions of the plate. The resultant field at any point P above the surface arises from the contributions of each element of charge on the plate, as shown in Figure 16.29. Those closest to P make the largest contribution because of the inverse square nature of the field around a point charge (← 16.2). Near the edge of the plate, the field becomes less uniform.

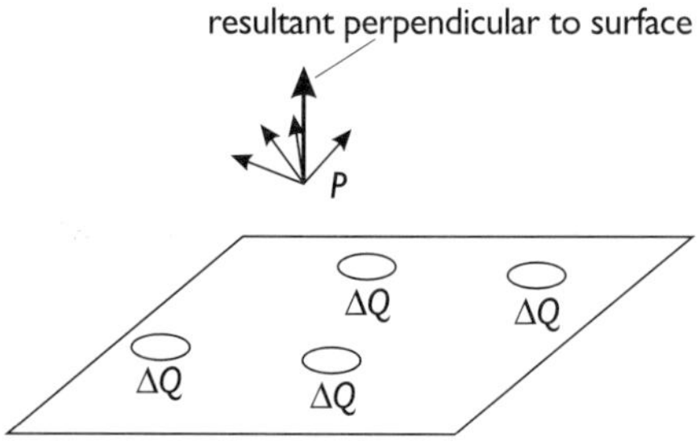

Suppose you add a second plate, this time carrying the same quantity of negative charge. When it is close to and parallel with the first, as shown in Figure 16.30, there is a relatively strong uniform field over the region between the plates, with a weak field outside the region (← 16.1). The fields due to each plate reinforce one another between the plates but have a cancelling effect outside.

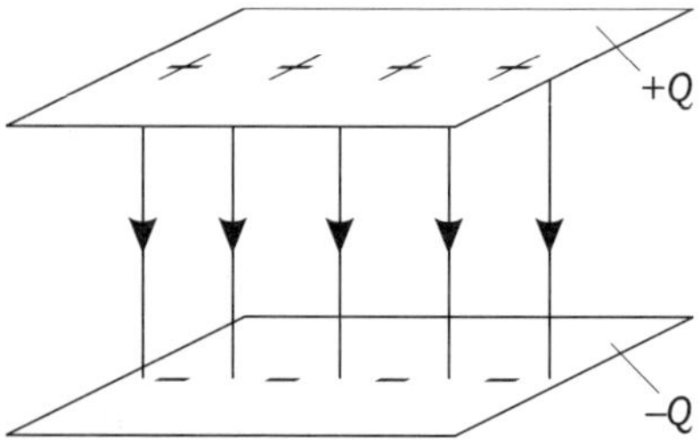

Figure 16.31 shows the charges $+Q$ and $-Q$ stored on the plates. When looking at charge and capacitance in 13.1, it was established that the charge Q displaced is proportional to the charging pd V. As long as the linear dimensions of the plate are large compared with the plate separation, the charge Q is proportional to the area A of the plates and inversely proportional to the plate separation d.

Figure 16.29 Electric field above a carpet of charge

Figure 16.30 Electric field between parallel plates

16 ELECTRIC FIELDS

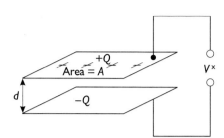

Figure 16.31 Charge on a pair of parallel plates

Summarising:

$$Q \propto V$$
$$Q \propto A$$
$$Q \propto 1/d$$

so $Q \propto VA/d$.

The actual charge stored depends upon the nature of the insulating medium, sometimes called the **dielectric**, between the plates. To change the above proportionality into an equation, you can introduce the permittivity ϵ of the dielectric. When the space between the plates is a vacuum, or 'free space', the constant ϵ_0 is used ($\leftarrow$ 16.2).

You can therefore write:

$$Q = \frac{\epsilon VA}{d}.$$

Question

16 In section 16.2, the value of ϵ_0 is quoted as being $8.854 \times 10^{-12}\,\text{F m}^{-1}$. Using equation 1 below, show that the unit F m^{-1} is equivalent to $\text{C}^2\text{N}^{-1}\text{m}^{-2}$.

We can therefore write:

$$Q = \frac{\epsilon VA}{d} \quad (1).$$

Re-arranging the equation gives:

$$\frac{Q}{V} = \frac{\epsilon A}{d}.$$

Since Q/V is equal to the capacitance C of the parallel plate arrangement we can write:

$$C = \frac{\epsilon A}{d} \quad (2).$$

To answer our original question as to the way in which charge and electric field are related, equation (1) may be written in another way:

$$\frac{Q}{A} = \frac{\epsilon V}{d}.$$

Since the electric field $E = V/d$ and Q/A is the charge surface density,

$$\sigma = \epsilon E \quad (3).$$

Or:

charge density on each plate = $\epsilon \times$ electric field between plates.

UNIFORM FIELD AND GAUSS'S LAW

You can use Gauss's law to derive an expression for the electric field between parallel plates, each of area A, one with charge $+Q$ and the other with charge $-Q$.

Gauss's law can be expressed simply by relating the flux Ψ emanating from a charge Q in a medium of permittivity ϵ by:

$$\Psi = \frac{Q}{\epsilon}.$$

Because electric field, **E** is the electric flux density, we can find the flux Ψ by multiplying **E** by the area A, through which it is passing. Because the sheet has two sides, each of area A, you can write:

$$E.2A = \frac{Q}{\epsilon}.$$

E is a vector so that, between the two plates the two electric fields, each of magnitude $Q/2A\epsilon$, add up to give:

$$E = \frac{\sigma}{\epsilon} \text{ as stated in equation (3)}.$$

Question

17 A capacitor is to be made from two thin strips of metal foil 30 mm wide, separated by paper 0.1 mm thick.
a How long should the strips be to produce a capacitor of 1 μF?
To make the capacitor easier to handle, the plates can be rolled.
b What precaution would you need to take before doing this?
c What other incidental advantage would this give?

EXPERIMENT 16.5 Factors affecting the charge on each plate of a parallel plate capacitor

You can use the arrangements shown in Figure 16.32 to measure how the charge on a capacitor, and therefore its capacitance, varies as the voltage, area and plate separation are changed. Since there are three variables, the experiment has three parts.

  i A graph of charge Q against V will show you whether or not they are proportional to each other.
  ii Keeping the charging voltage V constant, you can position the top plate to give different areas of overlap, as shown in Figure 16.33. You can then plot a graph of Q against A to see whether they are proportional to one another.

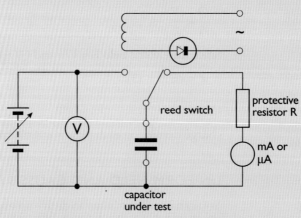

Figure 16.32 Circuit for experiment 16.5

16 ELECTRIC FIELDS

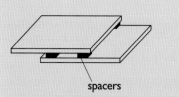

Figure 16.33 Changing plate area

iii Keeping the charging voltage V and the area A constant, you can obtain values of Q for various values of plate separation d. Plotting a graph of Q against d should indicate that inverse proportionality is likely and that Q should be plotted against $1/d$.

Knowing the plate separation and area, you will be able to calculate a value for the permittivity ϵ of the air between the plates.

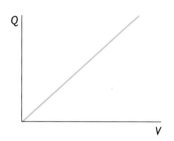
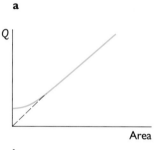
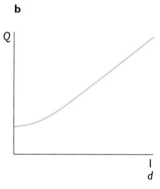

Figure 16.34 Dependence of charge on **a** voltage, **b** plate area and **c** separation

Questions

18 Figure 16.34a, b and c show sketches of three typical graphs obtained when doing experiment 16.5.
a Figure 16.34b suggests that Q is proportional to A as long as the area of overlap is not too small. Why do you think the experiment produced this result? If the experimenters had used pairs of plates of different areas, would they have been more likely to find Q proportional to A over a bigger range of values of A?
b Figure 16.34c shows that Q is proportional to $1/d$ for small values of d. When the separation becomes large, the charge becomes constant, regardless of the separation. Why do you think this might be?

19 a A pair of parallel plates are connected to a 10 V d.c. supply. The plates are 250 mm × 250 mm and are separated by a distance of 2 mm.
  i Estimate the electric field E between the plates ($\leftarrow$ 16.1).
  ii Calculate the charge stored on each plate.
  iii Calculate the capacitance of the plates.
b The separation of the plates is doubled to 4 mm, keeping the battery supply connected. What effect will this have on your answers to **a**?
c The battery supply is disconnected and the plates are then moved back to a 2 mm separation. Calculate the values of the electric field between the plates, the charge and the capacitance.
d How does the energy stored in the system in situation **b** compare with that in **c**?

THE MOTION OF A CHARGED PARTICLE IN AN ELECTRIC FIELD

A charged particle of mass m and positive charge Q in an electric field **E** experiences a force $Q\mathbf{E}$ in the direction of the field, so:

$$F_E = Q\mathbf{E}.$$

If the charged particle is initially at rest, it will accelerate uniformly in a straight line in the direction of the field. A particle moving initially at right angles to the direction of a field, and with no other forces acting on it, will follow a parabolic path. In either case, using Newton's second law of motion ($\leftarrow$ 6.4), the particle accelerates uniformly in the direction of **E** with an acceleration a given by:

$$a = \frac{Q\mathbf{E}}{m} \quad (1).$$

When the charge is negative, as it is for an electron, the acceleration takes place in the opposite direction.

FORCES AND FIELDS

Parabolic path of an electron in an E-field

Using Newton's equations for uniformly accelerated motion (← 6.4), we can find the x and y coordinates of an electron's path, shown in Figure 16.8. In the x-direction, the electron's initial speed v is unchanged as there is no force in this direction, so

and $\quad \begin{aligned} x &= vt \\ t &= x/v \end{aligned} \quad (2).$

In the y-direction, the electron starts from rest and accelerates uniformly so that at time t after entering the E-field region it will have moved a distance y given by:

$$y = \tfrac{1}{2} at^2 \quad (3).$$

Eliminating t and a between equations (1), (2) and (3) and replacing Q by the electronic charge e and m by its mass m_e, gives:

$$\begin{aligned} y &= \tfrac{1}{2}(e\mathbf{E}/m_e)(x/v)^2 \\ &= \frac{e\mathbf{E}}{m_e v^2} \cdot x^2. \end{aligned}$$

This has the form $y = k.x^2$, which is the equation for a parabola.

SUMMARY

Electric fields arise from charges. The direction of an electric field is the direction of the force on a positive charge placed in the field. The magnitude of an electric field is equal to the force on a unit charge placed in that field.

Electric field may be measured in $N\,C^{-1}$. The strength of an electric field can also be expressed in terms of the potential gradient and is then measured in $V\,m^{-1}$.

Coulomb's law states that the force F between two point charges Q and q, separated by a distance r, is given by:

$$F = \frac{Qq}{4\pi\epsilon r^2}.$$

The permittivity ϵ of a medium is equal to the product of its relative permittivity ϵ_r and the permittivity of free space ϵ_0:

$$\epsilon = \epsilon_r \epsilon_0.$$

Field E near a point charge Q, distance r from its centre, is given by:

$$E = \frac{Q}{4\pi\epsilon r^2}.$$

Potential V_E near a charge Q distance r from its centre is given by:

$$V_E = \frac{Q}{4\pi\epsilon r}.$$

If there is an electrical potential V_E at a distance r from the centre of charge, then V_E joules of energy will be transferred as 1 coulomb of charge is brought from infinity to that distance r.

Capacitance C of a conducting sphere of radius r is given by:

$$C = 4\pi\epsilon r.$$

Capacitance C of a parallel plate capacitor of plate area A and plate separation d is given by:

$$C = \frac{\epsilon A}{d}.$$

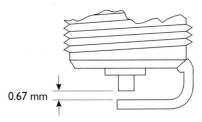

0.67 mm

Figure 16.35

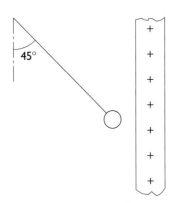

Figure 16.36

SUMMARY QUESTIONS

20 a In the sparking plug of a motor car engine there are two electrodes separated by a spacing of about 0.67 mm, as in Figure 16.35. If air begins to ionise when the electric field is about 3×10^6 volts per metre (at atmospheric pressure), roughly what p.d. must be applied across the electrodes to cause a spark in the air?
b The pd needed to cause a spark will depend on the gas pressure. Explain what effect you think a change of pressure will have on this pd. Is the gas pressure in a motor car engine greater or less than atmospheric pressure when the spark is needed?

21 A small, charged ball weighing $10^{-3}\,N$ is attracted to a charged plate as shown in Figure 16.36, and hangs at 45° to the vertical.
a Draw a diagram showing all the forces acting on the ball.
b Deduce the size of the electric force on the ball.

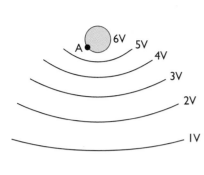

Figure 16.37

c If the electric field strength near the plate is 105 V m⁻¹, calculate the size and sign of the charge on the ball.

22 Figure 16.37 shows a full scale section of equipotentials at 1 V intervals between two conductors.
   On a tracing of Figure 16.37, starting from A on the upper conductor, construct a field line until it terminates on the lower conductor. By taking measurements, plot a graph of the variation of electric potential with distance from A along this line. How can the electric field strength at a point be deduced from this graph?

23 Figure 16.38 shows equipotentials drawn on a two-dimensional model representing a thundercloud over a churchyard. Assuming the charge is evenly spread over the base of the cloud, sketch lines of electric field between the cloud and earth on a copy of this diagram.
   Comment on the safety of the following locations during a thunderstorm:
   a on top of the church spire.   **c** leaning against the tree.
   b inside the church.   **d** in the car.

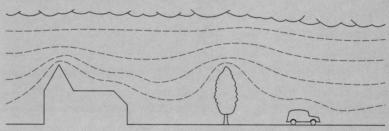

Figure 16.38

24 A 1 μF capacitor is to be made as follows. Long, 50 mm wide strips of thin metal foil, B and D, and of insulating paper 0.1 mm thick, A and C, are arranged in a sandwich as shown in Figure 16.39, and then rolled up to make a cylinder. The relative permittivity of the paper is 2.

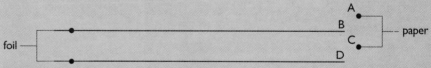

Figure 16.39

a At what point in the manufacture does the need for the top sheet of paper, A, become obvious?
b About how long a sandwich would be needed to get a final capacitance of 1 μF?
   Figure 16.40 shows a small area of the rolled up cylinder in cross-section. Now as well as capacitance between 'plates' B_1 and D_1 of the same layer, there is also capacitance between 'plates' D_1 and B_2 of adjacent layers, allowing extra charge to be stored on the foils for the same pd.
c By roughly what factor will the charge stored increase when the cylinder is completely rolled up?
d How does this affect your answer to **b** – the total length needed?
e Estimate roughly the diameter of the rolled up cylinder. You might then compare its size with that of a commercial 'paper' capacitor of about the same value.

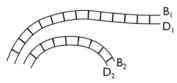

Figure 16.40

25 Figure 16.41a shows a section through a capacitor microphone; Figure 16.42b shows a circuit with which the microphone is used.

a The switch is initially closed. If it is opened, and the diaphragm pushed slightly inwards, explain what would happen to
  i the capacitance of the microphone, and
  ii the pd between B and C.

b Explain what would happen if, with the diaphragm still pushed in, the switch were closed.

c Why is the instrument constructed so that the diaphragm is as close to the first plate as possible?

d What is the time constant of the circuit in Figure 16.42b?

e Assuming that the switch is closed, state the changes of pd between B and C that you would expect to occur if a compression wave moved the diaphragm inwards in a time which was:
  i short compared with the time constant of the circuit – say about 10^{-5} s.
  ii long compared with the time constant of the circuit – say about 1 s.

f Two sources of sound, one of frequency 10 kHz, the other 50 Hz, are each found to produce the same amplitude of mechanical vibration in the diaphragm.
  i Why is the amplitude of the resulting variations of pd across BC smaller for the 50 Hz vibrations than for the 10 kHz?
  ii Explain what change you could make in the circuit to bring the amplitude of the electrical output from the microphone, when responding to the 50 Hz note, nearer to that produced by the 10 kHz note.

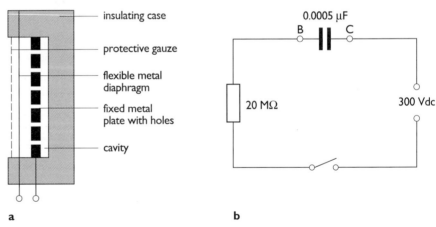

a   **b**

Figure 16.41

17 MAGNETIC FIELDS

You have probably known from an early age that magnets attract pieces of iron, but what causes this attraction? What is the nature of a magnetic field?

Hans Oersted and André Ampère (see Figure 17.1) investigated the magnetic effect of an electric current in the early part of the nineteenth century. Their work provided the basis for our present understanding of the role of electricity within magnetism. Magnetic fields, and the interactions between them and electric currents, give rise to effects that are truly amazing. These effects have been revolutionary in electric motors, meters and electrical technology generally.

Figure 17.1 **a** Ampère and **b** Oersted

17.1 MAGNETS AND MAGNETIC FORCES

The magnetic properties of some materials were known to the ancient Greeks, who found that a type of iron ore called lodestone, shown in Figure 17.2, was strongly magnetic and attracted other pieces of iron. They also found that a piece of lodestone pivoted about its centre always turned until it pointed north-south. This formed the basis of the magnetic compass, which has been used by Chinese, Arabs and Europeans since about 1500 BC. The effect led to the useful concept of a **magnetic pole**, the end of the lodestone or magnet pointing north being called the north-seeking or N pole. It was noticed that like magnetic poles repel one another and unlike poles are attracted.

D FORCES AND FIELDS

Figure 17.2 A piece of lodestone

Until the early part of the nineteenth century, magnetic forces were regarded separately from gravitational and electric forces. They were seen to be generally very much larger than gravitational forces on lumps of matter and, unlike electric forces, seemed to be a fundamental property of some types of matter.

Following the inverse square laws of gravitational and electric forces put forward by Newton and Coulomb, scientists looked for an inverse square law of force due to magnetism. A problem was that they could not completely isolate the north and south poles of a magnet. Breaking a magnet in half did not produce a north and a south pole; it simply produced two magnets, each with their own north and south poles. However, scientists found that when a pair of **long** magnets were brought together, end to end, the force between them did follow an inverse square law. The effect of the other poles was negligible, because they were far away from the poles under investigation.

Question

1 When you bring two **short** bar magnets towards each other, end to end, the force between them increases rapidly as they approach each other. In fact, the force is proportional to $1/\text{separation}^4$ in this situation. Why do you think this might be?

Figure 17.3 The GM Sunraycer electric car

Permanent magnets

In many aspects of physics and, indeed, in everyday life, there is a need for good permanent magnets. Devices using permanent magnetic materials range from door strips in refrigerators and small powerful motors in the motor manufacturing industry to body scanners in medical diagnosis, which require large regions of uniform field.

The materials used to make permanent magnets have improved considerably over the past century. The term **energy product** is used to describe the effectiveness of a magnetic material. This represents the energy transferred per m^3 in the magnetisation process, and is measured in $J\,m^{-3}$. The carbon-steels used at the turn of the century have energy products of less than $10\,kJ\,m^{-3}$. The addition of cobalt, and later aluminium and nickel, improved the energy product by a factor of about four in the 1930s. Samarium-cobalt compounds were introduced about 20 years ago. They possess energy products more than five times larger than for Fe-Ni-Al-Co compounds, but are expensive. More recently, in 1983, it was found that compounds of neodymium-iron-boron, which could be produced at relatively low cost, gave another significant rise in energy product of the order of $300\,kJ\,m^{-3}$. Figure 17.3 shows the GM Sunraycer, which uses Nd-Fe-B magnets in a brushless motor.

Another class of materials, ferrites, became available in the early 1950s. Ferrite powder can be incorporated into rubber or plastic, and can easily be bonded to create magnets of any desired shape. The material does not conduct electricity, an advantage in a number of applications. **Magnadur** magnets, often used in laboratory experiments, are based on ferrites. A pair of these magnets with adjacent poles facing each other produce a reasonably uniform field over a region of around $100\,cm^3$ of space.

17.2 MAGNETIC FIELDS AROUND CURRENTS

The link between magnetism and electricity was established in 1819, when Oersted brought a magnetic compass needle close to a wire carrying an electric

17 MAGNETIC FIELDS

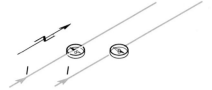

Figure 17.4 Oersted's experiment

current. He found that with the wire placed in a north-south direction and the needle placed over the wire, the needle swung at right angles to the wire when the current was switched on. His experimental arrangement is shown in Figure 17.4. Reversing the current direction, or placing the compass underneath the wire, caused the needle to deflect in the opposite direction.

> **Question**
>
> 2 Oersted carried out the experiment on the compass needle shown in Figure 17.4 on several occasions. However, he only noticed an effect due to the electric current during his later observations. Why was this?

Immediately following Oersted's discovery, Ampère showed that placing two parallel wires, each carrying a current, close together caused a force to be exerted between them. This was an attractive force when the currents were in the same direction and a repelling one when the currents passed in opposite directions.

The effect on the magnetised compass needle in Oersted's experiment illustrated the link between magnetism and electricity. Ampère showed the magnetic effect could be derived from electric currents on their own. His experiment forms the basis of the present SI definition of the ampere ($\rightarrow$ 18.3).

EXPERIMENT 17.1 The shapes of magnetic fields around conductors carrying a current

By setting up the arrangement in Figure 17.5a, using about half a metre of PVC covered copper wire, about 0.5 mm diameter, you can find the shape of the magnetic field around a straight wire.

To find the shape of the magnetic field around a coil, construct a flat coil of about ten turns on white card (Figure 17.5b). Remember that the Earth's field will modify the strength and direction of the field produced.

The magnetic field inside a solenoid can be found by making up a helical coil of about eight or ten turns on a card as shown in Figure 17.5c. Note the similarities between the field patterns created by this coil and those of a bar magnet.

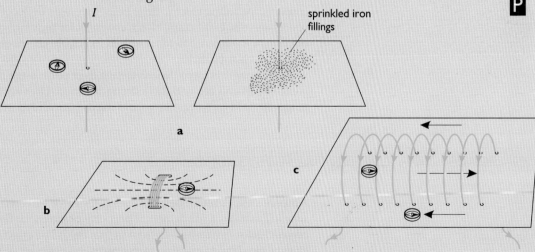

Figure 17.5 Finding the shape of the field around **a** a straight wire carrying a current, **b** a flat coil, **c** a solenoid

D FORCES AND FIELDS

MAGNETIC FIELD PATTERNS AND DIRECTION

Experiment 17.1 shows that, for a long straight wire carrying a current, the magnetic field 'circulates' around the wire becoming weaker the further away from the wire (→ 17.4). The following is a useful **rule for finding the field direction** of a given current. Hold the conductor in your right hand with the thumb pointing along the wire in the same direction as the conventional current flow. The manner in which the fingers curl around the wire indicates the field direction, as in Figure 17.6.

Using the shape of the field produced by a straight wire, it is a simple matter to predict the field around a flat coil, shown in Figure 17.7a, and the field in and around a solenoid, shown in Figure 17.7b. When carrying a current, a solenoid has a field pattern very much like that produced by a bar magnet. For a given current direction, you can predict which end of the solenoid behaves as a north- and which as a south-seeking pole, as in Figure 17.7c.

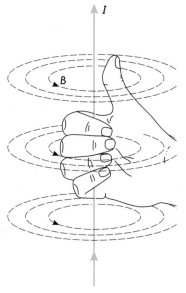

Figure 17.6 Field direction around a current

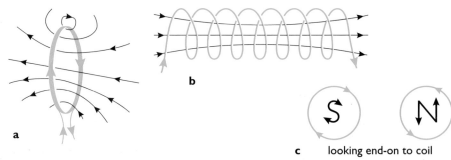

Figure 17.7 Field around **a** a flat coil, **b** a solenoid, **c** north and south poles

ELECTROMAGNETS

The strength of the magnetic field in Figure 17.5c can be enhanced considerably by winding the coil on to a rod of soft iron. The presence of the iron dramatically increases the magnetic flux in the system (→ 18). Other factors remain unchanged and so electromagnetic devices, such as motors, generators, transformers and relays, tend to be quite massive.

Electromagnets are made of coils of perhaps many hundreds or thousands of turns of copper wire wound on a soft-iron core. They are used in a variety of contexts, including electric door bells, magnetic deflection systems of mass spectrometers, giant crane electromagnets for moving scrap cars, and the production of the intense fields required in bubble chambers to investigate high-energy nuclear particles.

Electromagnets have the advantage over permanent magnets that they can be switched on and off, and their strength varied. However, large rates of change of magnetic field can create large and maybe unwanted induced voltages (→ 18.1).

17.3 FORCE ON A CURRENT IN A MAGNETIC FIELD

Ampère's experiment, demonstrating the force between two current-carrying wires, illustrates a fundamental property of magnetic fields: they can exert forces on conductors carrying a current in that field. The effect is analogous to that of gravitational fields exerting forces on masses or electric fields exerting forces on charges.

Suppose you place a wire in a steady and uniform magnetic field, provided by, say, two magnadur magnets, as shown in Figure 17.8a. When a current passes

17 MAGNETIC FIELDS

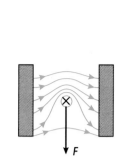

Figure 17.8 Poles of an electromagnet **a** A current-carrying conductor in a magnetic field **b** end view of current showing the field direction **c** end view showing force on conductor due to catapult field

through the wire, it produces its own field which circulates around the wire, as shown in Figure 17.8b. This field interacts with the field due to the magnets to produce what is sometimes called a **catapult field**. Above the wire, the fields are in the same direction and reinforce each other but below it, the fields cancel the effect of each other. The wire carrying the current experiences a downward force at right angles to both the steady field in which the wire is placed and the current flowing in the wire (see Figure 17.8c).

THE DIRECTION OF THE FORCE

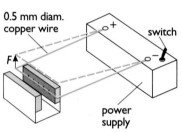

Figure 17.9 Arrangement to observe the force on a wire carrying a current in a magnetic field

To see the direction of the force due to the interaction of a magnetic field with an electric current, connect a length of copper wire of about 0.5 mm diameter across the terminals of a low voltage dc power supply that is capable of delivering a high current. Place a pair of strong magnets, such as magnadur magnets, around the wire so that the wire is in a steady field, as in Figure 17.9. In this arrangement, the wire deflects upwards when the current is switched on and the wire returns to its equilibrium position when the current is switched off.

It is sometimes convenient, and certainly quicker, to use the following rule when you want to find the direction of the force on a current in a magnetic field.

Fleming's left-hand rule

Place the thumb, first finger and second finger of the left hand at right angles to each other, as shown in Figure 17.10. The current flow is represented by the second finger, the direction of the field in which it is placed is represented by the first finger and the direction of the force on the wire is represented by the thumb.

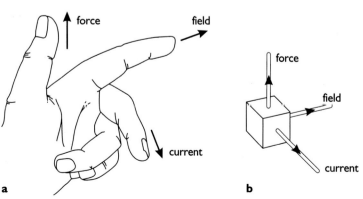

Figure 17.10 Left-hand rule

FORCES AND FIELDS

The usual conventions of magnetic poles (N = north-seeking) and current (conventional, opposite to electron flow) apply.

MAGNITUDE OF THE FORCE

More quantitative measurements can be obtained relating to the force on a conductor carrying a current in a magnetic field using a **current balance**. One type consists of a rectangular loop of wire pivoted on a support formed by two razor blades, as shown in Figure 17.11. A force F is exerted on the current in the arm AB of the balance when it is in a horizontal field. The magnitude of the force can be estimated by attaching to the arm AB small known masses, such as short lengths of small diameter wire, until they counteract the effect of the magnetic force and restore the balance.

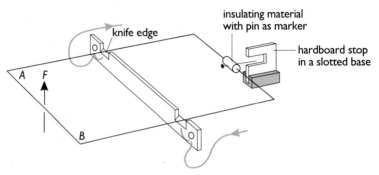

Figure 17.11 A current balance

Question

3 a Which of the situations in Figure 17.12 will cause the arm of the current balance to be deflected?
b Is there any force on the wires supplying a current to the arm AB of the balance? Bear in mind that the fields, in each case, are not always uniform over the whole region occupied by the balance.

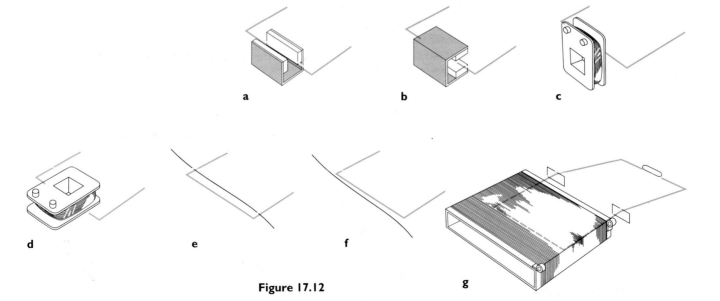

Figure 17.12

17 MAGNETIC FIELDS

EXPERIMENT 17.2 Force on a wire carrying a current in a magnetic field

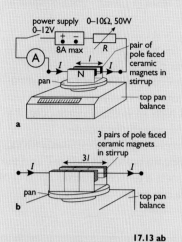

a Effect of changing current I

Using the apparatus shown in Figure 17.13a, you can record the current flowing in the wire. Repeat this with two, three and four weights, recording the value of the current I in each case.

WARNING: ensure that the current does not become so large that it over heats the resistance and wire in the circuit.

b Effect of changing length *l* of conductor

Keeping the current fixed you should add extra pairs of magnets of the same orientation as shown in Figure 17.13b so doubling and trebling the length of conductor in the field. In each case you need to record the force indicated on the top pan balance.

How could you find the effect on the force of orientation, or angle θ of the conductor in the magnetic field?

Figure 17.13 Apparatus for experiment 17.2

STRENGTH OF A MAGNETIC FIELD

Experiment 17.2 shows that the force F on a current placed at right angles to a magnetic field is dependent upon the magnitude of the current I and the length l of conductor in the field, for a given magnetic field strength. Careful measurements in such experiments show that:

$$F \propto I$$
$$F \propto l.$$

Placing the conductor at various angles θ, as shown in Figure 17.14, we find that:

$$F \propto \sin\theta$$

combining all three proportionalities gives:

$$F \propto Il.\sin\theta.$$

So $\dfrac{F}{Il.\sin\theta}$ = a constant.

This constant is called the **magnetic flux density B**, which is effectively the **strength of the magnetic field**. The equation

$$F = BIl.\sin\theta$$

is used to define the strength of a magnetic field.

When a conductor is at right angles to the magnetic field, $\sin\theta = 1$ and

$$B = \dfrac{F}{Il}.$$

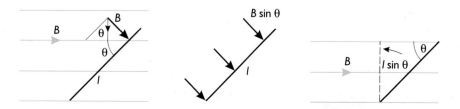

Figure 17.14 Effect of orientation

FORCES AND FIELDS

So, the magnetic field strength or flux density B can be defined as the force on a wire, placed at right angles to the magnetic field per metre of length and per ampere of current in the wire. The unit of B is the **tesla** (T), and 1 tesla is equal to $1\,\text{N}\,\text{A}^{-1}\,\text{m}^{-1}$.

Because B is the **force** on unit current and length, it is a **vector** quantity. It is similar to gravitational field strength g, which is the force per unit mass, and electric field strength E, which is the force per unit charge. For this reason we often refer to a magnetic field as a **B-field**. The direction of a **B**-field at a particular point is the same as the direction in which a compass needle points, when placed there. More strictly, it is the direction in which an isolated north pole of a magnet would move – assuming that a pole could actually be isolated.

The tesla is a large unit, so you will normally express fields in millitesla (mT). For example, the strength of the Earth's field in Britain is around 0.07 mT and the field between a pair of attracting magnadur magnets a few centimetres apart is of the order of a millitesla. **B**-fields of around 1 tesla can be produced by some magnetic alloys.

> ## Questions
>
> **4** A wire of length 80 mm and carrying a current of 5 A is placed at right angles to a steady magnetic field of 0.02 T. Calculate the force that acts on the wire and sketch a diagram showing its direction.
>
> **5 a** The Earth's magnetic field in the British Isles has a vertical component of 4.0×10^{-5} T and a horizontal component of 1.8×10^{-5} T. Show that the Earth's field is directed at an angle of about 66° to the horizontal. Calculate the strength of the **B**-field.
> **b** A conductor of length 0.5 m carries a current of 10 A.
>   **i** Sketch a diagram showing how it should be placed in the Earth's field for maximum force to be exerted.
>   **ii** In which direction would that maximum force act?
>   **iii** Could you detect the vertical component of this force using, say, a top-pan balance?

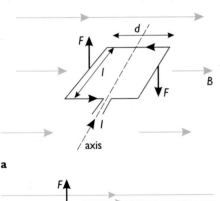

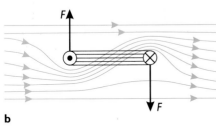

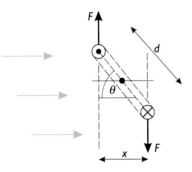

Figure 17.15 a Coil carrying a current I in a uniform **B**-field **b** Sectional view of field pattern around the coil **c** Angle θ between the plane of the coil and the field

TORQUE ON A COIL IN A MAGNETIC FIELD

Consider a rectangular coil carrying a current I in a uniform **B**-field, as in Figure 17.15a. There is an upward force on the left-hand arm of the coil and a downward force on the other tending to rotate the coil about its axis. Figure 17.15b shows a sectional view of the field pattern around the coil.

The force on each arm of the coil remains constant irrespective of the orientation of the coil to the field. Therefore, the couple ($\leftarrow$ B) on the coil is the product of the force F and the perpendicular distance x between lines of action of the forces. This decreases as the angle θ between the plane of the coil and the field increases, (see Figure 17.15c), and:

$$x = d \sin \theta$$

where d is the width of the coil.

Each arm of the coil has a length l and the force F on each arm is BIl. The resulting couple produces a torque T, which is equal to Fx or $BIlx$. So, the torque T on the single turn coil is given by:

$$T = BIld \sin \theta.$$

17 MAGNETIC FIELDS

The product ld is the area A of the coil. If the coil has N turns, the torque will be N times as great, and so:

$T = BANI \sin \theta$.

As it turns, the coil will experience a decreasing torque until its plane is at 90° to the **B**-field, when it will be zero.

The torque on a coil is used to good effect in moving coil meters and electric motors.

MOVING COIL METER

At the heart of a moving coil meter is a coil which rotates against a hair spring (see Figure 17.16a). This provides an opposite torque to the one provided by the unknown current through the coil. The torque due to the current causes the coil to rotate until it moves to a position where the coil is in equilibrium and the two torques are equal and opposite.

The torque due to the hair spring is proportional to the angle of deflection, assuming it is an elastic spring obeying Hooke's Law. However, in a uniform **B**-field, the torque provided by the current decreases as the coil rotates, giving a non-linear relationship between current and coil deflection. To overcome this problem, a **radial field** is used, as shown in Figure 17.16b. This produces a force that is always at right angles to the side of the coil and results in a steady couple of $BANI$.

The angle of deflection of the coil to the radial field is, therefore, proportional to the current in the coil – a feature of moving coil meters.

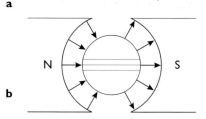

Figure 17.16 **a** Moving coil meter movement, **b** radial field in the moving coil meter

DC MOTOR

To enable the coil in a dc motor to rotate continuously, a system called a **commutator** is used (see Figure 17.17a). This both allows current to be delivered to a rotating coil and reverses the current in the coil every half cycle of rotation of the coil. Figure 17.17b shows the current being supplied to the coil through brushes in contact with a split ring commutator. This reverses the current through each side of the coil, so ensuring that the forces on the coil maintain the direction of the turning effect as the coil passes the vertical.

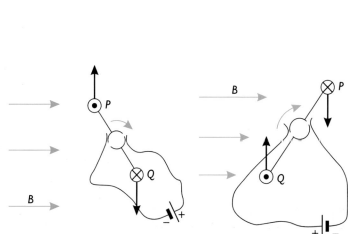

Figure 17.17 **a** Commutator system in a small dc motor **b**

FORCES AND FIELDS

Figure 17.18 Commutator of a car starter motor

A series of coils are used in commercially manufactured motors, such as that in Figure 17.18. So many pairs of contacts are required in the commutator, the two brushes at any given time making contact with just one pair of terminals. This provides greater torque and smoother running.

> **Question**
>
> **6** Explain in detail the currents and the forces acting on the armature of a simple motor, similar to the one shown in Figure 17.17, during a complete cycle of its rotation.

17.4 FORCE ON A MOVING CHARGE IN A MAGNETIC FIELD

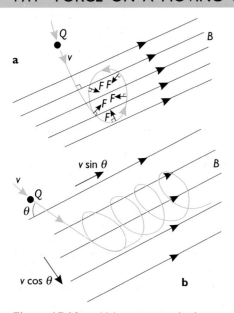

Figure 17.19 a Velocity perpendicular to the magnetic field – particle moves in a circle
b Velocity at angle θ to the magnetic field – particle moves in a spiral

When electrons pass along a wire and the wire is placed at right angles to a magnetic field, a force is exerted on the wire (← 17.2). What happens when the electrons are not confined within a wire but move along in a stream from an electron gun, as in a cathode ray tube (→ 35.6)?

In general, when a particle carrying a charge Q is moving at a speed v at right angles to a magnetic field B, there will be a force F on the particle given by:

$$F = BQv.$$

The direction of F is perpendicular to B and v, and can be found using the left-hand rule (← Figure 17.10). When electrons are being considered, it is important to remember that the direction of the current flow is opposite to that in which the electrons are travelling since they carry negative charge.

The force F causes the particle to follow a circular path and is thus a centripetal force (see Figure 17.19a). For a particle of mass m and path of radius r:

centripetal force, $F = BQv$,

so $\dfrac{mv^2}{r} = BQv.$

When the particle is travelling at angle θ to the direction of the **B**-field, only the component of the velocity perpendicular to the field, $v \sin θ$, causes a centripetal

Calculating the force on a moving charge in a magnetic field

You can use $F = BIl$ to find the force on a stream of charges moving at right angles to a magnetic field B. We relate the current in the stream to the flow of individual charges and the length of the stream to their speed.

Consider Figure 17.20 showing particles of charge Q, travelling at speed v within the cylinder XY. Suppose that in time Δt, all the particles leave and are replaced by more particles following on behind them in the stream. The length l of this cylinder of charge is given by:

$l = v.\Delta t$.

Suppose the number of charged particles within the cylinder is N. The current I is the total charge flowing per second in the stream, so:

$I = \dfrac{NQ}{\Delta t}$.

Applying $F = BIl$ to the stream and replacing l and I gives:

$F = \dfrac{B.NQ.v\Delta t}{\Delta t}$

or $\quad F = BNQv$.

This is the total force on N particles within the cylinder XY. The force on a single charge Q moving at a speed v at right angles to a **B**-field is therefore given by

$F = BQv$.

When the charge is moving at an angle θ to the field, the equation becomes

$F = BQv. \sin \theta$.

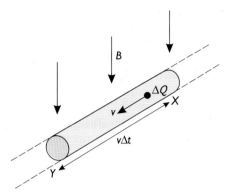

Figure 17.20 Section of wire in a magnetic field

force. The other component, $v \cos \theta$, is in the direction of the **B**-field. So, the particle follows a spiral path, as shown in Figure 17.19b.

The interaction between a magnetic field and moving charged particles produces many interesting effects, such as the auroras of the Northern and Southern Lights ($\to$ 35.4). It also has many applications such as the magnetic deflection of the electron beam in a television tube ($\to$ 35.6).

THE MOTION OF AN ELECTRON IN A MAGNETIC FIELD

As shown in Figure 17.21, the force on a charged particle Q moving in a **B**-field is:

$F_B = BQv \sin \theta$.

The force on a beam of electrons is opposite in direction, of course. This is due to the negative charge on an electron $e = -1.602 \times 10^{-19}$ C.

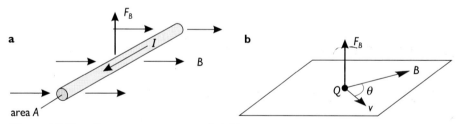

Figure 17.21a Force on a current in a **B**-field
b Force on charge Q moving with velocity v at θ to a **B**-field

Questions

7 a How fast is an electron travelling when it leaves an electron gun, having been accelerated through 50 V? (e, m_e)
b Calculate the force acting on the electron as it travels at right angles to a uniform **B**-field of 5×10^{-4} T.
c Show that
 i the electron travels in a circular path; draw a sketch to help your explanation
 ii the radius of that path is about 50 mm.

8 In one particular proton synchrotron ($\to$ 40.4), 4 MeV protons are 'injected' into a **B**-field so that the beam follows a circular path of radius 10 m. Calculate the strength of the **B**-field in tesla. (1e, mp)

HALL EFFECT

Consider a piece of conducting material, such as copper, connected into a circuit so that a steady current flows through it, as shown in Figure 17.22a. Charges, such as electrons, drift through the conductor at an average speed v. Suppose that a **B**-field is applied to the slice of conductor, normal to its flat surface, as shown in Figure 17.22b. Each electron in the magnetic field will experience a force F_B equal to Bev causing it to deflect towards the back of the slice. The left-hand rule gives the direction of the force, taking conventional current direction.

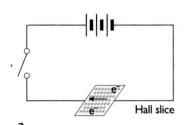
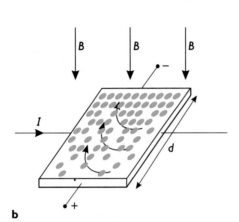

Figure 17.22 Hall effect
a a slice of conductor in a circuit
b applying a **B**-field to the slice
c a Hall probe

Equilibrium occurs when the accumulation of negative charge at the back of the slice repels the arrival of further electrons, which have built up the **Hall voltage V_H** across the slice. There then exists an electric field E equal to V_H/d, where d is the width of the slice. The force on the electrons in this field is:

$$F_E = eE$$
$$= eV_H/d.$$

In equilibrium:

$$F_E = F_B$$

or

$$\frac{eV_H}{d} = Bev,$$

giving $V_H = Bvd$.

Reversing the field direction causes the polarity of the Hall voltage to change. The magnitude of the Hall voltage is directly proportional to the applied **B**-field for a given conducting material carrying a given current. This forms the basis of a useful method for measuring **B**-field. To measure field values directly, the Hall circuit and probe must be calibrated beforehand, for example, using a current balance in a given field.

In an ordinary metallic conductor, there is a high density of electrons available for conduction, but the drift velocity of electrons is very small indeed ($\leftarrow$ 12.1). The link between current and drift velocity is given by $I = nAev$. So, the Hall voltage is difficult to measure unless high currents and large **B**-fields are used.

In p- and n-type semiconductors, fewer charges are available for conduction but charge carrier drift velocities are very much larger. A slice of p- or n-type material can therefore be used in place of metal. The Hall voltage V_H may then be measured using a mirror galvanometer or a digital millivoltmeter. The slice can be mounted on the end of a plastic probe and then used to sense the **B**-field in different situations, as shown in Figure 17.22c.

Hall effect in slices of p- and n-type semiconductors

In *n*-type material, electrons are mainly responsible for electrical conduction. Pure germanium does not conduct very well because it has few free electrons to contribute to conduction. 'Doping' it with antimony provides extra electrons and creates *n*-type germanium. In *p*-type material, holes, which can be thought of as positive charge carriers, are mainly responsible for conduction. Doping germanium with indium atoms creates *p*-type germanium (→ 27.6).

In Figure 17.23a, a **B**-field applied across a slice of *n*-type material causes electrons to accumulate at the back edge of the slice. This leaves the front edge at a more positive potential than the back edge. In Figure 17.23b holes move towards the back edge of the *p*-type slice creating a Hall voltage opposite in polarity to that in *n*-type material. The current and field directions remain unchanged.

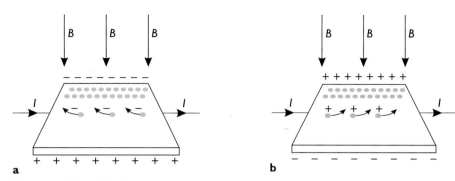

Figure 17.23 Hall effect in **a** *n*-type material **b** *p*-type material

Question

9 The Hall voltage V_H is given by Bvd, and the current I through a conductor is $nAQv$.

a Show that $V_H = \dfrac{BId}{nAQ}$.

b Why is the Hall voltage greatest in materials with a low density of charge carriers n, other factors being equal?

c Why is the Hall voltage greater the thinner the slice? Bear in mind that the area of the slice = width × thickness.

EXPERIMENT 17.3 The Hall effect

The circuit shown in Figure 17.24 contains a ready prepared slice of *n*-type material with a millivoltmeter connected to the Hall voltage contacts. Adjusting the potentiometer so that, on the most sensitive range of the meter a reading of zero volts is obtained, brings the two Hall contacts to the same potential in the absence of a magnetic field. You can then try to explain the effects of

- bringing the south pole and then the north pole of a magnet up to the slice,
- applying a stronger field by placing magnets on both sides of the slice,
- repeating the experiment using a prepared slice of *p*-type germanium.

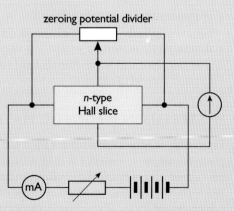

Figure 17.24 Circuit for experiment 17.3

FORCES AND FIELDS

> *Regulating the speed of electric motors*
>
> Motor control circuits regulate the speed of electric motors extremely accurately. A Hall effect generator, fitted to the motor, is used for monitoring the speed of the motor's shaft. This information is passed to the feedback circuit, which adjusts the current supplied to the motor so as to keep the speed constant. Such motors are often expensive, so their use is confined to sound and picture recording and replay systems, where accurate speed control is vital.
>
> Practical, everyday applications for the Hall effect almost always make use of semiconductors in which the effect is greatest. The detector chip includes further circuits for amplifying the Hall voltage and translating the signal into a form suitable for the particular application.

SUSCEPTIBILITY

Imagine that you place a magnetic material into a field of magnetic flux density B_0. The result is an increased **B**-field of flux density B. The amount of the increase B', is caused by the magnetisation of the material so that:

$$B = B_0 + B'.$$

The increase, B', as a ratio to B_0 is called the susceptibility X:

$$X = B'/B_0.$$

Now, because the relative permeability μ_r of the material is B/B_0, you can see that:

$$\mu_r = B/B_0 = \frac{B_0 + B'}{B_0}$$
$$= 1 + X$$

where X is the **susceptibility** of the material. X is defined as B'/B_0 and is the value of the relative permeability μ_r in excess of 1.

Like μ_r, X is a pure number.

For **ferromagnetic materials**, like iron, it has a very large value, often as high as 10 000, (although this reduces as saturation is reached when all the domains are aligned with B_0). With such high values it is generally unnecessary to distinguish X from μ_r.

For a completely non-magnetic material, X would be zero. However, this is very rare. All materials are very slightly magnetic and fall into two classes:

- **paramagnetic materials** where X is very small and positive. Such materials behave similarly to extremely weak ferromagnetic materials.
- **diamagnetic materials** where X is very small and negative. Here the value of B' is negative and the materials tend to align themselves perpendicularly to the applied **B**-field, B_0.

HYSTERESIS IN MAGNETIC MATERIALS

Hysteresis is responsible for the increase in internal energy which occurs as a piece of magnetic material is magnetised first one way, then the other. In the process of magnetisation **domains** tend to line up with the applied magnetic field. These domains act like tiny magnets. They consist of tiny crystals within which closed loops of electron currents are maintained (see Figure 17.25). As they realign, energy is transferred to the atomic structure of the material. When the material is magnetised in the opposite direction, the domains tend to turn again

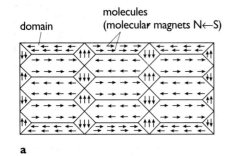
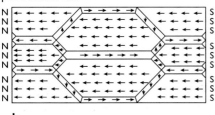

Figure 17.25 Magnetic domains in **a** unmagnetised, **b** magnetised ferromagnetic material

17 MAGNETIC FIELDS

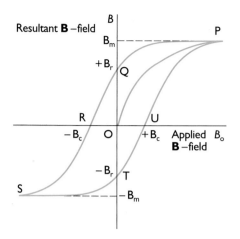

Figure 17.26 A Hysteresis Loop

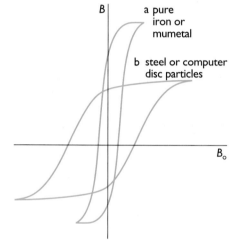

Figure 17.27 Hysteresis loops for
a pure iron or mumetal
b steel or computer disc particles

in that direction. So the internal energy, and therefore the temperature, of a specimen continues to increase as it is remagnetised in the opposite direction. This is particularly important when designing transformers for alternating current where the material of the core is remagnetised in opposite directions every 1/50 second.

The curve relating B and B_0 is shown in Figure 17.26. At point P, a large value of B is produced in the same direction as B_0. When B_0 drops to zero, B falls, slightly, to B_r as at point Q. This is referred to as the **remanence**.

To cancel this field, a reverse field B_c needs to be applied, to bring the material to point R. (B_c/μ_0 is known as the **coercivity** of the material.) In reality, it is quite difficult to apply exactly the right reverse field and the material tends to 'slip past' point R towards point S.

From this point, removal of the applied field and remagnetising the material in the opposite direction is the reverse of the above process but from S to P via T and U.

The area enclosed by the curve, known as a **hysteresis loop**, represents the energy needed to reverse the magnetisation of the material.

Hysteresis losses

Hysteresis of the core material is the main reason for energy loss in transformers. To cut down on hysteresis losses, the cores of transformers and other inductors are made of a special soft alloy called 'mumetal'. A hysteresis curve for this type of alloy is shown in Figure 17.27a. Since the area enclosed by the curve is small, the energy loss in reversing the domains is small.

For permanent magnets, and the magnetic particles in computer discs, audio and video tapes magnetisation needs to be retained. A hysteresis loop like the one shown in Figure 17.27b is required. The large area enclosed by this curve shows that a large amount of energy is needed to reverse the direction in which the domains are magnetised. If this energy is not available to the domains, the magnet will remain magnetised.

Question

10 a How does the domain model account for the properties of permanent magnets?
b How could you magnetise a bar of soft iron? How could you demagnetise it?
c Why are soft iron 'keepers' used to maintain the alignment of the domains and the strength of the magnet?

SUMMARY

Long ago, magnetic fields around permanent magnetic materials were seen to produce 'action at a distance' forces but these were thought to be quite distinct from those occurring in gravitational or electric fields.

Hans Oersted established the link between current electricity and magnetism when he observed that a magnetic field 'circulated' around a current flowing in a wire. This magnetic effect of an electric current is used in a variety of situations where control over the strength of a magnetic field is required.

Magnetic materials tend to retain their magnetisation due to hysteresis. When this effect is large, there will be a large energy change involved in reversing a magnetic field inside the material. This energy is represented by the area of the hysteresis loop.

A current *I* flowing in a wire of length *l* placed at right angles to a magnetic field *B* experiences a force *F*, where:

$F = BIl$.

The direction of the force is given by Fleming's left-hand rule. When the wire is inclined at an angle θ to the field, the force is given by:

$F = BIl \sin \theta$.

Magnetic field strength is measured in tesla (T); $1\,T = 1\,NA^{-1}\,m^{-1}$.

The force on a current carrying conductor is utilised in moving coil meters and motors.

A charge *Q* moving at speed *v* at right angles to a **B**-field experiences a force given by:

$F = BQv$.

The direction of the force is again driven by Fleming's left-hand rule, where the direction of *v* is the direction in which a positive charge moves.

The Hall effect arises when a **B**-field is applied across a conductor carrying a current. If the charge carriers move at a speed *v* in a slice of material of width *d*, the Hall voltage V_H will be given by:

$V_H = Bvd$.

The Hall voltage is easily measurable in *n*- or *p*-type semiconductor materials. The Hall effect in a small thin slice of semiconductor can be used to explore steady magnetic fields in various situations.

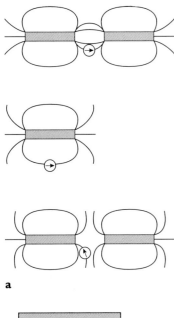

SUMMARY QUESTIONS

11 a Figure 17.28a shows field patterns around arrangements of permanent magnets and some field directions, indicated by a compass needle placed in various positions. Identify the poles on each magnet.
b Figure 17.28b shows the field between a pair of magnadur magnets. What is unusual about these magnets compared with the generally more common bar magnets?

12 Design a simple ammeter using a small plotting compass and a length of copper wire. What drawbacks would this ammeter have over more conventional types?

13 You have a strip of aluminium foil 50 mm by 10 mm by 0.1 mm lying on a bench, and a means of passing a steady current along the length of the strip.
a Calculate the strip's weight. (Density of aluminium = $2\,700\,kg\,m^{-3}$.)
b When a **B**-field of 0.05 T is applied, the foil just starts to lift off the bench. Calculate the current in the strip. Sketch a diagram to show the direction of the **B**-field.
c Calculate the value of the minimum current that would have to flow for the foil to be lifted in the Earth's magnetic field. Sketch a diagram to show the orientation of the current in the strip to the Earth's field. (The Earth's field strength = $4.5 \times 10^{-4}\,T$, dipping down towards the Earth's magnetic pole in the north at an angle of 66° below the horizontal.)

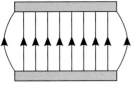

Figure 17.28

17 MAGNETIC FIELDS

14 An electron entering a magnetic field at right angles moves in a circular path. However, when it enters the field at any other angle, it moves in a spiral path around the field direction. Why is this?

15 When a Hall probe is placed with its thin germanium slice at right angles to a **B**-field, a Hall voltage of 100 mV is detected.
a The probe is adjusted so that the angle its plane makes with the **B**-field is 30°. What is the value of the Hall voltage now?
b The angle is changed to **i** 60° **ii** 0°. What is the value of the Hall voltage in each case?

16 This question illustrates why it is difficult to demonstrate the Hall effect in a piece of aluminium 50 mm long by 50 mm wide by 0.1 mm thick. (The mass number of aluminium is 27 and its density is 2 700 kg m^{-3}.)
a Given that there are 6.0×10^{23} atoms mol^{-1}, show that there are 6×10^{28} atoms per cubic metre.
b Using $I = nAQv$, estimate the drift speed v of charge carriers in the aluminium foil. Assume that one electron per atom is available to contribute to conduction and that a current of 5 A is flowing, as in Figure 17.29.
c The **B**-field across the slice is provided by an electromagnet and equals 0.1 T. Calculate the Hall voltage across the slice.

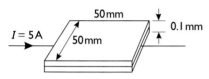

Figure 17.29

18 ELECTROMAGNETIC INDUCTION AND MAGNETIC FLUX

The technological society in which we live depends totally on the efficient generation and distribution of electricity. The principles of electricity generation were established by Michael Faraday (see Figure 18.1). In 1831, he succeeded in inducing a current to flow in one circuit by a change of current in another. Ideas about the nature of electromagnetic induction and magnetic flux have led to the design of more efficient electromagnetic machines, and are now of great importance to humankind.

Figure 18.1 Michael Faraday giving a Royal Institution children's lecture

18.1 ELECTROMAGNETIC INDUCTION

A conductor carrying a current at an angle to a magnetic field experiences a force at right angles to both the current and the field directions. Its direction is given by the left-hand rule (← 17.3).

Suppose that, instead of supplying a current, you move the conductor in the magnetic or **B**-field. You can investigate the effects quite easily using a strong permanent magnet and a connecting lead of about a metre in length attached across the input terminals of a sensitive centre zero galvanometer (see Figure 18.2). Moving the conductor parallel to the field in direction X, or 'through' the field in direction Z, produces no effect on the galvanometer. However, moving the conductor in direction Y produces a deflection on the galvanometer, but only while the conductor is moving. The emf, or voltage **induced** in the conductor

18 ELECTROMAGNETIC INDUCTION AND MAGNETIC FLUX

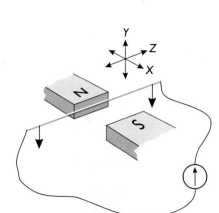

Figure 18.2 Moving a conductor in a magnetic field

drives a current through the galvanometer. The polarity of the induced voltage depends on whether the conductor is moved in the +Y or the −Y direction. Its magnitude depends upon the strength of the field and the speed at which the conductor is moved. In a conductor of length l moving at speed v perpendicularly to its length and a magnetic field of flux density B (← 17.3), the **induced voltage** v is given by:

$$V = Blv.$$

Question

1 Show that the quantity Blv can be measured in volts.

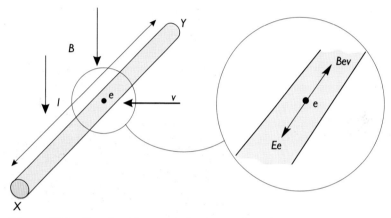

Figure 18.3 Electric and magnetic forces on a free electron

Origin and magnitude of induced voltages

The conductor in Figure 18.3 is moving through a **B**-field at speed v. A free electron e in the conductor will be travelling from right to left in the diagram, at right angles to the **B**-field. It will therefore experience a force tending to push it towards end Y of the rod. You can check this by applying the left-hand rule. Electrons, being negative, produce a current in the opposite direction to that of their travel.

The force arises for all free electrons in the rod, causing an accumulation of charge at the Y end of the rod. An electric field E is caused by this uneven distribution of charge. As a result, a force F_E acts on each charge. This is balanced by the magnetic force F_B on the moving charge. Note the similarity to the Hall effect (← 17.4). We can write:

$$F_E = F_B$$

or $$Ee = Bev.$$

Now $$E = \frac{V}{l},$$

giving $$V = Blv.$$

Moving the rod in the other direction changes the polarity of the induced voltage, while moving it faster increases the magnetic force and creates a bigger induced voltage.

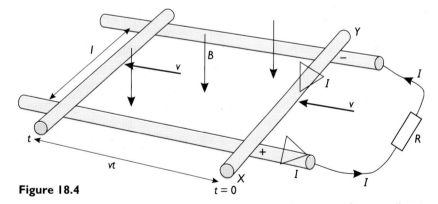

Figure 18.4

Questions

2 In this question, you can arrive at the relationship $V = Blv$ by considering energy transfer.

Suppose the conducting rod in Figure 18.3 moves at speed v along a pair of rails. The induced voltage V drives a current I through a circuit including the rod, connected to the pair of rails, as shown in Figure 18.4. In t seconds, the rod moves a distance vt.

a A current I is flowing in the rod at right angles to the **B**-field. Calculate the magnetic force on the rod. Using the left-hand rule (← 17.3), in which direction does it act?

FORCES AND FIELDS

Figure 18.5 Jumbo jet in level flight

b The rod is being moved against this force. Find an expression for the energy transferred by the work done in moving the rod in time t.
c Energy is transferred to the circuit as a result of the work done. How much energy, in the form of an expression, is transferred when an induced voltage V drives a current I for t seconds?
d Equate your answers to **b** and **c** to show that $V = Blv$.
e Note that the force $F = BIl$ opposes the motion. What is the effect on this force when the rod is
  i moved more quickly?
  ii moved more slowly?
  iii moved in the opposite direction at speed v?

3 Figure 18.5 shows a jumbo jet flying at cruising speed in level flight over the UK. Travelling due west, it is moving at right angles to the vertical component of the Earth's magnetic field. A voltage is induced across the wing tips.
a Could you demonstrate that this voltage exists? Could it be made use of?
b What is the value of the induced voltage? Estimate any quantities you think are necessary. (The **B**-field of the Earth in the UK is about 4.2×10^{-5} T orientated downwards towards the north magnetic pole at approximately 66° to the horizontal.)

4 The suspension of sensitive moving coil meters is often protected when the meter is in transit by short-circuiting the terminals. This may be done with a switch on the front panel of the meter, or often simply by connecting a copper wire across the meter terminals. Why does this offer protection and how does it work?

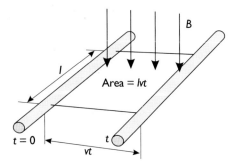

Figure 18.6 A conductor sweeps an area lvt in t seconds

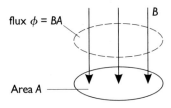

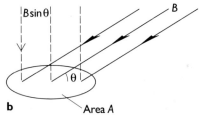

Figure 18.7a Flux, flux-density and area
b Effect of orientation

INDUCED VOLTAGE AND MAGNETIC FLUX

The concepts of **magnetic flux** ϕ ($\leftarrow$ 14.2) and flux density B are useful when considering the induced voltage arising in a conductor as it moves through a magnetic field.

During time t, the rod in Figure 18.6 sweeps out an area A perpendicular to the **B**-field, where

$$A = lvt.$$

However the induced voltage $V = Blv$, so

$$V = \frac{BA}{t}.$$

The quantity BA is equal to the magnetic flux, ϕ enclosed by the area A normal to the **B**-field (see Figure 18.7a):

$$\phi = BA.$$

So $\quad V = \dfrac{\phi}{t}.$

Here, the flux is assumed to be constant over time t. However, in general, it may be changing and, at any instant:

$$V = \frac{d\phi}{dt}.$$

This equation suggests, and experiment 18.1 can be used to confirm, that moving

18 ELECTROMAGNETIC INDUCTION AND MAGNETIC FLUX

Effect of orientation in a **B**-field

Suppose the **B**-field is inclined at an angle θ to area A, as in Figure 18.7b. The component of **B**-field normal to the area is $B \sin\theta$, and the equation $\phi = BA$ becomes:

$\phi = BA\sin\theta$.

a conductor through magnetic flux and changing the magnetic flux around the conductor, both induce a voltage. Induced voltages arising from conductors moving in magnetic fields form the basis of generators whilst those arising through the changing flux around a fixed conductor are at the heart of transformer action.

The SI unit for magnetic flux is the **weber (Wb)**, where 1 weber is the flux passing through an area of 1 square metre perpendicular to a **B**-field of 1 tesla. Because of the relationship $V = \phi/t$, the weber is sometimes referred to as the **volt-second**.

The strength of a magnetic field B is measured in **tesla (T)**. However, from the definition of the weber, B is also known as the flux density, with the unit of the **weber per m² (Wb m⁻²)**. So

$1\,\text{T} = 1\,\text{Wb m}^{-2}$.

Questions

5 The semiconductor slice of a Hall probe has dimensions 4 mm × 4 mm. Calculate the total flux passing through the slice when it is placed normally to a **B**-field of 0.01 T.

6 Calculate the total magnetic flux 'swept-out' per second by the aircraft wings in question **3**. Is there a relationship to the voltage?

EXPERIMENT 18.1 Observing induced voltages

The apparatus shown in Figure 18.8a enables you to measure the voltages induced as a conductor moves through a magnetic field.

Using a coil of 5 and then 10 turns wound around a matchbox, and connecting the ends to the galvanometer as shown in Figure 18.8b, you can investigate how the number of turns in a coil affects the induced voltage.

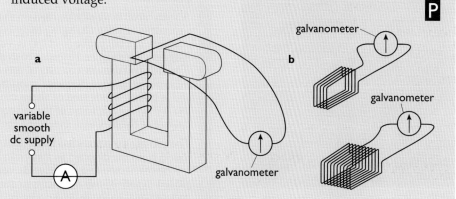

Figure 18.8a A 'demountable transformer kit', **b** coil and galvanometer

18.2 LAWS OF ELECTROMAGNETIC INDUCTION

Experiment 18.1 shows that the induced voltage in a coil is larger the more turns are used. The induced voltage per turn at any instant is given by the rate of change of flux ϕ through the coil, $d\phi/dt$. For a coil of N turns, the total flux linkage at any instant is $N\phi$, and so the induced voltage V is given by

$$V = \frac{d}{dt}(N\phi)$$

$$= N \cdot \frac{d\phi}{dt}.$$

This equation is a symbolic statement of **Faraday's law**, which states that the induced voltage is equal to the rate of change of flux linkage, that is, the total magnetic flux passing through the coil's turns.

> ### Questions
>
> 7 A rectangular coil of five turns has dimensions 50 mm by 20 mm. It is placed between the pole pieces of an electromagnet, as in experiment 18.1, and the field through the coil is increased steadily from zero to 0.1T in 10 seconds. Calculate:
> a the initial flux within the coil,
> b the flux passing through the coil after 10 seconds,
> c the induced voltage per turn of the coil,
> d the induced voltage across the ends of the coil,
> e the value of the induced voltage across the coil when the field is reduced steadily from 0.1 T to zero in 5 seconds,
> f the value of the induced voltage across the coil when the field is reduced to zero in 0.01 seconds.
>
> 8 Suppose the coil in question 7 is a circular one of diameter 20 mm and has 3 000 turns of very fine insulated copper wire. Explain the differences you would find in **a** to **e** of question 7.

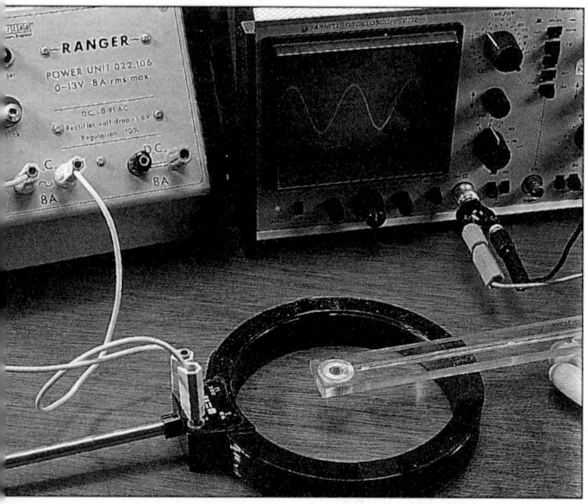

Figure 18.9 Search coil

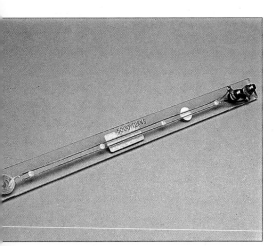

Figure 18.10 Using a search coil to investigate a changing field

Search coil

A search coil is a coil mounted on the end of a probe (see Figure 18.9). It is used to analyse **rapidly changing** magnetic fields, such as those due to alternating currents. A Hall probe (← 17.4) is a convenient way of analysing **steady** magnetic fields in and around current carrying coils and permanent magnets. A search coil usually consists of several thousand turns of insulated wire, 10 millimetres or so in diameter, mounted at the end of a plastic probe. When the coil is placed in a region of alternating flux, an alternating voltage is induced in the coil. The peak value of the induced voltage is proportional to the rate of change of the flux and, therefore, the peak value of the **B**-field within the coil.

The coil can be connected to the Y input of an oscilloscope and the peak voltage found from the display, as shown in Figure 18.10. Alternatively, the search coil can be connected to a sensitive dc voltmeter, as long as a suitable diode is placed in series with the meter to detect a dc signal. The meter reading will be proportional to the **B**-field under investigation.

POLARITY OF THE INDUCED VOLTAGE

In question **2**, you derived the equation $V = Blv$ by considering the energy transfers taking place when an induced voltage drives a current through a circuit. A force is set up on the moving conductor. This force **opposes** the motion that causes the induced voltage in the first place.

18 ELECTROMAGNETIC INDUCTION AND MAGNETIC FLUX

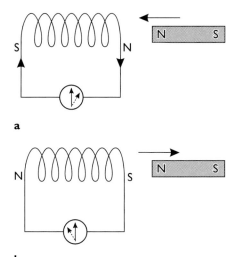

Figure 18.11 Magnet **a** entering and **b** leaving a solenoid

Suppose the north pole of a magnet enters a solenoid connected to a galvanometer, as in Figure 18.11a. The magnetic flux inside the solenoid increases and induces a voltage in the solenoid coil. This voltage drives a current through the circuit, creating its own magnetic flux. The end of the solenoid at which the north pole enters becomes a north pole and repels the north pole of the magnet. You have to do work against this force to put the north pole into the coil. So energy is transferred to the circuit and heating takes place. If you remove the magnet (see Figure 18.11b), you will induce a south pole at the end of the coil. Again, you need to do work to remove the magnet. In summary:

the polarity of the induced voltage is such that, when a complete circuit exists, it will drive a current that creates a flux that opposes the flux change causing it.

This is **Lenz's law**. When there is no circuit connected to the coil, there will be an induced voltage accompanying the change of flux. But there will be no resulting induced current and, therefore, no energy transferred within the system.

Essentially, Lenz's law is stating that energy within a system has to be conserved. It can be shown by a negative sign in Faraday's relationship:

$$V = -N \cdot \frac{d\phi}{dt}.$$

The negative sign indicates that the current due to the induced voltage produces an opposing change in flux. The induced voltage is an electromotive force, or emf, often denoted by the symbol E or E.

DIRECTION OF THE INDUCED CURRENT

When a straight conductor cuts through lines of flux, you can predict the direction of the induced current using **Fleming's right-hand rule**.

Place the thumb, first and second fingers of the right hand at right angles to one another. The thumb represents the direction in which the conductor is moved, the first finger indicates the field direction and the second finger, the direction of the induced current.

You can satisfy yourself that the rule works by applying it to Figure 18.4.

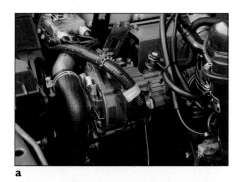

Figure 18.12 Generators of electricity **a** in a car, **b** in a bicycle, **c** in a power station

Generators

Generators work on the principle of electromagnetic induction as summarised by Faraday's and Lenz's laws as applied to the production of induced voltages arising from conductors moving in magnetic fields. They range from bicycle dynamos, used to provide a few watts for lighting, to power station generators delivering about 500 MW for distribution through the national grid. In between are car alternators, which can provide up to several hundred watts to charge the battery (see Figure 18.12).

In all types of electromagnetic generator, a voltage is induced as a result of a change in magnetic flux which links the turns of a coil. The resulting current delivered to the external circuit produces within the coil magnetic flux that opposes the change in flux causing it. Work has to be done to turn the generator against these electromagnetic forces and those of friction.

It follows that the more current is drawn from a generator, the harder it is to turn it and the greater the power required to maintain the generator's output voltage. Power stations use more coal when demand for electricity is heavy; you have to pedal harder when your bicycle dynamo is delivering current to the lights: and a car engine has to work harder, and therefore uses more fuel, when driving the alternator under load.

FORCES AND FIELDS

Question

9 Figure 18.13 shows diagrams of a dc and an ac generator. The dc generator is identical in construction to a dc motor.
a Describe the main differences between the two types of generator.
b Explain, in each case, why the output is as shown.

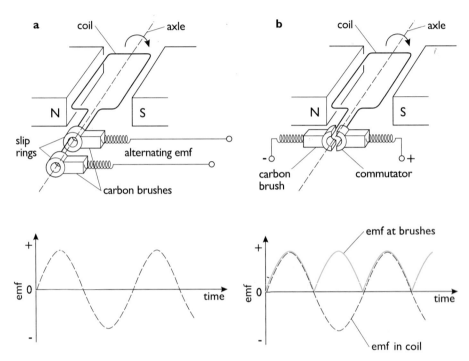

Figure 18.13 a An ac generator, **b** a dc generator

Experiment 18.2 shows an investigation of electromagnetic induction in changing fields. If you try it, you will be changing a number of variables to investigate the effect that each of them has separately.

EXPERIMENT 18.2 Electromagnetic induction in changing fields

a Effect of number of turns

Using the arrangement shown in Figure 18.14, you can look at an oscilloscope display of the induced voltage. This allows you to see the variation in the induced voltage with time as you change the number of turns in the coil.

b Effect of changing maximum flux

Keeping, say 10 turns of wire wound around the iron core, you can set the current in the 300-turn coil to convenient values and measure the peak voltages induced in the 10-turn coil.

c Effect of rate of change of flux

With the output frequency of the signal generator at 50 Hz, you can use the oscilloscope's Y_2 trace to find the voltage across the rheostat, and so the current in the 300-turn coil and the flux linking the 10-turn coil (see Figure 18.15a).

Adjusting the rheostat resistance to keep the peak current in the 10 Ω resistor and

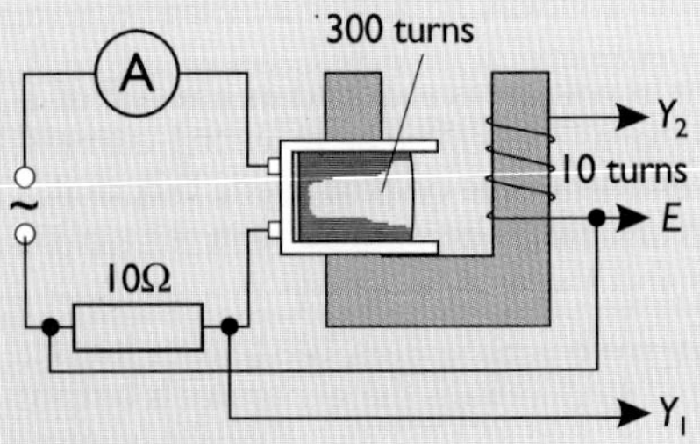

Figure 18.14 Circuit for experiment 18.2

18 ELECTROMAGNETIC INDUCTION AND MAGNETIC FLUX

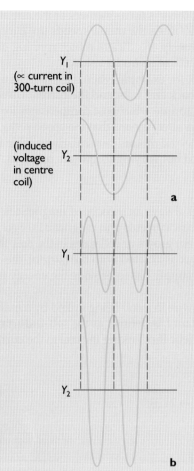

therefore the flux the same, you can increase the output frequency of the signal generator to, say 100 Hz, and note the change in the variation of induced voltage (see Figure 18.15b).

d Effect of iron

You can find the effect of iron by placing a yoke across the top, so completing a magnetic circuit.

Observe the effect, on the oscilloscope traces, of adding the iron yoke. Note that, in this case, the change in the oscilloscope traces may be so great that one or both of the traces may disappear from the screen. If this happens, try changing the 'y gain' of the oscilloscope before the iron yoke is added.

Figure 18.15 Effect of frequency on induced voltage: **a** at 50 Hz, **b** at 100 Hz

---

Question

10 This question refers to experiment 18.2.
a In Figure 18.15, why are the two traces displaying the variation of flux (Y_1) and induced voltage (Y_2) 90° out of phase?
b Explain carefully the reasons for the Y_2 trace in Figure 18.15b.
c Why is it twice the amplitude of the Y_2 trace in Figure 18.15a?
d What is the effect of completing the ring of iron in part **iv** of the experiment? Why do you think iron has this effect?

BEHAVIOUR OF A DC MOTOR

When a steady supply voltage V is connected to the armature coil of a d.c. motor ($\leftarrow$ 17.3), a current flows leading to a BIl force on the coil, causing it to rotate. However, as the armature coil moves through the surrounding magnetic field, an induced voltage E is set up. This opposes the applied supply voltage, and so is sometimes called a **back emf**. The faster the coil turns, the larger the value of E and the smaller the current I drawn by the motor.

If the armature resistance is R, the current I flowing in the coil is given by:

$$I = \frac{V - E}{R}.$$

At the instant the motor is switched on, there is no movement of the armature. So $E = 0$ and the current is a maximum at V/R. As the motor speeds up, E increases and I decreases to some final steady value. In an efficient motor running freely under no load, this may be well under 10% of its initial value.

When the motor is placed under load, for example when it lifts a mass against gravity, its speed tends to drop. So the back emf falls, and the current rises. The BIl force increases, providing increased torque which just compensates for the increased load placed on the motor. This effect means that the speed reduction is less than we might expect and makes electric motors particularly suitable for the many applications where constant speed but varying torque are required.

INDUCTANCE

When there is a current in a coil, there is always an associated magnetic field through the coil (see Figure 18.16). Electric currents and magnetic flux go together like links in a chain. If the current grows or collapses in the coil, the magnetic flux will too. The larger the flux associated with the current, the slower the changes will be. This is because the supply has to provide the energy to build up both the flux and the current.

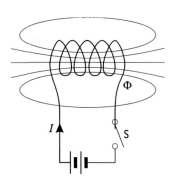

Figure 18.16 Self-induction

Suppose N is the number of turns of wire in a coil being linked by a magnetic flux ϕ through the coil, where the constant L is defined as the **self-inductance** of the coil. The equation relating the magnetic flux linking the coil to the current in the coil is given by:

$$N\phi = LI.$$

The unit of inductance is the **henry**, **H** ($\to$ 21.3).

From Faraday's law we know that:

$$V = N \cdot \frac{d\phi}{dt}.$$

If we **apply** a voltage V across the ends of a coil, the current will increase from zero. The flux through it will increase at a rate proportional to the voltage V across the coil. Since

$$N\phi = LI,$$

then

$$N \cdot \frac{d\phi}{dt} = L \frac{dI}{dt},$$

and we can write the following for the magnitude of this **applied** voltage:

$$V = L \cdot \frac{dI}{dt}.$$

The equation for the **induced** voltage with increasing current is the same but with a minus sign, from Lenz's law.

Suppose there is another coil near the first one so that flux generated by the first links the second coil, as in Figure 18.17. As the current I changes in the first coil, the changing flux induces a voltage V in the second. The magnitude of the induced voltage in the second coil can be written

$$V = M \cdot \frac{dI}{dt},$$

where the constant M, again measured in henry, is called the **mutual inductance** of the system. It depends upon factors including the number of turns, the proximity, and the nature of the medium in and around the coils. Mutual inductance is one of the basic principles on which **transformers** operate ($\to$ 22.3).

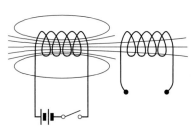

Figure 18.17 Mutual induction

18 ELECTROMAGNETIC INDUCTION AND MAGNETIC FLUX

Question

11 a On what factors does the self-inductance L of a coil depend?
b We usually apply the concept of self-inductance to a coil of wire. Does the concept only apply to coils? Does a short length of wire have self inductance? Explain your answers.

18.3 MAGNETIC FLUX AROUND ELECTRIC CURRENTS

In 17.1, we looked at the nature of magnetic fields around electric currents and their effects on a compass and iron filings. We then defined magnetic field strength and flux and found ways of measuring them. So you can now investigate and measure the variation of **B**-fields around currents. This leads to ways of maximising the magnetic flux within a system, and hence creating more efficient electromagnetic machines.

EXPERIMENT 18.3 The variation in **B**-fields around electric currents

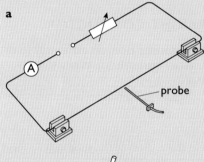
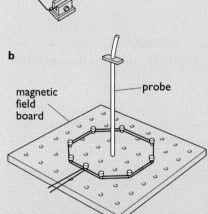
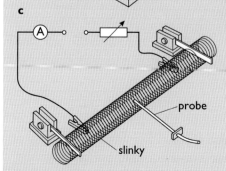

You can use either a search coil or a Hall probe to measure the **B**-field around different shapes of conductor carrying an electric current.

a Field around a long straight wire

Using the arrangement shown in Figure 18.18a and a suitable supply and probe, you can pass a steady current of 4–5 A to produce a detectable field several centimetres away. By taking readings, you can plot graphs to show that:

- the **B**-field at a given distance r from the wire is proportional to the current I in the wire and
- the **B**-field is inversely proportional to the distance r from the wire for a given current.

b Field around a flat coil

The arrangement in Figure 18.18b includes a coil of around 150 mm in diameter, made from about 10 turns of wire, mounted securely on to a flat board. It enables you to investigate the field at the centre of the coil.

c Field inside a solenoid

Using a 'slinky spring' as a solenoid (see Figure 18.18c), you can see how the **B**-field varies:

- across the area and along the length of the solenoid
- when the density of turns is doubled
- with the current passing.

Using another slinky of different area of cross-section, you can see whether the field depends on a solenoid's area.

Figure 18.18 Investigating fields near currents in **a** a long straight wire, **b** a flat coil, **c** a solenoid

FORCES AND FIELDS

r/mm	10	15	20	25	30	35
peak voltage/V	0.62	0.40	0.31	0.24	0.19	0.17

a

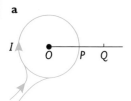

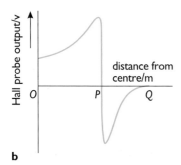

b

Figure 18.19

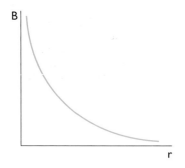

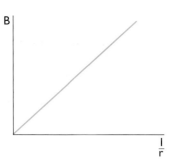

Figure 18.20 Field around a long straight wire

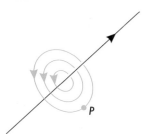

Figure 18.21

Question

12 This question refers to experiment 18.3.
a Figure 18.19a shows the readings of the peak voltage V induced in a search coil placed at varying distances r from the straight wire connected to a supply giving a steady ac output voltage. Plot a suitable graph to test whether or not the field varies according to a $1/r$ relationship.
 What are the main problems with the design of experiment **a** as a means of investigating the **B**-field around a straight wire?
b Figure 18.19b shows a graph of the variation of a Hall probe output voltage as the probe is moved along a radius from the centre of a flat coil to a point about 250 mm from the centre. Explain the main features of the graph.
c It is found from experiments similar to 18.3 **c** that the **B**-field at the centre of a solenoid is independent of the area of cross section A (for solenoids that are otherwise identical and have diameters small compared with their length). Explain qualitatively why you think this might be.
 If two similar solenoids, carrying the same current, have areas in the ratio two to one, how will the total flux passing through them compare?

MAGNETIC FIELD AROUND A STRAIGHT CONDUCTOR

You find experimentally that the **B**-field at a distance r from an extremely long and thin straight wire is proportional to $1/r$ (see Figure 18.20). When the wire carries a current I, B is proportional to I and so:

$$B \propto \frac{I}{r}.$$

The actual value of the **B**-field depends upon the nature of the medium around the conductor and so must the constant of proportionality. This constant is $\mu/2\pi$ where μ is called the **permeability of the medium** surrounding the conductor:

$$B = \frac{\mu I}{2\pi r}.$$

Accordingly, the value of the ampere determines the value of μ. The permeability **in a vacuum, or free space**, has the value $4\pi \times 10^{-7}$ N A^{-2} and the symbol μ_0.

Question

13 Figure 18.21 shows the magnetic field around a straight wire.
a Using the equation for this arrangement show that the unit of μ_0 is N A^{-2}.
b The value of μ_0 is usually given in H m^{-1}. Is this set of units consistent with N A^{-2}?

The definition of the ampere and the value of μ_0

The ampere is one of the base units in the SI system and many derived units depend on it. Its definition is based upon the force between two current carrying conductors.

18 ELECTROMAGNETIC INDUCTION AND MAGNETIC FLUX

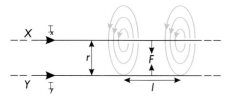

Figure 18.22 Force between two long wires

Consider two long parallel wires, X and Y, in a vacuum and carrying currents I_X and I_Y (see Figure 18.22). The **B**-field over a length l of Y due to the current in X is given by:

$$B = \frac{\mu_0}{2\pi r} \cdot I_X,$$

where r is the separation of the two conductors. The force on this length l is therefore given by:

$$F = BI_Y l.$$

Substituting for B:

$$F = \frac{\mu_0}{2\pi r} \cdot I_X I_Y l.$$

Newton's third law and symmetry lead to the same result for the force on a given length of X. Figure 18.22 shows the direction of the force.

From this equation the **ampere** is defined as:

that constant electric current which, when maintained in two straight parallel conductors, of infinite length and neligible cross-section, placed 1 metre apart in a vacuum, produces a force between the conductors equal to 2×10^{-7} newton per metre of length.

So, in the equation above,

$F = 2 \times 10^{-7}$ N, $l = 1$ m, $I_X = I_Y = 1$ A and $r = 1$ m. Substituting these values gives:

$$\mu_0 = 4\pi \times 10^{-7} \text{ N A}^{-2}.$$

Having established a value for μ_0, measuring the force between two currents would allow us to estimate the current flowing in a conductor, using the relationship between force and currents above. This is the principle behind a form of current balance used to make absolute measurements of current. An unknown current is passed through coils, causing the equilibrium to be upset. The force restoring balance can be measured accurately. Then, with a knowledge of the **B**-field created from the geometry of the coils, the current flowing can be estimated to a very high degree of accuracy.

FIELD INSIDE A LONG SOLENOID COIL

Experiment 18.3 **c** shows that

- the **B**-field inside a solenoid remains reasonably uniform along the length of the solenoid, dropping fairly quickly to a half this value at the ends,
- the field at the centre of a solenoid is directly proportional to the current and the number of turns, but inversely proportional to its length,
- the total flux at the centre of a solenoid is proportional to area. This means that the **B**-field is independent of area.

These relationships are summarised in Figure 18.23.

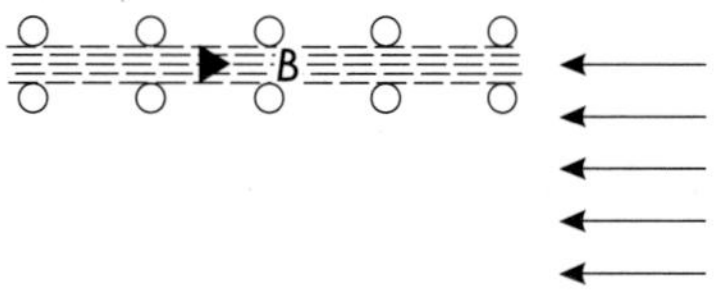

(a) N turns, length l, current I, field B area A, flux Φ
(b) $2N$ turns, length l, current I, field $2B$ area A, flux 2Φ
(c) N turns length $2l$, current I, field $B/2$ area A, flux $\Phi/2$
(d) N turns length l, current $2I$, field $2B$ area A, flux 2Φ
(e) N turns length l, current I, field B area $2A$, flux 2Φ

Figure 18.23 Field inside a solenoid

You might find it helpful to think of the magnetic flux ϕ as 'flowing' through the solenoid, driven by the current flowing through the turns of wire. There is, of course, no evidence that magnetic flux is a material substance which does actually flow!

FORCES AND FIELDS

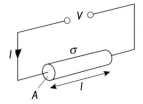

Figure 18.24 Electrical conduction

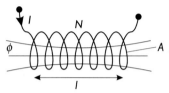

Figure 18.25 The 'flow' of magnetic flux

Now consider a source of voltage in a circuit driving a current through a length l of resistive material of area cross-section A (see Figure 18.24). You can write:

voltage = current × resistance

or $\quad V = I \times \dfrac{l}{\sigma A}$,

where σ is the electrical conductivity of the material ($\leftarrow$ 12.3). You can derive a similar relationship for the flux inside a solenoid (see Figure 18.25).

Figure 18.23 suggests that it is the combination of the number of turns and the current known as '**current-turns**' NI - that drives the flux along the coil. The 'resistance' of the region inside the solenoid is often called **reluctance** and is equal to $l/\mu A$ where μ is the permeability of the material inside the coil. It is sometimes called the **magnetic conductivity** because of the similarity to the electrical situation. When the core is air or a vacuum, μ is μ_0, and has the value $4\pi \times 10^{-7}$ H m^{-1}. When a material such as iron is inside the solenoid, μ can be found from:

$$\mu = \mu_r \mu_0$$

where μ_r is the **relative permeability** of the material in the core of the coil. For example, iron which is 99% pure has a relative permeability of 250; iron with a 2% silicon content has a value of μ_r of 1 000; and some nickel-iron, or NiFe, alloys have values of μ_r up to 10^5.

You can now write the equation for the flux inside a solenoid as

current-turns = flux × reluctance

or $\quad NI = \phi \cdot \dfrac{l}{\mu A}$.

Since $\phi/A = B$, the **B**-field inside a solenoid is given by:

$$B = \dfrac{\mu NI}{l}$$

or $\quad B = \mu n I$

where n is the number of turns per metre.

The concept of reluctance explains why the induced voltage in Experiment 18.2d is larger when a complete loop of iron links the 300-turn coil and the 10-turn coil. The reluctance of the completed magnetic circuit is small and the flux through the circuit relatively large, giving a large rate of change of flux within the smaller coil. The concept also explains why electromagnetic machines such as motors and generators are heavy: they contain iron which has the effect of reducing reluctance and increasing flux for given currents.

Question

14a Calculate the **B**-field inside a slinky coil of 40 turns carrying a steady current of 3 A and stretched out to a length of 500 mm. Assume that $\mu_r = 1$ for air. (μ_0)
b Given that 2 A is the current limit of the available wire for an air-cored solenoid, calculate the number of turns, current, and length required to give a field of 10 mT at the solenoid's centre.

18 ELECTROMAGNETIC INDUCTION AND MAGNETIC FLUX

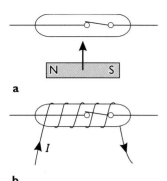

Figure 18.26 A reed switch controlled by **a** a bar magnet **b** a solenoid

Reed switch

The **reed switch** is a device for switching a current on or off depending on the magnetic field it is in. It is often used in burglar alarm systems to detect whether a door is open or closed. In this case the switch is usually fitted into the door frame. The magnet which triggers it is fitted into the thickness of the door, so that it is near to the switch when the door is closed. Other uses include some food-processors and electric toothbrushes, where a conventional switch could get wet (a moveable magnet activates the dry reed switch).

One type of reed switch consists of a pair of contacts encapsulated in a glass envelope which can be closed by a magnetic field aligned along the length of the contacts. A magnetic field can be provided by bringing a bar magnet to the switch, as shown in Figure 18.26a. An alternative method of controlling the switch is to enclose it by a solenoid, as in Figure 18.26b. When a current flows, a field is produced. A sufficiently strong field will close the contacts. The required field to operate the switch is usually given by manufacturers in terms of 'ampere-turns'. Typically, 50 ampere-turns will generate sufficient flux density to close the contacts. So, 20 mA through 2 500 turns will operate the switch in this case.

Question

15 a How would you calculate the reluctance of a complete loop of iron?
b What effect on the reluctance of the loop will a small air gap have?

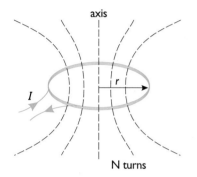

Figure 18.27 Field around a flat coil

FIELD AROUND A FLAT COIL

Experiment 18.3**b** shows that the field at the centre of a flat coil is proportional to the number of turns of wire in the coil and the current flowing, and inversely proportional to the radius of the coil (see Figure 18.27). Putting this in symbolic form:

$$B \propto \frac{NI}{r}.$$

The constant of proportionality is equal to $\mu/2$, so the **B**-field at the centre is given by:

$$B = \frac{\mu.NI}{2r}.$$

The Biot-Savart Law

The **B**-field near a conductor is affected by a number of factors: the distance from the conductor, the length, the current flowing, and the angle to the conductor.

Experimental work shows that the contribution of each of these to an element ΔB of the **B**-field due to an element Δl long is as follows:

$$B \propto I$$
$$\Delta B \propto \Delta l$$
$$\Delta B \propto \sin\theta$$
$$\Delta B \propto 1/r^2.$$

So: $\Delta B \propto \dfrac{I l \sin\theta}{r^2}.$

The constant of proportionality, r^2, turns out to be $\mu/4\pi$, so:

$$\Delta B = \frac{\mu I l \sin\theta}{4\pi r^2}.$$

Magnetic field inside a coil

At the centre of the coil shown in Figure 18.27, the flux density is due to the current I in each turn of the coil. Imagine that the circumference of each turn is made up of small lengths, Δl.

The contribution to the **B**-field of the current in Δl is given by the Biot-Savart law:

$$\Delta B = \frac{\mu I \Delta l \sin\theta}{4\pi r^2}.$$

Since θ is 90°, $\sin\theta = 1$, and:

$$\Delta B = \frac{\mu I \Delta l}{4\pi r^2}.$$

FORCES AND FIELDS

For the complete circumference, the value of the **B**-field is:

$$\Sigma B = \frac{\mu I}{4\pi r^2} \cdot \Sigma \Delta l$$

$$= \frac{\mu I}{4\pi r^2} \cdot 2\pi r$$

$$= \frac{\mu I}{2r}.$$

So, for N turns, $B = \frac{\mu N I}{2r}$.

The field along the axis of the coil falls quite rapidly with distance from the centre of the coil. However, two identical coils with a common axis and a separation equal to their radius can produce a highly uniform field. The coils are connected in series so that they both carry the same current in the same sense, as in Figure 18.28. A uniform field exists between the two coils. Such an arrangement is called a **Helmholtz coil** system ($\rightarrow$ 35.4).

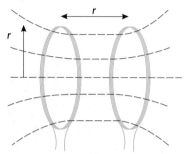

Figure 18.28 Helmholtz coils

> ### Question
>
> **16** A uniform electric field can be produced between a pair of parallel plates connected to a high voltage dc supply. What similarities and differences exist between this arrangement and the Helmholtz coil shown in Figure 18.28?

SUMMARY

When a conductor of length l moves at speed v at right angles to a **B**-field of strength B, the voltage induced in the conductor is equal to Blv.

The product BA is equal to flux and is measured in webers (Wb) when B is in teslas (T) and A in metre^2. The **B**-field is therefore a measure of magnetic flux-density.

The induced voltage in a coil of N turns is proportional to the rate of change of flux $d\phi/dt$ through it and on the number of turns N.

The direction of the induced voltage is such that it opposes the change that causes it.

μ_o is the permeability of free space. Its value is $4\pi \times 10^{-7}$ N A^{-2}, a value fixed by our choice of definition of the ampere.

μ_r is the relative permeability of a material medium. It is defined by:

$$\mu_r = \frac{\mu}{\mu_0}.$$

The **B**-field at distance r from a long straight wire carrying a current I is given by:

$B = \mu I / 2\pi r$.

The **B**-field at the centre of a solenoid of length l containing N turns and carrying a current I is given by:

$B = \mu N I / l$.

18 ELECTROMAGNETIC INDUCTION AND MAGNETIC FLUX

SUMMARY QUESTIONS

To answer the following questions, you will need to do some research, using whatever resources are available.

17 A petrol engine requires a high voltage across the gap of a spark plug to ignite the petrol-air mixture in the cylinder. There are usually four cylinders in a car engine. The sparks at each of the four plugs need to be generated in the right sequence and at the right time to maintain the smooth running of the engine.

How are such voltages obtained from the 12V dc battery supply and how are these distributed efficiently to each of the spark plugs?

18 Induction heating is an important industrial process. Describe the physical principles behind the process and give an example of its use.

19 A particular design of electric cooker-hob consists of flat coils embedded into a ceramic surface. The coils are supplied with a high-frequency alternating current of around 20 kHz. Electric currents in a pan placed on top of a coil cause the pan to get hot and so cook the food inside.
a What are the principles behind such a cooker?
b Why is a high frequency current necessary?
c Why is it important to choose a pan with just the right electrical resistance?

20 The SI definition of the ampere, in use for the last three decades, is based upon the force between two conductors carrying a current.

How was the ampere previously defined and why was the present definition found to be more acceptable?

UNIT E

OSCILLATIONS

OSCILLATIONS

Oscillations can be very fast or very slow. They range from the very rapid but tiny vibrations of a quartz crystal at the heart of a watch to the very slow and relatively large amplitude oscillations of the tides. Oscillations arise in many situations, but are always mechanical or electrical in nature. Sound waves (→ 29.4) falling on a microphone cause the diaphragm to vibrate, or oscillate. The same pattern of oscillation is then transferred to the electric charge that flows in the wires leading from the microphone.

For oscillations to occur in a mechanical system, there must be inertia and a degree of springiness. This is as true for a mass suspended by a vertical spring and displaced, as for a tall chimney driven into oscillation by the wind.

The same mathematics can be used to model mechanical oscillations and those of current and voltage in electrical systems. By gaining an appreciation of the behaviour of diverse oscillating systems you can begin to predict the behaviour of any oscillating system, such as the bridge in Figure 19.1.

Figure 19.1 The collapse of Washington State's Tacoma suspension bridge in 1940, caused by wind-induced oscillations

19 SIMPLE MECHANICAL OSCILLATIONS

An oscillation occurs when a particle or a system of particles moves backwards and forwards along the same path, passing through a rest or **equilibrium position**. One of the simplest vibrating systems consists of a mass suspended by a spring from a fixed point, as in Figure 19.2. When the mass is displaced and released, the energy of the system is continuously transferred between the potential energy in the spring due to elastic strain and the kinetic energy of the moving mass. Drag created by the movement through the air causes energy to be transferred from the system into the surroundings. So eventually the mass comes to rest at its equilibrium position. Such a motion is said to be **damped**.

Any real mechanical oscillation is damped unless there is a source of energy available to maintain it. The pendulum of a grandfather clock soon stops when the slowly falling weight ceases to supply the energy required to maintain a constant amplitude of swing. Similarly, a vibrating guitar string decays in amplitude as energy is dissipated to the surroundings.

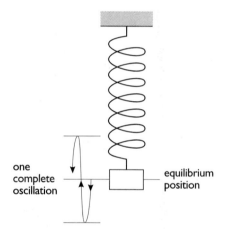

Figure 19.2 Oscillatory motion of a mass on a spring

> **Question**
>
> 1 A long simple pendulum with a massive bob goes on swinging for a much longer time than a short pendulum with a light bob. There are two main reasons for this. What are they?

19.1 SIMPLE OSCILLATORS

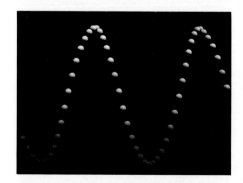

Figure 19.3 Motion of a pendulum bob from above illuminated by a strobe flash. The camera is moved horizontally at a uniform speed. The pattern corresponds to a displacement-time graph of the bob

The time for one complete oscillation is called the **period T** of the motion. This is the time it takes to complete one up-down motion of a mass on a spring or one to-and-fro motion of a pendulum bob (see Figure 19.3). The **frequency f** is the number of oscillations or cycles completed in one second. The unit of frequency is the **hertz (Hz)** or s^{-1}. Frequency and period are inversely related:

$$f = \frac{1}{T}.$$

The maximum displacement of the mass or pendulum bob from its equilibrium position is called the **amplitude A** of the motion (see Figure 19.4). You need to specify the value of each of these quantities to describe completely a particular simple oscillation.

By varying how far you move a mass on a spring before releasing it, you can vary the amplitude of the oscillation. The initial movement is the source of energy for the free oscillations that follow. Once released, the mass oscillates at a

OSCILLATIONS

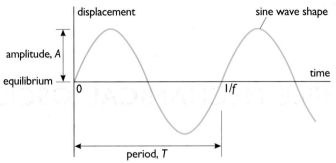

Figure 19.4 Sinusoidal displacement-time graph

particular frequency, called the **natural frequency**. This depends only on the **inertia** (← 6.4), or mass, of the system and the **stiffness** of the spring (← 4.1). The larger the mass, the lower the frequency, and the stiffer the spring, the higher the frequency. You can investigate the relationship between them further in experiment 19.1.

For an oscillatory motion to occur, there must be both an inertial and an elastic element within the system. There can then be a continuous interchange between kinetic and potential energy. When you displace a mass on a spring, you create a **restoring force** directed towards the equilibrium position and increase the potential energy of the system. When you release the mass, this potential energy is transferred into kinetic energy as the mass passes the equilibrium position and so the motion is maintained.

> ### Question
>
> 2 For the mass-spring system in Figure 19.2, at which points in the motion are
> a the restoring force and
> b the kinetic energy
> a maximum?

EXPERIMENT 19.1   Period of a mass-spring system

By hanging a small known mass from a single spring, and measuring the extension x, you can use Hooke's law, $F = kx$, to calculate k, the stiffness of the spring (← 4.1).

You can find how the period T of small vertical oscillations changes when you:

a double the stiffness by adding a second spring in parallel
b halve it by adding a second spring in series
c change the mass attached.

You can then verify that $T = $ a constant $\times \sqrt{m/k}$.

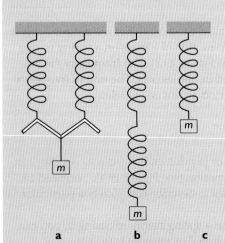

Figure 19.5 Arrangement of springs

19 SIMPLE MECHANICAL OSCILLATIONS

You can develop a mathematical model (→ 19.2) which shows that the value of the constant in experiment 19.1 has a value of 2π. So:

$$T = 2\pi\sqrt{m/k}$$

and $\quad f = \dfrac{1}{2\pi}\sqrt{k/m}.$

Definitions of the second

Until the beginning of the twentieth century, the second was defined as a fraction of the period of revolution of the Earth on its axis. When greater accuracy was required, the second was based on the period of the Earth's revolution round the Sun.

When you refer to the accuracy of a clock, you are really measuring the consistency of the clock or of one clock with another. Until the 1930s, the best pendulum clocks could keep in step, or **in phase**, for a longer period than any other type of clock, they differ by only 1 s in about 100 days. These were later replaced by quartz crystal clocks, which have an accuracy of 1 part in 10^8, or 1 s in 1000 days. A quartz crystal vibrates at a natural frequency determined by the dimensions of the crystal. By the **piezoelectric effect**, electric charge builds up and decays on the surfaces of the crystal in time with the mechanical vibrations. It is therefore possible to stimulate and sustain the vibrations by electrical means, as in Figure 19.6. The very constant frequency of the vibrations acts as an accurate regulator of an electric oscillatory circuit (→ 22.5), which is then used to drive an electric clock or digital watch, etc.

In the 1950s, an atomic clock was developed with an accuracy of 1 part in 10^{11}. This means that it would take 3000 years for two atomic clocks to disagree by 1 second.

Question

3 The natural frequency of a mass-spring system depends on the values of m and k. What do you think are the quantities on which the natural frequency of a simple pendulum depends?

OSCILLATORS AND TIME

The period of the mass-spring system in experiment 19.1 is independent of its amplitude: the decay of the amplitude due to damping does not change the period, as long as it is not too great. Any oscillator with a constant period is said to have **isochronous** motion and can be used as a clock. Mass-spring systems, pendulums and atomic lattice vibrations are all constituents of devices used to measure the passage of time.

The ultimate standard of time until the 1960s was based on astronomical measurements. In 1964, the second was redefined in terms of the period of a particular electromagnetic wave emission from caesium-133 atoms. The clock, based on this atomic energy change (→ 37), consists of a small radio transmitter tuned to match the frequency of the caesium atoms. The fundamental properties of the atoms are unaffected by all external factors except the tuned magnetic field that 'drives' them. So the clock is reproducible and consistent, making it the most suitable on which to base our definition of time.

Question

4 The second is defined as the duration of 9 192 631 770 periods of oscillation of a particular spectral line: the radiation corresponding to the transition between two hyperfine levels of the ground state of caesium-133.

What are **a** the frequency and **b** the period of oscillation of this radiation?

Time is related to length through a fundamental constant of the universe c, the velocity of light in a vacuum. The second and the defined value of the velocity of light are used together to define the standard metre (→ 32.3). So, the time for radar or laser pulses to be reflected from an object, such as an aircraft or even the Moon, can be used to measure its distance. In this way, the Earth-Moon distance has been measured to an accuracy of a few centimetres.

Question

5 Taking the Moon to be 4.0×10^5 km from the Earth, calculate the time interval between sending and receiving a light pulse bounced off the Moon. (c)

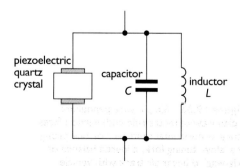

Figure 19.6 Constant frequency LC oscillator circuit controlled by a piezoelectric crystal

E OSCILLATIONS

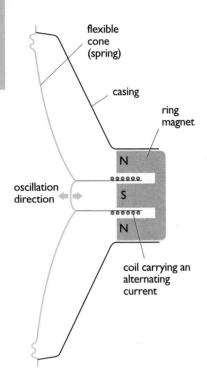

SOURCE OF WAVE MOTION

Any oscillator that can pass its energy to the medium surrounding it, causing an oscillation there, acts as a source of a wave motion (→ 29). For example, a vibrating loudspeaker cone creates the compressions and rarefactions of the air known as a sound wave. The moving coil and cone together are equivalent to a mass-spring system (see Figure 19.7). The loudspeaker must be driven continuously. This **forced vibration** (→ 20.2) replaces the energy lost to the surrounding air.

Figure 19.7 A cross-section of a moving-coil loudspeaker

SOME FREE OSCILLATORS

Experiment 19.2 allows you to study the motion of a variety of free oscillators.

EXPERIMENT 19.2   The motion of some free oscillators

Various oscillator systems are shown in Figure 19.8a–d. By setting up each in turn, you can investigate the effect of changing the mass and the stiffness on the period of oscillation of each system. Oscillators a–e are all sinusoidal with varying degrees of damping, as in Figure 19.9.

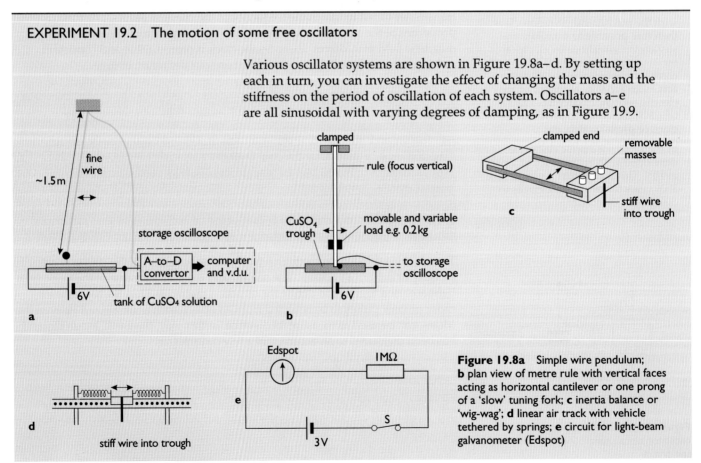

Figure 19.8a Simple wire pendulum; **b** plan view of metre rule with vertical faces acting as horizontal cantilever or one prong of a 'slow' tuning fork; **c** inertia balance or 'wig-wag'; **d** linear air track with vehicle tethered by springs; **e** circuit for light-beam galvanometer (Edspot)

19 SIMPLE MECHANICAL OSCILLATIONS

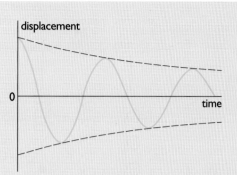

Figure 19.9 Lightly damped simple harmonic motion curve

You should be able to show that $T \propto \sqrt{m/k}$ in each case by using a computer as a storage oscilloscope or just an oscilloscope.

Question

6 How can you check that the damping is exponential (→ 41.3) after measuring successive amplitudes of the displacement-time graph in Figure 19.9?

All of the displacement-time motions in experiment 19.2 are **sinusoidal** – the graph is a sine or cosine curve. We call the class of oscillators with this property **simple harmonic oscillators**. The ball rolling in the V-shaped track in Figure 19.32d and the time-base sweep of an electron spot across the face of an oscilloscope screen are two examples of oscillators that are not simple harmonic. However, *any* oscillatory motion can be broken down into a sum of harmonic components with different amplitudes in a process called **Fourier analysis**. So we can consider simple harmonic motion as the basic component of all oscillatory motions.

Question

7 Which of the following have harmonic motions:
a a windscreen wiper,
b the pendulum of a grandfather clock,
c a diver flexing the end of a diving board, whilst remaining in contact with it?

19.2 SIMPLE HARMONIC OSCILLATORS

In a real system, resistive forces usually reduce the amplitude of oscillations. The oscillator has to work against these forces transferring energy to its surroundings. In an **ideal system**, there are no dissipative forces and no losses of energy.

The air-track vehicle connected between two springs, shown in Figure 19.8d, is close to an ideal system. The two springs act together in parallel as a single

OSCILLATIONS

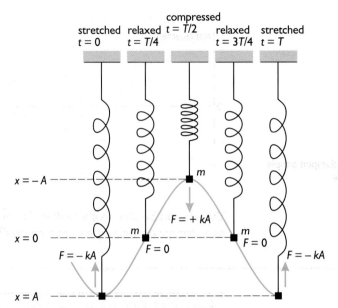

Figure 19.10 Mass m attached to a spring of stiffness k moving on a frictionless surface

spring of stiffness k, obeying Hooke's law (← 4.1). When the vehicle of mass m is displaced a distance x from equilibrium, it experiences a force in the opposite direction (see Figure 19.10):

$F = -kx.$

The negative sign occurs because the force F is measured towards the equilibrium point while the displacement x is measured away from it, and k is a positive constant.

Newton's second law of motion (← 6.4) leads to the equation:

$F = ma$

or $a = \dfrac{F}{m}.$

Replacing F in this equation by $-kx$ gives:

$a = -\dfrac{k}{m}x.$

This is the **equation of motion** of a simple harmonic oscillator. With initial values for such quantities as the position and amplitude of the oscillation, you can trace the mass's displacement from the equilibrium position at any particular **time**. You can do this using numerical methods or by finding an analytical equation as a solution (→ 41.3).

An equation to describe simple harmonic oscillations

Experimentally, it is found that the displacement x varies sinusoidally with time.

The motion of any object is simple harmonic if the restoring force, and therefore the acceleration, is proportional to the displacement x of the object from equilibrium, as in Figure 19.11. The acceleration is given by:

$a = \dfrac{d^2 x}{dt^2}.$

So the equation of motion is a second-order differential equation:

$\dfrac{d^2 x}{dt^2} = -\dfrac{k}{m}x.$

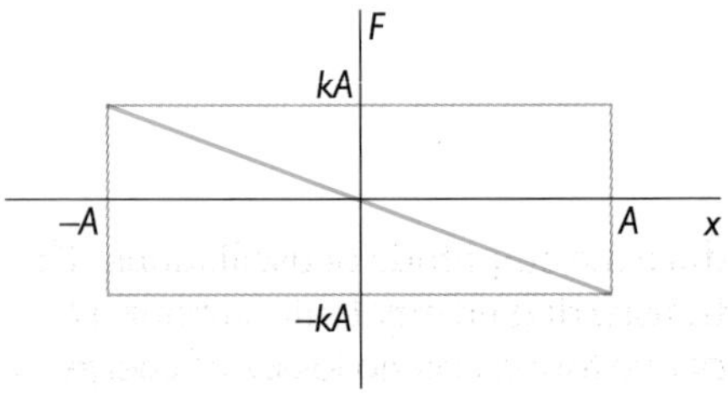

Figure 19.11 Force-displacement graph for a mass-spring system

19 SIMPLE MECHANICAL OSCILLATIONS

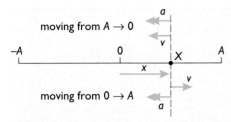

Figure 19.12 Mass is at X in its oscillation along the line between ±A. Above the line, the vectors represent the motion to the left and below the line, motion to the right

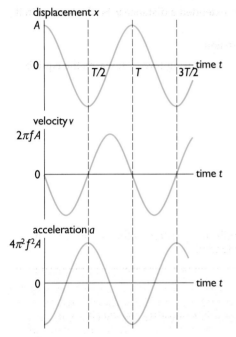

Figure 19.13 Relationships between displacement x, velocity v and acceleration a

Suppose that the motion is started by displacing the mass m by a distance A and then releasing it. Oscillations will occur with displacement x at a time t given by:

$$x = A \cos \omega t$$

where $\omega = 2\pi f$, the angular frequency of the motion. Figures 19.12 and 19.13 show the relationships between these quantities.

The velocity v of the mass at time t is given by the gradient of the displacement-time graph, which is the differential of the displacement with respect to time:

$$v = dx/dt = -\omega A \sin \omega t.$$

The acceleration a is the gradient of the velocity-time graph, or differentiating again with respect to time:

$$a = -\omega^2 A \cos \omega t.$$

Now, $x = A \cos \omega t$, so:

$$a = -\omega^2 x.$$

Since $\quad a = -\dfrac{k}{m} x,$

then $\quad \omega^2 = \dfrac{k}{m}$

and $\quad \omega = \sqrt{\dfrac{k}{m}}.$

$\omega = 2\pi f = 2\pi/T$, so the frequency of oscillation f of such a system is given by:

$$f = \dfrac{1}{2\pi} \sqrt{\dfrac{k}{m}}$$

and the period T by:

$$T = 2\pi \sqrt{\dfrac{m}{k}}.$$

Because the equation $a = -\omega^2 x$ does not include the amplitude A, you can see that frequency and therefore period are independent of the amplitude of such an oscillation.

Figures 19.12 and 19.13 show the relative values of displacement, velocity and acceleration at particular times. Note the phase differences between:

$$\text{displacement } x = A \cos \omega t$$
$$\text{velocity } v = -\omega A \sin \omega t$$
$$\text{acceleration} = a = -\omega^2 A \cos \omega t.$$

A MASS-SPRING SYSTEM

For a mass-spring system, the natural frequency f and the period of the oscillation T are given by:

$$f = \dfrac{1}{2\pi} \sqrt{\dfrac{k}{m}}$$

$$T = 2\pi \sqrt{\dfrac{m}{k}}.$$

The periods of the mass-spring systems measured in experiment 19.1 agree with this equation. Gravity produces a constant vertical force on the mass but does not alter the frequency. In fact, the frequency or period of a harmonic oscillator is **independent of its amplitude**.

OSCILLATIONS

Question

8 In Figure 19.12, at which point(s) in the motion does the mass have:
a maximum velocity to the left
b maximum potential energy
c maximum acceleration to the right
d zero acceleration?

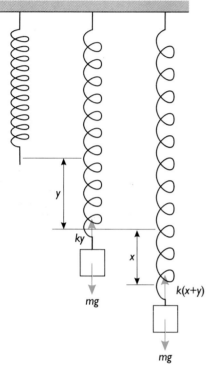

Figure 19.14 Forces on a mass-spring system extended by distance y and then displaced

Oscillations in an extended vertical mass-spring system

In Figure 19.14, a spring in equilibrium is extended a distance y. So the tension in it is given by:

$$\text{tension} = mg = -ky = -\text{stiffness} \times \text{extension}.$$

The mass is then displaced a distance x. The restoring force F towards the equilibrium position is given by

$$F = \text{tension} - mg$$
$$= -k(y + x) - mg.$$

Since $mg = -ky$, we can write:

$$F = -kx$$

and $$a = -\frac{k}{m}x$$

as before.

So the mass oscillates about the new equilibrium position of the extended spring in exactly the same way as when it is not extended.

Question

9a How can you find the period of a mass oscillating vertically on a spring when the only measurements you know are the static extension of the spring and the mass hanging on it?
b Suppose it is taken to the Moon where the gravitational field at the surface is one-sixth of that on Earth. How will the period of this particular mass-spring system change?

In experiment 19.2, we use an A-to-D converter and a computer to show that the displacement-time graph of an oscillating system. Experiment 19.3 uses a much less sophisticated method, requiring only a ticker-timer.

EXPERIMENT 19.3 The motion of a mass-spring oscillator

Figure 19.15 shows a trolley runway, raised at one end to compensate for friction, with a trolley connected to fixed supports by springs. Using this you can measure the force constant of the system k in N m^{-1} by pulling the trolley with a spring balance.

Using a ticker-timer to tape half an oscillation allows you to plot a graph of displacement against time, as in Figure 19.16. You can compare this experimental graph with:

$$x = A \cos \sqrt{k/m}\, t.$$

19 SIMPLE MECHANICAL OSCILLATIONS

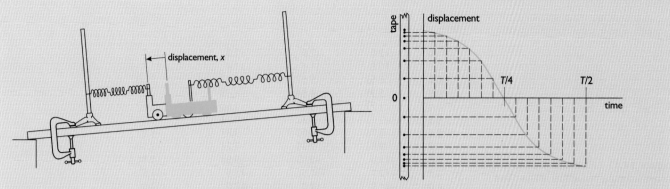

Figure 19.15 Trolley tethered between springs for experiment 19.3

Figure 19.16 Displacement-time graph from ticker-tape

Question

10a How could you use a velocity-time graph from experiment 19.3 to find the acceleration of the system at any *displacement* from equilibrium?
b Draw the shape of the acceleration-displacement graph.

Describing simple harmonic motion numerically

We can find a numerical solution to the equation

$$a = -\frac{k}{m} x$$

using an iterative method ($\rightarrow$ 41.3).

We assume that the oscillator's acceleration remains constant for a time Δt. We calculate a new value for the displacement and, using this, find the new acceleration and, hence, the new speed, etc. We then repeat these four steps. The equations for the steps of the nth set of calculations are:

$$a_n = -\frac{k}{m} x_{n-1}$$

$$v_n = v_{n-1} + a_n \Delta t$$

$$x_n = x_{n-1} + v_n \Delta t$$

$$t_n = t_{n-1} + \Delta t.$$

Suppose we take $k = 16$ N m^{-1} and $m = 1.6$ kg, so the value of k/m is 10 s^{-2}. At the beginning, we could take the values

$$n = 1$$
$$x_{n-1} = x_0 = A$$
$$v_0 = 0$$
$$t_0 = 0.$$

We could then take an interval of a tenth of a second:

$$\Delta t = 0.1 \text{ s}.$$

The table in Figure 19.17 shows the results of numerical analyses for different values of Δt and compares these to the results obtained using the equation

$$x = A \cos \sqrt{10}\, t$$

for an amplitude of 12 cm.

E OSCILLATIONS

t(s)	x(cm) calculated using			
	$\Delta t = 0.1$ s	$\Delta t = 0.01$ s	$\Delta t = 0.001$ s	$12 \cos \sqrt{10} t$
0	12.0	12.0	12.0	12.0
0.1	10.8	11.4	11.4	11.4
0.2	8.52	9.57	9.67	9.68
0.3	5.39	6.84	6.98	7.00
0.4	1.72	3.43	3.60	3.62
0.5	−2.13	−0.32	−0.14	−0.11
0.6	−5.76	−4.03	−3.87	−3.83
0.7	−8.81	−7.35	−7.21	−7.18
0.8	−11.0	−9.93	−9.83	−9.81
0.9	−12.1	−11.5	−11.5	−11.5
1.0	−11.9	−12.0	−12.0	−12.0

Figure 19.17 Values of displacement x, calculated using numerical analysis and equation

19.3 CIRCULAR MOTION AND SIMPLE HARMONIC MOTION

There is a clear correspondence between uniform circular motion and simple harmonic motion. This is shown in experiment 19.4.

EXPERIMENT 19.4 Oscillation and circular motion

Figure 19.18a shows a variable speed motor with a short rod and small ball at its end, attached at right angles to its spindle and a mass mounted on a vertical spring beside it. Using this, you can displace the mass to make vertical oscillations of amplitude equal to the length of the rods.

Figure 19.18 Apparatus for experiment 19.4 to compare projected circular motion and simple harmonic motion. The shadows stay in step as each oscillator's amplitude dies away

By mounting the small ball on a rotating turntable, as in Figure 19.18b, you can adjust the turntable speed until the motion of the ball appears identical to the motion of an oscillation – that of a tethered air-track vehicle, or the bob of a swinging pendulum (see Figure 19.18c).

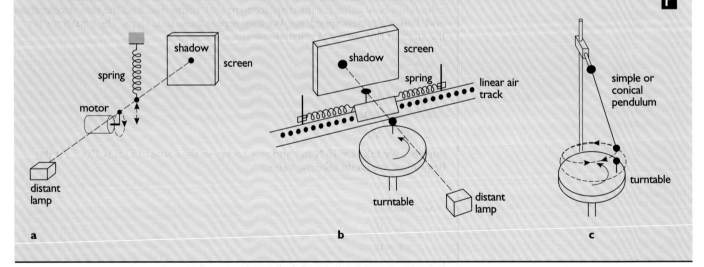

By projecting the motion of an object in a circular path on to a diameter of the circle, it can be seen as a motion in a line – simple harmonic motion. In Figure 19.19, imagine P to be moving round the circumference of a circle, radius A, with constant angular speed ω, measured in radians per second. Suppose that P is at X at time $t = 0$. Then at time t, the angle $POX = \theta = \omega t$. The projection of OP on the

19 SIMPLE MECHANICAL OSCILLATIONS

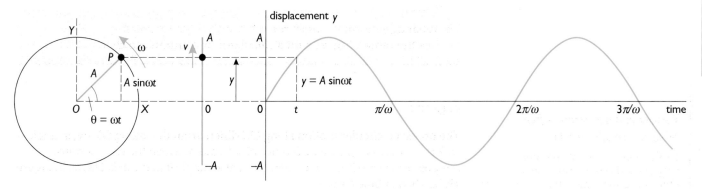

Figure 19.19 Steady circular motion at angular speed ω projected on to a vertical line

axis OY is given by:

$$y = A \sin \omega t.$$

The graph of the variation of y with time is a sine curve, showing that the projection is simple harmonic motion. Notice also that the amplitude A of the motion equals the radius of the circle.

P has the angular speed ω, which is the angle OP turns through in one second. For each revolution, $\theta = 2\pi$ radians. In 1 second there are f revolutions, so:

$$\theta = 2\pi f \text{ rad}$$

and $\omega = 2\pi f$.

For the simple harmonic motion along the diameter, ω is called the **angular frequency**, or **pulsatance**.

The time period of the motion T is given by:

$$T = \frac{2\pi}{\omega}.$$

In the above, we have projected on to the diameter along OY. We could have chosen any diameter, such as that along OX. The projection of OP on OX is given by:

$$x = A \cos \omega t,$$

the equation for displacement in oscillatory motion. This leads to the general equation of simple harmonic motion:

$$a = -\omega^2 x$$

just as for oscillatory motion.

Figure 19.20

> ### Question
>
> **11** A simple pendulum has a length l from the point of suspension to the centre of mass of its bob, which has a mass m (see Figure 19.20). It is displaced by a small angle θ.
> **a** By considering the forces acting on the bob, show that, when it is released, it oscillates about its equilibrium position with simple harmonic motion and that its time period T is given by:
>
> $$T = 2\pi \sqrt{\frac{l}{g}}$$
>
> where g is the acceleration due to gravity.

E OSCILLATIONS

b What does 'a small angle' mean? Are 1°, 5° or 10° small? Use the radian function on a calculator or computer to check.

Velocity and acceleration from projected circular motion

Expressions for simple harmonic velocity and acceleration can be obtained using Figure 19.21. The linear velocity of P is $A\omega$, as ω is the angle swept out by P in each second. So, the component of velocity v along OX in Figure 19.21b is given by:

$$v = -A\omega \sin \omega t.$$

The negative sign indicates that v is directed towards O.

The acceleration of P, the centripetal acceleration, is $A\omega^2$ ($\leftarrow$ 7.1) directed towards the centre of the circle, as in Figure 19.21c. The component of acceleration along OX is:

$$a = -A\omega^2 \cos \omega t.$$

PHASES

The projected circular motion along OX differs from that along OY by an angle of $\pi/2$ rad. There is a **phase difference** of $\pi/2$ rad between the two motions. Suppose time $t = 0$ is chosen when OP is at an angle θ to the axis OX, as in Figure 19.22. Then at time $t = 0$:

$$y = A \sin \phi.$$

At a later time, when P reaches P':

$$y = A \sin(\omega t + \phi).$$

The angle ϕ is called the **phase constant**. It is the **initial phase angle**, or the phase angle when time $t = 0$.

In experiment 19.4, when the oscillator and rotor do not start together there is an initial phase angle between them. As the speed of the rotor or turntable is adjusted to become equal to the angular frequency of the oscillator, the phase angle changes continuously. It is only when the two motions have the same angular frequency that the phase angle is constant. The combination of two simple harmonic motions where the frequencies are slightly different results in a phenomenon called **beats** ($\rightarrow$19.5).

Question

12 Look back at Figure 19.13.
a Write down the phase differences between the displacement and **i** the velocity and **ii** the acceleration of a simple harmonic oscillator at the same instant.
b What time interval, in terms of whole cycles of oscillation of period T, corresponds to a phase difference of **i** $\pi/2$? **ii** π? **iii** 2π?

PHASORS

The link between circular motion and simple harmonic motion leads to a simple geometrical method of representing an oscillation. It also provides a simple way

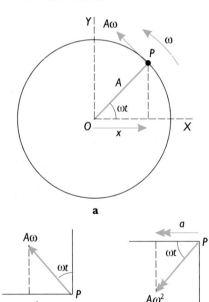

Figure 19.21a Projected circular motion, **b** component of velocity along OX, and **c** component of acceleration along OX

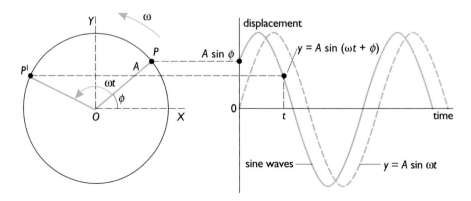

Figure 19.22 Projected circular motion showing phase difference and phase constant

19 SIMPLE MECHANICAL OSCILLATIONS

of combining oscillations that have the same frequency but different amplitudes and relative phase angles.

A **phasor** represents the amplitude and phase angle of an oscillation as an anticlockwise rotation. Its length represents the amplitude and its angle to the *x*-axis represents the phase angle relative to another phasor or to its value at $t = 0$. **Phasor diagrams** allow combinations of simple harmonic motions of the same frequency, but with different amplitudes and phase angles, to be added easily. They are particularly useful in the study of diffraction and interference of waves ($\rightarrow$ 31.2 and 33.2) and of the oscillations in electrical circuits ($\rightarrow$ 21 and 22).

All wave motions are generated by oscillations. To detect the resultant oscillation at a point in space, we measure the sum of the displacements caused by each wave at that point using a suitable detector. For example, a microphone is used for sound waves and an aerial for radio waves. The detected oscillation depends on the frequencies, amplitudes and relative phase angles of the wave oscillations at that point.

Adding oscillations using phasors

Suppose there are two oscillations at a point in space in the same direction. They have the same frequency but differ by a phase angle of $\pi/2$ rad. We can represent one, symbolically, by

$$y_1 = A_1 \sin \omega t$$

and the other by

$$y_2 = A_2 (\sin \omega t + \pi/2) = A_2 \cos \omega t.$$

The two oscillations can be represented either by the two phasors of Figure 19.23a or by the sine waves of Figure 19.23b. The resultant oscillation at the point is:

$$A \sin (\omega t + \phi),$$

where $A = \sqrt{A_1^2 + A_2^2}$ and $\phi = \tan^{-1} A_2/A_1$.

This can be seen directly from the phasor diagram, by vector addition. The resultant represents the amplitude and phase of the disturbance at the point.

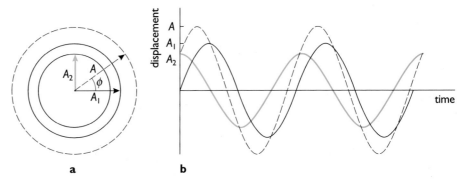

Figure 19.23 Phasor addition of oscillations

> ### Question
>
> **13a** Show that for two oscillations of equal amplitude A_o with a phase difference of $\pi/2$, the resultant amplitude is $\sqrt{2} A_o$ and the phase angle ϕ is $\pi/4$ rad. Suppose the phase difference between the oscillations was π. What would the resultant amplitude be?
>
> **b** A radio aerial receives two sinusoidal signals from a transmitter, one directly and the other after reflection from a large building. What will the amplitude and phase of the resultant signal depend upon? What importance will this have in the correct choice and siting of the receiving aerial ($\rightarrow$ 32.2).

19.4 THE ENERGY OF A SIMPLE HARMONIC OSCILLATOR

Energy is needed to start a mass oscillating. Damping forces on the mass will cause the amplitude of the oscillations to decay eventually to zero. If there is no damping, the total energy of a harmonic oscillator will remain constant.

For the simple mass-spring system of Figure 19.10, the energy is stored as potential energy in the spring and kinetic energy of the moving mass. We can

OSCILLATIONS

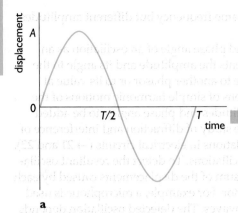

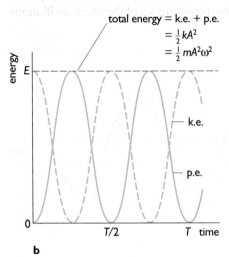

Figure 19.24 Variation of **a** displacement and **b** the corresponding kinetic, potential and total energy during one cycle, where $x = A \sin \omega t$

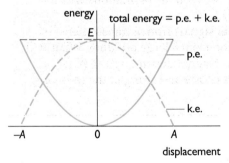

Figure 19.25 Variation of kinetic, potential and total energy with displacement

discover how the amount of the total energy is shared between them by:

- finding the variation of energy with time for one oscillation of the mass shown in Figure 19.24, or
- finding the variation of energy with displacement of the mass from the equilibrium position as in Figure 19.25.

Energy relationships in a harmonic oscillator

The energy stored within a harmonic oscillator, such as a mass-spring system, varies with time. The potential energy E_p stored in the spring at any instant is $\frac{1}{2}kx^2$ ($\leftarrow$ 4.3), where x is the displacement from equilibrium. We know that:

$$x = A \cos \omega t,$$

and $E_p = \frac{1}{2}kx^2$

so $E_p = \frac{1}{2}kA^2 \cos^2 \omega t.$

The velocity v is given by:

$$v = \frac{dx}{dt} = -A\omega \sin \omega t$$

and the kinetic energy by $E_k = \frac{1}{2}mv^2$,

so $E_k = \frac{1}{2}mA^2 \omega^2 \sin^2 \omega t.$

The total energy E of the oscillator is the sum of the potential and kinetic energies:

$$E = E_p + E_k$$
$$E = \frac{1}{2}kx^2 + \frac{1}{2}mv^2$$
$$= \frac{1}{2}kA^2 \cos^2 \omega t + \frac{1}{2}mA^2 \omega^2 \sin^2 \omega t.$$

We saw in 19.2 that

$$\omega = \sqrt{\frac{k}{m}}, \text{ so:}$$

$$E = \frac{1}{2}kA^2 \cos^2 \omega t + \frac{1}{2}kA^2 \sin^2 \omega t$$
$$= \frac{1}{2}kA^2 (\cos^2 \omega t + \sin^2 \omega t).$$

Since $\cos^2 \omega t + \sin^2 \omega t = 1$:

$$E = \frac{1}{2}kA^2 = \frac{1}{2}mA^2\omega^2.$$

These equations contain the following useful results:

- the potential energy has a maximum value $\frac{1}{2}kA^2$ when the spring is fully compressed or extended and the mass is at rest
- the kinetic energy is a maximum when the mass is moving at its maximum speed, ωA. This occurs at the equilibrium position when the potential energy in the spring is zero
- at any time, the sum of the kinetic and potential energies is constant at $\frac{1}{2}kA^2$
- the rate at which the energies interchange is twice the frequency of the oscillation, as shown in Figure 19.24.

The potential energy can be plotted directly against x, as in Figure 19.25 because:

$$E_p = \frac{1}{2}kx^2.$$

The kinetic energy can be plotted using:

$$E_k = E - E_p$$
$$= \frac{1}{2}kA^2 - \frac{1}{2}kx^2.$$

So, the kinetic energy curve can be plotted directly from the potential energy curve. From the equation above for the kinetic energy, we can also write:

$$v^2 = \frac{k}{m}(A^2 - x^2) = \omega^2(A^2 - x^2)$$

19 SIMPLE MECHANICAL OSCILLATIONS

or $v = \omega\sqrt{(A^2 - x^2)}$.

This is another expression for the speed of the mass, but in terms of its displacement from equilibrium. It shows clearly that:

- v is a maximum at $x = 0$, and
- v is zero at $x = +A$ and $x = -A$.

Questions

14 What is the average value of the kinetic or potential energy of a mass-spring oscillator over a complete number of cycles?

15 In Figure 19.25 at what displacement in an oscillation is the potential energy equal to the kinetic energy?

16 Look back at Figure 19.14. Show that the potential energy of the system increases by $\frac{1}{2}kx^2$ when the mass is displaced up or down a distance x from equilibrium (x is less than y).

The total energy E is given by:
$$E = \tfrac{1}{2}kA^2 = \tfrac{1}{2}mA^2\omega^2.$$

From this, you can see that **the total energy is proportional to the square of the amplitude of the oscillation.**

The eye and ear detect the *energy* of a light or sound wave, not the *amplitude* of the wave. Since the energy of a mechanical wave is carried by oscillations of the particles along the path of the wave, it is proportional to the square of the amplitude of the wave. So the eye and ear are called **square law detectors**.

When an oscillator is *lightly* damped, for example, through friction or viscous drag of the air, the energy dissipated per oscillation to the surroundings is found to be directly proportional to the total energy of the oscillator at that time. So the displacement against time curve shows an exponential decay of amplitude with time (see Figure 19.26). The amplitude decays towards zero as the energy in the system is transferred to the surroundings.

The frequency of the motion is effectively unchanged for light damping. So the natural frequencies measured in experiment 19.2 should be the same as the values calculated using the mathematical model of simple harmonic motion (← 19.3).

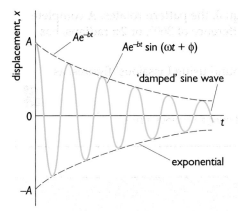

Figure 19.26 Displacement-time graph for a lightly damped oscillator

19.5 BEATS AND LISSAJOUS' FIGURES

The addition of two simple harmonic motions of the same frequency *in the same line* is relatively straightforward (← 19.3). However, if you add two harmonic motions *at right angles* to each other, you will generate a variety of patterns called **Lissajous' figures**. Examples are shown in Figures 19.27 and 19.28 (see experiment 19.5).

When the oscillations are in the same direction but at different frequencies, we observe **beats** and **modulation**. When two oscillations of slightly different frequency are added, the difference in frequency between the two sources is called the **beat frequency**. Figure 19.30 shows the resultant in the case where the amplitudes of the two oscillations are equal, giving beats with maximum modulation (see experiment 19.6).

E OSCILLATIONS

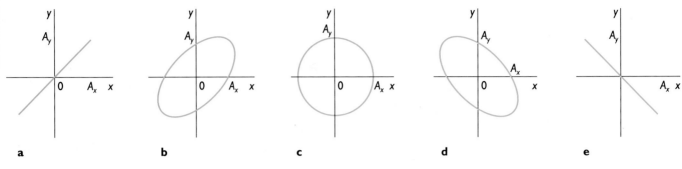

Figure 19.27 Lissajous' figures generated when horizontal and vertical frequencies are equal, and amplitudes have phase differences φ equal to **a** 0, **b** $\pi/4$, **c** $\pi/2$, **d** $3\pi/4$, **e** π

Figure 19.28 Lissajous' figures generated when and the vertical frequency is **a** twice, **b** three times, the horizontal frequency

EXPERIMENT 19.5 Lissajous' figures

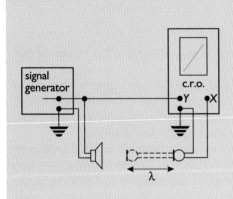

You can generate Lissajous' figures by connecting two signal generators to an oscilloscope, one to the Y-input and the other to the X-input. Using identical frequencies and amplitudes, you will notice that the shape of the Lissajous' figure only depends on the phase angle between the two oscillators (see Figure 19.27).

With the two frequencies slightly unequal, the pattern rotates. A complete rotation is completed when a phase difference of 360°, or 2π radians, has occurred between the oscillators.

The speed of sound ($\rightarrow$ 29.4) may be found using Lissajous' figures, as shown in Figure 19.29.

Figure 19.29 Measuring the speed of sound using a Lissajous' figure

Question

17a Explain the shapes in Figure 19.28a and b.
b If the X and Y inputs were interchanged, what shapes would you expect to see?

Equations of Lissajous' figures

Any Lissajous' figure can be described by an equation relating the x and y displacements at a particular instant in time, as in the following two examples.

Example a In Figure 19.27a, the x and y motions are of the same frequency and are in phase. Let:

$$x = A \cos \omega t$$

and $y = A \cos \omega t$.

Eliminating t gives:

$$y = x,$$

the equation of the straight line shown.

19 SIMPLE MECHANICAL OSCILLATIONS

Example b In Figure 19.27c, the x and y motions are of the same frequency, but are $\pi/2$ rad out of phase. Again, let

$$x = A \cos \omega t$$

and $\quad y = A \sin \omega t.$

Since $\cos^2 \omega t + \sin^2 \omega t = 1$, we can eliminate t by squaring and adding. This gives:

$$x^2 + y^2 = A^2,$$

which is the equation of a circle of radius A.

The shapes tend to move between a straight line and circle, unless the two frequencies are absolutely identical.

EXPERIMENT 19.6 Beats

Attaching a small mass to one prong of two identical tuning forks, sounded together, causes the intensity of the sound to fluctuate. Altering the position of the mass causes the beat frequency to change. Figure 19.30b shows the resultant displacement of the air when the tuning forks are sounded in front of a microphone connected to an oscilloscope.

Question

18 Two tuning forks, marked at 440 Hz, are sounded together. The sound fluctuates from maximum to minimum and back to maximum in 10 s.
a What is the value of the beat frequency?
b Calculate the difference in their two frequencies.
c One fork is correctly marked. What is the value of the percentage error in the frequency of the other?

BEATS

Two pendulums are displaced and released at the same time. Suppose one, A, has a time period of 5 seconds and the other, B, one of 4 seconds. Then, after an interval of 20 seconds, they will be back in phase. During this time, A will have completed four oscillations while B will have completed one extra oscillation.

Because the time period of the beats is 20 seconds, the beat frequency f is equal to $\frac{1}{20}\,\text{s}^{-1}$ or 0.05 Hz. But the frequency of A is $\frac{1}{20}\,\text{s}^{-1}$ or 0.20 Hz and the frequency of B is $\frac{1}{20}\,\text{s}^{-1}$ or 0.25 Hz. The difference in these two frequencies is equal to 0.05 Hz, the beat frequency.

In general, the beat frequency f resulting from two oscillators at frequencies f_1 and f_2, is given by:

$$f = f_1 - f_2.$$

Question

19 Give a generalised argument to show that the beat frequency f resulting from two oscillators f_1 and f_2 is given by:

$$f = f_1 - f_2.$$

OSCILLATIONS

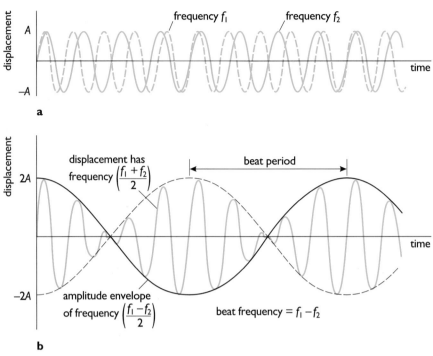

Figure 19.30 Displacement-time graphs for **a** the two oscillations, and **b** the resultant oscillation, in experiment 19.6

Mathematical analysis of beats

In Figure 19.30a, at time $t = 0$, we take the two oscillations of frequencies f_1 and f_2 to be of equal amplitude A and to be in phase. The displacements x_1 and x_2 are given by:

$x_1 = A \sin \omega_1 t$

$x_2 = A \sin \omega_2 t.$

The resultant displacement x is given by

$x = x_1 + x_2$

$\quad = A(\sin \omega_1 t + \sin \omega_2 t).$

Summing the two sines:

$$x = 2A \sin \frac{(\omega_1 + \omega_2)}{2} t \cdot \cos \frac{(\omega_1 - \omega_2)}{2} t.$$

Because $\omega = 2\pi f$, $x = 2A \sin 2\pi \frac{(f_1 + f_2)}{2} t \cdot \cos 2\pi \frac{(f_1 - f_2)}{2} t.$

The right hand side of this relationship consists of 4 factors:

2, A, $\sin 2\pi \frac{(f_1 + f_2)}{2} t$ and $\cos 2\pi \frac{(f_1 - f_2)}{2} t.$

Now sines and cosines only vary between −1 and +1. So when they are both +1 or −1 the resulting signal x has amplitude $2A$ and when they have opposite signs, x has amplitude $-2A$. The frequency, or number of times it varies between these extremes in each second, depends on the values of f_1 and f_2.

The sine term shows that the resulting signal has the average frequency of $(f_1 + f_2)/2$.

The cosine term shows that the resulting signal is **modulated** by the lower frequency $(f_1 - f_2)/2$. This modulation is heard as a **beat** whenever x is $+2A$ or $-2A$. So the beats occur with a frequency of $f_1 - f_2$.

19 SIMPLE MECHANICAL OSCILLATIONS

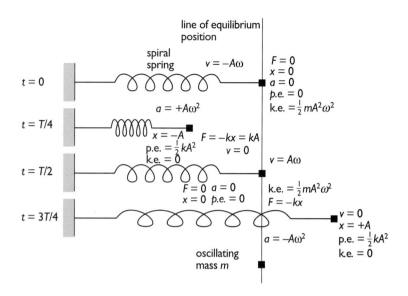

Figure 19.31 Motion of a mass-spring system

SUMMARY

An object moves in simple harmonic motion (see Figure 19.31) when one of the following conditions is satisfied:

- the restoring force F on the object is directly proportional to its displacement x from equilibrium;
- the acceleration a of the object is directly proportional to its displacement x from equilibrium, but is directed in the opposite sense

$$a = -\omega^2 x$$

- the displacement-time graph is a sine wave, $x = A \sin(\omega t + \phi)$.

The frequency f of a pure simple harmonic motion is independent of its amplitude A.

In practice, the motion of many systems, such as a simple pendulum, is only simple harmonic for small amplitude oscillations.

The period of a mass-spring system is:

$$T = 2\pi\sqrt{m/k}$$

and that of a simple pendulum is:

$$T = 2\pi\sqrt{l/g}.$$

The angular frequency is:

$$\omega = \sqrt{k/m},$$

where k is the stiffness, or force constant, of the elastic system of mass m.

If $x = A$ when $t = 0$, the relationships between x, v, a and t are:

$$x = A \cos \omega t$$

$$v = -A\omega \sin \omega t = \omega\sqrt{(A^2 - x^2)}$$

$$a = -A\omega^2 \sin \omega t = -\omega^2 x.$$

A phasor diagram gives the amplitude and phase angle of the resultant oscillation when two or more oscillations of the same frequency are added together in the same line.

OSCILLATIONS

The total energy of a harmonic oscillator is constant when there are no dissipative forces:

$$E = E_p + E_k = \tfrac{1}{2}kx^2 + \tfrac{1}{2}mv^2.$$

The total energy is proportional to the square of the amplitude of the oscillation:

$$E = \tfrac{1}{2}kA^2$$

and $\quad E = \tfrac{1}{2}m\omega^2 A^2.$

SUMMARY QUESTIONS

20 To decide whether an oscillatory motion is simple harmonic or not, we can use the following method:
  i Displace the object a distance x from its equilibrium position.
  ii Write down an expression for the restoring force in terms of x.
  iii Use Newton's second law of motion to see whether the equation is of the form:

acceleration $\propto -$ (displacement).

Which of the following do you think are simple harmonic motions (see Figure 19.32)?
a a hydrometer displaced vertically in a liquid
b a liquid oscillating in a U-tube
c a light, hard, conducting ball bouncing between charged capacitor plates
d a ball rolling along a V-shaped track
e a small magnet, suspended by a thread, oscillating in the centre of a current-carrying coil
f the vertical motion of a sewing machine needle
g an elastic ball bouncing vertically on a hard horizontal surface
h a positive electric charge moving along a line through the centre of a ring of negative charge.

21 For each of the systems **b, c, d** and **g** in question **20**, sketch graphs to show the variation with time, for at least one complete cycle, of the motion of
a the displacement,
b the velocity
c the acceleration of the oscillating object.

22 Each of the following can be used as time-keeping oscillators:
a a simple pendulum
b a mass suspended from a spring
c a torsional pendulum.
Which of these will have the same period when set up on the Earth and on the Moon?

23 A 0–10 N spring balance has a linear scale of length 50 mm. A mass is hooked on to the balance, which is suspended vertically. The mass oscillates with a frequency of 3.2 Hz.
a At what mark on the scale do you expect the pointer to come to rest?
b Suppose you attempted this experiment with a laboratory spring balance. Why would you be unlikely to observe such an oscillation?

24a When a loudspeaker, as in Figure 19.7, emits a pure note, its cone moves in simple harmonic motion. Taking the amplitude of the cone for a

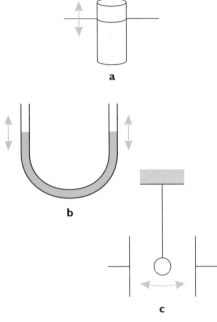
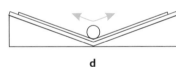
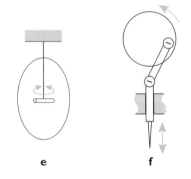
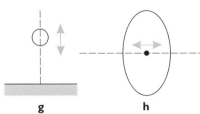

Figure 19.32

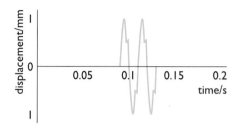

Figure 19.33 Displacement-time curve of loudspeaker cone

loud note at 1.0 kHz to be 1.0 mm, calculate the maximum speed and acceleration of the cone.
b The tip of a prong of a 440 Hz tuning fork has a maximum speed of 4.0 m s^{-1}. Calculate the amplitude and maximum acceleration of the tip.

25 A loudspeaker cone may oscillate as in the displacement-time curve of Figure 19.33 to produce a short pulse of sound. Draw the displacement-position graph of the air at time 0.2 s along a line perpendicular to the cone through its centre
a directly in front of it, and
b directly behind it.
Take the cone to be at the origin and the speed of sound to be 330 m s^{-1}.

26a The mass of the moving part of the loudspeaker in question **24a** is 0.015 kg. Calculate the maximum force exerted on the cone under the conditions in question **24a**.
b In a car engine, the pistons can be assumed to move approximately in simple harmonic motion. Taking the piston mass to be 0.5 kg and the amplitude of motion to be 0.04 m, calculate the maximum force that the piston experiences when the engine is running at 3 000 rpm.

Figure 19.34

27 A trolley of mass 1.0 kg is connected between two springs (see Figure 19.34). Each spring is of stiffness 20 N m^{-1} and is stretched by 40 mm. The trolley is pulled 40 mm to the right and released.
a At what points in the motion is the potential energy of the system
i a maximum? **ii** a minimum?
b Calculate
i the maximum and minimum potential energy
ii the maximum kinetic energy of the system.
c Draw graphs of the variation of potential, kinetic and total energy of the system
i with time
ii with displacement from equilibrium.

28 A second identical trolley is placed on the trolley of question **27**, whilst it is oscillating. There is no slipping. How does
a the period,
b the amplitude,
c the total energy, and
d the maximum potential energy
of the system change when the second trolley is attached
i at one end of the oscillation?
ii at the mid-point of the motion?
c Why does the total energy of the final system differ in **i** and **ii**?

29 The table of results in Figure 19.35 was obtained in experiment 19.2, using the inertia balance. It shows period T of oscillation against mass m added to the mass holder. The frequency f of oscillation of the system should obey the formula

$$f = \frac{1}{2\pi}\sqrt{\frac{k}{M+m}},$$

where k is the stiffness of the system and M is the mass of the holder.
a Show that the formula can be written:

$$m = (k/4\pi^2) \cdot T^2 - M.$$

m/kg	T/s
0	0.36
0.2	0.52
0.4	0.67
0.6	0.80
0.8	0.88
1.0	0.97

Figure 19.35 Table of results from experiment 19.2

E OSCILLATIONS

b Plot a suitable linear graph to show that the data satisfies the formula.
c Find the mass M of the holder and the stiffness k of the system.
d Is there a quite different method of measuring k to check your value? If so, describe it.

30 A constant circular motion is converted into a linear 'reciprocating' motion in many machines. An example is the domestic sewing machine shown in Figure 19.32f.
a Write down expressions for the displacement y, speed v and acceleration a of the needle point in terms of the frequency f of rotation and radius r of the driving wheel. Take the starting point of the motion to be when the needle is at its highest point.
b Estimate the maximum speed and acceleration of the tip of the needle by observing a sewing machine in action.

31 Show that the period of oscillation of the hydrometer in question **20a** is given by:

$$T = 2\pi\sqrt{d/g}$$

where d is the depth of immersion when the hydrometer is floating at rest.

32 A girl standing on the end of a diving board flexes it up and down. Assume that the end of the board executes simple harmonic motion of constant amplitude 0.20 m. Above its equilibrium position, the end of the board experiences a downward acceleration.
a Explain why the girl and board separate when the downward acceleration is greater than g.
b Find the highest frequency at which the girl and the board do not separate.

33 Imagine that a smooth surfaced tunnel is driven through the centre of the Earth. Assume that the Earth is a perfect sphere of uniform density. Newton's law of gravitation tells us that the force F on a mass m, a distance r from the centre, is given by:

$$F = GMm/r^2,$$

where M is the mass of the sphere of radius r 'beneath' the object (see Figure 19.36).
a Show that the mass m moves along the tunnel in simple harmonic motion.
b Find an algebraic expression for the period of the motion.
c Calculate this period. (G, g, E_R)
d A satellite just skims the surface of the Earth. Find an expression for the period of its orbit, and its value.

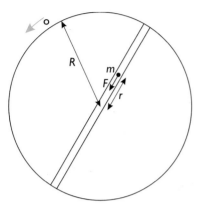

Figure 19.36

20 DAMPED AND FORCED OSCILLATIONS: RESONANCE

If you displace a child's swing from its rest position and then let it go, it will oscillate freely about its lowest point. If there were no friction at the support or no damping due to air resistance, the oscillations would carry on forever with constant amplitude (← 19). In reality, there is damping and the amplitude of the oscillation decays with time, as shown in Figure 20.1.

To maintain the amplitude of a real oscillation, we have to apply a periodic driving force. This transfers energy into the system to counteract the effects of damping. The resulting oscillation is called a **forced oscillation**. With little damping and a periodic driving force matching the natural frequency of the swing, large amplitudes can build up. This process is known as **resonance**.

Damping has different effects on free and forced oscillations. Its effects at resonance can be considered by varying the driving frequency and observing the resulting forced oscillations of a system.

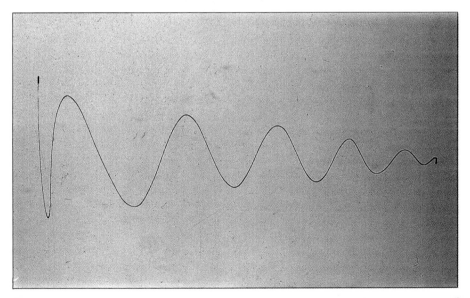

Figure 20.1 The trace against time of the decaying amplitude of an oscillation

20.1 THE EFFECT OF DAMPING

Plucking a guitar string sets it into free vibration. At each successive oscillation, the amplitude is reduced. This is because energy is transferred from the system in two ways: work done against the viscous forces of the air, and energy radiated away in the sound produced. For natural light damping, the amplitude often

OSCILLATIONS

Damped harmonic motion

For an undamped oscillation of a mass m on the end of a spring of stiffness k, the restoring force F is proportional to the displacement x:

$$F = -kx$$

($\leftarrow$ 19.2). Experimental work shows that, normally, the damping force F_d opposing the motion is proportional to the velocity v of the mass, and:

$$F_d = -bv,$$

where the constant b is the opposing force per unit velocity.

So the equation of the motion of the damped mass-spring system becomes

$$F - F_d = -kx$$

or $\quad F + bv + kx = 0$.

Since force equals mass × acceleration ($F = ma$):

$$a + \frac{bv}{m} + \frac{kx}{m} = 0$$

or $\quad a + \gamma v + \omega_0^2 x = 0,$

where the damping constant $\gamma = b/m$, and the natural angular frequency $\omega_0 = \sqrt{k/m}$. This equation can be solved numerically using a suitable modelling program or spreadsheet system with a computer ($\leftarrow$ 19.2 and $\rightarrow$ 41.3).

The solution to the equation is shown graphically in Figure 20.2. A general solution in the form of an analytical mathematical function can be found and has the following form:

$$x = Ae^{-(\gamma/2) \cdot t} \cdot \cos(\omega t + \phi),$$

where ω is the angular frequency of the damped oscillation given by:

$$\omega^2 = \omega_0^2 - \gamma^2/4.$$

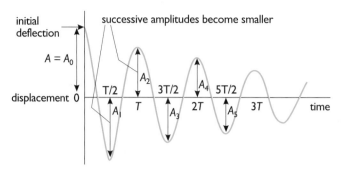

Figure 20.2 Damped oscillations

decays exponentially with time, as shown in Figure 20.2. The frequency of the oscillation hardly changes from the natural frequency. The ratio of adjacent amplitudes is a constant called the **decrement** δ, of the damping:

$$\delta = \frac{A_0}{A_1} = \frac{A_1}{A_2} = \frac{A_2}{A_3}, \text{ etc.}$$

Experimentally, for δ to be constant the damping force F_d must be proportional to the velocity of the system at any instant and oppose its motion. In a musical instrument, such as the guitar, the complete decay of the amplitude may take several seconds. This depends on the energy relationships in the system described by the Q-factor ($\rightarrow$ 20.4).

Questions

1 What fraction of the energy of an oscillating system has been dissipated when its amplitude has halved?

2 The damping force within a mechanical oscillating system depends upon the speed of the oscillating mass within a given fluid medium. For 'Newtonian' fluids, the damping force is proportional to the speed.

Devise an experiment you could perform to investigate the relationship between the damping force and the speed of movement through various fluids.

3a Using a suitable software package, set up the numerical model for an undamped harmonic oscillator ($\leftarrow$ 19.2). Satisfy yourself that the displacement of the oscillator varies sinusoidally with time and is of constant amplitude.
b Change the first line of the model to incorporate a damping term, as described in the box: Damped harmonic motion. Using $k = 16$ N m^{-1}, $m = 1.6$ kg and an initial amplitude of 12 cm, what value of b will cause the amplitude to decay to zero **i** within 5 oscillations? **ii** within 10 oscillations?
c Does the initial amplitude have any bearing on the time taken for the oscillations to decay to zero for a given degree of damping?

DEGREES OF DAMPING

When vibrations are undesirable, artificial heavy damping is designed into the system. For example, the suspension system of a car includes shock absorbers or dampers. In a simple shock absorber, shown in Figure 20.3, a piston moves in a cylinder filled with a viscous oil. After the suspension receives a sudden jolt,

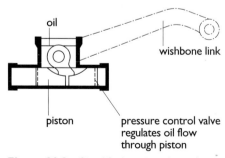

Figure 20.3 Simplified section through a car shock absorber

20 DAMPED AND FORCED OSCILLATIONS: RESONANCE

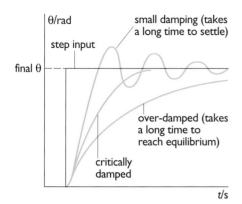

Figure 20.4 Deflection against time response for a moving coil meter with different degrees of damping

there is a smooth slow return to equilibrium without oscillation. This is an example of **over-damping**.

Some measuring instruments, such as a moving-coil ammeter, include electromagnetic damping, so the needle moves smoothly across the scale to a steady reading when the circuit is completed. Without this damping, the needle would oscillate many times, due to the inertia of the needle and coil, and the stiffness of the spring, before coming to rest. Such a situation is called **under-damping**. We normally want the meter system to reach a steady reading as quickly as possible. This requires **critical damping**, as shown in Figure 20.4.

The coil of a light-beam galvanometer on the direct range is almost undamped. We use this instrument in experiment 20.2 to show simple harmonic oscillations. It can also be used, with a variable resistor across its terminals, to demonstate a wide range of degrees of damping.

EXPERIMENT 20.1 Simple damping

You can investigate the effect of damping on the amplitude of a mass-spring oscillator by measuring the decrement δ and seeing whether or not the damping is constant. The damping can be introduced by:

a attaching card discs of varying diameter

b immersing the mass in water or oil in a dashpot, or cylinder, as shown in Figure 20.5. You can find out to what extent the amount of damping depends on how closely the mass fits the cylinder and on the viscosity of the liquid

c attaching a sheet of aluminium to the mass so that the sheet oscillates between the poles of two bar magnets, as in Figure 20.6. The distance between the magnets will alter the amount of damping.

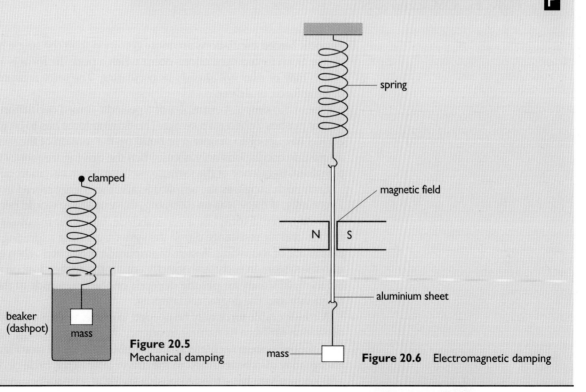

Figure 20.5 Mechanical damping

Figure 20.6 Electromagnetic damping

OSCILLATIONS

EXPERIMENT 20.2 Damping of a light-beam galvanometer movement

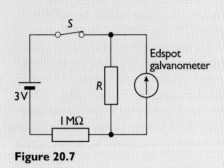

Figure 20.7

By setting up the circuit in Figure 20.7 and setting the galvanometer to 'direct' or undamped, you can measure the natural period of oscillation of the galvanometer. By reducing the value of the 1 MΩ resistor to give full-scale deflection and opening switch S, you can observe the oscillation of the galvanometer coil.

With R at, say, 3 000 Ω, you can observe lightly damped oscillation, as in Figure 20.1. Reducing R to, say, 100 Ω increases damping to its critical value. Further reduction in R causes the movement to be over-damped and the light spot will take longer to return to zero displacement (see Figure 20.4).

Questions

4 A moving-coil meter movement is often slightly under-damped. What advantage does this have?

5 An oscilloscope follows a step input very closely. Why do you think its response is so much faster than a moving-coil meter?

Buildings and other structures will oscillate to varying degrees when given a sudden push, such as a strong gust of wind. Artificial damping is built into each structure through its rigidity. It is important to consider the frequency of the gusts when predicting the subsequent behaviour of the structure.

20.2 FORCED OSCILLATIONS

Periodic forced oscillations are more common than the single impulse considered above. Such forced oscillations occur when a periodic force is applied to a system which may or may not already be oscillating. The force is usually called the driving force, or **driver**.

One of the simplest examples of a periodic forced oscillation occurs when a parent pushes a child on a swing. The driving force has to be repeated at the correct frequency to maintain or build up the motion of the swing. Large amplitude oscillations only occur when the driving frequency is close to the free or natural frequency of the swing's oscillations – a situation called **resonance**.

Each push increases the amplitude and the energy stored in the system. Eventually, if there was no damping, the amplitude would become infinite and the swing would complete a vertical circle. However the dissipative forces of friction and air resistance damp the system, creating a limiting value to the amplitude of the swing. A steady amplitude is reached when extra energy put into the system in each push exactly equals the energy loss on each oscillation. So, the final steady amplitude depends on the magnitude of the driving force, its frequency and the degree of damping.

A considerable force may be needed to cause, within a system, any significant oscillation far from its natural frequency. Most of the energy of the driver is used to make the system oscillate, so the amplitude is small (see Figure 20.8). The energy may also be returned to the driver later in the cycle of oscillation – as you will know if you have tried to push a swing at a frequency away from resonance.

20 DAMPED AND FORCED OSCILLATIONS: RESONANCE

Figure 20.8 Build up of displacement for different applied frequencies

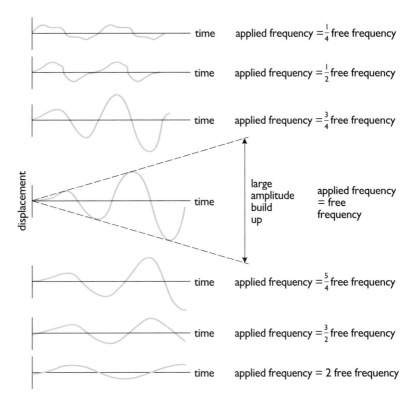

You can study a system similar to the child's swing in the laboratory using an arrangement called Barton's pendulums, as in experiment 20.3. This allows you to observe the effect of a *single* sinusoidal driving frequency on a set of oscillators having *different* natural frequencies, dependent on the lengths of the pendulums.

Body resonances

The cause of a series of helicopter crashes was traced to the forced oscillation of the pilots' eyeballs due to the vibration transmitted from the engine of the aircraft. The frequency of the vibration caused disturbance of the pilot's vision so that overhead power lines became very difficult to see.

Different parts of the body, such as the abdomen and other internal organs, possess different natural frequencies. These depend on their mass and the elastic constant of their supporting medium. Much lower forcing frequencies on the body can lead to motion sickness, while very high frequencies cause damage to the cells of the body (see Figure 20.9).

Figure 20.9 Effects of oscillations of various frequencies on the body

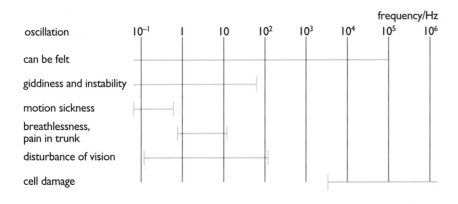
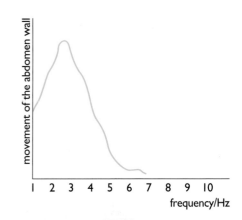

OSCILLATIONS

Question

6a On what factors does the natural frequency of oscillation of a pendulum depend?

b Relate these factors to the general expression for the natural frequency of a simple harmonic oscillator:

$$f_o = \frac{1}{2\pi} \sqrt{\frac{\text{force constant}}{\text{inertia}}}.$$

c Referring to **a** and **b**, estimate the natural frequency of oscillation of your own leg. What implications do you think this might have on the speed at which you walk?

EXPERIMENT 20.3 Barton's pendulums

Figure 20.10 shows Barton's pendulums – a set of pendulums of varying length with paper cone bobs at equal intervals along a string, driven by a heavy pendulum. If you pull the driver pendulum to one side so that it swings in a plane perpendicular to the diagram, you can observe what happens to the pendulums over a period of time. You can then repeat the exercise with the cones weighted.

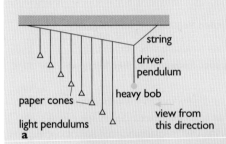

Figure 20.10a Barton's pendulums
b time exposure and instantaneous photographs of lightly damped oscillations

  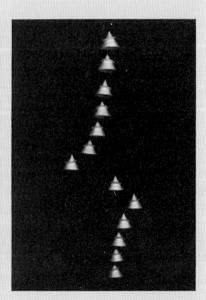

Question

7 Adding mass to the cones in experiment 20.3 affects the damping of the system. Why is this so and in what way is the degree of damping affected?

20 DAMPED AND FORCED OSCILLATIONS: RESONANCE

MOTION OF BARTON'S PENDULUMS

In experiment 20.3, the driving force of the heavy pendulum is applied to each light pendulum through the motion of the supporting string. The applied force varies sinusoidally with time. At first, the motion of each pendulum is very irregular. This **transient oscillation** has a frequency that combines the pendulum's natural frequency with the driver frequency. All the transient oscillations die away exponentially to leave all the pendulums oscillating at the driver frequency – **steady-state oscillation**. The time taken for this to happen depends on the degree of damping. The larger the effect of damping, the more quickly the steady state is reached.

The amplitude of oscillation of each pendulum depends on how close the driving frequency is to the natural frequency of the pendulum and also on the degree of damping. The pendulum with the maximum amplitude, the **resonating pendulum**, has a natural frequency equal to that of the driver. Maximum energy is supplied through the connecting string from the driver to this pendulum.

The pendulums do not all move in phase. The short pendulums, with natural frequencies higher than that of the driver, move in phase with the driver. The long ones, with lower natural frequencies, move in antiphase – 180° or π radians out of phase. The pendulum in resonance is one quarter of an oscillation behind the driver, lagging by a phase angle of $\frac{\pi}{2}$. Maximum energy is supplied by the forcing oscillator, the driver pendulum, when this force is in phase with the velocity of the resonating pendulum. This condition is called **energy resonance**.

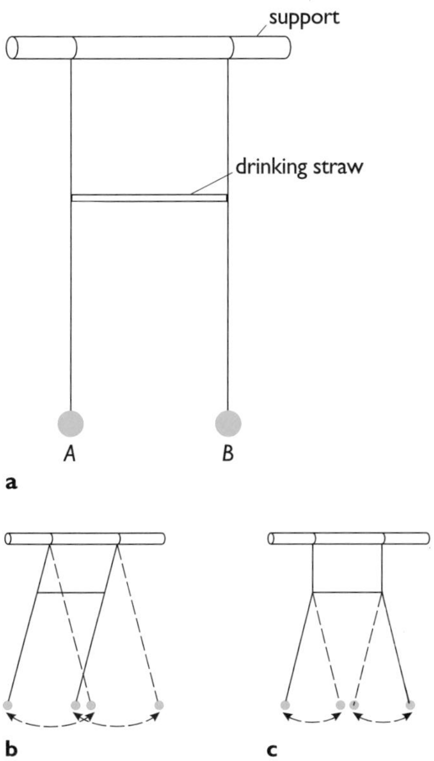

COUPLED OSCILLATORS

One of the major features of a resonating system, which makes many simple demonstrations difficult to perform, is the effect of the driven oscillator on the driver. This can be significant near resonance. In experiment 20.3, the inertia of the driving pendulum is much greater than that of the driven system, so the energy changes of the driver are insignificant. When two identical pendulums are coupled together, the situation is very different. The driving pendulum loses energy to the driven pendulum until the roles of the pendulums are reversed (see Figure 20.11). These are known as **coupled oscillators**.

If you set up a row of identical coupled oscillators, and displace and release the first one, the energy will be transferred progressively down the row from the first to the last and then reflected back. This model can be used to illustrate the propagation of energy of a wave pulse along a string or spring. In another model (→ 29.3) masses and springs are connected together as coupled oscillators to illustrate the propagation of a compression wave through a solid.

Another common laboratory example of coupled oscillations occurs when a mass oscillating on the end of a spring also exhibits a pendulum motion. The coupling is most evident when the periods of vertical oscillation and pendulum motion are in a simple ratio. The energy is clearly seen to transfer back and forth between the two modes of oscillation.

Figure 20.11 2 coupled oscillators

20.3 AMPLITUDE AND PHASE IN FORCED OSCILLATIONS

How an oscillating system behaves under the action of a periodic driving force has implications in many areas of science and engineering (→ 20.6). So it is of considerable practical importance to know about it.

Experiments 20.4 and 20.5 allow you to investigate forced oscillations in a single oscillator. You can apply a driving force over a range of frequencies

OSCILLATIONS

around its natural frequency, and vary the amount of damping within the system. Experiment 20.4 is more visually striking, while experiment 20.5 is better for making quantitative measurements of the phase difference between driver and driven, and for controlling the degree of damping.

EXPERIMENT 20.4 Forced oscillations of a mass on a spring

You can produce forced oscillations using a vibrator mounted vertically with a spring and a 0.05 kg mass hanging from it, as shown in Figure 20.12. The two straws indicate the motion of the vibrator and the mass.

Immersing the mass in water damps the oscillations. By varying the frequency of the signal generator between 1 Hz and 10 Hz, and maintaining a constant driver amplitude, you can plot a graph of the amplitude of the driven mass against frequency.

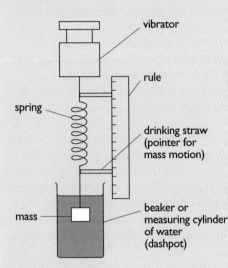

Figure 20.12

EXPERIMENT 20.5 Forced oscillations of a lightly damped light-beam galvanometer movement

If you connect the circuit in Figure 20.13a to an oscilloscope and a light beam galvanometer, as in Figure 20.13b, you can measure:

a the amplitude x_o of oscillations of the light beam
b the displacement x_1 between the light beam and the electron beam when it is at the centre of the screen.

Since the phase angle $\phi = \sin^{-1}(x_1/x_o)$, you can keep x_1 constant and plot graphs of x_o and ϕ against frequency for various shunt damping resistors. This will give you a set of curves similar to those in Figures 20.14 and 20.15.

Figure 20.13a Circuit and
b arrangement of apparatus for experiment 20.5

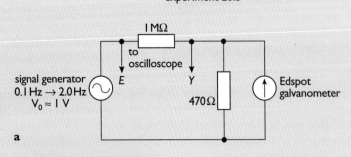
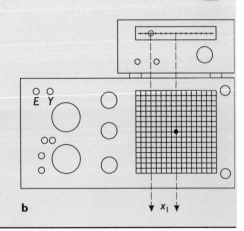

20 DAMPED AND FORCED OSCILLATIONS: RESONANCE

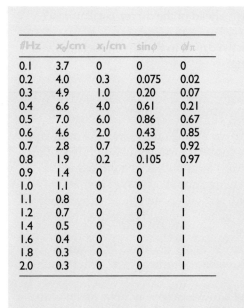

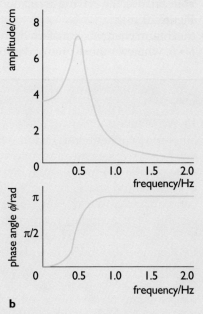

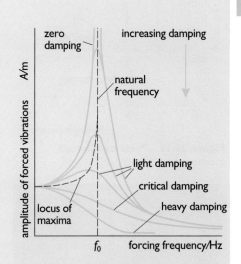

Figure 20.14a Results table and **b** graphs for $R = 470\ \Omega$

Figure 20.15 The amplitude response curves for a mechanical system subject to damping

Questions

8 In experiment 20.4, at what frequency do you observe resonance? Is this equal to the natural frequency:

$$f_o = \frac{1}{2\pi}\sqrt{\frac{k}{m}}?$$

9 Why does the extended wing mirror of a lorry often vibrate violently when the lorry is stationary with the motor idling, yet hardly oscillates when the lorry is moving?

The following are the more significant observations from experiments 20.4 and 20.5. (You might like to compare these with those from Barton's pendulums, experiment 20.3):

- the driven oscillator has the same frequency as the driver frequency f_f in the **steady state**. The larger the damping, the more quickly the steady state is reached. This is true irrespective of whether or not f_o, the natural frequency of the driven oscillator, and f_f are equal
- as the system approaches the steady state, the driven amplitude varies in the form of **beats** of frequency equal to the difference between the natural and the forcing frequency
- the amplitude of the steady state oscillations is a maximum when $f_f = f_o$. The system is in **resonance**. For a fixed driver amplitude, the driven amplitude at resonance depends entirely on the degree of damping
- the amplitude of the driven oscillations varies with the forcing frequency f_f as in Figure 20.15. As the damping increases, the amplitude becomes less at all values of f_f. The maximum amplitude occurs at a frequency slightly less than f_o as damping increases

E OSCILLATIONS

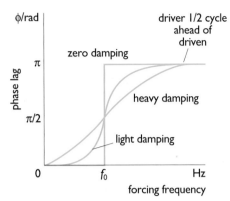

Figure 20.16 Phase lag in forced oscillations

- at resonance, the driving oscillation is $\frac{\pi}{2}$ ahead of the driven oscillator (see Figure 20.16)
- maximum energy is transferred from the driver to the driven oscillator when $f_f = f_o$, whatever the degree of damping.

Question

10 A girl hangs a mass of a few hundred grams on the end of a long, thin but strong rubber band. Holding the top of the rubber band quite still in one hand, she sets the mass into vertical oscillations. She finds that the mass oscillates with a natural frequency of about 1 Hz (see Figure 20.17).
a What would she observe if, instead, she moved her hand supporting the band up and down at a very low frequency, say, 0.1 Hz, and with an amplitude of a couple of centimetres?
b Suppose she moved her hand up and down very quickly at, say, a frequency of around 5 Hz with the same amplitude as in **a**? What would she observe?
c Suppose the frequency of her hand's movement was 1 Hz, again with the same amplitude as in **a**. What would happen?
d Describe the phase relationships between the motion of the suspended mass and the force exerted by the girl's hand in **a**, **b** and **c**.

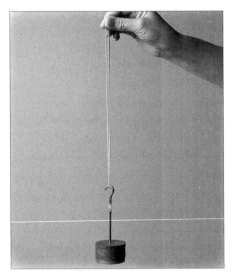

Figure 20.17

Look at the shapes of the amplitude curves shown in Figure 20.15. Far from resonance, the shapes can be explained simply from experiment 20.4. At very low frequencies, the spring in Figure 20.12 acts as a rigid rod. There is ample time for the mass to move in phase with the driver and with the same amplitude. The system is said to be **stiffness controlled**.

At frequencies well above resonance, the system is **inertia controlled**. The force in the spring is too small to accelerate the mass any distance before the direction of motion is changed. So the amplitude of the driven mass is close to zero and the motion is in antiphase with the driver.

20.4 THE QUALITY OR Q-FACTOR

The Q-factor is a means of expressing the energy lost by an oscillator. When a free vibration lasts a long time, the energy dissipated during each cycle of the oscillation due to frictional effects is small compared with the total energy of the oscillating system: its Q-factor is high.

The Q-factor is defined by:

$$Q = \frac{1}{\text{fractional energy loss per radian of oscillation}}$$

or $$Q = 2\pi \cdot \frac{1}{\text{fractional energy loss per oscillation}}$$

$$Q = 2\pi \cdot \frac{\text{total energy of the system}}{\text{energy loss in one oscillation}}.$$

The amplitude of a lightly damped oscillation decays exponentially and has a frequency slightly less than its natural frequency. The oscillator completes many cycles before the amplitude decays significantly. The Q-factor can be found experimentally by counting the number of oscillations n, before the amplitude

20 DAMPED AND FORCED OSCILLATIONS: RESONANCE

decays to $1/e$ of its initial value. Then:

$$Q = \pi n.$$

These definitions are difficult to apply when the degree of damping is large. An estimate of Q can be made by observing the total number of visible oscillations that occur before the oscillator comes to rest. Figure 20.18 lists typical values of Q-factor for some systems.

	Q-value
car suspension	1
trolley in Figure 19.16	10
low frequency electrical resonance circuits	100
pendulum in air	500
pendulum in vacuum	1 000
piezoelectric quartz crystal	10^5
atomic transitions	10^7

Figure 20.18 Table of Q-values

> **Questions**
>
> 11 What is the unit of the Q-factor?
>
> 12 Estimate the Q-factors for the following damped oscillations:
> a a guitar string giving a note of about 500 Hz
> b the balance wheel of a mechanical watch
> c a child swinging on a swing.
>
> 13 Show that Q is about 12 for an oscillator that loses half of its energy each cycle of its oscillation.

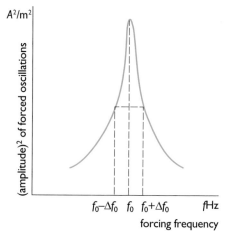

Figure 20.19 Band-width is defined as $2\Delta f_o$

Q AND THE SHARPNESS OF RESONANCE

In a forced oscillation, the sharpness of the resonance peak depends on the degree of damping. This determines the amplitude of the oscillation at resonance (see Figure 20.15). A lightly damped system has a high Q and a sharp resonance peak. In fact, the Q-factor of a system is one way of indicating the sharpness or selectivity of its resonance response to driving forces of various frequencies.

The Q-factor is used as a quantitative measure of the **band-width** of a resonance curve. It is a useful concept, for example, when considering the ability of the tuned circuit of a radio receiver to separate the chosen radio frequency from other nearby frequencies. A high Q means a narrow band-width and, therefore, good selectivity ($\rightarrow$ 22.5). The important quantity here is the energy at the selected frequency received by the aerial, which is proportional to the square of the amplitude of the oscillation. The band-width of the resonance curve is defined as the difference in frequency between the points at which the square of the amplitude is half the peak value (see Figure 20.19). Under light damping conditions, this peak is symmetrical. The Q-factor is given by:

$$Q = \frac{\text{resonant frequency}}{\text{width of resonance curve at half height}},$$

or $Q = \dfrac{f_o}{2\Delta f_o}.$

This is also known as the reciprocal of the relative band-width.

In practice, we quite often plot the amplitude resonance curve. Then:

$$Q = \frac{\text{resonant frequency}}{\text{range of frequencies at } 1/\sqrt{2} \text{ of maximum amplitude}}$$

$$= \frac{f_o}{\Delta f}.$$

Forced oscillations and the numerical method

The numerical method provides a way of analysing forced oscillations. The equation of motion of a damped mass-spring system is:

$$F + bv + kx = 0.$$

When driven by a sinusoidal driving force F_f, as shown in Figure 20.12, this becomes:

$$F + bv + kx = F_f.$$

Spreadsheets provide a method of investigating such systems ($\rightarrow$ 41.3).

20.5 RESONANCE IN CONTINUOUS MEDIA

In some systems, there is more than one frequency of resonance. In experiment

E OSCILLATIONS

20.4, we assume the inertia of the system is concentrated in the mass at the end of the spring and the restoring force is concentrated within the spring. There is only one natural vibrational frequency for this lumped or discrete system. However, if you increase the driving frequency of the vibrator by a factor of 100, as in experiment 20.6, you will find several related driving frequencies that give resonance (frequencies at which the energy of the system becomes a maximum). The mass will act as a fixed point and only the spring will vibrate. Each small part of the spring has inertia and stiffness, so the spring can be thought of as a collection of coupled oscillators.

EXPERIMENT 20.6 Resonances of a spring

By hanging a mass from a spring attached to a vibrator, as in experiment 20.4, and using driving frequencies of 100 and 200 Hz, you should observe resonances or nodes at particular frequencies, as in Figure 20.20. These are related by:

$$f_n = \frac{n}{2l}\sqrt{\frac{T}{\mu}},$$

where l is the length of the spring, T is the tension, μ is the mass per unit length and n is the number of vibrating sections.

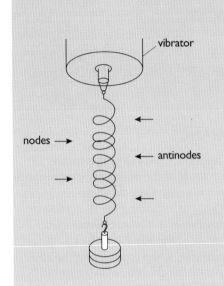

Figure 20.20 Nodal points observed in experiment 20.6

When resonance occurs in the spring, energy is trapped between points half a wavelength apart. These **nodal points** are permanently at rest. We usually call this phenomenon a standing or stationary wave (→ 30.6). The energy can be thought of as 'standing' in the spring, alternating between kinetic energy and potential energy. We can consider the system to be the superposition of two identical waves travelling in opposite directions with wavelengths matched correctly to the length of the spring.

Question

14 A motor car wheel, including the tyre, is unlikely to have its centre of mass exactly on its axis of rotation. When travelling at a certain speed, severe vibrations are sometimes felt through the steering or suspension of the car. To minimise these vibrations, wheels are usually 'balanced' when a tyre is replaced.
a Why do these vibrations occur?
b What do you think determines the actual speed at which these vibrations occur?
c Why would they not occur at speeds a little above or below this critical speed?
d With the help of Figure 20.21, explain what is done when a wheel is 'balanced'.

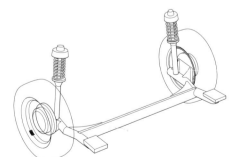

Figure 20.21 Wheel, balancing weight and suspension system of a car

20 DAMPED AND FORCED OSCILLATIONS: RESONANCE

20.6 EXAMPLES OF RESONANCE...

Resonance phenomena occur almost everywhere in nature. Some are useful and some inconvenient, as the following examples show.

... IN MUSICAL INSTRUMENTS

Musical instruments all rely on resonance for the amplification of sound. A tuning fork held over a closed tube sounds louder when the tube is of the correct length. So here two very lightly coupled oscillators, the fork and the air in the tube, result in a loud sound. In a violin or guitar, the strings are held taut over a hollow wooden box with a hole or holes in it. The air in the box vibrates, acting like a spring, and drives the air near the hole, which acts like the mass of the system. The note obtained by blowing across the top of a milk bottle is caused in the same way (see Figure 20.22). Such a system is called a **Helmholtz resonator**.

To amplify the sound from all the strings, the sounding box of a guitar or violin needs to respond over a broad band of frequencies. So, the system must have a low Q. This can be achieved by having a high degree of damping and by using several resonators coupled together to cover different ranges of the frequency spectrum. The interaction of these coupled oscillators on each other leads to many resonant frequencies in the system and broad band amplification occurs. The thin wood of the box gives the required damping and the complex shape leads to many resonant frequencies within the box. A sounding box is quite unlike a resonance tube, which has a high Q-factor ($\rightarrow$ 30.6).

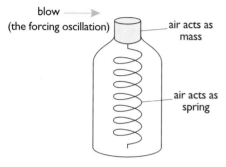

Figure 20.22 Milk bottle as a Helmholtz resonator

... IN ATOMS AND MOLECULES

Resonances occur at an atomic level where electromagnetic radiation causes atoms to vibrate. Usually the vibrations are very small, but near resonance the amplitude becomes significant. For example, NaCl (sodium chloride) crystals absorb radiation strongly in the infrared at a wavelength of 6.1×10^{-6} m (see Figure 20.23). The crystals can be thought of as formed of alternate Na$^+$ and Cl$^-$ ions held together by electric forces ($\rightarrow$ 26.4). Each ion behaves as a mass tethered by two springs whose stiffness ($\leftarrow$ 4.1) is determined by the interatomic forces.

Molecular structure is investigated by observing infrared absorption by a specimen, a technique known as **absorption spectroscopy**. The atoms of each molecule are driven in various modes of oscillation, as in Figure 20.24. The rate of absorption of energy is a maximum at any resonance ($\rightarrow$ 37.2).

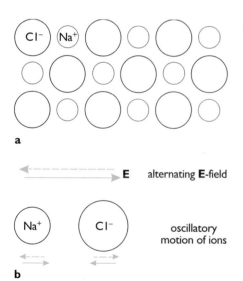

Figure 20.23a Model of NaCl crystal, **b** forced oscillations of ions by an **E**-field

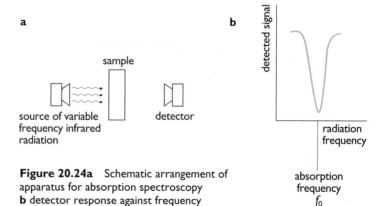

Figure 20.24a Schematic arrangement of apparatus for absorption spectroscopy **b** detector response against frequency

283

E OSCILLATIONS

> **Question**
>
> 15 The mass of a Na$^+$ ion is 3.8×10^{-26} kg. Estimate the 'spring constant' between a pair of ions, given that
>
> $f_0 = \dfrac{1}{2\pi}\sqrt{k/m}$. (c)

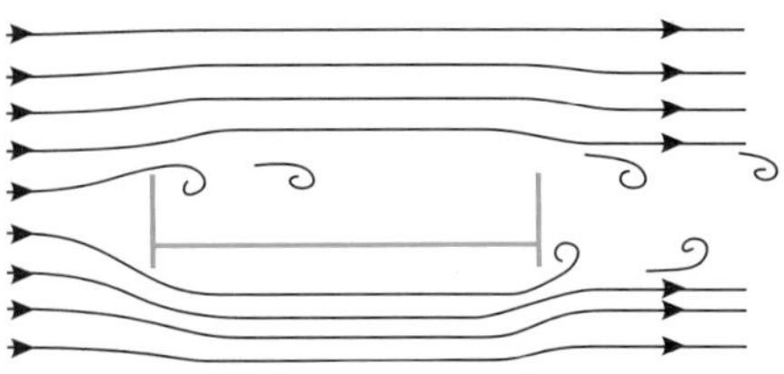

... IN STRUCTURES

Figure 19.1 shows the catastrophic failure of the suspension bridge across Washington State's Tacoma Narrows in 1940. This is a spectacular example of the resonant effect of wind-induced oscillations on a structure. The wind speed was quite low, only about 20 m s^{-1}, but it caused torsional and vertical oscillations in the bridge deck. Shackles between the cables and the deck were loosened, reducing the damping and allowing the oscillations to build up to destructive proportions. A similar failure had occurred to the Chain Pier at Brighton some 100 years before, but no research had been done in the intervening period to discover the cause of the oscillations.

Before building the Severn bridge, scientists studied the effects of the wind on the Tacoma bridge using scale models in a wind tunnel. They found that the problems could be avoided by replacing the solid sides of the bridge with lattice work or by making the deck section a more streamlined shape.

The Tacoma bridge collapsed because of a phenomenon called **vortex shedding** (see Figure 20.25). Wind blowing past a cylinder or a block-shaped object forms 'whirlpools' of air beyond called **vortices**. The fluttering of a flag or the trailing end of a sail in a breeze is also caused by these eddies. The vortices are shed first from one side of the object and then the other. Pressure changes in the air act as a regular oscillatory driving force on the object. This results in a forced oscillation with an amplitude and frequency dependent on the speed of the wind.

Most circular steel chimneys are fitted with helical fins called **strakes** (see Figure 20.26) to break up the regularity of the vortex shedding. This reduces the possibility of the build up of a resonance oscillation. For similar reasons, special dampers are fitted to long cables, such as overhead power lines, to minimise wind driven vibrations of the cables.

Figure 20.25 Vortex shedding over a bridge deck with solid sides

Figure 20.26 Strakes on a chimney

... IN MACHINES

It is important to avoid the build up of resonances between a machine and its base. For example, the accuracy of a sensitive weighing machine can be reduced by vibrations of the base on which it rests. A lathe or other rotating machine will vibrate on its mounts. If the mounting is too stiff, the machine may damage the floor; if too soft, it may sway around. The best anti-vibration mounting consists

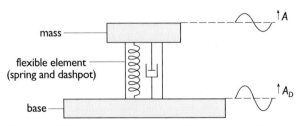

Figure 20.27 Schematic diagram of a simple anti-vibration mounting. For a system where the base drives the mass, the transmissibility $T = A/A_D$

20 DAMPED AND FORCED OSCILLATIONS: RESONANCE

of a flexible mount with a massive platform or machine so that any resonant frequencies are very low compared with the frequencies of the forcing vibration. Sufficient damping must be included so that any vibrations die away quickly. This arrangement can be modelled by the simple system in Figure 20.27. A car seat, for example, has both springiness and damping to reduce the oscillations of the passenger. On a larger scale, Figure 20.28 shows a model of a car suspension system used in computer analysis.

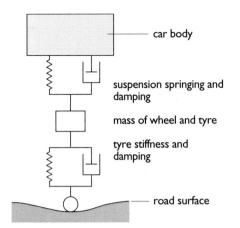

Figure 20.28 Model of a car suspension

Question

16 Referring to Figures 20.15 and 20.27, explain why the natural frequency of a system should be much less than the forcing frequency in order to achieve minimum amplitude.

SUMMARY

Any real oscillating system is damped. The potential energy and the kinetic energy transfer to other parts of the system and its surroundings. This increases their internal energies and, therefore, their temperatures, but usually only very slightly. When the damping forces are proportional to the velocity, the decay of the amplitude of oscillation is exponential.

A system, displaced from equilibrium, returns to equilibrium in the shortest time when it is critically damped.

When an external sinusoidal driving force, such as:

$$F = F_o \sin 2\pi ft,$$

is applied to a damped harmonic oscillator, the system will oscillate at the frequency of the driving force. The amplitude and phase of the driven oscillator will depend on:

i how close the frequency f of the driver is to the natural frequency f_o of the driven oscillator, and

ii the degree of damping.

When $f = f_o$, resonance occurs. There is maximum energy transfer between the driver and driven system. The driven oscillator lags $\frac{\pi}{2}$ radians behind the driver.

In a system where inertia m and stiffness k are concentrated, there is only one resonance at:

$$f_o = \frac{1}{2\pi} \sqrt{k/m}.$$

In a system where inertia and elasticity are spread uniformly, there is a series of resonances at related frequencies.

The quality or Q-factor is a numerical measure of the sharpness of resonance and the degree of damping. It is high for a lightly damped system, which will have a narrow resonance peak. One definition is:

$$Q = 2\pi/(\text{fractional energy loss per oscillation}).$$

Resonance can be both useful, for example, in musical instruments, and inconvenient, for example, in rotating machinery or wind-induced oscillations of large structures.

OSCILLATIONS

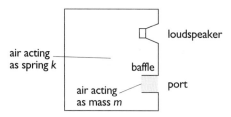

Figure 20.29 Bass reflex cabinet of a loudspeaker

SUMMARY QUESTIONS

17 What factors determine the amplitude, phase and frequency of the forced oscillations of the light pendulums in Barton's pendulums experiment?

18 The bass reflex cabinet of a loudspeaker behaves like a Helmholtz resonator (see Figure 20.29).
a Explain how the system helps to amplify low frequency sounds from the speaker.
b Suggest a suitable natural frequency for such a cabinet.

19 This question will help you to see the problems in using a model to study the behaviour of a real structure. An engineer makes a model of a bridge exactly to 1/8th scale. Assume, for simplicity, that you can use the expression:
$$f_0 = \frac{1}{2\pi}\sqrt{k/m}$$
for one natural frequency of oscillation of the bridge.
a How much smaller is the mass of the model than the real bridge?
b How much stiffer are the supports of the model than those of the real bridge?
c What is the value of f_0 for the model in terms of f_0 of the real bridge?

20a The undamped string of a piano has a Q-factor of 1 000. It is made to vibrate at 512 Hz, its fundamental frequency. Estimate how long it will oscillate before coming to rest.
b A mass of 0.2 kg, hanging on a vertical spring of stiffness 20 N m^{-1}, is set into oscillation. After 50 oscillations, the amplitude has fallen to 0.5 of the original displacement. Estimate the Q-factor of the system.

21 The 'spring constant' of the suspension system of a car is found to be 2×10^3 N m^{-1}, for small displacements. It is estimated that the mechanical power needed to keep the car rocking up and down with an amplitude of 10 mm is 10 W at the natural frequency of the suspension, 0.5 Hz.
a Suggest how you could measure the 'spring constant' of the suspension and the power to oscillate the car up and down.
b Estimate the Q-factor of this car.

22 A washing machine consists of a casing containing a motor and drum on a stiff but flexible mount.
a Why are the motor and drum mounted on springs?
b When clothes are badly loaded in the drum, the casing vibrates violently during spin drying, and the drum does not accelerate to maximum speed. When the clothes are re-arranged, the problems cease. Suggest an explanation.

23 A single trolley is tethered between two identical stretched springs, as shown in Figure 20.30a. It oscillates at a frequency f_0 when displaced in the line of the springs and released. A second trolley and third spring are then added to make the system of Figure 20.30b.
a Describe the motion of the two trolley system when the trolleys are released simultaneously with equal initial displacements:
 i both to the left
 ii one to the right and one to the left.
Calculate the oscillation frequencies of the trolleys in terms of f_0, in each case.

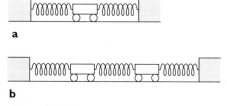

Figure 20.30

b One trolley is displaced whilst the other is held at rest. Both are released together. Describe their subsequent motion. State the frequency of the complete motion of the system, in terms of f_o, rather than the individual frequency of each trolley.
c Suppose you had a row of such trolleys connected by springs. What would happen if you displaced the end trolley and released it?

24 You can measure the acceleration, and so the velocity and displacement, of a moving vehicle by making observations on masses carried within the vehicle. Figure 20.31 shows the principle of one sort of device for measuring acceleration. Such devices are referred to as **accelerometers**. A mass m is free to move horizontally within a case, but is restrained by springs fixed to the case. A pointer on the mass can move over a scale fixed to the case. When the case and mass are at rest, the pointer is opposite the zero mark on the scale. When the pointer shows a displacement x from zero, the net force exerted by the springs is kx.

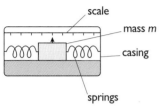

Figure 20.31

a Explain why the pointer is at zero when the mass and case are moving at constant velocity in the horizontal direction. Assume that the velocity has been constant for a long time.
b Explain why the pointer has a fixed displacement when the casing has been in a state of steady acceleration a to the left for a long time. Will the pointer's displacement be to the left or to the right?
c Give an expression for the magnitude of the displacement in **b**.
d Suppose the casing is displaced suddenly from rest by a sharp blow from a hammer and then held at rest. Describe the subsequent motion of the mass, if there is only a small amount of friction between it and the casing.
e It is suggested that, to measure varying accelerations, it would be good to have zero friction between the mass and the casing. Argue briefly for or against this idea.
f Suppose that, in use, appreciable changes of acceleration are expected to occur over times not exceeding time t. Give an argument to help decide whether the period T of natural oscillation of the mass and springs should be large or small compared with t.
g In designing an accelerometer for use in a car, a period T of $\pi/5$ seconds was chosen. It was assumed that accelerations up to 2 m s^{-2} should be measured. What would be the displacement at an acceleration of 2 m s^{-2}? (Note: the values of m and k are not needed.)
h It is decided that an accelerometer for use in a car accelerating at up to 2 m s^{-2} should have a period of 2π seconds. What problems would arise in designing this accelerometer?

25 This question is about the isolation of sensitive apparatus from floor vibrations. Isolation is achieved by mounting the apparatus and table of total mass M on a flexible suspension. The suspension may be modelled by a single spring. The floor vibrates at a frequency f and with an amplitude A_f. The natural frequency of the mass-spring system is f_o and the damping constant for the oscillation is ω_0/Q. The amplitude of vibration of M is given by:

$$A = \frac{\omega_o^2 A_f}{\sqrt{((\omega_o\omega/Q)^2 + (\omega^2-\omega_o^2)^2)}}.$$

a To analyse the effectiveness of isolation mountings, two dimensionless quantities, R and D, can be used. These are defined as:

frequency ratio $R = \omega/\omega_o$

damping ratio $D = 1/Q$.

Write down an expression for A in terms of R and D. Show that, as the floor frequency f is varied, the maximum amplitude of vibration of M occurs when:

$$R^2 = 1 - D^2/2.$$

Use $f/f_o = \omega/\omega_o$.

b Use your expression for A from part **a** to show that, for very small values of R, the amplitude of oscillation of M becomes equal to the amplitude of vibration of the floor.

Explain in physical terms how it is that, under these conditions, the spring behaves like a rigid rod.

c The ratio A/A_f is called the **transmissibility** T. Write down approximate expressions for T in terms of R and D for the cases when:
  i $R \rightarrow 0$
  ii $R = 1$
  iii $R \gg 1$.

Use these results to sketch graphs, on the same axes, showing how T varies with frequency for:
  iv $D = 0.1$
  v $D = 0.9$.

d For many situations, damping is small enough to be neglected, except near resonance. Write down an expression for T that would be valid everywhere except near resonance. Show that the isolation would be effective when $T < 1$ but only when $R > \sqrt{2}$.

e It is found that an analytic balance mounted on a slate slab cannot be used to its required accuracy because the floor is vibrating with an amplitude of 1.6 μm. From experience, it is known that this balance is only satisfactory when the amplitude of vibration is reduced to 0.2 μm. The main source of vibration is a nearby underground railway, and the predominant frequency is 22 Hz.
  i Calculate the natural frequency of the isolation system that would just enable the balance to operate satisfactorily.
  ii Given that the mass of the balance plus slab is 50 kg, what is the value of the required spring constant of the suspension?

21 ELECTRICAL OSCILLATIONS

When you first apply a voltage to a circuit, the voltage across each resistor, inductor and capacitor, and the currents within them, take time to become steady. Voltages and currents that are changing over a short period of time are called **transients**. Those involved when charging and discharging a capacitor through a resistor (← 13) are good examples. The behaviour of transients is important for producing pulses, for pulse shaping, and for analysing and designing counting circuits.

For electrical oscillations or sinusoidal signals, such as an alternating current, the electrical supply varies continuously, and there are no steady values of voltage and current. These signals can be processed – altered in frequency, amplitude or phase. Circuits can be designed, for example, to filter a particular range of frequencies from a complex signal; this process is used in hi-fi speaker systems, such as that in Figure 21.1.

Phasor diagrams (← 19.4) help you to understand how a.c. circuits work.

Figure 21.1 Hi-fi speaker and cross-over filter

21.1 RC CIRCUITS: THE TIME CONSTANT

A capacitor discharging through a resistor has less than 1% of its original charge when a time $5RC$ has elapsed (← 13.3). The exponential charge and discharge

OSCILLATIONS

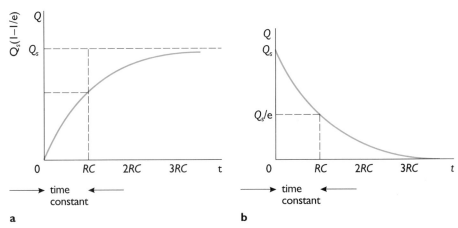

Figure 21.2a Charge on a capacitor during charging
b exponential decay of charge on a capacitor during discharge

curves are shown in Figure 21.2. Since charge $Q = CV$, the voltage and charge curves are identical in shape. The time for the charge on the capacitor to fall to $1/e$, or 0.37, of its initial value is called the **time constant** and is equal to RC.

The discharge curve is described by the equation:

$$dQ/dt = -Q/RC.$$

This is a differential equation with solution:

$$Q = Q_o e^{-t/RC}$$

or, when considering voltage, $V = V_o e^{-t/RC}$,

where Q_o and V_o are the charge on and the voltage across the capacitor at time $t = 0$.

The voltage across a capacitor when charged through a resistor varies with time according to the equation:

$$V = V_o(1 - e^{-t/RC}),$$

where V_o is the charging voltage.

EXPERIMENT 21.1 Plotting an exponential decay curve of charge stored on a capacitor

Using the circuit of Figure 21.3, you will find a time constant of about 50s. After measuring the initial voltage across the capacitor and current in the resistor before removing the flying lead, you can take readings of the current I at 10s intervals as the capacitor discharges.

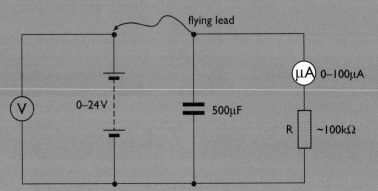

Figure 21.3 A CR discharge circuit

21 ELECTRICAL OSCILLATIONS

Question

1 The following is a typical set of results from experiment 21.1. Use it to plot a graph of current or charge against time in intervals of 10 seconds.

time/s	0	10	20	30	40	50	60
current/μA	90	73	59	48	39	31	25

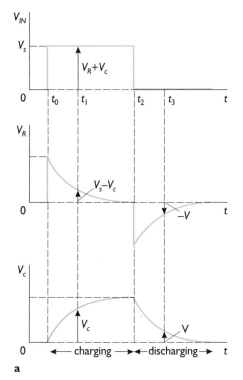

Suppose you apply a square-shaped voltage pulse V_S to a resistor and capacitor in series. The leading edge of the pulse causes an effect similar to that produced when switching on a battery supply to the circuit, as in Figure 21.4b. The falling edge has the same effect as short circuiting the supply (see Figure 21.4c). The voltage curves for a charge-discharge cycle are shown in Figure 21.4a. Note that, at any instant, the sum of the voltages across C and R is equal to the applied voltage.

Instead of a single square pulse, we can provide a continuous square wave input of a chosen frequency, as in Figure 21.5a and b, (see experiment 21.2). In a, the output is taken across C and in b, across R. The waveform of the output voltage depends on the ratio of the time constant of the circuit to the half period T of the input square wave. Note that here V_{IN} varies between $-V_S$ and $+V_S$.

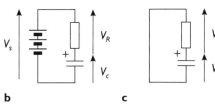

Figure 21.4a Voltages across input, R and C during charging and discharging of the capacitor,
b circuit with $V_S = V_R + V_C$ for $t_2 > t > t_0$,
c circuit with $V_R + V_C = 0$ for $t > t_2$

EXPERIMENT 21.2 Square pulses into a CR circuit

By supplying a square wave voltage input to circuits a and b in Figure 21.5 and observing the output on the oscilloscope screen, you can investigate the response of the circuit with $R = 1\,000\,\Omega$ and $C = 1\,\mu F$. You can slowly increase the frequency from an initial value of about 100 Hz to around 10 kHz, and then keep the frequency constant at, say, 1 kHz and vary the value of both R and C.

Figure 21.5 Applying a square wave to CR circuits: **a** an integrating circuit, **b** a differentiating circuit

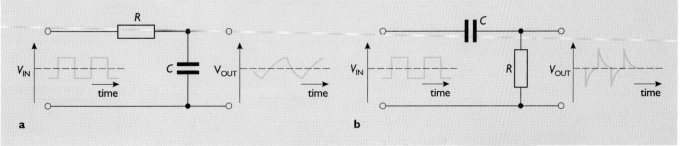

E OSCILLATIONS

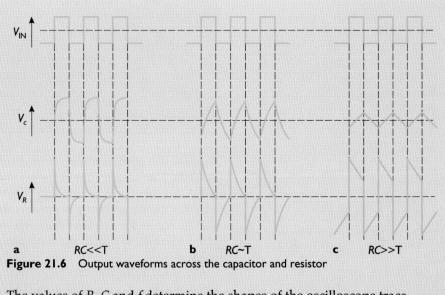

Figure 21.6 Output waveforms across the capacitor and resistor

The values of R, C and f determine the shapes of the oscilloscope trace shown in Figure 21.6.

Questions

2 In Figure 21.6, does the input voltage V_{IN} equal the sum of the voltage across the capacitance V_C and the voltage across the resistance V_R in each case?

3 Using Figure 21.4a, explain the changing shape of the output pulses as the pulse time T varies relative to RC.

21.2 INTEGRATING AND DIFFERENTIATING CIRCUITS

In this section, we look at two extremes: when RC is much less than, and much greater than T.

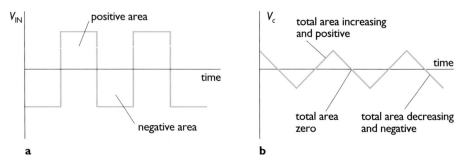

Figure 21.7a Input and **b** output waveforms for integration

292

21 ELECTRICAL OSCILLATIONS

When RC is much less than T, as in Figure 21.6a, we call the circuit a **differentiating** circuit. This is because the output across R just consists of positive and negative spikes with approximately the differential, or gradient, of the input square wave. Since the voltage across the capacitor cannot change instantly when charge flows, any sudden voltage change at the input causes the same change across the resistor, whatever the values of R, C and f. Consequently, the lengths of the vertical lines in Figure 21.6 are all the same, although the wave form has changed.

When RC is much greater than T, as in Figure 21.6c, we call the circuit an **integrating** circuit. This is because the output across C rises and falls almost linearly with time. The area under the input waveform also increases almost linearly with time (see Figure 21.7). So the output across the capacitor is roughly proportional to the integral of the input wave.

Questions

4 Using the fact that $V = V_o e^{-t/RC}$, $Q = V_C C$, $T = \dfrac{dQ}{dt}$ and $V = IR$, show that in the circuits of Figure 21.5:

$$V_C = \frac{1}{RC} \cdot \int V_R \, dt$$

and $V_R = RC \cdot \dfrac{dV_C}{dt}$.

5 Figure 21.6 shows that when RC is much less than T:

$V_C \sim V_{IN}$.

a Use the solution to question 4 to justify that:

$$V_{OUT} \propto \frac{dV_{IN}}{dt}.$$

b Justify that, when RC is much greater than T:

$$V_{OUT} \propto \int V_{IN} \, dt.$$

The RC combination appears extensively as a building block in electronic circuits with both square pulse and sinusoidally varying inputs. Square pulse inputs are used, for example, in the astable circuit and the relaxation oscillator ($\rightarrow$ 23.4 and 24.0).

21.3 INDUCTORS AND SELF-INDUCTANCE

When there is a current in a conductor, there is always a magnetic field, or **B**-field, around it ($\leftarrow$ 18.1). When the current changes, a voltage is induced by the **B**-field within the conductor itself. A conductor wound into the shape of a coil is called an **inductor** because in it such induced voltages are increased proportionally to the number of turns. An iron core inserted into the coil considerably enhances the inductive effects. These are most noticeable when a circuit is switched on or off because the rapid changes in **B**-field that occur induce large voltages.

OSCILLATIONS

EXPERIMENT 21.3 Behaviour of an inductor

If you adjust the circuit of Figure 21.8a so that both lamps have the same brightness, on switching on, the lamp in series with the inductor comes on later than the other lamp. The way the current grows can be investigated by connecting the circuit shown in Figure 21.8b. This circuit shows a make before break double pole switch. With the oscilloscope set on a very slow timebase, a trace similar to that shown in Figure 21.9a should be obtained.

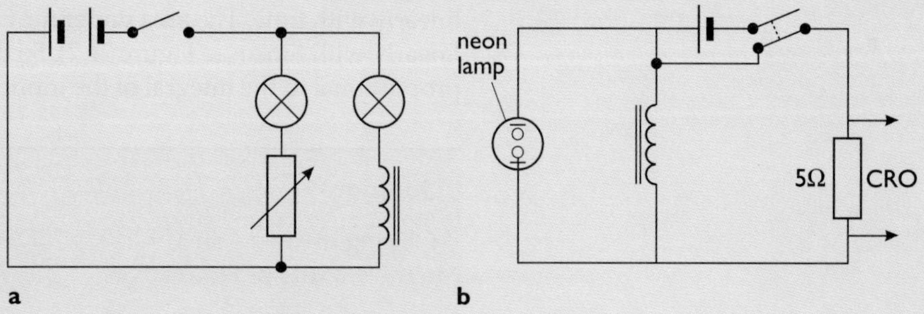

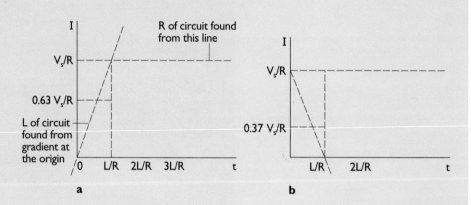

Figure 21.8 Circuits for investigating **a** decay and **b** growth of current

Figure 21.9a Growth of current in an inductor **b** decay of current and flux in an inductor when switched off

On opening the switch, you will notice that the spot drops very quickly to zero as the current ceases in the circuit. This causes the neon lamp to light if the 5 Ω resistance is removed. Without the iron core in the coil, the spot moves much more rapidly to reach the same final value as before.

Experiment 21.3 shows that a steady voltage V applied across an inductor causes the current I to rise initially at a steady rate so that V is proportional to the rate of change of current dI/dt. An increased current causes a larger voltage across the resistive parts of the circuit and, therefore, a lower voltage occurs across the inductive part. So the current does not go on increasing for ever. It levels out at a value that depends on the supply voltage, and the resistance in the circuit provided by the 5 Ω resistor, the internal resistance of the supply, the resistance of the coil of wire in the inductor and that of the connecting leads.

You can write:

$$V \propto \frac{dI}{dt}.$$

The ratio between the induced voltage and the rate of change of current is termed the **self-inductance L** of the coil. This is determined by the number of turns of wire in the coil, the shape and size of the coil, and the nature of the

21 ELECTRICAL OSCILLATIONS

Inductance and the rate of change of current

We know that:

$$N\phi = LI$$

We can consider a small time Δt, during which the current increases by ΔI, causing a magnetic flux increase of $\Delta\phi$, and:

$$N\Delta\phi = L\Delta I.$$

Dividing by Δt gives:

$$N\frac{\Delta\phi}{\Delta t} = L\frac{\Delta I}{\Delta t}.$$

In the limit as $\Delta t \to 0$, this becomes:

$$N\frac{d\phi}{dt} = L\frac{dI}{dt}.$$

These equations are the same if

$$V = N\frac{d\phi}{dt} \quad (\leftarrow 18).$$ So the voltage induced in each turn of a coil is given by:

$$\frac{V}{N} = \frac{d\phi}{dt}$$

and the induced voltage V is given by:

$$V = L\frac{dI}{dt}$$

as stated above.

So, a coil's self-inductance in H is given by the pd in volts across its terminals caused when the current in it changes at the rate of 1 ampere per second.

From the above equations:

$$1\text{ H} = 1\text{ Wb A}^{-1} = 1\text{ V s A}^{-1}.$$

medium inside the coil (air, iron or some other material). The magnetic flux ϕ in a coil of N turns is given by:

$$N\phi = LI,$$

so $$L = \frac{\phi}{(I/N)}.$$

So a coil has a self-inductance, in SI units, equal to the total magnetic flux for each ampere of current passing per turn of the coil.

The SI unit of self-inductance is the **henry, H**. 1 H is the inductance of a coil in which a magnetic flux of 1 weber is produced for each ampere of current passing per turn of the coil.

Question

6 In the experiment based on Figure 21.8b, a student observed that the spot rose initially at a rate of 0.75 V s^{-1} when a single 1.5 V cell was used.
a What was the initial rate of rise of current, dI/dt, in the circuit?
b What was the value of the self-inductance L?
c Suppose the total resistance in the circuit was 10 Ω, including that of the cell, the 5 Ω resistor and the inductor. What would the final steady current be?
d When the switch was closed, this current collapsed to zero in, say, 10 milliseconds. What average rate of change of current does this represent?
e What voltage across the terminals of the inductor would be generated by this collapsing current?
f Why does the neon light glow momentarily when switching off, although not when switching on the circuit?
g What implications might this have for the switches used to switch off d.c. currents in inductive components, such as motors, electromagnets, relay coils etc?

CURRENT GROWTH AND DECAY IN AN INDUCTOR

When the switch in Figure 21.8b is closed, a current starts to grow in the circuit. The changing magnetic field associated with the changing current limits the rate at which this can happen. The voltage of the supply V_S and the value of the inductance L determine the initial rate of current change:

$$V = L\,dI/dt.$$

There is no current initially when V_S is applied across the inductor, and $V_S = V_L$. As the current grows, its rate of growth decreases. So the voltage across the inductor V_L decreases and that across the resistance V_R increases. However, the voltage across the resistor and inductor must equal the voltage of the supply at all times by Kirchhoff's second law ($\leftarrow$ 12.2). This means that the voltage V_R across the resistor equals the difference between V_S and the voltage V_L across the coil:

$$V_R = IR$$
$$= V_S - V_L.$$

Finally the current reaches a constant value V_S/R with V_L becoming zero.

When the switch is opened, the magnetic flux linking the coil must fall to zero. The energy from the magnetic field during this collapse drives charge round the circuit until the field has fallen to zero. The changing flux tends to stop the current from collapsing – Lenz's law (← 18.2). As the flux collapses, so does the current. The change in current with time is shown in Figure 21.9b. There is an exponential decay of current, similar to that in the RC circuit, but with a time constant L/R.

Growth and decay of current in an inductor

At any instant in the growth or decay of current in an inductor, we know from Kirchhoff's second law:

$$V_S = V_L + V_R.$$

Substituting for V_L and V_R gives:

$$V_S = L\frac{dI}{dt} + IR.$$

The 'solution' to this differential equation gives the growth of current in the circuit as a function of time:

$$I = \frac{V_S}{R}(1 - e^{-Rt/L}),$$

which is the curve of Figure 21.9a.

When $V_S = 0$, corresponding to the short-circuiting of the RL circuit, the equation:

$$V_S = L\frac{dI}{dt} + IR$$

becomes $\quad \dfrac{dI}{dt} = \dfrac{-R}{L}I.$

This is similar to the equation of a mass-spring system:

$$\text{acceleration} = \frac{dv}{dt} = \frac{-k}{m}x \quad (\leftarrow 19.2).$$

Its 'solution' gives the exponential decay of current with time:

$$I = I_0 e^{-Rt/L}.$$

The value of L/R is the time constant in this case. The larger the value of L, the longer it takes for the steady state to be reached because any changes of current take place relatively slowly. The larger the value of R, the smaller the maximum current in the LR circuit and the shorter the time to grow to or decay from that small value.

DIFFERENTIATING AND INTEGRATING LR CIRCUITS

The effects of applying pulses to an inductor-resistor circuit are similar to those that occur with capacitor-resistor networks, as shown in experiment 21.4.

EXPERIMENT 21.4 Square pulses into an LR circuit

You can repeat experiment 21.2, with a high inductance coil replacing the capacitor, as shown in Figure 21.10. By using values of R up to 220 kΩ and input frequencies up to 1 kHz, you should obtain oscilloscope traces as in Figure 21.11.

21 ELECTRICAL OSCILLATIONS

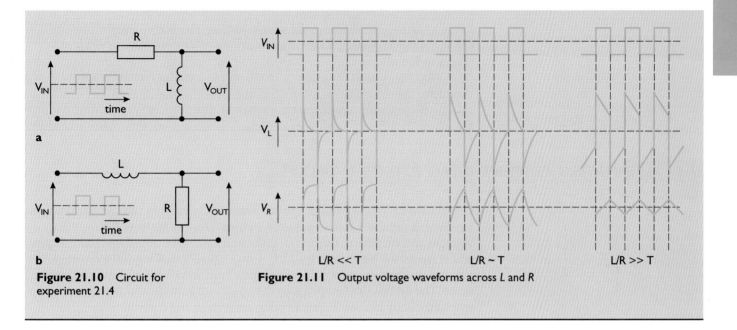

Figure 21.10 Circuit for experiment 21.4

Figure 21.11 Output voltage waveforms across L and R

The curves in Figure 21.11 are similar to those in Figure 21.6, but the V_R curves are interchanged in the two figures. Again V_S varies from $-V_S$ to $+V_S$.

The time constant of the circuit in Figure 21.10 is equal to the value of L/R. So you need to look at situations where the pulse length is long or short compared to L/R. With the iron core in place, L should have a value of about 10 H for a 1000 turn coil.

From Figure 21.11, you can see that:

- *for large values of resistance R*, the value of L/R is small compared to the half-period of the pulses T. So, the output voltage across L is large when the input is changing, but zero when it is steady. This causes the circuit output to be the *differential* of the input signal
- *for small values of R*, the value of L/R is large compared to T. The output across R changes steadily for a constant input voltage, and the circuit *integrates* the input signal.

From Kirchhoff's second law (← 12.2), the sum of the voltages across L and R must always equal the input voltage of the signal generator.

Questions

7 Show that the quantity L/R has the dimension of time.

8 Newton's second law of motion (← 6.4) may be written:

$$F = m\frac{dv}{dt}.$$

This is analogous to the situation where a steady voltage V across an inductor L causes the current I through it to increase at a steady rate:

$$V = L\frac{dI}{dt}.$$

E OSCILLATIONS

> a In what ways is this a useful analogy? Can it be misleading?
> b What factors limit the maximum velocity of a motor car under a steady driving force?
> c What factors limit the maximum current in a circuit containing an inductor?
> d Are these factors analogous to each other?

21.4 ALTERNATING CURRENT IN AN R, C OR L CIRCUIT

Experiment 21.5 allows you to apply a sinusoidally varying voltage signal across a resistor, a capacitor and an inductor, in turn. You can measure the resultant voltages and currents in two ways. You can measure the time-averaged voltages and currents, the rms values (→ 22.1) using digital or moving-coil a.c. meters. Or you can measure the peak values, and their variation with time, using an oscilloscope. The first method emphasises changes in magnitude, and the second emphasises any phase differences between voltage and current.

EXPERIMENT 21.5 Variation of current with frequency

You can connect, in turn, a resistor of about 500 Ω, a capacitor of about 50 nF and a coil of a few hundred turns with an air-core across the terminals XY of the circuit in Figure 21.12a. By varying the frequency at constant voltage, you can obtain graphs of current against frequency similar to those in Figure 21.12b.

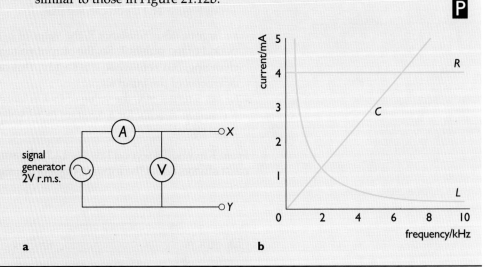

Figure 21.12a Circuit
b graph of current against frequency

The curves in Figure 21.12b show that:

- for a resistor, current is independent of frequency
- for a capacitor, current is directly proportional to frequency
- for an inductor, current is inversely proportional to frequency.

Question

9 Sketch the graphs of voltage V against frequency f you would expect to obtain in experiment 21.5 with constant current I in the circuit.

21 ELECTRICAL OSCILLATIONS

EXPERIMENT 21.6 Phase differences between applied voltage and current

Using the circuit of Figure 21.13a, observe the phase between voltage and current. The circuit contains a large capacitor made up of two electrolytic capacitors, a centre-zero high resistance dc meter and a frequency of 0.1 Hz. The pointers on the meters move 1/4 of a cycle out of phase.

Using the circuit of Figure 21™.13b, a double-beam oscilloscope screen should show you that I_C leads V_C by a phase angle of $\pi/2$ rad. So the voltage reaches its peak $T/4$, a quarter of a period, after the current, as in Figure 21.14.

If you replace the capacitor with an inductor, you should notice that V_L leads I_L by $\pi/2$ rad.

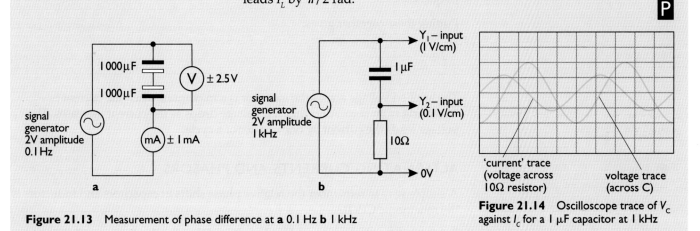

Figure 21.13 Measurement of phase difference at **a** 0.1 Hz **b** 1 kHz

Figure 21.14 Oscilloscope trace of V_C against I_C for a 1 μF capacitor at 1 kHz

EXPLAINING THE RESULTS OF EXPERIMENTS 21.5 AND 21.6...
... FOR A CAPACITOR

In the circuits of Figure 21.13, charge must be pumped round at a suitable rate to maintain the voltage across the plates equal to the changing voltage of the supply. At low frequencies, the charge flows on to and from the plates at a slow rate so only a small current is needed in the circuit. Should the frequency be doubled, the charge will have to reach the same value as before, but in half the time. So for a fixed value of V, the rate is proportional to the frequency of the supply:

$$I \propto f.$$

For a large capacitor, the change in charge must be larger in the same time. The current doubles when the capacitor doubles. So:

$$I \propto C.$$

Putting these together gives: $I \propto f\tfrac{1}{2}C$.

The charge changes at a maximum rate when the voltage across the plates is zero. This means that the current is a maximum and is decreasing when the voltage is zero and rising. So, I_C leads V_C by one quarter of a cycle (see Figure 21.14).

... FOR AN INDUCTOR

At low frequencies there is time for the magnetic field to grow as the current increases. The larger the inductance of a coil, the larger the magnetic field created for a given current. So it takes longer for the current to grow for a fixed voltage. Experiment 21.3 (← Chapter 18) shows that:

$$I \propto \frac{1}{L}.$$

At high frequencies, the current has to change direction before it has increased far from zero. Experiment 21.5 shows that:

$$I \propto \frac{1}{f}.$$

An inductor is frequently called a choke because it 'chokes' the current at high frequencies.

Putting these together gives:

$$I \propto \frac{1}{fL}.$$

The applied voltage across the inductor is a maximum when the current is increasing from zero. After this, the voltage drops as the current increases. The voltage leads the current by one quarter of a cycle.

ALTERNATING CURRENTS AND PHASORS

A simple way to remember the relative phase shifts in capacitors and inductors is the mnemonic **CIVIL**:

For capacitance C the current I leads voltage V but this leads current I for inductance L.

You can represent the phase relationship between voltage and current graphically by drawing phasor diagrams (← 19.3). The lengths of the phasors represent the peak values of the voltage and current. The projection of the phasor on to an axis represents the instantaneous value (see Figure 21.15).

There are two conventions to remember:

- the reference is always drawn at '3 o'clock'
- the phasors are always displaced anticlockwise.

REACTANCE OF INDUCTORS AND CAPACITORS

For both direct and alternating currents, the resistance in a resistor is given by the ratio of the voltage across the resistor to current in it at any instant (← 12.3):

$$V = IR.$$

But in inductors and capacitors, there is a phase difference of $\frac{\pi}{2}$ between the voltage and current when using alternating current. A more general equation that includes L and C is:

$$V = IX,$$

where X is the **reactance** of the component.

Reactance can be defined as the ratio of the peak voltage applied across the component to the peak current in it, even though the peaks do not occur at the same instant of time. As with resistance, reactance is measured in ohms. Reactance depends upon the frequency of the applied signal as well as on the component value (see experiment 21.5 and Figure 21.16).

21 ELECTRICAL OSCILLATIONS

For an inductor:

$$X_L = 2\pi f L$$

and for a capacitor:

$$X_C = \frac{1}{2\pi f C}.$$

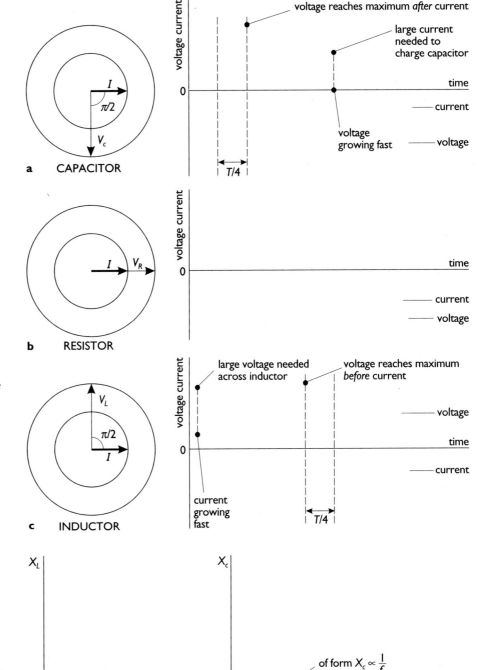

Figure 21.15 Phasor diagrams, and related current and voltage against time graphs, for a pure **a** capacitor, **b** resistor, **c** inductor

Relationships for reactances X_C and X_L

The defining equation for a capacitor is $Q = CV$ (← 13.1). So the current in the circuit I is given by:

$$I = \frac{dQ}{dt} = C\frac{dV}{dt}.$$

Suppose the applied sinusoidal voltage is $V = -V_o \cos \omega t$, as shown in Figure 21.15a. By differentiation and substitution, we find that:

$$I = \omega C V_o \sin \omega t.$$

This is of the form $I = I_o \sin \omega t$, where $I_o = \omega C V_o$.

So I leads V by $\pi/2$ rad and the reactance X_C is given by:

$$X_C = \frac{V_o}{I_o}$$

$$= \frac{V_o}{\omega C V_o}$$

$$= \frac{1}{\omega C} \text{ or } \frac{1}{2\pi f c}.$$

For an inductor, the relationship between V and I is:

$$V = L\frac{dI}{dt}.$$

Suppose current $I = I_o \sin \omega t$, to agree with Figure 21.15c. By differentiation and substitution, we find:

$$V = \omega L I_o \cos \omega t.$$

So V leads I by $\pi/2$ rad and V_o/I_o the reactance of the inductor is given by:

$$X_L = \omega L \text{ or } 2\pi f L.$$

Figure 21.16 Dependence of reactance on frequency

E OSCILLATIONS

> **Question**
>
> 10 Using the expressions for X_c and X_L, find the values of L and C in Figure 21.12b. The voltage was fixed at 2 V.

21.5 ALTERNATING CURRENTS IN RL AND RC CIRCUITS

Suppose you repeat experiment 21.6 (Figure 21.13b), but replace the 10 Ω resistor by a 1 kΩ variable resistor. You will find a change in the trace on the oscilloscope showing the relative amplitudes and phase angle φ between the voltage and current measured by the voltage across R_E. The phase difference will no longer be $\frac{\pi}{2}$ radians. The circuit is said to have **impedance** Z consisting of a significant resistive component as well as a purely reactive component. Impedance is measured in ohms.

IMPEDANCE AND PHASORS

You can represent the amplitude and phase of the current I and voltages across the resistor V_R and capacitor V_C in Figure 21.13b by a sequence of phasor diagrams for increasing R (see Figure 21.17). The supply voltage V_S is maintained at 2 V and the frequency at 1 kHz.

The Y_1-input of the oscilloscope is connected across the supply, not just the capacitor. When the resistance is very small, the voltages across the supply and capacitor are almost equal and in phase (see Figure 21.17a). However, as the resistance increases, the voltage across it increases although the current falls. The voltage across the capacitor falls and the phase angle φ between applied voltage and current decreases, as shown in Figure 21.17b.

Figure 21.17 Phasor diagrams for an RC combination with a fixed supply voltage V_s and increasing R of **a** 10 Ω, **b** 100 Ω, **c** 500 Ω, **d** 1 000 Ω

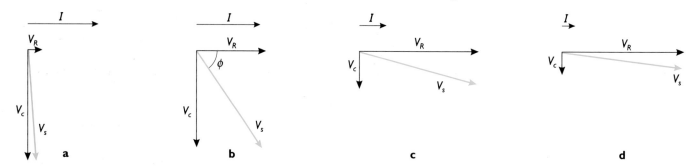

You can relate the amplitudes of the voltages in the RC combination by the expression:

$$V_S^2 = V_R^2 + V_C^2,$$

and V_S lags behind V_R by a phase angle $\phi = \tan^{-1}(V_C/V_R)$.

Connecting an a.c. voltmeter across each component in turn, as in Figure 21.18, will give meter readings indicating that V_S is not equal to $V_R + V_C$. Instead, they are related by the expression above. This is because the meter is measuring the time averaged rms (→ 22.1) value of the voltage, not its instantaneous value. So the phase angles must be taken into account.

It is always true, from Kirchhoff's second law, that

(instantaneous value of V_S) = (instantaneous value of V_R) + (instantaneous value of V_C).

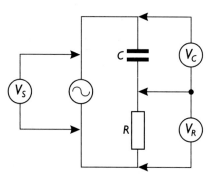

Figure 21.18 Voltage readings across resistance and capacitance in an a.c. circuit

21 ELECTRICAL OSCILLATIONS

This can be shown in a phasor diagram by projecting the phasors on to the *y*-axis (see Figure 21.19). The projections are the instantaneous values, so:

$$v_S = v_R + v_C.$$

Phasor diagrams show:

- the relative phases and amplitudes of the voltages in the circuit, and
- their instantaneous values, by rotating the diagram for the time required and then projecting the phasors on to the *y*-axis.

Impedance of an RC or RL combination

The current in a series circuit is the same in all components. So, for an *RC* circuit, instead of writing:

$$V_S^2 = V_R^2 + V_C^2,$$

you can write:

$$I^2Z^2 = I^2R^2 + I^2X_C^2$$
$$= I^2R^2 + I^2/\omega^2C^2$$

or $Z = \sqrt{R^2 + \dfrac{1}{\omega^2C^2}}$,

where *Z* is the impedance of the circuit. This leads to an equation relating supply voltage V_S and current *I*:

$$V_S = IZ$$

or $V_S = I\sqrt{R^2 + \dfrac{1}{\omega^2C^2}}$.

The current leads the voltage by a phase angle ϕ given by:

$$\phi = \tan^{-1} 1/\omega RC.$$

Similarly, for an *RL* circuit:

$$V_S^2 = I^2Z^2$$
$$= I^2(R^2 + X_L^2)$$
$$V_S = I\sqrt{R^2 + \omega^2L^2},$$

and V_S leads *I* by a phase angle ϕ given by:

$$\phi = \tan^{-1}(\omega L/R).$$

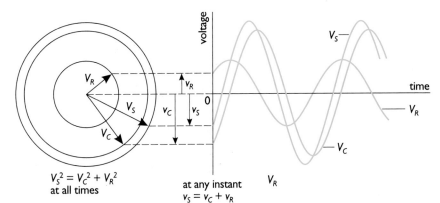

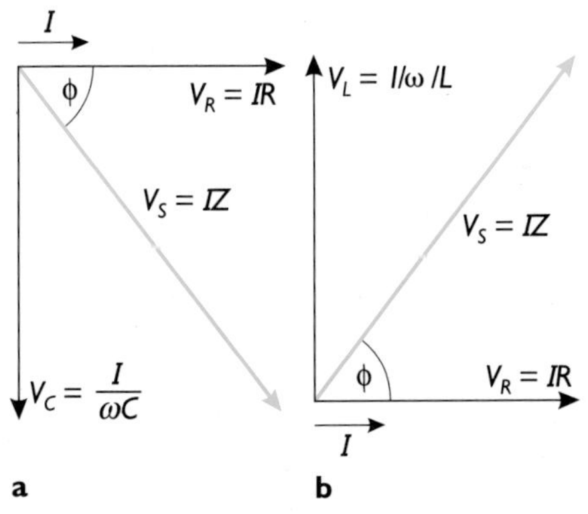

Figure 21.19 Relation between the phasor and instantaneous values of voltage in an *RC* circuit

Frequently, (← 21.4), the capacitor and inductor are considered to be pure components. However, all real components have some resistance. For example, the inductor is a coil of low resistance wire. When you do calculations, you should consider it to be a pure inductor in series with a small pure resistor. When you are working out the effects of real capacitors and inductors in a circuit, you should really replace *C* and *L* by phasor diagrams for *RC* and *RL* circuits, as shown in Figure 21.20. You can then calculate their true impedances.

The same is true of a wire-wound resistor. At high frequencies, its impedance increases noticeably because of its inductance. Capacitors have a very high resistance. This is not "infinite" due to slight charge leakage through the dielectric. This parallel resistance is normally too high to be a problem in most applications.

Filters

A capacitor and resistor in series act as a potential divider. The ratio of the voltage across the resistor to that across the capacitor is frequency dependent. This is because the reactance of the capacitor is inversely proportional to frequency. This frequency dependent response is used in filter circuits.

Figure 21.21a i shows a **treble cut** or **low-pass** filter. It is used for example in a hi-fi system to suppress the high frequency sounds, from scratches on a damaged disc. In it, V_{OUT} is equal to V_C. The circuit behaves very differently above and below the **break** frequency f_o which is given by:

$$f_o = 1/2\pi RC$$

at which $V_R = V_C$.

$$\text{Gain} = \frac{\text{output voltage}}{\text{input voltage}}$$

$$= \frac{V_{OUT}}{V_{IN}}.$$

Figure 21.20 Phasor diagrams **a** for *RC*, **b** for *RL*

OSCILLATIONS

The change in response is best shown by plotting a log-log graph (→ 41.3) of gain against frequency, as shown in Figure 21.21a ii. Below f_o, the output remains almost equal to the input: close to a gain of unity. Above f_o, the output falls at a rate f_o/f.

Taking the output across the resistor, so that $V_{OUT} = V_R$ as shown in Figure 21.21b i gives us a **high pass** or **base cut**, filter. This reduces any mains hum or motor rumble being amplified by the hi-fi system.

The log-log frequency response of the circuit shows the **attenuation** at low frequencies and constant gain at high frequencies. Figure 21.22 shows the phasor diagrams at multiple frequencies of f_o. Notice the similarities to Figure 21.17, where R rather than f is varied.

The ratio V_C/V_{IN} gives the gain for the low-pass filter and V_R/V_{IN} for the high-pass filter. A suitable choice of values of R and C causes cut off at the required frequency. Figure 21.23 gives a numerical example of a high-pass filter.

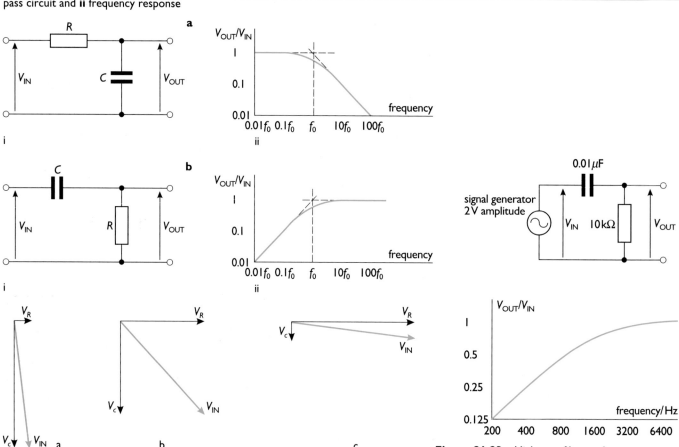

Figure 21.21 RC filters: **a i** low-pass circuit and **ii** frequency response, **b i** high-pass circuit and **ii** frequency response

Figure 21.22 Phasor diagrams at frequencies of **a** $0.1f_o$, **b** f_o, **c** $10f_o$

Figure 21.23 High-pass filter and response curve

Questions

11 Using the fact that, in a filter, $V_R = V_C$ at the break frequency f_o, show that $f_o = 1/2\pi RC$. Also show that, at f_o:

$$\text{gain} = V_{OUT}/V_{IN} = \frac{1}{\sqrt{2}}$$

12 Find the value of the break frequency directly from the graph of Figure 21.23 and check that it agrees with the calculated value.

21.6 MEASURING INDUCTANCE

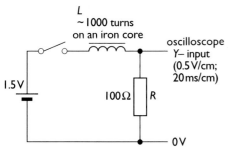

Figure 21.24 Circuit for measuring an inductance of about 5 H

Figure 21.24 shows a d.c. circuit containing an inductor and resistor. The growth of current in the circuit, from the instant of closing the switch, is shown in Figure 21.9a. The initial rate of rise of current depends on the value of the inductance, since all the applied voltage is across it. The initial current in the circuit is zero, so the voltage across the resistor is zero. The value of the inductance can be calculated from the gradient of the curve at the origin, since:

$$V = L\,dI/dt + IR, (\leftarrow 19.2).$$

At $t = 0$, $I = 0$, so $V_S = L\,dI/dt$ and:

$$L = V_S / (\text{gradient at } (0, 0)).$$

There are many ways of measuring inductance, two of which are suggested in experiment 21.7.

EXPERIMENT 21.7 Measurement of self-inductance L

By switching on the circuit in Figure 21.24, you can measure the rate at which the voltage across R initially increases on the oscilloscope screen. Given that $R = 100\,\Omega$, you can calculate the initial rate of rise of current in the circuit, dI/dt, measure V_s and calculate L. The curve also enables you to obtain a second estimate of L from the time constant L/R.

Alternatively, you can use experiment 21.5 and $V = I\omega L$. Plotting a graph of I against $1/f$ at low frequencies for a fixed voltage V should give you a straight line through the origin with a gradient $V/2\pi L$. The coil in Figure 21.24 has an inductance of about 5H and a d.c. resistance of 5 Ω.

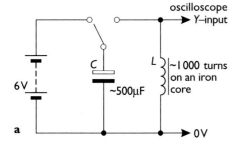

a

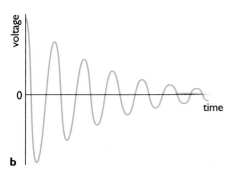

b

Figure 21.25a An *LC* circuit and **b** the free (resonant) oscillations of current in it

Questions

13 What do you think is meant by the term 'd.c. resistance'? How would you measure a coil's d.c. resistance?

14 Using the relationship $X_L = \omega L$, show that, in a coil of inductance 5 H and d.c. resistance 5 Ω, the effect of the resistance of an inductor is negligible in comparison with the reactance at a frequency $f = 100$ Hz.

A quick approximate method of measuring L is to use the resonant or natural frequency f_o of oscillation of an *LC* circuit (see Figure 21.25), given by:

$$f_o = \frac{1}{2\pi\sqrt{LC}} \quad (\rightarrow 22.5).$$

By measuring f_o from an oscilloscope trace, L can be calculated assuming C is known. The result has a large uncertainty due to the tolerance of the electrolytic capacitor.

Alternatively, the resonant frequency of the *LC* circuit can be found by driving the circuit with a signal generator. Then a much smaller value of C can be used to find L. This has the advantage that small capacitors tend to have smaller tolerances than large electrolytic ones.

E OSCILLATIONS

> **Question**
>
> **15a** Suggest a suitable graph to plot to find L, assuming you have a range of small capacitors available, a signal generator and an oscilloscope.
> **b** Why is the circuit with a large capacitor in Figure 21.25 suitable for finding L, whereas the same circuit with a small capacitor would not be?

One of the best and most direct methods of measuring L is to use the defining equation, $V = L\,dI/dt$, to calculate L. This can be done with an alternating supply connected to the inductor, as in experiment 21.8.

EXPERIMENT 21.8 Measuring self inductance by an ac method

The circuit of Figure 21.26a enables you to monitor both the voltage V and current I in the circuit at the same instant. With the resistance in the circuit small compared to the reactance, the rate of change of current, dI/dt, is a maximum when $I = 0$. The voltage V_L is also a maximum, equal to the peak voltage V_o of the supply.

As shown in Figure 21.26b, the amplitude of the V_L trace can be measured from the oscilloscope screen. You can find the maximum rate of change of voltage across the resistor by measuring the gradient of the V_R curve at $V_R = 0$. You can then calculate L from:

$$L = RV_L \frac{\Delta t}{\Delta V_R}.$$

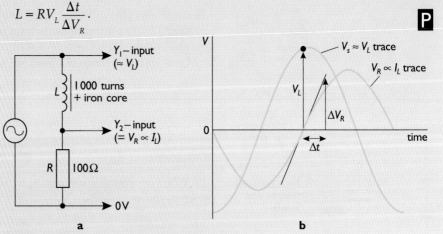

Figure 21.26a Circuit, and **b** oscilloscope traces for experiment 21.8

MUTUAL INDUCTANCE

When a changing magnetic flux from one coil links another coil, there is **mutual induction** (← 18.2). The induced voltage V generated across the secondary coil is proportional to the rate of change of magnetic flux through it. This is proportional to the current I in the primary coil at any instant. These quantities are related by the equation:

$$V = M\,dI/dt,$$

where M is a constant of proportionality called the **mutual inductance** of the coils. Its value depends on the product of the number of turns in each coil, the material linking the coils, and the shape and size of the core between the coils.

EXPERIMENT 21.9 Measuring mutual inductance

Placing two coils together without a linking iron core (see Figure 21.27) should give you oscilloscope traces, as in Figure 21.26b. You can measure mutual inductance in the same way as self inductance is measured in experiment 21.8. By changing the number of turns on one or both coils, you can show that the mutual inductance depends on the product of the turns.

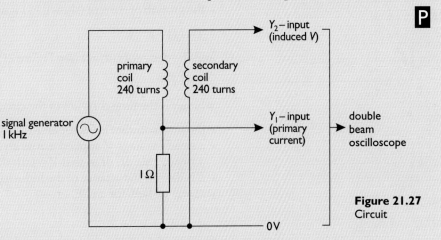

Figure 21.27 Circuit

Experiment 21.9 illustrates the principle of measuring mutual inductance, by using an air-cored system. This is because an iron core does not have linear magnetic properties, but undergoes hysteresis (← 17.5). The design of transformers has to take such non-linearity into account (→ 22.3).

Motor car petrol ignition system

In a conventional petrol engine, a distributor supplies a high voltage to the sparking plugs. The distributor is turned by the engine so that it connects a high voltage to the central electrode of each plug just when ignition is required.

In the older mechanical ignition, a square-shaped cam inside the distributor rotates as the engine turns and opens a mechanical switch called contacts or 'points'. Electronic ignition uses engine-triggered internal switching, which tends to be more reliable and requires no maintenance. In both cases, the switch is connected between the car's battery and the primary coil of a mutual inductor called the ignition coil. This has only a few turns but is capable of carrying a reasonably high current, which sets up a large magnetic flux in the core.

When the switch closes, current builds up relatively slowly in the primary due to its self inductance. When the switch has been closed for long enough, the current reaches a saturation level. As soon as the switch opens, the current drops quickly to zero causing the magnetic flux due to the primary to collapse and a high voltage to be induced in the many thousand-turn secondary by mutual induction. The distributor connects the secondary of the ignition coil to each sparking plug in turn at exactly the right instant. It supplies several kV about 3 ms before the piston inside the cylinder reaches the top of its stroke (known as top dead centre, or TDC).

Clearly, timing in this process is critical. For example, the switch must remain closed long enough for a sufficiently large flux to be established in the primary of the ignition coil.

OSCILLATIONS

SUMMARY

In a d.c. circuit containing an inductor L or a capacitor C in series with a resistor R:

- the currents and voltages reach their steady values after a time of at least $5L/R$ or $5CR$
- the maximum current equals V_S/R
- the growth of current (shown in Figure 21.28a) is expressed symbolically by:

$$I = I_0(1 - e^{-t/\tau})$$

and the decay of current (see Figure 21.28b) by

$$I = I_0 e^{-t/\tau}$$

where τ is the time constant of the circuit

- the time constant is CR for an RC circuit and L/R for an RL circuit
- a steady voltage V applied across an inductor L will cause the current to rise at a steady rate dI/dt so that $V = L\,dI/dt$
- L is the self inductance of the inductor, measured in henrys
- at any instant, for an RC circuit:

$$V_S = V_R + V_C = R\,\frac{dQ}{dt} + \frac{Q}{C}$$

for an RL circuit

$$V_S = V_R + V_L = L\,\frac{dI}{dt} + IR,$$

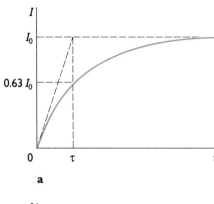

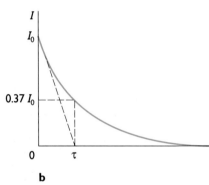

Figure 21.28 a Growth and **b** decay of current in an inductor or a capacitor

where V_S is the supply voltage or zero.

Figures 21.29 and 21.30 summarise the relationships between the alternating voltages and currents in the circuits shown. The current I is related to the frequency of the supply f by the equation:

$$I = I_0 \sin \omega t$$

so that at $t = 0$, $I = 0$.

Figure 21.29 a.c. voltage/current relationships

circuit	I for fixed V	V for fixed I	X or Z	V and I phasors	I leads V by phase angle ϕ
R	independent of f	independent of f	R	→ V_R ; → I	0
C	$\propto f$	$\propto 1/f$	$1/\omega C$	→ I ; ↓ V_C	$\pi/2$
L	$\propto 1/f$	$\propto f$	ωL	↑ V_L ; → I	$-\pi/2$

21 ELECTRICAL OSCILLATIONS

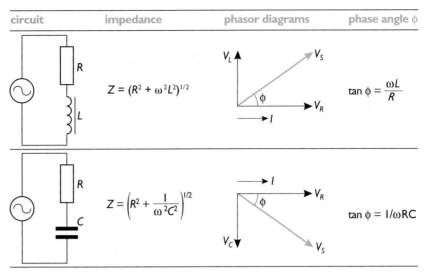

Figure 21.30 ac voltage/current relationships

SUMMARY QUESTIONS

16 The equation $dQ/dt = -Q/RC$ can be solved numerically by a repeated or iterative process as follows. Suppose in 10 s, 0.1 Q of the charge on the capacitor is lost. After 10 s, the charge remaining is 0.9 Q. In the next 10 s, 0.09 Q will be discharged leaving 0.81 Q remaining. Every 10 s, 0.1 Q is discharged by the capacitor.
a Use Figure 21.20 to show that the time constant CR of the circuit in this example is 100 s.
b Plot a graph of Q remaining on the capacitor against time at 10 s intervals, using the method described above.
c Compare your graph with the graph of $Q = Q_0 e^{-t/RC}$ and with that of question 1.
d Suppose, instead of 10 s intervals as in **b**, you had taken 1 s intervals with 0.01 Q being lost each interval. Your graph would then have been closer to the graph in **c**. Why?

17a Suppose the graph of Figure 21.31 represents the voltage across a capacitor.
 i At which times will the capacitor be fully charged?
 ii At which times will the current in the circuit be a maximum?
 iii During which time is the capacitor charging?
 iv During which time is it discharging?
b Suppose the graph of Figure 21.31 represents the voltage across an inductor.
 i At which times is the current in the circuit a maximum?
 ii During which times is the magnetic field through the inductor growing?
 iii During which times is the magnetic field collapsing?
c What does the mnemonic CIVIL mean in the context of this question?

18 What is meant by differentiating and integrating circuits?
a Draw one of each using **i** a CR combination and **ii** an LR combination.
b For each, must the input pulse time be long or short compared to the time constant of the circuit?

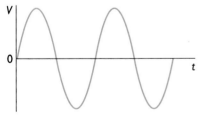

Figure 21.31

OSCILLATIONS

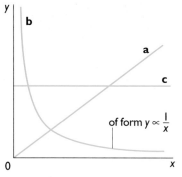

Figure 21.32

Figure 21.33

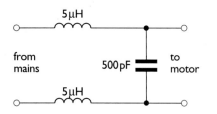

Figure 21.34

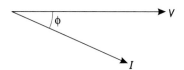

Figure 21.35

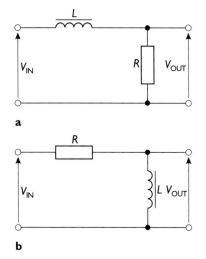

Figure 21.36

19 An alternating voltage $V_S = 32 \sin 100\pi t$ is applied across
a a resistor of 10 Ω
b a capacitor of 10 μF
c an inductor of 100 mH,
separately in turn. Write down expressions for the current in the circuit in each case.

20 Suggest at least *two* sets of suitable quantities from V, I, Z and f for the *x*- and *y*-axes of each of the curves **a**, **b**, and **c** in Figure 21.32. Sketch the experimental arrangement from which each of your choices would have been plotted.

21 A signal generator producing the oscillation shown in Figure 21.33 is connected across a capacitor. Copy the diagram and sketch the variation of the current in the circuit against time, on the same axes.

22 A 240-turn coil without an iron core has an inductance of about 0.4 mH.
a Neglecting the resistance of the wire, calculate its impedance at 1.0 Hz, 100 Hz, 10 kHz, 1 MHz.
b Explain why an inductor is sometimes called a choke.
c The circuit of Figure 21.34 is used in the supply to household electric motors, working off the mains supply at 50 Hz. It suppresses high-frequency oscillations which could interfere with television reception at 100 MHz. Explain how the inductor does not reduce the current to the motor but chokes the high-frequency oscillations.
d A telephone receiver has an inductance of about 10 mH. Explain why a person's voice has a different frequency response curve over a telephone than in direct speech.

23 Design a circuit that could cause the phase difference between supply voltage and current shown in Figure 21.35. Write down an equation relating the supply voltage to the current for your circuit.

24 What sort of filters are shown in Figure 21.36a and b? Write down an expression for the break frequency of each circuit.

25 Explain, with reference to high-pass and low-pass filters, why the circuit of Figure 21.37 is called a band-pass filter.

26 Explain how the circuit of Figure 21.38 acts as a cross-over filter, so that high frequency signals go to the high-frequency output (such as a 'tweeter' loudspeaker) and low frequency signals go to the other output, (such as a 'woofer' loudspeaker).

27 In a lighting circuit, an 80 W fluorescent tube is connected to the 230 V, 50 Hz mains supply in series with an inductor of 0.375 H. It behaves as a resistor of 20 Ω when lit.
a Draw a phasor diagram to show that the tube will function correctly.
b Calculate the current in the circuit.

28 a Draw phasor triangles for the supply voltage V_S in terms of the voltages V_R and V_C across AB and CD in Figure 21.39.
b The two arms of the circuit are in parallel, so when you place the two diagrams together with a common V_S, you should have a parallelogram with V_S as one diagonal. Show that the other diagonal is the output voltage V_{XY}.

21 ELECTRICAL OSCILLATIONS

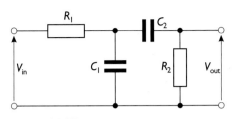

Figure 21.37

c At what frequency does your parallelogram become a square? Calculate the phase shift between the supply and output voltages at this frequency.
d Are the supply and output voltages always of the same magnitude, whatever the phase shift?
e What is a use for this circuit?

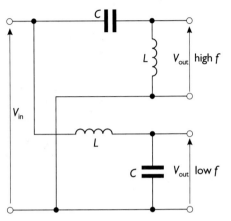

Figure 21.38

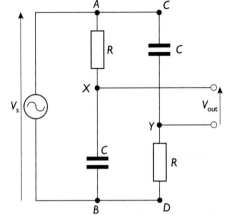

Figure 21.39

22 ENERGY AND POWER IN AC CIRCUITS

The electronic circuits used in many devices in the home, such as radios, hi-fi systems and computers, require stable low voltage dc supplies. Yet most of these devices operate from the 230 volt or 240 volt ac mains. They must have circuits built in to them that reduce the voltage and convert it to dc.

So why use ac in the electricity supply system? Why is electrical power carried by very high voltage cables across the country? Why not use dc at voltage levels that can be used directly? To answer these and other questions, you need to look at some of the properties of alternating voltages and currents. You also need to look at the energy transfers that take place in the circuits carrying them.

22.1 DESCRIBING ALTERNATING CURRENTS

When there is a direct current in a wire, charges move in one direction only whether the current is a constant or a varying one. The **average current** in amperes can be found from the total charge in coulombs passing a given point in the wire divided by the time taken. However, the average value of a sinusoidally alternating current over one complete cycle of oscillation is zero. The charge moves in one direction for the first half period and then returns during the second half of the cycle, as shown in Figure 22.1.

Clearly, you cannot use mean or average current as a way of describing the size of an alternating current. One possibility might be to use the **peak value** of the current, I_0. However, this occurs momentarily twice each cycle and so does not really represent the effective magnitude of the current.

When you connect a resistor to a source of electricity, energy is transferred from the source. Heating takes place in the resistor whether the current is direct or alternating. If, at any instant, the current through the wire is I and the voltage across it is V, then the energy transferred per second at that moment, or the power delivered P in watts, will be given by:

$$P = VI$$

($\leftarrow$ 12.2). Voltage, current and resistance are related by $V = IR$ ($\leftarrow$ 12.3). So you can also write the power transferred in the resistor as I^2R or V^2/R.

Note that the power is independent of the direction of the current or voltage. It is the product of the instantaneous current and voltage. Alternatively it depends on the square of the current in the resistor, or the square of the voltage across it. So power or energy transfer are used as a basis for describing the effective magnitude of an alternating current.

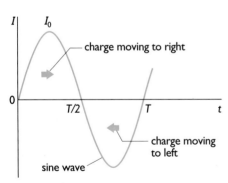

Figure 22.1 Charge flow for an alternating current

22 ENERGY AND POWER IN AC CIRCUITS

EXPERIMENT 22.1 Power in a resistive circuit

Using the circuit of Figure 22.2, you can adjust the ac and dc power supplies to the bulb until the light intensity is the same. By using an oscilloscope, you can compare the traces of the voltage across the bulb for each supply, as in Figure 22.3, and see that the steady dc voltage is 0.71 of the peak ac voltage.

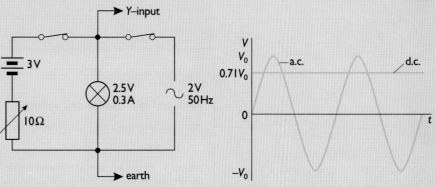

Figure 22.2 Circuit for experiment 22.1

Figure 22.3 Traces of voltage across bulb for ac and dc supply

RMS VALUE

Experiment 22.1 offers a way of finding the steady dc equivalent of an alternating voltage. This is known as the alternating voltage's **root mean square**, or **rms**, value:

$$V_{rms} = \frac{V_o}{\sqrt{2}}$$

$$= 0.71\, V_o,$$

where V_o is the peak voltage. Similarly, the rms value of an alternating current, I_{rms}, is equal to $I_o/\sqrt{2}$, where the peak current in a circuit is I_o:

$$I_{rms} = \frac{I_o}{\sqrt{2}}$$

$$= 0.71\, I_o.$$

In everyday life, alternating currents and voltages are always quoted as rms values. All the ac measuring instruments that you use, except the oscilloscope, are calibrated to display the rms value. For example, a 240 V mains lamp runs off a supply voltage with an rms value of 240 V. It actually peaks at $240\sqrt{2}$ V, or 339 V, positively and negatively. If the lamp is rated at 60 W, the rms current will be:

$$I_{rms} = 60/240$$

$$= 0.25 \text{ A}.$$

The peak current is $\sqrt{2}$ times this or 0.35 A. 60 W is the mean power dissipated in the lamp and is the product of the rms values of voltage and current – it is the power that would be transferred by a steady voltage of 240 V and current of 0.25 A. The peak power is 120 W, the product of the peak voltage and the peak current.

E OSCILLATIONS

The relationship between peak and rms values is different for other waveforms, such as square, triangular or sawtooth. It is only for a sinusoidal wave that the scaling factor between the rms and peak value is $\sqrt{2}$. Nevertheless, the rms ac value gives the same power dissipation as the dc value for all waveforms.

> **Question**
>
> **1** Find the rms and peak currents in a 240 V ac mains 3 kW electric kettle element.

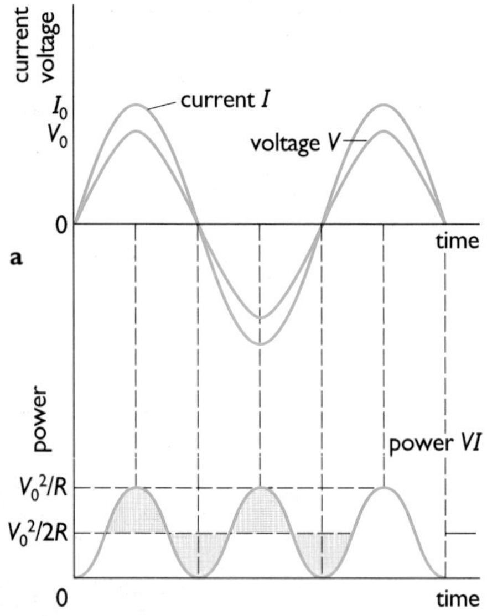

Peak and rms current and voltage

We can describe a sinusoidally varying voltage V by the expression:

$$V = V_0 \sin\omega t,$$

where V_0 is the peak voltage and ω the angular frequency is $2\pi f$. The current I delivered through a resistor R connected to the supply will vary so that:

$$I = I_0 \sin\omega t.$$

The voltage across and current in the resistor are in phase.

The instantaneous power dissipated in a resistor is given by:

$$VI = V_0 I_0 \sin^2\omega t$$
$$= \frac{V_0^2}{R} \cdot \sin^2\omega t.$$

Now $\cos 2\omega t = 1 - 2\sin^2\omega t$

so $VI = V_0 I_0 (1 - \cos 2\omega t)/2.$

Over a whole cycle, the average value of $\cos 2\omega t$ is zero. So the power delivered over a cycle is given by:

$$P_{av} = V_0 I_0 / 2 = V_{rms} . I_{rms}$$
$$= \frac{V_0}{\sqrt{2}} \cdot \frac{I_0}{\sqrt{2}}$$

So we see that $V_{rms} = V_0/\sqrt{2}$ and $I_{rms} = I_0/\sqrt{2}$.

Figure 22.4b is a graph of instantaneous power against time. It is another sinusoidal variation with a peak value of V_0^2/R and, over a whole number of cycles, an average value of $V_0^2/2R$. Note that the area under the power–time curve gives the energy dissipated in the resistor. This is equal to the area under the mean power line over a whole cycle. The value of the steady voltage V_{rms} that gives the same mean power of $V_0^2/2R$ is, therefore, $V_0/\sqrt{2}$.

Figure 22.4a Current and voltage, **b** instantaneous power against time, for a resistor. The shaded areas are equal

22.2 POWER IN REACTIVE AND RESISTIVE CIRCUITS

Voltage and current in a **resistive** circuit are in phase. So the rate of energy dissipation at any instant is equal to the product of the current and the voltage at that instant. The power is always positive or zero and we can calculate a mean power. For a resistor, this can be either $V_0 I_0/2$ or $V_{rms} I_{rms}$.

However, in **reactive** circuits, in capacitors or inductors, the voltage and current are not in phase (← 21.4). Energy may be taken from the supply over one part of the cycle but it will be returned later in the cycle. For example, energy is

22 ENERGY AND POWER IN AC CIRCUITS

taken from the supply when a capacitor becomes charged and returned when it is discharged. A similar argument applies to inductors. Because of this, the instantaneous power may take positive, zero or negative values and the mean power may be zero. So we need to distinguish between the **true** power – the mean power drawn from the supply that is actually dissipated in the circuit and not returned – and the **apparent** power – the product $V_{rms}I_{rms}$ in the circuit.

For capacitors or inductors, the apparent and true powers are quite different in value, but for resistors, they are the same.

EXPERIMENT 22.2 Power in a capacitive circuit

Using the circuit of Figure 22.5, with the switch connected to the bulb, the joulemeter reading increases by about 36 J every second. This is equal to the product VI taken from the meter readings. So the true power equals the apparent power for a resistor. With the switch connected to the capacitor, the product VI is slightly larger than for the bulb but the joulemeter reading is close to zero. So there is a large apparent power but the true power is very small.

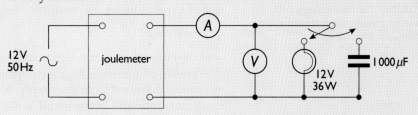

Figure 22.5 Circuit

Question

2 The capacitor in Figure 22.5 can be made up of two 1 000 μF electrolytic capacitors in series.
a Why must they be connected with reversed polarity?
b Why two 1 000 μF capacitors and not two 2 000 μF capacitors?

Figure 22.6 helps explain the difference between true and apparent power for a **pure** capacitor. There is a phase difference of $\pi/2$ rad between the current and applied voltage across a pure capacitor. During the first quarter cycle, the capacitor charges. Energy is transferred from the supply as the electric field between the plates is established. When the current falls to zero, the voltage across the capacitor is a maximum and the capacitor is charged. The capacitor discharges during the next quarter cycle returning the stored energy to the supply. In the next quarter cycle, the capacitor charges again but with opposite polarity. It discharges again in the last quarter cycle, and then the whole cycle repeats itself.

The power transfer at any instant is the product of the voltage and current. Notice that it takes equal positive and negative values, and goes through two cycles of change during one cycle of the supply. The area under the power curve from time zero to any instant tells us the energy stored in the capacitor at that instant. During the second quarter cycle, the area is negative, indicating the return of the energy to the supply. During half a cycle the capacitor has 'borrowed' and 'returned' all of its energy to the supply. This process is repeated every half cycle, so the mean or true power is zero. The apparent power is

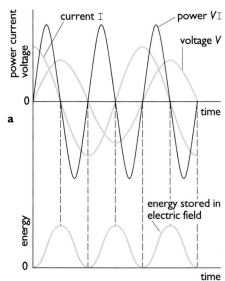

Figure 22.6a Current, voltage and instantaneous power, and **b** stored energy against time, for a capacitor

E OSCILLATIONS

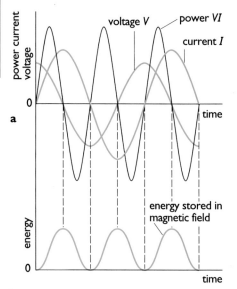

Figure 22.7a Current, voltage and instantaneous power, and **b** stored energy against time, for an inductor

positive and equal to the product of the rms values of voltage and current.

In the **real** circuit of experiment 22.2, the joulemeter reading for the capacitor is not exactly zero. This is because energy is dissipated in the resistance of the circuit containing the capacitor.

In a **pure** inductor, the phase shift between V and I is also $\pi/2$ rad but in the opposite sense to that in a capacitor (see Figure 22.7). The magnetic flux linking the inductor is in phase with the current. Maximum energy is stored when the current is a maximum and the voltage across the inductor is zero – unlike the capacitor circuit, where maximum energy occurs at zero current.

The phase difference between energy storage in a capacitor and an inductor is of fundamental importance in the behaviour of LC circuits ($\rightarrow$ 22.5).

REACTANCE, IMPEDANCE AND THE POWER FACTOR

The reactance X of a pure capacitor is $1/\omega C$ and that of a pure inductor is ωL ($\leftarrow$ 21.4). No net energy transfer takes place when such components are connected to an alternating supply.

However, when there is also a resistive component in the circuit, such as in the coil of a **real** inductor, the circuit has impedance Z given by

$$Z^2 = R^2 + X^2$$

and the phase angle ϕ between voltage and current is less than $\pi/2$ rad ($\leftarrow$ 21.5). The power-time curve is no longer symmetrical about the time axis and there is net power delivered from the supply. Note that power is dissipated only in the resistive component of the circuit, not in the reactive part.

The maximum permitted apparent power, rather than the real power, is quoted for components such as transformers. This is because the reactive component, as well as the resistive component, determines the current in the circuit. The current ratings of connecting wires, switches and fuses must take this into account. Similarly, the power supply must be capable of supplying both the true power and the energy stored in the magnetic field of the inductor. The maximum apparent power is measured in volt-amps (VA) to distinguish it from the true power measured in watts (W).

The ratio of the true to the apparent power is determined by the phase angle ϕ between voltage and current (see Figure 22.8) and:

true power = apparent power $\times$ cosϕ.

Cosϕ is called the power factor. It is 1 for a pure resistance and 0 for a pure reactance.

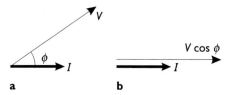

Figure 22.8a Apparent power VI, **b** true power $VI\cos\phi$

Fluorescent tube

The use of an inductor in a fluorescent lighting tube ($\rightarrow$ 35.3) illustrates a reactance in action. Suppose that the tube dissipates 80 W and acts as a pure resistance of 20 Ω when lit. Its rms current and power will be linked by:

22 ENERGY AND POWER IN AC CIRCUITS

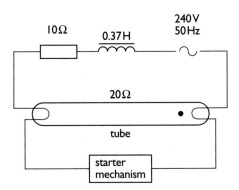

Figure 22.9 Circuit for fluorescent light tube

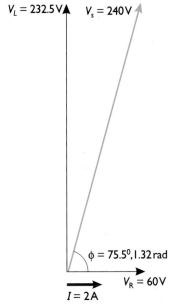

Figure 22.10 Phasor diagram for fluorescent tube circuit

$$P_{rms} = (I_{rms})^2 R$$

so, $(I_{rms})^2 = P_{rms}/R = 80/20 = 4 \text{ A}^2$

and $I_{rms} = 2 \text{ A}.$

It will draw this current whenever it is lit.

The tube (see Figure 22.9) is connected in series with an inductor of inductance 0.37 H and resistance 10 Ω, to the 240 V, 50 Hz mains supply. The starter mechanism, which only operates when the tube is cold, is shown as a 'black box'.

The true power drawn from the supply is:

$$P_{true} = I^2 R_{total}$$
$$= 4 \times 30$$
$$= 120 \text{ W}.$$

The apparent power drawn is:

$$P_{apparent} = I_{rms} \times V_{rms}$$
$$= 2 \times 240$$
$$= 480 \text{ W}.$$

So the power factor is:

$$\text{power factor} = P_{true}/P_{apparent}$$
$$= 120/480$$
$$= 0.25.$$

The phasor diagram of Figure 22.10 shows the relative phases of the voltages and current in the circuit.

Suppose we try using a series resistor R in place of the inductor. It will need to be chosen so that there is a voltage of 200 V across it and 40 V across the tube. So its power dissipation P_R will be:

$$P_R = V_R I$$
$$= 200 \times 2$$
$$= 400 \text{ W}.$$

This is five times as much as the tube dissipates, mostly in useful light output. The power factor with the resistor will be:

$$\text{power factor} = P_{true}/P_{apparent}$$
$$= 400/480$$
$$= 0.83.$$

With the inductor, the tube's **efficiency** η is:

$$\eta = \text{useful output power}/\text{total input power}$$
$$= 80/120$$
$$= 0.67 \text{ or } 67\%.$$

So, using an inductor both helps the tube to strike when the starter operates and allows most of the energy to be dissipated in the lamp.

With a series resistor, the total power dissipated is 480 W as opposed to 120 W: four times greater! The efficiency is:

$$\eta = \text{useful output power}/\text{total input power}$$
$$= 80/480$$
$$= 0.17 \text{ or } 17\%,$$

which is very low indeed. So, the resistor would cause considerable heating problems and four times the running costs for the same output.

E OSCILLATIONS

> **Questions**
>
> These questions relate to the box 'Fluorescent tube' (→ 22.2).
>
> **3** Show that the reactance of the inductor is 116 Ω. Satisfy yourself that all the figures quoted are correct.
>
> **4** Find out about the starter mechanism. How is the fluorescent tube lit from 'cold' (→ 35.3)?

22.3 THE TRANSFORMER

Alternating current can be stepped up or down in voltage easily by a **transformer**. This is one of the fundamental reasons for using ac when producing and distributing electrical power. Step-up transformers enable electrical power to be transmitted at very high voltages over large distances from generator to consumer, with low power losses in transit. Step-down transformers reduce the voltage to consumer levels where required. Whenever there is a need to change the voltage of an alternating supply, a transformer is likely to be used.

A transformer consists of two separate coils linked together by a 'magnetic circuit' of laminated soft iron (see Figure 22.11a). The alternating voltage V_p applied across the primary coil causes an alternating current in the coil and an associated alternating magnetic flux ϕ in the iron core linking the coil (← 18.1). The flux also links the secondary coil, inducing a voltage V_s across it. So the device is a **mutual inductor** (← 18.3).

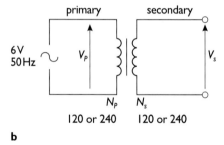

Figure 22.11a A transformer and **b** its circuit symbol

EXPERIMENT 22.3 The action of a transformer

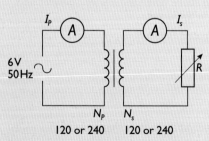

Figure 22.12 Circuit for transformer current measurement

Figure 22.11b shows a primary coil of N_p turns and a secondary coil of N_s turns. The coils are linked by a double iron C-core and clip. By connecting a 6 V ac supply to the primary coil and measuring the voltage across the secondary, you should find that the ratio of the voltages across the coils is the same as the ratio of their turns.

Similarly, using the circuit of Figure 22.12, you should find that the ratio of the currents in the two circuits is almost the inverse ratio of their number of turns, whatever the value of the load resistance. Disconnecting R should show you that the primary current is very small.

22 ENERGY AND POWER IN AC CIRCUITS

NUMBER OF TURNS, VOLTAGE AND CURRENT

When the secondary coil of a transformer is on open circuit, the primary coil can be considered as a self inductor. Then in the primary circuit, the voltage V_P is given by:

$$V_p = N_p \frac{d\phi}{dt} + Ir$$

where N_P is the number of turns in the primary coil and r is its resistance (← 18.2 and 21.2). In an ideal, or perfect, transformer you can assume that r is zero. In reality, the magnetising current I is very small, so the voltage across the resistance can be neglected.

In the open circuit of the secondary coil, a voltage $d\phi/dt$ is induced across each turn of the coil as long as very little of the alternating flux escapes from the iron core. In an ideal transformer, you can assume that all of the flux links both coils. Then the voltage V_s in the secondary coil is:

$$V_s = N_s \frac{d\phi}{dt}.$$

So the ratio of the primary to the secondary voltage is equal to the ratio of their number of turns:

$$\frac{V_p}{V_s} = \frac{N_p}{N_s}.$$

When a load resistance R is connected across the secondary coil (see Figure 22.12), there is a secondary current $I_s = V_s/R$. Again, we neglect the resistance of the secondary coil. The voltage across and current in the secondary coil are in phase. Power $V_s I_s$ is dissipated in the resistor R. The power output cannot exceed the power input. The alternating flux associated with the current I_s induces a current I_p in the primary circuit that is in phase with the primary voltage. So:

$$V_p I_p \geq V_s I_s.$$

The secondary circuit is dissipating power, which has to come from the supply. The supply only provides real power when V and I are in phase. So I_p must be in phase with V_p. For an ideal transformer, we can write:

$$\frac{V_p}{V_s} = \frac{I_s}{I_p}$$

$$= \frac{N_p}{N_s}.$$

In a step-up transformer, $V_s > V_p$ and $I_s < I_p$, both in the same ratio.

Transformer action and phasors

Consider the phase shifts in an ideal transformer connected to a pure resistive load. For real power to be drawn, voltage and current must be in phase. When the secondary coil is on open circuit, there is a $\pi/2$ rad phase shift between the voltage V_p and the magnetising current I and its associated magnetic flux in the primary circuit. The primary coil behaves as an inductor. The induced secondary voltage is a further $\pi/2$ rad out of phase with the magnetising current since it is induced by the changing flux. Figure 22.13 a shows the phasor diagram (← 19.3).

Addition of a load resistor causes a current I_s to flow in the secondary coil, in phase with V_s. The secondary flux caused by I_s changes the primary flux. So a current I_p is created in the primary coil by the changing secondary flux. This is π rad out of phase with the secondary current and in phase with the primary voltage. The primary flux must be maintained at the value it was when the secondary circuit was open.

The total primary current is the phasor sum of the magnetising current I and I_p. I is usually small compared to I_p. So the primary current is approximately I_p, in phase with V_p, as shown in Figure 22.13b. The real power drawn from the supply is $V_p I_p$.

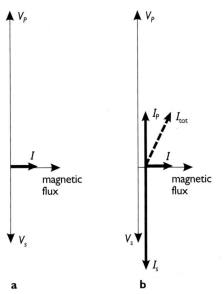

Figure 22.13 Phasor diagrams for an ideal transformer **a** on open circuit and **b** with a resistive load

Questions

5 Suppose you need to choose transformers for operating:
a a car headlamp bulb rated at 12 V, 60 W to be run from the 240 V mains via a transformer
b a 1 000 W electric iron, purchased in the USA and designed to run off a 120 V ac supply, again to be used from the 240 V mains via a transformer
c a 60 W table lamp designed to run off the 240 V mains but to be used in the USA, where the mains voltage is 120 V.
The choice of transformer in each case will be determined by the turns ratio and the currents in the primary and secondary coils. What are these?

E OSCILLATIONS

> What actual number of turns and wire thickness do you think would be appropriate for each of these transformers?
>
> **6** The secondary coil of a transformer has a resistance r. What effect does this have on the secondary voltage and current?
>
> **7** Explain why no power is consumed in an ideal transformer on stand-by when the secondary is on open circuit.

ENERGY LOSS

A real transformer loses about 5 %, or less, of its input power in two ways:

(i) in the windings of the coils because of their resistance. This is known as **Joule heating loss**
(ii) in the iron core through eddy currents and continual reversal of the magnetisation of the iron. This is known as **hysteresis loss** ($\leftarrow$ 17.5).

There is also a further small loss because not all of the flux from the primary links the secondary coil.

Figure 22.14 Components inside a transistor radio receiver

TRANSFORMERS IN ELECTRONICS

Most modern electronic equipment works at low direct voltages. A transformer is normally used to reduce the domestic mains supply to the required value before rectification and smoothing takes place ($\rightarrow$ 23.1).

Transformers also have other uses in electronic equipment. They can couple amplifier circuits together so that only the varying voltage signals are passed from one stage to the next. The steady state voltages in each stage are not affected by any other. Figure 22.14 shows **radio-frequency**, or **rf**, amplifier coupling transformers in a transistor radio. The aerial detection circuit also contains a tuned transformer (see Figure 22.36).

At the other 'end' of the radio, you may find a transformer connected to the loudspeaker. When connecting the output stage of an audio amplifier to a loudspeaker, we want to transfer the maximum power from the amplifier to the speaker. The amplifier behaves as a pure signal source with a series resistance called its **output impedance** (see Figure 22.15). The impedance of the loudspeaker may not equal the impedance of the power amplifier. Typically, a loudspeaker has a resistance of 4 Ω and a transistor amplifier has an output impedance of 2 kΩ. Maximum power is transferred from the amplifier to the loudspeaker when its output impedance equals the loudspeaker's resistance. The amplifier can be coupled to the speaker by a step-down transformer, called a **matching transformer**.

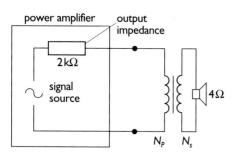

Figure 22.15 Use of a matching transformer: the power amplifier 'sees' a load resistance of 2 kΩ with $N_p/N_s = 22$

The amplifier's output needs to be in phase with the speaker's oscillations so that maximum power will be transferred. This means that the impedances need to be matched so that:

$$Z_{\text{primary}} = Z_{\text{output}}$$
$$Z_{\text{secondary}} = Z_{\text{speaker}}.$$

Since the signals are to be in phase, resistive terms can be used:

Z_{primary} we can call R_p

$Z_{\text{secondary}}$ we can call R_s.

22 ENERGY AND POWER IN AC CIRCUITS

We know that:
$$V_s = V_p N_s / N_p$$
and $I_s = I_p N_p / N_s$.
So $R_s = V_s / I_s$
$$= V_p / I_p \cdot N_s^2 / N_p^2$$
or $R_s = R_p \cdot N_s^2 / N_p^2$.

Therefore, you need to choose a matching transformer whose ratio of turns is equal to the square root of the ratio of the output and speaker impedances. Taking the typical values of 2 kΩ and 4 Ω, respectively:
$$N_p : N_s = \sqrt{2000:4} = 22:1.$$

Using a voltage-follower circuit ($\rightarrow$ 23.4) is another method of solving this matching problem.

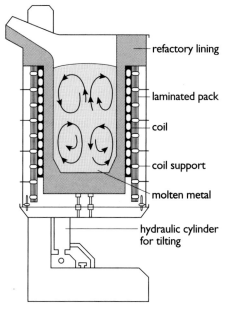

Figure 22.16 Section through a tiltable core-less induction furnace

> ### Induction heating
>
> **Eddy currents** are produced by electromagnetic induction ($\leftarrow$ 18). Some situations where eddy currents arise can be considered as single-turn secondary transformers. A solid metal plate or core acts as a single–turn secondary coil when placed close enough to a primary coil for the primary's changing magnetic flux to link the secondary plate ($\leftarrow$ 18). One design of electric cooker uses this principle. The primary coil operates at a frequency of about 18 kHz. This induces high eddy currents at low voltages in the base and sides of a saucepan placed on the coil. The resistance of the metal causes the required heating of the pan for cooking. If the system is not designed with appropriate safeguards, the eddy currents could be sufficient to melt the saucepan.
>
> Foundries use this effect in induction furnaces, such as that shown in Figure 22.16. The core of the furnace reaches a uniform temperature of 700–800 °C, high enough to melt the contents. The primary coil is water-cooled and surrounded by magnetic screening to avoid the metal casing of the furnace being melted by eddy currents too.

22.4 GENERATION AND TRANSMISSION OF ALTERNATING CURRENT

Electricity can be generated where it is needed, but it is cheaper to generate electrical power in only a few places and distribute it around the country to consumers through a grid. The distribution takes place at very high voltages to minimise power losses in the transmission cables. However, it is more economical and safer to generate electricity at lower voltages of between 10 kV and 30 kV. The consumer also needs safe power at low voltages. So step–up transformers are used between the generators and the distribution network. Step-down transformers are used from the distribution lines to the consumer (see Figure 22.17).

Transformers require alternating currents, so the electricity supply is an alternating one. Large generators and transformers are used because they are more efficient than small ones.

There is considerable fluctuation in demand for electricity, as shown in Figure 22.18. Yet the electricity industry has to ensure that the voltage at the consumer's power socket does not fluctuate by more than 5 % of its normal value. Many

OSCILLATIONS

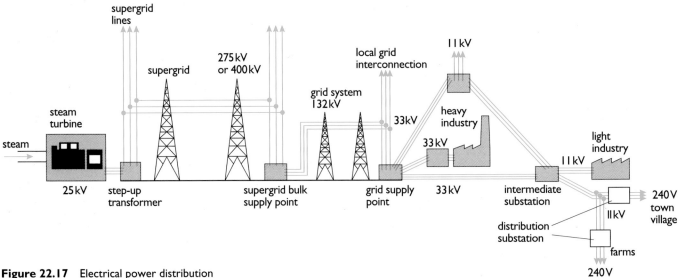

Figure 22.17 Electrical power distribution

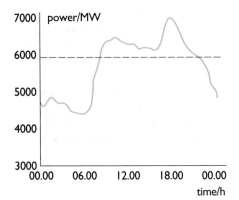

Figure 22.18 Typical power demand on a winter's day in MANWEB

industrial consumers close down at the time when domestic consumption increases, keeping demand reasonably constant for 14 to 16 hours a day. Power station controllers keep an eye on TV broadcast schedules so they can anticipate a surge when people make cups of tea, turn on lights and so on, at the end of popular programmes.

Electricity cannot be stored. However in the middle of the night, a time of low demand, some of the generators can be used to drive others which act as motors to pump water to high–level reservoirs. At peak demand time, the stored water is released to drive these 'pump' generators – a system know as **pumped storage**. The pumped storage system can be switched on almost instantaneously to respond to sudden surges in demand. For example, the pumped storage system at Dinorwig in North Wales can produce 1 800 MW for up to 5 hours.

Fluctuation in demand has a smaller effect on a large–scale system. The national supergrid is supplied by a fairly constant number of power stations working 24 hours a day. When demand increases in one part of the country, power can be switched from another region where demand is less. In this way fluctuations in demand can be smoothed out, without having to 'fire up' extra power stations. The demand shown in Figure 22.18 could be met from power stations providing a constant 6 000 MW over the 24 hour period plus the Dinorwig pumped storage system. Our national grid system also allows power stations to be sited close to fuel supplies, ports, industrial areas where demand is high, or places where hydro-electric schemes are possible.

There are over 100 power stations in Great Britain with a generating capacity of more than 50 GW, or 5×10^{10} W. All of the generators have to be synchronised so they rotate in phase with each other, to behave like a giant single supply.

TRANSMISSION CABLES

Overhead power cables are normally used in the transmission system rather than underground cables. This is partly due to the fact that soil is a poor thermal conductor. Extra electrical insulation is needed around an underground high voltage cable to avoid breakdown. This means that the cable is doubly thermally insulated. Any energy dissipated from the cable will be trapped causing a considerable rise in temperature unless the system is well designed.

A second problem arises because of the capacitance between underground cables and the ground. In an ac system, this leads to a charge-discharge current

22 ENERGY AND POWER IN AC CIRCUITS

Why are high voltages used for transmission?

Consider two supply lines of resistance R carrying a current I. The voltage drop across the cables is IR and the power lost is I^2R. Suppose the power delivered to the consumer at voltage V is P (see Figure 22.19). Then $P = IV$ and the power loss is:

$$(P/V)^2 . R.$$

The resistance of the cables is fixed, as is the power demand of the consumer. But the voltage drop across the cables and the power loss can both be reduced by increasing V. Doubling the voltage reduces the power loss to one quarter.

So, the higher the transmission voltage, the smaller are the power losses in the connecting cables. A high transmission voltage also reduces any fluctuation of the supply voltage as the demand increases. Each 400 kV supergrid conductor has a resistance of 0.017 Ω km^{-1}. At a current of 2 500 A, 1 000 MW of power is being delivered with a power loss of 210 kW km^{-1} from each pair of cables. This amounts to 0.021 % per km. If 40 kV cable was used to meet the same power demand, the power loss would be at least 100 times greater – and the cables might melt!

High voltages mean that more transformer substations are required. However the low power losses at high voltages more than compensate for the capital costs of the substations.

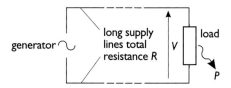

Figure 22.19 Power loss in transmission

(← 22.2), known as the **wattless power**, tens of times greater than that in overhead cables. So the ac generators have to provide more power, much of which cannot be used by the consumer but is just stored in the cable system and returned to the generator. The national grid needs a power factor as close to unity as possible for efficient transmission.

Wattless power is not a problem when using dc. The cross–Channel link, which has a maximum power rating of 2 000 MW, uses a dc voltage of 400 kV. This system has the added advantage that the British and French ac generating systems do not have to be synchronised.

Question

8 Suggest how a low voltage can be 'transformed' to 400 kV dc.

GENERATORS

The design and action of a single-phase generator is shown in Figure 22.20. The **rotor** is an electromagnet, supplied with a small direct current through slip rings. The coils on the rotor are wound non-uniformly so that the waveform generated in the stationary coils, or **stator**, is sinusoidal. With this arrangement, no moving contacts are needed and thick wires can be used for the stator coils so currents of thousands of amps may be generated.

The giant generators or **alternators** in power stations are driven at 3 000 rpm by steam turbines. However, in these alternators the rotor is driven past three separate sets of stator coils, spaced at angles of 120°, or $2\pi/3$ radians, to each other. The resultant output consists of three voltage waveforms spaced at one-third period intervals (see Figure 22.21).

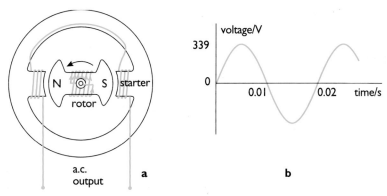

Figure 22.20a Single-phase generator and **b** output waveform

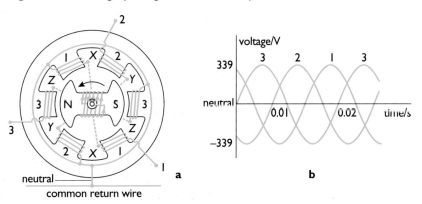

Figure 22.21a Three-phase generator and **b** output waveform

OSCILLATIONS

Electricity companies generate three phases because the three-phase alternator is smoother running and more efficient than a single-phase machine. It is also more economical, since only three wires plus a common return, or neutral, wire are needed to power three different loads. Six wires would be required for the three loads on a single phase supply. The current in the earthed neutral wire is small, so a thinner cable can be used.

The loads on the three phases need to be kept approximately equal. For this reason, all industrial users and factories are provided with all three phases. Machines such as large motors also run from all three to balance the loading. Such motors are more efficient and smooth running as a result.

Each domestic consumer receives a single-phase ac supply. Equal groups of private houses are supplied with each phase in the hope that all the phases will be equally loaded. Each of the phasors in Figure 22.22 is 240 V rms or 339 V peak value. The single-phase domestic supply of 240 V is obtained by connecting between one phase and the neutral point.

Connection between any two phases gives an rms voltage of 415 V.

Questions

9 With the rotor in the position shown in Figure 22.20, is the voltage being generated a maximum or zero?

10 Draw three resistive loads R connected to a three-phase alternator, and show that only four leads are necessary.

11 Using Figure 22.22, show that, for equal resistive loads, the current in the return wire is zero.

12 Using Figure 22.22, show that the rms voltage between two phases is 415 V.

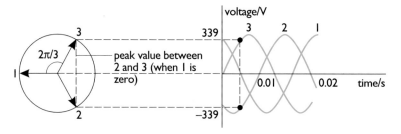

Figure 22.22 Phasor diagram for a three-phase 240 Vrms supply

22.5 RESONANCE IN ELECTRICAL CIRCUITS

The effect of applying a sinusoidally varying voltage to a resistor, capacitor or inductor separately, or to an RC or RL combination, is to offer a particular impedance to the signal frequency. This affects its magnitude and, sometimes, its relative phase (← 21.4 and 21.5). However, when both an inductor and capacitor are present, in LC or LCR combinations, oscillations can occur. These arise from the interchange of energy stored in the electric and magnetic fields. Experiments 22.4–22.6 allow you to find out how LC circuits behave when a sinusoidally varying voltage is supplied.

22 ENERGY AND POWER IN AC CIRCUITS

EXPERIMENT 22.4 A series LCR circuit

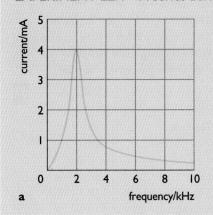

If you connect the circuit shown in Figure 22.23b, maintain the voltmeter reading at, say, 2 V rms and vary the frequency of the supply, you should find that the ammeter reading varies as shown in Figure 22.23a.

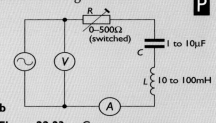

Figure 22.23a Current-resonance curve and **b** circuit for experiment 22.4

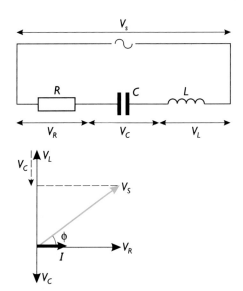

Figure 22.24 A series LCR circuit and its phasor diagram

SERIES RESONANCE

In experiment 22.4, each graph of current against frequency, such as that in Figure 22.23a, is called a **series-resonance curve**. It is similar to the forced oscillation curve obtained in mechanical vibrating systems (← 20.3).

At a particular frequency, the forced vibrations of charge in the circuit driven by the supply voltage have a **maximum** value. The maximum current is determined by the resistance of the circuit. In the case of experiment 22.4 this is the resistance of the inductor. The effect of resistance on the resonance curve from experiment 22.4 can be observed by adding a resistor of 5 Ω in series with the capacitor and inductor. The resonance curve becomes broader and the maximum current less.

At the resonant frequency f_o, the reactance of the inductor X_L is equal to the reactance of the capacitor X_C. Because L and C are in series, the currents through each are the same. This means that the voltages across each, V_L and V_C, have the same magnitude but are opposite in phase (see Figure 22.24). You can check this using the mnemonic CIVIL (← 21.4). The voltage across L and C together is, therefore, zero. The impedance Z of the circuit is a **minimum**, equal to the resistance of the circuit.

Since, at the resonant frequency, $X_L = X_C$, then:

$$2\pi f_o L = 1/2\pi f_o C \quad (\leftarrow 21.4)$$

and
$$f_o = \frac{1}{2\pi\sqrt{LC}}.$$

The impedance at resonance is purely resistive, so the current drawn and the voltage V_S across the supply are in phase at this frequency. This is also true for a **parallel resonance** circuit (see experiment 22.5). At frequencies below f_o, the reactance of the inductor becomes less significant and the circuit behaves like a capacitor, with I leading V_S. At frequencies above f_o, the reactance of the capacitor becomes less significant and the circuit behaves like an inductor, with V_S leading I (see Figure 22.25).

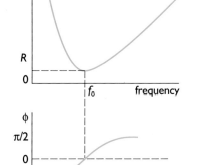

Figure 22.25 Variation of impedance Z and phase angle φ with frequency

Question

13 Check the resonant frequency of Figure 22.23a against previous measurements of L and C. (See Figure 21.12b and question **10** in 21.4.)

OSCILLATIONS

EXPERIMENT 22.5 Parallel resonance circuit

Connecting a capacitor and inductor in parallel, as in Figure 22.26, and maintaining the voltage across the supply at 2 V rms allows you to measure the currents I_L and I_C in each component and I through the supply, over a range of frequencies. You can then plot graphs of I_L, I_C and I against frequency, as shown in Figure 22.27.

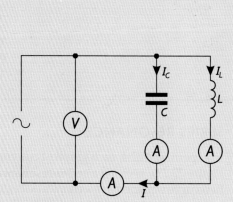

Figure 22.26 Parallel resonance circuit

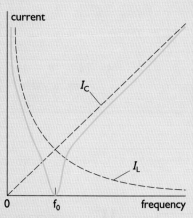

Figure 22.27 Current-resonance curve for a parallel LC circuit

PARALLEL RESONANCE

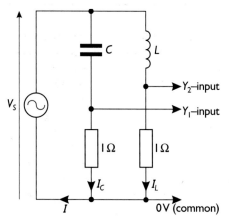

Figure 22.28 Circuit to investigate phase relationship between I_L and I_C

When L and C are connected in parallel to the supply, as in experiment 22.5, the graphs for I_L and I_C appear as in Figure 21.12b. The current I from the supply is approximately equal to the difference between I_L and I_C. At resonance the supply current is a **minimum** and the impedance of the circuit is a **maximum**. Contrast this with the series resonance circuit. This minimum supply current maintains a large current circulating between the capacitor and inductor. At resonance, V_S and I are in phase and the minimum current is determined by the resistance of the circuit. With no resistance, I would be zero.

Using the circuit of Figure 22.28, it is possible to compare the magnitudes and phases of the currents I_L and I_C on a double-beam oscilloscope screen. At resonance, $X_L = X_C$ as for the series resonance circuit. The voltage across L and C must be the same in magnitude and phase because the two components are in parallel. The magnitude of the current in each component must be the same when their reactances are identical. But, from the mnemonic CIVIL, we know that I_C leads V_C by $\pi/2$ and V_L leads I_L by $\pi/2$ radians. This means that I_L and I_C are π radians out of phase, which is often referred to as **antiphase**.

Figure 22.29b shows the current phasor diagram for the parallel resonance circuit. I_L and I_C are not in antiphase because resistances in the circuit lead to a small supply current I. For a circuit with only a small resistance, the resonant frequency is:

$$f_0 = \frac{1}{2\pi\sqrt{LC}}.$$

At frequencies below f_0, the circuit behaves like an inductor with V_S leading I. Above f_0, it behaves like a capacitor with I leading V_S. Experiment 22.6 shows a large current oscillating in the circuit at the natural frequency of the system. For small resistances, this is very close to the resonant frequency.

22 ENERGY AND POWER IN AC CIRCUITS

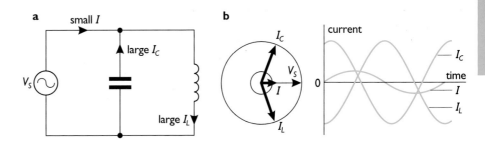

Figure 22.29a Current, **b** phasor diagram and current waveforms for parallel resonance circuit

EXPERIMENT 22.6 Natural oscillations in an *LC* circuit

If you charge the capacitor in Figure 22.30a and throw the switch, the circuit becomes that of Figure 22.23b but without a power supply. The oscilloscope shows you that the capacitor discharges and recharges, while energy is being transferred out of the circuit via the resistance. This causes the oscillations to die away. Notice that decreasing the capacitance or inductance causes the frequency of the oscillations to increase.

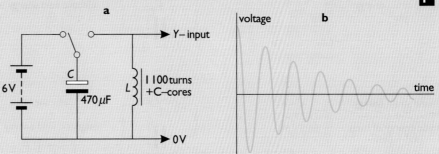

Figure 22.30a Circuit and **b** natural oscillations in experiment 22.6

The electrical oscillations seen in experiment 22.6 are analogous to free oscillations in a mass-spring system. The damping of the mechanical system is equivalent to the energy dissipation in the resistance. Figure 22.31 compares expressions for undamped oscillations in mechanical and electrical systems. The analogous mechanical and electrical quantities are listed in Figure 22.32.

In the mechanical system, as a spring is extended, potential energy is stored due to elastic strain in the stretched spring. This is transferred into kinetic energy when the mass is released and then back to potential energy. In the electrical system, energy is stored initially in the electric field between the capacitor plates. The current is zero. When the capacitor discharges, the energy is transferred into the magnetic field of the inductor, becoming a maximum when the current is a maximum. At this instant, no charge remains on the capacitor plates. The magnetic field then collapses, maintaining a decreasing current in the same direction and charging the capacitor plates in the opposite polarity to the initial state. The current falls to zero as the stored charge reaches its initial value.

At resonance, in a series circuit, the voltages across the capacitor and inductor are equal and opposite (see Figure 22.33). They are π rad out of phase with each other. The voltage across each capacitor and inductor is many times greater than the supply voltage when the resistance in the circuit is small. Similarly, in the parallel resonance circuit, the current circulating between the capacitor and inductor is many times the supply current. The supply current only needs to supply the energy lost in the resistance of the inductor and connecting leads (see Figure 22.24). Energy is stored in the electric field between the capacitor plates and the magnetic field of the inductor, and reaches a constant value.

OSCILLATIONS

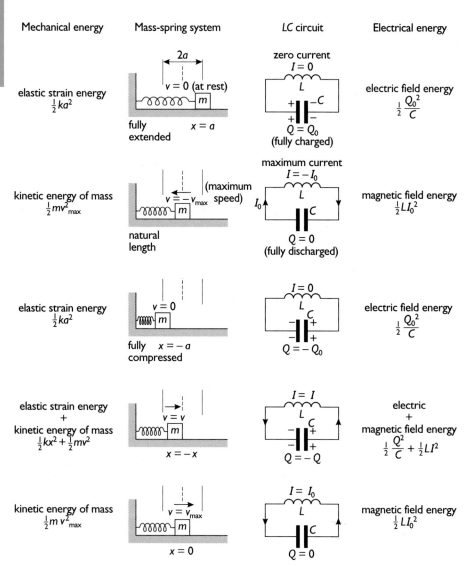

Figure 22.31 Comparison of mechanical and electrical oscillations

Mechanical		Electrical	
mass	m	L	inductance
displacement	x	Q	charge
velocity	$v = \dfrac{dx}{dt}$	$I = \dfrac{dQ}{dt}$	current
acceleration	$\dfrac{d^2x}{dt^2} = \dfrac{dv}{dt}$	$\dfrac{d^2Q}{dt^2} = \dfrac{dI}{dt}$	rate of change of current
force (Newton's second law)	$F = m\dfrac{dv}{dt} = m\dfrac{d^2x}{dt^2}$	$V = L\dfrac{dI}{dt} = L\dfrac{d^2Q}{dt^2}$	voltage (following from definition of L)
stiffness or spring constant	$k = F/x$	$1/C = V/Q$	
compliance	$1/k = x/F$	$C = Q/V$	definition of capacitance
elastic potential energy	$\tfrac{1}{2}kx^2$	$\tfrac{1}{2}Q^2/C$	electric field energy
kinetic energy	$\tfrac{1}{2}mv^2$	$\tfrac{1}{2}LI^2$	magnetic field energy
natural frequency of oscillation	$\dfrac{1}{2\pi}\sqrt{\dfrac{k}{m}}$	$\dfrac{1}{2\pi}\sqrt{\dfrac{1}{LC}}$	natural frequency of oscillation

Figure 22.32 Corresponding mechanical and electrical quantities

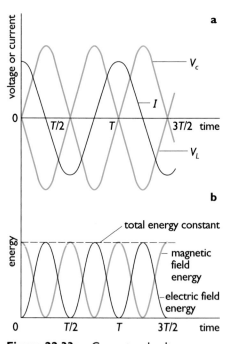

Figure 22.33 a Current and voltage, **b** stored energy against time, for a series resonance circuit

22 ENERGY AND POWER IN AC CIRCUITS

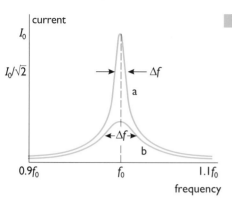

Figure 22.34 Current resonance curves for an *LCR* circuit with **a** $R = 10\,\Omega$, **b** $R = 30\,\Omega$

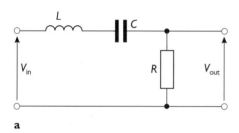

a

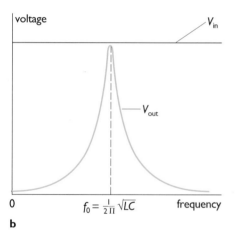

b

Figure 22.35a Circuit and **b** frequency response curve for a band-pass filter

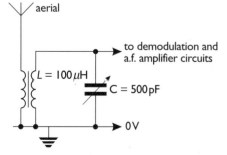

Figure 22.36 The rf section (aerial) circuit of a simple radio receiver

Tuned circuits: Q-factor and selectivity

The quality or Q-factor ($\leftarrow$ 20.4) of the coil determines the sharpness of the resonance curve in LCR circuits. The quality factor Q is given by:

$$Q = \frac{2\pi}{\text{fraction of energy of system lost per oscillation}}.$$

It represents the energy-storing ability of the system. Capacitors are almost free from energy losses. It is the resistance of the coil that determines the Q-factor.

The fraction of a system's energy lost per cycle is R/fL.

So: $Q = 2\pi fL / R = \omega L / R$.

At resonance:
$$V_S = V_R = IR$$
and $\quad V_L = IX_L = I.2\pi fL$
so $\quad Q = V_L / V_S$.

One way of measuring Q is to find the ratio of the voltage across the inductor or capacitor to the supply voltage at resonance (when the impedance is purely resistive). We can also define Q in terms of the band-width of the resonance curve. This is the ability of the circuit to select a particular signal from others at close frequencies, and is known as **selectivity**.

For a high Q series circuit, $Q = f_0 / \Delta f$, as shown in Figure 22.34. The band-width Δf is measured at the half power points, where $I = I_0/\sqrt{2}$. The resonant frequency f_0 of the circuit is 1.6 MHz, in the radio-frequency region of the electromagnetic spectrum. Suitable values for the capacitor and inductor of such a radio-tuning circuit are 100 pF and 100 μH.

From Figure 22.34, you can see that resonant circuits can be used as filters to select a small band or almost a single frequency from a mixture of signals. Figure 22.35a shows how an LCR combination can be connected to make a narrow **band-pass** filter. However, only certain ratios of component values will give the sharp response curve of Figure 22.35b.

Figure 22.36 shows the tuning circuit of the radio in Figure 22.14. The radio waves generate oscillations at many different frequencies in the aerial. In the tuned transformer circuit, any oscillations close to the resonant frequency build up to give a measurable alternating voltage across it. The sharpness of resonance or selectivity and the amplification factor are determined by the Q-factor of the circuit. The amplification factor is the ratio of the signal voltage across the capacitor to its value at the aerial. The selected frequency or frequency band is then demodulated and amplified ($\rightarrow$ 23.1 and $\rightarrow$ 23.3).

Questions

14 Refer to the box 'Tuned circuits' and Figure 22.34.
a Check that the two methods of measuring Q give the same value for Q in Figure 22.34.
b What would the resonance curve for $R = 0$ look like?
c What is the value of its Q-factor?
d Show that for half peak power, $I = I_0/\sqrt{2}$.

15 Outline the problems that might arise if two radio stations broadcast on frequencies close to one another.

OSCILLATIONS

SUMMARY

The rms and peak values of an alternating voltage are related by the expression:

$V_{rms} = V_o/\sqrt{2}$.

The rms value is the equivalent dc value, and is used for calculating power dissipation in ac circuits.

The impedance Z of the components of a circuit is given by:

$$Z = \frac{V_o}{I_o}$$

$$= \frac{V_{rms}}{I_{rms}}.$$

The reactance X of a component is the part of the impedance that draws wattless power. The power drawn by it from the supply is returned to the supply. The power drawn from the supply and not returned is dissipated in the resistance of the circuit.

In a perfect transformer:

$$\frac{V_p}{V_s} = \frac{I_s}{I_p}$$

$$= \frac{N_p}{N_s}.$$

Power dissipated $= V_p I_p = V_s I_s$,

where V and I are in phase.

The resonant and natural frequency f_o of a pure LC circuit is given by:

$$f_o = \frac{1}{2\pi\sqrt{LC}}.$$

At resonance V_{supply} and I are in phase.

At resonance, I is a maximum for the series combination and a minimum for the parallel combination. The resistance of the circuit determines the maximum or minimum current.

Electrical oscillations in an LC circuit are analogous to the mechanical oscillations in a mass-spring system. L is equivalent to M and $1/C$ is equivalent to k.

The Q-factor, $2\pi fL/R$ or $\omega L/R$, determines the selectivity – the sharpness of resonance – of an LC tuning circuit.

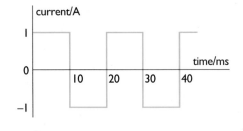

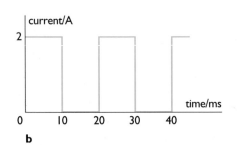

Figure 22.37

SUMMARY QUESTIONS

16 Figure 22.37 shows two square current waveforms.
a Find **i** the average value, **ii** the peak value and **iii** the rms value of each waveform.
b Find the relationship between the rms and peak value for each waveform.
c Calculate the power each current dissipates in a 10 Ω resistor.

17 When a lamp is connected to an ac supply, the average alternating voltage across it is zero. Why does the lamp light?

22 ENERGY AND POWER IN AC CIRCUITS

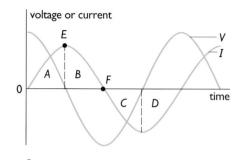

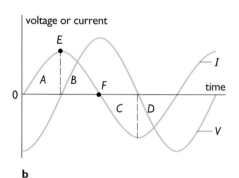

Figure 22.38

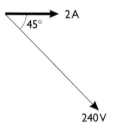

Figure 22.39

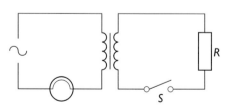

Figure 22.40

18 In Figure 22.38a and b, the current and voltage are in **quadrature** – $\pi/2$ rad out of phase.
a Which figure shows the V and I variations with time for **i** a capacitor, **ii** an inductor?
b For each of the sections, A, B, C and D, state whether the energy stored in the component is increasing or decreasing.
c At which point, E or F, is the energy stored a maximum in **i** a capacitor and **ii** an inductor?

19 This question shows that when you increase the size of a transformer its efficiency increases.
When you double the dimensions of a transformer:
a by what factor does **i** the number of turns, **ii** the length of the windings, **iii** the area of cross section of the windings, increase?
b by what factor does the resistance of the windings decrease?
c the inductance increases by a factor of 2. For the same flux through the core, by what factor does the current decrease?
d show that, under open circuit conditions, the resistive energy losses fall by a factor of 8 and, under load conditions, by a factor of 2.

20 RMS measurements across an alternating power supply are shown in the phasor diagram of Figure 22.39.
a What must the minimum, apparent power rating of the supply be?
b What is the value of the power factor of this supply?
c What is the ratio of the resistance to the reactance in the circuit?
d What real power is dissipated in the circuit?
e Would any of your answers change, if the measurements were of peak rather than rms values?

21 Explain the following observations. In the circuit of Figure 22.40, the switch S is initially open and there is no iron core linking the two coils. The lamp lights normally. Two C-cores are fitted together through the coils. The lamp goes out. Switch S is closed and the lamp lights again.

22 Consider the power station shown by one steam turbine in Figure 22.17. Assume **i** it generates 50 MW for heavy industry, **ii** the supergrid at 400 kV is used, **iii** each of the three sets of cables between the transformers has the same resistance of 20 Ω, and **iv** each of the three transformers loses 5 % of the incident power.
Calculate:
a the power available to heavy industry at 33 kV,
b the overall efficiency of the system.

23 Cable size is important in determining the cost for transmitting electricity over large distances. The minimum cost occurs when there is a current density of $5 \times 10^5 \, \text{A m}^{-2}$ in the cable.
a Calculate the total cross-sectional area of the supergrid cables, at 400 kV, carrying the current from a 2 000 MW power station.
b The cost of the power dissipated in the cables, for a fixed current value, is proportional to their resistance. How does the resistance depend on the radius of the line? Sketch a graph of the cost of power dissipation against cable radius.
c Suppose the total costs of building, maintaining, etc. a power cable system are proportional to the weight of the current-carrying cables. To your graph of part **b**, add a graph of capital costs against cable radius.

OSCILLATIONS

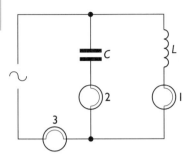

Figure 22.41

d From your two graphs, suggest how a minimum-cost cable size can be deduced.

24 Consider the action of a high-frequency electromagnetic induction cooker as a transformer (← 22.3). The metal saucepan acts as a secondary load resistance of about $0.1\ \Omega$.
a Calculate the power being drawn from the primary, when there is a current of 100 A in the saucepan base.
b Suggest why the chef will only be burnt rather than receive an electric shock, if she accidentally touches the metal of the saucepan.
c What would be the effect of having a saucepan with **i** too large or **ii** too small a resistance?

25 Explain the following observations. In the circuit of Figure 22.41, the frequency of the supply voltage is varied whilst its amplitude is kept constant. At low frequencies, lamps 1 and 3 are alight. At some middle frequency, lamps 1 and 2 are alight. At high frequencies, lamps 2 and 3 are alight.

26 A signal generator of constant rms output voltage 5 V is connected in series to an inductor, capacitor and resistor, as shown in Figure 22.24. The total circuit resistance is $25\ \Omega$. At the resonant frequency, $X_L = X_C = 250\ \Omega$.
a Calculate the current in the circuit.
b Calculate the voltages across **i** R **ii** C and **iii** L.
c Explain why V_S is not equal to $V_R + V_C + V_L$.
d Calculate the quality factor Q of the circuit.
e How could you increase Q by a factor of 10?

23 ANALOGUE ELECTRONIC SYSTEMS

Amplifiers are electronic systems which process continuously varying electrical signals such as an audio signal, a sound level, a pH level or a radio signal. **Integrated circuits** have revolutionised the design and construction of complicated electronic circuits. It has affected digital electronics ($\rightarrow$ 24) as well as amplifiers and other **analogue** circuitry. The term analogue comes from analogue computing where particular circuits act as analogues or representations, of real phenomena, devices or systems.

To use an analogue IC you only need to understand how it affects the input signal and produces a particular output. The internal workings of a chip can remain unknown: they are only of real consequence to the designers and manufacturers of such circuits. Complex electronic systems can be constructed using such basic circuits as black boxes, or building blocks. This is often called the **systems approach** to electronics. For example, a pn junction diode made of say silicon, can be treated purely as a building block which is part of a system. There is no need to consider how the charge carriers move within the p- and n-type materials ($\rightarrow$ 27.6).

An important building block is the **operational amplifier**, or **op-amp**. Its name stems from the original use of such amplifiers to perform mathematical operations in analogue computing. Op-amps are complete amplifying systems and are used in many circuits, for example, commercial electronic circuits designed for audio, radio and television systems. Various symbols are used to represent these.

23.1 THE P-N JUNCTION DIODE

An ideal diode allows charge to flow freely in one direction in a circuit but presents an infinite resistance in the other.

EXPERIMENT 23.1 The current-voltage characteristic of a silicon diode

By connecting the circuit of Figure 23.1a adjusting the variable resistor and measuring the voltage and current, then with the diode connections reversed or by connecting the circuit of Figure 23.1b, using an oscilloscope, you can obtain the characteristic shown in Figure 23.1c. Alternatively, you can connect an oscilloscope using the circuit of Figure 23.1b to obtain a display of Figure 23.1c on the screen.

E OSCILLATIONS

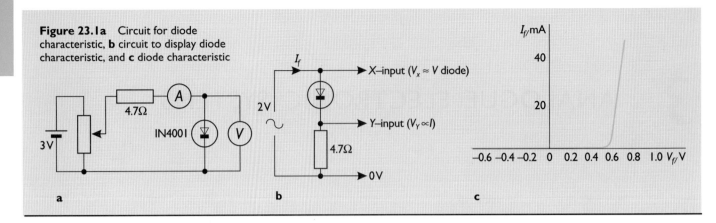

Figure 23.1a Circuit for diode characteristic, **b** circuit to display diode characteristic, and **c** diode characteristic

The characteristic in Figure 23.1c shows that to make a typical silicon diode conduct, a potential difference of 0.6 V must be applied across the diode in the forward direction. The forward resistance is then very small. The current increases from 1 mA to 100 mA for an increase of only about 0.2 V. So large-current, high-voltage diodes have characteristics that approximate very closely to the ideal. Germanium diodes, which are used for low power applications, require only about 0.2 V for conduction.

Question

1 Sketch the current-voltage characteristic of an ideal diode.

RECTIFIER CIRCUITS

Most dc power supply units first transform the ac mains to the required voltage and then rectify the ac using diodes to give dc. A simple **half-wave rectifier** circuit is shown in Figure 23.2a. The input and output voltage waveforms are shown in Figure 23.2b. An improved circuit, called a **full-wave rectifier**, can be constructed from a bridge network of diodes (see Figure 23.3a) or by using a centre-tapped transformer and two diodes (see Figure 23.3b). Although the whole ac waveform is then used, the resultant dc still consists of a series of pulses (see Figure 23.3c).

Most electronic equipment requires smooth, unfluctuating dc. To achieve this, a reservoir capacitor C is connected across the rectifier output, as in Figure 23.4. The capacitor charges nearly to the peak value of the ac supply. The diode acts as a switch, allowing charge to the capacitor only when the transformer voltage exceeds that across the capacitor. The diode disconnects the power supply from the capacitor when the transformer voltage falls below that across the capacitor,

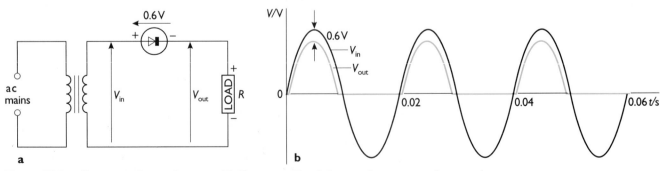

Figure 23.2a Power supply transformer and half-wave rectifier, **b** input and output waveforms

23 ANALOGUE ELECTRONIC SYSTEMS

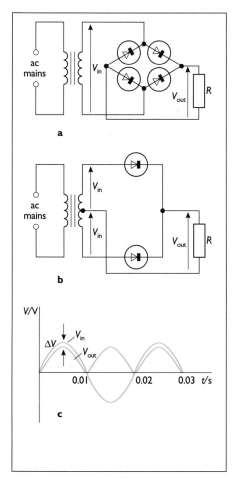

Figure 23.3a Full-wave bridge rectifier, **b** bi-phase full-wave rectifier, **c** input and output waveforms

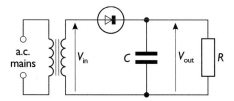

Figure 23.4 Half-wave rectifier with reservoir capacitor

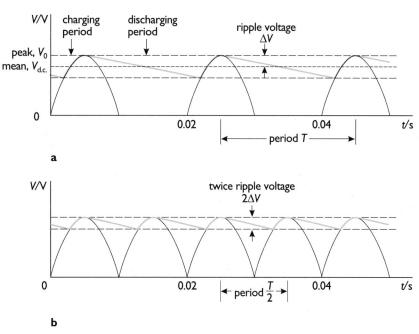

Figure 23.5 Output waveform for **a** half-wave rectifier, **b** fullwave rectifier, with reservoir capacitor

which then discharges through the load. So for most of the time, the current I in the load is provided by the capacitor.

The capacitor behaving as a reservoir smooths out the **ripple voltage** in the rectifier output. The ripple voltage is the fluctuation in the output waveform about its mean value (see Figure 23.5). Without the load resistor, the filtering action is perfect and the output voltage is constant, close to the peak value V_o. The ripple voltage is smaller the larger the value of C and the higher the frequency f of the supply.

Note that, for a full-wave rectifier, the ripple frequency is twice that of the supply.

Questions

2 Trace the current path through the two full-wave rectifier circuits in Figure 23.3a and b over one cycle of the ac input. Is the ripple voltage ΔV in Figure 23.3c the same for both circuits?

3 What happens to the ripple voltage ΔV as the value of the load resistor R, changes? Why is this?

4 Give some examples of the uses of circuits giving a fairly smooth dc output, such as that in Figure 23.4.

LIGHT EMITTING DIODES

Light emitting diodes or LEDs, are used in a variety of electronic systems, often to indicate whether the system is off or on. An LED behaves in a similar way to a conventional diode, conducting in one direction only. To conduct, an LED requires a larger forward voltage across its terminals than the 0.6 V for a silicon diode. Typically it requires a voltage of the order of 2 V with a current of around 20 mA. When it is conducting, it emits light of a well-defined frequency. The forward voltage and the precise frequency are determined by the nature of the

OSCILLATIONS

semiconductor material in the diode. Red, yellow, green and blue LEDs are available.

LEDs are more susceptible to damage than ordinary diodes, particularly from reverse-bias voltages, the maximum allowed being about twice the normal forward voltage. It is important that the current through an LED does not reach too high a value: typically it is 20–30 mA. So, a protective resistor of 200–300 Ω is usually connected in series with it: often internally within the LED package.

As with liquid crystals, LEDs are frequently manufactured in the form of an array. The most common form is probably the seven-segment display (see Figure 23.6). This is manufactured with either a common anode or common cathode. Switching on a combination of its segments can generate, for example, all digits between 0 and 9.

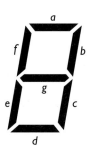

Figure 23.6

Question

5 Figure 23.6 shows the arrangement of segments a, b, c, d, e, f, g in a common cathode seven-segment display. Which segments should be switched on to display each of the digits 0 to 9?

EXPERIMENT 23.2 Rectifier circuits

By connecting the circuits of Figure 23.7a and b, with red and green light emitting diodes (LEDs), you will be able to identify the current directions in a full-wave rectifier circuit. In the circuit of Figure 23.7c, you will notice that the LEDs flash until S is closed and then one remains on. Repeating this last experiment with an ordinary diode, at a higher frequency of, say, 300 Hz, and an oscilloscope across the load resistor, enables you to see a waveform similar to Figure 23.5a with a 0.6 V drop across the diode.

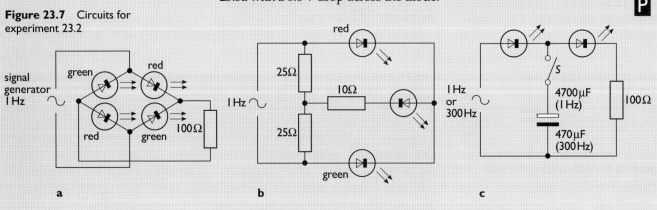

Figure 23.7 Circuits for experiment 23.2

THE DIODE AS A DETECTOR

The diode is also used with a capacitor as a filter in the **detection** or **demodulation** of an amplitude modulated (am) radio signal (see Figure 23.8). Circuits can be designed and used to detect and select a 1 MHz electromagnetic wave (see Figure 22.36). This radio frequency wave carries an audio signal. The audio frequency range lies between 20 Hz and 20 kHz. The audio wave **modulates** the amplitude of the **carrier** radio wave.

The half-wave rectifier circuit of Figure 23.4 would, in principle, separate the audio signal from the carrier wave. The transformer would be replaced by the signal source. The capacitor would be very much smaller at about 10 nF and the

23 ANALOGUE ELECTRONIC SYSTEMS

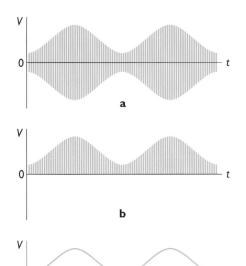

Figure 23.8a am waveform b effect of diode c with filter

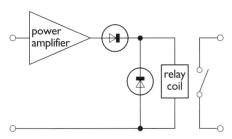

Figure 23.9 Diodes protecting a power amplifier output

resistor much larger at about 10 kΩ. In such a detector circuit, the diode 'cuts' the signal in half, as shown in Figure 23.8b and the time averaged signal is no longer zero. The time constant RC of the circuit must be short compared with the period of the audio signal so that the voltage across the capacitor keeps in step with the changes in the audio signal. However, RC must also be long compared with the period of the carrier wave so that little ripple is introduced. The output voltage across the load resistor R is then a faithful reproduction of the audio signal.

Question

6 Show that the values of R and C stated for the detector circuit above give a suitable RC for the am wave demodulation circuit.

Diode protection

Diodes are often used as safety devices. For example, the circuit of Figure 23.9 includes two diodes to protect the output of a power amplifier. Current only passes through the relay when the output voltage is positive. When the output from the amplifier is negative, the circuit is switched off. Then the collapsing magnetic flux (← 14.2) in the relay would produce a large negative 'voltage spike' but for the presence of the parallel diode. This conducts as soon as the voltage is more negative than –0.6 V, so protecting the amplifier.

Moving coil meters are sometimes protected from overload by a diode across their terminals. The diode is chosen so that it does not conduct for applied voltages within the design range of the meter. An excessive pd across the terminals causes the diode to conduct and act as a shunt across the meter, reducing the possibility of damage to the moving coil.

23.2 THE IDEAL OP-AMP AND FEEDBACK

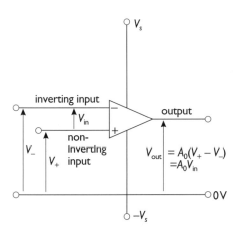

Figure 23.10 Symbol for an op-amp showing $V_{out} = A_o V_{in}$

The op-amp is an amplifier with a very large voltage gain A_o, typically about 10^5. It is a **differential** amplifier, which means that it amplifies the voltage difference **between** the two inputs. The inputs are:

- the **non-inverting**, labelled (+), and
- the **inverting**, labelled (−).

The output voltage V_{out} is given by:

$$V_{out} = A_o (V_+ - V_-)$$

where V_+ and V_- are the voltages at the two inputs, relative to the 0 V level.

The plus and minus signs refer to the sign of the output voltage rather than that of the input voltage. So when the voltage applied to the inverting (−) input is more positive than that to the non-inverting (+) input, the output is a negative voltage (see Figure 23.10). To provide negative as well as positive voltages at the output, the amplifier requires positive and negative balanced power supplies, often termed a 'dual rail supply'. In Figure 23.10, these are labelled V_S, 0 V and $-V_S$. In all circuit diagrams, the positive and negative power supply connections

OSCILLATIONS

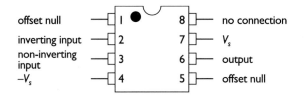

parameter	ideal op-amp	741 op-amp	TL 081 op-amp
open-loop gain	infinite	10^5*	10^5
input resistance	infinite	$2\,M\Omega$	$10^{12}\,\Omega$
output resistance	zero	$75\,\Omega$†	$100\,\Omega$
slew rate	infinite	$10^6\,Vs^{-1}$	$10^7\,Vs^{-1}$

* The open-loop gain decreases as the frequency increases, becoming 1 at 1 MHz.
† The output is protected against short circuits. The maximum current is 25 mA.

Figure 23.11 Pin connections (top view) for the 741 and TL081 op-amps, and performance parameters

to the op-amp are omitted for simplicity. They must, of course, be present in any practical arrangement. Input and output voltages are always given relative to the 0 V level.

Two popularly used op-amps are the 741 and TL081. Figure 23.11 shows their pin connections and their performance parameters contrasted with those of an ideal op-amp.

The ideal op-amp has **infinite gain**, which means that **any** difference in potential between the inputs drives the output to saturation – close to V_S or $-V_S$. **Infinite input resistance** means that there is zero input current so there is no loading of, or taking of current from, the input circuit. The ideal op-amp also has **zero output resistance**, so it can drive any load efficiently. The output responds instantaneously to any input change, and so is said to have a very high **slew rate**. You can use an op-amp as a means of comparing two voltages as shown in Figure 23.12. It is then referred to as a **comparator**. It works for voltages of between 1 mV and 10 V or so. However when V_+ is exactly equal to V_-, the output should be zero, but in practice, it is extremely difficult to achieve because of the high **open loop gain**, A_0. This causes the output to swing rapidly to $+V_S$ when V_+ is only slightly bigger than V_- or to $-V_S$ when the opposite is true.

EXPERIMENT 23.3 Op-amp comparator circuit

By setting up the circuit of Figure 23.12 you can measure the input and output voltages with high resistance voltmeters.

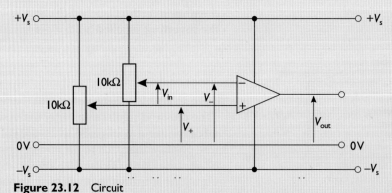

Figure 23.12 Circuit

23 ANALOGUE ELECTRONIC SYSTEMS

FEEDBACK

The op-amp has a very large gain A_o, normally called the **open-loop** gain. This makes its behaviour unstable so it is rarely used without a **feedback loop**. The feedback is produced by adding a fraction of the output voltage to one of the inputs. The resultant output voltage is created from the combined effect of the applied input voltage and the output voltage.

In **negative feedback**, part of the output is fed back so as to reduce the input change. Such an arrangement is used in amplifiers. It reduces considerably the gain of the circuit, but it also reduces distortion, improves stability and increases the band-width. This means that the whole range of frequencies used are amplified by the same factor (see Figure 23.13). In **positive feedback**, the fraction of the output fed back assists the change at the input. This arrangement is used in oscillator circuits and where rapid switching is required.

The **closed loop gain**, A, is the overall gain with feedback applied as shown in Figure 23.14:

$$A = \frac{V_{out}}{V_{in}}.$$

With a negative fraction β of the output fed back to the input this is related to the open loop gain by the equation:

$$A = \frac{A_0}{1 + \beta A_0}.$$

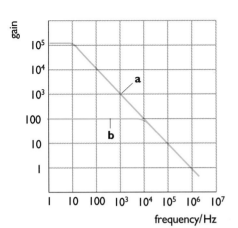

Figure 23.13 Frequency response of the 741 op-amp **a** open-loop gain A_o **b** closed-loop gain $A = 100$

Feedback equation

In Figure 23.10, the amplifier output $V_{out} = A_o V_{in}$. By adding a negative feedback loop, as shown in Figure 23.14, a fraction of the new output voltage is fed back to the input. This fraction is called the **feedback factor** β and:

$$V_{out} = A_o (V_{in} - \beta V_{out}).$$

Rearranging gives:

$$V_{out} = \frac{A_o}{1 + \beta A_o} \cdot V_{in}.$$

The closed-loop voltage gain of the amplifier is given by:

$$A = \frac{V_{out}}{V_{in}}$$

$$= \frac{A_o}{1 + \beta A_o}.$$

Normally, $\beta A_o \gg 1$, so A is approximately equal to $1/\beta$.

Notice that the gain of the amplifier is independent of the op-amp's open-loop gain as long as A_o is very large. So the mathematical model applies for any op-amp with a large A_o.

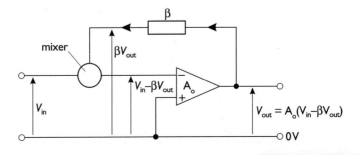

Figure 23.14 Schematic representation of negative feedback in an amplifier

E OSCILLATIONS

In section 23.3, we consider two basic op-amp circuits using negative feedback: the inverting and non-inverting amplifiers. In section 23.4 we describe two circuits using positive feedback: the Schmitt trigger and the relaxation or square-wave oscillator.

23.3 AMPLIFICATION AND NEGATIVE FEEDBACK

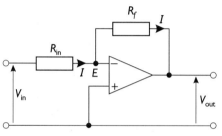

Figure 23.15 The inverting amplifier

The properties of the ideal op-amp given in 23.2 lead to two rules to use when analysing any circuit containing negative feedback. These are:

rule 1 the current into either op-amp input is zero,

rule 2 the output voltage changes in such a way as to maintain zero voltage between the inputs, as long as the amplifier is not saturated.

> ### Question
> 7 Show that the two rules above follow from the properties of the ideal op-amp.

THE INVERTING AMPLIFIER

The circuit of the inverting amplifier is shown in Figure 23.15. The feedback loop is R_f. The non-inverting (+) input and point E are held at 0 V. This is the function of the feedback loop (rule 2). We, therefore, refer to point E as a **virtual earth**.

The current in R_{in} must be the same as that in R_f since there is no current into the inverting (−) input (rule 1). So when V_{in} is positive, V_{out} must be negative. The two resistors behave as a voltage divider with the junction between them at 0 V, as in Figure 23.16.

The **closed-loop** gain A is the ratio V_{out}/V_{in}, which equals the ratio of the resistances R_f and R_{in}. When the output reaches the supply voltage V_S, saturation occurs. The mathematical model is then no longer applicable, because point E moves away from 0 V.

Note that the behaviour of the circuit in Figure 23.15 depends only on the values of the resistors. A typical value of A is 10 to 100 compared to a value for A_o of greater than 10^5. The ideal value of A_o is infinite. If you use a real op-amp with a sufficiently high gain and input resistance, it will give a closed-loop gain very close to the ideal value. Experiment 23.4 investigates the behaviour of an op-amp with negative feedback, configured in various ways.

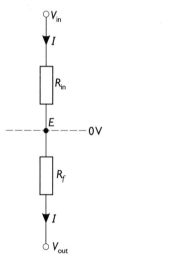

Figure 23.16 Resistors R_{in} and R_f acting as a voltage divider

Closed-loop gain of inverting amplifier

In Figure 23.16:
$$V_{in} = IR_{in}$$
and $V_{out} = -IR_{out}$
So $A = V_{out}$
$$= -R_f/R_{in}$$

> ### Question
> 8 Calculate the feedback factor β for an inverting amplifier with $R_{in} = 10\,k\Omega$ and:
> a $R_f = 10\,k\Omega$
> b $R_f = 100\,k\Omega$,
> c $R_f = 1\,M\Omega$.

EXPERIMENT 23.4 Behaviour of an op-amp with negative feedback

a By plotting a graph of output voltage against input voltage you can obtain the **transfer characteristic** of the circuit in Figure 23.17a or b. This is shown in Figure 23.18. If you reduce the feedback resistor R_f from 100 kΩ to 10 kΩ, you will find that the gain of the circuit is 1. If you increase it to 1 MΩ, the gain is 100. By connecting a double beam oscilloscope to the input and the output of the circuit, you will see the phase shift and the effect of saturation on the output signal.

b Adding one or more input resistors at the inverting input (see Figure 23.19a) produces a summing amplifier. Each input requires a 10 kΩ potentiometer to vary the input voltage, as in Figure 23.17a. You should investigate the effect of making R_1, R_2 and R_3 all equal to the feedback resistor. If you replace the potentiometers by low frequency ac signals from a low voltage transformer or a signal generator, you can investigate how the output signal changes as the frequency varies.

c You can produce a steadily changing voltage from a steady input voltage with the ramp generator circuit of Figure 23.20. With $R = 1$ MΩ and $C = 1$ μF the time constant RC is 1 second. If you adjust V_{in} to +1 V or −1 V and briefly discharge the capacitor, you should observe a steady change in V_{out} on a meter or oscilloscope until saturation is reached close to ±V_S. With $V_{in} = 0.1$ V, you should find that V_{out} takes 10 times as long to reach saturation.

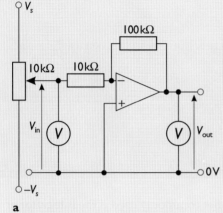

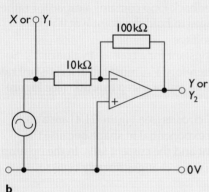

Figure 23.17 Circuits for experiment 23.4a

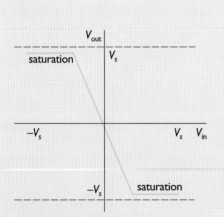

Figure 23.18 Transfer characteristic of an inverting amplifier

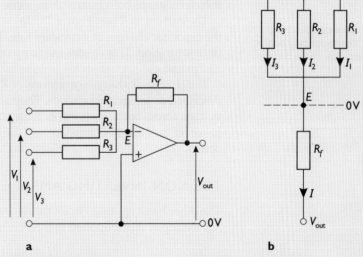

Figure 23.19a Summing amplifier, **b** Kirchhoff's laws applied to a summing amplifier

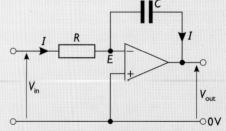

Figure 23.20 Circuit of integrator or ramp generator

OSCILLATIONS

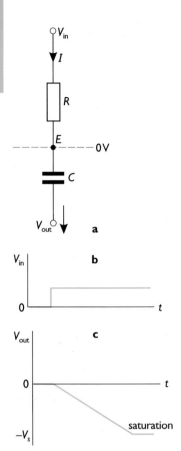

Figure 23.21a Voltages and current path, **b** input, and **c** corresponding output, of a ramp generator

The integrator

From Figure 23.21a:

$$V_{in} = IR$$

and $V_{out} = \dfrac{-Q}{C}$

where Q is the charge stored on the capacitor. Since:

$$I = \dfrac{dQ}{dt}$$

$$Q = \int_0^t I\,dt$$

and $V_{out} = \dfrac{-1}{RC}\int_0^t V_{in}\,dt,$

giving $V_{out}/t = -V_{in}/RC.$

So, for a constant V_{in}, the gradient of the ramp, or the time to reach saturation V_s, depends on R, C and V_{in}.

A **summing amplifier** is constructed by adding one or more resistors to the inverting input, as in Figure 23.19a. The output becomes:

$$V_{out} = -R_f\left(\dfrac{V_1}{R_1} + \dfrac{V_2}{R_2} + \dfrac{V_3}{R_3}\right).$$

The values of the various input resistors weight the importance of the input voltages. For example, with $R_1 = R_2 = R_3 = R_f$:

$$V_{out} = -(V_1 + V_2 + V_3).$$

Questions

9a Use Figure 23.19b and Kirchhoff's first law (← 12.1) to verify the formula for V_{out} of the summing amplifier.
b Show how to add an inverting amplifier to one input to produce a subtractor with output **i** $V_1 - V_2$ and **ii** $10V_1 - V_2$.
10 Sketch graphs of V_{out} as in Figure 23.21c from 0 to 30 s for:
a $RC = 1$ s and **i** $V_{in} = 1$ V, **ii** $V_{in} = 0.5$ V,
b $V_{in} = 1$ V and **i** $RC = 2$ s and **ii** $RC = 3$ s.
Take $V_s = \pm 9$ V.

Another variation of the inverting amplifier commonly used is the **integrator** or **ramp generator** (see Figure 23.20). The feedback loop contains a capacitor rather than a resistor. As before, the output changes to hold the point E at 0 V. It, therefore, keeps both inputs at the same voltage.

The current in R (see Figure 23.21a) is constant for a constant value of V_{in}. So the capacitor charges at a constant rate. Therefore V_{out} must fall at a constant rate, until saturation. This creates the ramp or integral of a step function (see Figure 23.21b and c).

The charge-discharge waveforms of RC circuits (see Figure 21.4) are also described as integrating circuits. They are exponential in shape because the voltage across both components is constant and the current falls. In the op-amp however the current is constant and the voltage changes linearly until saturation occurs.

THE NON-INVERTING AMPLIFIER

The other basic op-amp circuit configuration that makes use of negative feedback has the input signal applied to the non-inverting input. This circuit is sometimes

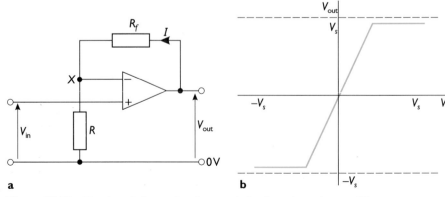

Figure 23.22a Circuit and **b** transfer characteristic of non-inverting amplifier

23 ANALOGUE ELECTRONIC SYSTEMS

Closed-loop gain of non-inverting amplifier

In Figure 23.22a, the output voltage V_{out} changes to maintain point X at the input voltage V_{in} as long as the amplifier is unsaturated (rule 2). There is no current into the amplifier (rule 1). So the resistors form a potential divider, as shown in Figure 23.23, and:

$$V_{out} = I(R_f + R)$$
$$V_{in} = IR.$$

The closed-loop gain A is given by:

$$A = \frac{V_{out}}{V_{in}} = \frac{R_f + R}{R}$$
$$= \frac{R_f}{R} + 1.$$

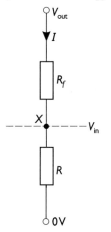

Figure 23.23 Resistors R_f and R acting as a voltage divider

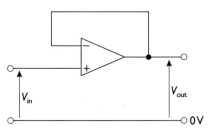

Figure 23.24 Buffer or voltage follower

called the **non-inverting amplifier** (see Figure 23.22a). It has an extremely high input resistance and a very low output resistance. Its transfer characteristic is shown in Figure 23.22b. The output changes in the same sense as the input and the closed-loop gain of the amplifier is $(R_f + R)/R$.

A particularly useful and simple version of this circuit is the **buffer** or **voltage follower** shown in Figure 23.24. This has 100 % negative feedback, so $\beta = 1$ and the gain of the circuit is unity. For the ideal op-amp, the input resistance is infinite and the output resistance is very small.

Buffer or voltage follower

This circuit can act as a current source to drive a device such as a moving-coil meter without loading the input circuit. Suppose, for example, the 100 μF capacitor of Figure 23.25a is charged to a voltage V and the voltmeter has a resistance of 100 kΩ. When the switch S is thrown, the capacitor discharges to 0.37 V in 10 s. The time constant RC for the arrangement is:

$$RC = 10^5 \,\Omega \times 10^{-4}\,F$$
$$= 10 \text{ s}.$$

When the same experiment is repeated with a voltage follower between the capacitor and the meter (see Figure 23.25b), the capacitor does not discharge significantly and the meter reading is steady. This is due to the high input resistance of the amplifier, which is about 10^{12} Ω for a TL081 op-amp.

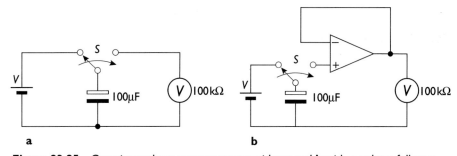

Figure 23.25 Capacitor voltage measurement **a** without, and **b** with, a voltage follower

Question

11a Show that the gain of a non-inverting amplifier with $R_f = 100$ kΩ and $R = 10$ kΩ is 11.
b Calculate the gain when $R_f = R$.
c Show that the gain of the voltage follower is unity. Note that R is infinite and $R_f = 0$.

MAXIMISING SIGNAL TRANSFER

You can couple several op-amp circuits together to obtain maximum voltage transfer between circuits and minimum signal loss. For the final stage of such a system, you require a power amplifier to drive a transducer such as a loudspeaker, a relay or motor. This is because such a device has a low resistance, requiring a large current to drive it. However, when the output of the system is only monitored by an oscilloscope, which is a high resistance load, no power amplifier is needed. Consider a two-stage amplifier system with input and

OSCILLATIONS

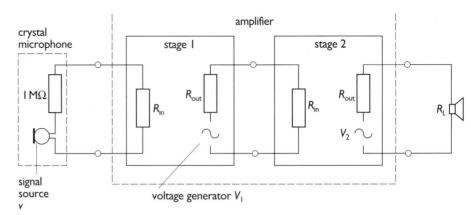

Figure 23.26 Equivalent circuits of an amplifier to illustrate impedance matching

output connections, as shown in Figure 23.26. The two stages can be thought of as black boxes. The voltage across the input to the second stage is

$$\frac{R_{in}}{R_{out}+R_{in}} \cdot V_1,$$

where R_{in} and R_{out} are the input and output resistances of the amplifier stages. The output of the first stage behaves like a supply of voltage V_1.

The condition for maximum transfer to the second stage is: $R_{in} \gg R_{out}$.

At the output to the second stage is a load R_L, such as a 4 Ω loudspeaker. For maximum power to be transferred to this load: $R_{out} = R_L$ (← 12.3).

At the other end, the crystal microphone across the input to the first stage may have a resistance of 1 MΩ. For maximum signal transfer into the amplifier: $R_{in} \gg 1$ MΩ.

In general, we require an amplifier to have a very **high input impedance** and a very **low output impedance**. Note that the term impedance is used rather than resistance. This is because every system has capacitance and inductance as well as resistance.

Questions

12a Show that the output current of a voltage follower is $I = V_{in}/R_L$ when a load resistor R_L is connected across the output.
b Trace the path of this current through the system.

13 Show that the voltage across the first stage of Figure 23.26 is given by:

$$\frac{R_{in}}{R_{in} + 1} \cdot V,$$

where R_{in} is measured in MΩ.

Impedance matching is of vital importance in all signal transfer processes (← 9.3). A transformer can also be used for impedance matching (← 22.3) where a high voltage and low current are transformed into a low voltage and high current. However, because of losses, the power gain of the transformer system is always less than 1. With the voltage follower circuit, the power supply provides the source of power. Although the voltage is unchanged with a voltage gain of 1, the current is increased, causing a power gain of greater than 1.

23 ANALOGUE ELECTRONIC SYSTEMS

EXPERIMENT 23.5 A simple am radio receiver

If you construct the circuit in Figure 23.27, you will have a simple radio receiver (← 22.5 and 23.1). You can tune to a particular radio signal by varying C_1. Having chosen a strong signal, you can follow it through the circuit by connecting an oscilloscope between points W, X, Y and Z and earth, 0 V.

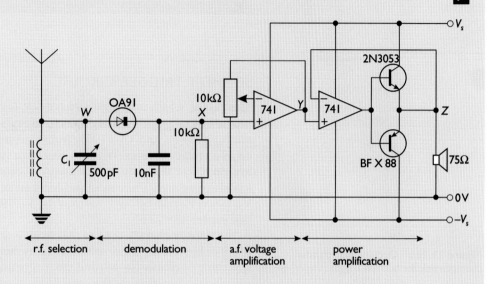

Figure 23.27 Simple radio receiver

23.4 POSITIVE FEEDBACK CIRCUITS

Where the feedback loop is between the output and the non-inverting input, an increase in the input causes an increase in the output. This, in turn, causes the input to increase further until the output voltage is driven rapidly to the limit, close to $+V_S$ or $-V_S$. This is **positive feedback**.

> *Feedback in control systems*
>
> **Servo mechanisms** in control systems use feedback to maintain steady conditions or smooth out any required changes. The feedback always reduces the change being made, so the feedback must be negative.
>
> Any time delay in the response of the servo mechanism acts like a phase shift and may cause the feedback to become positive. The system may then go into oscillation. Resistors can be used in a resonance circuit, as in Figure 23.34, to damp down any possible oscillations. Friction or viscous forces damp oscillations in mechanical systems in a similar way (← 20).

Question

14a Which of the following two statements demonstrates negative feedback and which demonstrates positive feedback?
  i The suggestion in the 'money markets' that the pound sterling is overvalued can lead to speculators selling sterling in favour of other

OSCILLATIONS

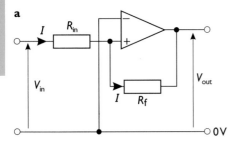

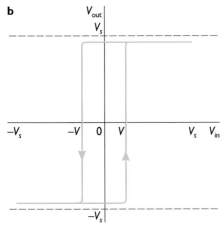

Figure 23.28a Circuit and **b** transfer characteristic for Schmitt trigger

currencies. Widespread selling reduces confidence in the currency and further reduces its value causing a 'run' on the pound.

ii When the body's temperature rises due to exercise, perspiration is secreted by the sweat glands of the skin. Evaporation of the fluid causes cooling, which reduces the body's temperature.

b Give one more example of each type of feedback, chosen from any context other than electronics.

SCHMITT TRIGGER CIRCUIT

This circuit, which uses positive feedback, is shown in Figure 23.28a. It can provide rectangular shaped pulses from ragged input waveforms. Suppose the output is at $-V_S$, and V_{in} is rising from a negative to a positive value. R_{in} and R_f act as a voltage divider. When V_{in} reaches a value of (R_{in}/R_f). V_S, the non-inverting input is at 0 V. Any further rise in V_{in} causes the non-inverting input to be at a positive voltage. The output suddenly switches to positive saturation close to V_S. When V_{in} falls below $V = -(R_{in}/R_f)$. V_S, the output switches back close to $-V_S$ (see experiment 23.6).

The transfer characteristic of the circuit is shown in Figure 23.28b. The loop is a result of **hysteresis**. Other examples of hysteresis loops in physics occur in the stretching of rubber (← 4.3) and the magnetisation of iron (← 18.0).

EXPERIMENT 23.6 Schmitt trigger circuit

This circuit is shown in Figure 23.28a. With $R_{in} = 10\ \text{k}\Omega$ and $R_f = 100\ \text{k}\Omega$, you should connect the X-input of an oscilloscope across V_{in} and the Y-input across V_{out}. By varying the value of V_{in}, you will find that V_{out} suddenly changes from negative to positive when $V_{in} = V_{out}/10$ and back to negative when V_{in} equals $-V_{out}/10$.

Using a low voltage transformer or signal generator, you can display V_{in} and V_{out} on a double-beam oscilloscope, as in Figure 23.29. You can also try connecting the output of the circuit to a ramp generator, as in Figure 23.30, with its output connected back to the Schmitt trigger input. You should find that the waveforms at X and Y are respectively, square and triangular in shape.

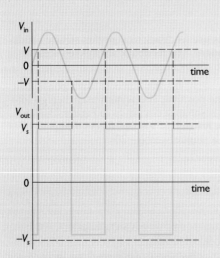

Figure 23.29 Squaring a sine wave

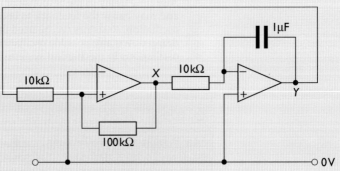

Figure 23.30 Square- and triangular-wave oscillator

23 ANALOGUE ELECTRONIC SYSTEMS

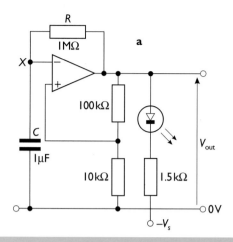

THE RELAXATION OSCILLATOR

There are many ways of generating electrical oscillations. One that uses a single op-amp is called a **relaxation oscillator** (see Figure 23.31). To see how this works, consider the circuit when the capacitor is charging through the resistor. At this instant, the output of the op-amp is close to V_S. The capacitor charges until the voltage across it is greater than $V_S/10$. The output of the op-amp is switched suddenly to about $-V_S/10$, because the voltage at the inverting input has become more positive than the non-inverting input. The capacitor then discharges completely and recharges with the opposite polarity until the voltage across it is less than $-V_S/10$. The output is switched again and the cycle restarts. Figure 23.31b shows the voltages at point X and at the output.

The exponential charge-discharge curves for the capacitor are nearly linear over the range considered.

23.5 ANALOGUE COMPUTING

In analogue computing, each mathematical process is performed by an electronic circuit which behaves in an analagous way. For example, an integrator circuit mimics mathematical integration.

To illustrate this, consider a single wire connecting the output of an integrator circuit back to the input. The voltages at the input and output become the same (see Figure 23.32a) just as the exponential function e^x, when integrated or differentiated, is the same. The capacitor is charged initially from a separate supply V_o. When it is reconnected to the output, the waveform is an exponential decay with a time constant CR (see Figure 23.32b). So the input and output voltage waveforms of this circuit vary exponentially with time.

Similarly, a series of such circuits, suitably set up with the correct applied voltages produces an output which is the solution of differential equations such as those that describe oscillatory motion (← 22.5) and radioactive decay (→ 39.4).

The digital techniques used in digital computers (→ 24) have now superseded many analogue computing systems.

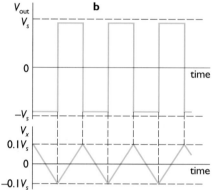

Figure 23.31a Relaxation oscillator and **b** voltages at point X and output

Exponential integration

The input and output voltages of an integrator (see Figure 23.32a) are related by:

$$V_{out} = \frac{1}{RC} \int_0^t V_{in}\, dt.$$

The feedback loop makes $V_{out} = V_{in} = V$. Putting the integrator equation in differential form gives:

$$\frac{dV}{dt} = \frac{-V}{RC}$$

with solution $V = V_o e^{-t/RC}$ for the input or output waveform (see Figure 23.32b).

There is another function that, when differentiated or integrated **twice**, is proportional to itself – the sine or cosine function. Sinusoidal oscillations occur in simple harmonic motion (← 19).

You can set up the differential equation for simple harmonic motion using two integrators and an inverter with the output fed back directly to the input, as shown in Figure 23.33a. The voltages at Y, X and W represent the displacement, velocity and acceleration, respectively, of the oscillating quantity, such as mass or charge. Each integrator produces a $\pi/2$ rad phase shift and the inverter, a π rad phase shift of the input signal. So the input at W is in phase with the output at Z (see Figure 23.33b). Oscillations begin once the charging battery V_o is removed.

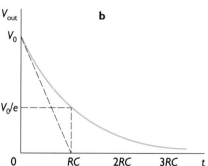

Figure 23.32a Integrator as an analogue computer, **b** its output voltage

OSCILLATIONS

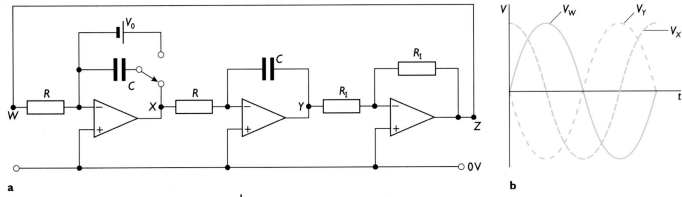

Figure 23.33a Circuit to solve simple harmonic oscillator differential equation and **b** waveforms at points W, X, Y and Z ($V_W = V_Z$)

Simple harmonic motion integration

The equations relating the voltages at W, X and Y in Figure 23.33 are

$$\frac{dV_X}{dt} = -\frac{V_W}{RC}$$

$$\frac{dV_Y}{dt} = -\frac{V_X}{RC}$$

$$V_Z = -V_Y = V_W = V.$$

Eliminating V_X and V_Y gives:

$$\frac{d^2V}{dt^2} = -\frac{V}{(RC)^2},$$

with solution $V = V_0 \sin 2\pi f t$, where $f = \frac{1}{2\pi RC}$.

EXPERIMENT 23.7 Solving differential equations

Using the circuit of Figure 23.32a and starting with $R = 1\,M\Omega$ and $C = 1\,\mu F$, you can investigate what happens for different values of R and C. By switching C from the cell to the output V_o, you can observe the exponential decay of the output on an oscilloscope.

The circuit of Figure 23.33a, with $R = R_I = 1\,M\Omega$ and $C = 1\,\mu F$, should provide you with spontaneous oscillations. Reducing the value of R and C to obtain a convenient frequency enables you to observe waveforms at W, X, Y and Z, using a double beam oscilloscope. You can also compare the phase shift between each op-amp output.

The circuit of Figure 23.33a can be forced to oscillate at frequencies other than its natural frequency by driving the first integrator from the output of a signal generator (see Figure 23.34). The amplitude of the driven oscillations rises to a maximum and resonates close to the natural frequency $f_o = \frac{1}{2\pi RC}$. All of the expected phenomena associated with resonance can be observed, such as the change of phase difference with frequency, between the driving oscillator and the driven system. The feedback loop through resistor R_D is included to change the damping of the amplitude of the oscillations. Free oscillations occur in the circuit of Figure 23.33a by ensuring there is a phase change of 2π radians between the input and output.

Real op-amps do not behave ideally, especially at frequencies of about 1 kHz and above. A small extra capacitance, about 2 pF, may be needed in the inverter

23 ANALOGUE ELECTRONIC SYSTEMS

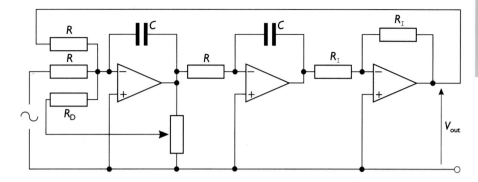

Figure 23.34 Forced oscillation circuit

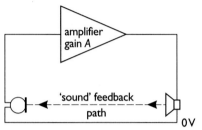

Figure 23.35 'Sound path' acting as positive feedback loop

circuit in parallel with the feedback resistor. This will adjust the phase shift to make the circuit oscillate.

A time delay in a circuit can also have the effect of changing the phase of a signal. This is demonstrated dramatically by bringing together a microphone and loudspeaker, connected to the same amplifier (see Figure 23.35). The 'sound path' acts as a positive feedback loop for certain frequencies, resulting in a whistle or howl being emitted from the speaker. By altering the path length between the microphone and speaker, the frequency of the whistle can be changed.

Questions

15 Add an inverter to Figure 23.32a to set up the equation

$$\frac{dV}{dt} = \frac{V}{RC}.$$

Solve this equation to obtain the value of the output voltage.

16 Compare the electrical and mechanical equations of simple harmonic motion and check that they agree with the table in Figure 22.32.

17 Why are only certain frequencies amplified in the circuit of Figure 23.35?

SUMMARY

A silicon diode requires 0.6 V in the forward direction to conduct.

Diodes are used in rectifier and voltage doubler circuits. Other applications include demodulating radio signals, clipping waveforms and as safety devices to suppress unwanted voltage spikes.

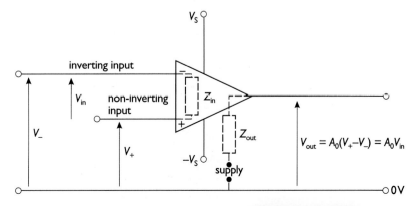

Figure 23.36 Behaviour of an op-amp

OSCILLATIONS

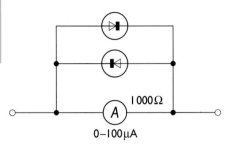

Figure 23.37

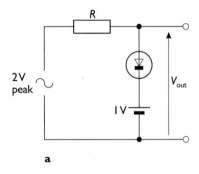

a

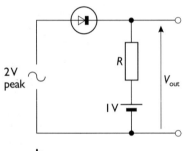

b

Figure 23.38

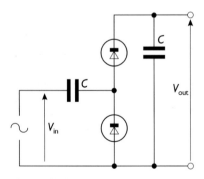

Figure 23.39

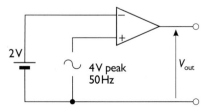

Figure 23.40

The operational amplifier is a differential amplifier with high input impedance, Z_{in}, and low output impedance, Z_{out}. It has an open loop gain A_o of about 10^5. For an ideal op-amp, the input resistance is infinite and the output resistance zero, with infinite open-loop gain.

Once saturation is reached, when V_{out} is approximately equal to V_S, the simple amplifier model no longer applies.

The amplifier is unstable without negative feedback. The closed loop gain A is normally between 1 and 100.

Positive feedback is used in oscillator circuits and the Schmitt trigger, which is a fast switching circuit.

There are two types of amplifier, inverting and non-inverting.

For inverting ones: $\qquad A = -R_f/R_{in}$

and for non-inverting ones: $\quad A = (R_f + R)/R$.

The voltage follower is a non-inverting amplifier with $A = 1$. It is used to drive a load requiring a large current, without loading the input circuit.

An amplifier requires a high input and low output impedance for maximum voltage transfer. The load impedance must match the output impedance for maximum power transfer.

The coefficients in differential equations can be represented by particular values of resistors and capacitors in integrating circuits. This is the basis of analogue computing.

SUMMARY QUESTIONS

18 Two diodes can be used to protect a moving-coil meter from overload, as in Figure 23.37. The meter shown has a resistance of 1 kΩ and a full-scale deflection current of 100 μA. Explain how the circuit works.

19 Figures 23.38a and b show two clipping circuits. For each, sketch the output waveform V_{out} when the input is a sine wave of peak value 2 V.

20 Figure 23.39 is a voltage doubler circuit. Explain how the circuit works.

21 The performance parameters given for a diode are the power rating, the maximum forward current and the peak inverse voltage. The peak inverse voltage V_{rrm} is the maximum voltage that a diode can withstand in the non-conducting direction before breakdown.
a Why are each of these quantities important?
b What is the value of V_{rrm} for each diode in the rectifier circuit of:
i Figure 23.2a, **ii** Figure 23.3a and **iii** Figure 23.3b?
Assume $V_{in} = V_o \sin 100\pi t$.
c Does the addition of a smoothing capacitor to a rectifier circuit make any difference to the required V_{rrm} rating?

22 Draw a circuit diagram of
a an inverting amplifier
b a non-inverting amplifier
with closed-loop gain of 50.

23 The op-amp is being used as a **comparator** (← 23.2) in Figure 23.40. Sketch a graph of V_{out} against time.

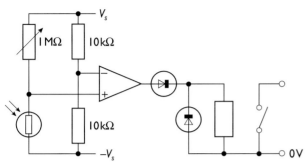
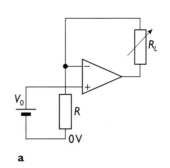
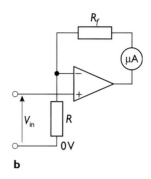

Figure 23.41

Figure 23.42

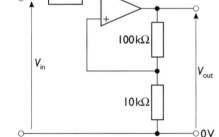

Figure 23.43

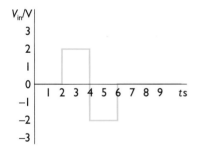

Figure 23.44

24 The op-amp is being used as a switch in Figure 23.41.
a Explain how the circuit works and suggest a use for it.
b Suggest other devices that might replace the LDR and the use to which each resultant circuit would be put.

25a Explain how the circuit of Figure 23.42a acts as a constant current source in R_L, whatever the value of R_L.
b What is the magnitude of the current?
c Explain how a high impedance voltmeter, connected in parallel with R_L, acts as a linear ohm-meter to measure the value of R_L.
d Explain how the circuit of Figure 23.42b acts as a high impedance voltmeter.

26 Figure 23.43 is a differential amplifier circuit. Show that the closed-loop gain $A = -R_f/R_{in}$.

27 Figure 23.44 is an inverting Schmitt trigger circuit. Six of the circuits are contained in a single IC, the TTL 7417 or CMOS 40106 ($\rightarrow$ 24.1).
a Explain how the circuit works.
b At what values of V_{in} does it switch?
c Sketch its voltage transfer characteristic.

28 The input V_{in} to the ramp generator of Figure 23.20 is shown in Figure 23.45. Sketch the output V_{out} against time for $R = 1\ M\Omega$ and $C = 1\ \mu F$.

29a Explain how the circuit of Figure 23.46 solves the differential equation of radioactive decay ($\rightarrow$ 39.4):

$$\frac{dN}{dt} = -\lambda N.$$

b Sketch the variation of the output against time.

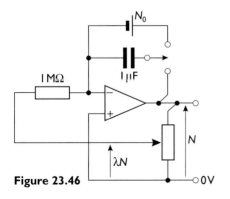

Figure 23.45

Figure 23.46

24 DIGITAL ELECTRONIC SYSTEMS

Many modern devices, such as computers, calculators, compact disc systems, watches and clocks, contain **digital** circuits. Such circuits generate and process **pulses** rather than continuously varying analogue signals (← 23).

The word 'digital' refers to counting, as on the digits, or fingers of each hand. When information is sent in digital form, it is a series of 'on' or 'off' signals. It has to be coded into this form for transmission and then decoded at the other end.

Digital signals have a big advantage over analogue signals. Interference accumulated by all signals during transmission can be removed, leaving signals that are 'clean' – identical to the original ones.

There is a vast array of names in use for the various types of digital circuit, including LSI (large-scale integration), VLSI (very large scale integration) and microprocessor. All except the very simplest are contained on a silicon chip or other integrated circuit (IC). Each has a slightly different meaning and the circuit it describes may have a simple or complex function. A microprocessor may be used in a kitchen toaster to time its toasting action or in a video to make sure that the user cannot eject a videotape while it is still threaded around the heads.

The names are often used as marketing ploys. For example, a video or toaster may be advertised as 'microprocessor controlled'. Some cars have been described as having 'multiple computers'! Many children's toys also contain simple microprocessors or equivalent ICs. So do compact disc and laserdisc players, NICAM Stereo, Teletext and Fastext TVs, traffic light, washing machine and dishwasher control systems, water-saving systems and many more.

The inclusion of microprocessors has enabled a huge increase in reliability of many devices. A burglar alarm's control system is now able to be programmed to check the type of signal from each sensor. If it is consistent with a cat opening a door, or a spider spinning a web inside an infrared sensor, the alarm will not be triggered.

24.1 BASIC DIGITAL CONCEPTS

When the voltage level of a pulse is small or zero it is said to be **low** and has a value of logic **0**. Above a specified value, it is said to be **high** and has a value of logic **1**. These levels are usually referred to as **0** and **1** or, sometimes, as **off** and **on**.

The shape of a pulse may become distorted as it passes through an electronic system. But, unless the distortion is very large, it can still be detected as either **1** or **0** (see Figure 24.1). A circuit called a **Schmitt trigger** (← 23.4) is often used to 'sharpen up' such signals. It does this by creating steep sides to the pulses, so that each pulse switches on or off in less than a nanosecond or so.

24 DIGITAL ELECTRONIC SYSTEMS

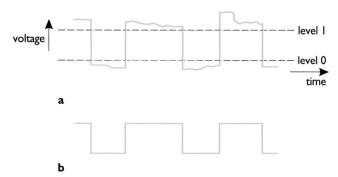

Figure 24.1 Input signal **a** is read by circuit as signal **b**

24.2 LOGIC GATES

Logic gates are the basic electronic building blocks at the heart of all digital electronic systems, however complex. They perform the five basic logic functions **NOT, AND, OR, NAND** and **NOR**. They are represented by the symbols shown in Figure 24.2. Their output for different inputs can be displayed in a **truth table** as shown for the NOT gate below.

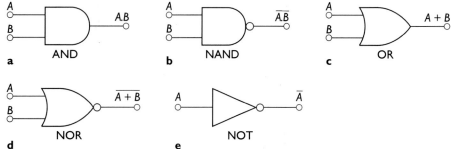

Figure 24.2 Gate symbols **a** AND **b** NAND **c** OR **d** NOR **e** NOT

LOGIC GATES AND TRUTH TABLES

The NOT gate changes a high input to a low one and vice versa – it changes a 0 to a 1 and a 1 to a 0. An input A producing an output Q gives the following truth table:

A	Q
0	1
1	0

This can be written as Q is NOT A. A more convenient and meaningful notation uses a short bar over a quantity to indicate NOT, so that:

when $Q = 1$ (high), $\overline{Q} = 0$ (low)

and, conversely,

when $Q = 0$ (low), $\overline{Q} = 1$ (high).

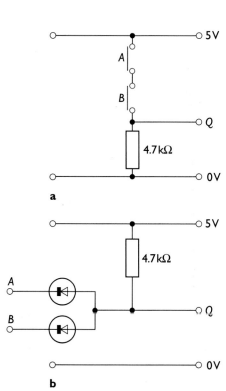

Figure 24.3 Two simple AND gates. Logic 1 is 5 V, logic 0 is 0.7 V.

Figure 24.3 shows two simple ways of producing the **AND** function. To express this function in words, we can write:

An AND gate has output Q high if input A AND input B are both high, otherwise output Q is low.

353

OSCILLATIONS

Logic gates and Boolean algebra

Boolean algebra is used in describing the complex operations of logic gates. In it '+' stands for OR and '.' for AND. So:

'$A.B$' means 'A AND B'

and '$A+B$' means 'A OR B'.

Boolean algebra is a very powerful tool for solving logic problems. However, for many problems the truth table is a straightforward and adequate method.

The behaviour of the output of the **NAND**, or **NOT AND** gate, is opposite to that of the AND gate:

A NAND gate has output Q low if input A AND input B are both high, otherwise output Q is high.

However, this expression is rather clumsy. Using the notation above, where a short bar indicates NOT, we can write for a NAND gate:

$\overline{Q} = 1$ if $A = 1$ AND $B = 1$.

The outputs for all of the possible inputs for each gate can be shown by truth tables. Figure 24.4 shows the truth table for each type of gate.

inputs		output, Q				
A	B	AND	NAND	OR	NOR	NOT (input A)
0	0	0	1	0	1	1
0	1	0	1	1	0	1
1	0	0	1	1	0	0
1	1	1	0	1	0	0

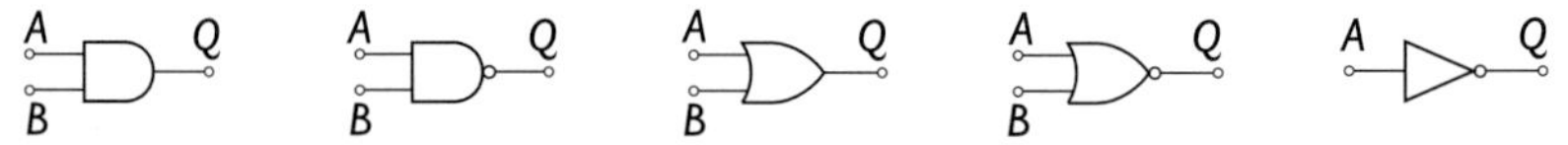

Figure 24.4 Truth tables for two-input gates

Questions

1 Show that each circuit of Figure 24.3 acts as an AND gate.

2 Design circuits for
a a NAND and
b an OR gate
using two switches and a resistor, where the output is measured across the resistor.

3 Write down the operations OR and NOR in words.

4 We use the logical terms AND and OR in everyday language. For example, to obtain cash from a bank cash dispenser, you need a valid card AND a valid personal identification number.

In a burglar detection system, when the pressure switch under the stair carpet OR the infrared sensor in the hall is triggered, the alarm will sound.

Suggest two more uses for each of the terms AND and OR.

EXPERIMENTING WITH DIGITAL SYSTEMS

Digital systems can be constructed by connecting logic gates together. It is normal in diagrams to omit the power connections to logic devices and show only the power supply rails where the inputs or output to the gate are connected to them.

Although the currents associated with integrated circuits (ICs) are small, the input current to an IC is particularly small compared with its output. This is because the current at the output comes from the power supply, while the input, is fed by a voltage signal across a very high input resistance.

A digital IC is not able to provide a large output current. As a **source**, it can usually provide enough to drive a single light-emitting diode (LED) as shown in Figure 24.5a.

It can act as a **sink** using the arrangement shown in Figure 24.5b or as a

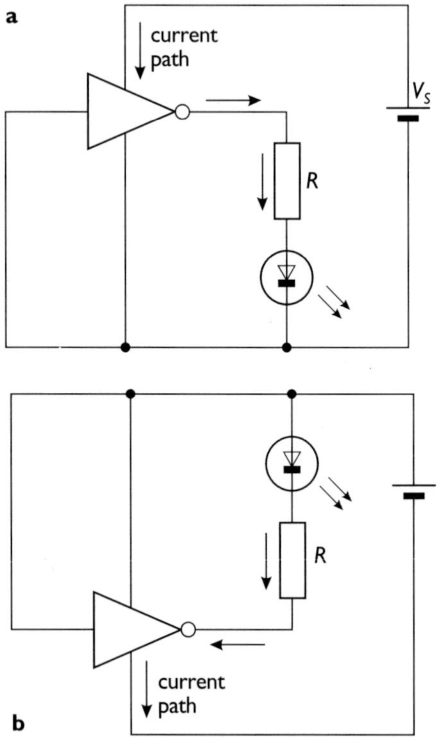

Figure 24.5 NOT gate as **a** a source of current, **b** a sink of current

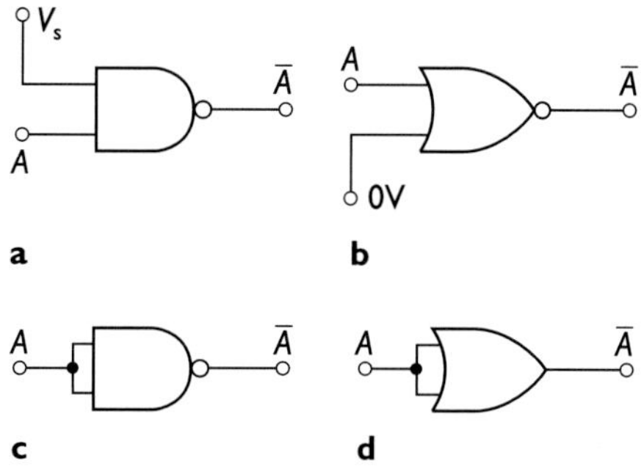

24 DIGITAL ELECTRONIC SYSTEMS

Figure 24.7 NOT gates **a** $\overline{A \cdot 1} = \overline{A}$, **b** $\overline{A + 0} = \overline{A}$, **c** $\overline{A \cdot A} = \overline{A}$, **d** $\overline{A + A} = \overline{A}$

AND
$A \cdot 0 = 0$
$A \cdot 1 = A$
$A \cdot A = A$
$A \cdot \overline{A} = 0$

OR
$A + 0 = A$
$A + 1 = 1$
$A + A = A$
$A + \overline{A} = 1$

Figure 24.8 Some logic identities in Boolean notation: A is 1 or 0

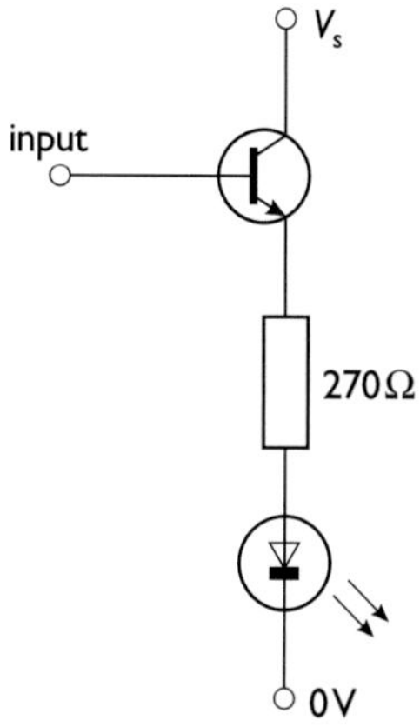

Figure 24.6 Voltage follower circuit used as a buffer to indicate logic levels

buffer coupled to the output of a circuit, to reduce the load on the output. A higher power transistor connected as a voltage follower can be used as a buffer as shown in Figure 24.6. Note that it acts as a NOT gate. You should use a circuit such as this to indicate logic levels in practical investigations of logic circuits, unless an alternative is shown.

NOR and NAND gates can be converted to NOT gates by connecting their inputs as shown in Figure 24.7. In fact, any logic gate can be constructed using only NAND gates (→ 24.3).

Figure 24.8 gives some useful identities written in Boolean notation. You can easily verify these by looking at the truth tables for the gates or by experiment. In some applications, three or four input gates may be needed. These can always be constructed from combinations of two-input gates.

EXPERIMENT 24.1 Two-input logic gates

By connecting the circuit shown in Figure 24.9 with inputs A and B to 0 V (0), or V_s (1), you should satisfy yourself that its behaviour is consistent with the truth table shown in Figure 24.4. Use each type of gate in turn.

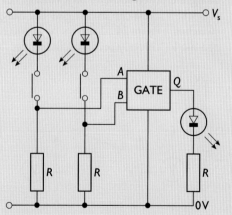

Figure 24.9 Logic gate circuit

Questions

5 Write truth tables for **a** a three-input and **b** a four-input AND gate. Design circuits for each of these gates using two-input AND gates.

6 Write a truth table for a three-input NOR gate and design a circuit for the gate using three two-input NOR gates.

OSCILLATIONS

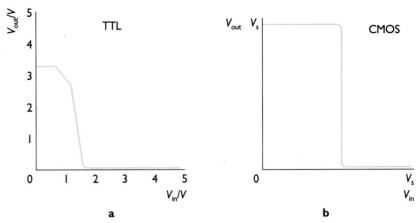

Figure 24.10 Transfer characteristics

The output voltage of a gate depends on its input voltage and its **transfer characteristic** (see Figure 24.10). The transfer characteristic of any device tells you how the input and output are related under a wide range of conditions. These conditions are usually defined by the voltages and currents applied to the device and the size of the input signal.

In Figure 24.10, the V_{in}/V_{out} transfer characteristics of a TTL, or 'Transistor–Transistor Logic' device are compared to those of a CMOS, or 'Complementary Metal Oxide Semiconductor' device.

We need to measure the voltage levels at which the output of a gate is at logic 1 and at logic 0. We will then know whether a particular input voltage will switch the gate from one state to the other.

EXPERIMENT 24.2 The transfer characteristic of a NOT gate

Using the circuit of Figure 24.11, you can vary the input voltage to the gate using the potentiometer as a voltage divider. By plotting a graph of values of V_{out} and V_{in} you can investigate the range of input voltage over which the gate switches. Figure 24.12 shows how you can display the characteristics of V_{out} against V_{in} directly on an oscilloscope.

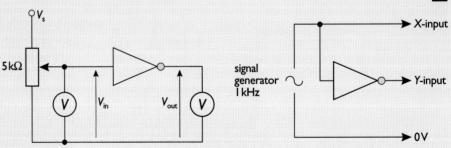

Figure 24.11 Circuit for measuring input and output voltages of a NOT gate

Figure 24.12 Circuit for displaying the transfer characteristics

COMBINING GATES

Two gates can be coupled together so that the output of one is fed to the input of the other. Then it is important that V_{out} logic 1 of the first gate is a higher voltage than the minimum V_{in} logic 1 of the second gate. Also V_{in} logic 0 of the second gate must be a higher voltage than the maximum V_{out} logic 0 of the first gate. For

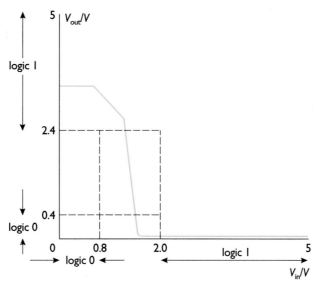

Figure 24.13 Logic levels for a TTL NOT gate

example, the dotted lines in Figure 24.13 indicate the logic levels for a typical gate. When these conditions are not met, the first gate may not control the second one correctly.

EXPERIMENT 24.3 Simple control circuits

Figure 24.10 shows that a NOT gate output switches from high to low over a small range of input voltage. By connecting the circuit of Figure 24.14a you can illustrate the principle of a simple control circuit such as a 'security light': the LED indicator is lit when the light-dependent resistor (LDR) is covered.

If you replace the NOT gate by a NAND gate, as shown in Figure 24.14b, you will find that the circuit only functions when the switch is in position A. This is a feature called an **enable** incorporated in multiplexer circuits (→ 24.3).

By connecting the circuit shown in Figure 24.14c you can use a single gate as a **latch** (→ 24.4). Note that once the LED indicator is lit, it cannot be switched off until the **positive feedback** (→ 24.4) connection between the output and one input of the OR gate is broken.

Figure 24.14a Simple 'parking light', **b** with enable, **c** with latch

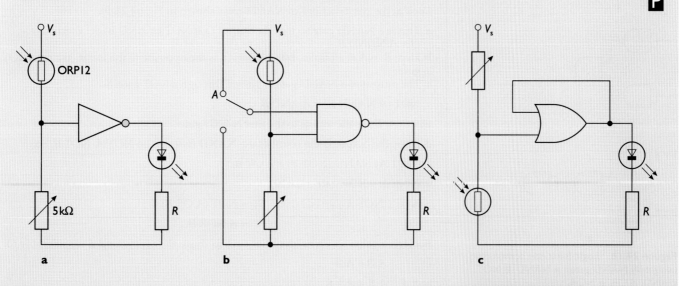

E OSCILLATIONS

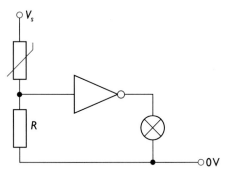

Figure 24.15

> ### Question
>
> **7** A thermistor is to be used in a frost warning device to sense when the air temperature falls below 2°C. Then a warning indicator light is to switch on. The resistance of the thermistor is known to be 5 kΩ at 2°C.
>
> The circuit to be used is shown in Figure 24.15. The NOT gate requires a maximum voltage of 0.8 V in order to produce a high output.
> **a** Explain how the circuit functions in the way described above.
> **b** What value of R is required?
> **c** Why would it be better to use a variable resistance rather than fixed resistor at R when setting up the circuit?
> **d** Suppose a circuit is required that switches on an indicator when the temperature **exceeds** a certain value. You could modify the circuit in two possible ways to achieve this. Describe and explain the two modifications.

24.3 COMBINATIONAL LOGIC: ADDITION

In a computer or microprocessor, there may be thousands or even millions of logic gates connected together. Any block of logic gates in which the outputs at any time are determined entirely by the state of the inputs is called a **combinational logic circuit**. Combinational logic circuits often contain **memory** blocks that provide pre-programmed data.

Sometimes combinational logic circuits are made using only one type of gate, such as NAND or NOR gates. This does not necessarily increase the number of gates required in the circuit very much and it has the advantage of reducing manufacturing costs. For example, all five logic functions can be constructed from combinations of NAND gates, as shown in Figure 24.16.

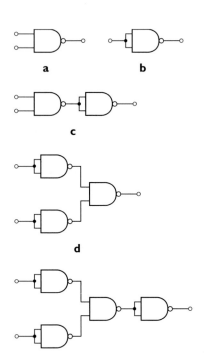

Figure 24.16 Logic functions constructed using only NAND gates: **a** NAND, **b** NOT, **c** AND, **d** OR, **e** NOR

> *Combinational logic and Boolean algebra*
>
> Look at the combinations of NAND gates shown in Figure 24.16.
>
> **a** This is a NAND gate so that the output Q is NOT (A AND B) or, alternatively, $Q = A$ NAND B.
> **b** Here $B = A$ at all times, but the output is the opposite to the input, giving a NOT gate.
> **c** The output of a NAND gate is passed through a NOT gate. So:
>
> NOT (A NAND B) = A AND B
>
> where A, B are the inputs to the first NAND gate.
>
> **d** Here, the output is the result of the NAND function on NOT A, NOT B. So:
>
> (NOT A) NAND (NOT B) = A OR B.
>
> **e** From **d**, a NOR gate must be:
>
> NOT ((NOT A) NAND (NOT B)) = NOT (A OR B)
>
> $\qquad\qquad\qquad\qquad\qquad\qquad = A$ NOR B.
>
> Boolean algebra shortens identities such as those shown in **c** and **d** considerably to:
>
> **c** $\qquad \overline{\overline{A.B}} = A.B$
>
> and **d** $\quad \overline{\overline{A}.\overline{B}} = A + B$.

24 DIGITAL ELECTRONIC SYSTEMS

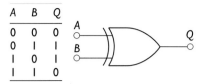

Figure 24.17 Exclusive OR gate

This last equality is an example of de Morgan's theorem: to obtain the inverse of a Boolean function, invert each variable and exchange 'AND' for 'OR'. Using this, we can write:

NOT (A OR B) = (NOT A) AND (NOT B)

or $\quad \overline{A + B} = \bar{A}.\bar{B}.$

This shows that a NOR gate can be made by inverting the inputs to an AND gate.

The exclusive OR gate (EOR or XOR) written in Boolean notation is:

$$Q = A.\bar{B} + \bar{A}.B$$

or, more simply $\quad Q = A \oplus B.$

Question

8a Draw a circuit showing how an AND gate can be made from two NOT gates and a NOR gate.
b How does this demonstrate de Morgan's theorem?

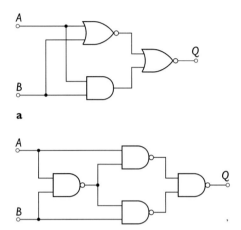

a

b

Figure 24.18 Examples of the exclusive OR gate

EXCLUSIVE OR GATES

This useful gate, often written as EOR or XOR, is defined by the truth table of Figure 24.17. The output is high when **either** input is high but, low when **both** inputs are high or low.

The gate can be constructed from combinations of gates in many different ways. Two versions are shown in Figure 24.18. Note that the one in b, uses only one type of gate, the NAND gate.

EXPERIMENT 24.4 Making all logic functions from one type of gate

After checking that the combinations in Figure 24.16 have the correct truth tables, you can connect similar combinations of NOR gates to make all five logic functions. You can then look for the symmetry between NAND and NOR.

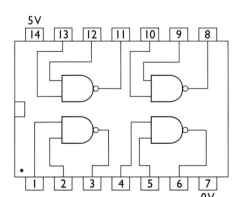

Figure 24.19 Pin connections for TTL 7400 quad two-input NAND (top view)

Questions

9 How does the EOR function differ from the OR function?
10 The **parity** or **coincidence** gate is the inverse of the EOR gate. Write down its truth table and design a circuit for it.

The IC shown in Figure 24.19 contains four gates so the EOR gate can be made on one IC chip.

The EOR gate is a major component of the simplest binary arithmetic circuit, the **half adder**. The rules of binary arithmetic are shown in the truth table for the half adder (see Figure 24.20). The 'sum' column is the EOR function and the 'carry' column is the AND function. The arrangement shown in Figure 24.18a can act as a half adder where the output from the AND gate is taken as the carry signal. A computer can add two binary numbers together using such methods.

E OSCILLATIONS

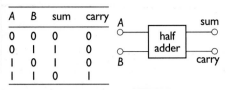

A	B	sum	carry
0	0	0	0
0	1	1	0
1	0	1	0
1	1	0	1

Figure 24.20 The half adder

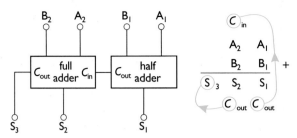

Figure 24.22 Two-bit addition

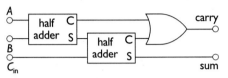

Figure 24.21 The full adder

However, for binary additions, a circuit called the full adder is required (see Figure 24.21). This is because a carry digit must be included from the previous addition. The block diagram for adding two binary numbers $A_2A_1 + B_2B_1$ is shown in Figure 24.22.

EXPERIMENT 24.5 Binary arithmetic

You can construct a half adder using either of the arrangements of Figure 24.18. You can then add another to make the full adder of Figure 24.21.

You can also investigate how two two-bit binary numbers can be multiplied together using a full adder and four AND gates.

MULTIPLEXERS

A **multiplexer** circuit is a means of sending several signals down the same cable. An **enable** signal selects each input in turn and passes the chosen signal to the output. Using this circuit, many input lines of data can be channelled into a single output line. The signals are sent down the cable sequentially. At the far end of the cable, a **demultiplexer** uses a synchronised enable signal to supply the data to each of the output lines, as shown in Figure 24.23. Figure 24.24 shows the arrangement for a two-input multiplexer and demultiplexer.

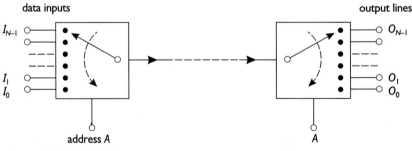

Figure 24.23 Multiplexer system

Figure 24.24a Two-input multiplexer, **b** two-input demultiplexer

Question

11 In the two-input multiplexer of Figure 24.24, satisfy yourself that
a when A is at logic 1, the output is the same as I_1
b when A is at logic 0, the output is the same as I_0.

24 DIGITAL ELECTRONIC SYSTEMS

24.4 SEQUENTIAL LOGIC: THE LATCH AND POSITIVE FEEDBACK

The most basic unit that can store digital information in a circuit is the **bistable** or **latch**. As the name suggests, it has two stable states.

The simplest latch is an OR gate with its output connected to one input, as in Figure 24.25a. However, once the input S goes to logic 1, the output Q is latched to logic 1 and cannot be reset to logic 0 at the input.

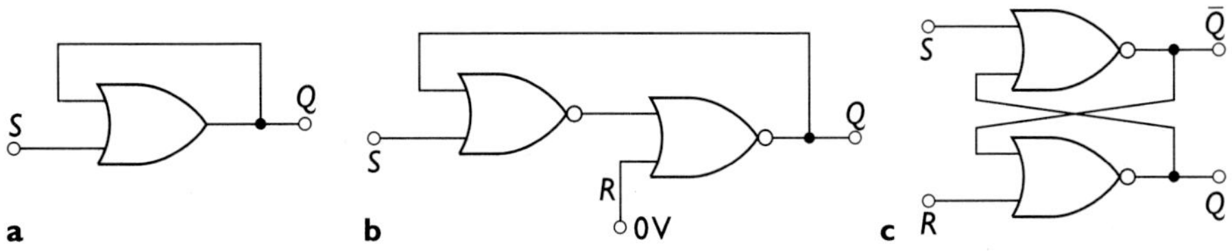

Figure 24.25 Simple latch to bistable: **a** OR gate; **b** NOR, NOT gates; **c** SR bistable

To make a better latch, consider the OR gate constructed from two NOR gates with the second acting as a NOT gate (see Figure 24.25b). The grounded input R can be used to reset the output Q to logic 0 by raising it to logic 1. The final arrangement is shown in Figure 24.25c. This has been drawn symmetrically to show two inputs S and R and two outputs Q and $\overline{Q}$.

The output of each gate is fed back to the input of the other. Both gates are latched. This means that they are held in particular logic states when the inputs are connected to logic 0. If

$$Q = 1 \text{ then } \overline{Q} = 0$$

or if $\quad Q = 0 \text{ then } \overline{Q} = 1.$

The switch between the two stable states happens very quickly.

This is an example of **positive feedback**, where any change at the output causes the input to **increase** the change at the output. The states of the outputs can be described by the table of Figure 24.26a. A voltage pulse to the **set** or S input of the bistable switches the output to logic 1. Further pulses to this input have no effect on the output states.

A pulse must go to the **reset** or R input to change Q and $\overline{Q}$. This bistable arrangement is known as a **Set-Reset** or **SR** bistable. Another SR bistable design is shown in Figure 24.27.

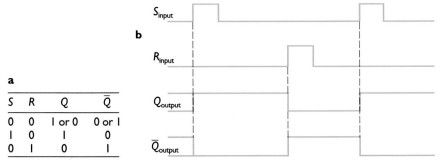

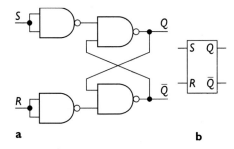

Figure 24.26 The SR bistable **a** logic states table, **b** voltage pulse sequence

Figure 24.27a NAND gate SR bistable and **b** its symbol

EXPERIMENT 24.6 The SR bistable

By constructing the arrangements of Figure 24.25c and Figure 24.27, you can show that the sequence of voltage input pulses change the outputs Q and $\overline{Q}$ as shown in Figure 24.26b.

E OSCILLATIONS

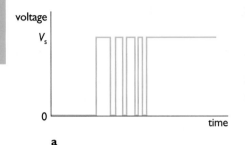

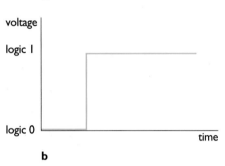

Figure 24.28 Voltage-time pulses of **a** 'chattering' switch, **b** a debounced switch

> ### Questions
> **12** What happens when both *S* and *R* are set to logic 1 by being given voltage inputs at the same instant?
> **13** Devise a circuit around the bistable in Figure 24.27b that will operate as a burglar alarm, triggered by a pressure switch and reset by a press switch, inaccessible to a burglar.

DEBOUNCED SWITCHES AND CLOCK PULSES

Most mechanical switches have voltage-time pulses as in Figure 24.28a. They are said to 'bounce' or 'chatter'. The arrangement of Figure 24.29 gives a **Q** output that is a single **on** pulse, as in Figure 24.28b. It is known as a **debounced switch**.

We can modify the inputs to Figure 24.27 to use the spare NAND (NOT) gate inputs as enables (see Figure 24.30). The signals applied to the *S* and *R* inputs can only change the outputs when the enable input is high. The enable input is often referred to as the **clock** input. This is because sequential logic systems often require a regular train of **clock pulses**, where each positive pulse enables signals to be passed from one set of gates to the next. In most ICs, the outputs change on the leading edge of the clock pulse. In this way we can ensure that the many ICs in a computer circuit, for example, all respond to their input signals at the right time and in the correct sequence.

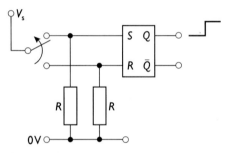

Figure 24.29 Circuit for a debounced switch

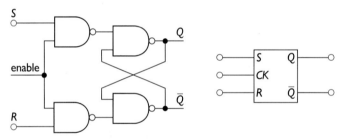

Figure 24.30 A clocked SR bistable

24.5 PULSE GENERATORS

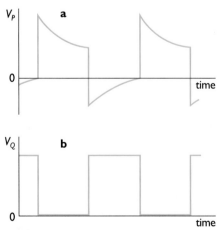

Figure 24.31 Voltage waveforms in a TTL astable circuit

Any square-wave oscillator or signal generator can be used to produce the clock pulses for counter circuits. A useful one is the **astable multivibrator** where the output continually oscillates between two states. The output signal is fed back to the input through an *RC* network, which causes a time delay and makes the oscillation happen. This is an example of **positive feedback** (← 23.4).

To explain this, consider what happens when applying square pulses to *RC* networks (← 21.1). Figures 21.5b and 21.6a show that the leading edge of a square voltage input pulse, switching from 0 to V_S, generates a positive voltage spike across the resistor *R*. The falling edge of the pulse (V_S to 0) produces a negative spike. This exponential waveform is only a spike when the product *RC* is small compared with the pulse length. For a value of *RC* that is not small, the output voltage has the shape of Figure 24.31a. When we feed this voltage into a NOT gate, as in Figure 24.32, the output voltage is a square wave like that in Figure 24.31b. When this square wave is fed into an identical circuit, and the output from this is fed back positively to the input of the first circuit (see Figure 24.33), oscillations due to **astable** action are produced.

24 DIGITAL ELECTRONIC SYSTEMS

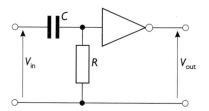

Figure 24.32 Circuit for squaring the waveform

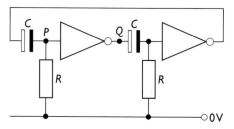

Figure 24.33 The astable circuit

How counters work

When an astable multivibrator is connected into a clocked bistable, the output of the bistable is at half the frequency of the multivibrator. This is a **divide-by-two** counter. One type of bistable, known as the **data** or **D-type** bistable, is shown connected as a divide-by-two device in Figure 24.34a.

The clocked SR bistable shown in Figure 24.30 can be made into a D-type bistable by adding a NOT gate, as shown in Figure 24.35. If D changes while CK (the clock input) is high, Q changes. So while CK is high, Q can 'see through' D and the latch is said to be transparent. On the other hand, the tandem arrangement, known as a **master-slave** data latch, shown in Figure 24.36, is opaque. This is because the logic state at D is not passed to Q except for the brief moment when the leading edge of each clock pulse triggers it.

An **asynchronous** counter is also known as a **ripple through** counter because the clock pulse triggers the first bistable which then triggers the second, etc., to the end. In the count-up device, each bistable must trigger the next on the trailing, or falling, edge of the pulse waveform. Such a count-up device, constructed from D-type bistables, is shown in Figure 24.37a. The least significant bit, or **lsb**, output Q_A, is obtained from the output of the first bistable. Its change in state is triggered on successive rising clock inputs. Q_B, the output of bistable B, changes every second rising clock input. The output Q_C occurs every fourth and output Q_D, the most significant bit, or **msb**, every eighth rising clock input. The voltage waveforms of the outputs of such a four-bit asynchronous counter, relative to the input clock pulse, are shown in Figure 24.37b.

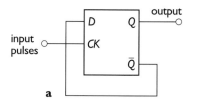

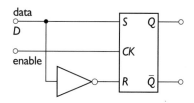

Figure 24.35 Transparent D-type bistable

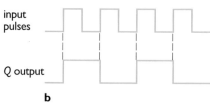

Figure 24.34a Bistable connected as a divide-by-two counter, **b** waveforms

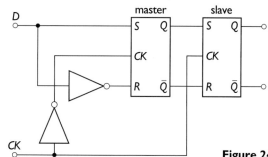

Figure 24.36 Master-slave data latch

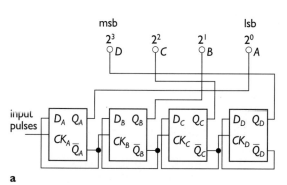

Figure 24.37a Four-bit binary counter, **b** time-related waveforms

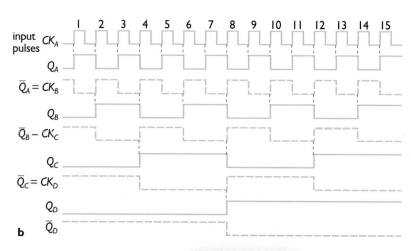

OSCILLATIONS

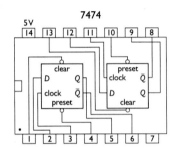

CK	Q_D	Q_C	Q_B	Q_A	$\bar{Q}_D$	$\bar{Q}_C$	$\bar{Q}_B$	$\bar{Q}_A$
0	0	0	0	0	1	1	1	1
1	0	0	0	1	1	1	1	0
2	0	0	1	0	1	1	0	1
3	0	0	1	1	1	1	0	0
4	0	1	0	0	1	0	1	1
5	0	1	0	1	1	0	1	0
6	0	1	1	0	1	0	0	1
7	0	1	1	1	1	0	0	0
8	1	0	0	0	0	1	1	1
9	1	0	0	1	0	1	1	0
10	1	0	1	0	0	1	0	1
11	1	0	1	1	0	1	0	0
12	1	1	0	0	0	0	1	1
13	1	1	0	1	0	0	1	0
14	1	1	1	0	0	0	0	1
15	1	1	1	1	0	0	0	0

Another way of displaying this information is in the form of a logic states table, as in Figure 24.38. Notice that when each bistable is triggered on the leading, or rising, edge of the pulse, by interchanging connections to Q and $\bar{Q}$, the counter counts down.

A counter based on four bistable elements, a four-bit counter, can be reset at any number less than 15, or binary 1111. This is done by using the **clear** facility on each IC. For example, the left hand side of Figure 24.39 shows the pin arrangement for the TTL 7474 IC. This contains two D-type bistable circuits. Figure 24.40 shows the logic diagram for two 7474 ICs connected as a decade counter. Without the clear facility, the counter would count input pulses from 0000 to 1111 in binary as indicated by the states of the outputs at DCBA, A being the least significant bit. However, the NAND gate resets the outputs to zero at binary 1010. This happens because B and D are connected to the NAND gate inputs. When they are both at logic 1, the NAND gate output falls to logic 0, causing the counter to restart from zero. The four inputs can be decoded to display a figure as shown in Figure 24.41.

Figure 24.38 Logic states table for four-bit binary counter

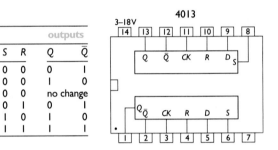

Figure 24.39 A dual D-type bistable (top view)

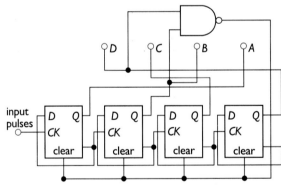

Figure 24.40 Decade counter

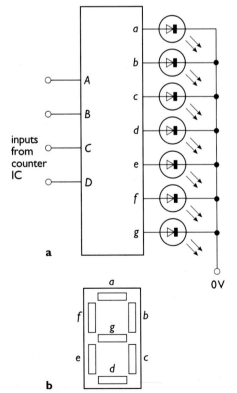

Figure 24.41a A 7-segment display circuit, **b** display

Questions

14 Figure 24.37a shows a D-type bistable counter which can be used to count up: an up-counter. How could you rearrange the connections to make it into a down-counter by subtracting one from the output at each pulse?

15 How can the up-counter be used as a divide-by-four or divide-by-eight circuit?

16 How can the counter of Figure 24.40 be modified to act as a die so that it resets at 7 and displays 1? (You need a three-input AND or NAND gate.)

24 DIGITAL ELECTRONIC SYSTEMS

EXPERIMENT 24.7 An astable circuit

An astable multivibrator can conveniently be made from NOR gates, acting as NOT gates, by constructing the circuit shown in Figure 24.42. The LEDs will flash at a rate of 1 Hz when you press the switch.

Reducing C to $1\,\mu F$, you can observe the waveforms at P and Q on an oscilloscope and see to what extent they resemble those of Figure 24.31.

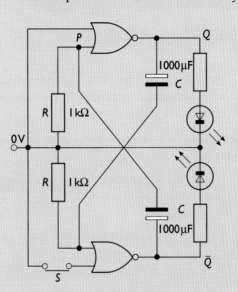

Figure 24.42 Astable circuit using NOR gates

THE MONOSTABLE

A **monostable** circuit is used to produce a single pulse. It appears like a combination of half a bistable with half an astable circuit (see Figure 24.43a).

The output pulse $\overline{Q}$ is negative and always the same length, whatever the length of the input pulse at A (see Figure 24.43b). The output must be taken at Q to produce a positive pulse.

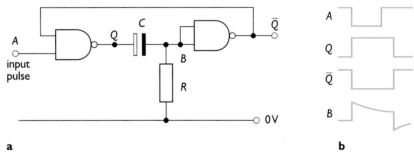

a

Figure 24.43a A monostable circuit, **b** voltage waveforms

Questions

17 What happens when the values of R and of C in Figures 24.33 and 24.42 are not the same?

18 Explain why the output pulse at $\overline{Q}$ in Figure 24.43 is of constant length.

E OSCILLATIONS

EXPERIMENT 24.8   A monostable circuit

Construct the circuit of Figure 24.43a to produce a single pulse. You can then investigate how its length varies when you adjust R and C.

24.6   USING ELECTRONIC BUILDING BLOCKS

By investigating the behaviour of various logic gates, a binary adder, a counter, an astable and a monostable, you are in a position to understand the behaviour of complex digital systems.

EXPERIMENT 24.9   Electronic system design

Below are the requirements for five circuits a – e. Design each circuit on paper first and then test that it meets the specifications using a suitable electronics kit.

a An indicator, flashing with a frequency of 1 Hz, flashes a total of ten times when a switch is pressed.
b A circuit sounds an audible alarm when the temperature goes below, say, 10 °C, and lights an indicator lamp when that temperature is exceeded.
c A circuit converts a 50 Hz sine wave into **i** 50 Hz and **ii** 100 Hz square pulses.
d A thermometer, suitable perhaps for a blind person, gives an audible tone whose frequency increases with increasing temperature.
e A frequency counter displays the frequency of a square wave in binary, up to 15 Hz.

COMMUNICATION SYSTEMS

In an analogue signal, the information is carried by the amplitude of the signal. Interference often alters the amplitude and so distorts the signal. Once this has happened, it is almost impossible to remove the interference and obtain the original signal again.

To produce a digital signal, the amplitude of the signal is **sampled** and a series of pulses produced. This is known as **pulse code modulation**. Normal levels of interference will only alter the size and shape of these pulses. It takes a very high level of interference to make them impossible to distinguish from the background. So a digital signal can usually be 'cleaned up' ready for further processing by a Schmitt trigger circuit (← 23.4).

Headphones and loudspeakers must receive analogue signals, since they are analogue devices. The same goes for almost every other system where digital signals are used; even a computer monitor usually produces an analogue signal to feed to the electron gun. So when the digital signal has been converted from its original analogue form, amplified, transmitted and cleaned up, it is converted back to analogue. It is then no different from an analogue signal that has been transmitted as analogue. It has simply suffered less from interference. The final analogue signal must, as far as possible, match the original.

24 DIGITAL ELECTRONIC SYSTEMS

ANALOGUE SIGNALS AND DEVICES

Most audio and visual signals are generated by analogue devices, such as microphones and need to be fed to analogue output devices, such as loudspeakers. To process or store these signals digitally, they need to be converted into digital signals, processed, then converted back into analogue signals. This is carried out by ICs: the first by an **ADC** (analogue to digital converter) and the latter by a **DAC** (digital to analogue converter).

ADCs operate by sampling the analogue voltage signal at different times, and then representing the value of the sampled signal as a sequence of highs and lows (see Figure 24.44). The DAC has the opposite task of interpreting these digital sequences and outputting a voltage similar to the original input voltage. Each ADC and DAC operates in a different way, producing different speeds and types of sampling error. In general, providing the ADC and the DAC are compatible with each other, the resulting signal will be far better than a signal processed and stored in analogue form.

In laser-scanned CD systems, numbers of bits and 'oversampling rates' of DACs are often quoted. All CDs are recorded to an international standard. So these factors are likely to have less effect on sound quality than features such as the laser's mounting and guidance system. The disc is a true digital record with the logic 1 signals as the flat surface of the disc and the logic 0 signals as the tiny pits in the surface. The laser scanner traces a spiral path across the disc, picking up the two voltage levels from the scattering of the laser light reflected by the disc surface. This is shown in Figure 24.45. The distortion is reduced by a factor of at least 100 over analogue plastic discs, and wear is non-existent.

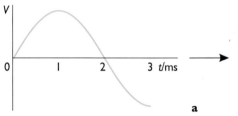
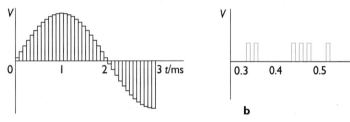

Figure 24.44a Principle of analogue to digital conversion, **b** section of four-bit binary, pulse code modulated signal

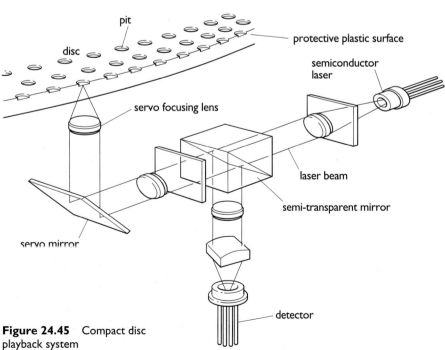

Figure 24.45 Compact disc playback system

PULSE CODE MODULATION: PCM

In PCM, information is carried by groups of pulses in a binary code. The voltage signal to be transmitted is divided into short time intervals. Its value is sampled, and the amplitude of each sample is converted into binary code. Each number corresponds to a particular voltage level, as shown in Figure 24.44. The binary coded digital signal is transmitted as a series of logic 0 and logic 1 voltage levels. The received signal is decoded from the binary numbers back into an analogue output signal identical to the original.

PCM has many advantages over analogue or linear systems. These include reduced distortion, absorption and noise. Noise is caused by random fluctuation of voltage levels picked up in transmission. As the signal only has two constant voltage levels, the noise can be filtered out by the receiver. For sound transmission, the sampling frequency must be at least twice the highest sound frequency for good reproduction of the sound waveform.

Radio, television and telephone systems all use digital signals. The BBC, for example, has a PCM system for distributing radio programmes to its transmitters. Digital techniques are used to provide the teletext systems on television. Several of the 625 transmission lines are now used for such digital signals. Each page of information is stored in the memory of the teletext circuit inside the receiver. The viewer selects and displays any page for as long as required using the remote control. British Telecom has developed a similar system, Prestel, using the telephone network. **Modems**, or **mod**ulator-**dem**odulators, computers and VDUs are used in this process.

SUMMARY

In **combinational** logic systems, the output states depend entirely on the input states of the logic gates at the same instant of time. The five basic logic functions and corresponding gates are AND, OR, NAND, NOR and NOT. The NOT gate function can be produced by a NAND or NOR gate.

One input of a gate can be used as a control, or enable device. This means that the signal on the other input is passed to the output when the control input is in a particular logic state: for example AND with the control input at logic 1.

For binary addition, a half adder can be constructed from an EOR and an AND gate. Two half adders with an OR gate make a full adder.

In **sequential** logic systems, the output states may depend on the previous values of the inputs, enabling data to be stored. The bistable is the simplest such memory circuit.

Positive feedback is used in the bistable or latch circuit to produce a device with two stable states.

A clock or enable pulse is used to control or synchronise the changes of the bistable output states.

An asynchronous or ripple counter is constructed from a string of bistable circuits, connected as divide-by-two circuits. The counter may be reset at any number. If it is reset at 10, it will act as a decade counter.

An astable multivibrator circuit, a simple square-pulse generator, consists of two NOT gates coupled together by positive feedback loops containing RC time-dependent voltage dividers.

A monostable circuit, when triggered, produces a single pulse of length determined by the RC values.

As well as producing and controlling all the functions in a computer, digital electronics is being used increasingly in communication systems. PCM signals produce higher quality, lower noise transmissions than analogue-modulated carrier waves.

24 DIGITAL ELECTRONIC SYSTEMS

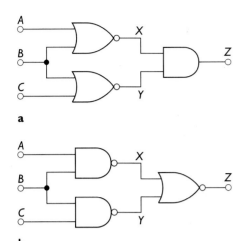

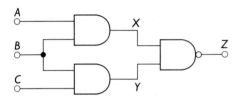

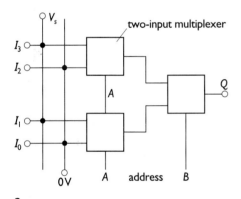

Figure 24.46

Figure 24.47a Four-input multiplexer, **b** truth table

A	B	Q_1	Q_2
0	0	0	1
1	0	1	1
0	1	0	0
1	1	1	0

SUMMARY QUESTIONS

19 Show that the logic diagrams a, b and c of Figure 24.46 are three-input NOR, AND and NAND gates, respectively.

20 Show that an EOR gate can be used as a **controlled inverter**. This means that:

if input $A = 1$ then output $Q =$ input $\overline{B}$ and

if input $A = 0$ then output $Q =$ input B.

21a The logic diagram of Figure 24.24a is an electronic double-pole-single-throw, or DPST, switch: a **2-to-1 line multiplexer**. The circuit can be placed between the sections of the four-bit counter, as shown in Figure 24.37. This way, the counter can be used as either an up- or down-counter, depending on the logic state of the address A. Show how this can be done.
b The four-input multiplexer of Figure 24.47a is being used as a **PROM**, or programmable read only memory.
  i Show that the four addressable locations have the values Q_1 shown in the truth table of Figure 24.47 b.
  ii Reprogram the PROM to give the outputs Q_2.

22 Design a circuit that will give a decimal number output for a two-bit binary number input. Such a device is called a **binary-decimal decoder**. Figure 24.48 shows the truth table for the device.

23 Design a circuit, called a **comparator**, to compare two one-bit binary numbers, A and B. It has three outputs X, Y and Z, where $X = 1$ when $A>B$, $Y = 1$ when $A = B$, and $Z = 1$ when $A<B$.

24 Design a voting system for three people using three AND, two OR and one NOT gate. One LED light shows a majority in favour and another shows a majority against.

25 The component T in Figure 24.49 is a thermistor. Its resistance decreases with increasing temperature.
a Explain how the circuit functions and suggest a purpose for it.
b How would you include a relay to activate a power circuit?
c How would you include an alarm circuit?
d Suggest a suitable circuit to produce an audible sound.

inputs		outputs			
A_2	A_1	Q_1	Q_2	Q_3	Q_4
0	0	1	0	0	0
0	1	0	1	0	0
1	0	0	0	1	0
1	1	0	0	0	1

Figure 24.48 Truth table for a binary-decimal decoder

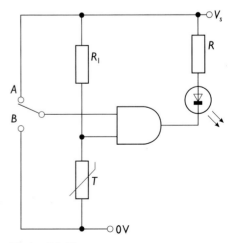

Figure 24.49

OSCILLATIONS

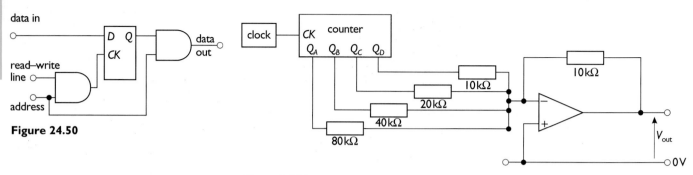

Figure 24.50

Figure 24.51

26 A **RAM**, or random access memory, is an array of one-bit read/write cells, such as the one in Figure 24.50. A particular cell or row of cells is chosen by its address. One bit can be written or read in each cell when the address or clock is high. Explain
a how the read-write line operates
b the action to store and retrieve data from the cell. Does the cell retain its information once it has been read?

27a Design an astable circuit, as shown in Figure 24.42, using NAND gates.
b Design a monostable circuit, as shown in Figure 24.43a using NOR gates.

28 Explain how the circuit of Figure 24.51 can act as a ramp generator. How could you make the changing output voltage smoother so that it changes in smaller steps?

29a Draw the timing diagram for a three-bit counter (see Figure 24.37). Shade in the areas where $\overline{Q_C}$, $Q_A.Q_B$ and $Q_C.\overline{Q_A.Q_B}$ are at logic 1.
b These shaded areas can represent the times at which a set of traffic lights are at red, amber and green respectively. For how long is each light on?
c Design a traffic light control circuit using the outputs to a three-bit counter, suitable gates and LEDs.

30 Design a reaction timer so that when one person pushes and holds down button *A*, an LED lights and a millisecond counter starts, and when another person pushes button *B*, the counter stops.

UNIT F

MATERIALS: THEIR PROPERTIES AND STRUCTURE

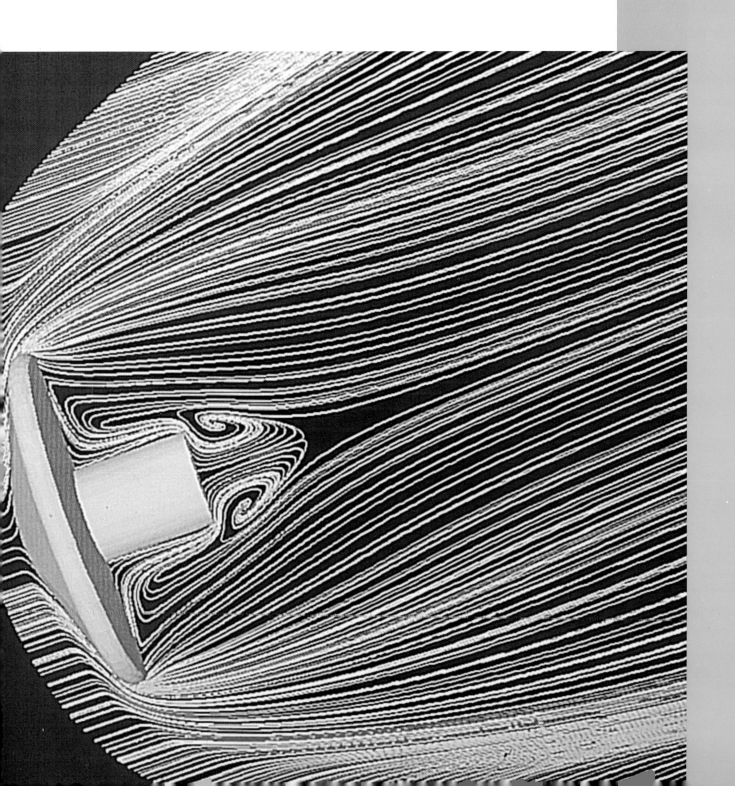

MATERIALS: THEIR PROPERTIES AND STRUCTURE

When you wrap a piece of copper wire round a terminal, how do you know whether it will retain a kink or whether it will break when you straighten it out? You need to understand enough of the structure of metals to predict what force will deform the wire plastically. You can interpret the properties of metals and any other materials on a large scale in terms of their atomic structure. In this unit, we begin by exploring a model for the structure of solids and go on to investigate the properties of fluids.

25 NATURE OF SOLIDS

All matter is made up of particles known as atoms or molecules. Yet the properties of a particular substance are not solely dependent upon the nature of the particles that make it up. The arrangement and energies of those particles are also very important.

Humphrey Davy

Over a century ago, in the Italian city of Florence, Humphrey Davy focused the Sun's rays on to a small diamond in a sealed glass globe. The diamond burnt with a brilliant light leaving no solid residue. Analysis of the residual gas in the globe showed that the diamond was pure carbon. Yet the soft, shiny black crystals of graphite are also almost pure carbon. So, the hardest and one of the softest solid materials known to us are made of the same element. The explanation for these observations must lie in the way carbon atoms are arranged within the samples.

Various models exist for the microscopic structure of solids. Investigating these allows us to understand their macroscopic, or bulk, properties, such as density, elasticity and ultimate tensile, or breaking, stress. A broad classification of the structures of solids divides them into four groups, crystalline solids, amorphous solids, glasses and polymers.

25.1 MACROSCOPIC PROPERTIES

When force is plotted against extension for a solid, such as a steel wire, the graph comes to an end when the sample breaks ($\leftarrow$ 4). From the graph you can find the Young modulus E of the material. Figure 25.1 shows values of the **specific stiffness** E/ρ of a variety of materials, where ρ is the density of the material. You can see that there is only a small variation in the value of E/ρ across very

material	specific stiffness/MN m kg^{-1}
steel	26
aluminium	26
earthenware pottery	20
electrical porcelain (used for insulation)	25
soda-glass	30
glass reinforced polyester	28
perspex	25
oak (hardwood)	22
carbon fibre/epoxy resin	200

Figure 25.1 Table of specific stiffness of materials

F MATERIALS: THEIR PROPERTIES AND STRUCTURE

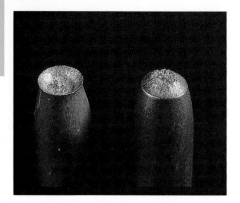

Figure 25.2 A cup and cone fracture

different materials. As you build up a picture of different materials and use different models to interpret their behaviour, this lack of variation may seem less strange.

Another important property of a material is its **mode of fracture**. Some materials such as soda-glass are **brittle**. When they break they do so with little or no warning. They do not deform much and little energy is released on fracture. Metals such as copper, on the other hand, exhibit **ductile** fracture. When copper is drawn out, its area of cross-section reduces until a characteristic cup and cone separation occurs, as shown in Figure 25.2. Many mild steels fail by ductile fracture at room temperature. However, at low temperatures, where cracks can be propagated through the metal, they break by brittle fracture. Figure 25.3 shows a welded ship that broke in two in cold arctic waters. Another way fractures can occur is through fatigue (→ 25.5 and Figure 25.4).

Figure 25.3 Brittle fracture in a welded ship

Figure 25.4 Fatigue failure surface of a drive shaft

25.2 METALS

Commercial metal samples are **poly-crystalline**, or composed of many crystals. The crystals are usually randomly oriented. In Figure 25.5, the crystal structure on the surface of a sample of copper is clearly visible. Within a single crystal, the metal atoms are arranged in a regular **lattice**. This arrangement could be imagined by stacking piles of polystyrene spheres.

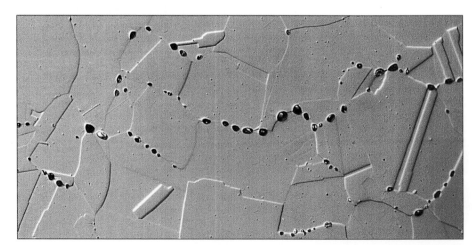

Figure 25.5 Microstructure of copper metal

25 NATURE OF SOLIDS

EXPERIMENT 25.1 Looking at the microstructure of metals

Using a microscope with a magnifying power of 100 and the ability to illuminate the specimen from the top, you can observe fine detail on the polished and etched surface of a metal object. Figure 25.6 shows a simple arrangements for this experiment.

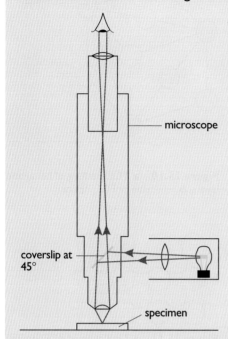

Figure 25.6 Microscope illuminator

Imagine a solid made out of a large number of identical spheres, each representing an atom of the solid. To get a high density, you must pack these spheres in as tightly as possible. In a two-dimensional array, the **hexagonal packing** shown in Figure 25.7 creates the maximum density of spheres. In three dimensions, there are several ways of packing the spheres that agree with structures found in different materials. Three of the common arrangements are the **hexagonal close pack (HCP)**, the **face-centred cubic (FCC)** and the **body-centred cubic (BCC)** structures.

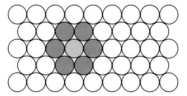

Figure 25.7 Two-dimensional hexagonal packing

Three-dimensional structures

Starting with the hexagonal layer of Figure 25.7, you can add a second layer by placing each sphere over, and touching, a group of three spheres on the original layer, as in Figure 25.8. This second layer has hexagonal symmetry identical to the original layer. Further hexagonal layers can be added to make up a three-dimensional structure. When alternate layers are vertically over each other, as shown in Figure 25.9, the lattice is called a **hexagonal close pack** or **HCP** crystal. The layers are stacked in sequence $A B A B A B \ldots$. Metals with HCP crystal structures include magnesium, zinc and cadmium.

Figure 25.8 Two hexagonal layers

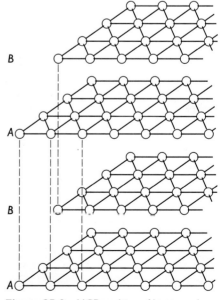

Figure 25.9 HCP packing of hexagonal layers

MATERIALS: THEIR PROPERTIES AND STRUCTURE

Using the same hexagonal layers as for the HCP lattice, you can construct other arrangements by changing the order in which the layers are stacked. After placing the base layer *A* and second layer *B*, the third layer *C* can be added as in Figure 25.10a rather than vertically over *A*. This arrangement, *A B C A B C A B C* ... is called the **face-centred cubic** or **FCC** structure. The relation of this structure to the **unit cube** is shown in Figure 25.10b. The hexagonal layers are the diagonal planes through the unit cube, which has one sphere at each corner and one in the centre of each face. Copper, aluminium and nickel have FCC structures.

The unit cube of the **body-centred cubic** or **BCC** structure has a sphere at each corner and one at the centre of the cube (see Figure 25.11a). Slices through the lattice show layers with a square arrangement in which the spheres do not actually touch neighbours in that layer. The layers repeat *A B A B A B* ..., as shown in Figure 25.11b. Iron at room temperature is a BCC structure, as are sodium, chromium and tungsten.

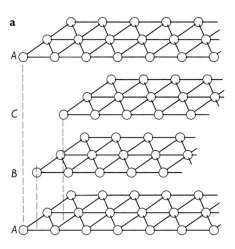
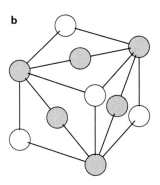

Figure 25.10 **a** FCC packing of hexagonal layers, **b** unit cube of FCC lattice

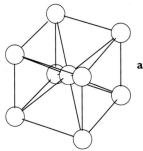
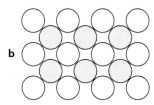

Figure 25.11 **a** Unit cube of BCC lattice, **b** layers of BCC lattice

Questions

1 Iron exists in different structures at different temperatures. It is BCC below 1 183 K and above 1 673 K, and FCC in between these temperatures. Predict one effect you might observe when an iron specimen is heated from below 1 183 K to above 1 183 K.

2 The molar volume (volume of 1 mole, → 35.2) of copper is 7.1 cm³. Estimate values for
a the diameter of a copper atom
b the spacing of the close pack planes in a copper crystal.

LOOKING AT ATOMIC STRUCTURES

To obtain direct evidence of these structures, we need to be able to 'see' objects of atomic dimensions. Using an optical microscope, we can see the crystals making up the metal. With an electron microscope, we can study smaller structures such as groups of atoms. The field ion microscope (→ 35.5) uses more massive particles than the electron, enabling us to look at structures right down to the atomic scale. Figure 25.12 shows a photograph of helium ions scattered from the tip of a tungsten needle. It certainly looks as though there is an ordered system scattering the ions.

To investigate structures with dimensions of order 10^{-10} m requires waves of a similar wavelength – X-rays. Most of the quantitative data on lattice structures comes from X-ray diffraction studies. A beam of X-rays interacts with a solid as in Figure 25.13, so that the X-rays are scattered in all directions. When the atoms have a regular arrangement, the scattered X-rays interfere coherently, and reinforcement and destruction occur in various directions (→ 32.2 and 33). This produces a pattern on the film. In an amorphous substance the atoms have no

Figure 25.12 Field ion micrograph of the surface of a specimen of tungsten

Figure 25.13 Experimental arrangement for X-ray diffraction

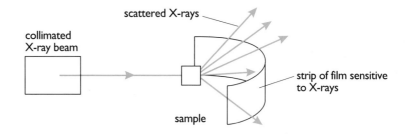

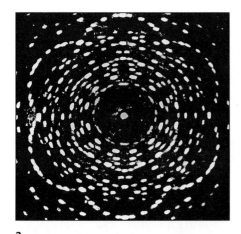

order, so there is a diffuse fogging of the film. Figure 25.14 shows two photographs taken with metal specimens. Compare these with X-ray diffraction photographs of other materials (→ Figures 25.25 and 25.32).

DISLOCATIONS IN CRYSTALS

What happens to a metal's crystal structure when the metal is deformed? By considering the lattice model, you can imagine that pulling two complete layers further apart would require a substantial force. It would be much easier to slide one layer over another – a process called **slip**. In a poly-crystalline metal specimen, there will be crystals whose slip planes are not perpendicular to the applied force and it is these that will slip (see Figure 25.15).

Figure 25.14 X-ray diffraction photographs of **a** a single metal crystal, **b** a polycrystalline metal

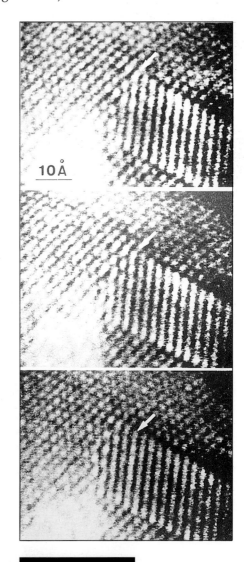

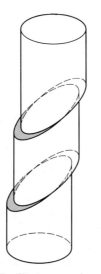

Figure 25.15 Slip in a metal specimen

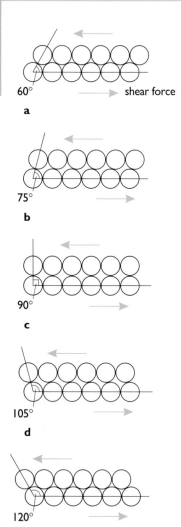

Figure 25.16 Sliding layers of atoms

Slip at an atomic level is illustrated in Figure 25.16. The layers separate as shown in (a) to (c) and then go back together again as shown in (c) to (e), shearing one layer over another. The shear stress τ, to produce this deformation is $G\Gamma$ where G is the shear modulus of elasticity and Γ the shear strain, or angle of shear ($\leftarrow$ 4.2). In positions (a) and (e), the layers are in stable equilibrium and the shear stress is zero. In position (c), the shear stress is also zero but the layers are in unstable equilibrium.

The maximum shear stress τ_m to initiate plastic deformation will occur about position (b), where the angle Γ corresponding to the slip between the layers is about 0.25 radians, or 15°. The measured value of G for mild steel is 60 GPa, predicting a value of τ_m of 15 GPa. The actual value of the limiting stress to initiate slip is 1 to 10 MPa – a factor of 10^3 to 10^4 times smaller than predicted. So metals are much weaker than predicted. In 1934, G.I.Taylor suggested that this is because commercial metal specimens are not perfect crystals.

One defect in crystals is a **dislocation**. Figure 25.17a shows an edge dislocation, a partial extra layer of atoms inserted between two complete layers. Looking at it in one plane, as in Figure 25.17b, shows that when a shear stress is applied, the lattice can slip one row at a time and the incomplete row or dislocation moves through the lattice. This could be expected to take place at a much lower stress than that required to slip complete layers. Experimental evidence for dislocations can now be obtained from electron micrographs (see Figure 25.17c). The English physicist, William Bragg investigated dislocations using a two-dimensional analogue – a raft of bubbles. You can do this too in experiment 25.2.

Figure 25.17 a An edge dislocation, **b** movement of a dislocation, **c** photograph of a dislocation

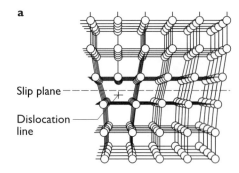

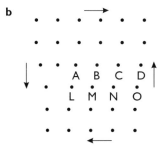

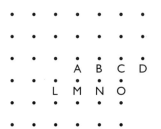

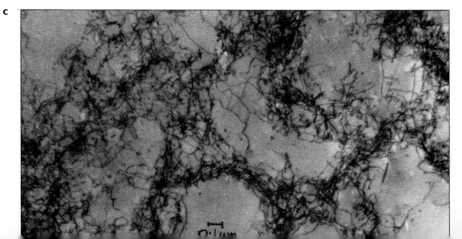

25 NATURE OF SOLIDS

EXPERIMENT 25.2 The bubble raft

Using a fine jet connected to a gas or hydrogen supply, you can blow a single layer of small bubbles about 2 mm diameter over the surface of water in a large dish. The water should contain a few drops of detergent and be moved about at a constant rate to blow bubbles of a constant size. Viewed over a small area a raft of bubbles will appear to be a hexagonal close pack. Looking more closely, you should be able to identify areas of homogeneity, or crystals, and positions where defects occur. By deforming the bubble raft between two blocks, as shown in Figure 25.18, you can see the effect of shearing, compressing and expanding the raft.

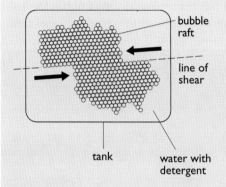

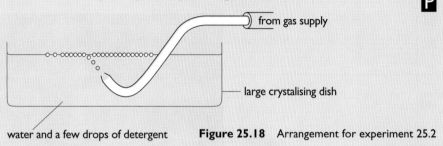

Figure 25.18 Arrangement for experiment 25.2

The movement of dislocations helps explain many of the properties of metals. Dislocations move very easily and are present in very large numbers. In a typical annealed metal, there may be kilometres of dislocations per cm^3. A single crystal wire of zinc, 1 mm in diameter, is so weak that it may be drawn out to two or three times its original length by a tensile force of less than 10 N. To strengthen a metal, you need to hinder the movement of dislocations. Methods include **mechanical working**, **heat treatment** and **alloying**.

In experiment 25.2, you may have noticed small elastic deformations occurring in the bubble raft. The dislocations produced move back again to their original positions when the stress is removed. However, under large stresses, the dislocations may move to the edge of the 'crystal' and be lost, or may meet other dislocations and be annihilated or trapped.

Deforming a metal in a plastic way, for example through rolling, extruding or hammering it, is known as **work hardening**. It tends to break up the crystals into smaller crystals, trapping dislocations in the crystal or grain boundaries and tangling them up within the crystal (see Figure 25.19). Reheating the metal to

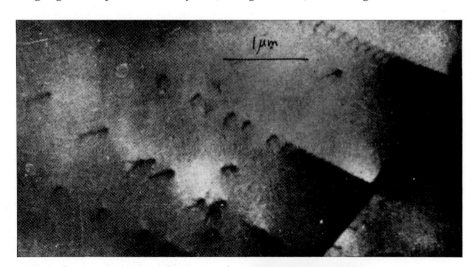

Figure 25.19 Tangled dislocations in a work-hardened metal

bright red heat, with the temperature greater than 1 100 K, usually allows the dislocations to untangle and anneals the metal into its soft state again.

COPPER

Copper is elastic for very small strains of about 0.1 % and then becomes ductile (see Figure 25.20). If the wire is unloaded in the ductile region and then reloaded, it maintains its original stiffness for greater stresses. This is another example of work-hardening.

The copper wire shown in Figure 25.21 has not broken cleanly but has formed a 'neck'. At the neck, the true stress is very large because the cross-sectional area

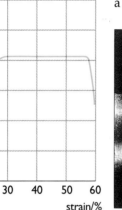

Figure 25.20 Stress-strain graph for copper

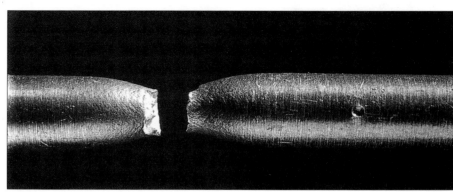

Figure 25.21 Broken ends of copper wire showing 'neck'

of the wire is small. The stress on the graph is less because it is calculated using the original cross-sectional area of the wire. Copper has a high conductivity, both of heat and of electricity. Its low yield strength has led to its extensive use in tubing for plumbing installations. Otherwise, copper is not an important engineering material. But it has a fairly simple structure, and an understanding of its behaviour can lead to an understanding of other metals and alloys that are important constructional materials, such as steel.

EXPERIMENT 25.3 Work hardening copper wire

By holding at one end a fresh piece of thin copper wire of, say, 1 mm diameter and about 200 mm long, you can see how much it bends under its own weight. If you do this again after winding each end round a pencil and pulling the wire until it deforms plastically, you will find that it bends much less. It has become work-hardened. If you then heat the wire to bright red in a bunsen flame and quench it in cold water, you should find that it has been annealed and is soft again.

STEEL

Steel is an alloy of iron and carbon, usually containing less than 1.5 % carbon by weight. Its properties change remarkably with the amount of carbon added (see Figure 25.22). Increasing the carbon content from 0.22 %, for mild steel, to 1.04 %, for high-carbon steel, increases the strength and reduces the ductility. Further increase in carbon content has the opposite effect. Steel's properties also depend on the heat treatment that the steel has undergone during its manufacture as shown in experiment 25.4.

25 NATURE OF SOLIDS

EXPERIMENT 25.4 Heat treatment of steel

Warning: Eye protection must be worn for this experiment.

You can heat different steel sewing needles and then produce:
a **quenched steel** by plunging one red-hot needle into cold water
b **annealed steel** by **slowly** cooling another red-hot needle
c **tempered steel** by warming a previously quenched and cleaned needle above a flame until a coloured layer appears.

By snapping each with two pairs of pliers, you should find that quenched steel is brittle, annealed steel is flexible and ductile, and tempered steel is tough and elastic. Snapping an untreated one, should allow you to determine the condition of a steel needle.

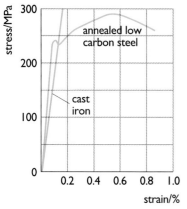

Figure 25.22 Stress-strain graphs for iron and steel

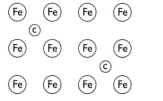

Figure 25.23 Interstitial carbon atoms in an iron lattice

ALLOYS

The presence of impurities in a metal leads to the crystal lattice being distorted locally by atoms of different sizes. The distortions inhibit the movement of dislocations and strengthen the metal. In such mixtures, or **alloys**, of metals, small proportions of impurity atoms may be **interstitial atoms**. These are atoms squashed in between the layers of the lattice (see Figure 25.23), like the carbon atoms in carbon steel which is iron containing up to 1 % carbon.

Alternatively, the alloying atoms may displace host atoms and occupy lattice sites. This occurs, for instance, in α-brass which has up to 30 % zinc in copper. When very pure metal crystals are grown in a clean controlled environment, very few dislocations occur. Such crystals are known as **metal whiskers** and have strengths that approach theoretical predictions.

Questions

3 Dislocations are important in helping crystals to grow. Find out about the different sorts of dislocations and explain why this is so.

4 Read about twinning, another process by which metals deform plastically. Write a paragraph explaining this process to a fellow student.

25.3 CERAMICS AND GLASSES

Ceramics and glasses are families of materials that have been made and used by people for thousands of years. They are generally hard, brittle, thermal and electrical insulators, and have a high resistance to corrosion. The majority exhibit elastic behaviour up to the point of fracture. Like metals, but for a different reason, their measured strength is orders of magnitude smaller than the predicted strength.

CERAMICS

The structure of a typical ceramic material is based on ionic bonding between oxygen and elements such as silicon, magnesium and aluminium. The basic building blocks of oxygen-based ceramics are the oxides: silica, magnesia and

F MATERIALS: THEIR PROPERTIES AND STRUCTURE

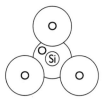

Figure 25.24 Silica tetrahedra

alumina. Silica (SiO₂) forms tetrahedra, as shown in Figure 25.24. These can be arranged in different ways: in a rigid three-dimensional assembly to give quartz, in long silicate chains to give asbestos, and in sheets of tetrahedra to give mica.

Clay, the raw material for pottery ceramics, contains silica and alumina combined to form small 'plates' of kaolinite. The plates slide over each other when wet but interlock to form a rigid material when water is driven off by heating.

GLASSES

Glasses also contain silica mixed with other oxides to alter their softening temperature, colour and other properties. But, unlike ceramics, glasses have no long-range order in the arrangement of their silica tetrahedra. The glassy state is that of a **supercooled liquid**. When the liquid cools, the viscosity of glass (→ 27.4) becomes very large. So unless left for centuries, its silica tetrahedra do not form a highly ordered, crystalline, structure.

As molten glass cools, it gets more and more 'treacle-like' until flow can no longer be detected. There is no definite solidification temperature, no change in volume as it solidifies and no latent heat associated with solidification. This is quite unlike most other materials. Figure 25.25 shows the X-ray diffraction pattern of glass. It has no structure, indicating a random arrangement of the silica tetrahedra. Quite near the softening temperature, some of the metal ions move. (← 12.5)

A freshly drawn thin fibre of glass has a surface free from scratches and is quite strong. But when glass comes into contact with other solids, its surface becomes covered in many fine scratches. These make the glass weak in tension.

Experiment 25.5 shows how easily a crack will spread from a scratch in a glass surface when it is in tension. Resistance to the spreading of cracks is called **toughness**. A tough material is one that requires a relatively large amount of energy to break it. Toughness can be measured from the strain energy stored, which is equal to the area under the force-extension graph (← 4.3).

Figure 25.25 X-ray diffraction photograph of glass specimen

Question

5 Optical fibres made of glass (→ 34.2) are being used increasingly in electronics and communications. Although quite strong, glass is very susceptible to brittle fracture. Find out how engineers overcome this problem.

EXPERIMENT 25.5 Effect of cracks in the surface of glass

Warning: Eye protection and a safety screen must be used.

a Breaking After making a scratch across the top of a glass rod and placing this over a matchstick, try pressing down gently on both ends of the rod (see Figure 25.26a). You should find that the rod breaks easily at the scratch point.

b Bending After heating a 3 mm diameter soda glass rod and drawing out a thin fibre, you will find that the fibre can be bent considerably without breaking, as shown in Figure 25.26b. If you touch the inside of the curved fibre with another glass rod, it will remain intact, but if you touch it on the outside of the curve, it will snap.

Figure 25.26 Cracks and **a** breaking, **b** bending

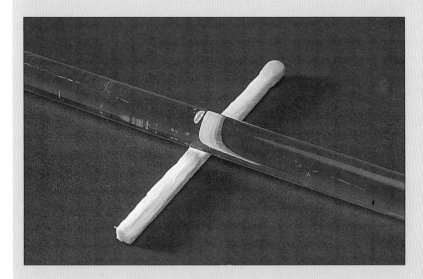
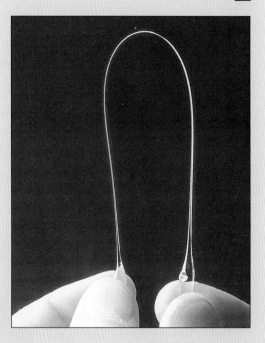

Stopping cracks

Toughened car windscreens are made by blowing cold air jets on to them, so that the outside surfaces cool and contract quickly. When the inside cools and contracts, it holds the outer surfaces in compression, making it difficult for cracks to propagate.

Another method of stopping cracks is to use thin fibres. A crack in one fibre will stop at its boundary and not affect the others. For flexible materials, the fibres can just be twisted together, for example, as the hemp in ropes and the steel in cables for cable cars and suspension bridges. If the material is brittle, the fibres have to be embedded in another substance called the **matrix**, making a **composite** material. One such composite is glass-reinforced plastic (GRP), which consists of glass fibres in a polymer matrix. GRP is light, strong and flexible. Carbon fibres in a polymer matrix makes a weaker but stiffer composite.

GRIFFITH CRACKS

Experiments on glass (← 4.2) show that, although glass has a high value of the Young modulus, the fibres break at very small strains. The stress necessary before the fibre breaks is different on each occasion. However, the less handling a fibre is subjected to, the larger the breaking stress.

In 1920, A.A. Griffith suggested that the discrepancy between the observed and predicted strengths of glasses and ceramics was due to the presence of minute cracks, both on the surface and in the bulk of the material. He argued that such cracks would distort the stress pattern (in a similar way to that shown in Figure 25.27), giving rise to stress concentrations at the roots of the cracks. When loaded, the stress in the area at the root of the crack would rise to the predicted

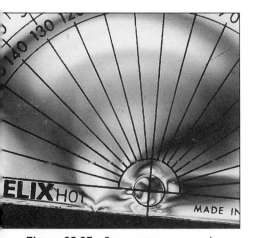

Figure 25.27 Stress patterns around cracks in a plastic protractor

MATERIALS: THEIR PROPERTIES AND STRUCTURE

value to separate the planes of ions. Then the crack would spread through the material. So the more cracks, the smaller the breaking stress of the sample. Griffith showed that the **minimum stress** σ for a crack to grow is given by:

$$\sigma = \sqrt{\frac{E\Gamma}{l}},$$

where E is the Young modulus of the glass, Γ is the **surface energy** and l is the length of the crack.

The low observed breaking stresses of ceramics and glasses suggest, using this relationship, that the cracks are very small. The minutest surface scratch or contamination is enough to produce an enormous reduction in strength. Only when the fibre is just formed, or its surface has been remelted by flame polishing, will l be small enough for the fibre's strength to approach the predicted maximum.

Metals also have surface scratches like glasses and ceramics. But metal atoms can slip, reducing the stress at the end of the crack. So it never reaches the atomic breaking stress that would allow the crack to deepen.

Question

6 For a typical glass, the breaking stress, which propagates a crack right across the sample, is 70 MPa. The free surface energy is 0.2 J m^{-2} and E is 70 GPa.
a Use these values with the relation proposed by Griffith to estimate the initial length of crack to cause failure.
b How many atoms long is this crack?

EXPERIMENT 25.6 Observing Griffith cracks

Slowly heat a small crystal of lithium nitrate in the centre of a microscope slide so that it dehydrates and melts. Then, after allowing the slide to cool, wash it off and, under a microscope, you should see the crack system in the slide's surface.

In the table of Figure 25.1, one material stands out as stronger than the rest – carbon fibre. When correctly made, carbon fibre behaves as a crystalline material with few cracks and very few dislocations. It is used extensively in aircraft manufacture and there is continuing interest in its commercial development.

25.4 POLYMERS

Polymers are long chain molecules. They can be divided on structure into **glassy**, or amorphous polymers and **crystalline** polymers. Glassy polymers, as their name suggests, are a random jumble of long chains. They include poly vinyl chloride (PVC) and perspex at room temperature. Crystalline polymers have some long range order from chains being folded in parallel sections, and because adjacent chains are laid down parallel to each other. Examples of crystalline polymers are polyethene and nylon 6-6 at room temperature. The strength of the

25 NATURE OF SOLIDS

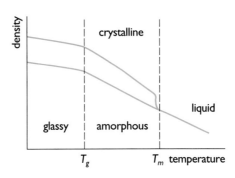

Figure 25.28 Variation of density with temperature for polymers

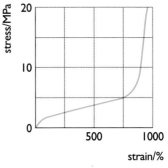

Figure 25.29 Stress-strain graph for rubber

resulting polymeric material depends on how the chains interlock and how much cross-linking there is between chains.

Unlike metals and ceramics, the properties of a polymer are very sensitive to its structure, the temperature at which the measurements are made, and the way in which the tests are carried out. So it is very important to change only one variable at a time when testing polymers.

Any given polymer may exhibit glassy or crystalline behaviour depending on the temperature of measurement. Two characteristic temperatures are the glass temperature T_g for the glassy to crystalline transition and the melting temperature T_m for a crystalline polymer. Figure 25.28 shows how the density changes with temperature for a typical sample.

RUBBER

Rubber is unusual in being elastic for very large strains. It can be extended by several times its original length and still return to its original size. Rubber stretches easily at first, then becomes harder to stretch (see Figure 25.29). The cross-sectional area of a rubber band decreases considerably when it is stretched. Since nominal tensile stress is calculated using original cross-sectional area (← 4.2), the measured stress is rather less than the actual stress rubber experiences when stretched. So rubber is rather stiffer than it would appear.

Apart from tyres and other obvious applications, rubber is used, for example, in place of coil springs in the suspensions of heavy lorries, railway carriages, and at the top of the columns that support motorway bridges (see Figure 25.30).

Figure 25.30 Photograph of rubber suspension in **a** a railway carriage and **b** motorway bridge supports

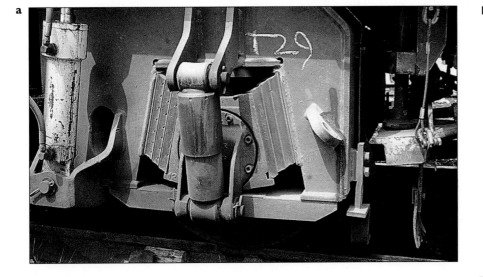

MATERIALS: THEIR PROPERTIES AND STRUCTURE

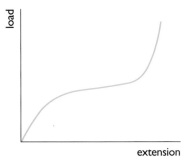

Figure 25.31 Load-extension graph for rubber

> **Question**
>
> **7** Figure 25.31 shows the loading graph for a rubber band. Rubber is an elastomer, which is a substance with long chain molecules coiled up in a zig-zag fashion when unstressed. Explain the form of Figure 25.31 in terms of changes in the shape and arrangement of the long chain molecules. Use Figure 25.32 to help you.

Figure 25.32 X-ray diffraction photographs of **a** unstretched rubber, showing no structure; and **b** stretched rubber, showing some order at molecular level

EXPERIMENT 25.7   Static fatigue, or creep, in polyethene

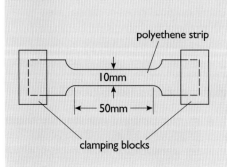

By hanging various loads from a set of polyethene strips of the shape shown in Figure 25.33, you can plot graphs of extension against both load and time, and record the time for the specimen to fail. You will then be in a position to ask questions about the mechanical properties of polyethene. By raising the temperature around the specimen and repeating the experiment, you can see whether temperature changes affect these properties. Because polyethene bags are stretched parallel to their length during manufacture, you may find differences between specimens cut parallel and perpendicular to the tube from which they have been taken.

Figure 25.33 Shape of polyethene strips for experiment 25.7

MECHANICAL PROPERTIES OF POLYMERS

These are difficult to predict because of the variation in a polymer's properties with temperature, time and mode of experimenting. Figure 25.34 shows the shape of the stress-strain relationships for polymers at various temperatures. None of them obeys Hooke's law, even at constant temperature. The value of the Young modulus varies with temperature much more rapidly for a polymeric material than for a metal or a ceramic (see Figure 25.35). Most amorphous polymers go through an elastomeric region when heated from the glassy to the liquid state. The response of these polymeric materials to sudden changes in load is particularly variable, as shown in Figure 25.36.

25 NATURE OF SOLIDS

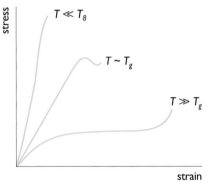

Figure 25.34 Shapes of stress-strain graphs for polymers at various temperatures T relative to glass temperature T_g

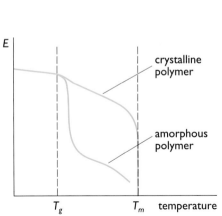

Figure 25.35 Variation of Young modulus E with temperature for polymers

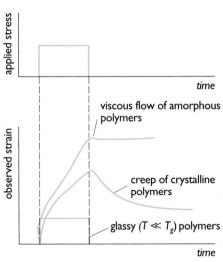

Figure 25.36 Time response of various polymers to stress

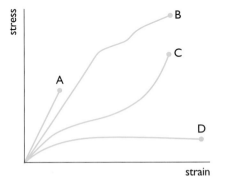

Figure 25.37 Stress-strain graphs for various materials

Question

8 Referring to the stress-strain graphs in Figure 25.37, which material
a is the strongest?
b is the stiffest?
c is brittle?
d has the highest yield stress?
e is the most ductile?

25.5 MEASURING OTHER MECHANICAL PROPERTIES

Many types of tests have been devised to quantify the mechanical properties of materials, such as hardness and fatigue.

Hardness indicates the resistance of a material to plastic deformation under load. For some materials, it is related to tensile strength. Hardness tests on metals usually involve pressing a steel ball or diamond into the metal surface with a particular force for a given time. This results in a small indentation for a hard metal, and a large one for a soft metal. The area of the indentation is measured. The hardness is then quoted as a number, such as the Brinell hardness number which is calculated from the load per unit area.

Fatigue is a common cause of failure when a material is subjected to continuously changing stresses well below the breaking stress. For example, if you bend a piece of copper wire repeatedly at one point, it will eventually break. Many plastic materials behave in a similar way. One method of testing fatigue uses a piece of the material with a loaded ball race on one end. The other end is connected to a rapidly spinning motor. The load causes the test piece to be flexed in opposite directions once each revolution. A counter records the number of revolutions until the specimen breaks (see Figure 25.38).

The number of cycles (times a load is applied and removed) to produce failure of the component varies with the load applied. If the load causes a stress equal to the ultimate tensile stress (UTS) (← 4.2), then the component fails at the first cycle. For some metals, such as aluminium, the number of cycles to failure increases as the load decreases, but it never becomes infinite. For other materials, such as steel and cast iron, there is a value of stress below which fatigue will

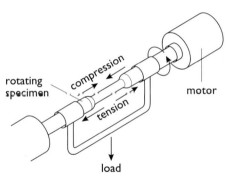

Figure 25.38 Testing metal fatigue

When a metal fails by fatigue, it appears that the oscillating stress sets up a crack. This gradually spreads through the metal until the stress in the remaining cross-section reaches the UTS and the component fails by ductile fracture. Since fatigue failures are initiated by cracks, the presence of surface cracks and scratches has a very significant effect on the fatigue life. This means that the history of a specimen before it is tested is very important. It can lead to very different fatigue lives for apparently identical specimens.

Another measurement engineers need to know is the **density** of the material – the mass in kilograms of unit volume, or 1 m^3, of the material. Figure 25.39 shows the densities of some constructional materials.

material	density/kg m^{-3}
steel	7700
aluminium alloy	2800
wood: oak	720
softwood	400 to 600
chipboard	450 to 1300
brick	1500 to 1800
concrete	2200 to 2400
glass (windows)	2500
glass reinforced polyester	1600

Figure 25.39 Densities of some constructional materials

The Comet disasters

In the mid-1950s, Comets became the first pressurised jet passenger aircraft to fly international routes. Several were inexplicably lost with all passengers and crew while flying at high altitude. Immediately following a crash in April 1954, the airworthiness certificate of all Comet aircraft was withdrawn and a detailed investigation followed. In one incident, debris was spread over more than 200 km^2 in deep water off the coast of Elba in the Mediterranean. Much of the wreckage was recovered and re-assembled at the Royal Aircraft Establishment at Farnborough.

It was deduced from the wreckage that the fuselage had split along the row of portholes and the number of rivet holes around the portholes had caused the failure. Each time the aircraft took off and climbed to high altitude, the cabin was pressurised and the skin of the fuselage was put under tension. Stresses were concentrated at the corners of apertures in the skin. After a large number of flights, fatigue had so weakened the metal that it failed at the corner of a porthole, where the stress was greatest.

Many tests were subsequently carried out with the fuselage in a tank of water. Forces were applied to the wings by hydraulic rams and the fuselage was put through a pressurisation cycle repeatedly. The fuselage failed in the same way, but it was possible to investigate the mode of failure in more detail because the fuselage, being in water, had not disintegrated completely. Lessons learnt from the metal-fatigue investigations on the Comet have led to the design and manufacture of much safer aircraft. They have also pointed to the need for regular and careful monitoring of aircraft performance.

SUMMARY

Solids can be grouped into classes: metals, ceramics, glasses and polymers.

The bulk properties of materials, such as their mode of fracture, can be related to their microstructure.

Metals are crystalline assemblies of ions. Their mechanical properties are dominated by the number and movement of dislocations. It is the presence of dislocations that makes metals so weak.

Ceramics are also crystalline but do not have movable dislocations. Glasses are supercooled liquids – a jumble of silica tetrahedra and other oxides with no ordered arrangement. Ceramics and glasses break by the propagation of surface cracks through the material.

Polymeric materials can have both crystalline and glassy properties.

SUMMARY QUESTIONS

9 Failure by metal fatigue came to public prominence with the crashes of two Comet airliners. Explain, giving examples, what is meant by 'fatigue failure'. (← 4)

10 Some very small drill bits are made entirely from a ceramic, silicon carbide. However, most large masonry drill bits consist of a small tip of silicon carbide brazed to a steel shank, rather than a complete silicon carbide bit. Why do you think this might be?

11 Graphite consists of planes of carbon atoms with strong bonding forces between atoms within the plane. The forces between atoms in neighbouring planes are relatively weak. How does this explain the behaviour of graphite and its everyday uses, such as in pencils and for lubrication?

12 Answer the following for **each** of the materials steel, glass and polyethene.
a Where might it be used?
b What physical properties make it suitable for each application?
c How are the physical properties explained in terms of the structure of the material?

26 INTER-PARTICLE FORCES AND ENERGIES

Some forces between particles attract, while others repel. Steel cables supporting the large weight of the road deck of a suspension bridge are under tension. This is because strong forces of attraction exist between the particles that make up the cable, keeping them together. On the other hand, the vertical steel pillars that support the roof of a warehouse are under compression. Strong repulsive forces exist between the particles, preventing the pillars from collapsing under the weight of the roof.

Models of the structures of solids, developed by physicists and mathematicians and used by civil and structural engineers, assume the existence of such attractive and repulsive forces (← 25). Using these inter-particle forces and their corresponding potential energies, we can build up a model of structure on an atomic scale.

26.1 FORCES BETWEEN PAIRS OF PARTICLES

Consider an isolated pair of particles. When they are far apart, as in a gas, they exert negligible forces on each other. When they approach each other, there is a resultant force of attraction between them. As they get closer, this force increases, reaches a maximum and then decreases to zero. For the particles of a solid or liquid, the force becomes repulsive as the particles get very close. This happens when an attempt is made to compress a solid or liquid. The **equilibrium separation** occurs when the resultant force is zero. A force-separation graph in Figure 26.1 shows these changes.

Question

1 Figure 26.2 shows two vehicles on an air track linked by an elastic cord and with two magnets arranged so that they repel each other. The vehicles are shown in equilibrium.
a What happens when the vehicles are pushed **i** a little closer together?
ii a little further apart?
b In what ways is this a good model of the forces between neighbouring particles in a solid material? In what ways is the model inaccurate?

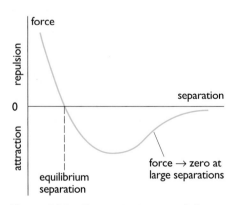

Figure 26.1 Force-separation graph for a pair of particles

The forces between particles are all electrical in origin and such forces decrease in magnitude with increasing separation of the particles. From Coulomb's Law (← 16.2) the force between individual pairs of charges is inversely proportional to the square of their separation.

26 INTER-PARTICLE FORCES AND ENERGIES

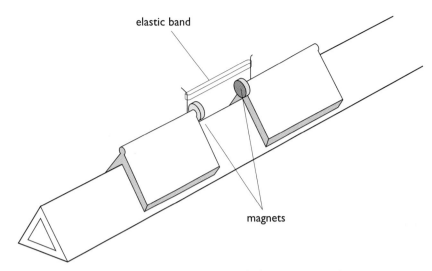

Figure 26.2 Attraction-repulsion between vehicles on an air track

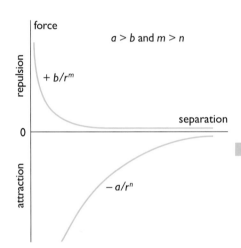

Figure 26.3 Components of the force-separation graph

An inverse power law for the variation of force with distance gives a negligible force at large separations. The higher the power in the law, the faster the force will change at small separations. The curve in Figure 26.1 could be made by summing two separate curves, each of which extends from zero separation to a large distance. In Figure 26.3, the repulsion-force curve changes much more rapidly at small separations than the attraction-force curve, and must, therefore, have a higher power in the inverse power law.

Force between particles

The force F between particles of separation r is given by:

$$F = -(a/r^n) + (b/r^m),$$

where a and b are constants. The first term is negative because it represents the force of attraction between the particles, while the second term represents the repulsive force.

When index m is greater than index n, and constant a is greater than constant b, the expression closely matches what we observe experimentally between, for example, molecules or nucleons. In fact, it applies to any particles held together by an interplay of closely repulsive and distantly attractive forces.

Question

2a Using the equation in the box 'Force between particles', write down a relation between a and b at the equilibrium separation r_o when $m = 13$ and $n = 7$.
b Sketch a graph of force against separation over the range $0.7\,r_o$ to $2.0\,r_o$.

26.2 POTENTIAL ENERGY BETWEEN A PAIR OF PARTICLES

The existence of forces between a pair of particles implies that work has to be done on or by the particles to separate them or push them together. There is,

MATERIALS: THEIR PROPERTIES AND STRUCTURE

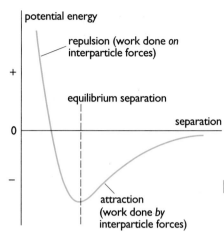

Figure 26.4 Potential energy-separation graph

therefore, a potential energy associated with the particles. This is taken to be **zero** when the particles are at **infinite separation** (← 16.3).

Because they attract each other at large distances, particles must have work done on them to separate them. So, to increase with increasing separation, the potential energy must be negative. To reduce the separation below its equilibrium value, work also has to be done on the particles. So the potential energy increases and becomes large and positive. Figure 26.4 shows a curve fitting these conditions. The equilibrium or stable position is that of minimum potential energy. The minimum of the energy-separation graph occurs at the separation for zero force, when $r = r_0$.

Potential energy and force between particles

An increase in potential energy ΔU is the energy $F\Delta r$ transferred as work is done **on** the particles. So for work done **by** the force F between the particles:

$$F = -\frac{\Delta U}{\Delta r}$$

or, in the limit as Δr tends towards zero:

$$F = -\frac{dU}{dr}.$$

The value of the force at any separation is equal to the negative of the gradient of the potential energy-separation graph. So, when a force F obeys an inverse nth power law, the potential energy obeys an inverse $(n-1)$th power law.

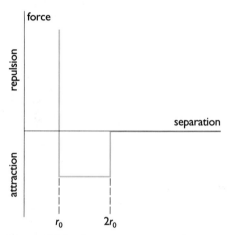

Figure 26.5 Force-separation graph

Questions

3 Sketch force-separation and potential energy-separation graphs for two hard spheres that exert no forces on each other until they collide. Assume they do not deform when they collide.

4 Sketch the potential energy-separation graph for two particles whose inter-particle force changes with separation as shown in Figure 26.5.

26.3 FITTING THE MODEL TO MATERIALS

A solid is not a system of isolated pairs of particles. We need to allow for the forces and energy terms between all the particles in a solid. Since force is a vector quantity, a particle at equilibrium in the solid has zero resultant force acting on it. However, the energy terms between each pair of particles can be added as scalars to arrive at a total potential energy for the particles (← 16.3). To provide quantitative scales for the graphs of inter-particle forces and energies, we must relate them to measured macroscopic properties, such as elasticity. The concept of a potential energy well helps us to do this.

POTENTIAL ENERGY WELLS

The shape of the potential energy-separation graph is often called a potential energy well (← 15.3 and 16.3).

A gravitational situation illustrates this. Imagine you are in a shallow well in

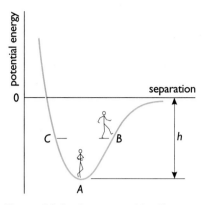

Figure 26.6 Gravitational 'well'

the ground, as in Figure 26.6. At the bottom of the well, A, you will be in a stable situation with no resultant force trying to move you. To escape from the well, you will need to climb a height h to reach the rim. You will have to supply extra energy equal to mgh. At point B you will still require extra energy to escape. This, however, will be less than from A. Released from B you could slide, with no friction, down to A and up to C, oscillating between B and C. In the case of particles in a liquid or solid, the well depth ΔU is the energy required to separate the particles completely. Such a process results in the sublimation of the solid or the evaporation of the liquid. In both processes, the energy input is the relevant latent heat energy. This increases the potential energy of the particles by moving them many particle diameters apart.

As an example, we will estimate the depth of the potential energy well for water molecules. The molar latent heat of vaporisation of water is 40 kJ mol^{-1}. One mole of water contains N_A molecules, where N_A is the Avogadro constant ($\rightarrow$ 35.2). So the latent heat of vaporisation per molecule is:

$$\frac{\text{molar latent heat}}{N_A} = \frac{40 \times 10^3}{6 \times 10^{23}} = 7 \times 10^{-20} \text{ J, approximately.}$$

Each water molecule within the water touches about 10 neighbours. So, to evaporate it, the energy of 10 bonds must be overcome. But, each bond links a pair of molecules and we must avoid counting each bond twice. So the estimate for the bond energy or well depth in water is one fifth of the latent heat energy per molecule, or about 1×10^{-20} joules.

Questions

5a The molar latent heat of vaporisation of water is 40 kJ mol^{-1}. How much energy is needed to change 1 kg of water into 1 kg of steam at the boiling point?
b An electric kettle containing 1 kg of water fails to switch off when the water reaches boiling point. Estimate how long it will be before the kettle boils dry, assuming the power of the electric element is 3 kW. What factors would you need to take into account to improve the accuracy of your estimate?

6 If a breeze blows against your body as you get out of a swimming pool, you will feel chilled, even on a warm day. How do you explain this at a microscopic level?

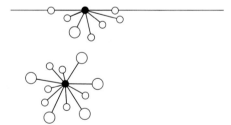

Figure 26.7 Molecules at the surface and inside a liquid

To check the estimate for water, you can use it to predict another property. Consider a molecule at the surface (see Figure 26.7). This molecule lacks neighbours above and, hence, its energy is different from that of a molecule well below the surface. The difference in energy is called the **free surface energy** per molecule. This is equal to the sum of the bond energies overcome when bonds are broken and not reformed as a molecule moves from inside the liquid up to the surface layer ($\rightarrow$ 27.2).

The molar volume, or volume of 1 mole ($\rightarrow$ 35.2), of water is 18 cm^3, since the molar mass of water is 18 and its density is approximately 1 g cm^{-3}, or 1 000 kg m^{-3}. If we assume each molecule occupies the space of a cube of side 3×10^{-10} m, then 1 m^2 of water surface will contain approximately 10^{19} molecules. Imagine that each of these has two fewer neighbours than those in the bulk liquid. This gives a predicted free surface energy of 0.2 J m^{-2}. The estimate is about three times the measured value but does suggest the model is worthy of further study.

MATERIALS: THEIR PROPERTIES AND STRUCTURE

> **Questions**
>
> **7** If we had used data for liquid argon rather than water in the prediction of free surface energy, the agreement would have been much better. Suggest reasons why.
>
> **8** The molar latent heat of vaporisation of argon is 6.5 kJ mol^{-1}.
> **a** Estimate the well depth for argon.
> **b** Would you have expected the well depth for argon to be less than that for water? Give your reasons.
>
> **9** The molar latent heat of atomisation of diamond is 715 kJ mol^{-1}. Estimate the bond energy, or well depth, for carbon, given that each carbon atom in diamond is bonded to four others.

Elasticity and bond stiffness

Consider a simple lattice of particles with mean separation r (see Figure 26.8). One square metre of the surface of this lattice will contain $1/r^2$ particles. The stiffness k of the bonds between the particles is the gradient of the force-separation graph. When the lattice is stretched, the mean separation of the particles increases by Δr. So:

$$\text{strain produced} = \frac{\Delta r}{r}.$$

Since the force to stretch one bond is $k\Delta r$, the force per unit area, or stress, is given by:

$$\text{stress} = \frac{k\Delta r}{r^2}.$$

The Young modulus ($\leftarrow$ 4.2) for the material is given by:

$$\begin{aligned}\text{Young modulus} &= \frac{\text{stress}}{\text{strain}} \\ &= \frac{K\Delta r/r^2}{\Delta r/r} \\ &= \frac{k}{r}.\end{aligned}$$

This argument cannot be applied easily to metals because their mechanical properties are mainly determined by defects such as dislocations.

PREDICTING THE DETAILED SHAPE OF FORCE – AND ENERGY – SEPARATION GRAPHS

When many crystalline solids are subjected to small stresses, they obey Hooke's law ($\leftarrow$ 4.1), and the Hooke's law constant k is the same for compression and tension. This suggests that near the equilibrium separation of the particles, the force-separation graph is a straight line and the energy-separation graph is a parabola.

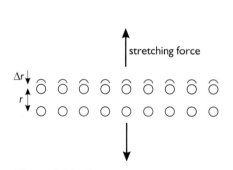

Figure 26.8 Stretching a crystal lattice

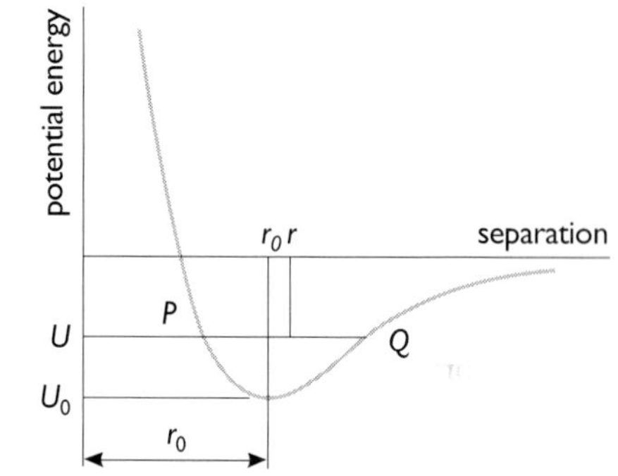

Figure 26.9 Asymmetric potential well

When their temperature increases, materials expand. This implies that the equilibrium separation increases with increase in temperature. In a solid, the particles have potential and kinetic energies, both of which increase as the temperature rises.

In Figure 26.9 a particle with energy U_0 at the bottom of the well is at a very low temperature and rests at an equilibrium separation of r_0. A particle with energy U is at a much higher temperature and is confined between P and Q with an equilibrium separation of r. The property of expansion requires that r is greater than r_0, making the shape of the well assymmetric.

To arrive at a detailed shape, we need information on the type of bonding between the particles that make up the solid.

BONDING IN SOLIDS

Ionic solids such as sodium chloride, are crystalline. The particles making up the lattice are positive and negative ions (see Figure 26.10a). The force between any

26 INTER-PARTICLE FORCES AND ENERGIES

Origin of van der Waals forces

A quantitative description of van der Waals forces involves quantum mechanics. However, it is possible to imagine how they arise. Some molecules are polar – they have equal positive and negative charges whose centres of charge are separated and, therefore, possess an electric dipole moment. In a solid, polar molecules will be arranged positive to negative for minimum potential energy and will attract each other (see Figure 26.10d). This dipole-dipole attraction was proposed as the origin of van der Waals forces by W.H. Keesom in 1912.

Just as a bar magnet will attract an unmagnetised piece of iron by induction, a polar molecule may induce a dipole moment in a non-polar molecule and attract it. This dipole-induced dipole attraction was suggested by Peter Debye in 1920.

In 1930, H. London proposed what have come to be known as **dispersion forces**. He argued that even when all the molecules are non-polar, weak forces of attraction must be present to explain the existence, for instance, of solid neon and argon. The varying configurations of the electron cloud around an atom could result in that atom having an instantaneous electric dipole moment. That instantaneous dipole moment could induce a dipole moment in a neighbouring atom and give a momentary attractive force. Averaged over time this would give a net force of attraction to hold the solid together.

All three types of van der Waals force lead to an attraction potential energy term obeying an r^{-6} law. The repulsion term follows an r^{-12} law, giving an energy-separation graph represented by

$$U = -(a/r^{-6}) + (b/r^{-12}),$$

where a and b are constants for a given solid. The use of U rather than E_p for molecular potential energy is only a convention.

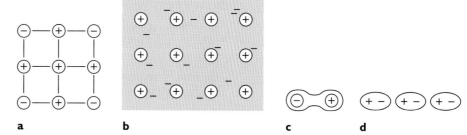

Figure 26.10 Forces in **a** ionic solids, **b** metals, **c** covalently bonded solids **d** molecular solids

two ions obeys an inverse square law. As long as the attractions from pairs of unlike charges exceed the repulsions from pairs of like charges, a stable solid results (→ 26.4).

In **metals** most atoms contribute one electron to a 'gas' of free electrons that can move through the lattice. The lattice, therefore, consists of positive ions held together by the 'gas' of free electrons, as shown in Figure 26.10b. Metallic bonding is much more complex than that of ionic solids and requires a quantum model (→ Unit H) to analyse it.

Covalently bonded solids The covalent bond between two atoms involves the two atoms sharing a pair of electrons (see Figure 26.10c). The resulting molecule is neutral and is not strongly bonded to its neighbours. Covalently bonded crystals such as diamond are effectively macromolecules, requiring very large energies to atomise them.

Molecular solids formed from neutral molecules or atoms, are bound by so-called **van der Waals forces**. They are mechanically weak and have small latent heats of sublimation. The simplest examples are solid noble gases, such as neon and argon.

STOPPING SOLIDS COLLAPSING

If the only forces between particles were forces of attraction, then matter would condense to a point. Our model suggests that, at very small separations, the forces of repulsion increase rapidly and swamp the forces of attraction. An exception occurs when the gravitational force is extremely large, such as when a **black hole** is formed.

Measurements of the radii of atoms by different methods give different values, showing that the electron cloud defining the atom does not have a sharp edge. When a solid is compressed, the separation of the atoms or molecules is reduced. This distorts the electron clouds and the distortion gives rise to the repulsion forces. With all types of bonding, the repulsion force increases much more rapidly than the attraction force as the particles move together.

26.4 SODIUM CHLORIDE – AN IONIC SOLID

Sodium chloride (NaCl), an ionic crystal, can be imagined to be made up of equal numbers of sodium and chloride ions in a simple cubic lattice. Much experimental work has been done on sodium chloride, so predictions of the model can be

MATERIALS: THEIR PROPERTIES AND STRUCTURE

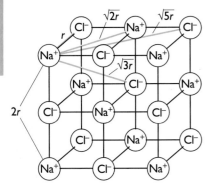

Figure 26.11 Sodium chloride lattice

Effect of adjacent ions

Figure 26.11 shows the cubic lattice of sodium chloride. The nearest neighbours to each Na$^+$ ion are six Cl$^-$ ions, each a distance r away. All six contribute to the attraction potential energy. Further away, separated by $\sqrt{3}r$, are another eight Cl$^-$ ions. Still further away, at $\sqrt{5}r$, are 24 Cl$^-$ ions and so on through the lattice. There are also repulsion potential energies between nearby sodium ions. Each Na$^+$ ion has 12 Na$^+$ ions $\sqrt{2}r$ away, six ions $2r$ away and so on.

Adding up the contributions from all the ions in the lattice is best done on a computer. Such calculations show that the total electrical potential energy of an ion pair in a crystal of sodium chloride is 1.75 times that of an isolated pair. The factor 1.75 is called the **Madelung constant** and is a different number for different arrangements of the ions in a lattice. The predicted value of U was -8.27×10^{-19} J. Adding in the effect of all other ions changes the prediction to:

$$U = -1.45 \times 10^{-18} \text{ J}.$$

This larger value is closer to the measured value.

Repulsion energy

Let us assume that the remaining discrepancy between measurement and prediction is due to the energy of repulsion between ions. The model predicts a repulsion energy of $+1.5 \times 10^{-19}$ J. This has the correct sign and represents a little over 10% of the attraction energy at the equilibrium separation. So, our predicted value of U is now given by:

$$U = -1.45 \times 10^{-18} + 0.15 \times 10^{-18} \text{ J}$$
$$= -1.30 \times 10^{-18} \text{ J}.$$

checked from the results. We know that:

- from X-ray diffraction studies, the equilibrium separation r of sodium and chloride ions is 2.8×10^{-10} m,
- each Na$^+$ and Cl$^-$ ion carries a charge of magnitude 1.6×10^{-19} C
- the energy required to separate one mole of NaCl crystal into Na$^+$ and Cl$^-$ ions a long way apart – the lattice energy – is 780 kJ mol^{-1}.

Since 1 mol of NaCl contains 1 mol of Na$^+$ Cl$^-$ bonds, the energy of one bond is given by:

energy of one bond = the energy of 1 mol of bonds divided by the Avogadro constant ($\rightarrow$ 35)

$$= (7.8 \times 10^5)/(6.0 \times 10^{23})$$
$$= 1.30 \times 10^{-18} \text{ J}.$$

This is the energy needed to separate two ions. Work is done on the bond to separate the ions, giving U a negative sign, so:

$$U = -1.30 \times 10^{-18} \text{ J}.$$

We can now find the value of U predicted by the model and compare it with this measured value.

The potential energy U between two ions of charge Q is given by:

$$U = \frac{Q^2}{4\pi\epsilon_0 r}$$

($\leftarrow$ 16.3). Substituting the data for sodium chloride above gives:

$$U = -8.27 \times 10^{-19} \text{ J}.$$

The predicted value is about two-thirds of the measured value. This is a good estimate for a first attempt – many models in physics are much poorer predictors. However, the model can do better if we also include neighbouring ions, and the repulsion energy terms (see box).

Questions

10 For sodium chloride, the total energy can be expressed by the relation: $U = -(a/r) + (b/r^9)$.
a From the data in the main text, find a and b.
b Plot the graph for separations up to 1 nm.

11 Using the formula for the potential energy between two ions given in question 10, calculate the attraction and repulsion energies for sodium and chloride ions within a distance $2r$ of a given ion.

SUMMARY

The structure of solids can be understood in terms of the forces between the constituent particles. These forces are attractive at large distances and repulsive at small distances. At the equilibrium separation, the resultant force is zero and the potential energy between the particles is a minimum.

In certain simple cases, such as sodium chloride, the force-separation and energy-separation relationships can be investigated quantitatively. The general shape of these relationships is the same whatever the nature of the bonding between the particles.

26 INTER-PARTICLE FORCES AND ENERGIES

SUMMARY QUESTIONS

12 Explain why the potential energy of particles that attract each other is said to be negative.

13a One mole of aluminium (Al) metal has a volume of 10 cm^3. Use this to make an estimate of the mean separation of centres of Al ions in Al metal.
b The energy of atomisation of Al metal is 330 kJ mol^{-1}. Estimate the 'well depth' in aluminium in J and in eV.

14 Graphite consists of layers of carbon rings. Within the rings, the carbon atoms are linked by strong covalent bonds. Between the rings are van der Waals bonds. Use this information to explain the use of graphite as a solid lubricant.

27 FLUIDS AND TRANSPORT

Liquids are characterised by their high density, viscosity and extremely low compressibility. The values of these quantities fall between those of solids and gases. Solids are like liquids but they have a high rigidity and so cannot flow. In contrast, gases have no rigidity at all, are easily compressed and expand to fill the volume available.

In a few liquids, such as in liquid crystals, a short-range order occurs but it is much less extensive than in solids. The ability of atoms, molecules and conglomerations of these to move around the bulk of the liquid by flowing or by diffusion is similar to gases. The concepts of inter-particle forces and energies ($\leftarrow$ 26) and simple properties of electrical conductors ($\leftarrow$ 12) help us to understand why liquids behave as they do.

27.1 A MODEL FOR A LIQUID

When a solid melts to form a liquid, there is a small change in volume. The density of the liquid is only slightly different from that of the solid. So the number of neighbours in contact with a particle in a liquid cannot differ much from that in the corresponding solid. The inter-particle bonds in the liquid may have the same general form as those in a solid. However, there is a major difference – they break and reform. This allows the liquid to flow, and particles to diffuse through the liquid.

The average kinetic energy of the molecules making up a substance determines its temperature ($\leftarrow$ 10.5). The nature of a liquid is due to its particles constantly joining each other and then separating during collisions. So the average kinetic energy per molecule is comparable to the average potential energy, or **bond energy** ($\leftarrow$ 26.3), per molecule.

The shape of the energy-separation graph for the particles of a liquid depends on the type of bonding, as it does for a solid. You can investigate this most easily by boiling the liquid. The strength of the bonding, or **'well depth'** ($\leftarrow$ 26.3), can be deduced from the value of the molar latent heat of vaporisation of the liquid. An analysis ($\leftarrow$ 26.3) gives a value for water of about 1×10^{-20} J. This value is not very different from that for a solid, yet it is different enough to give liquids very different properties.

Question

1 The value of the well depth for water ($\leftarrow$ 26.3) is larger than that for most liquids. Why might this be?

27 FLUIDS AND TRANSPORT

27.2 FREE SURFACE ENERGY AND SURFACE TENSION

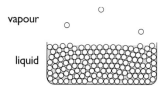

Figure 27.1 Molecules in a liquid and a vapour

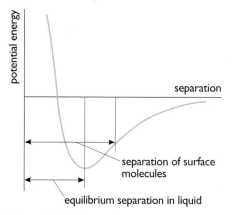

Figure 27.2 Potential energy-separation graph for a pair of molecules

One easily observable property of liquids is the presence of surface effects. Some liquids curl up into drops, while others spread out over a surface. Work has to be done to move an object through a liquid surface.

Near to its surface, a liquid's density decreases over the last few molecular diameters. This is because the molecules at the surface do not touch quite as many neighbours and are slightly further apart (see Figure 27.1). So their potential energies are higher than those of molecules in the body of the liquid (see Figure 27.2). The difference between this higher potential energy and that of molecules lower down, summed over all the molecules in the surface, is called the **free surface energy** of the liquid (← 26.3).

The existence of free surface energy explains why small drops of liquids and bubbles are usually spherical. A sphere is the shape that has the minimum surface area for a given volume. A liquid will always flow until it has the lowest possible potential energy. So the surface area reduces until the surface becomes spherical.

Animals and surface tension

If you look at the surface of a liquid, you will see that it appears to be in tension, rather like a stretched sheet of rubber. Figure 27.3 shows two small animals that rely on surface effects to survive. The pond skater in a deforms the surface but does not sink through it. The spider in b carries its air supply with it under water as a bubble. The apparent tension in the surface arises from the net attractive forces, which act sideways between the molecules in the surface.

At equilibrium, the separation of the molecules in the bulk of a liquid is such that the attractive and repulsive forces are equal and opposite, and so the net force is zero. The average separation of molecules near the surface is greater than this **equilibrium separation**. So any pair of molecules in the surface attract each other. These attractive forces give rise to **surface tension**.

a

b

Figure 27.3 a Pond skaters, **b** a water spider

Question

2 Surface tension may be defined in terms of the force required to pull a rod through a liquid surface. Show that this definition is consistent with surface tension having the unit N m^{-1}, given that the free surface energy has the unit J m^{-2}.

DYNAMIC EQUILIBRIUM

The molecules in a liquid are continually changing places: some of those in the bulk of the liquid move into the surface; some in the surface evaporate; some of those in the vapour condense; and molecules in the surface move into the bulk of the liquid. This situation is known as a **dynamic equilibrium**, and involves very large numbers of particles.

When dynamic equilibrium is established near the surface of a liquid, as shown in Figure 27.4, the rate of evaporation from the surface equals the rate of condensation, so, no net evaporation occurs. Similarly, at equilibrium, the rate of diffusion up to the surface equals the rate of diffusion away from the surface

MATERIALS: THEIR PROPERTIES AND STRUCTURE

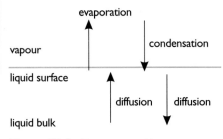

Figure 27.4 Dynamic equilibrium near a liquid-vapour surface

($\leftarrow$ 10.7). The populations of molecules in the surface layer and the vapour do not change but the particular molecules making up those populations do. The vapour above the liquid is saturated and exerts a **saturated vapour pressure**.

> ### Question
> 3 Think of several other examples of dynamic equilibrium. Write some notes identifying the various processes that compete to make up the equilibrium in each case.

WETTING OF SURFACES

Figure 27.5 shows two common observations on water. In a, the water in the glass tube has its surface or **meniscus** curved up the tube. In b, the water on the waxed surface curls up into drops and does not spread or **wet** the surface. In both cases, the free surface energy of the water is the same. The difference lies in the equilibrium established in the layer a few molecules thick where solid, liquid and vapour all meet.

To understand the shape of the surface, consider the dynamic equilibrium set up where the meniscus meets the solid (see Figure 27.6). Vapour molecules can form a thin film on the impermeable surface of the wall of a container, a process known as **adsorption**. Additionally, molecules in the liquid can adhere to the wall. When more energy is liberated on adsorption than is required for evaporation, there is energy available to spread the liquid over the surface. The final shape of the liquid surface will be that giving the minimum potential energy for the system. There are four potential energy terms involved: the free surface energies of the liquid-vapour, liquid-solid and solid-vapour interfaces; and the gravitational potential energy.

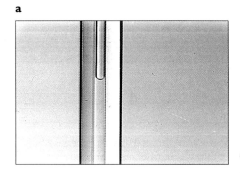

Figure 27.5 Water **a** wetting, **b** not wetting, surfaces

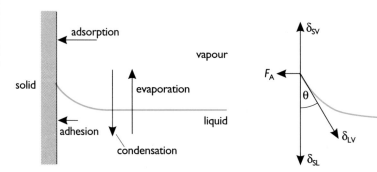

Figure 27.6 Interactions near a liquid-vapour-solid interface

Figure 27.7 Surface forces at a liquid-vapour-solid interface

Alternatively, you can consider the surface tension forces associated with the interfaces (see Figure 27.7). At equilibrium, the resultant force is, of course, zero. The surface tension forces γ_{LV} (liquid-vapour), γ_{SV} (solid-vapour) and γ_{SL} (solid-liquid) are all tangential to their respective surfaces, while the adhesion force F_A is perpendicular to the wall. The angle θ depends on the relative magnitudes of these forces. It is called the **angle of contact**.

Figure 27.8 shows the angle of contact in some common cases. The wetting of a surface depends on this angle:

- when $\theta = 0$ the liquid wets the surface. Spreading is limited by things other than surface energy
- when $\theta < 90°$ the liquid wets the surface and the meniscus is concave

interface (liquid – solid)	angle of contact (degrees)
water – glass	0
ethanol – glass	0
mercury – glass	140
paraffin oil – glass	26

Figure 27.8 Table of angles of contact between some liquids and solids

- when θ > 90° the surface is not wetted. The meniscus is convex and the liquid curls up into drops on the surface.

Agents, such as silicones, can increase the angle of contact. This is the usual method of 'waterproofing' a fabric. When water wets the fibres, rain falling on the fabric spreads over the surface and capillarity (→ 27.2) encourages water to seep through. When the fibres are coated so that the angle of contact with the fibre is above 90°, the water tends to form drops on the surface and the fabric is waterproof.

EXPERIMENT 27.1   Measuring surface tension

The apparatus for this experiment is shown in Figure 27.9a. If you pull the wire ring supported on cotton threads horizontally through the surface, you will notice that the reading on the balance drops and then recovers to its former value. From this increase in force ΔF and the length of wire L pulled through the surface, you can calculate the surface tension γ, since both 'sides' of the wire break the surface, $\gamma = \Delta F / 2L$.

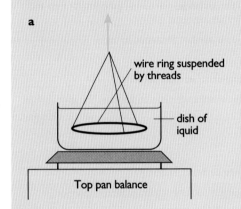

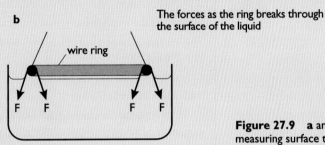

Figure 27.9 a and b Apparatus for measuring surface tension forces

As the ring in experiment 27.1 breaks through the surface, it experiences a maximum force of γ multiplied by the total perimeter of the ring (see Figure 27.9b). This force is measured by the change in balance reading multiplied by g. For example, a ring, 5 cm in diameter immersed in water with an angle of contact of zero, experiences a force of 0.02 N, producing a change in balance reading of about 2 g.

EXPERIMENT 27.2   Measuring the angle of contact

A perfectly clean microscope slide, set up as in Figure 27.10, can be adjusted until its angle is such that the liquid appears to meet the solid surface without a curved meniscus. Checking this using a low-power microscope, you can note the angle of contact.

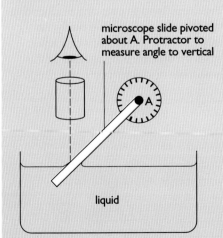

Figure 27.10   Apparatus for measuring the angle of contact

MATERIALS: THEIR PROPERTIES AND STRUCTURE

BUBBLES AND DROPS

The pressure inside a soap bubble is greater than the atmospheric pressure outside it. The excess pressure is given by:

$$\text{excess pressure} = 4 \times \frac{\text{surface tension}}{\text{radius of bubble}}.$$

Excess pressure in a bubble

a Inside a liquid

Imagine, within a liquid, a rising air bubble of radius r, volume V, surface area A, and with an excess pressure inside of P. As it rises, the radius of the bubble increases by Δr because the pressure of the liquid surrounding it is reducing. Its surface area increases too, as does its surface energy. The extra surface energy comes from the work done by the gas expanding inside by an amount ΔV.

Energy transferred = surface energy per unit area × increase in area

$$P\Delta V = \gamma \Delta A$$

where γ is the surface energy per unit area – the surface tension.

The surface area of a sphere is $4\pi r^2$. So for small increases in radius, the increase in volume is given by:

$$\Delta V = 4\pi r^2 \Delta r$$

and $\Delta A = 4\pi(r+\Delta r)^2 - 4\pi r^2$

$\qquad = 8\pi r \Delta r$ (neglecting Δr^2 and higher orders).

Substituting these expressions in the equation for energy transferred gives:

$$P \times 4\pi r^2 \Delta r = \gamma \times 8\pi r \Delta r$$

or $\qquad P = \dfrac{2\gamma}{r}.$

b Inside a floating soap bubble

A soap bubble has two liquid-vapour interfaces, one inside and another outside. So, its excess pressure P is twice that for a bubble in a liquid:

$$P = \frac{4\gamma}{r}.$$

The presence of this excess pressure is of major importance in the formation of small bubbles and droplets, including raindrops. The smaller the radius of the drop, the greater the excess pressure. So tiny drops tend to evaporate rather than grow. To form, raindrops initially require the presence of suitable nuclei on which they can grow. These nuclei need only have the dimensions of a few micrometres. Modern rain-making techniques involve seeding regions of the atmosphere containing water vapour with suitable nuclei so that water droplets grow and fall as rain.

Questions

4a Calculate the excess pressures in bubbles of water of radii **i** 5 μm, **ii** 50 μm and **iii** 0.5 mm. Take γ to be 0.07 J m^{-2}.
b Calculate the same excess pressures as fractions of 1 atmosphere, which is approximately 100 kPa.

27 FLUIDS AND TRANSPORT

Figure 27.11 Rise of liquid in a capillary tube

> **5** Cloud chambers used to detect α-particle tracks make use of the observation that droplets grow preferentially on electrically charged particles. The tracks show up as lines of droplets growing on particles charged by ionisation caused by the α-particle. Explain why a charged droplet might grow more quickly than an uncharged one.
>
> **6** An excess pressure is involved in creating bubbles in a liquid. When you watch and listen to a kettle boiling, small bubbles nucleate around the sides and bottom, grow and float to the surface. It continues boiling steadily but later becomes irregular and explosive. This is called bumping. Try to explain these observations.

CAPILLARITY

Next time you look at the surface of a glass of water, observe that the surface is flat and horizontal except for a narrow meniscus round the edge. The liquid surface remains horizontal even when the glass is tilted. However, when a narrow tube such as a **capillary tube** is placed in the water, the water rises up the tube (see Figure 27.11). Touching water with the edge of a piece of cloth leads to water spreading through the cloth. Similarly, when a sugar cube touches the surface of coffee, the coffee creeps up through the pores of the cube. All these effects suggest that surface energy can be transferred into gravitational potential energy.

We can explain these observations using the pressure difference across a curved surface, which is a feature of bubbles and drops. The smaller the radius of the capillary tube, the greater the rise. The rise also depends on the liquid's surface tension, angle of contact θ and density and on g. It is given by:

$$\text{rise of liquid} = 2 \times \frac{\text{surface tension} \times \cos\theta}{\text{radius of tube} \times \text{density} \times g}.$$

Measuring the height of capillary rise is one of the best ways of experimentally determining the surface energy. A separate experiment has to be done to measure θ.

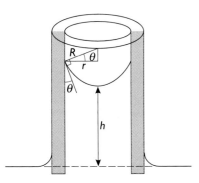

Figure 27.12 Capillary rise, showing that $r = R \cos\theta$

Capillary rise formula

Figure 27.12 shows a capillary tube of radius r with its lower end immersed in liquid. The liquid rises up the tube a distance h, ending with a spherical meniscus of radius R. Assuming that the radius of the tube is small enough for gravity not to distort the meniscus, there will be a pressure difference P across the meniscus given by:

$$P = \frac{2\gamma}{R}.$$

The liquid is pulled, by surface adhesion between the liquid and the tube, to a height that provides atmospheric pressure at the bulk liquid surface. So the pressure just below the meniscus surface is less than atmospheric pressure by the same amount, P.

The pressure due to a column of liquid of height h and density ρ is $h\rho g$ (→ 27.3). Substituting this in the equation above gives:

$$\frac{2\gamma}{R} = h\rho g.$$

MATERIALS: THEIR PROPERTIES AND STRUCTURE

The geometry of Figure 27.12 shows that the radius R of the meniscus is related to the radius r of the tube by $r = R \cos \theta$, where θ is the angle of contact. So the equation for h becomes:

$$h = \frac{2\gamma \cos \theta}{r \rho g}.$$

Questions

7 Mercury and glass have an angle of contact of 140°. What do you think happens to the meniscus and the equilibrium level in a glass tube while it is being plunged into a container of mercury?

8 Some years ago, the Parker company marketed a pen that stored ink in a cluster of capillaries. Estimate the diameter of the capillaries for the ink to rise to the top of a capillary 10 cm long when the nib is dipped in ink.

9 Describe the part played by surface energy effects in the action of the wick of a simple spirit burner or lamp.

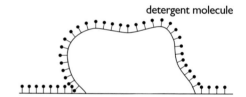

Figure 27.13 Detergent molecules clothing a grease particle

Soaps and detergents

One of the problems in washing dishes is that grease, and other materials not wetted by water, adhere to the dishes and can only be removed by mechanical abrasion. The purpose of detergent is to solve the wetting problem and let grease 'float off' the surface of the dishes.

Soaps and detergents have carbon chains with, typically, 10 to 30 carbon atoms in each molecule. The two ends of the chain are composed of different functional groups. One end attracts water molecules and is known as **hydrophilic**. The other bonds well to grease and non-polar molecules and is said to be **hydrophobic**, or water hating.

When a greasy dish is put in a solution of detergent, the detergent molecules surround the grease particles with a layer whose outer ends are hydrophilic (see Figure 27.13). This solves the wetting problem, and water can then penetrate between the dish and the grease and float it off.

Soap bubbles

When you shake pure water, bubbles form, but they collapse almost immediately when you stop shaking. The addition of quite small amounts of soaps or detergents reduces the surface energy by a factor of two, enabling you to blow bubbles with lifetimes of many tens of seconds (see Figure 27.14a).

The soap molecules are not distributed uniformly throughout the solution. They congregate at the surfaces with the hydrophobic end of the molecule outwards, as in Figure 27.14b. This soap molecule layer reduces evaporation of the water and gives the film stability.

Figure 27.14a Blowing bubbles,
b diagrammatic section through a soap film

27 FLUIDS AND TRANSPORT

27.3 PRESSURE IN FLUIDS

The kinetic model of the ideal gas interprets pressure in terms of bombardment of the walls by the particles of the gas (← 10.4). The same sort of model can be used for pressure in liquids. At any given point, pressure is not directional. **Pressure** is defined as **force per unit area**. The force resulting from the pressure is always perpendicular to the area on which it acts.

Simple demonstrations, such as Pascal's vases (see Figure 27.15), suggest that liquids flow until the pressure is the same at all points on the same horizontal line. This assumes, of course, that they are in a uniform gravitational field, such as on the surface of a planet, and are in containers open to the atmosphere.

Figure 27.15 Pascal's vases

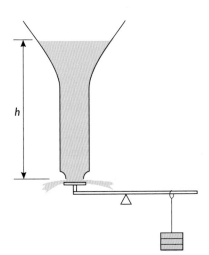

Figure 27.16 Apparatus for liquid pressure experiment

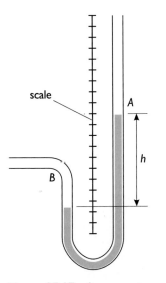

Figure 27.17 A manometer

Question

10 What do you think happens when Pascal's vases are
a closed vessels?
b in a space vehicle with the gravitational field not constant in time or position?

Pressure increases with depth. The experiment shown in Figure 27.16 shows that the pressure causing sufficient force to unseat the valve at the base depends only on the depth of liquid and not on the shape of the container of liquid. The increase in pressure with depth is given by:

increase in pressure = density × g × increase in depth

$$\Delta P = \rho g \Delta h.$$

When using this relation, note that it gives the *increase* in pressure rather than the absolute pressure.

The manometer shown in Figure 27.17 enables the pressures on two sides to be compared. With side A open to the atmosphere, the pressure P_B in side B is given by:

$$P_B = P_A + h\rho g,$$

where P_A is atmospheric pressure.

MATERIALS: THEIR PROPERTIES AND STRUCTURE

Question

11 Standard atmospheric pressure supports a column of mercury of height 760 mm. Look up the density of mercury in a data book and use the relation between pressure and depth to calculate atmospheric pressure in pascals. Show that, at normal atmospheric pressure, air exerts a force of about 10 N on each square centimetre of surface.

Relation between pressure and depth

Consider a rectangular block of liquid, height Δh and area of cross-section A, in the body of the liquid (see Figure 27.18). The force on the top is PA, where P is the pressure at the level of the top of the block. At the bottom of the block:

pressure = $(P + \Delta P)$
force   = $(P + \Delta P)A$.

Because the block is in equilibrium, the force on the top plus the weight of the block must balance the force on the bottom. The weight of the block is its volume multiplied by its density ρ and the gravitational field strength g. So:

$(P + \Delta P)A = PA + A\Delta h \rho g$

giving $\Delta P = \rho g \Delta h$.

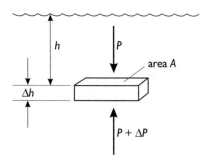

Figure 27.18 Rectangular block of liquid

A car braking system

When a car driver brakes, force exerted by the foot is applied to the piston of the master brake cylinder, shown in Figure 27.19. This increases the pressure in the **hydraulic** system. The increased pressure results in forces on the pistons of the slave cylinders. The pressure in this enclosed fluid system is the same at all points, and:

force = pressure × area.

So the forces on the various pistons are proportional to their areas. The distribution of braking effort between the wheels depends on the relative areas of the pistons and on any changes introduced by the system's pressure-reducing valves or any advanced braking system.

To provide a large braking force, as in **disc** brakes, the piston areas of the slave pistons must be larger than that of the master cylinder. Liquids are incompressible, so the volume of liquid moved must equal the slave piston areas multiplied by their movements. In disc brake, large pistons are used so the movement of the brake pad is less than 1 mm.

To provide the larger piston movement needed in **drum** brakes, the piston areas of the slave pistons are smaller than that of the master cylinder. The ratio of the forces is the ratio of the piston areas, so stopping a fast or heavy vehicle still requires a large force on the master cylinder piston. In most modern cars, the force from the driver's foot is supplemented by a force from a servo mechanism.

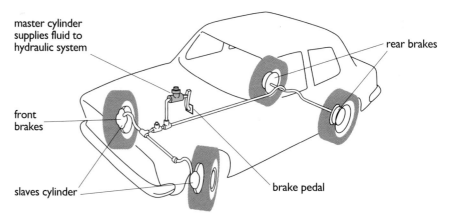

Figure 27.19 The hydraulic brake system of a car

Questions

12a Calculate the energy transferred when work is done in moving a foot brake pedal 100 mm with an average force of 10 N.
b Suppose this energy is all transferred to a car's disc brakes through four

27 FLUIDS AND TRANSPORT

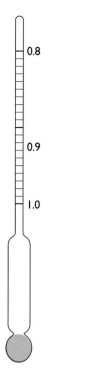

Figure 27.20 An hydrometer

50 mm × 50 mm brake pads. What forces do the four pads exert on the discs, assuming they each move 1 mm?

13 Find out:
a with reference to thermal effects, why most cars have disc brakes at the front and drum brakes at the rear
b with reference to mechanical effects, why rear brake drums are generally of the 'one leading, one trailing shoe' design.

FLOTATION

The imaginary block of liquid in Figure 27.18 was assumed to have the same density as the rest of the liquid. Otherwise it would not have been in equilibrium. The difference between the forces on the top and bottom of the block equalled the weight of the fluid block.

If the fluid block is replaced by one of different density, there will no longer be equilibrium. The difference between the forces on top and bottom will still equal the weight of the original fluid block. This weight of fluid will be displaced. So, the new block will either sink or float according to whether its density is greater than or less than that of the original fluid. This is summed up in **Archimedes's principle**, which states that:

when a body is wholly or partially immersed in a fluid, it experiences an upthrust equal to the weight of fluid displaced.

Questions

14 Figure 27.20 shows a hydrometer used to measure the density of liquids. The bottom is weighted so that it floats upright and the stem is calibrated to enable density to be read directly. Explain why the scale is non-linear and has the highest density at the bottom of the scale.

15 Describe what you would expect to observe when a cork is pushed below the surface of a bucket of water and released
a on the surface of the Earth
b in an orbiting spacecraft.

16 Explain the following observations.
a Fat people may find it easier to float in water than thin people.
b To pick an object off the bottom of a swimming pool, it may help to breathe out before ducking under the surface.

17 Cargo is stacked on the deck of a pontoon being towed along a canal as shown in Figure 27.21. The pontoon has to get under a low bridge. Explain whether more cargo should be added to sink the pontoon lower in the water, or cargo should be unloaded.

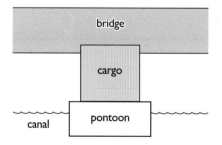

Figure 27.21

27.4 FLUID FLOW

STREAMLINED FLOW

A pressure difference inside a fluid, whether a liquid or gas, causes it to flow. When one end of a trough of water is lifted, the water flows to the lower end until the pressure difference has been removed and the surface is again horizontal (see Figure 27.22).

Consider the simple closed cooling system of a modern water-cooled motor car engine or a microbore central heating system. In both, a pump provides the

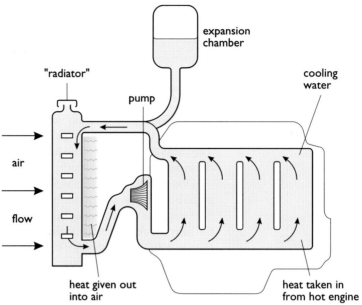

Figure 27.23 Car cooling system

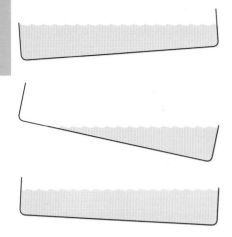

Figure 27.22 Water in tipped trough

pressure difference to circulate the hot water to radiators and back, as shown in Figure 27.23. Assume that the water filling such systems behaves as an **ideal fluid**. This means that it:

(i) is **incompressible**
(ii) has **no viscosity** or internal friction
(iii) has a laminar or **streamlined flow**.

Assumption (i) implies that the volume per second of fluid crossing an area perpendicular to the flow is the same at all points around a circuit. This is analogous to the conservation of charge in electrical circuits (← 12.1). Assumption (ii) implies that all the work put into the pump is used to circulate the water, in which there are no 'losses'. This is not a bad assumption for water at low speeds. Assumption (iii), streamlined flow, only occurs at low speeds when the water follows flow lines with no turbulence or mixing between the flow lines. Figure 27.24a shows some ideal streamlined flow patterns around various shapes, while Figure 27.24b shows some computer simulated flow patterns, indicating where turbulence tends to occur.

For a given volume of water to circulate per second, the water must move faster in a smaller bore pipe than in a larger pipe. To accelerate the water from the larger pipe into the smaller pipe, a pressure difference is needed. The mathematics of this situation was worked out by Daniel Bernoulli in the eighteenth century. **Bernoulli's equation** can be written for a given system as:

$$P + \tfrac{1}{2}\rho v^2 = \text{a constant},$$

where P is the pressure, ρ the density of the fluid and v the velocity of the fluid.

The equation predicts that when the velocity of flow is increased, the pressure is lower where the fluid is flowing faster. As the kinetic energy increases, the potential energy falls. Figure 27.25 shows a simple demonstration confirming this prediction.

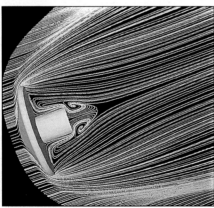

Figure 27.24a Idealised streamlined flow patterns, **b** computer simulation of flow patterns produced during aerobraking around a spacecraft designed to be used on a proposed NASA manned Mars mission

27 FLUIDS AND TRANSPORT

Bernoulli's equation

Imagine a section of tube of cross-sectional area A_1 narrowing down to an area A_2 (see Figure 27.26). The velocity of the fluid crossing A_1 is v_1, and crossing A_2 is v_2. The volume passing along the tube each second is the same in each section:

volume per second = $A_1 v_1$
  = $A_2 v_2$.

The change in kinetic energy of fluid per second is given by:

kinetic energy per second = $\frac{1}{2}$(mass per second) (velocity of fluid)^2
  = $\frac{1}{2}(\rho A v) v^2$.

The change in kinetic energy per second equals the energy transferred per second as work is done by the pressure difference. The energy transferred per second by the fluid in a tube is also given by:

energy transferred per second = pressure × volume per second
  = PAv.

Equating these gives:

energy transferred per second = change in kinetic energy per second

$$P_1 A_1 v_1 - P_2 A_2 v_2 = \tfrac{1}{2}(\rho A_2 v_2) v_2^2 - \tfrac{1}{2}(\rho A_1 v_1) v_1^2.$$

The volume per second is contant, so we can rearrange the equation like this:

$$P_1 + \tfrac{1}{2}\rho v_1^2 = P_2 + \tfrac{1}{2}\rho v_2^2.$$

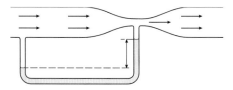

Figure 27.25 Pressure drop in a constricted tube

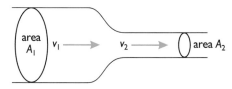

Figure 27.26 A narrowing tube

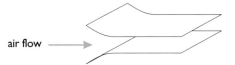

Figure 27.27

Figure 27.28 An aerofoil

Questions

18 What assumptions are made in arriving at the equation:

$$P_1 + \tfrac{1}{2}\rho v_1^2 = P_2 + \tfrac{1}{2}\rho v_2^2?$$

(See box 'Bernoulli's equation'.)

19 Explain the following in terms of Bernoulli's equation.
a When you blow between two pieces of paper held as shown in Figure 27.27, the pieces come together rather than separating.
b When air flows smoothly past an aerofoil of the shape shown in Figure 27.28, the pressure above the aerofoil is less than the pressure below.

VISCOUS FLOW

When a real fluid flows, even in the streamlined case, there are drag, or **viscous, forces** present between the layers of fluid moving along adjacent streamlines. Work has to be done against these forces to keep the fluid flowing. This causes heating and represents a 'loss' of energy, and therefore efficiency, of the fluid transport system.

Figure 27.29 shows some flow lines for a liquid flowing past a solid boundary. The lengths of the arrows represent the velocities of the various streamlines. The

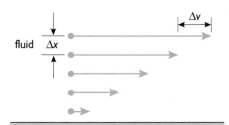

Figure 27.29 Velocity gradient in a flowing liquid

liquid in contact with the boundary is at rest. There is a velocity gradient $\Delta v/\Delta x$ with the velocity v increasing as the distance x from the boundary increases. When this velocity increase with distance is linear, we refer to the liquid as a **Newtonian liquid**. The viscous force F causes one layer of liquid to slip relative to the next. The **coefficient of viscosity** η of the liquid is defined by:

$$\eta = \frac{F/A}{\Delta v/\Delta x}$$

where F/A is the viscous force per unit area parallel to the direction of motion. This relation does not apply to turbulent flow.

EXPERIMENT 27.3 Viscous flow of water in capillary tubes

The apparatus is shown in Figure 27.30. Using tubes of several different lengths l and different diameters d, and varying the height h, you can investigate the relationships between the pressure difference ΔP, the volume flowing per second, the length of tube and its diameter.

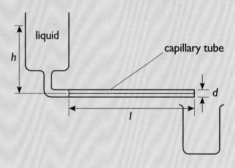

Figure 27.30 Apparatus for experiment 27.3

Figure 27.31 displays some typical results from experiment 27.3. Analysis of them shows that

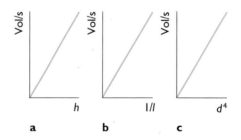

Figure 27.31 Graphs from experiment 27.3

(a) for a given tube, the volume flowing per second is proportional to the pressure difference,
(b) for a given pressure difference and diameter of tube, the volume flow rate is inversely proportional to the length of the tube,
(c) for a given pressure difference and length of tube, the volume flow rate is proportional to the fourth power of the diameter.

These can be combined into a single equation, known as Poiseuille's equation:

$$\text{volume per second} = \frac{k\, d^4\, \Delta P}{l},$$

where k is a constant. The value of k is found practically to be about $1/40\eta$ and was found theoretically by Poiseuille to be $(\pi/128\eta)$. So, from experimental measurements, you can calculate a value for η.

Because d^4 is proportional to the square of the area of cross-section A, Poiseuille's equation can be rearranged in the form:

$$\text{volume per second} = \frac{A^2}{8\pi\eta} \cdot \frac{\Delta P}{l}$$

This is one example of a **transport equation**. The rate of transport of water is proportional to (the area of cross-section of the transport system)^2, and to the pressure gradient $\Delta P/l$ transporting the water.

The viscosity of liquids varies rapidly with temperature, decreasing as the temperature increases. Treacle becomes much less viscous when warmed. Even 'viscostatic' oils become less viscous as the temperature rises, though the change is not as dramatic as with other oils. Figure 27.32 shows how the viscosity of water changes with temperature, and Figure 27.33 is a table of the viscosities of common liquids. The SI unit for viscosity is the kg m^{-1} s^{-1}. It has not yet been given its own name.

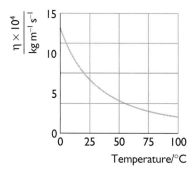

Figure 27.32 Variation of viscosity of water with temperature

liquid	viscosity $\times 10^4$/kg m^{-1} s^{-1}
water	9
mercury	15
ethanol	11
(glycerol)	
propane 1,2,3-triol	4000
machine oil	~1000

Figure 27.33 Viscosities of some liquids at 298 K

> **Question**
>
> **20** Calculate the pressure drop along a hypodermic needle of length 40 mm and a bore, or internal diameter, of 0.25 mm, used for injecting a liquid of viscosity 9×10^{-4} kg m^{-1} s^{-1} at a rate of 0.1 cm^3 s^{-1}.

Note: throughout the discussion on viscosity, we have assumed streamlined flow. Analysis of turbulent flow leads to a quite different set of relationships.

27.5 TRANSFERRING ENERGY BY THERMAL CONDUCTION AND CONVECTION

This kind of energy transfer requires a temperature difference, and so is caused by heating. It obeys a transport equation similar to that for liquid flow. Experimental measurements show that the rate of transfer P of energy through a material is proportional to its cross-sectional area A and the temperature difference ΔT. It is inversely proportional to its thickness or length l, assuming no loss of energy from the sides of the specimen.

In symbolic form the rate of transfer of energy can be written:

$P = dQ/dt$,

where Q is the energy transferred by temperature differences ($\leftarrow$ 11.2) and

$$P \propto \frac{A\Delta T}{l}.$$

A constant of proportionality can be introduced. This is a negative constant because energy is transferred from hot to cold, the direction in which the temperature decreases. So the thermal transport equation becomes:

$$P = -\lambda \frac{A\Delta T}{l}.$$

The constant of proportionality λ is called the **thermal conductivity** of the material. The unit of λ is W m^{-1} K^{-1}.

F MATERIALS: THEIR PROPERTIES AND STRUCTURE

Question

21 The transfer of energy by thermal conduction through a thermally conducting material, such as a pane of glass, can be thought of in terms of the equation:

driving force = transfer rate × resistance.

a What corresponds to driving force in this situation?
b What is being transferred and in what unit is it measured?
c What is the resistance of the system?

EXPERIMENT 27.4 Measuring thermal conductivity

a Searle's method for good conductors. The apparatus is shown in Figure 27.34a. The bar is so well lagged that energy transferred from the sides may be neglected. One end of the bar is heated and the other end is cooled by a stream of water.

b Lees's method for poor conductors. The apparatus is shown in Figure 27.34b. To maximise the energy transfer, the temperature difference is applied across a thin disc of comparatively large cross-sectional area.

Once a steady state has been reached, in either case, you can calculate λ from:

- the rate of transfer of energy through the material calculated from the rise in temperature of the cooling water and its flow rate
- the temperature gradient calculated from the temperature difference between two points in the material or across it, and
- the area of cross-section of the material.

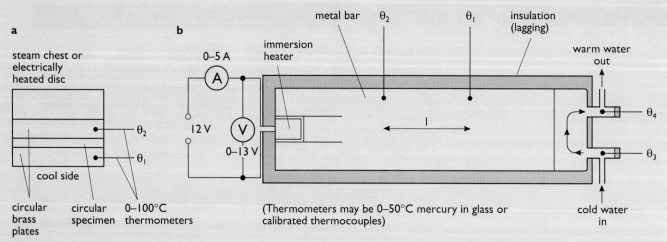

Figure 27.34 Apparatus for measuring the thermal conductivity of **a** a poor conductor, **b** a good conductor

MECHANISM OF THERMAL CONDUCTION

Temperature is directly related to the average kinetic energy of particles of matter. The transfer of energy by conduction involves the transfer of energy by internal vibrations and by collisions between particles. Figure 27.35 shows values for the thermal conductivity of some materials.

material	thermal conductivity/ W m^{-1} K^{-1}
copper	385
aluminium	238
iron	80
earthenware	2
wood	0.1– 0.4
perspex	0.2
plastic foam	0.02

Figure 27.35 Thermal conductivities of various materials

In **gases**, most of the internal energy is kinetic energy of translation of the particles. When one group of particles is heated, their kinetic energy increases. This extra energy is shared with other particles in collisions. Since the density of gases is low, the amount of energy transferred is small. So gases are very poor thermal conductors.

A similar process is responsible for thermal conduction in **liquids**. Because the density is much higher, the collisions are much more frequent and the inter-particle forces are larger. So, the thermal conductivities of liquids are much higher than those of gases.

Solids show a wide variation of thermal conductivities. Metals are good conductors, while ceramics and glasses are poor conductors, and molecular solids are very poor. The variation in thermal conductivity over different types of solid parallels that of electrical conduction. The good electrical conduction of metals is interpreted in terms of a free electron 'gas' which can move through the metal lattice. These free electrons also make the major contribution to thermal conduction, transferring kinetic energy and sharing it out in their collisions with the lattice.

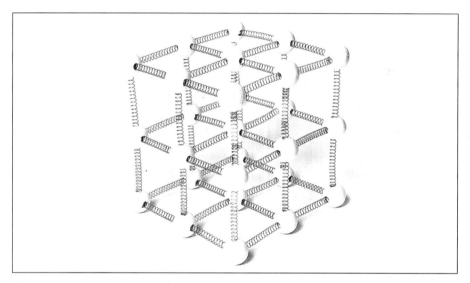

Figure 27.36 A model solid

The other mechanism for energy transfer in solids is through vibrations of the lattice itself. Imagine a lattice of particles as shown in Figure 27.36. Each particle has kinetic and potential energy of vibration. When the plane of particles at one end is heated, energy will be transferred through the bonds linking the particles until it is shared out between all. The stronger the bond, the more effective is this method of energy transfer. Diamond is an electrical insulator because it has no free electrons. However, it has a very strongly bonded lattice and so is an excellent thermal conductor. By the same reasoning, molecular solids, such as plastics, with weak Van der Waals bonding are very poor thermal conductors.

CONVECTION

Most fluids, whether liquids, gases or vapours, are poor thermal conductors. Convection is the dominant means of exchanging internal energy between parts of fluids at different temperatures. Less dense parts of the fluid experience a net upthrust in the Earth's gravitational field and tend to rise, setting up a convection current. Gravity is required to cause the upthrust. Without it, natural convection cannot occur.

F MATERIALS: THEIR PROPERTIES AND STRUCTURE

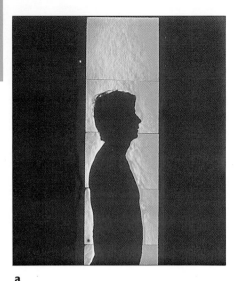

a

Convection is a very effective method of transferring energy by heating. This can be seen in examples as diverse as the atmosphere (see Figure 27.37a) and a beaker of water (see Figure 27.37b). In both these cases the source of heating is at the bottom. This is necessary for convection to take place in fluids where the density decreases as the temperature rises.

Convection can be minimised by heating the fluid from the top or trapping the fluid in pockets too small for significant currents to be set up. Using cellular materials for **thermal insulation** has an added advantage. To transfer energy from solid to fluid requires a temperature difference so that the fluid is always cooler than the hot solid and warmer than the cold solid. The greater the number of interfaces between solid and fluid, the smaller the proportion of the temperature difference available to set up convection currents.

Questions

22 Would you expect convection currents to occur inside an orbiting spacecraft's atmosphere when it is travelling freely, assuming that one part of the atmosphere is hotter than another?

23 The process of convection can best be explained in terms of the bulk properties of a fluid. How would you attempt to explain convection using the kinetic model of a fluid?

24 The density of water increases as the temperature rises from 0 °C to 4 °C. What convection currents will be set up in a pond when:
a ice first forms on the surface?
b all the water is below 4 °C?

Keeping warm in winter

Reducing the transfer of energy from inside to outside plays a major part in energy management schemes for the home.

Energy is conducted through a wall because of

- the temperature difference across the wall,
- the temperature differences between the air in the room and the wall, and
- the temperature differences between the outside of the wall and the air outside.

This energy transfer occurs due to the temperature difference between inside and outside. Most practical methods of reducing it rely on incorporating pockets of air in the walls and then subdividing these so that convection becomes negligible.

The industrial unit for the rate of thermal energy transfer is the U value. This is equal to λ/l, so the thermal transport equation becomes

$$P = -UA\Delta T.$$

The U value of a sheet of material is the rate of transfer of energy per square metre per kelvin temperature difference across it. Building regulations require that walls have a U value of less than $1 \text{ W m}^{-2}\text{K}^{-1}$.

The thermal conductivity of brick is about $1.2 \text{ W m}^{-1}\text{K}^{-1}$. So a brick wall two bricks thick, each of standard thickness 230 mm, has a U value of about $2.5 \text{ W m}^{-2}\text{K}^{-1}$. This high value can be reduced to an acceptable value by building the wall with a cavity between the two layers of brick. It may be reduced further by filling the cavity with foam whose effective thermal conductivity is over 100 times less than that of brick. Figure 27.38 shows the U values of various materials. Most good thermal insulators are cellular in construction, such as expanded polystyrene foam. Air has a very low thermal conductivity

b

Figure 27.37 Convection currents in **a** the atmosphere, **b** a beaker of coloured water

27 FLUIDS AND TRANSPORT

and the air trapped by the cells ensures that the material is an exceptionally good insulator.

The 'thermal resistance' of rendering, plaster and other wall surfacing may make a larger contribution to the insulation than the poor thermal conductivity of the brick. Crumpled aluminium foil can have a low U value because, although aluminium is a good thermal conductor, crumpled foil has many surfaces, each of which has a high thermal resistance.

material	U value/W m^{-2} K^{-1}
120 mm-thick brick wall	6
250 mm-thick brick wall with cavity	1
20 mm-thick soft wood	7.5
4 mm-thick glass	250
4 mm-thick glass including surface effects	5
double-glazing window including surface effects	2.5
100 mm-thick foam lagging	0.2

Figure 27.38 U values of building materials

Questions

25 Energy inside a house may be lost to the outside by thermal conduction through walls, windows, doors, the roof and the floor. Given the U values of each of the above, how would you attempt to calculate the total energy transferred when the inside of the house is maintained at a constant temperature above the surroundings?

26 Estimate the rate of energy transfer by thermal conduction into a refrigerator, from the following data:
Surface area of refrigerator walls is 4 m^2.
Walls are of thickness 5 cm and of thermal conductivity 0.01 W m^{-1} K^{-1}.
Room temperature is 17 °C and temperature inside refrigerator is 5 °C.

27 A room has 2 m^2 of window area singly glazed with glass 4 mm thick. Estimate the energy transferred per second across the window due to a temperature difference of 12 °C between the room and outside air. The thermal conductivity of glass is 1 W m^{-1} K^{-1}. Is your estimate realistic?

28a Explain why cavity-filled walls help to reduce energy loss.
b Is the effect of adding the thermal resistances of each wall and the filled cavity the same as adding their U values?

29 Make a list of the uses of cellular materials for thermal insulation, such as string vests, survival bags.

27.6 THE FLOW OF ELECTRIC CHARGE

In an electrical circuit, charge is transported down an electrical potential gradient. Experiments suggest that the current I through a material is proportional to its cross-sectional area A and the voltage V across it. It is inversely proportional to the length l. This leads to another transport equation, which can be written:

$$\text{charge per second} = \text{current} = \frac{-\sigma A V}{l},$$

where σ is the **electrical conductivity** ($\leftarrow$ 12.3).

MATERIALS: THEIR PROPERTIES AND STRUCTURE

> **Question**
>
> **30** The transport equation for the flow of charge along a conductor due to potential difference is similar to that for the flow of energy through a material due to temperature difference. Each equation takes the general form:
>
> driving force = rate of flow × resistance.
>
> **a** In each case:
>   **i** what is the 'driving force' associated with the flow?
>   **ii** what is it that is flowing?
>   **iii** what is the resistance associated with the flow?
> **b** Look at the equation for the **B**-field inside a long solenoid (← 18.3). What similarities does magnetic flux share with the above situation? What are the differences?
> **c** Can the general equation above be adapted to cater for the flow of water along a pipe?

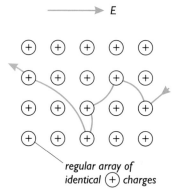

Figure 27.39 Typical path of a free electron in a metal

CONDUCTION IN METALS

A simple model of a metal is a lattice of positive ions held together by a 'gas' of free electrons (← 26.3). The free electrons have random speeds, known as **thermal speeds**, of the order of 10^5 m s^{-1} and may travel 50 or more ionic diameters in between collisions with the lattice. The Hall effect (← 17.4) shows that electrons carrying negative charges are the main charge carriers in metals.

When a potential difference is applied across a piece of metal, an electric field is set up. The electrons accelerate in a direction opposite to the applied field. The path of a typical electron can be imagined as similar to that in Figure 27.39. In a collision, the electron retains its high thermal speed but, on average, loses the extra component of velocity obtained from the electric field. In a steady state, therefore, the application of a field results in a small drift velocity of about 1 mm s^{-1} in the free electron gas (← 12.1).

CONDUCTION IN SEMICONDUCTORS

A crystal of pure silicon or germanium is a fairly good insulator because it has very few free electrons available for conduction. Semiconductors become much better conductors as their temperatures rise (← 12.5). This increase in conductivity occurs because the increased internal energy at higher temperatures releases more free electrons from the silicon or germanium lattice. A semiconductor's increase of conductivity with temperature is exponential (→ 28.4).

Commercial semiconductors are always **doped** by carefully adding controlled amounts of impurities to the pure base material. Measurements of the Hall effect in semiconductors suggest that in some materials, known as *n*-type materials, the charge carriers are predominantly negative. In *p*-type materials, the Hall effect indicates that the carriers are predominantly positive.

Normally such a positive charge carrier is explained as the absence of an electron, or **hole**, which travels in the opposite direction to electrons in an electric field. When a hole is filled by an electron from a neighbouring atom, the vacancy moves to the neighbouring atom. This process continues in sequence, so that the vacancy moves through the lattice, giving the impression of a moving positive charge. To make this model work quantitatively requires quantum mechanics which describes how energy quanta interact between themselves and with

27 FLUIDS AND TRANSPORT

matter. It is a highly exacting discipline providing numerous mathematical models that explain physical phenomena at a fundamental level.

Silicon is in Group IV of the periodic table and has a diamond-type structure. The addition of a Group V impurity, such as phosphorus or arsenic, dopes the structure with atoms possessing an extra electron to give *n*-type material. These extra conduction electrons increase the conductivity considerably. Doping the structure with Group III atoms, such as gallium and indium, introduces atoms deficient in electrons and creates *p*-type material. Similarly, the number of holes is greatly enhanced and the conductivity is increased.

> ### Question
>
> **31** Sketch diagrams to illustrate how electrons moving from one vacancy to the next in sequence can give the impression of a positive vacancy or hole moving in the opposite direction to the movement of electrons.

CONDUCTION IN LIQUIDS

Free electrons do not exist to any great extent in most liquids, apart from mercury. In liquids, the charge carriers are ions. Those that do not possess ions are known as **non-polar liquids** and are very good insulators. Experiments 27.5 and 27.6 illustrate the part played by ions in conduction in liquids.

EXPERIMENT 27.5   The electrical resistance of liquids

By setting up the circuit shown in Figure 27.40a you can find the resistance of a liquid between two carbon rod contacts. With pure, or de-ionised, water, you should find that the current is too small for the lamp to light. However, the lamp lights with dilute sulphuric acid or barium hydroxide solution, showing that they are good conductors.

If you add small amounts of dilute sulphuric acid to a solution of barium hydroxide in the beaker, you will see that, as barium sulphate solid is precipitated, the lamp dims. Adding further acid causes the lamp to glow again. Data such as that in Figure 27.40b enables you to find how the resistance of the solution depends on the number of ions present.

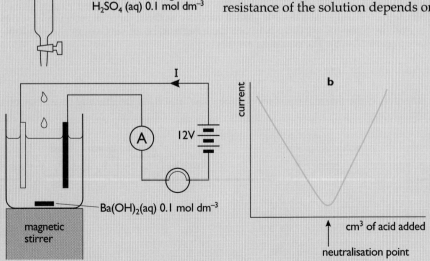

Figure 27.40a Apparatus for, and **b** results from experiment 27.5

Analogy between electrical and thermal conduction

Many problems concerning the transfer of energy by thermal conduction can be solved by analogy to the flow of charge in a simple electrical circuit containing a battery and resistors (← 12).

The basic equation for the charge passing through a resistor per second is:

$$V = IR = \frac{Q}{t} \cdot R.$$

For a uniform resistor of conductivity σ, length l and area of cross-section A:

$$R = \frac{l}{\sigma A}$$

so

$$\frac{Q}{t} = \frac{\sigma A V}{l}.$$

The basic equation for the energy transfer per second, Q/t, through a slab of material by thermal conduction is similar:

$$\frac{Q}{t} = \frac{\lambda A \theta}{l}$$

for a slab of thickness l, area of cross-section A and thermal conductivity λ, with a temperature difference θ between its surfaces.

Comparing the two equations shows that voltage or potential difference is equivalent to temperature difference. It also shows that we can define a **thermal resistance** given by:

$$R_{th} = \frac{l}{\lambda A}.$$

This analogy can be used to solve, for example, the problem of thermal conduction through the wall of a house containing, say, one window and a door. The wall acts like three resistors **in parallel**, one for the brick R_b, one for the glass R_g and one for the wood R_w as in Figure 27.41a. Knowing the area, thickness and conductivity of each material, we can calculate the total thermal resistance R_T of the wall from:

$$\frac{1}{R_T} = \frac{1}{R_b} + \frac{1}{R_w} + \frac{1}{R_g}.$$

We can then find the energy transfer using:

$$\frac{Q}{t} = \frac{(\theta_1 - \theta_2)}{R_T}.$$

Similarly, we can treat a double glazed window as three thermal resistors **in series** to find the temperature drop between the inner and outer surfaces of the window (see Figure 27.41b). The equation for this is:

$$(\theta_1 - \theta_2) = \frac{Q}{t} \cdot (2R_g + R_a)$$

where R_g is the thermal resistance of the glass, and R_a is the thermal resistance of the air gap.

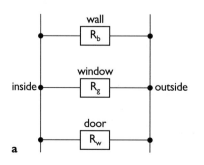

a

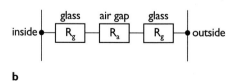

b

Figure 27.41 Thermal resistors **a** in parallel, **b** in series

27 FLUIDS AND TRANSPORT

EXPERIMENT 27.6 Migration of ions in a liquid

By using the circuit shown in Figure 27.42, you can see the migration of coloured ions and complexes. With the filter paper wet with ammonia solution, and two small crystals of, say, copper (II) sulphate and potassium manganate (VII) placed as shown, the blue colour of the copper ammonium complex will move towards the negative contact, and the purple of manganate (VII) towards the positive.

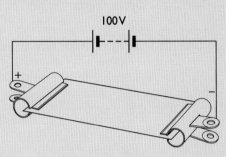

Figure 27.42 Circuit for experiment 27.6

SUMMARY

The properties of liquids are intermediate between those of solids and gases.

Liquids have densities comparable with solids but can flow like gases. They possess free surface energy and surface tension, and can adhere or adsorb on to solid surfaces.

Because the particles of a liquid are not fixed in a lattice, its properties are discussed in terms of dynamic equilibria between competing processes.

The pressure in a liquid varies with the depth of immersion h and is given by:

$\Delta P = h\rho g$,

where ρ is the density.

The flow of an ideal fluid may be modelled by Bernoulli's equation:

$P + \frac{1}{2}\rho v^2 =$ a constant.

Both air and water at low speeds approximate to ideal fluid behaviour.

Viscous forces oppose the flow of fluids. To overcome them, energy must be transferred to the fluid. This energy is dissipated, and increases the internal energy of the fluids and their surroundings.

The transport of energy by temperature differences through materials involves thermal conduction and, in fluids, convection.

Electrical conduction is the transport of charge. In solids, the only moving charges are electrons. In liquids, ions with both signs of charge may move.

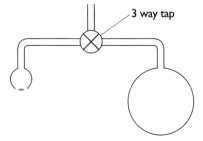

Figure 27.43

SUMMARY QUESTIONS

32 A T-piece tube has different sized soap bubbles blown on each arm, as in Figure 27.43. Discuss what will be observed when the tap is turned to connect the two bubbles.

MATERIALS: THEIR PROPERTIES AND STRUCTURE

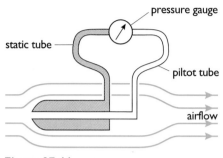

Figure 27.44

33 The density of solid polystyrene is 980 kg m^{-3}. A student measures the density of a block of polystyrene foam and arrives at a value of 20 kg m^{-3}. Explain his result.

34 Figure 27.44 shows a type of air-speed indicator called a pitot-static tube. When mounted in the airflow around an aeroplane, the pressure gauge reads the pressure difference between the static and moving air. Use Bernoulli's equation to calculate the pressure difference on the gauge when the aeroplane is moving at 120 m s^{-1} in air of density 1.4 kg m^{-3}.

35 A car aquaplaning on a wet road is uncontrollable because the large frictional force between the tyre and the road is replaced by the tiny shear force on the water film between the tyre and the road. Calculate this shear force to slide a tyre with an area of 0.025 m^2 resting on a film of water 0.1 mm thick when a car aquaplanes at 20 m s^{-1}. Take the viscosity of water to be 1.0×10^{-3} kg m^{-1} s^{-1} and assume that water behaves as a simple Newtonian fluid.

36 Figure 27.45 shows the temperature distribution near a double glazed window of a room maintained at 20 °C when the outside temperature is 0 °C. Both panes of glass are of thermal conductivity 1.4 W K^{-1} m^{-1}. Calculate the rate of heat loss through a window of area 2.5 m^2 under these conditions.

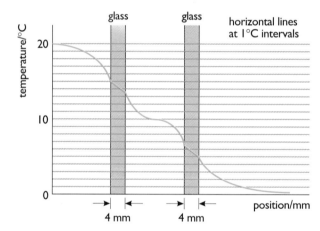

Figure 27.45

28 THERMODYNAMICS

Thermodynamics grew out of studies of the conservation of energy in mechanical and thermal systems is the mid-nineteenth century (← 9 ← 11). It deals with systems, which may be as large as the Universe, or as small as a simple cell.

A system has properties such as pressure and temperature and all its thermodynamic processes may be observed.

Thermodynamics is concerned with the thermal and mechanical properties of matter in bulk, not atoms and molecules. Its predictions are independent of the microscopic model used. For example, we cannot ascribe a meaning to the temperature of a single molecule, or the pressure exerted by a single atom. Quantities such as temperature and pressure only have a meaning for large assemblies of particles.

Knowing about kinetic and potential energy (← 9.2) and familiarity with the thermal properties of matter and the model of an ideal gas (← 10) are very useful in studying thermodynamics.

28.1 ENERGY AND TEMPERATURE

The transfer of energy from one system to another forms the basis of the **zeroth law of thermodynamics**:

when systems A and B, and B and C are in thermal equilibrium with each other, systems A and C are in thermal equilibrium.

No **net** transfer of energy takes place when two systems are in thermal equilibrium with each other. However, energy may be freely exchanged between the systems.

The zeroth law gives rise to the concept of **temperature**. Two systems are said to be at the same temperature when they are in thermal equilibrium. Thermometers depend upon this being the case. When there is a net rate of energy transfer from one system to another, the system from which energy leaves is at the higher temperature.

You can change the temperature of a system by adding or removing energy from the particles that make it up, so altering its internal energy. It is also possible to increase the internal energy of a system by doing **work** on it. Two surfaces of an unlubricated bearing rubbing against one another cause an increase in the internal energy of the particles that make up the bearings. Work done on the system causes this increase (← 9.1).

Energy ΔE, can be transferred to or from a system by heating or by doing work (← 11.2). A sign convention distinguishes the two directions (see Figure 28.1).

Figure 28.1 Sign convention for energy transfer by heating or doing work

F   MATERIALS: THEIR PROPERTIES AND STRUCTURE

Traditionally, because the origins of thermodynamics lie in designing steam engines, the sign convention used is:

energy **supplied to** the system through **heating** is denoted by **Q**
energy **output by** the system through heating is denoted by **−Q**
energy **output by** the system as useful **work** is denoted by **W**
energy **supplied to** the system as **work**, is denoted by **−W**.

Imagine a volume of air trapped in a closed cylinder by an air-tight piston. The air in the cylinder is heated so that an amount of energy Q transfers into the gas. The air expands and does work, transferring energy W. This means that the piston and the mass attached to it are raised to a higher potential energy in the Earth's gravitational field. This increase in potential energy, ΔE_p, is W (see Figure 28.2).

The remainder of the energy Q is given to the gas as increased internal energy ΔU of the particles that make up the gas:

$$Q = \Delta U + W.$$

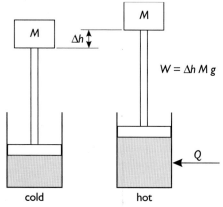

Figure 28.2 Increase in W by doing work in Earth's gravitational field

This is a statement of the **first law of thermodynamics**, which follows from the conservation of energy (← 11.3). Note that, in this statement of the first law:

Q is the net energy transferred **into** the system by means of temperature differences alone – by a **heating** process
ΔU is the **increase** in internal energy of the system
W is the net energy transferred out as **work** is being done **by** the system.

By the sign convention above, the values of Q, ΔU and W here are positive. They will be negative in the following circumstances:

- when a system **loses** a net amount of energy by conduction, convection or radiation, Q is negative,
- when a net amount of work is done **on** the system, W is negative
- when the internal energy of the system **falls**, ΔU is negative.

In the case of freezers or refrigerators, for example, all three quantities are generally negative.

28.2 HEAT ENGINES AND REVERSIBILITY

Heat engines do useful work while transferring energy from a hot system to a colder one (see Figure 28.3). Heat engines take many forms, including steam, petrol, diesel and stirling engines (see Figure 28.4). They operate by taking a gas through a cycle of changes which is initiated by transferring energy into the gas. In the case of a petrol engine, energy is transferred in through the combustion of a petrol-air mixture.

Seebeck and Peltier effects in semiconductors

The thermoelectric effect was discovered by Thomas Johann Seebeck in 1821. When two dissimilar metals are joined together a small p.d. exists when the joints are at different temperatures. This effect occurs also in semiconductors and is the basis of thermoelectric generators.

The number of free electrons per cubic metre in a material depends on both the temperature and the composition of the material. Thus, pieces of different semicon-

28 THERMODYNAMICS

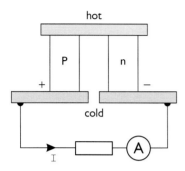

Figure 28.3 The Peltier Effect

a

b

c

Figure 28.4a A stationary steam engine, **b** a petrol car engine **c** a diesel car engine

ductor in isothermal contact will have different electron densities and hence will exhibit a small p.d. which is a function of temperature.

When a current is passed through a pair of such junctions (a thermo-couple) a temperature difference is produced between the junctions. This effect was noted by Jean Peltier in 1834 and today banks of thermo-couples are used as heat pumps or thermoelectric generators. A practical arrangement uses p-type and n-type semiconductors as the couple. Many couples are connected electrically in series and thermally in parallel. The device generates a p.d. when a temperature difference is maintained across the slab (Figure 28.5). When a current is driven through the slab it transfers thermal energy from the cold to the hot side. The heat sink on the hot side has to remove the energy transferred plus the energy dissipated in the slab. A typical slab of size 30 mm × 30 mm × 6 mm contains 31 couples and is capable of pumping 18 W when a current of 9 A is passed through it. Such Peltier effect heat pumps are used to cool sensitive electronics.

A refrigerator is a form of heat engine 'in reverse' (see Figure 28.5). Energy is transferred from a cold reservoir to a hotter one, with work being done on the system to achieve this.

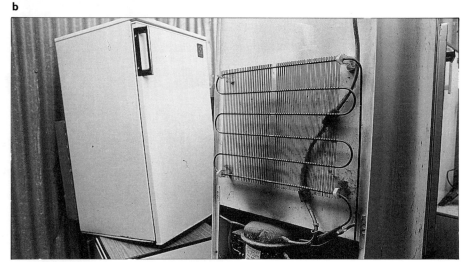

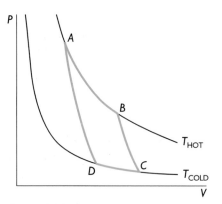

Figure 28.5a Block diagram, and **b** photograph of, refrigerators

Figure 28.6 Pressure-volume diagram for the Carnot cycle

Carnot cycle

The Carnot cycle is a simple, idealised example of a heat engine. Imagine taking a fixed mass of an ideal gas around a closed loop so that the final state of the gas is the same as its initial state. Assume that the changes occurring in the system do so **reversibly**. This means that they take place slowly, in infinitesimally small steps, so that the system never departs from an equilibrium state. The closed loop in Figure 28.6 shows the state of the gas at any instant. The engine operates between two temperatures T_{HOT} and T_{COLD}.

In **section A-B**, starting at point A the gas expands at constant temperature T_{HOT} to B. The internal energy of the gas remains constant at constant temperature. Energy Q_1 is absorbed from the hot source and external work by the engine transfers this energy to the engine output.

Section B-C is an **adiabatic** expansion (← 11.3). So, overall, no energy enters or leaves the gas. Its internal energy is reduced by the energy needed to do the work in expansion.

On **section C-D** the gas is compressed **isothermally** (← 11.3) at T_{COLD}. During this section, work is done on the gas, rejecting energy into the outside world. The internal energy produced, Q_2, is removed from the gas to the cold reservoir.

The **section D-A** is an adiabatic compression in which work is done on the gas, transferring energy to increase its temperature from T_{COLD} to T_{HOT}. This restores the gas to its original state at A.

So, at the end of a cycle, the gas has returned to its original state, its net internal energy is unchanged and ΔU is 0. Writing the first law of thermodynamics in the form:

$$Q = \Delta U + W,$$

it is clear that the useful output energy W is equal to the energy Q supplied. Since

$$Q = Q_1 - Q_2,$$

the output energy is the difference between the energy supplied by the hot source and that rejected to the cold reservoir. The efficiency η of a heat engine is defined by:

$$\eta = \frac{\text{energy output by the engine in doing work}}{\text{energy supplied to the engine}}$$

$$= \frac{Q_1 - Q_2}{Q_1}.$$

28 THERMODYNAMICS

The energy transfer in an isothermal expansion or compression is **proportional to the kelvin temperature**. So Q_1 and Q_2 are proportional to T_{HOT} and T_{COLD} respectively, and the efficiency is given by:

$$\eta = \frac{T_{HOT} - T_{COLD}}{T_{HOT}}.$$

EFFICIENCY

Heat engines are necessarily inefficient. Their **efficiency** η is related to the kelvin temperatures of the hot and the cold reservoirs and is given by:

$$\eta = \frac{T_{HOT} - T_{COLD}}{T_{HOT}}.$$

This relation was first suggested by Sadi Carnot in 1824 (see box on the **Carnot cycle**). Note that:

(i) the only way to increase the efficiency of an ideal heat engine is to increase the temperature difference across it, and
(ii) the efficiency will only approach 100 % when the temperature of the cold reservoir approaches the **absolute zero of temperature**, 0 K.

Real heat engines, such as steam, petrol and diesel, operate around different cycles and have efficiencies less than that of the Carnot cycle engine. A steam engine follows the Rankine cycle; petrol the Otto cycle; and a diesel engine, the Diesel cycle.

A steam turbine in a large power station has an inlet steam temperature T_{HOT} of 900 K and a condenser temperature T_{COLD} of around room temperature, 300 K. Its Carnot cycle efficiency is, therefore:

$\eta = (900\ K\ to\ 300\ K) / (900\ K)$

$= 0.67\ or\ 67\ \%.$

This is much larger than its actual efficiency of around 0.4. It is possible to engineer considerable improvements in the design of heat engines but the Carnot cycle efficiency provides an upper limit.

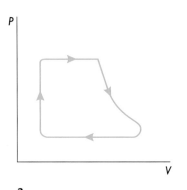

a

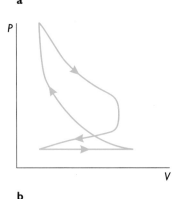

b

Figure 28.7 Pressure-volume diagrams for **a** the Rankine cycle for steam engines, **b** the Otto cycle for petrol engines

Questions

1 Figure 28.7 shows pressure-volume diagrams for the Rankine cycle of steam engines and the Otto cycle of petrol engines. Describe the various stages of each cycle, identifying where energy is supplied and absorbed, and where energy is transferred as work is done on and by the gases in the cylinder.

2 When the working gas is taken round the Carnot cycle in the opposite direction to that in Figure 28.6, the engine behaves as a heat pump or refrigerator.
a Explain how this happens.
b A heat pump extracts energy from a reservoir at 7 °C and delivers it at 67 °C. From the definition of the efficiency of a heat engine given above, express the energy transfer through work input to the heat pump as a fraction of the energy delivered.

MATERIALS: THEIR PROPERTIES AND STRUCTURE

SECOND LAW OF THERMODYNAMICS

In 1849–51, three people separately put forward a second law of thermodynamics to follow on from the first law and bring in Carnot's ideas. The three were Rudolph Clausius, William M. Rankine and William Thomson (who later become Lord Kelvin). There are many ways of stating the second law, three of which are:

(i) No heat engine can operate on a closed cycle whose **only** result is to transfer the internal energy of the working fluid, by doing work, into potential or kinetic energy.
(ii) No heat pump can transfer internal energy from a cold reservoir to a hot reservoir without an external work input.
(iii) If the first law says 'you cannot win', then the second law states that 'you cannot break even except at absolute zero'.

The second law allows us to give a numerical value to temperature from which the efficiency of the ideal heat engine is defined. This temperature scale has an absolute zero but no upper limit. It is important that this thermodynamic scale, known as the **Kelvin Scale**, is identical to the ideal gas temperature scale (← 10.2), and it turns out to be so.

28.3 CHANCE AND PROBABILITY

The laws of thermodynamics are not dependent on any particular microscopic model of matter. However, such a microscopic model helps us to understand the underlying concepts.

In the ideal gas model a gas is imagined as a large assembly of particles in rapid random motion. One of the model's assumptions is that the kinetic energies of these particles are distributed in a particular way known as the **Maxwell-Boltzmann distribution** (← 10.3). The model extends these ideas to ask questions about the distribution of energy and the chance of events happening.

IRREVERSIBLE PROCESSES

Everyday, you observe some processes to be irreversible, or one-way. Dough placed in the oven and baked becomes bread. The reverse, bread becoming dough, is never observed. Again, when you drop a stone it falls to the ground. The reverse, a stone spontaneously jumping up from the ground and lodging in your hand, is not observed to occur. When a film of such events is played backwards it is a source of amusement because it obviously conflicts with common experience. However, were you to film the pendulum of a clock swinging to and fro, you would have much more difficulty in deciding whether the film was running forwards or backwards. In these, and all other cases, energy is conserved. So energy alone is not a sufficient criterion to decide what takes place.

DIFFUSION: A ONE-WAY PROCESS

You can build a **statistical model** to predict the direction of one-way processes, starting with a simple example of **diffusion** (← 10.7). Figure 28.8 shows a gas diffusion experiment with bromine vapour. A drop of liquid bromine at the bottom of the tube evaporates and, after some time, the brown vapour fills the tube. No matter how long you wait, you would never observe the vapour condense back into a drop at the bottom again.

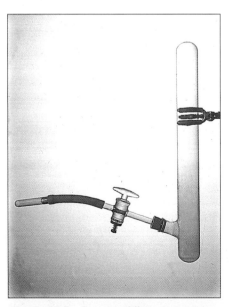

Figure 28.8 Bromine diffusion experiment

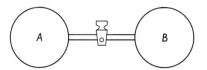

Figure 28.9 Two identical containers A and B connected through a valve

28 THERMODYNAMICS

Question

3 A person wearing a distinctive perfume enters a room. Describe the process of diffusion by which the scent quickly spreads throughout the room. Is it possible for the process to work in reverse? Is it probable?

To model the process of diffusion, imagine two identical containers A and B, as shown in Figure 28.9, connected by a tube with a valve. At the start all the gas is in A, and B is evacuated. Opening the valve allows gas to spread into B. The question is: 'how many particles end up in A and how many in B?'. The equilibrium is a dynamic one so, although the actual populations change, the average remains the same. In fact, bulk measurements show that the pressures in the two containers are the same. When both are at the same temperature, you might predict that, on average, they contain the same number of particles.

Suppose there are only two particles shared between the two containers A and B. The only ways of arranging two particles are shown in Figure 28.10a. Assume that all four ways are equally likely to occur and that each occurs in a random fashion. Then the chance of **both** being in container A is $(\frac{1}{2})^2 = \frac{1}{4}$. In a cine-film of the particles in the containers, you would see, as you would expect, both particles in A on one quarter of the frames.

Figure 28.10b extends the argument to three particles in A and B and Figure 28.10c to four particles. With three particles, all will be in A on $(\frac{1}{2})^3 = \frac{1}{8}$ of the occasions. With four particles, all will be in A on $(\frac{1}{2})^4 = \frac{1}{16}$ of the occasions. As the number of particles increases, the chance of finding all of them in one container rapidly becomes very small indeed. With 10 particles, the probability has shrunk to one in 1024. With one sixth of a mole, containing 10^{23} particles, the probability is only one in 2 raised to the power 10^{23}. This is a very large number indeed, which, in standard form, is 3×10 to the power 10^{22}. As a figure, it is 3 followed by ten thousand million million million zeros!

How many times would you expect to find **equal** numbers of particles in the containers? A look at Figure 28.10 shows that equal numbers, or only one different, occur in 2 out of 4 cases when there are two particles, 6 cases out of 8 with three particles, and 6 cases out of 16 with four particles. So the probability of finding equal, or almost equal, numbers of particles is much greater than that of finding all particles in one container.

When the number of particles has risen to 10^{23} the number of ways leading to equal numbers in each container is very large. The number of ways leading to almost equal numbers in each container is enormous. With probabilities this big, we are really dealing with certainties.

It is assumed that in the model all arrangements of particles are equally likely, and occur at random.

Question

4a Estimate the number of particles of air in a telephone kiosk.
b How many ways are there of arranging them so that a particle can go either in the top half or the bottom half of the kiosk?
c Of this number of ways, how many will place all of the particles in the bottom half?

Using an ordinary calculator, it is not possible to evaluate part **b** of the question. To obtain some idea of the scale of the problem, imagine there are just 100 particles in the telephone kiosk.

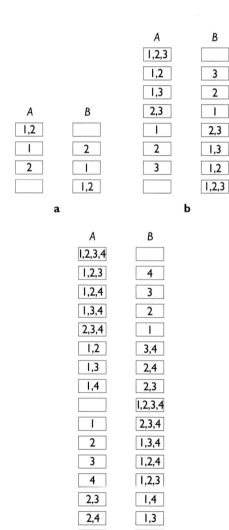

Figure 28.10 Ways of arranging between containers A and B **a** two particles, **b** three particles, **c** four particles

F MATERIALS: THEIR PROPERTIES AND STRUCTURE

> **d** What now is your answer to **b**?
> **e** What now is your answer to **c**?
> **f** Suppose the particles re-arrange themselves into a new position every microsecond. How long might you have to wait before there were no particles in the top half of the kiosk?

The model leads to two very important predictions:

(i) we will never observe an event requiring an arrangement of particles that can only occur in one way because the chance of that one critical way occurring is infinitesimally small, and

(ii) we will observe an event requiring an arrangement of particles that can arise in many ways because the probability of one of those ways occurring is finite.

We never observe all the gas in one container and a vacuum in the other. Equal pressures and equal amounts of gas in the containers occur when both are at the same temperature. Of course, there are only approximately equal numbers of gas particles in each container. Dynamic equilibrium exists between the containers and so there is continuous interchange of molecules between them.

28.4 THE STATISTICAL MODEL AND ENERGY

In the ideal gas model we assume a continuous distribution of all possible energies for the particles, called the Maxwell-Boltzmann distribution (← 10.3). Evidence from **spectra** (→ 37) suggests that the energies of atoms and molecules are **quantised** (→ 37) electrons in atoms and molecules can only exist in states that have well-defined energies. So the distribution of energies is a histogram rather than a smooth curve.

The height of each column of the histogram in Figure 28.11 represents the number N_p, of atoms or molecules with a particular total energy. That total energy is made up of a number p of **energy quanta**. The statistical model assumes that these energy quanta are shared out and shuffled randomly between the atoms or molecules. When this is done, we find a simple relationship between the heights of adjacent columns on the histogram of the more energetic atoms or molecules.

The ratio of the number of atoms or molecules with p quanta to those with one less quantum, $p-1$, is a constant for a given number of energy quanta being shared amongst the atoms or molecules at a given temperature T. This ratio is an exponential term known as the **Boltzmann factor**.

The Maxwell-Boltzmann distribution of energies assumed in the ideal gas model contains the same form of exponential term. At room temperature, most properties depend on those particles with the most energy. For example:

- evaporation occurs when there are molecules with enough energy to escape from a liquid
- electrons with enough energy to become 'free' enable electrical conduction to occur in semiconductors
- molecules with enough energy to break up on collision cause chemical reactions.

All of these are found by experiment to vary exponentially, the exponential term being the Boltzmann factor.

The Boltzmann factor

When N_p and N_{p-1} are the numbers of atoms or molecules with p energy quanta and with $(p-1)$ energy quanta, we find that:

$$N_p = N_{p-1}\, e^{-E/kT}.$$

The term $e^{-E/kT}$ is called the Boltzmann factor. E is a characteristic energy for the system concerned, k is the **Boltzmann constant** and T is the temperature in K, or kelvin. The Boltzmann constant has the value 1.38×10^{-23} J K^{-1} h (← 10.5).

The number of particles with p or more energy quanta, or the number N_E with energy greater than or equal to E, is related to the total number of particles N_T by a similar equation:

$$N_E = N_T\, e^{-E/kT}$$

or $\ln(N_E) - \ln(N_T) = -E/kT.$

28 THERMODYNAMICS F

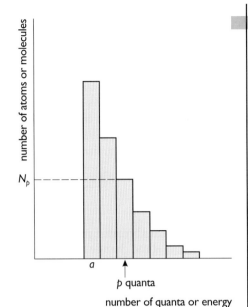

Figure 28.11 Histogram of numbers of atoms or molecules with a particular energy, against energy

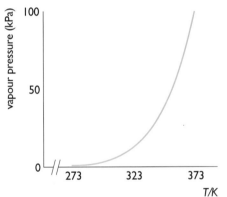

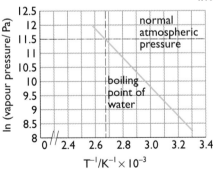

Figure 28.12 Vapour pressure of water against temperature

Vapour pressure and the Boltzmann factor

The vapour pressure of a liquid in equilibrium with its vapour increases rapidly as the temperature increases. It reaches the external pressure when the liquid boils. An empirical equation derived purely to fit the experimental data is:

ln (vapour pressure) = $-\Delta H_{vap}/RT$ + (a constant)

where R is the universal gas constant, or gas constant per mole, and ΔH_{vap} is the molar latent heat of vaporisation of the liquid.

ΔH is a change in enthalpy of a system at constant pressure: $\Delta H = \Delta U + P\Delta V$, ($\leftarrow$ 11.2) here a change of state on vaporisation.

Figure 28.12 shows a graph of the vapour pressure of water against temperature and of ln [vapour pressure] against $1/T$. The line corresponding to the equation above is drawn in. The graph show that water fits this empirical equation well.

The vapour pressure equilibrium occurs when the number of molecules escaping per second equals the number condensing per second. To escape, a molecule must have an energy greater than or equal to E, where E is the energy necessary to break all the bonds to neighbouring molecules. So the vapour pressure should be proportional to N_E, the number of molecules with energy greater than or equal to E. Comparison of the empirical equation with the defining equation for the Boltzmann factor shows that both have the same form:

$E = \Delta H_{vap}/N_A$

where N_A is the Avogadro constant, and the constant is a function of the total number of molecules. The Boltzmann factor enables you to predict a variation of vapour pressure with temperature that agrees with observation.

Questions

5a Use the Boltzmann factor to predict the proportion (N_E/N_T) of water molecules that have enough energy to escape at 100 °C and at 90 °C, given the molar latent heat of vaporisation of water = 40 kJ mol^{-1}.
b The saturated vapour pressure of water at 100 °C is 101 kPa, and at 90 °C it is 70 kPa. Are your answers to **a** compatible with the vapour pressure data for water?

6 The resistance of semiconductors decreases exponentially with increase in temperature ($\leftarrow$ 12.5). This is due to the number of conduction electrons increasing as the temperature rises ($\leftarrow$ 27.6). Explain this in terms of the Boltzmann factor.

7 Many chemical reactions obey a rate equation of the form:

rate at temperature $T = B\, e^{-A/RT}$

where A and B are constants. Some double their rate of reaction for every 10 °C rise in temperature around 300 K. For these reactions, estimate the constant A in the rate equation.

28.5 THE STATISTICAL MODEL AND THE SECOND LAW

The statistical model interprets heating a body as the transfer of energy quanta. Changing the distribution of quanta results in a change in the number of ways

MATERIALS: THEIR PROPERTIES AND STRUCTURE

W of arranging those quanta between the atoms or molecules of the body. To link the model to the second law of thermodynamics, you need to equate the addition of quanta to the energy transfer due to temperature differences and to define a bulk or macroscopic quantity related to W.

When energy Q is transferred, the number of ways of arranging the quanta changes from W before to W^* after. The Boltzmann factor also enters into statistical model calculations of the number of ways of arranging quanta. So W and W^* are related by the same form of equation as that relating the numbers of atoms or molecules with given numbers of quanta. In its logarithmic form, the relationship is:

$$\ln(W^*) - \ln(W) = \frac{Q}{kT}.$$

The change in $\ln(W)$ is related to the change in the thermodynamic property **entropy S**. Entropy is a bulk property of the system, and the change in entropy ΔS is defined by the equation:

$$\Delta S = \frac{Q}{T}.$$

Compare this with the equation for $\ln(W)$. You can see that ΔS equals $\Delta \ln(W)$ multiplied by the Boltzmann constant k.

You will find values for entropy changes in a data book. They are calculated from measurements of Q and T. ΔS cannot be measured directly, nor can we measure the change in $\ln(W)$ because of the enormous numbers of atoms or molecules and quanta in real systems.

Figure 28.13 Boltzmann's gravestone in Vienna

USING ENTROPY

The link between entropy and the number of ways of sharing energy is important in helping us to understand the concept of entropy. Boltzmann thought it so important that he asked for it to be put on his gravestone (see Figure 28.13).

The statistical model predicts that the events you observe are those that are most probable – those that can arise in the largest number of ways. Events happen spontaneously when they lead to an increase in the number of ways of arranging quanta. An increase in the number of ways is equivalent to an increase in entropy. So you will always observe events for which the total entropy of the system increases.

The entropy of some individual component may decrease, but not that of the system as a whole. This can be applied to the Universe as the ultimate system. The entropy of a system can decrease but the entropy of the Universe as a whole must increase. Outside any system the entropy increases more than any decrease within the system. These ideas on entropy give us another statement of the second law:

in any change, the total entropy of the system cannot decrease.

$\Delta S = 0$ corresponds to an idealised reversible system. In a real system, the entropy change ΔS is always positive.

Questions

8 By considering the Carnot cycle (← 28.2) follow the cycle shown in Figure 28.6 and work out the entropy change for each of the four sections. Show that, for the complete cycle, $\Delta S = 0$.

28 THERMODYNAMICS

9 When sodium chloride dissolves in water, the water cools slightly.
a Why is there a temperature change in the water?
b Why does this suggest a decrease in the entropy of the water?
c The second law of thermodynamics states that in a closed system, the entropy will increase. What has also changed in the above situation to cause an increase in entropy that more than offsets the decrease due to the change in temperature?

28.6 THIRD LAW OF THERMODYNAMICS

This law simply states that:

it is impossible to attain the absolute zero of temperature, 0 K.

This is because the energy required to extract any chosen amount of internal energy from a system becomes larger and larger as 0 K is approached.

SUMMARY

The first law of thermodynamics extends the conservation of energy to include thermal systems. Conservation of energy alone is not enough to predict whether a machine will transfer energy or what its efficiency will be.

The second law of thermodynamics states that energy is transferred spontaneously from higher to lower temperature bodies that are in thermal contact. It also defines the maximum efficiency of any useful energy transfer.

The quantity entropy is defined by the equation:

$$\Delta S = \frac{Q}{T}.$$

It provides an additional means of stating the second law of thermodynamics: for any real observable change, the total entropy of the Universe must increase.

A model's predictions about the behaviour of a system must be the same as those of thermodynamics for the model to be acceptable.

The third law of thermodynamics states that it is impossible to attain the absolute zero of temperature, 0 K.

Summary Questions

10 Calculate the maximum thermodynamic efficiency of a heat engine burning fuel at 1100 K and having an exhaust temperature of 500 K. Give reasons why its overall efficiency will be much less than this.

11a Find the value of E/kT for which the Boltzmann factor has values of
i 10^{-1},
ii 10^{-3} and
iii 10^{-6}.
b Estimate the value E needs to have so that the Boltzmann factor is 10^{-6} at room temperature.

UNIT G
WAVES: TRANSFERRING ENERGY THROUGH SPACE

WAVES: TRANSFERRING ENERGY THROUGH SPACE

Whether out at sea in a boat or standing by the side of a pond or river, most of us at some time or another have observed the phenomena of waves.

A number of questions may arise. What are waves? How are they caused? What do they do? What are their properties?

Answers to questions of this sort enable you to have a better understanding of sound and light. They also are the key to the understanding of the nature of matter and even space itself.

29 NATURE OF WAVES

When a stone is dropped into a still pond, some water is disturbed by the stone. This disturbance is transmitted across the surface of the water. The result is a set of circular ripples of steadily increasing radii. Clearly, water itself does not travel outwards: a leaf floating on the water surface will just bob up and down as the disturbance passes by. However, energy is carried by the ripples as they spread outwards from the source of the disturbance.

The basic characteristics of waves – in wires, springs, on the surface of water, in air or in space – can be explained by force and motion.

29.1 TRANSVERSE AND LONGITUDINAL PULSES

Pulses are very short bursts of wave motion. Imagine a pulse of short duration along a stretched spring, such as a slinky. There are essentially two types of pulse that could be sent along the stretched spring. The first is a **transverse** pulse, where the free end of the spring is displaced rapidly to the side and then back again. With transverse pulses, particles of the medium move at **right angles** to the direction of travel of the pulse. The second is a compression or **longitudinal** pulse, where the free end of the spring is compressed rapidly and then returned to its original position. So the particles move **parallel** to the direction of travel (see Figure 29.1).

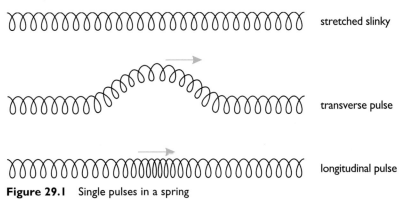

Figure 29.1 Single pulses in a spring

In either case, the pulse travels down the spring as a result of the elastic and inertial properties of the spring. The disturbance of a coil in the spring causes the next coil to experience a force which, in turn, causes it to accelerate. In this way a force is transmitted along the length of the spring.

WAVES: TRANSFERRING ENERGY THROUGH SPACE

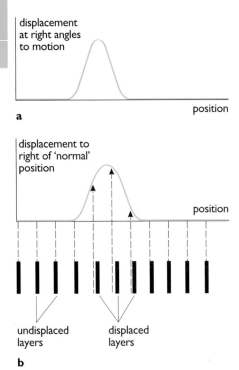

Figure 29.2 Displacement-position graph for **a** a transverse pulse **b** a longitudinal pulse

Figure 29.3 Pulse in a rope

Question

1 A slinky is stretched out along the floor and fixed at one end. A transverse pulse is sent along the spring by momentarily displacing the free end, and the speed of the pulse is observed. In answering the following questions, think about the forces acting within the spring and the effect of these forces on the mass of each coil of the spring.
a A pulse of larger amplitude is sent along the spring. Do you think the speed of the pulse will now be greater than, smaller than or the same as before?
b The spring is stretched by another metre or so. Will the pulse speed now be greater than, smaller than or the same as that of the original pulse?

You can compare the behaviour of transverse and compression pulses by plotting displacement-position graphs as in Figure 29.2. Clearly, the shape of the graph in Figure 29.2a looks like the pulse itself. It is sometimes called the **wave profile**. In Figure 29.2b, the displacements are measured parallel to the pulse direction, with displacements to the right indicated as positive.

The maximum displacement of a particle, the **amplitude**, can be related to the energy stored within the pulse. The larger the amplitude, the greater the stored energy for a given system. Studies on oscillations indicate that the energy of an oscillator is proportional to the square of its amplitude (← 19.4). This suggests that the energy stored in a pulse might also be proportional to the square of its amplitude. When a mechanical pulse passes along a spring, energy is dissipated through frictional losses as the pulse travels along the spring and so the amplitude decreases in size.

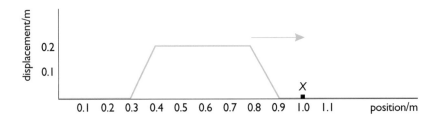

Question

2 Figure 29.3 shows an idealised sketch of a pulse in a rope. The pulse is travelling from left to right at a speed of 0.5 m s^{-1}. A scale has been superimposed on the sketch to indicate position and amplitude of the pulse. The diagram represents the situation at $t = 0$.
a Sketch a graph of displacement against time for the point marked X on the rope as the pulse passes by.
b Sketch a graph of the velocity of X against time as the pulse passes by.
c In what way is this an 'idealised' sketch?

29.2 TRAVELLING WAVES

Suppose you were to oscillate the free end of a spring continuously to produce a steady train of positive and negative transverse pulses. When the oscillation is

29 NATURE OF WAVES

type	how caused	effect on medium	application
water waves	earthquakes or winds at sea	oscillation of water particles at right angles to wave direction, i.e. transverse oscillations	waves at sea, tidal waves, river bores
waves in strings	bowing or plucking of stretched string	oscillation of string at right angles to wave direction, i.e. transverse oscillations	musical instruments, e.g. guitar, violin, cello, harp
sound waves in air	vibration of an object, e.g. loudspeaker cone, bell, reed	oscillation of layers of air parallel to wave direction, i.e. longitudinal oscillations	musical instruments, e.g. flute, trumpet, trombone; loudspeaker; siren
ultrasonic waves	oscillation of a quartz crystal	oscillation of particles of medium parallel to wave direction, i.e. longitudinal oscillations	testing of materials; sonar; medical screening
seismic waves	earthquakes, controlled explosions	longitudinal in the Earth and transverse oscillations between the surfaces inside the Earth	earthquakes and geological surveys

Figure 29.4 Examples of mechanical waves

simple harmonic (← 19.2), you have effectively set up a continuous wave, which looks like a sine wave, travelling along the spring. Some other examples of waves are given in Figure 29.4.

Figure 29.5 shows a transverse wave demonstrated using a wave machine. The wave is travelling from left to right but, of course, the photograph represents just an instant of the motion. A similar effect could be obtained using a slinky spring. Note that, when the wave is allowed to continue, reflections from the fixed end cause a change in the appearance of the travelling wave and so create a more complex waveform (→ 30.2).

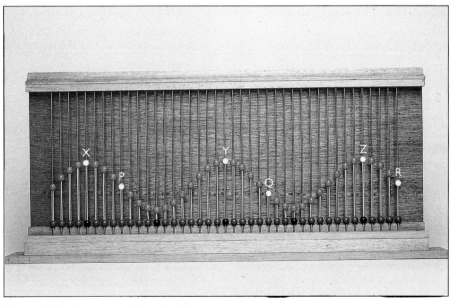

Figure 29.5 A wave machine displaying a transverse wave

WAVES: TRANSFERRING ENERGY THROUGH SPACE

Points X, Y and Z in Figure 29.5 are of particular interest. They are each at maximum displacement from their equilibrium position and, as the wave is travelling, they move in step with each other. They are said to be **in phase**. Points P, Q and R lie at the equilibrium position and, again, move in phase with each other. But they are one quarter of a cycle, or $\pi/2$ radians, out of phase with X, Y and Z. In any wave, two adjacent points vibrating in phase are separated by **one wavelength**. Suppose we denote the wavelength of the wave in Figure 29.5 by λ. Then $\lambda = XY = YZ = PQ$, and so on.

Someone sitting on the wave experiences an up-and-down movement as the wave passes. This motion is repeated regularly. The time for one cycle is known as its **period, T**. The number of cycles per second is known as the **frequency, f** and is equal to $1/T$. Its unit is the second^{-1}, or **hertz (Hz)** ($\leftarrow$ 19.1). So the wave velocity v is given by:

$$v = f\lambda.$$

Wave velocity does not usually change much with frequency or wavelength, if at all. So, usually, the higher the frequency of a wave, the smaller its wavelength.

Figure 29.7 shows how these ideas apply to longitudinal waves.

AMPLITUDE, POWER AND INTENSITY

Amplitude is the maximum displacement in a wave, such as that of particles from their equilibrium positions shown in Figure 29.8. A wave of small amplitude will deliver less energy in a given time than a wave of larger amplitude. As with pulses, the energy delivered per second, the **power**, is proportional to amplitude^2 rather than amplitude. **Intensity** is the power delivered per unit area by a wave. So, it is also proportional to the square of the amplitude.

Wavelength, frequency and velocity of a wave

Figure 29.6 shows a travelling wave from the source S. Suppose that in time t it reaches a point P, a distance of vt. In this time, ft cycles will have been produced, each of wavelength λ.

Distance travelled by wave = no. of cycles × length of each cycle

so $\qquad vt = ft \times \lambda$

or $\qquad v = f\lambda$.

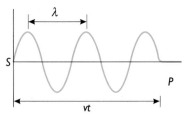

Figure 29.6 The wave equation

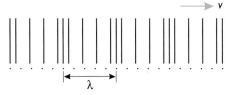

Figure 29.7 Wavelength of a longitudinal wave

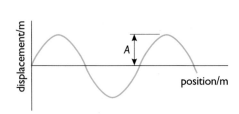

Figure 29.8 Amplitude A of a wave

Question

3 A transverse wave travels down a long spring. At a particular moment $t = 0$, the wave appears as in Figure 29.9. The wave is travelling from left to right, and a horizontal scale has been drawn. The wave frequency is 2 Hz.
a What is the wavelength of the wave in m?
b Estimate the amplitude of the wave.
c What is the speed of the wave in m s^{-1}?
d Sketch graphs of:
i the displacement of point P with time. On the same axes, repeat for points Q and R
ii the velocity and the acceleration of P, plotted on the same axes.

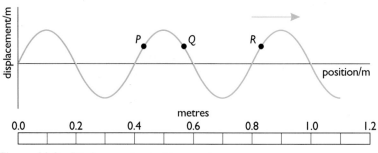

Figure 29.9 Displacement – position graph

29.3 SPEED OF MECHANICAL WAVES

Pulses and waves travel through a medium as a result of the elastic properties of the medium, but what are the factors that affect wave speeds?

Imagine a row of trolleys linked by compression springs, as in Figure 29.10a. The masses in the system are clearly defined, as are the forces acting on each of the masses. This model can be used to represent pulses of sound travelling in a medium as longitudinal oscillations of the molecules. When the end trolley is displaced, a force is transmitted through the springs to the next trolley in the line. This, in turn, causes the third trolley to move and so on down the line.

When the stiffness (← 4.1) of the springs is increased, the trolleys accelerate more quickly for a given displacement. This causes the disturbance to be passed to neighbouring trolleys more rapidly, so increasing the pulse speed. However, when the masses of the trolleys are increased, their accelerations are reduced and the pulse speed decreases. So, the effect each spring has on the next trolley depends upon the stiffness, k of the spring and the mass m of each trolley. Using an analytical model, you can show that, when the separation between trolley centres is x, the speed of a compression pulse v is given by:

$$v = x\sqrt{k/m}.$$

Speed of a mechanical wave

Figure 29.10 shows a pulse passing along a line of trolleys linked by compression springs. In b, the pulse has just reached trolley Q. A little later, in c, the pulse has just reached R.

Suppose you regard Q as a simple trolley-spring arrangement. In b, it is at the equilibrium position. In c, it has reached the right-hand extreme position of its oscillation. So, the time for the pulse to be passed on from trolley Q to the next trolley R, a distance x, is around one quarter of the natural period of oscillation of Q. Note that x is the distance between neighbouring trolley centres. A closer and more detailed analysis shows the time to be $\frac{1}{2}\pi$ of an oscillation.

Time period, T, of $Q = 2\pi\sqrt{m/k}.$

Time to cover distance $x = T/2\pi$

$\qquad = \sqrt{m/k}.$

So the speed of the pulse v is given by:

$\qquad v = x/\text{time interval}$

or $\quad v = x\sqrt{k/m}.$

You can use this equation to calculate v in experiment 29.1.

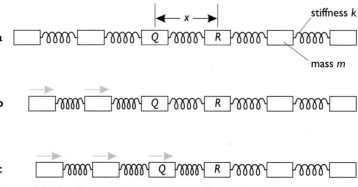

Figure 29.10 Compression pulses in a trolley-spring system

WAVES: TRANSFERRING ENERGY THROUGH SPACE

EXPERIMENT 29.1 Speed of a compression pulse

By setting up six or seven trolleys as shown in Figure 29.11, with pairs of compression springs used to link adjacent trolleys for stability, you cause a compression pulse to be generated when the trolleys hit the block. The pulse travels to the free end of the row and is reflected back. When it is reflected again at the block, the trolleys lose contact with it.

a Calculating the speed From the trolley separation, the spring constant of a pair of springs and the average mass of a trolley (found by weighing several of them), you can predict the speed by substituting into $v = x\sqrt{k/m}$.

b Measuring the speed By measuring length l (see Figure 29.11) and finding the time between the trolleys hitting and leaving the block, you can calculate the velocity from $v = 2l/t$.

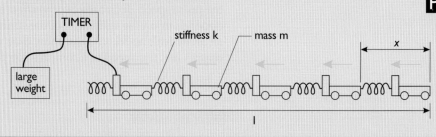

Figure 29.11 Measuring the speed of a pulse

Question

4 This question refers to experiment 29.1.
a Obtain a predicted value for the speed of the pulse, using the following data: trolley spacing = 0.35 m, trolley mass = 0.75 kg, stiffness of springy links between trolleys = 75 N m⁻¹.
b If the trolleys are pushed more gently against the block, will this affect the time of travel and so the speed of the pulse?
c If you doubled the mass of each trolley, what effect would this have on the pulse speed? Is your answer confirmed by experiment?
d Suppose you doubled the spring constant by connecting adjacent trolleys with four springs in parallel. How would this affect the pulse speed?

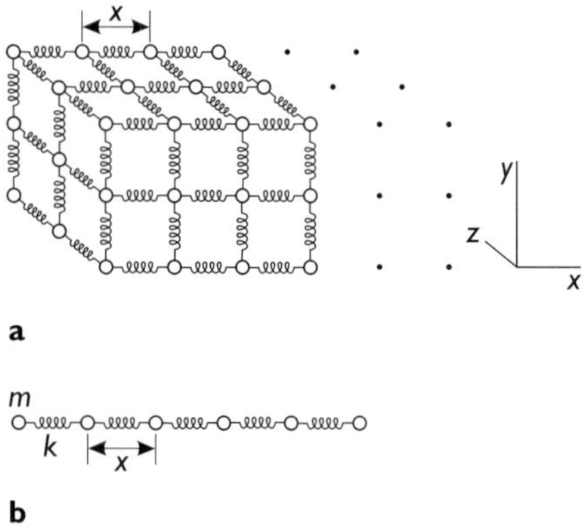

SPEED OF A PULSE ALONG A METAL ROD

The trolley-spring model can be used to predict the speed of a sound pulse along a metal rod. Imagine the rod to be a long row of atoms linked by bonds in a three-dimensional array, as in Figure 29.12a. When a compression pulse is applied in the positive x direction, a disturbance travels along the length of the rod. Since there is little propagation in either the y or z directions, you can simplify the model to a single row of atoms linked by their bonds, as in Figure 29.12b.

Figure 29.12a Crystal model, **b** simplified model of metal rod

Question

5 This question uses the fact that trolleys linked by springs can be compared to atoms linked by their bonds.

29 NATURE OF WAVES

Speed of sound in a metal rod

The model of a solid shown in Figure 29.12a can be used, with limitations, to obtain an expression for the Young modulus E of a metal (← 26.3)

$$E = \frac{k}{x},$$

where k is the spring constant of the bonds and x is the atomic separation. The volume occupied by a single atom is x^3, so the density ρ is given by:

$$\rho = m/x^3,$$

where m is the mass of an individual atom.

We know that $\quad v = x\sqrt{\dfrac{k}{m}},$

$$k = Ex,$$

and $\quad m = \rho x^3.$

So $\quad v = x\sqrt{\dfrac{Ex}{\rho x^3}}$

$$v = \sqrt{\dfrac{E}{\rho}}.$$

> The trolleys used in experiment 29.1 have a typical mass of 0.75 kg. Suppose these are replaced by individual atoms with a typical mass of 10^{-25} kg, and their separation reduced to 3×10^{-10} m, the interatomic separation in steel. The force constant of the springs and the interatomic force constant is very similar at 75 N m^{-1}.
> **a** Given that the speed of a pulse along the row of trolleys and springs is about 3.5 m s^{-1}, predict the pulse speed along a row of atoms in a steel rod.
> **b** In what ways is the trolley-spring model **i** a sensible one, **ii** an unsatisfactory one?

The Young modulus E (← 4.2) of a metal is related to the stiffness of the atomic bonds. The density of the metal ρ is closely related to the mass of the atoms and their separation. By looking at these macroscopic and microscopic properties, we can show that the speed of a pulse of sound along a metal rod is given by:

$$v = \sqrt{E/\rho}.$$

The speed of a pulse of sound in a metal rod turns out to be very much greater than the pulse speed along a row of trolleys and springs. This is due mainly to the large difference in mass between the atoms and the trolleys. Surprisingly, perhaps, the elastic constant of the springs and the bonds is similar.

Experiment 29.2 suggests a way of measuring the speed of sound in a metal rod. For a mild steel rod, a typical time might be 0.4 ms, which is 10 cycles of the 25 kHz signal. So:

$$\text{speed of pulse} = 2 \text{ m}/0.4 \times 10^{-3}\text{s}$$
$$= 5000 \text{ m s}^{-1}.$$

EXPERIMENT 29.2   Measuring the speed of sound in a metal rod

By dropping one end of a 1m steel rod on to a solid metal block, you produce a compression pulse which passes up the rod and is reflected at the top end. When the pulse returns, the rod briefly loses contact with the block.

Because the time of contact is only about 10^{-4} s, you need to set up the arrangement shown in Figure 29.13. When the rod strikes the block, the 25 kHz signal is connected to the Y input. By counting the number of cycles appearing on the oscilloscope, you can estimate the time of contact.

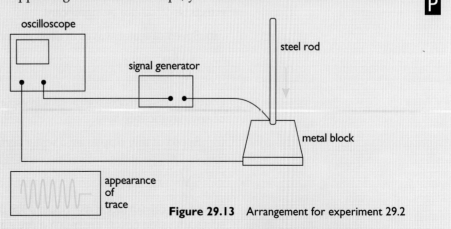

Figure 29.13 Arrangement for experiment 29.2

WAVES: TRANSFERRING ENERGY THROUGH SPACE

> **Question**
>
> **6** This question refers to experiment 29.2.
> **a** The main source of error in the experiment is the measurement of the pulse's time of travel. Suppose that, when using the oscilloscope, the time base is set to 0.1 ms cm^{-1} and the uncertainty in the measurement of the number of cycles is ±1. What is the uncertainty in the final value for the speed of the pulse?
> **b** Could it be that separation occurs after the pulse has travelled through the metal block and back? How could you investigate this and demonstrate that this is not the case?
> **c** The value of the Young modulus for steel is 2×10^{11} N m^{-2} and its density is 7 700 kg m^{-3}. Using the equation $v = \sqrt{E/\rho}$, calculate the predicted speed. Is this in good agreement with the measured speed in the experiment?
> **d** Suppose you tried the same experiment using a 1 m long aluminium rod. What time of travel of the pulse, in s, would you predict?
> ($E_{Al} = 7 \times 10^{10}$ N m^{-2}, $\rho_{Al} = 2 700$ kg m^{-3}.)

OTHER WAVE SPEED FORMULAE

Other wave speed formulae have a pattern similar to those for trolleys linked by springs or sound along a rod, as shown in the following examples.

(i) Speed of a transverse pulse along a stretched wire or spring:

$$v = \sqrt{\frac{T}{\mu}},$$

where T is the tension in the wire and μ is the mass of unit length of the wire. Note that T is a force within the wire or spring and is due to the interaction between it and the support at each end.

(ii) Speed of waves in shallow water:

$$v = \sqrt{gh},$$

where g is the gravitational field strength, or force per unit mass, and h is the depth of water. This equation applies when the wavelength is greater than the depth of water, and when the amplitude is small compared with the depth, for example waves in a ripple tank. When the depth is very small, surface tension effects ($\leftarrow$ 27.2) can be significant.

(iii) Speed of water waves in deep water:

$$v = \sqrt{\frac{g\lambda}{2\pi}}.$$

Gravity is responsible for these waves. Surface tension effects are too small to be of any significance here.

(iv) Speed of tiny ripples on a water surface:

$$v = \sqrt{\frac{2\pi\gamma}{\lambda\rho}}$$

where γ is the surface tension and ρ the density. When the wavelength is small, surface tension effects are much more significant than the pull of gravity on the

29 NATURE OF WAVES

tiny crests of water. So the speed is dependent upon the wavelength.

Note that each expression involves a quantity involving force or an elastic modulus, and a mass or density term.

> ### Question
>
> 7 The formula for wave speed in (i) on the previous page gives the correct unit for speed, as the following shows. For a stretched spring:
>
> $$v = u\sqrt{\frac{T}{\mu}}$$
>
> unit of $u\sqrt{T/\mu}$ is $(N/kg\,m^{-1})^{0.5} = (kg\,m\,s^{-2}/kg\,m^{-1})^{0.5}$
> $= (m^2 s^{-2})^{0.5}$
> $= m\,s^{-1}$,
>
> the unit of speed.
>
> Satisfy yourself that each of the formulae in (ii) – (iv) also gives the correct unit for speed.

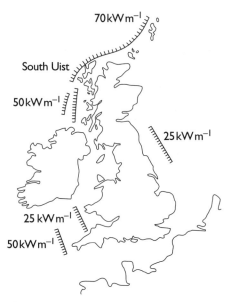

Figure 29.14 Available wave power around the coasts of Britain

WAVE ENERGY

Waves are a means by which energy may be transferred from one place to another. The rate at which a water wave delivers energy to a point in its path is dependent not only upon the amplitude of the wave but on its frequency and also its velocity.

Over recent years, many studies have taken place on the viability of extracting energy from water waves. Figure 29.14 illustrates the average power available per metre of frontage at various positions around the coast of Britain. As the shoreline is approached, the water depth decreases and becomes more suitable for the location of sea-bed mounted devices. However, the power available also decreases. Average power levels in 50 metres depth range typically from 40 – 50 kW m^{-1}. Up to 30 GW of power might be available, if all possible locations were exploited.

Various devices have been developed to capture wave energy. Salter's duck in Figure 29.15 is designed to oscillate as a wave passes. It is shaped in such a way as to transfer maximum energy from the wave, over 90 % in laboratory trials. The rocking motion of a series of these 'ducks' can drive a generator to produce electricity which, amongst a number of possibilities, could be transmitted to the shore.

At the moment, water waves are not thought to be a viable energy source in this country. Other countries, notably Norway, are continuing their research into it. One advantage of this method of electricity production is that it does not contribute to the greenhouse effect (→ 36.1).

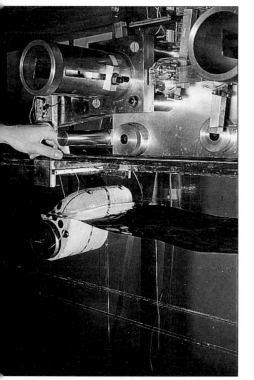

Figure 29.15 Salter's Duck

> ### Question
>
> 8 An object floating on a water surface moves up and down with simple harmonic motion (SHM) as a water wave passes. Its maximum kinetic energy E_{Kmax} occurs as it moves through the mid-point of its oscillation. The value of E_{Kmax} is $\frac{1}{2}mv^2$, where v is the maximum velocity of the object and m its mass.

WAVES: TRANSFERRING ENERGY THROUGH SPACE

a For simple harmonic motion (← 19.2) the displacement s of the object from its equilibrium position is given by:

$$s = A \sin 2\pi ft$$

where A is the wave amplitude and f its frequency.
Show that the maximum velocity v of the object is given by:

$$v = ds/dt = 2\pi fA.$$

b Show that E_{Kmax} of the oscillator is $\frac{1}{2}m(4\pi^2 f^2 A^2)$.
c Explain why the power delivered by the wave is proportional to $f^2 A^2$.
d Suppose the wave velocity is doubled but the frequency and the amplitude remain unchanged. What will happen to the power delivered?
e How will the power in the wave be dependent upon the length of the wavefront?

29.4 SOUND IN GASES

Discovery of the mechanism of sound transmission

One of the first experiments on the transmission of sound was carried out by Otto von Guericke (1602–86). Following his invention of the vacuum pump, he investigated the effect of removing air from a chamber in which a bell was being struck. As the air was gradually removed, the sound became fainter. The experiment was repeated by Robert Boyle (1627–91) shortly afterwards with the same results. They concluded that sound is transmitted as a result of a to-and-fro movement of the air, without the air as a whole moving in the direction of transmission of the sound.

Sounds in air are produced by vibrating objects such as a tuning fork, a loudspeaker or guitar string. The vibrations cause the layers of air to oscillate and this to-and-fro motion is passed on to other layers. The ears or other sensors respond to the tiny movements of the air reaching them.

The human ear can generally respond to sound frequencies between about 20 Hz and 20 kHz. Older people tend to have a narrower range, though, finding high frequencies particularly difficult to detect. Exposure to loud sounds can also cause a reduction in the ear's audible range, especially when the exposure is frequent or prolonged.

The longitudinal oscillations of the air produce regions of compression where the air is more dense. Lower density, and therefore lower pressure, regions are known as **rarefactions** (see Figure 29.16). For barely audible sounds, the pressure variation may be as little as 10^{-5} Pa above and below a normal atmospheric pressure of 10^5 Pa. Displacements of the air resulting from such sounds may be of the order of 10^{-11} m, less than the diameter of a single atom! Very loud sounds may have pressure amplitudes of 20 Pa with displacements of around 10^{-5} m.

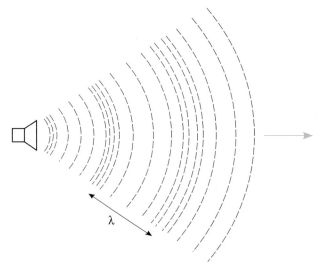

Figure 29.16 Compressions and rarefactions in air

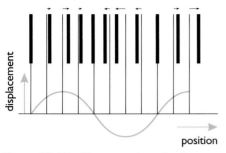

Figure 29.17 Diagram and graph, of displacement of layers of air as sound wave passes

29 NATURE OF WAVES

Question

9 Figure 29.17 shows the variation in displacement of layers of air at different points as a sound wave passes.
a From the diagram, note where the pressure is normal, above normal and below normal. Copy the graph and, on it, sketch the variation in pressure of the air.
b Calculate the phase difference between the displacement graph and the pressure graph.

Amplitude has been defined as the maximum displacement of a medium. For a sound wave, it can also be defined as the maximum pressure.

ULTRASONIC WAVES

These are longitudinal waves, similar to audible sound waves but of higher frequency. The frequency of ultrasonic waves ranges from the upper end of the audible spectrum at about 20 kHz to several hundred MHz. Beyond this frequency, ultrasonic waves travel only very short distances before their energy is absorbed by the medium through which they pass.

Ultrasonic radiation is normally produced, and detected, by **piezoelectric transducers**. Certain types of crystal, notably natural quartz, contract when a pd is applied across them. This is known as the **piezoelectric effect**. High-frequency alternating pd's cause the crystal to oscillate and act as a transmitter of ultrasonic waves. Such a crystal can also act as a receiver. Ultrasonic waves falling on it cause it to oscillate mechanically and so produce a voltage across it.

An ultrasonic wave's high frequency means that its power will be quite large even at small amplitudes. Lower frequency audible sounds would need to have a very large amplitude to travel over the same distance (← question 8). Ultrasonic waves are used for underwater imaging, the non-destructive testing of materials, the destruction of bacteria and in sonar (→ 30.2).

Underwater imaging

Ultrasonic radiation is particularly useful for undersea communication and detection systems. High frequency radio waves, used in radar, travel just a few centimetres in water, whereas highly directional beams of ultrasonic waves can be made to travel many kilometres.

SPEED OF SOUND IN GASES

Sound travels along a metal rod at about 5 000 m s^{-1} (← 29.3). This is fairly typical of the speed of sound in solids and contrasts with its much lower speed in liquids. Water-borne sound waves, for example, travel at about 1 400 m s^{-1}. In air, sound travels at even lower speeds. This is why the first sign of an approaching train at a railway station is often heard through the rails rather than through the air. The difference in speed is due mainly to the elastic nature of the different media. Much larger forces act on layers displaced by a given amount in a metal rod than in air and so the disturbance is transmitted more rapidly.

MEASURING THE SPEED OF SOUND

Using measured values of f and λ, the equation $v = f\lambda$, ($\leftarrow$ 29.2) can be used to calculate the speed of sound.

The phase difference, ϕ in radians between two signals that have a path difference x is given by:

$$\phi = 2\pi . \text{(fraction of a cycle between signals)}$$

$$= \frac{2\pi . \text{(path difference)}}{\text{(wavelength)}}$$

$$= \frac{2\pi . x}{\lambda}.$$

Experiment 29.3 shows two methods of finding λ and f. A phase shift of 2π radians occurs when the distance moved by the microphone is one wavelength λ of the sound wave. A typical wavelength is 33 centimetres.
So, the speed of sound = $0.33 \text{ m} \times 1000 \text{ s}^{-1} = 330 \text{ m s}^{-1}$.

EXPERIMENT 29.3 Speed of sound in air

Using the apparatus in Figure 29.18 with a frequency of 1 kHz and the microphones placed side by side, you can try either of the following methods:

a Wave-form comparison Connect each microphone to one of the y-inputs of the double-beam oscilloscope to produce two in-phase traces on the screen (see Figure 29.19).
b Lissajous' figures Connect each microphone to the x-input or y-input of the oscilloscope as in Figure 19.30. Adjust the oscilloscope input gain and adjust the distance between the microphone and loudspeaker to produce Figure 19.28a on the oscilloscope.

In either case, the distance you have to move one of the microphones away from the speaker until the signals are again in phase provides you with one wavelength. Finding the distance for, say, 10 or 20 complete cycles or wavelengths enables you to calculate v from f and λ.

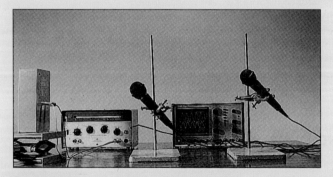

Figure 29.18 Apparatus for experiment 29.3

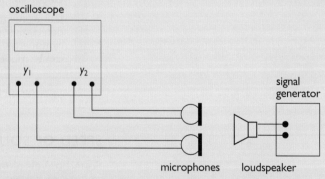

Figure 29.19 Arrangement and oscilloscope output for wave-form comparison

29 NATURE OF WAVES

Speed of sound in a gas

The speed of mechanical waves in any medium takes the form:

$$\text{speed} = \sqrt{\frac{\text{elastic modulus}}{\text{density}}}$$

(← 29.3). In the case of gases, it is appropriate to use the **bulk modulus of elasticity**. This is equal to γP for a rapid change in pressure, where P is the pressure of the gas and γ is the ratio of the specific heat capacity at constant pressure to that at constant volume for the gas involved (← 11.3). So, we can write the speed of sound as:

$$\text{speed} = \sqrt{\frac{\gamma P}{\rho}}$$

Effect of temperature on the speed of sound

For one mole of gas of mass M at temperature T, $PV = RT$, where P is the pressure, V the volume and R is the universal gas constant (← 10.2). So for one mole:

$$\rho = \frac{M}{V}$$

and $P = \frac{RT}{V}$.

Substituting these in the equation for the speed of sound gives:

$$v = \sqrt{\frac{\gamma P}{\rho}}$$

$$= \sqrt{\frac{\gamma RT}{M}}.$$

Since, M is the mass of one mole of gas, it is a constant for a particular gas. So:

$$v \propto \sqrt{T}.$$

Question

10 This question refers to experiment 29.3.
a Sketch the appearance of the traces when the microphone has been moved a distance of **i** half a wavelength, **ii** quarter of a wavelength.
b What is the value of the phase difference between the two signals when the path difference between the two microphones is **i** half a wavelength, **ii** quarter of a wavelength?
c Suppose the path difference between the two microphones is s. What is the phase difference, in radians, between the signals received by the two microphones?

The equation for the speed of sound in air is:

$$\text{speed} = \sqrt{\frac{\gamma P}{\rho}}.$$

For air, at normal atmospheric pressure, about 10^5 Pa, the density is 1.3 kg m^{-3} and γ takes the value 1.4. So, the speed of sound should be 330 m s^{-1}, which agrees well with the value calculated above from measurements found by experiment 29.3.

The wavelength of sound in air will clearly depend upon the frequency. At the lowest audible frequency, 20 Hz, the wavelength will be 340 m s^{-1}/20 Hz, using the equation $v = f\lambda$. This comes to 17 metres. At the upper limit, 20 KHz, the wavelength is only 17 mm. The wavelength is of importance in explaining the behaviour of sound waves (→ 30).

Question

11 The bulk modulus of elasticity for sea water is 2.3×10^9 N m^{-2} and its density is about 1 000 kg m^{-3}. Calculate the speed of sound in sea water.

TEMPERATURE AND THE SPEED OF SOUND

One might expect from the equation above that an increase in air pressure would mean an increase in the speed of sound. However, as the pressure increases, the density also increases. The ratio:

$$\frac{\text{pressure}}{\text{density}}$$

is constant, predicting that the speed of sound is unaffected by air pressure changes alone.

When the temperature changes, the ratio alters and then the speed does change. In fact, the speed of sound in a gas is proportional to the square root of the absolute temperature T of the gas.

Question

12 Air temperatures fall linearly with height from sea level until, at an altitude of about 10 km, they reach a fairly steady value of about 215 K (or –58 °C). The speed of sound in air at 273 K is 331 m s^{-1}.

G WAVES: TRANSFERRING ENERGY THROUGH SPACE

> **a** What is the speed of sound in m s^{-1} at an altitude of 10 km, the average cruising height of a passenger jet?
> **b** Sketch a graph of the variation of the speed of sound with height above the surface of the Earth.
> **c** Suppose an aircraft is travelling very close to but not quite at the speed of sound. Describe the effect on the air immediately in front of the aircraft. What happens as the aircraft passes through the speed of sound?

29.5 SOUND AND HEARING

Having looked at some aspects of the nature of sound, consider the response of the ear to sound.

Sounds can be very broadly categorised as being either musical or unmusical. **Unmusical** sounds are rough, irregular, unpleasant and without any definite pitch. **Musical** sounds, by contrast, are smooth, regular, generally pleasant and of definite pitch.

The human ear

The human ear consists of three main parts (see Figure 29.20):

- the **outer ear**, including the pinna, which is the visible or external part. Some animals are able to move the pinna in order to sense the direction from which sound is coming. Human pinnae no longer perform this function!
- the **middle ear**, which is separated from the outer ear by the ear drum. The middle ear is completely enclosed except for a connection to the throat, the eustachian tube. This tube is normally closed but opens during swallowing. It enables equal pressures to be maintained on either side of the ear drum. When pressure on the outside of the drum suddenly changes, the imbalance of pressure prevents the ear drum from oscillating freely. You will have noticed your ears popping, for example, when taking off in an aircraft or entering a tunnel in a high-speed train. Swallowing usually restores the balance.
- the **inner ear** which consists mainly of the semi-circular canals and the cochlea. Attached to the ear drum is the first of three tiny bones, known from their shape as the hammer, anvil and stirrup. The stirrup is attached to the oval membrane that separates the middle ear from the inner ear. Sound arriving at the ear causes the drum to oscillate with the same pattern as the source of sound. The three bones transmit the vibrations to the inner ear. This converts the tiny movements into electrical impulses which then travel to the brain.

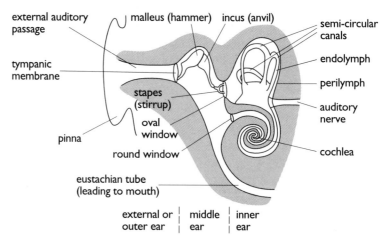

Figure 29.20 The human ear

PITCH AND FREQUENCY

The term **pitch** describes whether a sound is high or low in the musical scale. A link between pitch and frequency has long been established. Originally, this was done by placing card on the rim of a rotating toothed wheel in front of a siren. The faster the wheel turned, the higher the frequency and the higher the pitch of the sound perceived by the ear.

The ear has a non linear response to frequency. As the pitch of a series of notes is heard to rise by equal intervals, the frequency increases by equal ratios. This means that there is an exponential relationship between pitch and frequency. Frequencies of 128 Hz, 256 Hz, 512 Hz, 1024 Hz are perceived by the ear to rise uniformly in pitch, a doubling of frequency being equal to one octave.

AMPLITUDE, INTENSITY AND LOUDNESS

The **intensity** of a sound is the power crossing unit area normal to the direction of propagation of the wave. It is measured in W m^{-2} and is proportional to the

29 NATURE OF WAVES

square of the sound wave's pressure amplitude. When a sound's pressure amplitude is doubled, it is four times as intense. **Loudness** is a subjective sensation, depending both on the intensity of the sound and on the sensitivity of the listener's ear. This, in turn, depends upon the frequency of the sound wave.

There is a lower limit of intensity below which a sound is inaudible and its loudness is zero. This lowest value, the **threshold of audibility**, is 10^{-12} W m^{-2} for a sound of about 1 000 Hz. It corresponds to a sound pressure amplitude of 2×10^{-5} Pa. A sound of intensity 10 times greater than this, at 10^{-11} W m^{-2}, is said to have an intensity of 10 **decibels**, or 10 **dB**. An increase by a factor of 100 or 10^2 to 10^{-10} W m^{-2} corresponds to an intensity change of 20 dB.

A sound may also be too intense to be heard comfortably. This limit is called the **threshold of feeling** (or **pain**). At 1 000 Hz, sounds painful to the ear are about 10^{12} times as intense as those that are barely audible. This corresponds to a range of audible sound intensities of 120 dB. Generally, sounds of above 100 dB are uncomfortable to the ear. The graph in Figure 29.21 shows how the thresholds of audibility and feeling vary with frequency.

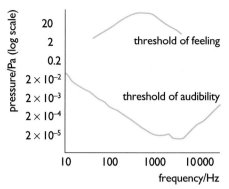

Figure 29.21 Sensitivity of the human ear

Decibel scale

This is a linear scale designed to cover the vast range of intensities with which the ear can cope and the logarithmic nature of its response to sound. The zero on the scale coincides with the threshold of audibility.

Suppose the threshold of audibility is assigned an intensity I_0. A sound of intensity I will be given a sound level of:

$\log_{10} I/I_0$ bels,

or

$10 \log_{10} I/I_0$ decibels

since there are 10 dB in a bel. Decibels are usually used because they are much more convenient in size than the bel.

An increase from I_0 to $10 I_0$ will correspond to an increase in intensity of $10 \log_{10} 10$ or 10 dB. A doubling of intensity will cause the level to increase by:

$10 \log_{10} 2$ dB = 10×0.3010 dB

= 3.01 dB

= 3 dB (approximately).

If P and P_0 are the pressure amplitudes causing intensities I and I_0, then:

sound level = $10 \log_{10}(I/I_0)$

= $10 \log_{10}(P/P_0)^2$

= $20 \log_{10}(P/P_0)$.

The decibel is not an SI unit. However, it is very useful in situations, such as those found in acoustics, where measurements range over many powers of ten. In electronics, amplification and attenuation ratios are often expressed using a decibel scale for similar reasons.

Question

13 The intensity of sound at the threshold of feeling is 10^{12} times that at the threshold of audibility when the pressure amplitude is 2×10^{-5} Pa. What is the pressure amplitude of the air in Pa at the threshold of feeling?

Again, our response to changing intensity of sound varies exponentially with intensity. The ear is very sensitive to small changes in intensity at low levels of intensity, but much less sensitive at high levels. So you may hear a pin drop in a very quiet room but not when the room is full of people talking.

SOUND LEVELS

The ear is particularly sensitive to the middle frequencies of its range and less to the low and high ends. Instruments measuring sound levels may respond differently to the ear. Filters may be built into them to make them less sensitive to the low and high ends. There is also an **A scale**, expressed in **dB(A)**, which gives extra weight to the middle frequencies. Figure 29.22 gives levels on the A scale of some common sounds and noises.

source	sound level/dB(A)	pressure amplitude/Pa
threshold of pain	130	63
jet taking off	125	36
thunder	105	3.6
average city street	70	6.3×10^{-2}
conversation	60	2×10^{-2}
whisper	20	2×10^{-4}
threshold of audibility	0	2×10^{-5}

Figure 29.22 Some typical sound levels

The effect on a person's hearing from industrial noise is dependent not only on the intensity level of the noise but also the time for which a person is exposed to that noise. Personal **noise monitors** display the product of the sound level and the time of exposure. Results from such monitors often suggest where improvements might be made to working environments so that damage to hearing can be reduced.

QUALITY OF A MUSICAL NOTE

Suppose a note of given pitch is played in turn on two different instruments at the same loudness. You will detect a difference and be able to identify the instrument by assessing the **timbre**, or **quality**, of the sound. This is because, generally, musical sounds do not consist of pure sinusoidal notes.

When a tuning fork is allowed to oscillate near a microphone connected to an oscilloscope, a sinusoidal waveform appears on the screen (see Figure 29.23b). A violin playing the same notes will appear as in Figure 29.23c and an oboe as in Figure 29.23d. The notes emitted by the violin consist of a **fundamental** of relatively large intensity and of the same frequency as the tuning fork. To this are added smaller intensity and higher frequency notes, called **overtones**. These have frequencies that are whole number multiples of the fundamental frequency. The number and intensity of the overtones determine the quality of the note from the particular instrument ($\rightarrow$ 30.6).

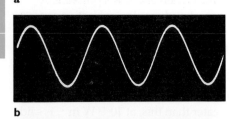
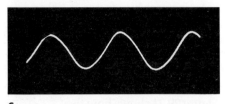
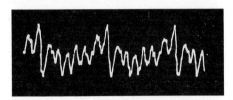
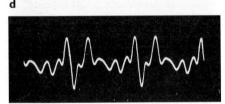

Figure 29.23 Quality of note from **a** a Pure tone, **b** a Tuning fork, **c** a Violin, **d** an Oboe

SUMMARY

Pulses and waves carry energy from one place to another. They are not material entities but disturbances in an elastic medium.

Waves may be either longitudinal or transverse. Longitudinal waves are characterised by oscillation of particles of the medium along the direction of travel of the wave. Transverse waves, are associated with a movement of particles in the medium at right angles to the direction of travel.

The following terms describe the characteristics of a wave, whether it is transverse or longitudinal:

- amplitude: the maximum value of the displacement of the wave
- frequency: the number of complete cycles of the wave passing a given point per second
- period: the time for a particle in the medium to undergo one complete cycle of oscillation
- wavelength: the minimum distance between two particles in the medium oscillating with the same phase. For example, the distance between two neighbouring crests of a transverse wave or compressions of a longitudinal wave
- phase: particles in a travelling wavefront moving in the same direction with the same displacement are said to be in phase.

The velocity, wavelength and frequency of a wave are related by the formula:

velocity = frequency $\times$ wavelength.

So, for a given wave velocity, the higher the frequency, the shorter the wavelength.

A wave of twice the amplitude of another carries four times the energy.

For mechanical waves, the density and the elasticity of the medium are important factors in determining their velocity.

Sound is a longitudinal wave. Its velocity in a gas is of the order of a few hundred metres per second; in liquids, around 1 000–2 000 metres per second; and in solids, the order of a few thousand metres per second.

Ultrasonic waves are sound waves with frequencies outside the audible range. Because of their high frequency, they can carry considerable energy at small amplitudes.

The ear does not respond equally to sounds of different frequencies. It is sensitive to sounds from around 20 Hz to 20 kHz, with a maximum sensitivity at about 3. 5 kHz.

The decibel is not an absolute measure of sound intensity level but effectively measures sound intensity ratios. The level in decibels of a sound of intensity I is defined as $10 \log I/I_0$, where I_0 is the intensity at the threshold of hearing. A scale, expressed in dB(A) corrects for the bias in the ear's response to different frequencies.

SUMMARY QUESTIONS

14 Figure 29.24 shows a sketch of an idealised pulse at a particular instant of time, $t = 0$. The pulse is travelling at 2 m s^{-1} along a rope.
a On a copy of the diagram, show the position of the pulse 2 seconds later.
b On separate axes, show the variation of displacement of P with time.
c On the same axes, show how the displacement of Q varies with time. Label clearly the time scale for parts **b** and **c**.

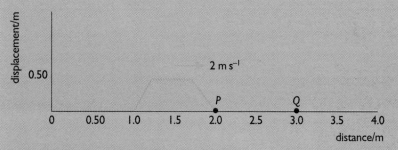

Figure 29.24 Displacement-distance graph for a pulse

15 Ultrasonic waves travel at 1 400 m s^{-1} in water and 340 m s^{-1} in air.
a Calculate the wavelength of ultrasonic waves of frequency 1 MHz **i** in air and **ii** in water.
b A bat, flying by sonar, emits ultrasonic waves of frequencies ranging from around 50 kHz to 0.2 MHz. Dolphins communicate using ultrasonic waves of around 0.1 MHz. Compare the wavelengths in the two situations.

16a The ear is sensitive to sounds in air with frequencies between about 20 Hz and 20 kHz. To what range of wavelengths does this correspond?
b An individual is able to sense the direction from which a sound is coming by detecting the difference in phase of the sound arriving at each ear. Draw a diagram and explain, in principle, how this works. Why is it much more difficult for the ear to detect the direction of high-frequency sounds than low-frequency ones?

17 Earthquakes radiate two types of wave through the Earth: a primary or P-wave, which is a compression wave similar to sound, and a shear or S-wave, which is transverse in nature. The P-wave travels at 7 800 m s^{-1}, while the S-wave travels at 4 200 m s^{-1}.

A seismograph at a recording station receives a P-wave 100 seconds before the S-wave. What distance, in km, is the recording station from the centre of the earthquake? How, with two or more recording stations, could you locate the earthquake?

18 The speed v of a longitudinal wave on a stretched slinky spring is given by $v = \sqrt{(F/\mu)}$ where F is the tension in the spring and μ is its mass per unit length.
a A slinky, of mass 0.5 kg and spring constant 1.5 N m^{-1}, is stretched to a length of 4 m. What is the value of the speed of the pulse and how long does it take to travel along the length of the slinky? Assume that the slinky has negligible unstretched length.
b Repeat the calculations in **a** for a stretched length of **i** 3 m and **ii** 6 m. Explain your answers by describing how the pulse is propagated along the spring.

19 The speed v of a wave along a stretched wire, such as that on a violin, is equal to $\sqrt{T/\mu}$, where T is the tension in the wire and μ its mass per unit length.
a Show that the wire will oscillate with a lowest possible frequency f given by $f = (\frac{1}{2}L)\sqrt{T/\mu}$, where L is the stretched length of the wire.
b Suppose the wire is to vibrate at a natural frequency of 650 Hz. Using the formula, calculate a value for the tension T in the wire as it vibrates at this frequency. To do this, you will need to estimate values for L and μ.
c What is the value of the uncertainty in your calculated value for T?

20 The speed v of a transverse pulse on the surface of water in a ripple tank is equal to $\sqrt{gh}$, where g is the gravitational field strength and h the depth of water. Sketch a graph showing how the speed of the pulse varies with depth using values for h from 1 cm up to 10 cm. What problems might arise if you tried to estimate the speed using values for h outside this range?

21a An increase in noise level of 3 dB increases the sound intensity by about a factor of 2. About how many times more intense is:
i a jet taking off at 125 dB than the sound of thunder at 105 dB?
ii conversation at 60 dB than a whisper at 20 dB?
b Why are noise levels measured on the decibel scale, which is a logarithmic scale? What is meant by a level of 0 dB on this scale?

22a There is evidence that light is a type of wave motion whose average wavelength is about 550 nm. What frequency of sound would have a similar wavelength in **i** a rod of steel, where it travels at about 6 000 m s^{-1}, **ii** water, where it travels at about 1 500 m s^{-1}?
b Acoustic waves have been generated with frequencies up to a few gigahertz. They travel reasonably well through solid materials but are rapidly attenuated in liquids or gases. Why might this be?

30 WAVE BEHAVIOUR

Frequency, wavelength, velocity, amplitude and phase are all terms used to describe waves (← 29). More important is the manner in which waves behave in various circumstances.

The nature of the reflection, refraction, diffraction and interference of waves is of particular interest to us. A doctor may use reflection of ultrasonic waves to investigate the well-being of an unborn baby (see Figure 30.1). Trawlers are able to locate shoals of fish also using reflection of high-frequency sound waves. Television aerials receive signals from the transmitter by diffraction, even though there may be no clear line of sight between the receiving aerial and the transmitter.

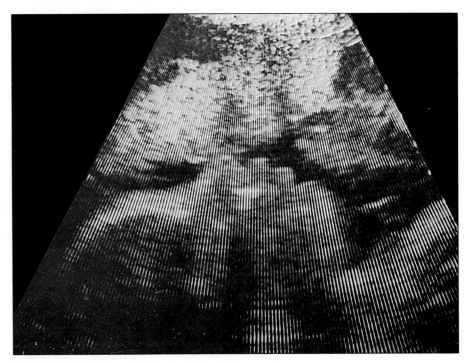

Figure 30.1 Ultrasound scan of a foetus in the womb

30.1 PRINCIPLE OF SUPERPOSITION

When a handful of pebbles is thrown into a still pond, many sets of circular ripples are produced. What happens to the water at a point where two or more ripples meet?

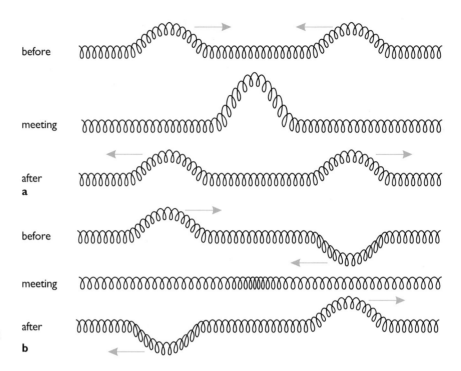

Figure 30.2a Two positive pulses nearing and **b** two opposite pulses meeting

To help answer this, you can simplify the problem. Figure 30.2a shows two positive transverse pulses approaching each other on a stretched spring. Where they meet, the displacement of the spring is larger than either of the individual pulse amplitudes. A short time later, the two pulses seem to have passed through each other and continue along the spring. The pulses may be temporarily modified in shape as they meet but afterwards, they continue as though nothing had happened. Figure 30.2b shows what happens when opposite pulses meet. The positive pulse momentarily cancels the negative pulse.

Question

1 In Figure 30.2b, energy is stored within each of the pulses before and after they meet. What becomes of this energy at the instant they meet and cancel each other?

So, at a point in the pond where ripples from a number of pebbles meet, the water is displaced by an amount equal to the sum of the displacements due to each of the individual waves at that instant. The same is true of waves from any source.

Everyday experience helps to confirm this conclusion. Sound waves emerging from a number of different instruments in an orchestra are able to pass through the air without permanently affecting each other. At any moment, the air at a given point will be displaced by an amount equal to the sum of the displacements due to all the separate waves at that instant. If sound waves could not pass through each other, it would be impossible to listen to music and hold a conversation at the same time!

This is summarised in the **principle of superposition**: at any instant, the resultant disturbance due to two or more waves is the algebraic sum of the disturbances arising from each separate wave. When waves meet, the propagation of each is completely unaffected by the others.

30 WAVE BEHAVIOUR

> **Question**
>
> **2** List some of the consequences if waves did permanently affect each other when they met.

30.2 REFLECTION OF WAVES

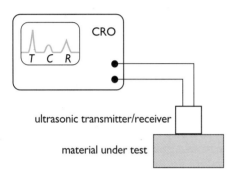

T : transmitted pulse
C : pulse reflected from flaw
R : pulse reflected from back of object

a

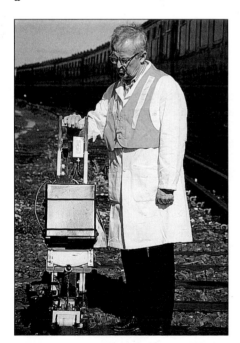

b

Figure 30.3a Ultrasonic flaw detection, **b** detector testing railway track

When you shout or clap your hands near a wall of large surface area, you hear an **echo** of the sound. Waves coming from the source of the sound are **reflected** at the wall. The time between the clap and hearing its echo will depend on the distance the wave travels and the speed of the wave.

> **Question**
>
> **3** A girl finds that, when she repeatedly claps her hands in front of a large brick wall, she can time each clap to coincide with the echo from the previous clap. Using a stopwatch, she finds she is clapping at a frequency of 3 Hz. Given that sound travels at a speed of 330 m s^{-1}, estimate how far away she is from the wall.

> *Ultrasonic flaw detection*
>
> An echo-sounding technique using ultrasonic waves (← 29.4) can check for flaws or cracks in materials. A big advantage of this method is that it is non-destructive — the material to be inspected does not have to be cut up or affected in any way. It is particularly useful for inspecting railway tracks and pipes.
>
> A piezoelectric crystal (← 29.4), able to generate ultrasonic waves of frequency around 1 MHz, is placed in contact with the object under test. Pulses of radiation are transmitted through the object. Reflections take place from the back surface of the object and the pulses return to the transmitter, which also acts as a detector. Any cracks or flaws within the material cause reflected pulses to return sooner. When the received pulses are displayed on an oscilloscope, their shape gives an indication of the size and position of the cracks or flaws (see Figure 30.3).

Reflection may be investigated using a long spring, such as a slinky. When a transverse pulse is sent along a stretched slinky, reflection occurs at the fixed end and the pulse returns along the spring.

Notice that, on reflection, there is a change in phase of the pulse and at one instant during reflection, the spring is actually straight. Figure 30.4 explains how this arises. As the pulse arrives at the fixed end a an interaction occurs so that equal forces are exerted on the support and on the spring in the downward direction at that point. This causes the positive pulse to be reflected as a negative pulse b. This pulse begins to be formed as soon as the leading edge of the incident pulse reaches the fixed end. In c, the spring is momentarily straight as the trailing edge of the incident pulse cancels the leading edge of the reflected pulse. This is what one would expect from the principle of superposition. The reflected pulse then returns along the spring in d and e.

WAVES: TRANSFERRING ENERGY THROUGH SPACE

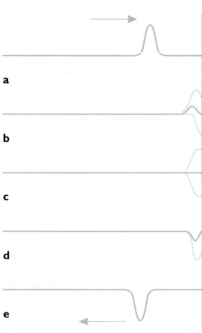

Figure 30.4 Formation of a reflected pulse

Waves are, in effect, a train of positive and negative pulses, and so behave in much the same way as they arrive at a reflecting boundary.

Question

4 When a wave is sent along a spring that is not held or fixed at the far end, the pulse reflects but without a phase change. Using the principle of superposition, try to explain this.

REFLECTION OF WATER WAVES

The reflection of two-dimensional waves can be investigated using a ripple tank. This is basically a large shallow container filled with water to a depth of about a centimetre. Waves on the surface of the water focus light from a small source above the tank on to a white sheet of paper beneath the tank (see Figure 30.5). The ripples may be continuous waves or single pulses and circular or straight wavefronts can be used. The effect of various barriers on the waves may also be observed.

Figure 30.6 shows photographs taken using a ripple tank. In a, circular waves originating from a dipper strike a straight barrier. After reflection, the waves remain circular in shape and appear to come from a point which is as far behind the barrier as the dipper is in front. In b, a straight wavefront meets a barrier inclined at an angle to the incoming wave. In c, straight waves meet a curved barrier, and are reflected and focused to a point.

Ripple tank patterns provide a means of understanding wave processes. Sound waves through air are similar in some ways to ripples on water, but there are important differences. Sound waves are longitudinal, and wavefronts from a small source are spherical in shape and travel three dimensionally.

Questions

5a What experiments would you perform to demonstrate the effects shown in Figure 30.6? Explain how you would obtain the equivalent of circular and straight waves.
b What experiments would you devise to demonstrate that light consists of waves?

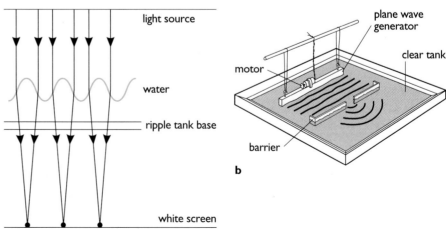

Figure 30.5a Ripple tank **b** Ripple tank

30 WAVE BEHAVIOUR

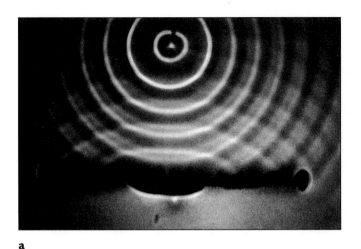

a

b

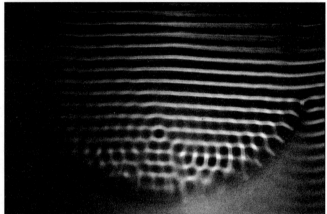

c

Figure 30.6 Waves in a ripple tank:
a circular waves reflected at straight barrier,
b straight waves meeting inclined barrier,
c straight waves meeting curved barrier

Question

6 This question refers to Figure 30.7.
a How long does it take for the sound to reach the sea bed and back?
b The speed of the sound waves is 1500 m s^{-1}. How deep is the water?
c What range of depths can be measured by this system?
d If the time delay can be measured to the nearest 0.1 ms. What will be the uncertainty in the calculated value of depth?

Navigation and ranging using ultrasonic waves

The ease with which bats fly and feed at night is due largely to the manner in which they emit and receive ultrasonic waves. Bats of the Vespertilionidae family emit through their nostrils ultrasonic waves of frequency up to 110 kHz and receive echoes through their highly developed ears. While cruising, they emit pulses of ultrasound of about 3 ms duration and 70 ms apart. Echoes received 60 ms after a pulse is sent out would correspond to reflection from an object 10 m away because the speed of the wave is 330 m s^{-1}. As a bat nears an obstacle, the duration of its emitted pulses decreases to about 0.3 ms and they are sent out at about 5 ms intervals. So objects only 50 cm away can be detected.

Some other species of bat appear to locate objects by sensing the intensity of the echo. When an object is small compared to its distance from the bat, the echo intensity received by the bat varies inversely as the fourth power of the bat's distance from the object. Changes in echo intensity indicate whether the bat is approaching or receding from an object.

G WAVES: TRANSFERRING ENERGY THROUGH SPACE

> ### Sonar
> Sonar, or **so**und **na**vigation and **r**anging, is a system used, amongst its many applications, to measure the depth of the sea bed. A typical system sends pulses of 40 kHz ultrasonic waves from a boat to the sea bed, where they reflect back to a receiver on the boat. Pulses of 1 ms duration might be sent every 10 ms. The transmitted and received signals can be displayed on an oscilloscope.

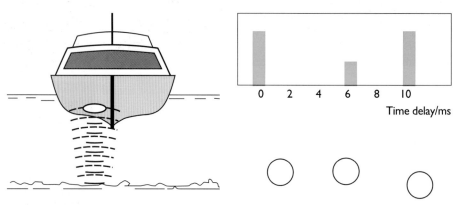

Figure 30.7 Measuring the depth of the sea bed

30.3 REFRACTION OF WAVES

When waves passing from one region to another undergo a change in speed, the waves **refract**.

> ### Question
> **7** Figure 30.8 shows a pulse travelling along a spring and into another, heavier one.
> **a** Compare the speeds of the pulses in the two springs.
> **b** Is all of the pulse energy transmitted into the heavier spring?

Refraction may also be observed in a ripple tank. To change the speed of the ripples, you can get them to travel from relatively deep water to rather shallower water. A sheet of thick glass or perspex on the floor of the tank will produce this effect, the edge of the glass marking the boundary between the two regions.

Figure 30.9a shows what might be observed. The ripples are meeting the boundary normally and pass into the shallower region without any change in direction but with a decrease in wavelength. The frequency remains unchanged.

Figure 30.9b shows a similar set of ripples meeting a boundary at an angle. Note that the wavefront changes direction as it enters the shallower region and slows down. Figure 30.9c illustrates that waves of higher frequency are refracted less than those of lower frequencies.

Christian Huygens, a Dutch physicist (1629–98), established a principle for predicting wave behaviour in circumstances such as those described above ($\rightarrow$ 31.2).

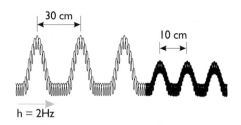

Figure 30.8 Change of wave speed in springs

30 WAVE BEHAVIOUR

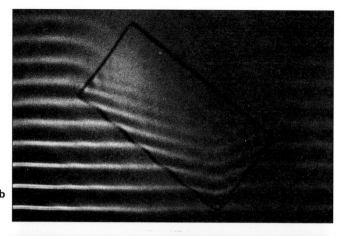

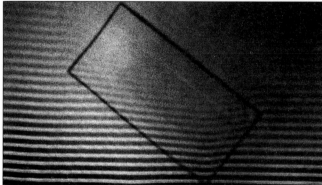

Figure 30.9 Refraction in a ripple tank: **a** change of speed of water waves, **b** plane waves meeting inclined boundary, **c** refraction of higher frequency waves

Question

8 What will happen to straight waves in a ripple tank when they pass from a region of shallow water to one where it is deeper? What happens when you gradually increase the angle between the incidence wavefront and the boundary between the shallow and deep water?

REFRACTION OF SOUND

You may have noticed that sounds do not travel far in the middle of a hot sunny day but carry much further in the evening when it is much cooler. Sound waves travel more rapidly in warm air than in cool air.

During the day, the ground is hot and the air cools as you rise above ground level. So, the lower edge of a wavefront travelling near to the ground moves faster than the upper edge of the same wavefront. The result is that the wavefront is deviated upwards, missing observers on the ground. In the evening, the temperature gradient reverses, and the air above is hotter than that in contact with the ground. The upper edge of the wavefronts travel faster and the sound waves deviate towards the ground (see Figure 30.10).

Figure 30.10 Refraction of sound waves

30.4 DIFFRACTION OF WAVES

If you shout to a friend out of your direct line of sight, say, on the opposite side of a high brick wall, he or she is likely to hear you quite clearly. This will be the case even if there are no reflective surfaces nearby to provide a path for the waves. Sound waves seem to spread out as they pass over and around the wall. This spreading effect is called **diffraction**. The amount of diffraction is determined by the size of the diffracting obstacle relative to the wavelength of the waves.

The wavelengths of sound vary from a couple of centimetres at the high frequency end to 10 m or so at the low frequency end of the audible spectrum. This range is of the **same order of magnitude** as the dimensions of many sources, and receivers, of sound: doorways, window openings, rooms, buildings, the human ear and the mouth. So diffraction effects are very important in acoustics.

Light seems to behave in a very similar way but the size of the effect is generally very much smaller ($\rightarrow$ 31).

DIFFRACTION OF WATER WAVES

Again, diffraction effects may be conveniently demonstrated using water waves in a ripple tank. Figure 30.11a shows a typical pattern obtained when waves are allowed to pass through a gap in a barrier. Notice that the waves spread into the 'shadow' of the barrier.

With a smaller gap but the same wavelength, as in Figure 30.11b, the spread of the pattern is greater. Note the secondary maxima and minima. Increasing the wavelength, as in Figure 30.11c, has the effect of increasing the amount of diffraction. Figure 30.12 shows the effect of water waves passing around the edge of a harbour wall.

Note that, a sharp 'shadow' of the gap, edge or obstacle does not occur.

Figure 30.11 Diffraction of water waves: **a** at a large aperture, **b** at a small aperture, **c** of longer wavelength

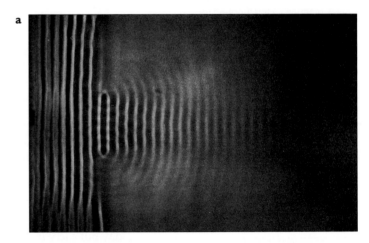

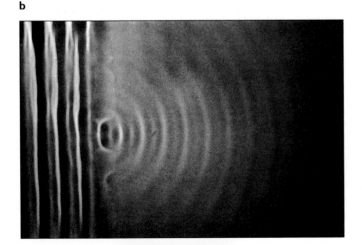

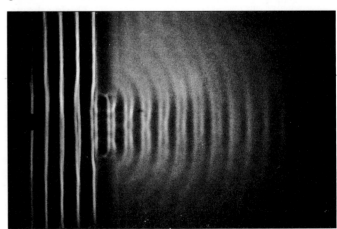

30 WAVE BEHAVIOUR

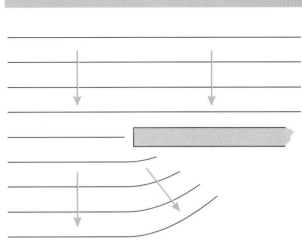

Figure 30.12 Diffraction of water waves around the edge of a harbour wall

Questions

9 If you sit directly in front of the speaker when listening to music on the radio, the sound will usually be clear and crisp. But if you sit to one side, the sound may become rather muffled and lacking in the higher frequencies. Explain this in terms of diffraction by considering the cone of the loudspeaker as a diffraction aperture.

10 Both light and radio signals seem to possess wave properties. Radio signals can be detected by an aerial even though there may be no direct line of sight between the transmitter and the aerial. However, light does not seem to spread as it passes through windows and other openings. What does this suggest about the relative wavelengths of light and radio waves?

30.5 INTERFERENCE OF WAVES

In 30.1, we looked at the superposition of two pulses travelling in opposite directions along a spring. What happens when two sets of circular waves superpose can be seen in Figure 30.13a. The two sources S_1 and S_2 produce circular waves of the same frequency and phase, and so are described as **coherent** sources ($\rightarrow$ 33.1).

Figure 30.13b illustrates what is taking place at any instant. With transverse

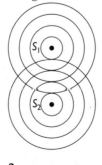
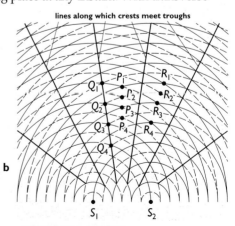

Figure 30.13a Coherent circular waves, **b** Interference pattern produced

WAVES: TRANSFERRING ENERGY THROUGH SPACE

waves, such as those on the surface of a ripple tank, waves from both S_1 and S_2 cause the water at P_1 to move up and down as they pass. The two waves meet at P_1 in phase because it is equidistant from S_1 and S_2. The waves superpose constructively to produce a large amplitude of movement at P_1. Points P_2, P_3, P_4, etc. experience similar large amplitudes of movement.

At Q_1, however, the two waves arriving at any instant are half a wavelength out of step, that is, they have a phase difference of π. So the waves superpose destructively and the water has very little movement. Points Q_2, Q_3, Q_4 and so on, experience similar effects.

At R_1, the waves are again in step because those from S_1 have travelled a whole wavelength further than those from S_2. So constructive interference takes place at points R_1, R_2, R_3, etc.

The result is a stationary pattern of superposition in which the water moves with large amplitudes in some places and very little in other places. This is known as an **interference pattern**.

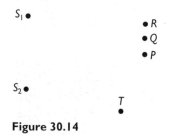

Figure 30.14

Question

11 In Figure 30.14, S_1 and S_2 are two coherent sources. Points P and R are successive maxima and Q is a minimum.
a What is the value of the phase difference between the two waves arriving i at P? ii at Q? iii at R?
b What is the path difference in wavelengths between the two waves arriving i at P? ii at Q? iii at R?
c T is a general point. What is the path difference in wavelengths between the two waves arriving at T?
d Write down the phase difference at T in terms of the path difference. (A path difference of one wavelength corresponds to a phase difference of 2π radians.)
e What general condition needs to apply for T to be i a maximum? ii a minimum?
f Suppose you arranged for the sources S_1 and S_2 to produce waves of the same frequency but out of phase with one another. Would you still expect to see a pattern? If so, how would it differ from that of coherent sources?

INTERFERENCE OF WATER WAVES

In a ripple tank, two dippers driven by the same motor can produce wave sources that are coherent and of about the same amplitude. Figure 30.16a shows the interference pattern produced with a separation of the dippers of about 5 wavelengths. If this distance is reduced, the pattern will spread out further as in b, where the dipper separation is about 2 wavelengths.

Similar results are obtained when straight waves fall on two gaps in a barrier in a ripple tank. The two sets of diffracted waves produced superpose to produce an interference pattern, as shown in Figure 30.15c.

Interference can also take place with other waves, such as sound waves (see Figure 30.15 and experiment 30.1).

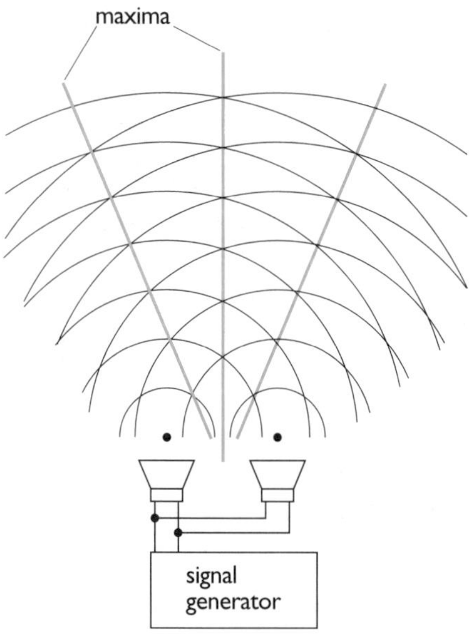

Figure 30.15 Interference of sound waves

Question

12a This question refers to Figure 30.16. Why is the interference pattern in b broader than that in a?

30 WAVE BEHAVIOUR

b Why does increasing the wavelength also cause the pattern to become broader?

c Suppose the gaps in c were increased, keeping their separation the same. Why might you not then see an interference pattern?

Figure 30.16 Interference of water waves: **a** of small source separation, **b** of large source separation, **c** after diffraction

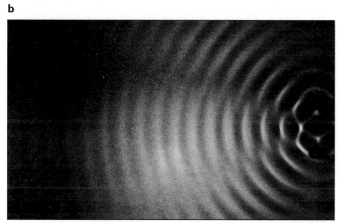

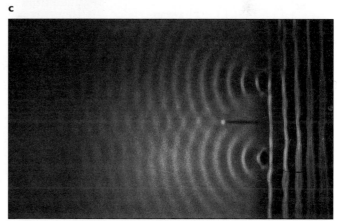

EXPERIMENT 30.1 Interference of sound waves

Two small loudspeakers driven by the same oscillator set at, say, 3 kHz, will give you two coherent sources of sound of equal amplitude. The speakers should be away from reflective surfaces, such as out of doors or in an anechoic room. If you place the speakers a few metres apart and walk slowly between them, you should hear the sound changing in intensity as you move. If you carry a microphone connected to an oscilloscope, as in Figure 30.17, you can see corresponding changes in amplitude.

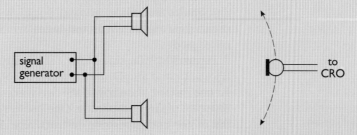

Figure 30.17 Arrangement for experiment 30.1

WAVES: TRANSFERRING ENERGY THROUGH SPACE

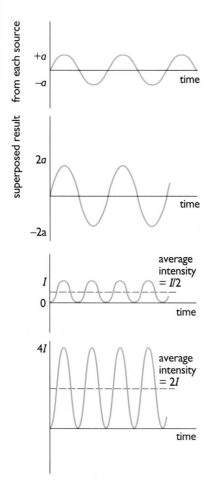

Figure 30.18 Energy and superposition

ENERGY AND INTERFERENCE

When waves from two sources interfere, the wave amplitude at a position of maximum is the sum of the wave amplitudes from each source — it is twice as large as either of the sources. Since the intensity of the wave is proportional to amplitude squared (← 29.2), the intensity of the waves at a maximum is four times that due to either of the two sources on their own. At minima there is little sound intensity. Averaging the intensity over the pattern, the intensity due to the presence of the two sources is twice that due to either on their own, as one would expect.

Figure 30.18 illustrates this. Suppose the amplitude and intensity due to each source is a and I, respectively. Then the result of superposition at a maximum produces a wave of amplitude $2a$ and intensity $4I$. As a result of superposition, the energy is not spread uniformly throughout the pattern but redistributed.

Questions

13 Suppose the connections to one of the speakers in Figure 30.18 were reversed.
a What effect would this have on the value of the frequency of sound emitted from each speaker?
b What effect would it have on the value of the phase difference between the two sources?
c Why would there still be an interference pattern produced?
d How would it be different from the original pattern?

14 A large piece of card has a hole cut in it. The size and shape of the hole is such that it matches the cone of a small loudspeaker. The speaker is connected to an audio frequency oscillator set at 1 kHz. With the card placed over the speaker, so that just the cone is visible, the sound is louder than when the card is not there. Suggest an explanation for this.

30.6 STATIONARY WAVES

Two dippers in a ripple tank can produce a stationary pattern of interference between two circular travelling waves (← 30.5). Along the line joining the two dippers, transverse waves are travelling in opposite directions and superposing. They form a **stationary wave**. There are points in it where the water is moving with large amplitude and others where the water is still.

A similar effect can be produced by sending a transverse wave along a slinky spring, fixed at one end. The wave reflects with a phase change and the reflected wave superposes with the incidence wave. At certain frequencies, a stationary wave will be produced. Again, parts of the spring will be oscillating with large amplitudes and other parts will be nearly stationary. The points with the biggest oscillations are known as **antinodes**, while those that are virtually stationary are called **nodes**.

Figure 30.19 illustrates how a stationary wave arises. Two waves of the same frequency and amplitude are travelling at the same speed in opposite directions. At one instant, shown in a, the two waves superpose constructively to give a resultant of large amplitude. One-quarter of a cycle later, in b, the two waves are out of step and superpose destructively. After half a cycle, the two waves are again in step and superpose constructively, as in c. Three-quarters of a cycle later

30 WAVE BEHAVIOUR

Figure 30.19 Formation of a stationary wave

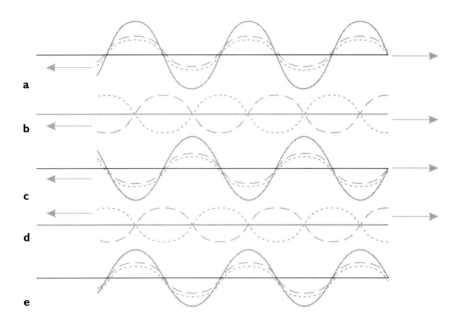

there is a resultant of zero amplitude, as in d. Finally in e, after one cycle of the wave, they are in step as they were in a.

So, although there are two travelling waves present, a stationary pattern is produced. The distance between successive nodes, or antinodes, is half a wavelength.

EXPERIMENT 30.2 Investigating stationary waves: Melde's experiment

Figure 30.20 shows a 1 m or 2 m length of rubber cord, thin latex rubber tubing or thin piano wire, stretched between a pulley and a vibrator driven by a signal generator. Using this arrangement, you can generate a standing, or stationary, wave by starting at a low frequency and slowly increasing it until you find a frequency at which the string oscillates with a large amplitude, as in Figure 30.20a. Another stationary wave occurs at double this frequency, as shown in Figure 30.20b. Similarly, one occurs at treble and quadruple the frequency as shown in Figure 30.20c and d. By observing these patterns with a stroboscopic light flashing at a frequency close to that of the vibrator, the nature of the stationary wave can be seen clearly.

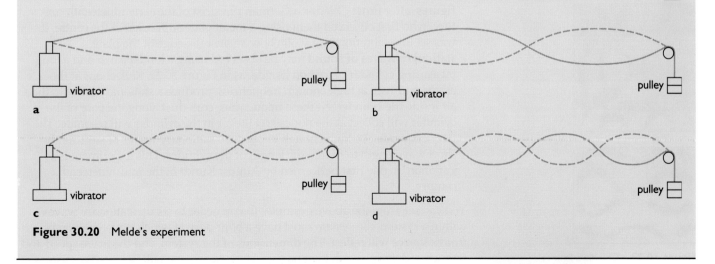

Figure 30.20 Melde's experiment

WAVES: TRANSFERRING ENERGY THROUGH SPACE

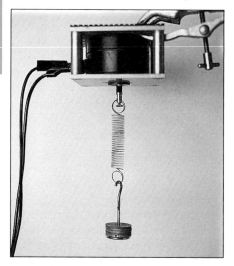

Figure 30.21 Producing longitudinal standing waves

Experiment 30.2 also shows that the distance between successive nodes or antinodes is a half a wavelength. In any given loop, all parts of the rubber cord move in phase, although with different amplitudes. Maximum amplitude occurs at the antinode. In adjacent loops, the oscillations of the cord are in antiphase, or π radians out of phase. Superposed incident and reflected waves will only produce a stationary wave pattern at certain frequencies. The time taken for the wave to travel along the length of the cord has to be an integral number of periods of the oscillator producing the wave. Suppose that the wave takes 1/15 second to travel down to the fixed end and back to the source. If, in this time, the vibrator has gone through one cycle of oscillation, the wave returns in phase with the source. This causes the stationary wave amplitude to build up so that one stationary wave loop is produced. The wave will return in phase with the source when the period of oscillation of the vibrator is: 1/30 second for two loops, 1/45 second for three loops, 1/60 second for four loops and so on.

Question

15a The stretched length of the rubber in Figures 30.21a to d is 1.5 m. What is the value of the speed of the wave along the rubber cord used in the experiment?
b Suppose the tension in the cord is increased so that the wave speed in it doubles. What frequencies of oscillation of the vibrator will be required to produce stationary waves in a 1.5 m length of the cord at its new tension?

Stationary waves may be observed in many situations:

- **longitudinal standing waves** can be set up along a stretched spring by means of a vibrator, as in Figure 30.21. As the frequency is adjusted, the reflected wave combines with the transmitted wave and a stationary wave pattern appears. Parts of the spring become nodes while others parts oscillate longitudinally with large amplitudes. The number of nodes depends on the frequency of oscillation.
- the same vibrator can be used to send waves across a flat horizontal steel plate. At certain frequencies, quite complex stationary wave patterns occur as a result of wave reflections from the edge of the plate. They may be observed by sprinkling salt or sand on the surface of the plate. The grains collect along the nodes of the pattern (see Figure 30.22). Such patterns are known as **Chladni's figures** after Ernst Chladni, a German physicist of the early nineteenth century, who first observed them using a metal plate supported at the centre. It was set into vibration by a violin bow against the edge of the plate.
- **stationary waves of sound** may be observed using a loudspeaker and a long measuring cylinder placed on its side, as in Figure 30.23. Reflections at the closed end will, at high enough frequencies, produce a stationary wave in the air inside the cylinder. At these frequencies, cork dust along the base of the cylinder will collect at the nodes and the air in the cylinder will **resonate**. The frequency must be high enough to produce stationary waves of sufficiently short wavelength to fit into the length of the cylinder. This experiment is a variation on one first performed by August Kundt in the mid-nineteenth century.

You can see from the above examples that in order to set up stationary waves within a system, the system must have a boundary at which travelling waves from a source will reflect. The dimensions of the system, and the wave speed and frequency determine whether or not a stationary wave will 'fit' into the system.

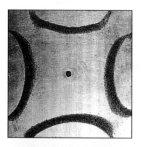

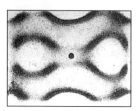

Figure 30.22 Chladni's figures

MUSICAL INSTRUMENTS

A note played on an instrument, such as a guitar, consists of a fundamental and overtones (← 29.5). The fundamental corresponds to the first mode of vibration (the lowest frequency mode) of the guitar string. The overtones' frequencies are multiples, or **harmonics**, of the fundamental frequency. If you clamp a plucked string by touching it very lightly at its mid-point with a piece of card, or a finger you will remove the fundamental from the note. You will then hear quite clearly the second harmonic, which is twice the frequency of the fundamental.

Stringed instruments

Experiment 30.2 illustrates that a string stretched between two fixed supports will vibrate in a number of possible modes. The first five modes are shown in Figure 30.24. The first mode is the fundamental or first harmonic, whose frequency is f. There is an antinode in the middle with nodes at each end. The second mode is the first overtone, or the **second harmonic**, and has a frequency of $2f$. There is a node in the middle with antinodes one quarter and three quarters of the way along the string. The other modes can be described in a similar way. For example, the fifth mode is the fourth overtone, the **fifth harmonic** and has a frequency of $5f$. There are five antinodes present in this mode of vibration.

When a string is plucked, transverse waves travel to both ends and their reflections set up a stationary wave pattern. Vibrations of the string generate longitudinal waves in the surrounding air. Those frequencies producing waves that fit into the string will be present. The note heard is dominated by the fundamental frequency, but the presence of the second, third and higher overtones contributes to the quality of the note from the instrument. The harmonic quality depends on resonances in the body of the instrument and the method of causing the vibration.

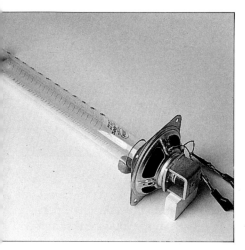

Figure 30.23 Stationary sound waves

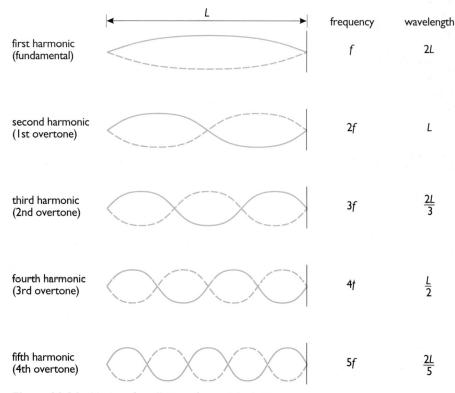

Figure 30.24 Modes of oscillation of a stretched string

WAVES: TRANSFERRING ENERGY THROUGH SPACE

> **Question**
>
> **16** The speed of a transverse wave along a stretched string is given by $v = \sqrt{F/\mu}$, where F is the tension and μ the mass per unit length of the string. For a string of length L:
> **a** what is the wavelength, in terms of L, of the transverse wave in the string when it is vibrating in its fundamental mode?
> **b** what is the frequency of the fundamental, in terms of T, μ and L?

Wind instruments

Wind instruments, such as a recorder, a flute and an organ pipe, depend upon vibrations of a column of air similar to those produced by the loudspeaker in the measuring cylinder in Figure 30.24.

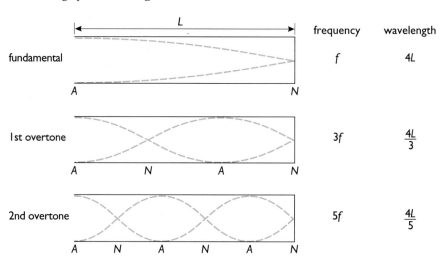

Figure 30.25 Modes of oscillation of a closed pipe

Consider first a pipe closed at one end. The first three modes of vibration are shown in Figure 30.25. The first mode, producing the fundamental frequency, has a node at the closed end with an antinode at the other. So there is a large amplitude of longitudinal vibration at the open end, decreasing to zero amplitude at the closed end. The effective length L of the pipe is the distance between the node and antinode. This is equivalent to a quarter of the wavelength, or $\lambda/4$. In practice, the air slightly beyond the open end of the pipe will oscillate. This means that the effective length of the air column is greater than the physical length of the pipe.

Since $v = f\lambda$ ($\leftarrow$ 29.2), the fundamental frequency f for a closed pipe is:

$f = v/4L$,

where v is the velocity of sound in air.

The second mode, or first overtone, has an extra node and antinode. The length of the pipe is then $3\lambda/4$ and the wavelength, $4L/3$. The first overtone has a frequency $3v/4L$, or $3f$, and so is the third harmonic. Similarly, the third mode, or second overtone, has a frequency of $5f$, and is the fifth harmonic.

Now consider a pipe open at both ends. Stationary waves will still occur but there must be antinodes at each end. The first three modes of vibration are shown in Figure 30.26. You can see that the fundamental has a wavelength of $2L$ and a frequency of $v/2L$. The first overtone has a wavelength of L and a frequency of v/L. This is twice the fundamental frequency and so is the second

30 WAVE BEHAVIOUR

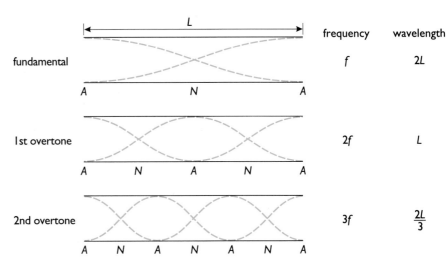

Figure 30.26 Modes of oscillation of an open pipe

harmonic. The second overtone will have a frequency of three times that of the fundamental.

From the above, you can see that only odd harmonics are possible for a closed pipe while for an open pipe all harmonics are possible. As a result, open pipes usually give a more satisfying sound than closed pipes. Note also that the fundamental frequency for an open pipe is twice that for a closed pipe of the same length. In a wind instrument, different notes are produced by changing the effective length of the pipe by opening holes, or stops. Alternatively, players can produce higher frequencies by blowing harder.

Various methods are used to set the air in a pipe into vibration but all make use of the high-frequency **vortex shedding** (← 19) that occurs when there is a steady air flow across, for example, a hole or a reed. The vortex shedding around the hole or the reed causes vibration of the air inside the pipe.

30.7 DOPPLER EFFECT

The pitch of a note from a car engine or a train appears to change as it passes you. This apparent change in frequency as the source of sound passes is known as the **Doppler effect** (see Figure 30.27). It occurs whenever the wave source, or observer or both are moving relative to each other.

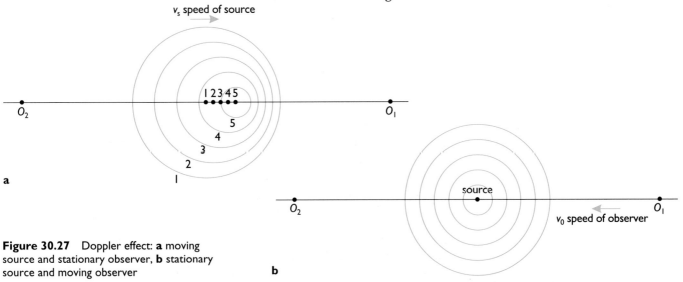

Figure 30.27 Doppler effect: **a** moving source and stationary observer, **b** stationary source and moving observer

WAVES: TRANSFERRING ENERGY THROUGH SPACE

Apparent frequency change due to the Doppler effect

The apparent frequency of a source due to the Doppler effect depends on whether it is the source, the observer or both that are moving.

a Moving source and stationary observer

Suppose wavefronts emerge from a stationary source and travel at speed v to two observers O_1 and O_2. Clearly, they both hear the same pitch of note of frequency f. The wavelength $\lambda = v/f$.

Suppose now, that the source moves towards O_1 at speed v_s, as in Figure 30.27a. Wavefront 1 leaves the source when it is at position 1. A short time later, wavefront 2 leaves the source when it is at position 2 and so on to wavefront 5, which leaves the source when it is at position 5.

Along the direction S_1 to O_1, f wavefronts leave the source in 1 second. The distance occupied by this number of wavefronts is $v - v_s$ because v is reduced by an amount v_s. So the new wavelength λ' becomes $(v - v_s)/f$. The apparent frequency f' is given by:

$$f' = \frac{\text{speed of wave}}{\text{new wavelength}}.$$

The speed of the wave is determined only by the medium through which it travels. So it stays constant at v and:

$$f' = \frac{f.v}{(v - v_s)}.$$

A similar argument leads to the apparent frequency at O_2 becoming:

$$f' = \frac{f.v}{(v + v_s)}.$$

b Stationary source and moving observer

In Figure 30.27b, observer 1 detects f wavefronts arriving from the source in 1 second. Observer 2 is moving towards the source at a speed v_o, so that more than f cycles in 1 second are detected. In fact, v_o/λ more cycles are detected per second. So the apparent frequency is given by:

$$f' = f + \frac{v_o}{\lambda}$$

$$= f + \frac{v_o f}{v}.$$

When the source of sound is moving towards the observer, the distance between the compressions and rarefactions in the air in the direction of travel is reduced. So the wavelength detected by the observer is reduced and the apparent frequency of the sound increases. As the source passes the observer and recedes into the distance, the wavelength is increased and the apparent frequency falls. Similar effects are obtained when the observer is moving relative to a stationary source.

When the source and observer are approaching one another, the apparent frequency f' is given by:

$$f' = \frac{f.(v + v_o)}{(v - v_s)},$$

where v is the speed of the sound wave, v_s the speed of the source, v_o that of the observer and f is the true frequency. When they are separating, the apparent frequency is given by:

$$f' = \frac{f.(v - v_o)}{(v + v_s)}.$$

These equations assume that both source and observer are moving in the same straight line. Note also that the apparent frequency remains constant while the source and observer approach one another. It falls to a lower value when the two pass each other and then remains at that new lower value.

Questions

17 A whistle is attached to a 1 m long rubber tube. A person blows through the tube, causing a note of frequency of 4 kHz to be emitted. The person then spins the whistle around on the end of the tube, taking 2 seconds to make one complete rotation. Calculate the highest and lowest frequencies of the note heard by an observer standing nearby.

18 The equations describing the Doppler effect were derived on the assumption that the medium carrying the sound was stationary and all speeds were measured relative to the medium. Describe qualitatively the effect of a steady wind on the apparent frequency heard by an observer.

19 Horseshoe bats make use of the Doppler effect to detect that they are flying towards or away from an object. Explain, in principle, how they might do this.

Using the Doppler effect to measure blood flow

The Doppler effect has been used quite successfully to monitor blood flow through major arteries (see Figure 30.28). Ultrasonic waves of frequency 5–10 MHz are directed towards the artery and a receiver detects the back-scattered signal. The apparent frequency of the received signal depends on the velocity of flow of the blood. Air-solid interfaces cause excessive reflections and reduce the sensitivity of the system. So gel is used to ensure good acoustic coupling between the transmitter and the skin, and the efficient transfer of the ultrasound's energy into the body.

This method is particularly effective for detecting thrombosis, since this causes a marked change in the speed of flow of the blood. It is also relatively inexpensive to perform and provides less discomfort for the patient than more traditional methods.

$$f' = \frac{f.(v + v_o)}{v}.$$

When the observer is moving away from the source, the apparent frequency f' is:

$$f' = \frac{f.(v - v_o)}{v}.$$

c Both observer and source moving

When both source and observer are moving, the equations are combined to give:

$$f' = f(v + v_o)/(v - v_s)$$

for source and observer moving towards each other, and:

$$f' = f(v - v_o)/(v + v_s)$$

for source and observer moving away from each other.

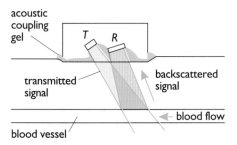

Figure 30.28 Measuring blood-flow rates using the Doppler effect

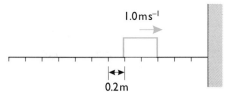

Figure 30.29

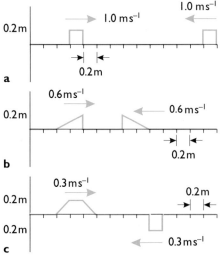

Figure 30.30

30 WAVE BEHAVIOUR

SUMMARY

When two or more waves meet at a point, the net displacement of that point is equal to the sum of the individual displacements caused by each wave as it arrives at that point. This is known as the principle of superposition.

When a wave meets a boundary between two media, some of the wave's energy may be reflected off the boundary and not pass into the new medium. Waves that cross the boundary undergo refraction.

Diffraction occurs when waves pass through an aperture or meet an obstacle. Waves undergoing diffraction spread into the shadow area of the object: the shorter their wavelength, the greater the diffraction.

Interference patterns occur when two or more wavefronts superpose. Waves arriving in phase interfere constructively while those out of phase superpose destructively.

The superposition of waves may also produce a stationary wave pattern. This consists of nodes, where the cord is stationary, and antinodes, where the cord is vibrating with maximum amplitude. Nodes are half a wavelength apart. Stationary waves are utilised in musical instruments to produce harmonics of a fundamental frequency and give character or quality to the sound from a particular instrument.

Sound emitted by a source will appear to be of higher frequency when a source and observer are closing on each other. When they are moving further apart, there is a shift towards the lower frequencies. These changes in frequency are, known as the Doppler effect.

SUMMARY QUESTIONS

20 Figure 30.29 shows the position at $t = 0$ of a wave pulse travelling along a rope fixed to a rigid post. Using the principle of superposition, draw a sequence of diagrams at 0.2 s intervals showing the appearance of the pulse during reflection at the post.

21 Figure 30.30 shows the appearance of three idealised wave pulses on a stretched spring. The graphs are of displacement against distance and show the situation at time $t = 0$. Use the principle of superposition to draw, in each case, the wave profile at $t = 1$ second.

22 Images of unborn babies can be produced using pulses of ultrasonic waves. A transmitter generates ultrasound pulses of about 2 MHz frequency. The pulses are of very short duration and are transmitted at a rate of about 1 kHz. The crystal used to generate the ultrasound rocks rapidly and steadily to and fro so that the whole region around the baby receives the ultrasonic radiation. Echoes can be detected after reflection at various parts of the baby's body and these can be displayed on an oscilloscope to give an image of the baby.
a What is the wavelength, in m, of 2 MHz waves in water, assuming their speed to be $1\,500\text{ m s}^{-1}$?
b To obtain narrow beams of ultrasound, a high frequency must be used. Lower frequency waves spread out too much. Why is this?
c Suppose the transmitter swings steadily through 50° in 0.1 s. How many of the 1 kHz pulses are produced in this time?

d There needs to be good acoustic coupling between the transmitter and the mother's skin for the waves to penetrate into the mother's body. How is this achieved?

e What determines the strength of the reflected pulse? How might the varying strengths be displayed on the oscilloscope screen?

f How might the system be improved so that finer details could be resolved? What are the system's limitations?

23 Railway lines can be tested quickly for flaws or cracks using ultrasonic waves (← 30.2). An ultrasonic probe, acting as both transmitter and receiver, is connected to an oscilloscope. Pulses of ultrasonic waves of very short duration are emitted every millisecond. The frequency of the radiation is 3 MHz and it travels at 6 000 m s^{-1} through the steel rail (see Figure 30.31a).

a When pulses are sent through a rail, a trace appears on the screen of the oscilloscope as in Figure 30.31b. Calculate from the trace the depth of the rail.

b At another place along the rail, an extra pulse appears on the screen of the oscilloscope, as in Figure 30.31c. Why might a flaw in the rail be the cause of this extra pulse? How far from the surface is the flaw?

c What is the value of the wavelength of the ultrasonic waves in steel? Why is it important to use high-frequency ultrasonic waves to detect small cracks in the steel?

24 A method for measuring the speed of sound in a metal rod is shown in Figure 30.32. When the hammer strikes the 1.5 m-long rod, the signal generator is connected to the oscilloscope. A trace appears on the screen for the short time that the hammer is in contact with the rod. The hammer loses contact with the rod when the pulse arrives back at the hammer after reflection at the free end of the rod.

The signal generator output is set at 20 kHz and 10 complete cycles are observed on the screen of the oscilloscope.

a How long is the rod in contact with the hammer?

b How fast does the pulse travel through the metal rod?

c To demonstrate that the contact time is due to the pulse travelling along the rod and not within the hammer, a student uses a rod of half the length. Sketch what you would expect to see on the oscilloscope screen.

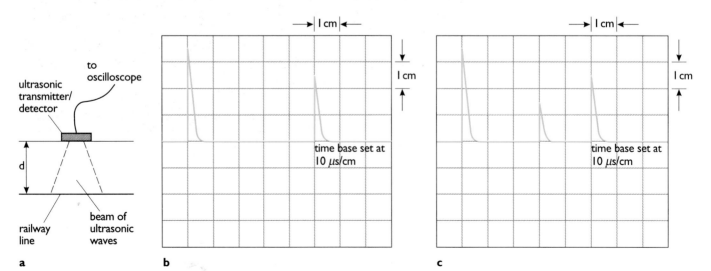

Figure 30.31 Ultrasonic probing of track: **a** arrangement, **b** and **c** traces on oscilloscope

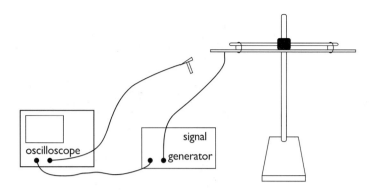

Figure 30.32 Measuring the speed of sound in a metal rod

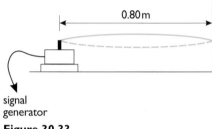

Figure 30.33

d If the signal generator frequency in c was increased to 30 kHz, what would you expect to see?

25 A rubber cord is clamped at one end to a rigid support and at the other to a vibrator. When the vibrator oscillates at 12 Hz, the string appears as in Figure 30.33.
a Sketch the appearance of the cord at **i** 24 Hz and **ii** 48 Hz.
b What is the wavelength of the wave, in m, on the cord at **i** 12 Hz, **ii** 24 Hz and **iii** 48 Hz?
c How fast do the waves travel along the cord?

26 A violin string has a mass of 4×10^{-4} kg m^{-1}. The wire is placed under a tension of 30 N.
a What is the value of the fundamental frequency of a note played on the string when its length is restricted to 35 cm?
b Sketch graphs to show the variation of the fundamental frequency against **i** string diameter and **ii** tension in the string, assuming other factors remain unchanged in each case.

27 A loudspeaker is placed just above the top of a measuring cylinder with an internal height of 40 cm. The loudspeaker is connected to a signal generator whose output frequency is slowly increased from 100 Hz to 1 000 Hz. Assume that sound travels at 340 m s^{-1}.
a Describe in as much detail as possible what you would expect to hear.
b The frequency is then fixed at 500 Hz and water is slowly poured into the cylinder until it is full. What would you hear?

28 An observer stands by the side of a road and listens to the horn of a car being continuously sounded as it approaches him at a steady speed of 30 m s^{-1}.
a Describe what the observer will hear as the car passes by and recedes into the distance. Assume that the speed of sound in air is 340 m s^{-1} and the frequency of the emitted note is 750 Hz.
b What would the observer hear if a 'supercar' moved away from the observer at the speed of sound?

31 WAVE NATURE OF LIGHT

The energy we receive from the Sun comes almost entirely as **electromagnetic radiation** ($\rightarrow$ 32). Light, mostly visible, infra-red and some ultraviolet, is predominant in this electromagnetic radiation. How does the light travel from the Sun to the Earth through the vacuum of space?

To help answer this, you can consider the possibility that light is a wave motion. In such a model, waves of light spread outwards from the Sun. They do so in much the same way as ripples travel outwards from a disturbance on the surface of a still pond.

Evidence that light has wave properties comes from the diffraction effects that occur as it passes through apertures and around obstacles, much as you can observe waves on the surface of water.

31.1 EVIDENCE FOR THE WAVE NATURE OF LIGHT

Waves, such as those in a ripple tank, can undergo reflection, refraction, diffraction and interference ($\leftarrow$ 30). If light consists of waves, it should behave in similar ways.

EXPERIMENT 31.1 Demonstrating the wave nature of light

Warning: take care when using razor blades close to the eye

Setting up the arrangement shown in Figure 31.1a allows you to observe wave behaviour with light. The slit can be made using two halves of a razor blade, as in Figure 31.1b. Best results are obtained when a lamp with a long vertical filament is used.

The pattern is a broad central patch of light with smaller 'fringes' on either side. You will see a clearer pattern with a green filter placed between the slit and your eye. Reducing the slit width will cause a broadening of the pattern (see Figure 31.2).

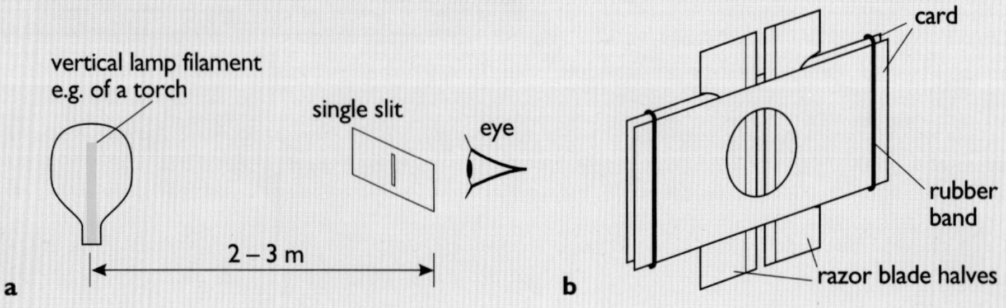

Figure 31.1 a Arrangement for observing diffraction of light at a single slit, **b** an adjustable single slit

31 WAVE NATURE OF LIGHT

The photographs below show single slit diffraction patterns made using a slit of width b and light of only one wavelength. In **a** $b \approx 20\lambda$, **b** $b \approx 5\lambda$, **c** $b \approx 3\lambda$, and **d** $b \approx \lambda$. These images were created using extremely precise equipment. Figure 31.2e shows the sort of image that you would expect to see using the apparatus described opposite.

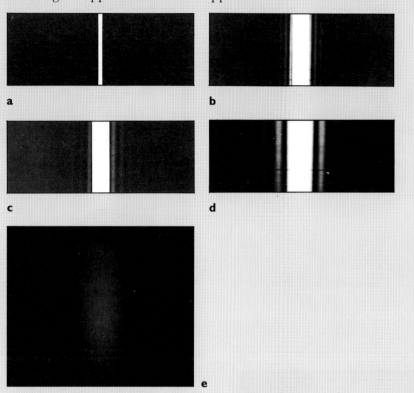

Figure 31.2 Single slit diffraction patterns using green light of $\lambda \approx 0.5$ μm and a parallel slit of approximate width: **a** 10 μm, **b** 2.5 μm, **c** 1.5 μm, **d** 0.5 μm, **e** typical experimental view

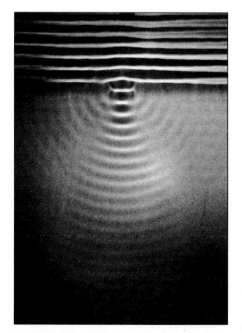

Figure 31.3 Diffraction of water waves passing through a gap in a barrier

Question

1 This question refers to experiment 31.1.
a Why should a green filter make the pattern clearer?
b Will it make any difference if you place the filter between the slit and the lamp, instead of between the slit and your eye?

Experiment 31.1 shows that light does not pass 'cleanly' around the edges of a slit, but seems to spread or **diffract** (← 30.4). Comparison of its behaviour (see Figure 31.2) with that of water waves in a ripple tank (see Figure 31.3) suggests that light does behave like a wave. Clearly, the scale of the two situations is quite different suggesting that the wavelength associated with the light waves is much smaller than that of the water waves in the ripple tank.

Diffraction does not only occur as light passes through an aperture. Figure 31.4a shows the shadow cast by an object, here a razor blade. The edges of the shadow are not very clearly defined and display fringes similar to those obtained with a single slit. Again, compare this with Figure 31.4b, illustrating water waves passing around a harbour wall. Figure 31.5, shows the shadow cast by a steel ball. The bright spot in the middle of the shadow shows that light is diffracting around the edges of the obstacle. These and other observations provide evidence for the wave nature of light.

475

G WAVES: TRANSFERRING ENERGY THROUGH SPACE

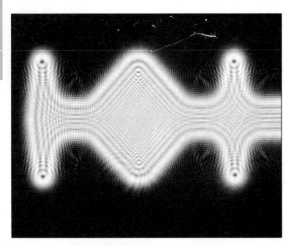

Figure 31.4 a Diffraction of light around a razor blade, **b** diffraction of water waves around a harbour wall

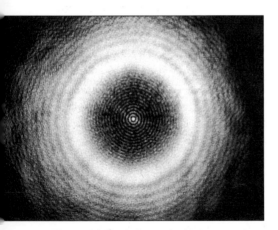

Figure 31.5 Diffraction of light around a steel bearing

Questions

2 To pass the driving test, it is necessary to be able to read a car number plate at about 20 metres. How might diffraction play a part in limiting our ability to see a number plate clearly at much greater distances?

3 What limitations might diffraction impose on the ability of a camera, installed in an orbiting satellite, to take clear photographs of human-sized objects on the Earth's surface? Would a small lens aperture be better than a large one for this purpose?

31.2 DIFFRACTION AT A SINGLE SLIT

To explain the observations in 31.1, it is convenient to make use of **Huygens' principle**. This states that every point on a wavefront behaves as though it were itself a point source emitting secondary waves, or **wavelets**.

To explain what happens when a plane wave from a light source meets a parallel slit, you can imagine a set of sources in the plane of the slit. These

31 WAVE NATURE OF LIGHT

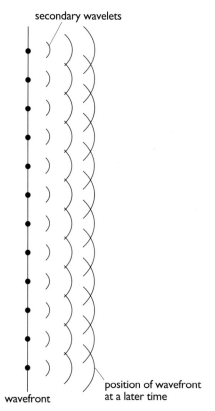

Figure 31.6 Huygens' principle

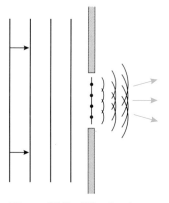

Figure 31.7 Wavelets leaving a single slit

Huygens' principle

At the end of the seventeenth century, the Dutch scientist, Christian Huygens, established a principle which enables us to predict the behaviour of a wavefront in a variety of situations. In his model for light, he suggested that all points on a wavefront act like secondary sources, each producing secondary wavelets which spread in all directions. The resulting wave is determined by the **envelope** of the secondary wavelets.

Figure 31.6 explains how the principle may be applied to a straight wavefront. Each point on the wavefront behaves as a secondary source and wavelets spread from these sources. A short time later, a new wavefront is formed by the envelope of the secondary wavelets. Huygens made the assumption that the amplitude of the secondary wavelets varies from zero in the backward direction to a maximum in the forward direction of the wave. This produces the forward propagation of the wave.

imaginary sources radiate waves outwards from the slit (see Figure 31.7). The way in which the waves from the sources superpose gives rise to a pattern of light typical of a single slit.

Figure 31.8a shows a plane wave of wavelength λ falling normally on to a single slit of width b. Huygens' principle suggests that, in the plane of the slit, points on the wavefront will act as secondary sources. Suppose there are, say, 16 sources across the slit, labelled a to p in the Figure 31.8. (The number chosen makes no basic difference to the argument.)

Along the forward direction, where $\theta = 0$, the waves are in step and add constructively to produce a large amplitude and so a bright patch of light. However, there will be an angle θ, shown in Figure 31.8b, where light from source a will add destructively with light from source i. In this direction, the path difference between light from these two sources is $\lambda/2$ and the waves are exactly out of step. Similarly, light from sources b and j will cancel, as will light from c and k, d and l, and so on. The path difference across the whole slit is one wavelength. So, in the direction θ, there will be a net amplitude of zero – no light.

There is also a greater angle θ where light from source a will cancel with light from source e one-quarter of the way across the slit (see Figure 31.8c). The path difference between light from these two sources is $\lambda/2$. Light from b and f, c and g, d and h, and so on will also cancel. There will be a path difference of two wavelengths across the slit and no light in this direction.

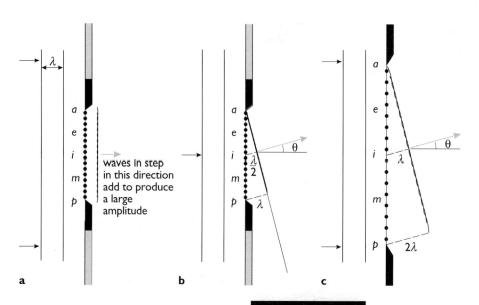

Figure 31.8 a Plane wave falling on single slit, b angle of first minimum in diffraction pattern, c angle of second minimum in diffraction pattern

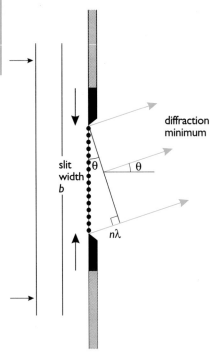

Figure 31.9 General condition for a diffraction minimum

In general, in any given direction, when the path difference between light from extreme edges of the slit is a **whole number of wavelengths**, there is **no light** in that direction. Figure 31.9 illustrates that a minimum occurs along a direction θ given by:

$$n\lambda = b \sin \theta,$$

where $n = 1, 2, 3$, etc. The intensity of a wave is proportional to the square of its amplitude. So a plot of intensity against sin θ would appear as in Figure 31.10. The model we have developed does not predict the shape or relative intensities of the maxima, only the positions of the minima. To find the intensities for a given angle, phasors can be used (← 19.3 and box).

If the slit width b is much greater than the wavelength of the light, minima will occur at fairly small angles. So sin θ will be very close to the angle θ in radians. This means that, approximately:

$$n\lambda = b\theta$$

and

$$\theta = \frac{n\lambda}{b}.$$

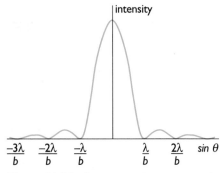

Figure 31.10 Intensity variation in a single slit diffraction pattern

Phasor treatment of diffraction

A single slit is illuminated by monochromatic light, as in Figure 31.11. For simplicity, imagine the slit to be composed of eight strips. Each strip will contribute a similar proportion of the total amount of light falling on a particular point on a distant screen. However, there will be a phase difference between light from each of the strips. The light from any given strip can be represented by a **phasor** (← 19.3). The resultant amplitude at that particular point on the screen may be found by summing the phasors, or joining them end to end. The angle between successive phasors is the phase difference between the waves they represent.

Figure 31.12 gives the phasor diagrams at points P, Q and R. At P, shown in a, the phase difference between light from each of the strips is zero and the phasors add together to produce a large amplitude. At Q, suppose there is a phase difference of π between the light from the first and last strips. Then the phase difference between light from successive strips is $\pi/8$, as in b. The resultant amplitude is smaller than before, although the total length of the phasors is unchanged. At R, there is a phase difference of 2π between the light from the first and last strips. The phase difference between light from successive strips is $2\pi/8$ or $\pi/4$, giving a phasor diagram as in c. The resultant amplitude is clearly zero.

Figure 31.13a – c show similar phasor diagrams but with the slit split into 80 strips. The phasor length is then much smaller, one-tenth in fact, because the amplitude of light from the narrower strips is much reduced. Also, for any given phase difference across the slit, the phase difference between light from adjacent strips is one-tenth what it was before. So the phasor diagrams appear as continuous curves but, otherwise, the net result is essentially the same as with eight slits.

Figure 31.13d – f show phasor diagrams for phase differences across the slit of 3π, 4π, and 5π radians. Suppose the total length of the phasors is equal to A. The resultant amplitude is represented by the straight line joining the beginning and end of the phasors. The first maximum is near to that shown in d and the second maximum is near to that shown in f. They actually occur at 2.86π and 4.92π.

From the resultant amplitude of light at each point on the screen, we can calculate the intensity variation. This is proportional to the square of the wave amplitude. Figure 31.14 shows the variation of amplitude and intensity across the screen. Compare the ripple tank pattern of single slit diffraction in Figure 31.3, with the amplitude variation in Figure 31.14.

Phasors provide ideal data for a microcomputer, enabling the pattern to be found precisely.

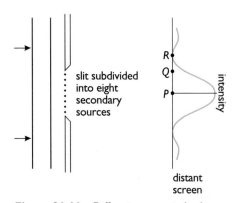

Figure 31.11 Diffraction at a single slit

31 WAVE NATURE OF LIGHT

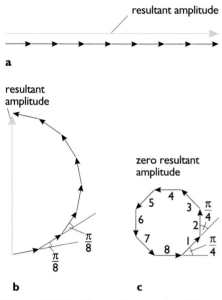

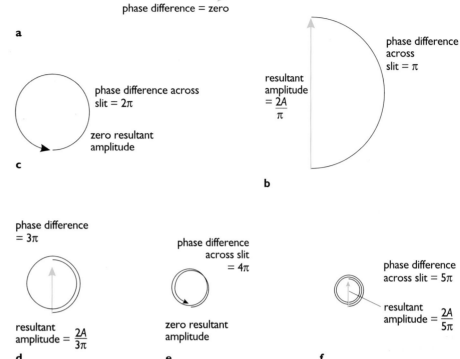

Figure 31.12 Phasor diagrams for eight strips: **a** at P, **b** at Q, **c** at R

Figure 31.13 Phasor diagrams for 80 strips: **a** at P, **b** at Q, **c** at R; and for phase differences of **d** 3π, **e** 4π and **f** 5π radians

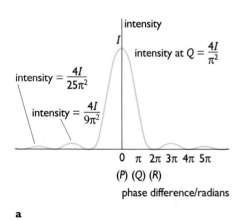

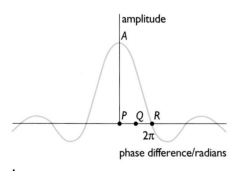

Figure 31.14 Intensity/amplitude pattern plotted against phase angle for a single slit

Question

4 The phasor diagrams in Figure 31.13a – c represent the situation at points P, Q and R in Figure 31.11.
a What is the path difference between light from the first and last strips at point R on the screen?
b What is the path difference, in m, between light from the first and last strips at point Q on the screen?
c At what angle is the first minimum R from the central maximum P? Assume that the angle is small and that the slit width is b.
d The resultant amplitude A of light arriving at P is the sum of the vector amplitudes from each of the strips. Calculate the resultant amplitude of the light arriving at point Q. How will the intensities at P and Q compare?

FINDING THE WAVELENGTH OF THE LIGHT

The equation $n\lambda = b \sin \theta$ predicts that the narrower the slit, the larger the spread of the diffraction pattern. The equation also predicts that shorter wavelengths will diffract less than longer ones. In fact, the ratio of the wavelength to the aperture width, λ/b, determines the angular spread of the pattern (→ 31.3). Note that the angular width of the central maximum is twice that of secondary maxima.

The link between the angular spread of a diffraction pattern and λ offers a way of finding the wavelength, as in experiment 31.2. Typical results from this show the separation of the first minima on either side of the central maximum to be 25 mm. So, the angular width of the central maximum is 25/2000 radians.

This means that the angle θ between the first minimum and the central maximum is just half that, and:

θ = 25/4000

= 6.25 × 10⁻³ radians.

Since θ is equal to λ/b, the wavelength of the light is given by:

λ = 10⁻⁴ × 6.25 × 10⁻³

= 630 nm to two significant figures.

Question

5 Light, of wavelength λ, falls on a single parallel slit whose width is 10 λ.
a Calculate the angular width of the central maximum in the diffraction pattern.
b Calculate the width of the central maximum on a screen placed 2.5 m away.

EXPERIMENT 31.2 Finding the wavelength of light from a diffraction pattern

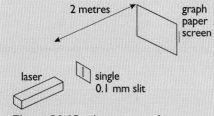

Figure 31.15 Arrangement for experiment 31.2

Warning: when using a laser, do not look directly into the beam

After setting up the arrangement shown in Figure 31.15, and checking that the slit is 0.1 mm using a wire of this diameter, you can observe the pattern produced. You should notice that the central maximum of the pattern is twice as wide as each of the secondary maxima. By measuring the width of the central maximum in mm, you can calculate the angle for the first minimum. Using the equation for single slit diffraction, you can then find the wavelength of the laser's light.

Question

6 In an experiment similar to 31.2, the slit width is 0.06 mm and the wavelength of light is 632 nm.
a What is the value of the angular separation between the first minima on each side of the centre?
b How far apart are the first minima when projected on to paper placed 2 m from the slit?

DIFFRACTION AT A CIRCULAR APERTURE

The equation λ = b sin θ for the first minimum applies to a narrow parallel slit. Where diffraction occurs at a circular aperture, the first minimum occurs at an angle θ given by:

1.22 λ = D sin θ,

where D is the diameter of the aperture.

You can observe diffraction at a circular aperture by stretching and breaking a piece of very fine copper wire of about 0.2 mm diameter, so that the end is work hardened (← 25.2). Pierce a hole in a piece of aluminium foil with the wire. Then, looking through the hole you have made, view a small torch lamp filament from

31 WAVE NATURE OF LIGHT

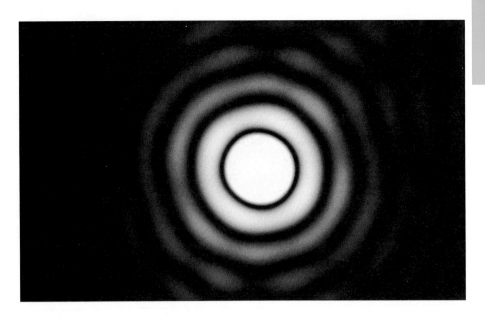

Figure 31.16 Diffraction through a narrow circular aperture

a metre or two away. Figure 31.16 shows the concentric rings of light, you can expect to see.

Questions

7 White light falls on to a parallel slit of width 0.1 mm. The resulting diffraction pattern is viewed on a screen 2 m away.
  a How far from the central maximum is the first minimum for
   i the red component of the light, with an average wavelength of 650 nm?
   ii green light, with an average wavelength of 550 nm?
   iii blue light, with an average wavelength of 400 nm?
  b Describe in detail the appearance of the diffraction pattern.

8 A tiny lamp filament is viewed with the naked eye from a distance of several metres.
  a Estimate the pupil diameter and the distance from the pupil to the retina.
  b Over what angle, in degrees, will most of the intensity fall after diffraction at the pupil of the eye?
  c Calculate the radius of the central maximum at the surface of the retina.

31.3 RESOLUTION OF SOURCES

Diffraction limits your ability to see things clearly, whether with the naked eye, through a telescope or any other optical instrument. This is because, in every case, light has to pass through an aperture.

If you look at a glowing lamp filament a couple of metres away, the focused image of the filament on the retina will not be perfectly sharp, even if your eyesight is very good. As light from the filament enters the pupil of the eye, a small amount of diffraction occurs. It is small because the pupil diameter is fairly large, of the order of 1 millimetre. So, normally, you will not be aware of any blurring of the image. But if, you attempt to read, for example, the number on the front of a bus some 100 m away, you will experience considerable difficulty.

Light coming from all parts of the number will diffract as it enters the pupil of your eye. The retinal image will consist of many diffraction patterns, resulting in a blurred image in which it is very difficult to distinguish detail. A number of factors may cause an inability to see detail, such as eye defects and light intensity, but diffraction is a significant one.

EXPERIMENT 31.3 Resolving sources

By viewing a twin filament light source (see Figure 31.17), through a **green** filter and a slit you can see diffraction patterns on either side of the filament, which, on gradually reducing the width of the slit, become so wide that it is only just possible to distinguish the filaments as separate. On replacing the **green** filter with a **red** one and, keeping the slit width constant, it is now impossible to see the separate lamp filaments clearly. Replacing the **red** filter with a **blue** filter, the filaments should now be easily distinguishable. Figure 31.18 summarises these observations.

Figure 31.17 A twin filament light source

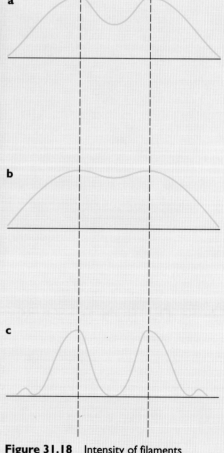

Figure 31.18 Intensity of filaments through a narrow slit with **a** a green filter, **b** a red filter, **c** a blue filter

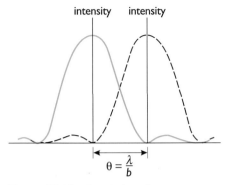

Figure 31.19 Resolution of two sources

RAYLEIGH CRITERION

Experiment 31.3 shows that, beyond a certain separation of diffraction patterns, two sources cannot be **resolved**, or seen as separate. It was suggested by Rayleigh (1842-1919) that two sources can just be resolved when the first minimum of one diffraction pattern coincides with the central maximum of the other. This criterion assumes that the light from each source is of equal intensity.

Suppose, for example, you view two lines through a rectangular aperture of width b. According to the criterion, the two images can just be resolved when the angle separating the images θ is greater than λ/b (see Figure 31.19). If the angle of separation is smaller, they will not be distinguishable.

RESOLVING TWO LINE-SOURCES OF LIGHT

Figure 31.20 shows the diffraction patterns formed by light, of wavelength λ, from two sources after passing through a narrow slit of width b. The angle θ, between the central maximum and the first minimum of each is λ/b, assuming b is considerably larger than λ ($\leftarrow$ 31.2). The central maximum of one source must fall on the first minimum of the second for the two sources to be resolved. So the angular separation of the two sources must be λ/b. When the separation of the two sources is s and the distance of the sources from the slit is x then:

$$s = x\theta.$$

When viewing through a circular aperture of diameter D, the angle is slightly larger, at $1.22\,\lambda/D$. Such an aperture could be the pupil of the eye or the objective lens of a telescope. For example, suppose you look at two lines drawn fairly close together on a sheet of paper. Light from the two lines will have an average wavelength of about 500 nm, and the pupil of the eye has a diameter of around 1 mm. The Rayleigh criterion suggests that, for the two lines to be just resolved, the angular separation of the two lines must be:

$$1.22\,\lambda/D = 1.22 \times 5 \times 10^{-7}\,\text{m}/10^{-3}\,\text{m}$$
$$= 0.0006 \text{ radians}.$$

So, to just be able to see clearly lines drawn 2 mm apart, you need to be as close as 3.5 m, assuming you have no eye defects. This suggests that diffraction plays an important part in determining whether or not you see things clearly.

Experiment 31.3 also looks at the effect of colour on diffraction. Red light diffracts more, or bends through a greater angle, than green because it has a longer wavelength. In fact with red, there is so much overlap of the diffracted images of the filaments that it is impossible to see them as separate sources.

In summary your ability to see things clearly as far as diffraction is concerned depends on:

- how far you are away from the object
- the wavelength of light emitted by the sources
- the size of the aperture through which you view the object.

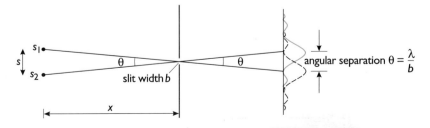

Figure 31.20 Angular separation of sources

G WAVES: TRANSFERRING ENERGY THROUGH SPACE

> **Question**
>
> **9** The vapour trails from the two tail-mounted engines of a jet aircraft are just distinguishable as being separate with the naked eye.
> **a** Assuming that the light from the trails has an average wavelength of 550 nm and that the pupil diameter is 1 mm, calculate the angular separation of the engines.
> **b** Assuming the engines are 4 m apart, how high is the aircraft?

When viewing an object under a microscope, your ability to resolve detail is limited by diffraction. Simply increasing the magnification will not improve the definition – it will just produce larger diffraction patterns. However, the diffraction depends on the diameter of the aperture through which light enters the microscope. It also depends upon the wavelength of the light in use. Clearer detail may be obtained by illuminating the object with light of shorter wavelength, or by using a microscope with a larger aperture at the objective. Electron microscopes use the wave properties of the electron which may have wavelengths as low as 0.1 nm.

31.4 DIFFRACTION OF RADIO WAVES

DIFFRACTION OF OTHER ELECTROMAGNETIC RADIATION

Diffraction can affect the emission and propagation of any waves from a source. The radar dish acts as a diffracting aperture to the radio waves generated at the focus of the dish. The larger the dish diameter, the narrower is the beam of radio waves. The transmitting dish on a communications satellite also needs to be of such a diameter that the radio wave is concentrated in a sufficiently narrow beam and is intense enough to be able to be detected by a ground-based aerial system.

There is evidence that radio waves possess wave properties very similar to light but have wavelengths very much longer. For example, radio waves are diffracted at relatively large apertures (see experiment 31.4).

EXPERIMENT 31.4 Diffraction of microwaves

By setting up the arrangement in Figure 31.21, you can investigate the diffraction of a beam of microwaves passing through a slit formed by two metal plates.

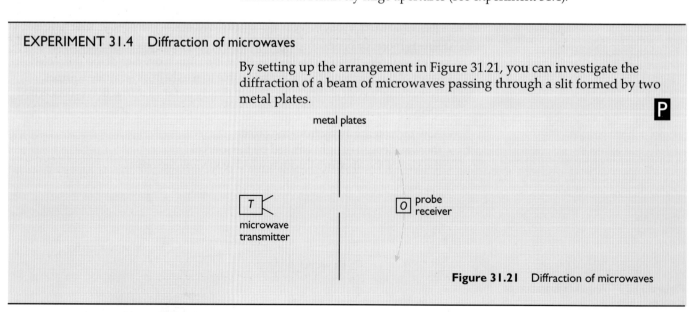

Figure 31.21 Diffraction of microwaves

31 WAVE NATURE OF LIGHT

> **Question**
>
> **10** Microwaves are a form of electromagnetic radiation similar to light, but of much longer wavelength. A parallel beam of microwaves of wavelength 3 cm passes through a single slit of width 6 cm.
> **a** At what angle, in degrees, to the forward direction will the first minimum appear?
> **b** For this angle, could you approximate $\sin\theta$ to θ radians?

Television signals sent out from a transmitter diffract around obstructions, such as buildings and hills, and also the horizon itself. Provided the transmitter is not too distant, the resulting signal detected at a receiving aerial is still of sufficient strength to obtain good reception.

In a radio telescope, radio waves from, say, a star are received through a circular aperture formed by the reflecting dish. Diffraction of the signals occurs and, as a result, they can be detected by the receiving aerial at the focus of the dish for quite large angles of movement. The Jodrell Bank radio telescope has a dish 80 metres in diameter. The characteristic wavelength λ of radiation emitted by hydrogen molecules in space is 0.21 metres. The diameter of the central maximum of the diffraction pattern of a source is twice the distance between the central maximum and the first minimum – 2 (1.22 λ/b). For the Jodrell Bank telescope, this works out at about 0.006 radians, and it is unable to distinguish radio sources with smaller angular separation. In fact, the eye is better by around a factor of ten at resolving detail than the Jodrell Bank telescope.

Increasing the diameter of the dish would reduce the angular width of the central maximum of the diffraction pattern and increase the resolving power. But clearly there is a maximum limit to the size of dish that can be constructed. There are other ways of improving the resolution of radio telescopes (→ 33.2).

> **Question**
>
> **11** Radio signals from space are a form of electromagnetic radiation. Figure 31.22 shows the reflector and aerial of the radio telescope at Dwingeloo in the Netherlands. The dish is 25 m in diameter and the aerial detects 21 cm radio waves emitted by the hydrogen in our own galaxy. Over what angle, in degrees, on either side of the central maximum position will the telescope detect an appreciable intensity of radiation?

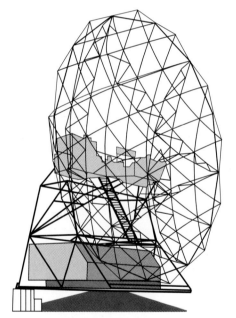

Figure 31.22 Radio telescope at Dwingeloo

SUMMARY

Evidence of the wave nature of light comes from diffraction. The wavelengths of light (around 550 nm on average) are much smaller than those of water waves. Tiny apertures are required to produce noticeable diffraction effects with light.

To explain the behaviour of waves as they pass through an aperture, Huygens' Principle can be used; this suggests that all points on a wavefront act like a secondary source, each producing secondary wavelets.

The angles at which minima occur as light passes through a single slit of width b are given by:

$n\lambda = b\sin\theta,$

where $n = 1, 2, 3, 4$, etc. When b is much greater than λ, then approximately:

$\theta = n\lambda/b.$

WAVES: TRANSFERRING ENERGY THROUGH SPACE

Either equation can be used to find the wavelength of light from measurements of its diffraction pattern.

Radio signals also have wave properties. They have much longer wavelengths than those of light, ranging from around a centimetre in the microwave region up to several kilometres in the long radio wave region (→32). Diffraction of radio waves occurs in and around buildings. This enables signals to be received even when there is no direct line of sight between the transmitting and receiving aerials.

When you view an object with the naked eye or through an optical instrument, you are looking through an aperture. Diffraction occurs and causes a blurring of the image.

The Rayleigh criterion states that two sources can just be resolved when they are separated so that the central maximum of the diffraction pattern of one falls on the first minimum of the other. When viewed through a narrow slit of width b, the angular separation of two sources must be greater than λ/b for them to be resolved. When viewed through a circular aperture of diameter D it must be greater than $1.22\,\lambda/D$.

Decreasing the wavelength used or increasing the aperture size of an instrument improves its resolution.

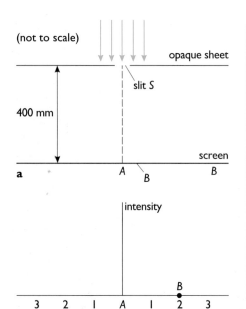

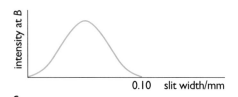

Figure 31.23

SUMMARY QUESTIONS

12 In Figure 31.23a, S is a narrow slit, the width of which can be varied. It is illuminated by a parallel beam of monochromatic light of wavelength 500 nm. A screen is placed 400 mm from S.
a Show that the intensity at B, a point on the screen 2.0 mm to one side of A, is a minimum when the slit width is 0.10 mm.
b On a copy of Figure 31.23b, draw two sketch graphs, each to the same scale, showing the intensity pattern close to A:
 i for a slit width of 0.10 mm,
 ii for a slit width of 0.05 mm.
Label the graphs clearly.
c Figure 31.23c shows how the intensity at B varies as the slit width is changed between zero and 0.10 mm. The intensity at B turns out to be greatest when the slit width is 0.05 mm.
Explain why the intensity at B diminishes when the slit width:
 i is decreased below 0.05 mm towards zero,
 ii is increased above 0.05 mm towards 0.10 mm.

(From Physics (Nuffield) Short Answer Paper, 1984.)

13 A radio telescope 80 m across detects radio waves of wavelength 0.21 m from two adjacent stars.
a Calculate their angular separation, given that the telescope can just resolve them.
b Assume that light travels at 3×10^8 m s^{-1} and that there are 3×10^7 seconds in 1 year. They are both 200 light-years from the Earth. How far apart are they?

14 An ultrasonic wave transmitter is about 2 cm across and emits waves of frequency 40 kHz.
a Over approximately what angle, in degrees, will the ultrasonic wave be emitted? Assume that the waves travel at 340 m s^{-1} in air.

b What changes would you need to make to the system to concentrate the energy into a much narrower beam?

15 Microwaves of 10 GHz frequency are generated at the focus of a parabolic dish of effective diameter 2 m.
a What is the angular spread, in degrees, of the beam as a result of diffraction at the dish's aperture?
b Estimate the area over which the beam will spread at a distance of 20 km from the transmitter.

16 A camera is mounted on a satellite orbiting the Earth at a height of 200 km.
a Calculate the minimum diameter its lens would need to have to resolve rocks on the Earth's surface 10 m apart. Assume the camera receives light of an average wavelength of 500 nm.
b What other factors might influence the resolving power of such a system?

17 The largest refracting telescope has an objective lens of about 100 cm in diameter. In what ways is this telescope better, optically, than one with a much smaller objective lens?

18 Two stars of angular separation 4×10^{-6} radians are viewed through a refracting optical telescope with an effective aperture of 250 mm. Can they be resolved? Assume that the average wavelength of the light used is 600 nm.

19 If the eye was sensitive to radiation ten times longer in wavelength than visible light, what problems might this pose?

20 The Jodrell Bank radio telescope, 80 m across, is sensitive to radio waves of 0.21 m wavelength. The eye, sensitive to light of average wavelength 500 nm, has a pupil diameter of about 2 mm. A Newtonian reflecting telescope with a mirror diameter of 20 cm is also sensitive to light of average wavelength 500 nm.
Compare quantitatively the abilities of each of these instruments to resolve astronomical details.

21 The divergence of a beam of laser light depends on the amount of diffraction at the laser output aperture.
a Calculate the angular divergence of the beam, assuming the output mirror of a helium-neon laser has an aperture with a diameter of 2 mm and the laser light has a wavelength of 632 nm.
b Calculate the diameter of the beam after it has travelled a distance of 100 m.

32 ELECTROMAGNETIC WAVES

There is much evidence to show that light possesses wave properties similar to those of sound and other mechanical waves (← 31).

Light is only one member of a whole family of waves. The Sun also emits X-rays, γ-rays and radio waves.

Unlike mechanical waves, these electromagnetic waves require no medium and are able to travel through the vacuum of space. For example, Figure 32.1 shows an image of the planet Neptune transmitted by radio signals back to Earth.

This chapter is concerned with the nature and properties of electromagnetic radiation and the many important uses to which it is put. To understand the ideas, you may need to refer to Chapters 29–31 which are concerned with the nature and behaviour of waves in general.

Figure 32.1 Image of Neptune taken by the Voyager spacecraft, transmitted by radio signals to Earth in August 1989

32 ELECTROMAGNETIC WAVES

32.1 GENERATING ELECTROMAGNETIC WAVES

In 1887, Heinrich Hertz discovered waves that were undoubtedly electromagnetic in origin and had properties similar to those of light. His discovery came about 15 years after the existence of such waves was predicted by James Clerk Maxwell.

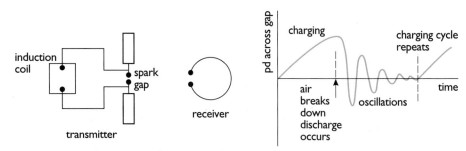

Figure 32.2 Hertz's spark transmitter

Hertz generated the waves by spark discharge across an air gap (see Figure 32.2). Two brass conductors were attached to two rectangular 50 cm × 50 cm brass plates. The plates were connected to a source of high voltage, an induction coil, which rapidly charged up the plates and caused a spark discharge between the two brass knobs. The discharge was oscillatory in nature, similar to that produced when a capacitor discharges through an inductance. However, because the capacitance of the plates and the self-inductance of the conductors were small, high-frequency oscillations were produced with an electric field across the air gap and a magnetic field around the discharge current. The result was the propagation of an electromagnetic disturbance.

Hertz detected the disturbance using a loop of wire broken by a similar spark gap. The magnetic flux (← 17) from the disturbance threading the loop induced an alternating voltage in the loop. With a loop diameter chosen so that the circuit resonated (← 22.5) at the frequency of the disturbance, he was able to see noticeable sparks across the gap caused by the large emfs induced. Amazingly, these sparks were quite visible when the spark transmitter and the receiving loop were separated by up to a couple of metres or so. When the sparking at the transmitter was stopped, no sparks were produced at the receiver.

He confirmed the wave nature of the disturbance by reflecting the transmitted waves from a large metal sheet and using the receiving loop to detect the nodes and antinodes in the resulting stationary wave (← 30.6). Later work confirmed that their speed was the same as that of light.

NATURE OF THE WAVES

The nature of Hertz's waves may be investigated more easily with a modified version of his apparatus (see Figure 32.3). An oscillator producing a signal of around 1 GHz replaces the high tension supply. It is connected to a pair of metal rods, a dipole transmitting **aerial**, mounted on an insulated stand. The receiver is a pair of similar rods connected to a sensitive galvanometer. A diode across the dipole receiver aerial ensures that its output is d.c. and registers on the meter.

A number of investigations may be carried out with this apparatus, as in experiment 32.1. All lead to the same conclusions about the nature of the disturbance leaving the transmitter. The disturbance seems to possess wave properties, exhibiting diffraction and superposition effects. Also, the receiving dipole only detects a signal when its plane lies parallel to that of the transmitting dipole. Rotating either dipole through 90° causes the received signal to drop to zero. Turning the second dipole to match restores the signal.

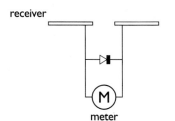

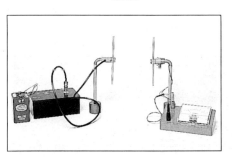

Figure 32.3 Apparatus for investigating Hertz's waves

WAVES: TRANSFERRING ENERGY THROUGH SPACE

EXPERIMENT 32.1 Investigating 1 GHz radio waves

Using the apparatus shown in Figure 32.3, you can check:

a through which materials 1 GHz radio waves pass,
b how the intensity of the waves decreases with distance,
c how the receiver must be orientated for maximum signal strength.

The arrangement of Figure 32.4 allows you to observe the effect of moving a large reflecting metal screen in the direction shown by the arrows (← 30).

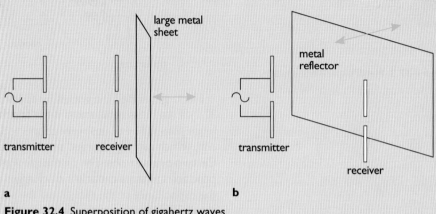

Figure 32.4 Superposition of gigahertz waves

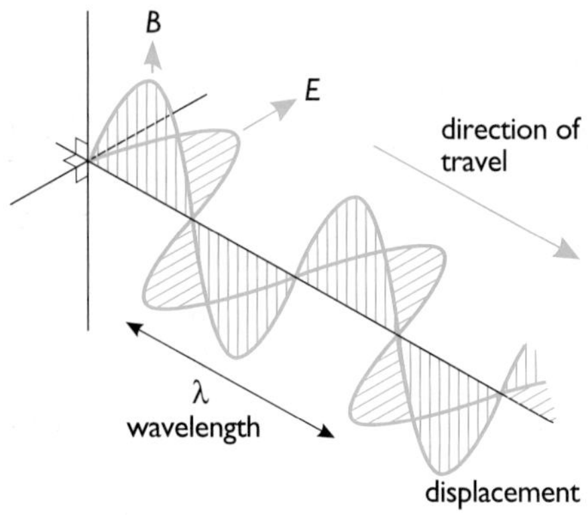

The experiment gives clues to the origin of the electromagnetic waves leaving the transmitter. Charge is continually flowing on and away from the metal rods of the dipole. So, currents in the rods are changing continuously, producing changing magnetic fields around them. These changing magnetic fields in turn induce changing electric fields (← 18). If the rods are each one-quarter of a wavelength long, or a multiple of this, the system will efficiently transfer energy from the oscillating charges in the rods into electromagnetic radiation.

From the variation of signal strength with receiver orientation, it is evident that the wave is **polarised** and so is transverse in nature (→ 32.4). It consists essentially of two components – an oscillating electric **E**-field and an oscillating magnetic **B**-field, in phase with each other. The **E**-field is parallel to the rods and therefore in the same direction as the electric field in the rods. The **B**-field is at right angles to the **E**-field and to the direction of travel of the wave (see Figure 32.5). The **E**-field and the **B**-field cannot be separated one from another. Each arises as a result of the presence of the other. Electromagnetic waves do not require a medium for their propagation and so are able to pass through a vacuum, where they travel at about 3×10^8 m s^{-1} (→ 32.3). This velocity c is linked to the frequency f and wavelength λ of the wave by the equation $c = f\lambda$ (← 29.2).

Figure 32.5 **E**- and **B**-fields associated with an electromagnetic wave

Questions

1 A 1 GHz electromagnetic wave is radiated by an aerial dipole when connected to an oscillator.
a What is the wavelength of the disturbance? (c)
b For the dipole to be as efficient a radiator as possible, how long should each rod be?

32 ELECTROMAGNETIC WAVES

2 Figure 32.4a shows a 1 GHz transmitter with its rods vertical placed close to a large metal reflector. The receiver, also with its rods vertical, picks up two signals, one direct and one after reflection by the metal sheet.
a As the receiver is moved along the line joining the transmitter and reflector, its output increases to a maximum, then decreases to a minimum and then back to a maximum. Explain this in terms of the wave properties of the signal (← 30).
b The distance between the maxima is found to be 15 cm. How fast are the waves travelling?
c The receiving dipole is rotated through a right-angle so that it is horizontal. What would you expect to observe?
d The arrangement is altered to that shown in Figure 32.4b. What would you expect to observe as the sheet is moved in the directions shown by the arrows?

32.2 THE ELECTROMAGNETIC SPECTRUM

The electromagnetic, or radio, waves discovered by Hertz share many of the properties of light. It is now known that these are just two members of a whole family of waves, all of which are transverse in nature, electromagnetic in origin and travel at the same velocity in a vacuum. More fundamentally, they all arise from electric charge moving in such a way that its rate of acceleration is changing. Electrons oscillating in an aerial produce radio waves, and high-energy electrons striking a solid object radiate X-rays because varying acceleration is occurring in both cases. The characteristics of electromagnetic waves are determined only by their frequency and wavelength, measured in a vacuum. Their frequencies lie along a **spectrum**, ranging from less than 50 Hz at one end to around 10^{24} Hz at the γ-ray end.

Question

3 What is the wavelength of:
a radio waves of frequency 200 kHz in km?
b microwaves of frequency 10 GHz in cm?
c visible light of frequency 5×10^{14} Hz in nm?
d γ-rays of frequency 10^{24} Hz in m? (c)

Figure 32.6 shows the relative positions of the various types of radiation in the electromagnetic spectrum. As is common when displaying a very large range of

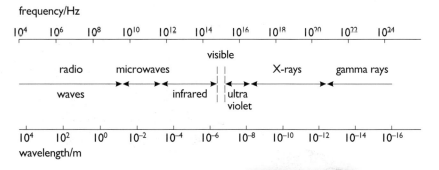

Figure 32.6 The electromagnetic spectrum

WAVES: TRANSFERRING ENERGY THROUGH SPACE

values, the frequencies are plotted on a logarithmic scale. One centimetre along the scale corresponds approximately to a change in frequency by a factor of 100.

Question

4 Suppose the frequency scale in Figure 32.6 was linear. How long would it be if, say, an interval of 1000 Hz occupied a distance of 1 mm? What would be the drawbacks of any linear scale?

Quantum nature of electromagnetic radiation

Although there is much evidence that electromagnetic radiation has wave properties, there is equally strong evidence that it has particle properties. A particle or **quantum** of electromagnetic radiation is called a **photon**. Its energy E is given by $E = hf$, where h is the Planck constant and f the frequency associated with the photon ($\rightarrow$ 36.1). The quantum nature of radiation becomes very evident at the high-frequency end of the electromagnetic spectrum. It is often impossible to consider the wave nature of high-frequency radiation without referring to its particle properties.

RADIO WAVES

Radio waves arise as a result of electrical oscillations that occur naturally or by design using oscillating circuits. Their frequencies lie on a spectrum from around 1 kHz up to 5 GHz, a range extensively used for radio communications and broadcasting. Figure 32.7 gives the frequency, wavelength and photon energy of different parts, along with some common applications.

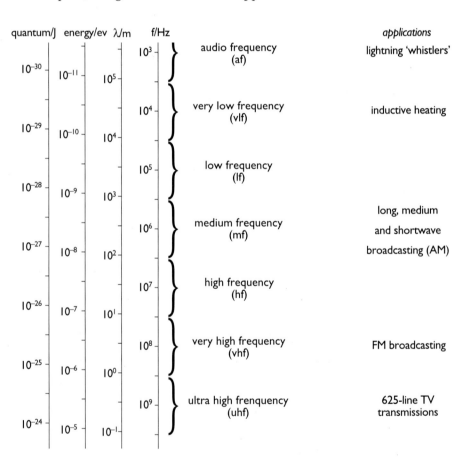

Figure 32.7 The radio wave spectrum

Question

5 Lightning in thunderstorms transfers a considerable amount of energy, much of it by electromagnetic waves. Some of these have frequencies in the audio range. Telephone lines of sufficient length can act as an aerial and waves having a frequency of the order of 10 kHz can cause electrons in the wires to oscillate.

a What effect do you think this might have on telephone conversations carried by the line?
b Calculate the wavelength of 10 kHz electromagnetic waves.
c Would a telephone line of 100 m be of 'sufficient length'?

Broadcasting and communication

For use in communication, radio waves must be modified or 'modulated' to carry the information to be transmitted (see Figure 32.8). In radio broadcasting, amplitude modulation (AM) is used in the low, medium, and high frequency bands. The variation in amplitude of the radio frequency carrier wave presents the information in the transmitted audio signal. Frequency modulation (FM) is used in the very high frequency band. Here, the information is stored in the pattern of changing frequency of the carrier wave. This system can convey much more information because the width of the band carrying the information is large (← 20.4 and 22.5). Music requires a bandwidth of around 20 kHz to encode frequencies between 30 Hz and 20 kHz. A bandwidth of only 3 kHz is sufficient to contain all the information in, say, a typical telephone conversation.

Any range of frequencies can therefore support only a limited number of broadcasting or communication 'channels'. Clearly, there can be few audio channels in the low frequency band, whilst on the uhf band, many channels are possible and the signal bandwidths can be very much greater. Television signals in Britain were originally broadcast on the vhf waveband and, to begin with, there was just one channel. With the demand for more stations and colour, more channels were required, each of greater bandwidth: a bandwidth of 5.5 MHz is needed for colour television whereas only 3 MHz is needed for black and white signals. Television signals were then moved to the uhf waveband.

By International agreement, specific channels have been allocated to broadcasting and other authorities for television and radio signals across the whole radio wave spectrum. For example, 47 television channels in the range 21 to 68 have been allocated in the uhf band at frequencies ranging from around 470 MHz to about 850 MHz. In the shf band, 40 channels have been allocated for satellite broadcasting in the 11–12 GHz range. In contrast, the UK has only one channel allocated on the lf band at a frequency of 198 kHz. Figure 32.9 summarises the limits of the broadcasting bands. There are many gaps in the bands. These are allocated to navigation for aircraft and shipping, the emergency services, the police, radio-controlled models, air traffic control, amateur radio and many others.

Despite the huge number of allocated uhf channels in the UK, the range and interference limitations prohibit us having more than six terrestrial broadcast channels. The problem is that the strong signal area around each transmitter is relatively small compared with the area of interference further out due to signals from other transmitters using the same or very similar frequencies. This severely limits re-use of channels.

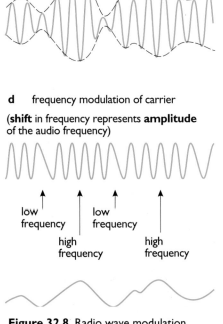

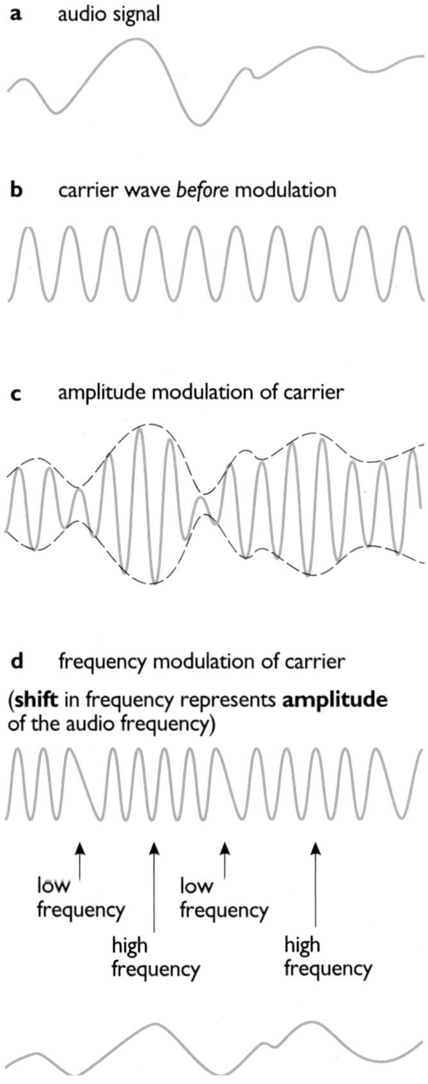

Figure 32.8 Radio wave modulation

very low frequency	50-100 kHz	6km - 3000 m	time signals
low frequency	150-285 kHz	2000 - 1053 m	AM radio: longwave
medium frequency	525-1 605 kHz	571 - 187 m	AM radio: mediumwave
high frequency	5 MHz-20 MHz	60 - 15 m	AM radio: shortwave
band II (vhf)	88-108 MHz	3.4 - 2.8 m	FM radio
band IV (uhf)	470-582 MHz	channels 21-34 television	
band V (uhf)	614-854 MHz	channels 39-68 television	
band VI (shf)	11.7-12.5 GHz	channels 1-40 DBS channels	

Figure 32.9 Limits of the broadcasting bands

WAVES: TRANSFERRING ENERGY THROUGH SPACE

AERIALS

Aerials of good design are essential for the efficient transmission and reception of radio signals across the whole of the radio frequency spectrum. Transmitting aerials connected to an oscillator emit energy carried by electromagnetic waves of the same frequency. Output power levels of up to 1 000 kW can be generated in the case of television transmissions. Receiving aerials intercept just a tiny part of the power within the wave and transfer it to the receiver as an electrical signal.

Aerials come in various shapes and sizes, depending on the wavelength of the transmitted or received signal. However, the main part of an aerial usually consists of a metal rod, half a wavelength long, known as a **half-wave dipole** (see Figure 32.10a). This is divided at its midpoint so that each length is a quarter of a wavelength of the transmitted wave.

Electromagnetic waves of all wavelengths are continually sweeping past a receiving aerial at the speed of light. The oscillating electric fields induce tiny currents to flow to and fro along the aerial. When the rods are parallel to the E-field, resonance causes relatively large currents to be produced over a fairly narrow range of frequencies. The bandwidth of the aerial has to be at least as big as the bandwidth of the signal being transmitted. Wide band aerials are available covering all the television channels in the uhf band. However, it is more usual for an aerial to cover a limited group of channels.

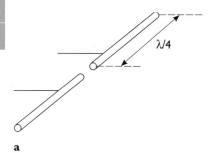
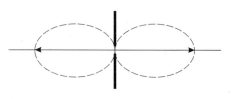

Figure 32.10a A dipole aerial and **b** its polar response curve

Question

6 A 100 MHz electromagnetic wave is to be picked up by a half-wave dipole.
a Calculate the length of rods required for the efficient reception of this wave.
b What dipole length would be required for a TV signal of 800 MHz?
c Compare the relative aerial sizes for vhf/FM radio and uhf/TV reception.

The response of a simple dipole aerial is shown in Figure 32.10b. This **polar**, or circular, plot gives the **sensitivity**, or relative response, of the aerial to a constant signal from all directions. The symmetry of the plot shows that the dipole has equal response in the forward and backward directions. To increase the overall sensitivity of the aerial, **director rods** may be placed in front of the dipole and a **reflector** placed a quarter of a wavelength behind the dipole. Such an arrangement is usually referred to as a **yagi array** (see Figure 32.11). The additions make the dipole aerial very sensitive in the forward direction and less so in others. Similarly, spectra become brighter and sharper when more slits are added to a difraction grating ($\rightarrow$ 33.3).

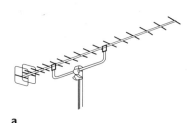
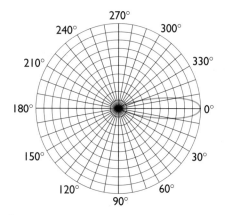

Figure 32.11a 17-element uhf yagi array and **b** response curve

Question

7 The reflector on a uhf aerial is placed a quarter of a wavelength behind the dipole. Electromagnetic waves undergo a phase change of π radians when reflected in this way.
a Calculate the phase difference between the main wave and the reflected wave at the dipole.
b Why does the reflector improve the gain of the aerial?

32 ELECTROMAGNETIC WAVES

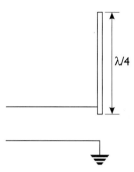

Figure 32.12 A quarter-wave element

Radio aerials

Aerials on portable FM radios or car radios tend to consist of just one rod of one-quarter of a wavelength long, which is about 75 cm for a 100 MHz signal. The second half dipole is formed by reflection into the ground, or the vehicle itself. The receiver aerial input is connected between the aerial and earth (see Figure 32.12).

For maximum received signal, the aerial must be aligned parallel to the electric field associated with the incoming wave. Most of the main TV transmitters give a horizontally polarised signal. So, the dipole has to be set in a horizontal position facing towards the transmitter. However, signals from relay transmitters feeding difficult reception areas are often vertically polarised.

On the low– and medium-frequency bands, radios usually pick up the broadcast signals with a ferrite rod aerial. Unlike the half-wave dipole, this senses the changes in the **B**-field associated with the electromagnetic wave. So for best reception of the signal, it needs to be parallel to the **B**-field of the incoming wave. The ferrite rod amplifies the **B**-field and tiny changing voltages produced in coils wound around the ferrite core vary according to the pattern of the broadcast signals. Other circuits in the radio select and amplify the desired signal. Ferrite rod aerials cannot respond to vhf signals because of the low inductances required in tuning circuits at such frequencies.

MICROWAVES

Figure 32.13 shows the frequencies and wavelengths associated with the microwave and infrared parts of the electromagnetic spectrum, and some of their

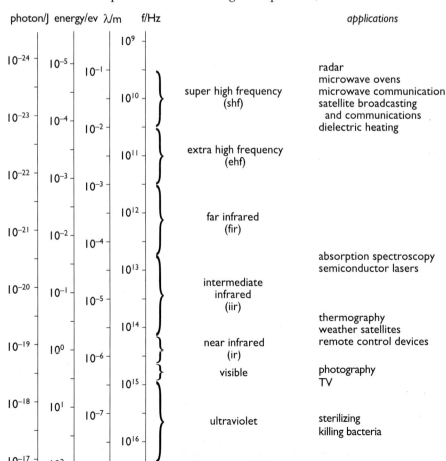

Figure 32.13 The microwave, infrared visible and ultraviolet spectrum

applications. Microwaves are very short wavelength radio waves. They are rather more difficult to produce, contain and transport along wires than waves of longer wavelength. This is because conductors of electromagnetic waves radiate more energy to the surroundings as the frequency increases. However, microwaves travel easily through the air and so offer great advantages as a means of communication compared to infrared waves.

> ### Question
>
> **8a** Calculate the wavelength of 10 GHz microwaves. (c)
> **b** Why is it that circuits will radiate more microwave energy to the surroundings than longer wavelength radio waves?

Because of the high frequency, many more telephone conversations and television channels can be carried using microwaves than longer radio waves. Long-distance links use microwave transmitters and receivers up to 70 km apart. Both consist of an aerial at the focus of a parabolic reflector. The narrow, well-defined beam produced is not diffracted as much as waves of longer wavelengths would be (← 31.2).

Their short wavelength and high frequency make microwaves ideally suited to use in communication and radar systems, but also make them quite difficult to produce. Electrical oscillators cannot be made to operate at frequencies above 3 GHz. Either the **klystron** or the **magnetron** can be used to produce high power microwaves. In these electron oscillators, electric and magnetic fields, respectively, cause oscillations in the motion of electrons inside an evacuated tube.

The microwaves produced are fed out via a short aerial in a tube, or **waveguide**. Over short distances, the waves can be confined within the waveguide as the maxima and minima of the electric and magnetic fields move down the guide like crests and troughs of a water wave. For efficient propagation of the wave, the cross-sectional dimensions of the waveguide must be of the same order of size as the wavelength of the microwaves.

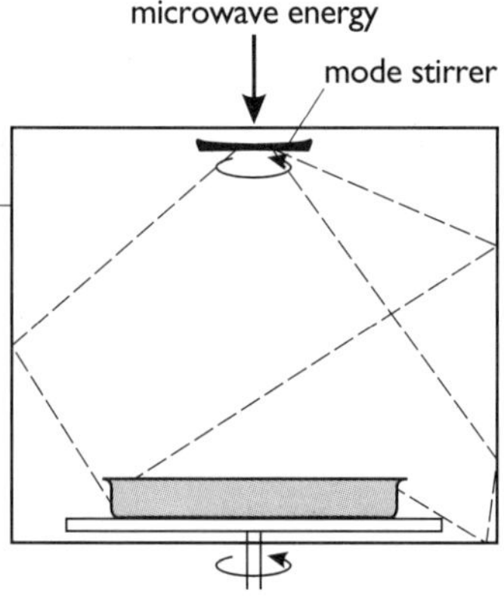

Figure 32.14 A microwave oven

Microwave ovens

When charges in a non-conducting, or **dielectric**, material have an electric field applied, they move a short distance but then stop, restrained by the binding forces in the material. When the field is switched off, the charges undergo a series of decaying oscillations before returning to their equilibrium position. This dissipates energy in the dielectric, much as a swinging pendulum dissipates energy to the surroundings. The energy appears as increased kinetic energy of the surrounding atoms, and the dielectric warms up.

An alternating electric field applied to a material transfers energy continuously to the internal energy of the material. The larger the amplitude of the electric field and the higher the frequency, the greater the rate at which energy is transferred. Water, with traces of salt, gains considerable internal energy when exposed to alternating electric fields of high enough frequency, such as microwaves. This is the basis of the microwave oven (see Figure 32.14). The moisture content of food placed in it gains enough internal energy from the microwaves to raise the temperature high enough to cook the food. Traditional cooking involves heating food in an enclosure maintained at a high temperature. This is a fairly inefficient process and there is often a danger of overheating the outer layers of the food and undercooking the centre. In microwave ovens only the food material itself is heated and the microwaves penetrate into the food by a few centimetres.

To conform to international regulations, electromagnetic waves of frequency 2 450 MHz are used in microwave ovens. These are generated by a magnetron,

32 ELECTROMAGNETIC WAVES

which is a thermionic valve (→ 35) incorporating cavities that resonate at the above frequency. Microwave powers of around 600 watts are typical. A waveguide feeds the microwaves into a metal enclosure where complex standing waves are set up, in much the same way as a metal plate vibrated at a high frequency will exhibit two-dimensional stationary waves (← 30.6). To ensure that food placed in the enclosure is uniformly heated, either a 'paddle' is rotated in front of the emerging microwave beam so that the standing wave pattern is continually changing or the food is rotated on a turntable. Some cookers employ both methods. Microwaves at large powers damage living tissue. So for safety, the doors are made of a microwave-reflecting glass and have switches that do not allow the magnetron to work when the door is open.

Question

9a In a normal oven, cooking time is largely independent of the amount of food being heated. However, small amounts of food heat up more quickly than large amounts in a microwave oven. Explain why this is, in terms of the way the food is heated.
b A microwave detector is set up in front of a microwave transmitter so that a large signal is received. A dry sponge is then placed just in front of the detector and doused with water. What would you expect to observe?

Direct broadcasting by satellite (DBS)

At Geneva in 1977, the World Broadcasting Satellite Administrative Radio Conference established a framework for television broadcasting by satellite in Europe and throughout much of the world. Forty channels were assigned in the shf microwave band ranging from 11.7 GHz to 12.5 GHz, five of which were allocated to the UK. Signals from a transmitter on Earth are sent to a satellite in geostationary orbit, which is about 36 000 km above the equator. The satellite receives the transmissions, converts them to another frequency, amplifies and re-transmits them to receiving aerials in the country of origin and to an area covering much of Europe (see Figure 32.15).

Satellites use frequency modulation (FM) of the television carrier wave. A dish receiving aerial of around 80 cm in diameter is required to pick up a sufficiently strong signal.

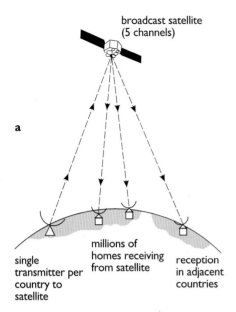
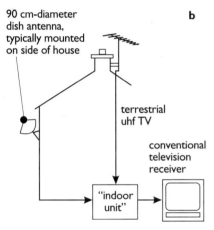

Figure 32.15a Direct broadcasting by satellite, **b** satellite reception in the home

INFRARED RADIATION

The wavelengths of infrared radiation lie between the microwave and visible parts of the electromagnetic spectrum (see Figure 32.13). Infrared radiation is invisible to the human eye, but objects exposed to it warm up as they absorb the radiation. This effect is used in **thermal infrared detectors**.

A thermal detector generates an emf when the infrared radiation falling on it raises its temperature. It may be **a thermopile**, which is a set of thermocouples connected in series with all their hot junctions close together. This can generate several volts. A **thermistor** can also be used as a thermal infrared detector. Changes in the intensity of incident radiant energy change its temperature causing a large change in its resistance. These can be measured using a bridge circuit (← 12.8).

The other main category of infrared detectors is the **quantum detectors**. They are made from materials whose properties change with the rate of arrival of photons upon their sensitive surfaces. The sensor on the front of a remotely controlled television is a quantum detector. It detects infrared emissions from a

WAVES: TRANSFERRING ENERGY THROUGH SPACE

Figure 32.16 TV remote control unit

remote control unit (see Figure 32.16). Photographic film is also a quantum detector. It is particularly effective at detecting infrared wavelengths around 1–2 micrometres (μm), the near infrared. The longer wavelengths of the far infrared do not have sufficient quantum energy to cause chemical changes in the film. Figure 32.17 shows an example of the output from a thermal imaging camera.

Thermal detectors have a reasonably uniform response across the infrared spectrum. But since they must wait for their own rise in temperature, their response time is generally much greater than that of quantum detectors.

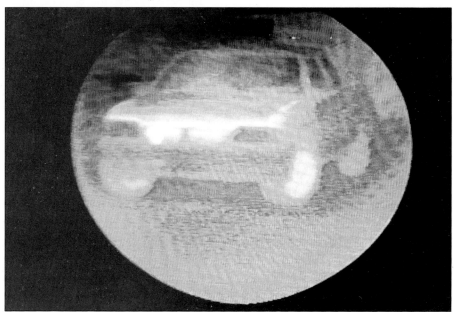

Figure 32.17 Output from a thermal imaging camera

Sources of infrared radiation

All substances emit infrared radiation, generated by the vibration and rotation of their atoms and molecules. At 0 K, there is no molecular activity and no infrared is emitted. As a substance gets hotter, activity increases and more infrared is emitted. The human body radiates on average around 200 W m^{-2} of its surface area. Most of this energy is emitted at 5–14 μm, with a peak at 9.3 μm which corresponds to a frequency of 32 000 GHz. The Sun emits about 120 000 W m^{-2} more than this. Indeed, more than 50 % of the energy emitted by the Sun is in the infrared part of the electromagnetic spectrum. It peaks at a wavelength of 500 nm in the visible part of the electromagnetic spectrum, a frequency of 600 000 GHz.

Question

10 Glass or transparent plastic transmits near infrared, but not long wavelength infrared.
a Explain why greenhouses can be maintained at quite high temperatures even though ambient temperatures may be quite low.
b Why is a 'full' greenhouse rather warmer than an empty one?

Tungsten lamps are excellent sources of relatively high-energy infrared, emitting about 7 × 10^5 W m^{-2} of filament surface at 2 700 K. The glass envelope around the filament restricts the range of wavelengths to below about 3 μm, the near

32 ELECTROMAGNETIC WAVES

infrared and visible part of the spectrum. (→ 36). Another important artificial source of infrared radiation is the semiconductor diode laser, such as that in a compact disc player (→ 37.4).

Weather satellites

Meteorological satellites are placed in either low **polar orbits**, at an altitude of about 900 km, or in a **geostationary orbit**. In the latter, the period of orbit is the same as that of the Earth on its own axis, so the satellite seems to be stationary above a point on the equator. Meteosat, built by the European Space Agency, is in a geostationary orbit and so is above the same part of the Earth all the time. The output from Meteosat shows clearly the moving cloud patterns over Europe, the Atlantic and Africa.

Polar-orbiting satellites, such as NOAA–7 and NOAA–8, are in solar-**synchronous** orbits. This means that the satellite keeps pace with the Sun throughout the day and crosses a given latitude at the same solar time each orbit so all of the Earth can be observed twice in one day.

The satellites have cameras that can view the Earth in both visible and infrared radiation. The infrared gives information about the temperature of the cloud and the ground surfaces. Figure 32.18 shows a typical output from a NOAA satellite.

Figure 32.18 NOAA–7 satellite, image of Hurricane Alicia

Questions

11 One polar-orbiting satellite ensures that a given part of the Earth may be observed twice in one day.
a How could two similar satellites give observations of a given location every six hours?
b For this to happen, why is it essential for the satellite to sense infrared emissions from the Earth's surface and not just visible radiation?

12 It would be very useful to have a weather satellite in geostationary orbit 900 km above the UK. Why is this not posssible?

WAVES: TRANSFERRING ENERGY THROUGH SPACE

LASERS

The word 'laser' is short for **Light Amplification by the Stimulated Emission of Radiation**. There are many laser systems giving a variety of wavelengths and powers in the visible, infrared and ultraviolet parts of the spectrum (→ 37.4). One very important aspect of a laser's output is that it is **coherent** (→ 33.1).

THE VISIBLE SPECTRUM

When sunlight or light from a white-hot filament lamp is viewed through a prism or diffraction grating, a **continuous** white light spectrum is observed. Analysis of the spectrum shows that wavelengths range from about 400 nm at the violet end to 700 nm in the red. There are so many possible excited states of the atoms in the lamp or the Sun that the whole range of visible frequencies are emitted. In contrast, a small number of excited states exist in a limited quantity of, for example, neon or sodium. The outer electrons of their atoms, after being excited into higher energy levels, return to their unexcited level and emit light characteristic of the substance. Neon has a characteristic red glow, mercury vapour, blue and sodium vapour, yellow. Such gases exhibit **line spectra** (→ 37.2).

Colour and the eye

The retina is the light-sensitive part of the eye (→ 36.4). It houses two types of receptors, rods and cones, each of which contains light-sensitive chemicals. These respond to incident light by generating electrical impulses in the nerve fibres leading to the brain. **Rods** enable you to see when light intensity levels are low. They do not provide information concerning colour of incident light and show maximum sensitivity at a wavelength of about 510 nm. **Cones** enable you to discern the colour of light. There are three types of cone in the retina, sensing the red, green and blue regions of the visible spectrum. Between them, they can sense wavelengths ranging from about 400 nm to about 750 nm. The relative absorption curves for the cones are shown in Figure 32.19.

Rods outnumber cones by about twenty to one. Their sizes are very similar, being just a few μm across. Rods are largely responsible for peripheral vision while cones predominate around the optical axis of the eye. Their outputs are scanned by the bundle of nerve tissues in the **optic nerve**, which relays the information to the visual cortex of the brain. There is still much discussion of how the retina and visual cortex decode the image. One problem facing researchers in this area is that neither retina nor cortex can be operated independently.

Question

13 Estimate the total number of rods and cones in a typical retina of diameter 15 mm, given that rods are about 0.01 mm apart.

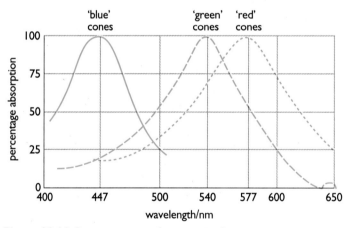

Figure 32.19 Response curves for cones in the eye

ULTRAVIOLET RADIATION

Electromagnetic radiation just beyond the violet end of the visible spectrum is known as ultraviolet radiation. Its wavelength range extends down to 10 nm, below which electromagnetic waves are regarded as soft X-rays (see Figure 32.20). Ultraviolet radiation is produced by relatively large energy transitions within excited atoms. It is emitted by ionised gases in a discharge tube, by electrical arc or spark discharge, or by incandescent objects of high enough temperature. The Sun's surface, at around 6 000 K, emits a considerable amount of ultraviolet radiation (see Figure 32.21). At the surface of the Earth, we receive only the near ultraviolet because the ozone layer in the atmosphere absorbs much of the shorter wavelength ultraviolet.

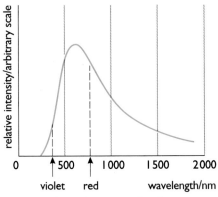

Figure 32.21 Solar energy distribution

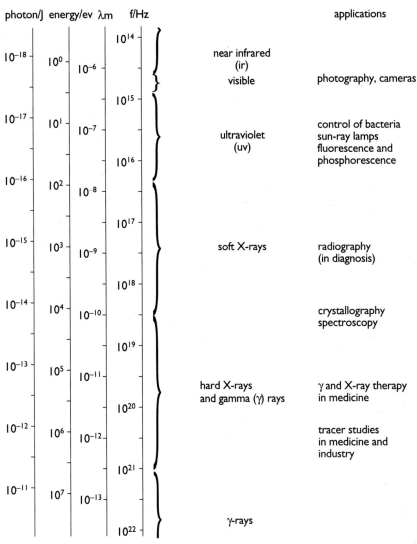

Figure 32.20 The high-frequency electromagnetic spectrum

A very effective source of ultraviolet is the **mercury vapour discharge tube**. Mercury vapour, excited by passing a current through it, emits visible and ultraviolet wavelengths. When the inside of the tube is coated with a fluorescent material, practically all of the ultraviolet emitted by the vapour is converted into visible light by fluorescence (→ 37.4). This is the basis of strip lighting. Blackening the tube to cut out visible radiation produces an ultraviolet source. The tube of such a lamp has to be made of a material that transmits ultraviolet, such as quartz, and not ordinary glass, which is opaque to such frequencies.

Ultraviolet wavelengths from about 320–400 nm have a beneficial effect on us, causing tanning and the production of vitamin D in skin cells. But prolonged exposure to shorter ultraviolet wavelengths can cause serious problems for eyes and skin. This has led to great concern over preserving the ozone layer in the Earth's atmosphere. Such wavelengths also have a lethal effect on micro-organisms, the most effective wavelength being 260 nm. Since mercury vapour emits strongly at 257 nm, low pressure mercury discharge lamps are frequently used to control bacteria.

Photographic emulsion provides a simple means of detection over the whole ultraviolet spectrum. Fluorescent and phosphorescent materials absorb incident ultraviolet radiation and re-radiate the energy at optical frequencies (→ 37.4).

WAVES: TRANSFERRING ENERGY THROUGH SPACE

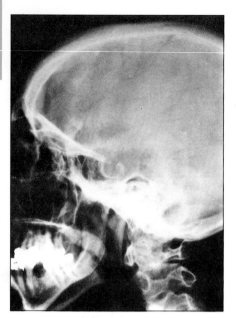

Figure 32.22 An X-ray photograph of a human skull

X-RAYS

X-rays range in wavelength from around 10 nm to about 10^{-12} m. They were discovered by Wilhelm Röntgen in 1895 (see Figure 32.22). He found that cathode rays, electrons, accelerated through a large voltage and striking the walls of a glass discharge tube, caused a strong luminescence of the glass itself. Superimposed on this continuous spectrum was the characteristic X-ray line spectrum of the tube target ($\rightarrow$ 37.5).

Medical uses of X-rays

X-rays for diagnosis are generated by directing a high-energy electron beam onto a target in an evacuated tube ($\rightarrow$ 37.5). Photographic film can be used to detect the rays, producing a **radiograph** (see Figure 32.24). Alternatively, the X-ray image is allowed to fall on a fluorescent screen, when a continuously changing image may be observed. This is important where a moving image is required, such as during investigation of the gastro-intestinal tract.

Bone has a sufficiently dense structure to produce radiographs with good contrast. For some materials, such as tumours and blood vessels, an agent is required to increase the contrast artificially. For example, barium sulphate in an aqueous suspension is frequently used when diagnosing problems associated with the intestinal tract.

Higher-energy X-rays are used in therapy to kill cancer cells. Clearly, normal cells must not be exposed to large doses of radiation in the process. So multiple X-ray beams are directed at the offending area, or a single beam is used and rotated about a point centred on the tumour. In either case, intervening tissue receives a much smaller dose of X-radiation than the tumour.

Question

14 X-rays are used at airports to produce 'shadow photographs' of the inside of closed packages and luggage. Discuss how, in principle, such systems might work.

X-ray crystallography

Visible light passing through a grating, or regularly spaced apertures, scatters and diffracts. If the aperture spacing is sufficiently small, a reasonably spread out diffraction pattern is observed ($\rightarrow$ 33).

A single crystal of, say, sodium chloride acts as a three-dimensional grating. The spacing of layers of atoms in crystals is of the order of 0.1 nm. For diffraction to occur, electromagnetic waves of approximately this wavelength are needed. Such soft X-rays may be obtained using a filter and an X-ray tube with a copper target ($\rightarrow$ 37.5).

X-rays have been used to investigate the structure of many crystalline materials and to find precisely the spacing of their crystal planes (see Figure 32.23). A series of spots is seen on film placed in the path of X-rays diffracted by a single crystal, as shown in Figure 32.24. A crystal crushed into the form of a powder presents to the X-ray beam a very large number of tiny crystals orientated at random. The result is a series of rings. Amorphous materials form no distinct pattern apart from broad and fuzzy rings ($\leftarrow$ 25).

During the course of this century, through the work of William Bragg and others, X-ray crystallography has played a very significant role in the investigation of the structure and behaviour of crystalline and polycrystalline materials. Another important technique is **electron diffraction**, which makes use of the

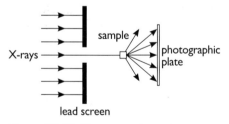

Figure 32.23 X-ray crystallography

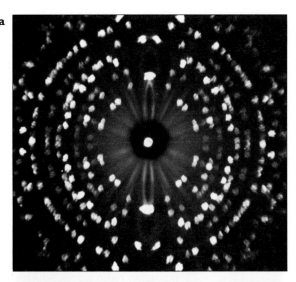

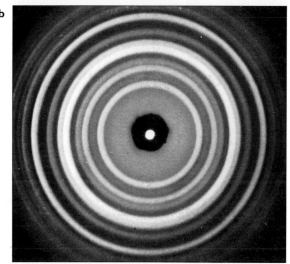

Figure 32.24 X-ray diffraction by **a** a single crystal, **b** a powder sample

wave properties of the electron (→ 37.6). It has the advantage over X-ray diffraction that a stream of electrons can be focused into a very fine diameter beam. This enables you to observe small changes in composition over different regions of a sample.

GAMMA-RAYS

Gamma (γ)-ray photons are identical to X-ray photons of similar energy. Only the way that they are produced differs. Gamma-radiation is emitted by the nucleus of an atom in an excited state, usually after emission of an alpha– or beta-particle (→ 38.5). It is analogous to the emission of a photon of visible light by an excited atom. Typical γ-ray frequencies range from around 10^{21} Hz upwards.

Question

15 Gamma-rays are used in the aircraft industry to check the internal parts of jet engines. Find out how this procedure is carried out. What advantages do γ-rays have over X-rays in this situation?

WAVES: TRANSFERRING ENERGY THROUGH SPACE

Atmospheric absorption of electromagnet radiation

The Sun and other objects in space emit all frequencies of electromagnetic radiation. When the radiation falls on the Earth, some frequencies penetrate the atmosphere and reach the Earth's surface. Others, however, are partially or totally absorbed, or are reflected (see Figure 32.25).

The absorption of ultraviolet radiation by the atmosphere results in a spherical shell of ionised air around the Earth called the **ionosphere**. It extends from about 40 km to over 1 000 km above the surface of the Earth. The ionosphere consists of several distinct regions or layers that may change in thickness between day and night, and also show latitude and seasonal variations. The C- and D-layers, lying approximately 40–90 km above the Earth, contain relatively small amounts of ionised air and reflect radio waves of low frequency. The E-layer lies about 90–150 km from the Earth and contains larger numbers of ions and electrons. It reflects medium-frequency radio waves. The highest layer, the F-layer, extends from about 150 km up to 1 000 km and contains the highest concentration of free electrons. It strongly reflects high-frequency radio waves, which aids long-range radio communication across the Earth's surface.

At night, when there is no solar radiation to maintain their ion concentrations, the D- and E-layers become inactive.

Electromagnetic waves with frequencies ranging from about 50 GHz down to about 15 MHz (wavelengths of approximately 6 mm to 20 m) pass straight through the ionosphere irrespective of the time of day. This **radio window** allows the use of satellites for intercontinental radio and television communication. It also enables Earthbound radioastronomers to investigate the radio emissions from stars and other objects in space.

Infrared radiation is particularly susceptible to atmospheric absorption. The atmosphere is composed of gases, liquids and solid particles, all of which tend to absorb or scatter infrared. Water vapour is probably the single most important attenuator of infrared radiation, with carbon dioxide and ozone also making a large contribution. **Infrared spectroscopy**, the analysis of the infrared absorption spectra of substances, provides scientists with a powerful tool for detecting very small quantities of certain substances.

Visible light is readily transmitted by the atmosphere, which is why most of our energy from the Sun is received in this form. Ultraviolet, X-rays and γ-rays are absorbed and scattered by the atmosphere.

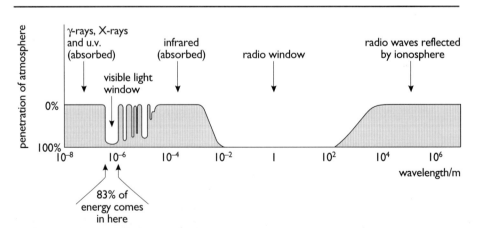

Figure 32.25 Atmospheric absorption of electromagnetic radiation

Question

16a What is meant by the 'greenhouse effect' ($\rightarrow$ 36.1)? How does the Earth's atmosphere have a greenhouse effect on solar radiation?

b In December 1901, Marconi transmitted the first radio signals across the Atlantic, from England to Newfoundland. Despite this, we still need microwave links and communication satellites. Why?
c Why does the deterioration of the ozone layer have potentially serious implications for life on Earth?

COSMIC RAYS

This radiation is received on the Earth from outer space. Cosmic rays consist of many different types of particles and electromagnetic waves. Some of the waves have considerably shorter wavelength than those emitted by the highest energy nuclear reactions that can be caused on Earth. These can only be due to extremely energetic processes taking place in stars.

32.3 VELOCITY OF ELECTROMAGNETIC WAVES

It was apparent to scientists as far back as the seventeenth century that light possessed a finite speed but that this speed was very high. They realised they would have to time light over very large distances to measure its speed. In 1676, Römer used the changing distance between the Earth and Jupiter to measure the speed of light. Nearly two centuries later, terrestrial methods were developed by people such as Fizeau and Foucault.

Methods for measuring the speed of light

In 1850, Foucault used a rotating mirror system, similar in principle to that shown in Figure 32.26, to measure the speed of light. His method involved sending pulses of light over a known distance and measuring the time taken. Light from a slit source s passed through a glass plate, on to a curved mirror R, which could be rotated about a vertical axis at high speed and then to a curved mirror M. From the speed of rotation of R, found using a stroboscope, and the displacement of the image d, Foucault calculated the time taken for the light to travel twice the distance between the two mirrors. He could only measure the displacement of 0.7 mm with limited accuracy. Nevertheless, he was able to obtain a value of 2.98×10^8 m s^{-1} for the speed of light.

The basis of direct measurements of the speed of light is to measure the time taken for a pulse of light to travel a known distance. In the nineteenth century, the pulses were long, so the distances had to be large. Fizeau used a toothed wheel to 'chop' the beam of light, Foucault, and later Michelson, used a mirror rotating at high speed. Even so they required distances of kilometres to observe a measurable time delay. Today, with electro-optic shutters 'chopping' the beam at 100 MHz, the spacing of pulses is only 3 m and measurements can be done on the laboratory bench.

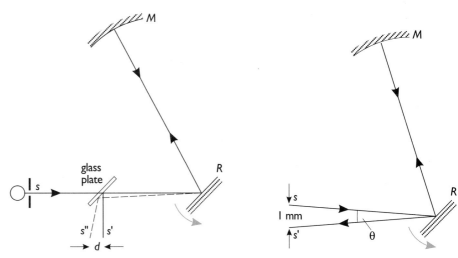

Figure 32.26 Apparatus for Foucault's experiment   **Figure 32.27**

Question

17 In an experiment similar to the one performed by Foucault, the rotating mirror is 2 m from the fixed mirror and an image displacement of 1 mm is required (see Figure 32.27).
a How long would a pulse of light take to travel from the rotating mirror to the fixed mirror and back again? (c)
b If the image is 50 cm from the rotating mirror, through what angle θ is the light beam turned after reflection from the fixed mirror?

WAVES: TRANSFERRING ENERGY THROUGH SPACE

> c Through what angle in degrees, does the rotating mirror turn?
> d Using your answer to **a**, calculate the rate of rotation of the mirror.
> e How could you arrange to obtain such a speed and how would you measure it?

In 1873, James Clerk Maxwell proposed a mathematical model that explained the observed properties of light. It also predicted the radio waves detected by Hertz in 1887. More specifically, it suggested that all electromagnetic waves, independent of frequency, travel at the same speed as light in free space. Maxwell's model shows that the speed c of such waves is dependent only on the electric permitivity ϵ_0 (← 16.2) and the magnetic permeability μ_0 (← 18.3), of free space. These are related by the expression :

$$c = \frac{1}{\sqrt{\epsilon_0 \mu_0}}.$$

From the definition of the ampere, μ_0 is defined as exactly $4\pi \times 10^{-7}\,\text{N A}^{-2}$. From measurements, ϵ_0 is $8.854 \times 10^{-12}\,\text{F m}^{-1}$. Combining these two quantities in the equation gives a value for c of $2.9979 \times 10^8\,\text{m s}^{-1}$.

> **Question**
>
> 18 Show that the equation $c = \frac{1}{\sqrt{\epsilon_0 \mu_0}}$ gives the correct unit for c by substituting the appropriate units for ϵ_0 and for μ_0.

CONSTANCY OF C AND THE DEFINITION OF THE METRE

A fundamental property of all electromagnetic waves is that they travel at exactly the same speed in a vacuum. The measured speeds given in Figure 32.28 confirm this. **So c has now been defined to be 299 792 458 m s⁻¹ in a vacuum.**

To avoid inconsistencies in our system of units, some change also had to be made to the definition of the second or of the metre. In 1983 the metre was redefined to be that length of path travelled by light in a vacuum during a time interval of 1/299 792 458 s. So, to measure distance all one needs is a clock! Any departure from the value of the metre could only mean that other measurements were inaccurately made — or that the speed of light varied with time. In the latter case, of course, the metre would have to be redefined.

Measuring the speed of light in a vacuum is now inappropriate. But measuring its value in any other medium can give useful information about the nature of that medium, such as its refractive index (→ 34.2) or its permittivity. Experiment 32.2 suggests a way of measuring the speed of light in a fibre-optic cable. The method is similar in principle to the one used by Foucault.

wavelength/m	speed/10^8 m s⁻¹
6.4	2.997 8 ± 0.000 3
1.8	2.997 95 ± 0.000 03
1.0	2.997 92 ± 0.000 02
1.0×10^{-1}	2.997 92 ± 0.000 09
1.2×10^{-2}	2.997 928 ± 0.000 003
4.2×10^{-3}	2.997 925 ± 0.000 001
5.6×10^{-7}	2.997 931 ± 0.000 003
2.5×10^{-12}	2.983 ± 0.015
7.3×10^{-15}	2.97 ± 0.03

Figure 32.28 Measured speeds of electromagnetic waves across the spectrum

EXPERIMENT 32.2 Measuring the speed of a pulse of light along a fibre-optic cable

Using the circuit in Figure 32.29a, you can detect light pulses with a photodiode at the end of a 10 cm fibre-optic cable. The pulses are sent from a light-emitting diode (LED) and modulated by a frequency of 1 MHz. The received and transmitted pulses should be similar to those shown in Figure 32.29b. The delay between pulses is introduced by the circuit components and the length of the cable. So, using a 20 m-length of cable causes the received pulse to be displaced further to the right, as in Figure 32.29c. Since the component delay does not change, the extra delay arises because of 19.90 m extra path length, enabling you to calculate the velocity of the pulses.

Note that the value obtained for the speed of light in this experiment is not 3×10^8 m s^{-1}, because light travels more slowly along optical fibre than in a vacuum. Typically, the speed along the cable is about 2×10^8 m s^{-1}.

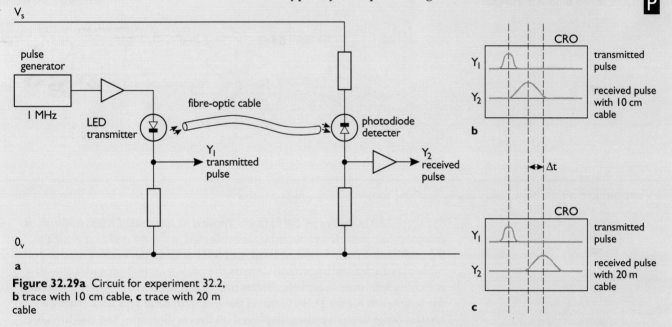

Figure 32.29a Circuit for experiment 32.2, **b** trace with 10 cm cable, **c** trace with 20 m cable

Question

19 In experiment 32.2, the time base was set to 0.1 μs cm^{-1} and the received pulse was displaced a further 1 cm to the right when the short length of cable was replaced by the 20 m length of cable.
a Calculate the time taken for the pulse to travel along the extra length of cable.
b Estimate the speed of the pulse of light in the cable.
c What is the value of the refractive index (→ 34.2) of the glass from which the fibre is made?
d Estimate the uncertainty in the value you calculated in **c**. Assume that the cable length is accurate to within 1 % and the displacement of the trace can be measured to within 5 %.
e What modifications would you make to improve on the accuracy of the results from this experiment?

G WAVES: TRANSFERRING ENERGY THROUGH SPACE

A similar experiment to 32.2 can be performed with an electrical signal transmitted along a conductor, such as a coaxial cable. A coaxial cable is often used to carry television and radio signals because it creates no external fields nor is it affected by them. It consists of two coaxial copper conductors separated from one another by a polyethene insulator.

An electrical pulse applied to one end of the cable will rapidly move along the cable. In the region occupied by the **E** and **B** fields associated with the pulse, charges will be set into motion as the pulse passes by and will come to rest again as the pulse moves on. The speed of individual charges is very slow, perhaps of the order of 10^{-6} m s^{-1}. The speed of the electromagnetic disturbance causing the movement is very fast, as demonstrated in experiment 32.3. The nature of the **E** and the **B** fields associated with the pulse and moving charges are shown in Figure 32.30. The fields are always at right angles to each other and to the direction of propagation of the pulse.

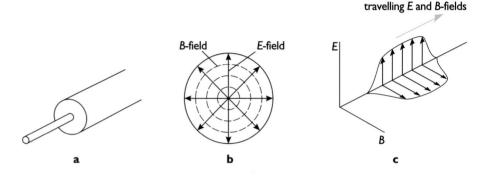

Figure 32.30 **E** and **B** fields in a coaxial cable

EXPERIMENT 32.3 Measuring the speed of a pulse along a coaxial cable

In Figure 32.31a pulses of 200 kHz are applied to one end of a 200 m drum of coaxial cable, pass down the cable, are reflected at the far end and return to the oscillator. Connecting a resistance of 50 Ω or 75 Ω, depending on the cable's characteristic impedance, across the cable's far end provides it with a matching termination and eliminates multiple reflections. You can measure the time taken for the pulse to travel the length of cable and back from the separation of the two pulses displayed on the oscilloscope. You can then find the speed of the pulse. Figure 32.31b shows the sort of traces you might obtain.

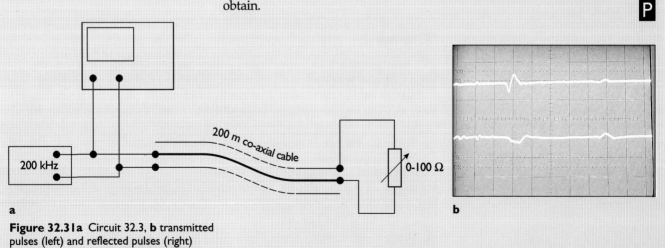

Figure 32.31a Circuit 32.3, **b** transmitted pulses (left) and reflected pulses (right)

32 ELECTROMAGNETIC WAVES

Question

20 This question refers to experiment 32.1.
a Estimate the separation of the pulses in Figure 32.31b and so find the time for the pulse to make a return journey. The time-base is set to 0.5 μs cm^{-1}.
b Calculate the speed of the pulses along the cable.
c Why is this value not 3×10^8 m s^{-1}? Estimate the relative permittivity of the space between the inner and outer conductors.
d If the frequency of the pulses was increased slightly, what effect would this have on the oscilloscope trace?
e Why is it important to match the input impedance to a television with the characteristic impedance of the aerial downlead?
f Pulses may be absorbed or reflected at the end of the cable with the same or with opposite polarity, depending on the termination resistance. What analogy may be drawn with transverse pulses on a slinky spring?

32.4 POLARISATION OF ELECTROMAGNETIC WAVES

Radio waves from a transmitting aerial are **polarised** (← 32.1). The **E**-field associated with the broadcast wave is parallel to the aerial dipole. Receiving aerials, on the other hand, pick up that component of the transmitted **E**-field parallel to their dipoles (← 32.2).

Microwaves from the horn of a transmitter are also polarised. A maximum signal is only picked up by the microwave receiver when it is orientated parallel to the incoming wave. Rotating the receiver through 90° around an axis parallel to the incoming wave causes the received signal to drop to zero. The same occurs when the receiver is orientated to the maximum signal and a **polarising filter** is rotated. Such a filter consists of a grille of parallel metal rods about 1 cm apart. When the grille is inserted between the transmitter and receiver, so that the grille is parallel to the **E**-field, no signal is detected. Rotating the grille until the rods are perpendicular to the field causes the received signal to grow to a maximum (see Figure 32.32).

When a transmitter and receiver are **crossed**, so that zero signal is detected, a grille inserted at 45° between them causes a signal to be detected. This is because only the **E**-field component perpendicular to the grille passes through. The detected signal is the component of this parallel to the receiver dipole. The plane of polarisation has been rotated through 90°.

In contrast, light from a lamp is unpolarised. There is no preferred plane for the oscillating **E**-field and all possible planes are equally likely. However, light can be polarised using a polarising screen or filter, similar to the grille used with microwaves. Light from a lamp falling on a sheet of **polaroid** contains all possible planes of oscillation of the electric vector. Microscopic crystals are aligned within the transparent plastic polaroid sheet by stretching during manufacture. They absorb light with a component of **E**-field parallel to their length and allow through light with a component of **E**-field perpendicular to their length. Light emerging from the sheet is therefore polarised. The lamp viewed through the polaroid will appear slightly dimmer due to absorption by the sheet. Another sheet placed on top of this, with its crystals aligned in the same direction, will cause a further reduction in intensity due to absorption. However, rotating the second sheet slowly causes the intensity to drop to near zero as the sheets become crossed. Further rotation causes the intensity to increase again to a maximum as the two sheets become aligned once more.

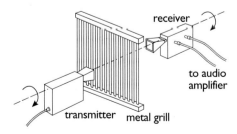

Figure 32.32 A polarising grille

WAVES: TRANSFERRING ENERGY THROUGH SPACE

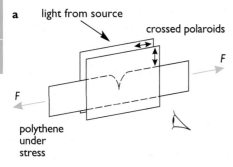

Questions

21 Look obliquely at a window, at the surface of water or a shiny table top through a piece of polaroid in, say, some sunglasses. You will notice that the intensity of light changes considerably as the polaroid is rotated. What does this tell you about the nature of reflected light? Explain why this happens. Why do many people regard polaroid sunglasses as better than 'ordinary' sunglasses?

22 A student views a lamp through two sheets of polaroid so that she can see the filament clearly. Then she places a third sheet of polaroid between the other two sheets. What will she see when she rotates the middle sheet slowly?

Photoelastic stress analysis

Structural engineers often make models of a structure to enable them to analyse the magnitude and distribution of stresses in it. Although computer models play an important part in this, the polarisation of light can also be used.

White light from a source falling on a sheet of polaroid becomes polarised. A second sheet placed so that the two pieces are crossed, ensures that no light will be transmitted (see Figure 32.33a).

When an object made from transparent materials, such as polythene or perspex, is put under stress and placed between the two sheets, it displays coloured fringes (see Figure 32.33b). Each coloured fringe corresponds to a particular value of the stress. Analysis of such fringe patterns in perspex models allows the stresses in a structure to be mapped out. This helps the engineer choose the most suitable materials, shape and dimensions for a particular structure.

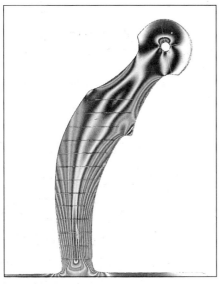

Figure 32.33a Observing stress patterns in polythene, **b** photoelastic stress patterns

POLARISATION OF LIGHT BY SCATTERING

Tiny particles placed in the path of unpolarised light scatter the light in all directions. The scattering is due to the oscillations of electrons in the particles in response to the incoming electromagnetic wave. The oscillating electrons themselves radiate electromagnetic wave energy.

Figure 32.34 shows unpolarised light travelling in the direction *OP*. Light is scattered in all directions by a particle at *P*. Consider two perpendicular directions, *PX* and *PY*, of this scattered light. Light emerging along *PX* can have its electric vector only in a vertical plane. No other plane from the original beam is possible along *PX* because it would imply oscillations of the **E**-field along the direction *PX* of the wave. Similarly, the **E**-field oscillations are in a horizontal

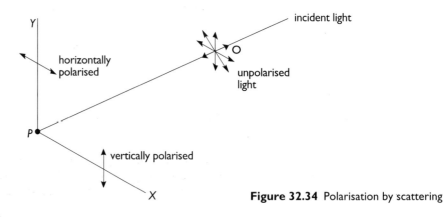

Figure 32.34 Polarisation by scattering

32 ELECTROMAGNETIC WAVES

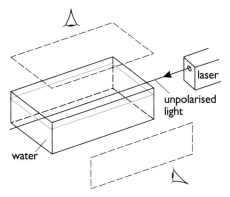

Figure 32.35 Observing polarisation by scattering

plane along *PY*. In this way, light emerging along each of these two perpendicular directions is plane polarised, with their planes at right angles to one another.

The effect may be observed using unpolarised light from a laser and a small transparent plastic tank of tapwater (see Figure 32.35). In a darkened room, the laser beam should be clearly visible as it passes through the water. To make it even clearer, a *tiny* droplet of milk can be mixed with the water (if you add too much, multiple scattering will occur and the effect will not be seen). The laser beam is viewed from vertically above the tank through a sheet of polaroid. When the sheet is slowly rotated, the beam disappears twice per rotation, showing that the emerging light is polarised. A similar effect occurs when the tank is viewed from the side.

You can also see the effect by viewing a clear sky through polaroid sunglasses. Look at the sky in a direction at right angles to that in which sunlight is coming and then rotate the polaroid. You will see the blue sky darken considerably as the polarised light is absorbed. Photographers frequently use polaroid filters on the front of their cameras to enhance colour photographs.

Question

23 The amount light is scattered seems to be very dependent on its wavelength. The shorter the wavelength, the greater the scattering. During a clear day, the sky away from the region of the Sun looks blue because small particles in the atmosphere scatter the blue light more effectively than red. Explain:
a why the Sun and sky often appear red at sunrise and sunset.
b what colour you would expect the 'sky' to appear to an astronaut on the Moon.
c what colour you would expect the Sun to appear from the Moon.

SUMMARY

Electromagnetic waves are produced by charged particles undergoing non-uniform acceleration. The waves travel at 2.998×10^8 m s^{-1} in a vacuum.

The waves are transverse and consist of travelling electric and magnetic fields. They require no medium for their propagation. The **E**-field is at right angles to the **B**-field, and both are perpendicular to the direction of travel.

The speed c, frequency f and wavelength λ of the waves are linked by the equation $c = f\lambda$.

The characteristics of a wave depend on its frequency. Radio waves appear at the low-frequency end of the spectrum, and progressively higher frequencies are associated with infrared, visible, ultraviolet, X-rays and γ-rays. Cosmic rays contain electromagnetic waves of even higher frequency.

The particle properties of electromagnetic radiation become increasingly important towards the high-frequency end of the spectrum.

Electromagnetic waves, being transverse, can be polarised. Radio waves emitted by an aerial are polarised, while light from, say, a lamp is unpolarised. Light may be polarised by reflection, scattering or by the use of a sheet of polaroid.

WAVES: TRANSFERRING ENERGY THROUGH SPACE

SUMMARY QUESTIONS

24 The following are some forms of electromagnetic radiation: radio waves, microwaves, infrared radiation, visible light, ultraviolet radiation, X-radiation and γ-radiation. For each type, describe
a how it is produced.
b how it might be detected.
c an important use or application for it.

25 A radio station broadcasts on a frequency of 95.4 MHz.
a What is the wavelength of this radiation, in m?
b Why, when travelling under a bridge in a motor car with the radio tuned to this frequency, is the reception not noticeably affected?
c Why are aerials used to receive such signals significantly larger than aerials used to pick up TV colour transmissions?

26 The Voyager spacecraft has completed its journey through the outer planets and is now beyond Neptune, some 5 000 000 000 km from the Earth. Figure 32.1 shows a photograph of Neptune sent from the craft back to Earth.
a How, in principle, are we able to obtain a photograph over such a large distance?
b How long does it take for a radio signal to travel that distance? (c)
c The pattern of signals received from the craft is exactly the same as that transmitted by the craft. What does that suggest about the way radio signals travel through the vast distances of space?

27 Over the past century or so, people's lives have been transformed by applications of the properties of electromagnetic radiation. Describe four such applications and for each explain how they have depended upon our understanding of electromagnetic radiation.

33 SUPERPOSITION OF ELECTROMAGNETIC WAVES

Interference patterns (← 30.5) and **diffraction** patterns (← 30.4) are frequently created when light interacts with itself and with matter. The patterns are made up of **fringes**, one fringe being an alternate light and dark stripe. Coloured patterns and fringes are often produced when **white light**, light consisting of a mixture of many wavelengths, is used.

Similar effects occur with other electromagnetic waves and much of the early evidence of the wave nature of light came from the observation of such patterns. The nature of each pattern can be explained using the principle of superposition (← 30.1). The effects produced by the superposition of electromagnetic waves and their applications are of considerable importance in communicatons, in entertainment and, increasingly, in art.

Young's experiment

In 1807, Thomas Young performed an experiment that demonstrated superposition effects with light. He used a pin-hole illuminated by the Sun and allowed light from this source to fall on two pin-holes, S_1 and S_2 (see Figure 33.1). On a screen placed some distance away, he saw a stationary pattern consisting of bright and dark bands of light. He explained it by suggesting that, in the space beyond the two pin-holes, the diffracted beams became superposed.

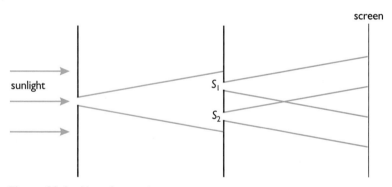

Figure 33.1 Young's experiment

33.1 INTERFERENCE OF ELECTROMAGNETIC WAVES

In Young's experiment, two light beams superposed in much the same way as water ripples from two dippers superpose in a ripple tank (← 30.5).

The two dippers must either by moving in phase or, at least, with a constant phase difference – they must be **coherent** sources. Similarly, to see superposition effects with light, we need coherent sources. The random nature of the emission of light from individual atoms in a source means that the light from two sources cannot have a constant phase relationship with each other.

We need to split the light from one source into two distinct parts to produce the effect of two mutually coherent sources. There are two methods of doing this: **division of wavefront** and **division of amplitude**. Using two slits is the simplest way of dividing the wavefront. Dividing the amplitude can be done using a semi-reflecting mirror, so that half of the light is transmitted and half of it is reflected. In either case, the beams produced are mutually coherent, containing

WAVES: TRANSFERRING ENERGY THROUGH SPACE

wave trains of the same phase mix and having a constant phase relationship, as shown in Figure 33.2.

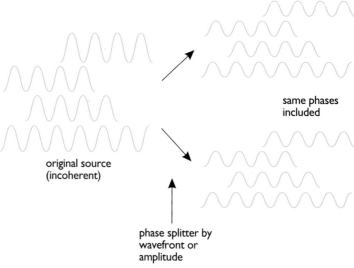

Figure 33.2 Division of amplitude and wavefront

EXPERIMENT 33.1 Observing Young's fringes

By observing a lamp with a vertical filament through a double slit close to the eye, as in Figure 33.3, you will see vertical fringes due to interference. The two slits need to be narrow enough, say 0.1 mm, to produce a reasonable amount of diffraction and close enough, about 0.3 mm, to allow the diffracted beams to superpose. The green filter placed in front of the light source allows through only a narrow band of wavelengths and gives a clearer interference pattern.

You might like to compare the pattern you see through the double slit with the diffraction pattern for a single slit (← experiment 31.1).

Figure 33.3 Observing interference fringes

Figure 33.4 shows a photograph of the pattern you see when you perform experiment 33.1. The main feature is the alternate fringes of light and dark of uniform separation. These are known as **Young's fringes** and show a strong similarity to the stationary patterns seen with water waves (← 30.5). The fact that we can obtain such a pattern is further evidence in favour of the wave model of light.

Question

1 Comment on the following statement:
'If it were not for diffraction, it would not be possible to see interference fringes in Young's experiment.'

FORMATION OF INTERFERENCE PATTERNS

Figure 33.5 helps to show how the pattern in Figure 33.4 might arise. The two waves originating from the two slits S_1 and S_2 are in phase. So, there is a bright

Figure 33.4 Young's fringes

33 SUPERPOSITION OF ELECTROMAGNETIC WAVES

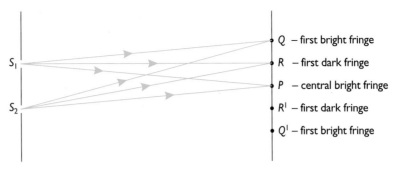

Figure 33.5 Why fringes occur

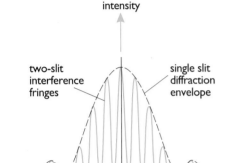

Figure 33.6 The single slit pattern and Young's fringes

patch at a point P on the screen, equidistant from the two pin-holes. At P, the two waves are exactly in phase and the waves from each source add constructively to produce a wave amplitude that is twice as large.

The next bright fringe occurs a little further along the screen at Q or Q'. Again, light arrives at this point in phase and adds constructively. For this to happen, light from S_2 has to travel **one wavelength further** than light from S_1. There are also on the screen, points such as R and R' where the waves from S_1 and S_2 are exactly out of step, or in antiphase. In the case of points R and R', the path difference is **half a wavelength**. The two waves add destructively to produce no light.

In Figure 33.4, there is a variation in intensity of the maxima over the pattern. This is caused by diffraction at each of the slits. The shape of the **envelope** of intensity of the fringes matches exactly the shape of the single slit diffraction pattern produced by each aperture (see Figure 33.6). If we could eliminate the effect of diffraction, the fringes would all be of equal intensity. Of course, without diffraction, there would be no fringes to be seen because the light from the two slits would not spread outwards and be able to interfere!

It is possible to predict where the maxima and minima of an interference pattern will occur. In Figure 33.7a, d represents the slit separation and λ the wavelength of the incident light. **Maxima** will occur on the screen in a direction θ from the central bright fringe, where:

$$n\lambda = d \sin \theta,$$

$n = 0, 1, 2, 3,$ etc. **Minima** will occur at angles θ where:

$$(n + \tfrac{1}{2}) \lambda = d \sin \theta.$$

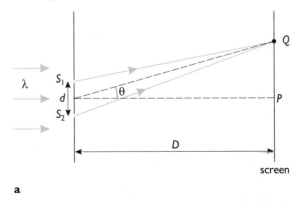
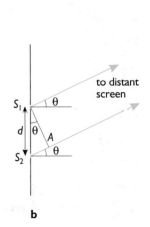

Figure 33.7a Arrangement for predicting the fringe pattern, **b** the situation close to the slits

Linking Young's fringes to the wavelength of light

When monochromatic light of wavelength λ falls on a double slit of separation d, the two diffracted beams superpose and a stationary interference pattern is formed. Suppose that, along a direction θ, the waves from each slit are in step and constructive interference takes place to give the first bright fringe at Q (see Figure 33.7). S_1 and S_2 are of the order of, say, 0.5 mm apart and the separation of slits and screen is about a metre. The distances S_1 to Q and A to Q are equal, and the extra path length travelled by light from S_2 is equal to S_2A. Since Q is the first bright fringe, the extra path length must also be one wavelength.

In the triangle S_2S_1A, the angle S_2S_1A is θ and:

$$\frac{S_2A}{S_1S_2} = \sin \theta.$$

As $S_1S_2 = d$, the slit separation, and $S_2A = \lambda$, the equation becomes:

$$\frac{\lambda}{d} = \sin \theta.$$

For the second bright fringe, $S_2A = 2\lambda$.
So, there will be a second bright fringe along a direction θ where:

$$\frac{2\lambda}{d} = \sin \theta.$$

Generally, for a bright fringe to occur:

$$\frac{n\lambda}{d} = \sin \theta$$

or $\quad n\lambda = d \sin \theta$,

where $n = 0, 1, 2, 3$, etc.

For a minimum in the pattern, the extra path length has to be $\lambda/2$ or $3\lambda/2$, $5\lambda/2$, etc. Generally, a minimum occurs along a direction θ where:

$$(n + \tfrac{1}{2})\lambda = d \sin \theta.$$

Note that when λ is much less than d, the angle of the first maximum is small and the equation becomes:

$$n\lambda = d\theta$$

where θ is in radians.

For the first bright fringe, $n = 1$ and:

$$\theta = \frac{PQ}{D} = \frac{y}{D} \quad \text{radians}$$

where y is the spacing between fringes. So:

$$\theta = \frac{\text{fringe spacing}}{\text{slits to screen distance}}.$$

For the first bright fringe, $\lambda = d\theta$, so the wavelength is given by:

$$\lambda = \frac{d.y}{D}$$

or **wavelength = slit separation** $\times \dfrac{\text{fringe spacing}}{\text{slits to screen distance}}$.

White light is composed of a range of wavelengths from about 400 nm to 700 nm. When white light falls on a double slit, there are a number of fringe systems superimposed upon each other. The central maxima is clearly white but subsequent maxima become less distinct and show coloured edges. The minima consequently become less well defined. It is not normally possible to observe more than four or five fringes within the central region of the pattern for white light. Fringes are much more clearly defined when a monochromatic source is used, for example, a sodium lamp or a laser.

COHERENCE OF ELECTROMAGNETIC RADIATION

Dippers driven by the same force in a ripple tank vibrate with the same frequency, and a stationary interference pattern forms on the water surface (← 30.5). The two sources are described as **mutually coherent sources**. When the dippers are in phase, maxima occur in the pattern. This happens wherever the path difference between waves from the two sources is a whole number of wavelengths. When the two sources are out of phase by 180° or π radians, minima occur.

Suppose the two dippers are not driven by the same vibrator, and so are independent of one another, and that they stop and start at random. Sometimes they will be in phase and sometimes completely out of phase. Normally there will be some phase difference between them. The resulting interference pattern will not be constant or stationary. The two sources are now **mutually incoherent**.

A light source, such as a gas discharge tube, consists of atoms which radiate as a result of electron transitions to lower energy levels (→ 37). Each transition can occur quite spontaneously and electromagnetic wave energy is radiated in the process for about 10 ns or so. Sometimes the transitions are stimulated to occur by a passing photon of exactly the same type. The radiation produced consists of a random collection of wave trains emerging from many different atoms within the source. The source is incoherent. When a double slit is placed directly in front of a perfectly monochromatic source, light from the two slits only has a constant phase difference over a time interval of 10 ns or so. Over longer times, the phase differences are changing randomly. Most detection systems will average out the resultant pattern over many 10 ns intervals and see no steady interference pattern. In particular, the eye has an 'averaging time' of 0.05 – 0.1 s, very much longer than the wave train duration of most sources. Only with sensitive detection systems, such as high-speed photomultipliers, have stationary interference patterns been observed over a period of 10 ns or less.

In addition, ordinary sources, even with the use of filters, are not perfectly monochromatic. They emit a range of spectral frequencies Δf around the dominant frequency f. As with dippers in a ripple tank, oscillations from opposite ends of the frequency range become out of phase in a time of $1/\Delta f$. This means that the pattern of independent sources will not be stationary for a time longer than about $1/\Delta f$. The sources are mutually incoherent.

So, two slits illuminated by an incoherent broad source of light form two mutually incoherent sources. Each slit is illuminated by different sets of wave trains from the incoherent source. Suppose, however, a single slit is illuminated and used as the source placed parallel to, and some distance away from, the double slit. Then, a constant relative phase is maintained between light from the two slits. Both slits can be thought of as being 'driven' by the narrow source (see Figure 33.8).

In **lasers**, atoms are stimulated to emit radiation in such a way (→ 37.4) that the majority emit photons in phase. Lasers can produce very long, high-amplitude wave trains for times of the order of a second, far greater than the 10 ns coherence time of many ordinary light sources. The light is highly monochromatic, 632.8 nm for the helium-neon laser. There is a very much smaller distribution of spectral frequencies around the dominant frequency than occurs with ordinary sources.

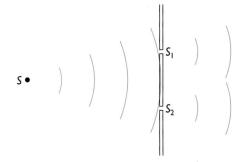

a S_1 and S_2 are driven by a single point source and maintain constant relative phase

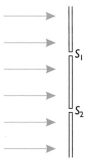

b S_1 and S_2 are now driven by independently radiating atoms

Figure 33.8 Mutually coherent sources

WAVES: TRANSFERRING ENERGY THROUGH SPACE

The waves emerging from a laser tube contain much the same phase components as those several metres away. The **coherence length** of a laser indicates the distance over which this is true, within specified limits. The phase components present at any one time in an emergent laser beam are similar to those present some fractions of a second later. So laser light is highly coherent, both in space and time, making it particularly useful in applications, such as holography (→ 34.5). You can investigate Young's fringes using laser light in experiment 33.2.

> **Question**
>
> **2a** Explain why two loudspeakers driven by the same signal generator can be regarded as coherent sources.
> **b** Two dippers in a ripple tank driven at the same frequency produce an interference pattern. What would be observed if they were each driven independently by oscillators of slightly different frequency?
> **c** A lamp filament consists of a large number of atoms which emit light when the filament passes a current. Why do the atoms form an incoherent set of sources?
> **d** Why is it not possible for superposition effects to be observed using two torchlamp filaments?

EXPERIMENT 33.2 Investigating Young's fringes

Using the arrangement in Figure 33.9a, with the two slits 0.06 mm in width separated by a distance of 0.25 mm, you should see a pattern similar to that in Figure 33.9b. You can investigate the nature of Young's fringes, measure the wavelength of the laser's light, note the brighter centre fringes, the outer diffraction envelope and the fringes within the secondary maximum of the single-slit diffraction envelope.

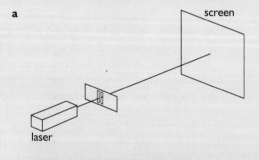

Figure 33.9a Obtaining fringes
b Young's fringes, and **c** single-slit pattern, with laser light

33 SUPERPOSITION OF ELECTROMAGNETIC WAVES

Question

3a In experiment 33.2, the fringe spacing was found to be 5.0mm for a slits-to-screen distance of 2m and a slit separation of 0.25mm. Use this data to obtain a wavelength for the light from the laser.
b What would be the effect on the fringes of:
 i reducing the separation of the slits?
 ii increasing the width of each slit?
 iii using light of longer wavelength?

INTERFERENCE OF MICROWAVES

Microwaves are a form of electromagnetic radiation with wavelengths of the order of a few centimetres. So, it should be possible to observe superposition effects with an arrangement similar to that in Young's experiment on light but with the much larger slit widths and separations, as in experiment 33.3.

EXPERIMENT 33.3 Superposition of microwaves

a The double slit shown in Figure 33.10a is made from two metal sheets and a strip of metal placed so that each gap is about 5 cm and their separation is about 10 cm. Microwaves falling on the double slit diffract and superpose, and you will see an interference pattern, similar to that in a ripple tank. Increasing the separation of the slits reduces the distance between maxima.

b In the arrangement of Figure 33.10b the receiver R is placed so that it picks up a direct beam from the transmitter T and one reflected from the metal sheet. Using this arrangement, you will observe a variation in intensity produced by the superposition between the two beams.

Figure 33.10a Arrangements for **a** experiment 33.3a, and **b** experiment 33.3b

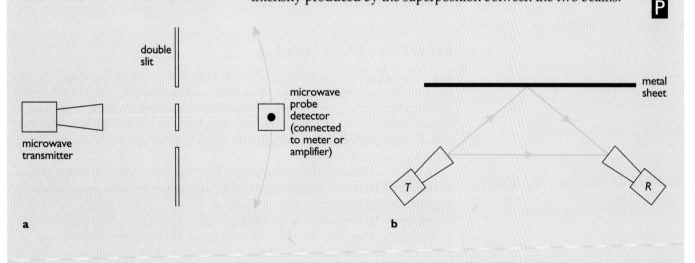

PHASE CHANGE ON REFLECTION

In experiment 33.3b, when the receiver is placed in contact with the reflector, a minimum occurs and not a maximum as one might expect. This is because microwaves undergo a **phase change of π on reflection**, much as pulses on

WAVES: TRANSFERRING ENERGY THROUGH SPACE

springs change phase on reflection (← 30.2). As a general rule, a phase change of π radians occurs when waves travelling in a less dense medium are reflected by a more dense medium.

When experiment 33.3b is performed with light, it is known as **Lloyd's mirror**. It simulates the two slit experiment by using the mirror to create a virtual image of a second slit. Figure 33.11 shows a possible arrangement that can be used with light. As with microwaves, light changes phase on reflection at the mirror surface. However, a dark fringe can only be observed when the screen is placed in contact with the end of the mirror.

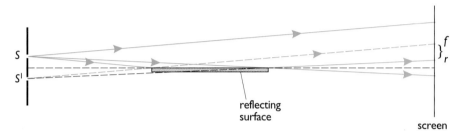

Figure 33.11 Interference by reflection – Lloyd's mirror

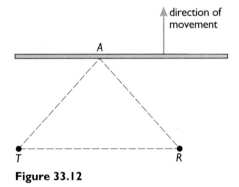

Figure 33.12

Question

4 Figure 33.12 shows a 10 GHz microwave transmitter T and receiver R placed close to a metal reflector. As the reflector is moved along the direction marked by the arrow, the received signal goes through a series of maxima and minima.
a Explain why this happens.
b The distance $TA = AR = 15$ cm and $TR = 24$ cm. Would you expect the receiver to detect a maximum or minimum signal in this position?

RECEPTION OF TV SIGNALS

Television signals on the 625-line BBC and ITV services are radiated by separate transmitting aerials but from the same location. For example, viewers in London receive television signals from the Crystal Palace transmitter. Each of the four stations is allocated a channel in the uhf waveband and so has its own separate broadcast frequency and wavelength (← 31.2). Viewers need only one receiving aerial to pick up all stations.

Reception difficulties can arise due to reflection of the signals from the ground or rooftops. The unwanted reflected signal may superpose with the wanted signal, resulting in partial cancellation. Since this effect is wavelength dependent, the reception on just one or two channels may be rather poor. Moving the aerial a small distance may cure the problem on one channel but introduce a similar problem on another channel. In some locations, it may be impossible to achieve satisfactory reception on all channels with any one aerial position. This problem can sometimes be solved by using an aerial that is more directional and strongly rejects the unwanted signal, such as a Log periodic aerial.

In some areas 'ghost' images may be seen on the screen. This is a separate effect caused by reflections from distant hills, buildings or other obstacles, where the extra path lengths are very much larger. A second aerial, in parallel with the first, can create a much more directional system and remove such 'ghost' images.

33 SUPERPOSITION OF ELECTROMAGNETIC WAVES

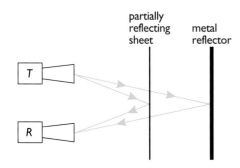

Figure 33.13

Question

5 Microwaves are directed at a partially reflecting hardboard sheet. Those that pass through are reflected by a metal sheet (see Figure 33.13). Two beams of microwaves return to the receiver R and superpose to give a resultant signal. As the metal reflector is moved 1.5 cm further away from the hardboard sheet, the received signal changes from a maximum, through a minimum and to the next maximum.
a What is the value of the wavelength of the radiation?
b For close separations of the reflectors, the signal detected at a minimum is very nearly zero. However, for large separations, the minimum signals are quite perceptible. Why is this?
c If the reflector is moved away from the receiver at 6 cm s^{-1}, at what frequency in Hz will the maxima occur?

Decca system of navigation

The Decca system designed for air, marine or land-based navigation, makes use of the interference pattern produced from a network of radio transmitters. These transmit radio waves in the low frequency band, 70–130 kHz. Fixed transmitting stations provide lines of maxima and minima, like those in interference patterns. Figure 33.14 shows the pattern produced by two transmitters, M the **master** and S the **slave** transmitter. Along line AB, and line CD,

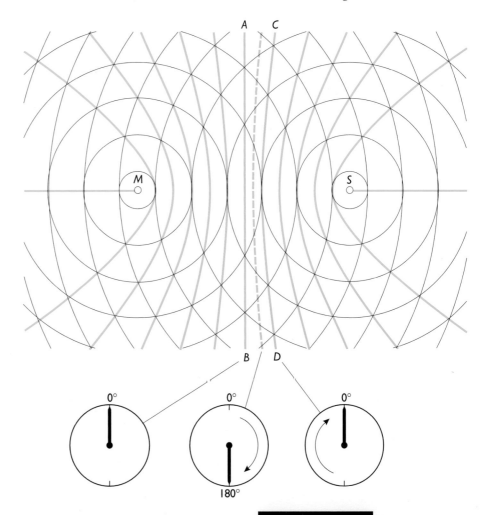

Figure 33.14 Decca lane pattern for master and slave transmitters

the signals received from the two transmitters are in phase. A radio receiver tuned to the correct frequency picks up a signal of large amplitude. The space bounded by two adjacent lines of maxima is referred to as a **Decca lane**. The radio receiver detects a minimum around the middle of a lane.

A chain of Decca navigator stations consists of a central master station and a number of outlying slave stations. The distance between stations is typically 100–200 km, and the range of the system is about 400 km. There are usually three slaves, known as red, green and purple from the colours printed on charts. The master has a control/monitoring function ensuring correct phase and frequency of transmitted signals. Users can fix the position of their ship, aircraft or vehicle using Decometers linked to the radio receiver. Automatic or computer-based methods are often used to obtain a fix of position. They can be plotted manually on a map or chart numbered in Decca lane units, as in Figure 33.15.

To distinguish between the red, green and purple lanes, harmonics of a fundamental frequency (← 30.6) are transmitted by the master and slave transmitters. Multipliers in the receivers bring the signals to a common frequency so that master and slave appear to be transmitting equal frequencies. The fundamental frequency f is about 14 kHz. In the red Decca lane pattern, the master transmits on a frequency of $6f$ and the red slave at $8f$. The receiver multiplies by 4 and 3 respectively to give a common frequency of $24f$, which is about 340 kHz.

The lanes are half a wavelength wide between transmitters. Red lanes are about 440 m wide since the wavelength of the signals is 880 m at 340 kHz. Figure 33.16 shows the frequencies and lane widths for one Decca chain.

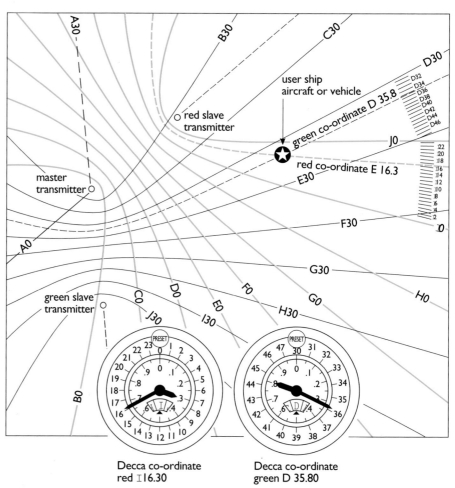

Figure 33.15 Decca lane chart

33 SUPERPOSITION OF ELECTROMAGNETIC WAVES

radiated frequencies

station	harmonic	frequency/kHz
master	6f	85.0000
purple slave	5f	70.8333
red slave	8f	113.3333
green slave	9f	127.5000

a

comparison frequencies and lane widths

pattern	harmonic	freq./kHz	lane/width
purple lanes	30f	425	352.1 m
red lanes	24f	340	440.1 m
green lanes	18f	255	586.8 m

b

Figure 33.16 Data on the Decca chain 5B, located in south-east England

Question

6 Look at Figure 33.16. Explain why is it necessary for each transmitter to radiate waves at different frequencies.

THIN FILM INTERFERENCE

When sunlight reflects off a thin film of oil floating on a pool of water, a brightly coloured pattern may be seen. Such patterns are produced by superposition effects of light and not by pigments, dyes or any other colouring material in the oil itself. Similar effects can be seen in soap films and in the feathers of certain birds, notably the peacock. They occur because light falling on a thin transparent film partially reflects at the first surface and partly at the second one (see Figure 33.17). The eye receives the reflected beams, one of which has travelled further than the other. The thickness of the film at a particular point determines which of the incident frequencies will undergo constructive interference. Interference as a result of reflections of a single wavefront is an example of interference by division of amplitude. Experiments 33.4 and 33.5 illustrate other examples of superposition of electromagnetic radiation.

When white light falls on a vertically mounted soap film, a series of coloured fringes is visible. As the soap film drains downwards, the thickness of the top part of the film becomes much less than the wavelength of the light and there is a negligible path difference between the two reflected beams. However, instead of a bright region arising, darkness occurs. The light undergoes a phase change of π radians on reflection at the air (less dense) to soap (more dense) surface. However, no phase change takes place at the interface between the soap-film and the surface of the air. So when the film is very thin, the two beams are π radians out of phase and superpose destructively.

Figure 33.17 Thin film interference

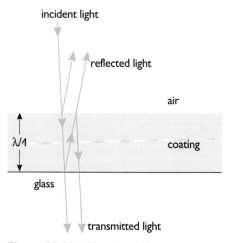

Figure 33.18 Blooming of lenses

Blooming of lenses

In camera lens systems, and other optical instruments, the reflection of light at the glass surface of lenses can reduce the amount of light transmitted by the system. To overcome the problem, the lens may be coated with a thin film of magnesium fluoride in a process described as **blooming**. The film is transparent and has a refractive index ($\rightarrow$ 34.2) between that of air and glass. So light striking either surface of the lens from the air undergoes a phase change of π radians (see Figure 33.18).

The thickness of the film is $\lambda/4$, giving a path difference of $\lambda/2$ between the two reflected rays. Since both rays have undergone a phase change of π on reflection, they superpose destructively, giving little reflected light. The transmitted rays also have a path difference of $\lambda/2$, but only one of them undergoes a phase change on reflection. So they superpose constructively.

The thickness can only be $\lambda/4$ for one particular wavelength of incident light. This is usually chosen to be in the green part of the visible spectrum. The red and blue extremes are still partially reflected giving the lens a purplish colour.

WAVES: TRANSFERRING ENERGY THROUGH SPACE

EXPERIMENT 33.4 Interference fringes in an air wedge

Using the arrangement shown in Figure 33.19 you can observe the superposition of monochromatic light reflected by an air wedge. By measuring the number of fringes in the wedge, and knowing the wavelength λ of the light, you can calculate:

a the angle θ of the wedge,
b the thickness of the paper, given that for n fringes the thickness of an air wedge is $n\lambda$.

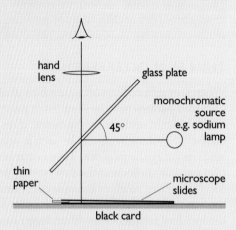

Figure 33.19 Arrangement for experiment 33.4

Question

7 This question refers to experiment 33.4.
  a Why are there interference fringes in the wedge?
  b Explain why $n\lambda$ = thickness of the wedge film for n fringes.
  c Suppose the glass is not perfectly 'flat' and varies (even microscopically) in its thickness. What effect will this have on the interference pattern?

EXPERIMENT 33.5 Superposition of vhf radio waves

After tuning a vhf/FM radio to a signal emitted by a fairly distant transmitter, place a large metal reflecting sheet in such a way that the radio is able to pick up both direct and reflected signals. If you move the radio slowly towards and then away from the sheet, as indicated in Figure 33.20, you should hear the output of the radio as a series of maxima and minima. By measuring the distance between maxima, you can calculate the wavelength and therefore the frequency of the received signal.

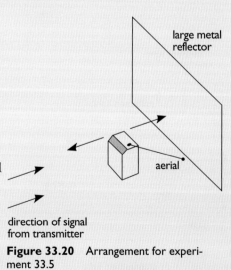

Figure 33.20 Arrangement for experiment 33.5

33 SUPERPOSITION OF ELECTROMAGNETIC WAVES

Question

8 This question refers to experiment 33.5.
a When repeating the experiment, the distance between two successive maxima is found to be 1.5 m. What is the wavelength, in m, of the received signal?
b The velocity of the transmitted radio waves is $3 \times 10^8 \, \text{m s}^{-1}$. Calculate the frequency of the wave. Is this consistent with the frequencies used in the vhf/FM band?
c Why is it necessary to tune into a weak signal rather than a strong one from a more local transmitter?

33.2 MULTIPLE-SLIT INTERFERENCE

Young's fringes are formed when light passes through two very narrow, closely separated slits. If there are three slits of the same width and the same distance apart, diffraction will take place as light passes through each of the slits. The three diffracted beams superpose and produce an interference pattern, which is very similar to the double slit pattern. However, the fringes are brighter, because more light reaches the screen, and the principal fringes are slightly sharper. When you add a fourth slit, the fringes increase in brightness further and become even sharper. With six slits, the effect becomes clearer still.

Figure 33.21a – d summarises the essential features of multiple-slit interference. Note that the fringe separation is the same in all cases, as the slit separation is constant. The double slit equation $\theta = \lambda / D$ ($\leftarrow$ 33.1) predicts the principal fringe separation in the multiple-slit pattern. Also, the fringes 'fit' within an outer diffraction envelope. The narrower each individual slit, the wider is the envelope. Although the fringes are dimmer, there are many more of them visible. Calculating overall amplitudes using phasors allows the pattern to be predicted for any number of slits.

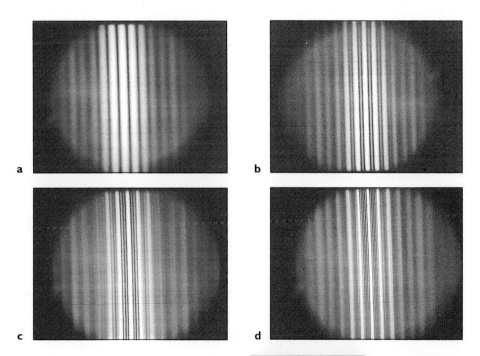

Figure 33.21 Light pattern produced by **a** two, **b** three, **c** four and **d** six slits

WAVES: TRANSFERRING ENERGY THROUGH SPACE

Predicting interference patterns using phasors

Light falling on a double slit forms an interference pattern on a distant screen (see Figure 33.22a). P is the central bright fringe, R the first minimum and T the first bright fringe. Q and S are points midway between the maxima and minima.

In Figure 33.22b, light is emerging from the double slit along an angle θ. The path difference between light from the two slits to a point on a distant screen is θd, since θ in radians is small enough to be considered equal to its sine or tangent ($\leftarrow$ 33.1). The fraction of a complete cycle that this represents is $\theta d / \lambda$. So the phase difference is given by:

phase difference = $(\theta d / \lambda) \cdot 2\pi$ radians.

Suppose that the amplitude of light from each of the two slits is A. Two phasors may be drawn representing this light with a phase angle between them of $(\theta d / \lambda) 2\pi$. In Figure 33.22c, the resultant amplitude is shown by the phasor drawn across them. The intensity of the light is proportional to the square of its amplitude.

Considering points $P - T$ on the screen:

- at P, $\theta = 0$ and two phasors are in phase. They add to produce a large resultant amplitude $2A$ (see Figure 33.23a).
- at Q, $\theta = \lambda / 4d$ and the resulting amplitude is $A\sqrt{2}$ (see Figure 33.23b). Squaring this gives an intensity half that at P
- at R, $\theta = \lambda / 2d$. Because the phase difference is now π, the two phasors add to give zero amplitude and intensity (see Figure 33.23c).
- at S, $\theta = 3\lambda / 4d$. The resulting amplitude on the screen along this direction is again $A\sqrt{2}$, giving an intensity half that at P (see Figure 33.23d)
- at T, $\theta = \lambda / d$ and the path difference between light from the two slits is one wavelength. The phase difference is 2π. The two phasors are now in step again, as in Figure 33.23 a, and add to give a resultant amplitude of $2A$.

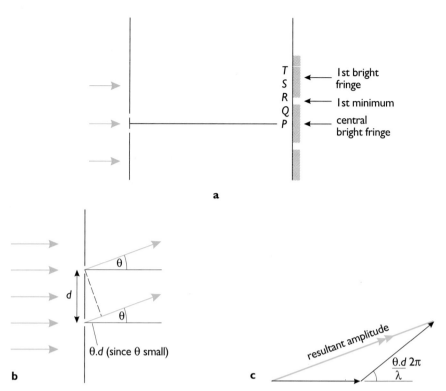

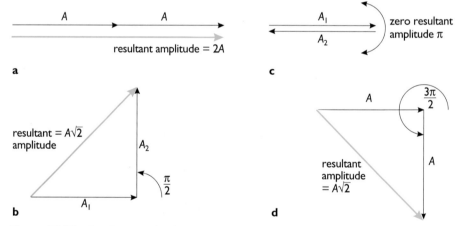

Figure 33.22 Double slit arrangement: **a** interference pattern on screen, **b** path difference of light, **c** resultant amplitude

Figure 33.23 Resultant amplitude at **a** P, **b** Q, **c** R, **d** S

Question

9a Draw phasor diagrams similar to those in Figure 33.23, but for four slits of the same constant separation.
b Ignoring the effect of diffraction at each slit, calculate the theoretical intensities at points P, Q, R, S and T.

STELLAR INTERFEROMETER

The ability of a radio telescope to resolve two sources is limited by its aperture ($\leftarrow$ 31.4). Even one with a dish 100 m across has much poorer resolution than the

33 SUPERPOSITION OF ELECTROMAGNETIC WAVES

The intensity pattern is shown in Fig. 33.24a.

This procedure predicts the correct angular separation of the fringes. However, the fringes decrease in brightness further from the central fringe because the diffraction envelope reduces the overall intensity as θ increases. The overall amplitude may be found by calculating the amplitude of the diffracted light in the direction θ before going through the above procedure. This is a very repetitive process, ideally suited to a microcomputer.

A two-slit intensity pattern predicted by the above technique is shown in Figure 33.24b. It is calculated for two slits whose separation is four times their width. The procedure may be extended to predict intensity patterns for any number of slits. For example, a six-slit pattern requires the addition of six phasors whose amplitudes in a given direction are determined by how much diffraction occurs at each individual slit.

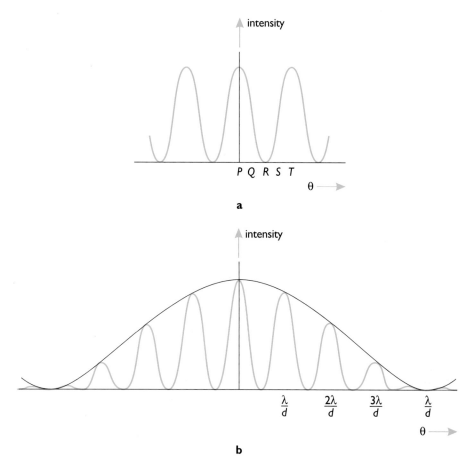

Figure 33.24a Intensity pattern using phasors **b** predicted intensity pattern using phasors

human eye. One way around this problem is to use two small telescopes placed a long way apart. Superimposing the two signals produces a superposition pattern with spacing finer than that in the diffraction pattern of one telescope on its own. Such a device, shown in Figure 33.25, is known as a **stellar interferometer**. The Mullard Radio Observatory in Cambridge contains one of these.

In Australia, a stellar interferometer has been constructed by placing an aerial array on top of a 100 m high cliff near the entrance to Sydney Harbour (see Figure 33.26). The system was designed for the study of 1.5 m wavelength radiation emitted by the Sun.

Figure 33.25 Stellar interferometer

WAVES: TRANSFERRING ENERGY THROUGH SPACE

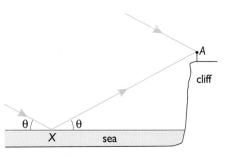

Figure 33.26

Question

10a Figure 33.26 shows the Sun's rays when the received intensity at A is a maximum. On a copy of the figure, mark a point Y on the direct ray from the Sun to A, so that $AX-AY$ is an odd number of half wavelengths. Note that the phase is reversed on reflection by the sea.

b As the Sun sets, the angle θ gradually changes.
 i Explain how and why the intensity at A varies as the Sun sets.
 ii Derive a formula for the path difference in terms of θ to predict the angles at which maxima occur.

33.3 DIFFRACTION GRATINGS

Figure 33.21a–d shows the intensity patterns for one to six parallel, closely separated slits. They all share the property that, due to diffraction at each slit, the intensity of the fringes decreases with distance from the central fringe. However, the sharpness of each fringe increases with the number of slits through which the light passes.

If each individual slit was made much narrower, the outer diffraction envelope would be much larger. There would be more fringes, with the same separation as before, within the central maximum (see Figure 33.27a). Suppose many more of these much narrower slits were used, keeping the same separation as before. The pattern would be similar but the fringes would be much sharper as in Figure 33.27b. If the separation of the slits was then decreased, sharp, widely separated fringes would be obtained, as in Figure 33.27c. Such an arrangement of slits is often described as a **diffraction grating**. It is possible to produce gratings with over 1 000 lines per mm.

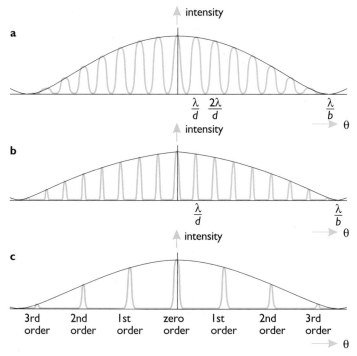

Figure 33.27 Intensity pattern of fringes when **a** slits are made much narrower, **b** many more narrow slits are used, **c** separation of slits is decreased

33 SUPERPOSITION OF ELECTROMAGNETIC WAVES

Using visible, monochromatic light, the outer diffraction envelope from a grating fills the whole field of view. The fringes, known as **orders**, appear as sharp lines in the grating pattern. The angle at which a given order appears depends upon the wavelength of light used. The diffraction grating disperses white light into its component colours, and is more useful than a prism in many applications.

EXPERIMENT 33.6 Using a diffraction grating

a With a laser light source. Placing the grating in the path of the laser beam 2–3 m from the screen, as in Figure 33.28a enables you to investigate the effect.

b Figure 33.28b shows the grating pattern obtained with a slit source illuminated by monochromatic light.

c With a white light source. Looking at a white light source, such as a torch filament, through the grating, you should be able to see first and second order spectra on either side of the undeviated, white zero order spectrum. Note that the short-wavelength violet light appears closer to the zero order than the longer-wavelength red light.

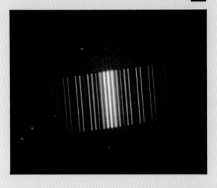

Figure 33.28 Diffraction grating used with monochromatic light: **a** from a laser, **b** from an illuminated slit

PREDICTING THE GRATING PATTERN

It is possible to predict the spacing of the grating pattern, given certain information about the grating and the wavelength of the incident light.

Figure 33.29a shows plane wavefronts of monochromatic light incident normally on a diffraction grating. The aperture spacing of the grating is d. Because the slits are very narrow, light waves emerging from each of the grating apertures diffract over a wide angle. In the forward direction, waves add constructively to form the zero order. Along an angle θ_1, as in Figure 33.29b, light

Figure 33.29 Forming **a** zero order, **b** first order, **c** second order, maxima in grating pattern

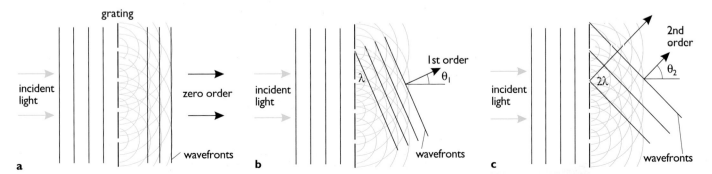

WAVES: TRANSFERRING ENERGY THROUGH SPACE

waves from each slit are again in phase and add constructively to produce another maximum, the first order line in the grating pattern. In this case, there is one wavelength path difference between light from adjacent slits. The second order line appears at an angle θ_2 to the forward direction, when the path difference between light from adjacent slits is two wavelengths, as in Figure 33.29c. There will be a maximum in the grating pattern whenever the path difference between adjacent slits is a whole number of wavelengths. The **nth order line** corresponds to a **path difference** between adjacent slits of $n\lambda$.

Figure 33.30 shows the situation at two slits, X and Y, a distance d apart. If, along a direction θ, light from Y travels n wavelengths further than light from X, there is an nth order line in this direction. The right-angled triangle XYZ gives the **grating equation**:

$$n\lambda = d \sin \theta.$$

The more slits the light passes through, the sharper are the maxima in the interference pattern. A grating may contain many thousands of slits and so the spectral lines in its pattern can be made to appear very sharp, with little energy transmitted at angles other than those for maxima.

Calculating the position of orders

Figure 33.31 illustrates the intensity distribution for light of wavelength 600 nm incident on a grating of 400 lines per mm. For such a grating, the slit separation d is 1/400 mm or 2.5×10^{-6} m. Each slit has a width b, which is typically around 10^{-6} m. The outer diffraction envelope arises from diffraction at individual slits and falls to minimum at an angle θ given by:

$$n\lambda = b \sin \theta. \quad (\leftarrow 31.2)$$

For the first minimum, $n = 1$ and:

$$\sin \theta = \frac{\lambda}{b}$$

$$= \frac{6 \times 10^{-7}}{10^{-6}}$$

$$= 0.6$$

giving $\theta = 37°$.

The value of $2\lambda/b$ is 1.2 which is greater than 1. A sine can never exceed unity, so a second minimum does not occur.

The angle of each order in the pattern is obtained from the equation $n\lambda = d \sin \theta$.

1st order: $1(6 \times 10^{-7}) = 2.5 \times 10^{-6} \sin \theta$
$\theta = 14°$.

2nd order: $2(6 \times 10^{-7}) = 2.5 \times 10^{-6} \sin \theta$
$\theta = 29°$.

3rd order: $3(6 \times 10^{-7}) = 2.5 \times 10^{-6} \sin \theta$
$\theta = 46°$.

4th order: $4(6 \times 10^{-7}) = 2.5 \times 10^{-6} \sin \theta$
$\theta = 74°$.

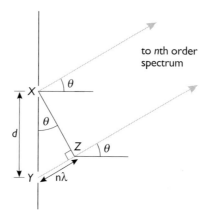

Figure 33.30 Deriving the grating equation

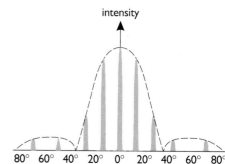

Figure 33.31 Intensity distribution of grating

Question

11 White light from a distant source ranges in wavelength from 400 nm in the violet to 700 nm at the red end of the spectrum. It falls on a grating of 300 lines per mm.
a Describe the appearance of the pattern of light seen through the grating.
b Calculate the angles for the first order **i** violet and **ii** red lines.
c What are the angles, in degrees, of the second order **i** violet and **ii** red lines?
d You should have calculated that the angle for the first order red line is less than that for the second order violet line. This means that the first and second order spectra do not overlap. **i** Is this true for the second and third order spectra? **ii** Is it also true for the third and fourth orders?

MEASURING THE WAVELENGTH OF LIGHT

A diffraction grating of known line spacing provides a very simple way of measuring the wavelength of light, as in experiment 33.7.

33 SUPERPOSITION OF ELECTROMAGNETIC WAVES

EXPERIMENT 33.7 Wavelength of light from a diffraction grating

The lamp in Figure 33.32a is placed 2–3 m from a grating of known line spacing, with the room darkened. If you look through the grating at the lamp filament, you should observe a pattern of spectra. With the help of a blue filter, you can mark the distances x of the blue light in the first order spectrum from the zero order and calculate the angle θ between the first and zero order. Similarly, you can measure θ for green and red light. Using the grating equation, you can then calculate the wavelengths of blue, green and red light. Some typical results are shown in Figure 33.32b.

Figure 33.32 Experiment 33.7: **a** arrangement, **b** typical results

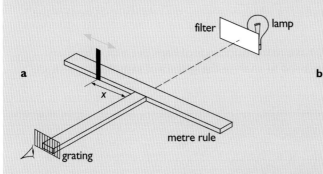

	red	green	blue
x/mm	199	164	130
$\tan \theta$	0.199	0.164	0.130
θ degrees	11.2	9.3	7.4
$\sin \theta$	0.195	0.162	0.129
wavelength/mm	$1/300 \times 0.195$ or 6.5×10^{-4}	$1/300 \times 0.162$ or 5.4×10^{-4}	$1/300 \times 0.129$ or 4.3×10^{-4}
wavelength/m	6.5×10^{-7}	5.4×10^{-7}	4.3×10^{-7}
wavelength/nm	650	540	430

Question

12 This question refers to experiment 33.7.
a What is the value of the uncertainty in the measurement of distance x? Do you think it is accurate to within 1 mm, 2 mm...?
b Bearing in mind your answer to **a**, do you think it would be reasonable, for the angles concerned, to make the approximation $\sin \theta = \tan \theta = \theta$ radians?
Would this simplify the calculations?
c Calculate the uncertainty in the final value for the wavelength. Assume that the grating has 300 lines per mm and that this value is accurate to within 1%.

SUMMARY

Superposition effects may be observed with light. The maxima and minima appear as light and dark bands, known as interference fringes. They can be produced by:

- division of wavefront, as in Young's double slits experiment and Lloyd's mirror, or
- division of amplitude, as in thin film interference.

Path differences required to produce superposition effects with light need to be very small because its wavelengths are very short. These superposition effects provide further evidence of the wave nature of light.

For a Young's slit arrangement, the separation of the fringes x is given by:

$$x = \frac{\text{wavelength of light} \times \text{slit to screen distance}}{\text{slit separation}}.$$

Other types of electromagnetic radiation also produce superposition effects.

Interference patterns produced by radio waves are used in navigation systems and telescopes, but can cause problems in the reception of some radio transmissions.

Light passing through a multiple-slit system, such as a diffraction grating, also exhibits superposition effects. The interference pattern produced consists of maxima that are much sharper and brighter than in a two-slit system.

The grating equation relates the wavelength λ, the line spacing d and the angle θ of a maximum in the pattern produced:

$n\lambda = d \sin \theta$

where n is the order (number of whole wavelengths relatively shifted) of the spectrum.

Diffraction gratings may have many hundreds of lines per mm. The sharp patterns they produce are used to analyse the spectra of samples in grating spectroscopes.

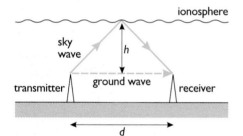

Figure 33.33 Transmitting in the mediumwave radio band

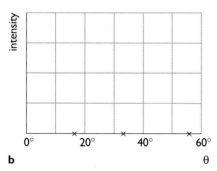

Figure 33.34

SUMMARY QUESTIONS

13 A beam of laser light, of wavelength 632 nm, is directed towards a double slit. A set of interference fringes is seen on a distant screen. The width of each slit is 0.04 mm and the separation of the slits is 0.20 mm. The distance between the slits and the screen is 2 m.
a Calculate the separation of the fringes seen on the screen.
b How many fringes will be observed within the central maximum of the diffraction envelope?
c The opaque material between the two strips is removed. What effect does this have on the pattern observed on the screen?

14 A student attempts to set up Young's experiment (← 33.1). He uses a lamp with a vertical filament and rules two slits with a pin on a glass microscope slide coated with aquadag, a soft, opaque, graphite-based substance. He places a translucent screen about 2 m from the lamp, and puts the double slit midway between the lamp and screen. Looking at the screen from behind, the student can see no fringes.
   Explain what the problem might be and suggest what he might do to obtain a good set of fringes.

15 In the mediumwave radio band, waves reach a receiver by two routes: a ground wave which travels direct, and a sky wave, which is reflected off the ionosphere (see Figure 33.33). Superposition occurs between them.
   At night, the sky wave is particularly strong, its amplitude being comparable with that of the ground wave. The receiver may get a strong signal, or almost none at all, depending on the effective height h of the ionosphere at that moment. As h varies over a few seconds, the signal rises and falls, a phenomenon called **fading**.
   Suppose a maximum signal is received when h is effectively 80 km, λ = 250 m and d = 120 km. Calculate the distance h would have to change in order for this signal to become a minimum.

16 A parallel beam of monochromatic light of wavelength 580 nm is incident normally on a diffraction grating having a large number of regular slits, each of width 0.70×10^{-6} m, as shown in Figure 33.34a.
   After passing through the grating, the light will have, as a result of

interference, intensity maxima in certain directions. Calculations predict that for the first, second and third order interference maxima, the values of the angle θ between the direction of the incident light and the directions of these maxima should be approximately 16°, 34° and 56° respectively.
a Calculate the value θ would need to have for the light diffracted by a **single** slit of width 0.7×10^{-6} m to have its first intensity minimum?
b Show, by drawing on a copy of Figure 33.34b, how the intensity of the light passing through the grating would vary over the range of θ from 0° to 60°. The angles 16°, 34° and 56° have been indicated by means of small crosses on the θ-axis.

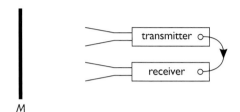

Figure 33.35

17 In a radar speed-measuring device, a transmitter emitting a radio wave of wavelength 30 mm is placed alongside a receiver, as shown in Figure 33.35. Some of the output of the transmitter is fed, by the wire shown, directly to the receiver and is then compared with the reflected signal.
a A sheet of metal, M, is held in front of the device as shown, so that the intensity detected by the receiver is a maximum. When the metal sheet is moved 7.5 mm towards the device, the intensity falls to a minimum. Explain, in words, why:
 i the receiver signal decreases,
 ii the minimum is not necessarily zero.
b The radiation is now directed at an approaching motor car which is moving directly towards the device at a steady speed, and the receiver detects a signal fluctuating in strength with a frequency of 2.0 kHz. Calculate the speed of the car.
c When the car, of mass 1 200 kg, has travelled a distance of 100 m in a straight line towards the device, the observed frequency is found to be 1.0 kHz. Calculate the average braking force applied, stating any assumptions made.

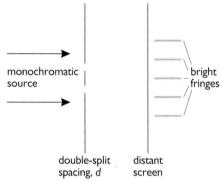

Figure 33.36

18 You may decide to carry out an experiment using monochromatic light passing through two narrow slits, as shown in Figure 33.36, which is **not** drawn to scale. The lines of light and dark fringes seen on a distant screen are shown at the right-hand end of the diagram.
a Assuming that the slit width and spacing are both known, state what you would need to measure in order to estimate the wavelength of the light.
b You replace the two-slit barrier by one with 100 slits, all with the same width and spacing as the two slits previously used. Nothing else is altered. With the aid of a labelled diagram, explain why the fringes on the screen would have the same separation for the 100 slits as for the two slits.
c State **two** ways in which the fringes formed by the 100 slit arrangement differ from those formed by the two slits.

19 A parallel beam of light of wavelength 7.0×10^{-7} m falls normally on to an opaque screen in which there are two parallel and identical slits S_1 and S_2, 3.5×10^{-4} m apart, as shown in Figure 33.37a.
 Imagine that slit S_2 is covered up. The light transmitted by S_1 varies in intensity with angle θ to the original direction producing a pattern on a screen several metres away. This pattern is represented by a graph of intensity against sin θ as shown in Figure 33.37b.

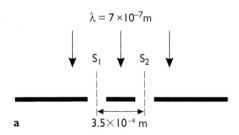

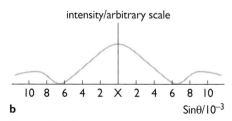

Figure 33.37

a Calculate the slit width, showing your working.
b i Imagine slit S_2 is uncovered. There is still a maximum at X, where $\sin \theta = 0$. Copy Figure 33.37b and mark with crosses other places where there are now maxima. Show how you calculate these positions.
ii Now show how the intensity of the light transmitted by both slits varies with $\sin \theta$, indicating also how the intensity scale may need to be changed.
c In what ways will the graph you have drawn in **b ii** change when the width of each slit is doubled? Assume that their distance apart remains unchanged. (From Physics (Nuffield) Short Answer Paper, 1978).

34 OPTICAL SYSTEMS

Visible light is a central part of the electromagnetic spectrum. As such, it exhibits all of the properties of waves and these are utilised in optical systems, such as cameras and telescopes. However, these very wave properties produce limitations in the systems. Optical microscopes, for example, cannot magnify indefinitely because they cannot resolve the detail of an object as small as the wavelength of the light being used. In this chapter, we look at some optical systems and relate their performance to the wave properties of light.

34.1 WAVEFRONTS AND RAYS

In Chapter 30, we looked at the behaviour of wavefronts, for example, as they met a reflective surface or changed speed during refraction. However, it is often more convenient to consider the direction of travel of wavefronts rather than the wavefronts themselves. Figure 34.1a shows wavefronts radiating from a source, carrying energy in all directions. They can be represented by a series of **rays** emerging radially from the source. Figure 34.1b shows how rays illustrate the direction of travel of a series of parallel wavefronts.

Figure 34.2 shows rays of light from a lamp striking a mirror.

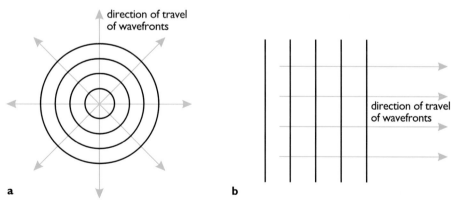

Figure 34.1 Rays and wavefronts **a** radiating in all directions, **b** parallel

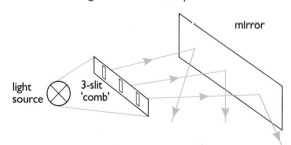

Figure 34.2 Ray-streaks reflecting off a mirror

WAVES: TRANSFERRING ENERGY THROUGH SPACE

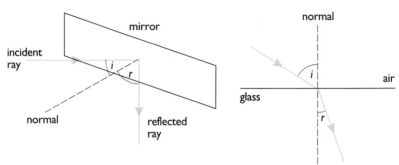

Figure 34.3 Reflection of a ray of light

Figure 34.4 Refraction of a ray of light

Figure 34.3 shows the **reflection** of light using the ray model. The angle of incidence, i is equal to the angle of reflection, r. It is usual to measure the angles of incidence and reflection with respect to the **normal** – an imaginary line drawn at right angles to the surface at the point of incidence.

Figure 34.4 shows a ray of light being **refracted**. The angles of incidence and refraction are those that the incident and refracted rays make with the normal. When light passes from air into glass, the ray is bent towards the normal. Light travelling in the reverse direction follows the same path and bends away from the normal.

Lenses and mirrors – the fundamental building blocks of optical systems – make use of the refractive and reflective properties of light.

34.2 REFRACTION

Light passing from one transparent medium into another undergoes a change in velocity, in much the same way as a water wave crossing a boundary between deep and shallow water (← 30.3). When the light ray meets the boundary at an angle other than 90°, the ray changes direction as it crosses the boundary (see Figure 34.4). In 1626, the Dutch mathematician, Snell, observed that, for a given pair of media, the ratio of the sine of the angle of incidence to the sine of the angle of refraction was constant. Over 50 years later, Huygens explained refraction by assuming that light changed its speed as it travelled from one medium to another. He concluded that:

$$\frac{\text{speed of light in medium}_1}{\text{speed of light in medium}_2} = \frac{\sin i}{\sin r}$$

$$= \text{a constant, } n,$$

which is known as **Snell's law** (see Figure 34.5). The constant n is the **refractive index** of medium$_2$ with respect to medium$_1$, or $_1n_2$.

When light travels from a vacuum into a given medium the constant n is called the **absolute refractive index** of the medium and:

$$n = \frac{\text{speed of light in vacuum}}{\text{speed of light in medium}}.$$

For most media transparent to visible light, the value of n increases with decreasing wavelength. In fact n varies considerably with wavelength outside visible light frequencies. This is known as **dispersion** (→ Question 3). The constant n is also a very non-linear function with discontinuities where the material ceases to be transparent. For this reason, absolute refractive indices are quoted for a particular wavelength – 589.0 nm, one of the yellow spectral lines of sodium.

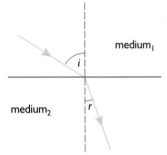

Figure 34.5 Snell's law

34 OPTICAL SYSTEMS

Velocity of light and refractive index

Imagine a wavefront AB moving at a speed v_1 (see Figure 34.7). It meets a boundary between two different media, 1 and 2, at an angle of incidence i. Huygens' construction (← 31.2) shows that, time t later, the wave, will be refracted, and will be moving at speed v_2 at an angle r to the boundary. From the diagram:

$$AD = \frac{v_1 t}{\sin i}$$

$$= \frac{v_2 t}{\sin r}.$$

So $$n = \frac{v_1 t}{v_2 t}$$

which means that $$n = \frac{\sin i}{\sin r}.$$

When medium$_1$ is a vacuum, this ratio is the absolute refractive index n of medium$_2$.

Question

1 Figure 34.6 gives the absolute refractive index of various media.
a Assuming that the speed of light in a vacuum is $3 \times 10^8 \, \text{m s}^{-1}$, what is the value of the speed of light in **i** air, **ii** crown glass **iii** water?
b A ray of light in a vacuum meets a crown glass surface at an angle of 30° to the normal. What angle, in degrees, does the refracted ray make with the normal?
c A ray travels from crown glass into a vacuum at an angle of incidence of 30°. Calculate the angle of refraction.
d Suppose, in **b**, the ray was travelling from air into crown glass. Would that make a significant difference to your answer?
e A ray travels from water into crown glass at an angle of incidence of 30°. What is the angle of refraction in degrees?

medium	vacuum	air	water	crown glass	flint glass
absolute refractive index (at 589 nm)	1	1.0003	1.33	1.517	1.650

Figure 34.6 Refractive index of some materials

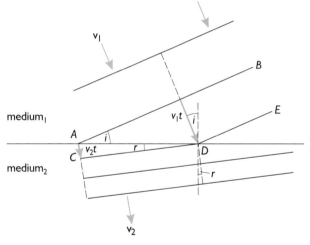

Figure 34.7 Huygens' and refraction

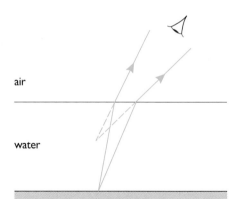

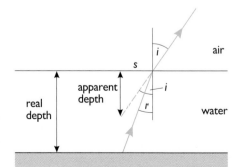

Figure 34.8 Real and apparent depth in a swimming pool

Questions

2 When looked at from above, a swimming pool appears less deep than it actually is.
a Use Figure 34.8a to show that this effect is caused by refraction.
b Explain why distance s in Figure 34.8b is equal to
i the apparent depth $\times \tan i$ and also **ii** the real depth $\times \tan r$.
c Since it is being looked at almost directly from above, the distance s is small and the tangents of the incident and refracted angles are not significantly different from their sines. Use this fact to show that:

$$n = \frac{\text{real depth}}{\text{apparent depth}}.$$

d The refractive index of water is 4/3. Calculate the apparent depth, when viewed from above, of a swimming pool that is actually 2 m deep.

WAVES: TRANSFERRING ENERGY THROUGH SPACE

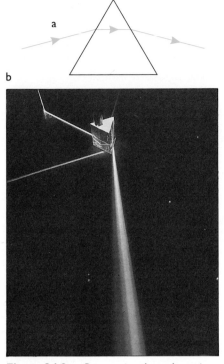

Figure 34.9a Ray passing through a prism
b a prism producing a spectrum from white light

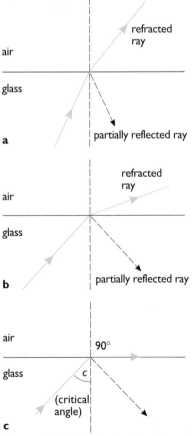

Figure 34.10 Critical angle and total internal reflection

3 Figure 34.9 shows the path of a monochromatic ray of light as it travels through a triangular glass prism. Smoke or dust particles in the air around the prism make the incident and emergent beams visible.
a Explain why the ray takes the path shown.
b When a white beam of light is used, the result seen in Figure 34.9b is obtained; white light is split into the colours of the spectrum. Explain this in terms of the way the velocity of light in a medium is dependent upon the frequency of the light.
c Some students need to increase the angle of dispersion, or spread, of the colours. One suggests using a glass with a higher refractive index. Another suggests choosing a material that has a large rate of change of refractive index with wavelength. Which suggestion do you think is correct and why?

TOTAL INTERNAL REFLECTION

Suppose a ray of light travelling through glass meets a boundary with the air. The refracted ray emerges as shown in Figure 34.10a, but some **partial reflection** occurs. When the angle is increased slightly, the angle at which the ray leaves the surface also increases as shown in Figure 34.10b. Increasing the angle still further, a point is reached when the refracted ray emerges at 90° to the normal, as in Figure 34.10c. Any further increase in the angle of incidence causes the ray to be **totally internally reflected** from the glass-air surface. The angle beyond which refraction ceases and the ray is totally reflected is known as the **critical angle** c. Total internal reflection can only take place when light enters a medium of lower refractive index – an optically less-dense medium.

Applying Snell's law to the ray of light in Figure 34.10c as it travels from **glass into air**:

$$\frac{\sin c}{\sin 90°} = \frac{1}{n}.$$

Since $\sin 90° = 1$, we can write:

$$c = \sin^{-1}(1/n).$$

In Figure 34.11, laser light is directed towards the side of a tank of water, with a tiny drop of milk added to scatter the light and to make the beam visible. As the angle i is increased beyond the critical angle for water, the light is totally internally reflected from the top surface of the water.

Question

4 The refractive index for glass is 1.5 and that for water is 1.33.
a Show that the critical angles for glass and water are about 42° and 49° respectively.
b Which medium must light be travelling through to be totally internally reflected at a glass/water boundary? Calculate the critical angle.
c Figure 34.12 shows a ray of light incident normally on one surface of a 90°, 45°, 45° prism. Draw a diagram of the prism and trace the path of the ray through it. Measure the angle through which the ray is turned.

Total internal reflection is made use of in a variety of instruments, such as 35 mm

34 OPTICAL SYSTEMS

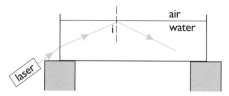

Figure 34.11 Observing total internal reflection

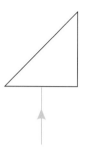

Figure 34.12

single lens reflex cameras. These have a pentaprism to direct light from the object to the eyepiece without reversing the image. The pentaprism also provides a long optical path between the focusing screen and the eyepiece, reducing eyestrain and the need for complicated wide-angle eyepiece optics. Prism binoculars use prisms for similar reasons.

Question

5 This question asks you to consider some familiar effects of reflection and refraction of light. Find out about the processes that take place in each of the following, and write a few sentences on each.
a A light shone towards a bicycle or car reflector from almost any angle is reflected back towards the source.
b The eyes of a cat seem to shine very brightly when illuminated by a source. A similar effect occurs with the **cats' eyes** used as lane markers on the road.
c When swimming underwater, we can see much better wearing goggles.
d Several 'ghost images' are seen when viewing the Moon through a double-glazed window.
e The rainbow consists of the colours of white light. With each primary rainbow, a secondary rainbow may be observed at a higher angle. The colour sequence in the secondary rainbow is opposite to that in the primary rainbow.
f Driving along a road on a hot day, a 'water mirage' can be seen. The surface of the road in the distance appears to be covered in water.

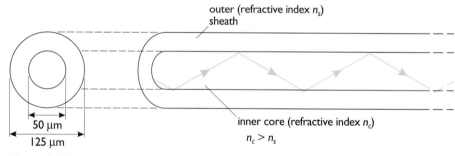

Figure 34.13 Optical fibre

Optical fibres

These are playing an increasingly important part in telecommunications and medicine. An **optical fibre** consists of a central glass core, clad in a sheath of glass of comparatively lower refractive index. Light is fed into the core at such an angle that much of the light travelling along the core undergoes total internal reflection as it meets the outer glass sheath (see Figure 34.13). As a result, the passage of the light is unaffected by any bends or kinks in the fibre.

In medicine, short optical fibres are used for internal medical investigations of the digestive tract and other parts of the body. They provide more information for diagnosis with less discomfort to the patient than traditional methods. The **fibre-optic endoscope** consists of a sealed shaft of about 10 mm in diameter and about 1.5–2.0 m in length. Contained within the shaft are a bundle of fibres that carry light from the viewing end to the tip of the endoscope. Another set of fibres carry the image from the lens at the tip back to the viewing end. All the fibres are arranged in an ordered fashion, forming a coherent bundle. Also contained within the endoscope are an irrigation channel through which water

WAVES: TRANSFERRING ENERGY THROUGH SPACE

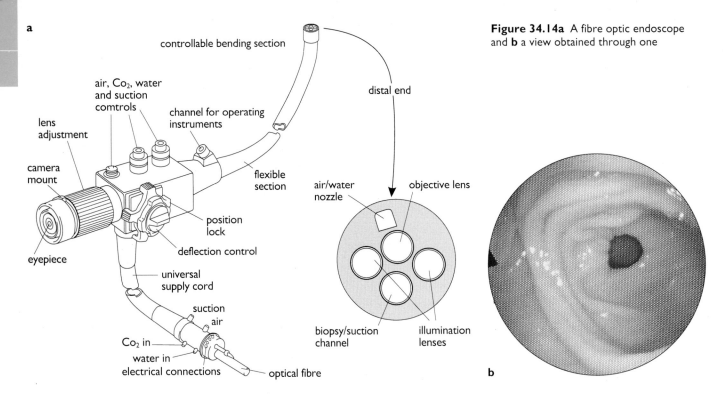

Figure 34.14a A fibre optic endoscope and b a view obtained through one

Figure 34.15 Optical cable and its electrical equivalent

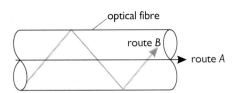

Figure 34.16 Light travelling by different routes in an optical cable

can be pumped; control cables; and a channel through which a tool may be inserted, allowing a variety of tasks to be performed (see Figure 34.14a).

Attenuation may occur in an optical cable due to the scattering and absorption of light by impurities in the fibre. Over short distances, this is not a problem because powerful light sources can be used.

Optical fibres are also employed in long-distance communication. The light used as the carrier signal has a very high frequency of about 10^{14} Hz. This allows very large numbers of channels of communication, such as telephone conversations or TV signals, to be carried by a single optical link. This fact, coupled with its lack of vulnerability to electrical interference, makes optical communications systems increasingly attractive (see Figure 34.15). Electrical cables typically have a 50 % power loss over about 3 km. Optical fibres have been developed that halve these losses.

Over very long distances, optical cable may need to be joined, or other equipment may need to be connected into the system. Such joints have to be carefully designed so that minimal amount of light is lost and the surfaces are kept free of contamination. A further problem is that the transmitted light pulses get 'smeared out' because light travelling along different paths arrives at different times. For example, in Figure 34.16, a pulse by route A arrives before a pulse by route B. To make the time by both routes the same, the pulse down the centre (A) must be slowed down. Therefore, the refractive index of the centre of the fibre must be greater than that at the edge. Such **graded index fibres** are essential for all long-distance applications.

34.3 SIMPLE OPTICAL SYSTEMS

Light-imaging systems are in everyday use. Waves play a fundamental part in any imaging system. The air-traffic controller uses short-wavelength radio waves, or radar, to create images of the air space (see Figure 34.17). Similarly, a doctor uses ultrasonic waves to observe the progress of

34 OPTICAL SYSTEMS

an unborn baby (← Figure 30.1). The materials scientist uses the wave properties of the electron to investigate the structure of new metallic alloys.

While all these imaging processes are very different, they do share similarities. In all of the examples, there are essentially two stages in the process. Firstly, there is an interaction between waves and an object: the object scatters the incident waves or radiates waves itself. Either way, waves are emitted that convey information about the object. Secondly, the imaging system gathers the emitted waves and recombines them to form a pattern, or **image** of the original object. All the processes also share a limitation: detail on the original object less than a wavelength across is not conveyed to the image. No matter how carefully an optical microscope is made, it cannot resolve detail smaller than the wavelength of light. Likewise, radar cannot detect an object smaller than the wavelength of the incident radio waves. So, the wavelength of radiation used to illuminate an object has to be matched to the detail required in the final image. Figure 34.18 shows the relationship between the size of objects that might be imaged and the wavelengths and frequencies of sound and electromagnetic radiation.

typical objects in size range	order of magnitude /m	frequency of sound waves with this wavelength/Hz	frequency of e-m waves with this wavelength/Hz
city	10^4		3×10^4
airfield	10^3		3×10^5
jumbo jet	10^2	3.3	3×10^6
house	10	3.3×10	3×10^7
child	1	3.3×10^2	3×10^8
sparrow	10^{-1}	3.3×10^3	3×10^9
bee	10^{-2}	3.3×10^4	3×10^{10}
daphnia	10^{-3}	3.3×10^5	3×10^{11}
amoeba	10^{-4}	3.3×10^6	3×10^{12}
chlorella	10^{-5}	3.3×10^7	3×10^{13}
typhoid bacillus	10^{-6}	3.3×10^8	3×10^{14}
smallpox virus	10^{-7}	3.3×10^9	3×10^{15}
turnip yellow virus	10^{-8}	3.3×10^{10}	3×10^{16}
vitamin A molecule	10^{-9}		3×10^{17}
carbon atom	10^{-10}		3×10^{18}
	10^{-11}		3×10^{19}
	10^{-12}		3×10^{20}
	10^{-13}		3×10^{21}
uranium nucleus	10^{-14}		3×10^{22}
hydrogen nucleus	10^{-15}		3×10^{23}

Figure 34.18 Sizes of objects and frequencies of waves with corresponding wavelengths

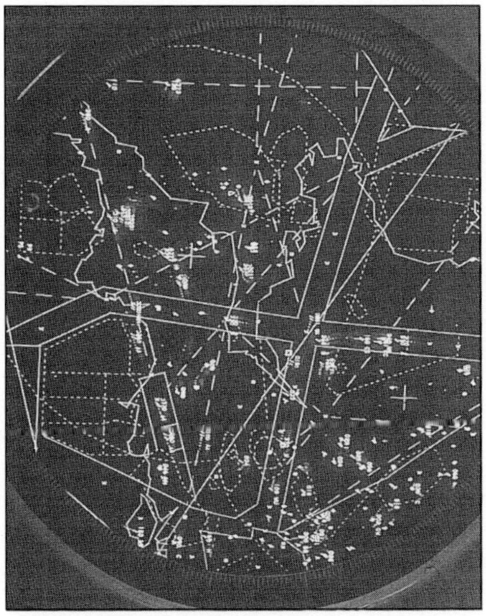

Figure 34.17 Radar display of an air traffic controller's air-space

Question

6a Why is it not possible to see individual atoms using an optical microscope?
b Air-traffic controllers are heavily dependent on radar when monitoring air space. What sort of wavelengths would be suitable? Why is visible light unsuitable?

PLANE MIRRORS

One of the most familiar, and probably the simplest, of imaging systems is the plane mirror. When you stand in front of a mirror, some of the light scattered by yourself and objects around you falls on the mirror surface and is reflected

WAVES: TRANSFERRING ENERGY THROUGH SPACE

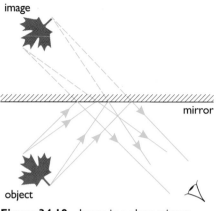

Figure 34.19 Image in a plane mirror

according to the laws of reflection (← 34.1). The eye intercepts a small amount of this reflected light, which appears to come from behind the mirror (see Figure 34.19). The image produced is called a **virtual** image, since no light actually passes through the image. Simple geometry shows that the image is the same size as the original object and as far behind the mirror as the object is in front.

CURVED MIRRORS

A slightly less familiar system is the curved mirror. Light falling on a **concave** mirror surface reflects according to the laws of reflection. As long as the beam is not too wide, parallel light arriving at a **spherical** concave mirror is reflected so that it passes through a point known as the **principal focus** F of the mirror. In practice, parallel rays are reflected through different points on the axis of the mirror, depending on how far from the axis they strike the mirror. This effect is known as **spherical aberration**. The problem can be overcome by using a mirror with a **parabolic** rather than a circular section. With a circular mirror, the problem can be reduced by using only the part of a mirror near to the axis.

Rays of light passing through F and on to the mirror will emerge as parallel rays. The distance between this point and the mirror is called the **focal length** f of the mirror (see Figure 34.20a). Figure 34.20b shows how a concave mirror can create an image of a distant object. Light scattered from any given point on the object falls on the mirror. With the object in this position, the reflected light recombines to form a **real**, inverted and diminished image.

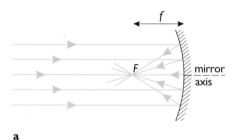

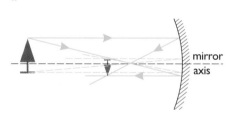

Figure 34.20a Reflection by a concave mirror, b a real image produced by a concave mirror

Questions

7a Why is the image in Figure 34.20b called a 'real, inverted and diminished image'?
b Copy Figure 34.21a and find, by completing the drawing, the type and nature of the image produced. Use the fact that rays travelling parallel to the axis of the mirror reflect through the principal focus F, and rays passing through F emerge parallel. Note that a ray travelling along a radius of the mirror will reflect along its own path because it meets the mirror surface along a normal. Note also that the mirror surface is drawn as a straight line to indicate that only a tiny part of the mirror is being used. This overcomes the problem of spherical aberration.
c In Figure 34.21b, the object is placed between the principal focus and the mirror. **i** Complete the sketch and explain why the image produced is a virtual one. **ii** Is the image larger or smaller than the object? Is it inverted or upright? **iii** Where might such a mirror be useful?

8 Convex mirrors are also able to produce images, as in Figure 34.22.
a Why will these images always be virtual, upright and smaller than the original object?
b Give two examples of where convex mirrors might be put to good use.

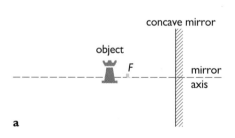

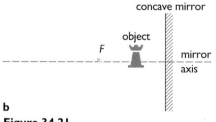

Figure 34.21

LENSES

The **pin-hole camera** is an imaging device, although not a very efficient one. In Figure 34.23a, some of the light scattered by a point on an object O_1 will pass through the pin-hole and form an image point I_1. The rest of the light scattered from that point of the object does not contribute to the final image. Light from many such points, $O_{2,3...}$, on the object form similar image points on the screen of the camera, so an inverted image is seen.

34 OPTICAL SYSTEMS

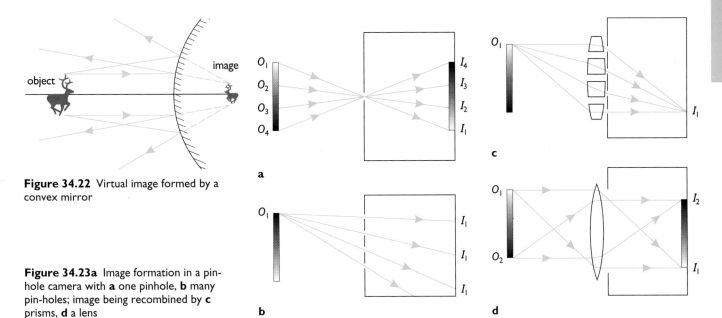

Figure 34.22 Virtual image formed by a convex mirror

Figure 34.23a Image formation in a pin-hole camera with **a** one pinhole, **b** many pin-holes; image being recombined by **c** prisms, **d** a lens

Figure 34.24 Focal length of a convex lens

If there are many pin-holes as in Figure 34.23b, there will be many images of point O_1 on the screen. These images could, in principle, be brought to a single point by a system of prisms of differing angles placed in front of the pin-holes, as in Figure 34.23c. Taking this idea further, the pin-holes could be replaced by one large hole, and the prisms by a suitable **convex lens**, which is effectively a collection of variable angled prisms. All scattered rays of light from O_1 reaching the lens will then be recombined to form a bright clear image I_1 on the screen. Other points on the object will be similarly recombined at different parts of the screen, as in Figure 34.23d.

As with curved mirrors, it is useful to describe a lens by its focal length. Parallel light falling on a convex or **converging** lens passes through the principal focus. The distance between the centre of the lens and the principal focus is its focal length (see Figure 34.24). There is an inverse relationship between **power** and focal length: the stronger the lens, the shorter its focal length. When the focal length of a lens is expressed in metres, the power in **dioptres D** is given by:

power = 1 / focal length.

Figure 34.25a–d shows the real images created by a thin lens when an object is

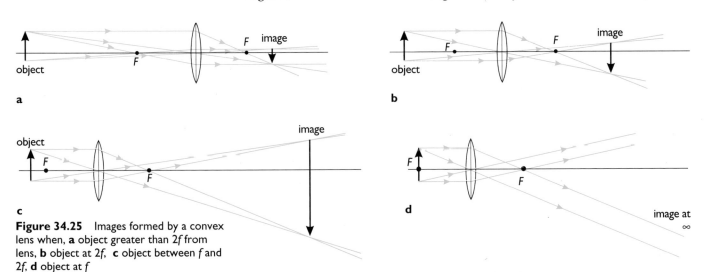

Figure 34.25 Images formed by a convex lens when, **a** object greater than 2f from lens, **b** object at 2f, **c** object between f and 2f, **d** object at f

placed at different distances from it. You can determine the nature and position of these images by following three simple rules:

i rays of light travelling parallel to the axis of the lens are refracted through the principal focus,
ii rays passing through the principal focus are refracted parallel to the axis of the lens, and
iii rays passing through the optical centre of the lens are not deviated.

Rule **iii** leads to the conclusion that the **linear magnification** produced by the lens is the ratio of the distance of the lens from the image to its distance from the object. When the object is closer to the lens than the focal length, a virtual image is formed (see Figure 34.26). The eye placed in the position shown will see the rays of light as though they came from *I* rather than from *O*. The image is upright and enlarged, and the lens is functioning as a **magnifying glass**.

Concave lenses are similar to convex mirrors in that they cannot on their own recombine light scattered by an object to form a real image. Thinner in the middle than at the edges, they diverge light and so are often known as **diverging** lenses. Parallel light falling on such a lens diverges, appearing to come from a point behind the lens a distance equal to its focal length, as shown in Figure 34.27.

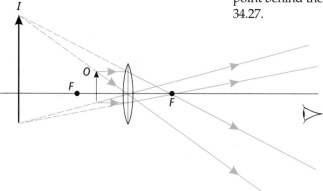

Figure 34.26 Convex lens as a magnifier

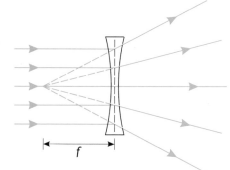

Figure 34.27 Focal length of a concave lens

Formula for a thin lens

Finding the image position for a given lens and object distance using a scale drawing every time is a little tedious. Instead, the position may be calculated from a simple formula.

In Figure 34.28, when the focal length of the lens is *f*, the distance between the object and the lens is *u*, and the image distance is *v*, the equation relating the three quantities is:

$$\frac{1}{u} + \frac{1}{v} = \frac{1}{f}.$$

The formula applies equally to converging or diverging lenses. When working with the formula, you need to follow a strict **sign convention**. There are two conventions in general use:

- the **new cartesian** convention, in which co-ordinates are taken from axes centred on the optical system (lens or mirror)
- the **real-is-positive** convention, in which distances to **real** objects and images are considered **positive** and distances to their **virtual** counterparts are considered **negative**. The focal length of a converging lens is taken to be positive whilst that of a diverging lens is negative.

In either case, object and image distances are measured from the optical centre of the lens to the object or image position.

34 OPTICAL SYSTEMS

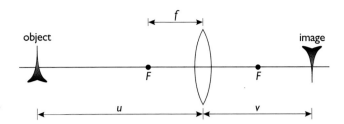

Figure 34.28 Relationship between object distance, image distance and focal length

Question

9a In Figure 34.28, the lens has a focal length of 20 cm. How far from the lens is the image when the object distance is **i** infinite? **ii** 100 cm? **iii** 40 cm? **iv** 20 cm?
b Calculate the linear magnification in **a ii** and **iii**.
c Plot a graph of the image distance against the object distance for a lens of power +10D for object distances ranging from 1 m down to 10 cm.
d Suppose the lens in Figure 34.27 has a focal length of 5 cm. Where will the image be when the object is placed 2 cm from the lens?

DEFECTS OF LENSES

The amount of refraction occurring when a ray of light passes from one medium into another is dependent not only on the media involved but also on the wavelength of the light (← 34.2). The focal length of a simple lens is also dependent on the wavelength of light: the smaller the wavelength, the greater the refraction and the shorter the focal length. The image of a white light source, such as a lamp

Figure 34.29a Chromatic aberration, **b** an achromatic lens

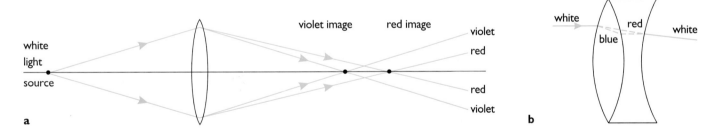

filament, produced by a convex lens will have coloured edges (see Figure 34.29a). The effect is known as **chromatic aberration**. The problem is overcome by using multiple element lenses, as in more expensive cameras. An **achromatic** lens system consists of two or more lenses made from different optical glasses, as shown in Figure 34.29b. The spacing and focal lengths of each are chosen so that their combined power for red light is the same as that for violet light.

Like curved mirrors, lenses display **spherical aberration**. A wide beam of light falling on a spherical lens is not brought to a clear focus. Parallel rays falling near the edge of a lens converge over a shorter distance than those near the centre of the lens (see Figure 34.30).

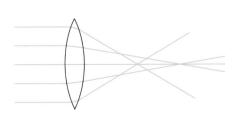

Figure 34.30 Spherical aberration

34.4 MORE COMPLEX OPTICAL SYSTEMS

HUMAN EYE

The human eye (see Figure 34.31) is sensitive to electromagnetic radiation ranging from about 400 nm to 700 nm. Light from a distant object entering the

WAVES: TRANSFERRING ENERGY THROUGH SPACE

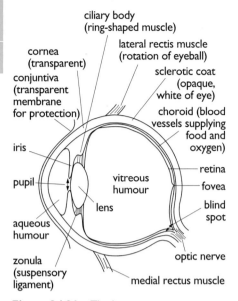

Figure 34.31 The human eye

eye is focused by the optical system of the eye to produce a real inverted image on the retina, the light-sensitive surface at the back of the eye. The amount of light falling on the retina is controlled by the iris. Bright light conditions cause the iris to constrict the pupil while low light levels cause it to dilate, or enlarge the pupil. Focusing is adjusted by the lens. Light is refracted as it enters the cornea, again as it passes into the aqueous humour, then into the lens and finally as it passes into the vitreous humour on its way to the retina. The aqueous and vitreous humours are transparent fluids that maintain the even shape of the eyeball.

Figure 34.32 illustrates possible paths of rays passing into an eye focused on a very distant object. The refractive index of each of the media is also shown. Most of the refraction takes place at the air-cornea boundary, which contributes a constant 45 dioptres to the refractive power of the optical system. Refraction at successive interfaces is not so marked due to the smaller changes in refractive index. The eye lens provides about 18 dioptres in its relaxed, or **unaccommodated** state. The total power of the unaccommodated eye is typically around 63 dipotres.

The ciliary muscles are able to change the curvature of the eye lens. In young people, this may increase the power of the lens to around 30 dioptres, enabling the eye to **accommodate** to objects as close as 250 mm. Elderly people are generally not so fortunate and have a much smaller range of accommodation.

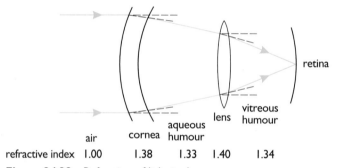

Figure 34.32 Refraction of light in the eye

Question

10 The eye in its relaxed state can be thought of as having a single lens of power 58 dioptres.
a Calculate the focal length of this lens.
b Is this consistent with the dimensions of the eye itself?
c The pupil of the eye is enlarged at low light levels. Will this cause blurring or sharpening of the image due to spherical aberration?

DEFECTS IN THE OPTICAL SYSTEM OF THE EYE

The normal eye is able to accommodate to objects as far away as infinity, the **far point**, and as close as 250 mm, the **near point**. Objects closer than this will not be clearly in focus, giving a blurred retinal image.

A person suffering from **long sight**, or hypermetropia, has an insufficiently powerful optical system to focus on near objects (see Figure 34.33a). The far point is still at infinity but the near point is further than the 250 mm for the normal eye. To correct this defect, the power of the optical system needs to be artificially increased. This can be done by placing a convex lens just in front of the eye

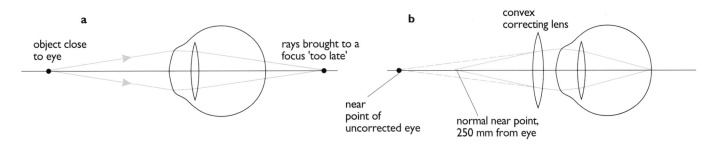

Figure 34.33a Long-sight and **b** its correction

which has the effect of moving the near point back to the normal 250 mm from the eye. The eye then has a full range of accommodation (see Figure 34.33b).

Question

11 One person with long sight has a near point 750 mm from the eye.
a Calculate the power of lens required to cause an object placed 250 mm from the eye to appear 750 mm away, where the person can see it clearly.
b A long-sighted person, without the aid of spectacles, may be able to read a newspaper when it is held at arms' length. Why is this?
c A long-sighted person without spectacles may be able to read the tiny print in a telephone directory when it is viewed through a pin-hole. Why?
d In young children, the eye lens can reach adult size before the eyeball. Why might this be a cause of hypermetropia?

Short sight, or myopia, is a condition in which the power of the optical system is too large. Distant objects are brought to a focus in front of the retina and appear blurred (see Figure 34.34a). So, the far point is closer than infinity and the near point is closer than 250 mm. To correct this condition, a diverging lens is placed in front of the eye. Rays emerging from an object placed at the far point of the eye then appear to come from infinity (see Figure 34.34b). As a result of the correction, the near point appears further from the eye.

Question

12a One person with short sight has a far point 1 m from the eye and a near point 150 mm away. Using Figure 34.34b, explain why it is necessary to use a correcting lens with a power of −1 dioptre.
b An object placed 150 mm in front of this person's uncorrected eye is seen to be in focus. A pair of spectacles containing a correcting lens moves the near point further from the eye. **i** Draw a diagram to show why this happens. **ii** Work out how far the near point is now from the eye.
c Why do short-sighted people often remove their spectacles for close work?

Figure 34.34a Short sight and **b** its correction

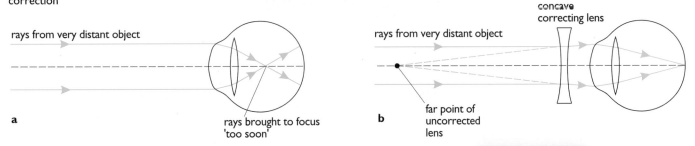

WAVES: TRANSFERRING ENERGY THROUGH SPACE

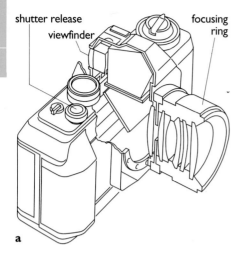

The ability of the eye to adjust itself to focus on objects at different distances reduces dramatically in middle age. Above around 50 years of age, the lens is unable to significantly change its power leaving the eye long-sighted. This condition is known as **presbyopia**. Some people with the condition use two pairs of spectacles for close and middle distance vision while others use **bifocal** lenses.

Question

13 Draw ray diagrams to show how bifocal lenses help people with presbyopia.

CAMERAS

The optical system of the camera shares some similarities with that of the eye, but also differs in important ways. A standard 35 mm camera consists of a multiple-element converging lens of focal length around 50 mm (see Figure 34.35a). Pressing the shutter button opens the shutter for a fraction of a second. During this brief time, the lens produces an image of a scene on film at the back surface of the camera. Before operating the shutter, focusing is carried out by rotating the lens on a threaded mount, which allows it to move towards or away from the film surface. In many cameras focusing is performed automatically by detection of reflected infrared or ultrasonic beams emitted from the front of the camera.

Because a film requires a predetermined amount of light for proper exposure, the aperture diameter and shutter speed have to be chosen carefully. In many cameras, this is again performed automatically. Shutter speed settings may range from a few seconds to perhaps 1/1000 second. The lens aperture diameter is usually expressed as a fraction of its focal length. For example, $f/8$ represents an aperture diameter for a 50 mm lens of 6.25 mm. The following apertures are often seen on a typical camera: $f/2, f/2.8, f/4, f/5.6, f/8, f/11, f/16$.

While the shutter is closed, light passing through the lens is diverted by a plane mirror mounted at 45° into a pentaprism. Total internal reflection (← 34.2) within the prism returns the image to an upright position for viewing through an **eyepiece**. On opening the shutter, the mirror is removed from the path of the light and light falls on the surface of the photographic film (see Figure 34.35b).

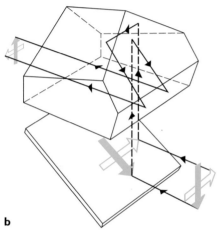

Figure 34.35a A 35 mm camera, **b** its prism and mirror action during focusing

Question

14a Calculate the distance between the film and the optical centre of a 50 mm camera lens when it is focused on objects at infinity. In what direction must the lens be moved so that objects 2 metres away are in focus?
b The **depth of field** on a photograph is a measure of the range of distance over which objects are reasonably in focus. Use Figure 34.36 to explain why the smaller the aperture, the greater the depth of field.
c i Find the ratio of the aperture diameter at $f/4$ compared with $f/8$.
 ii How do their areas compare?
 iii Why, for a given shutter speed, will an aperture of $f/5.6$ allow twice as much light on to the film as an aperture of $f/8$?
d The standard 50 mm lens is exchanged for a **telephoto** lens of 200 mm focal length. Explain why:
 i the telephoto lens is much larger in physical size than the standard lens,
 ii the long focal length lens produces a larger image but a smaller field of view on the film.

34 OPTICAL SYSTEMS

e Wide-angle lenses have shorter focal lengths than standard lenses, typically around 30 mm. Why do they produce smaller images on the film and cover a wider field of view?

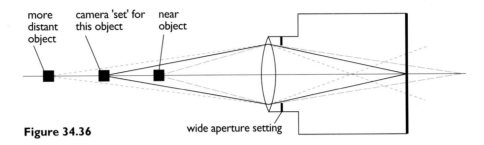

Figure 34.36

TELESCOPES

The telescope, first used some 400 years ago, is an optical system containing lenses or mirrors, which allows an observer to obtain a magnified image of distant objects on Earth or in outer space. The simplest form of **astronomical telescope** consists of just two lenses. Parallel light from a distant source falls on the first lens, the **objective**, and is brought to a focus at a distance f_o from the lens. The objective is a weak lens, with a focal length of possibly a metre or more, which produces an inverted real image. This is observed through the **eyepiece**, which has a much shorter focal length and acts as a magnifying glass. The eyepiece is usually positioned a distance f_E from the real image so that a virtual, magnified image is produced at infinity (see Figure 34.37). This ensures that the eye is in a relaxed state and the observer is able to continue viewing for long periods of time without undue strain.

To find the **angular magnification** of a telescope, suppose θ_1 is the angle subtended at the eye by the distant object and θ_2 the angle subtended by the virtual image seen through the eyepiece. The angular magnification is the ratio θ_2/θ_1. This ratio is dependent on the powers of the two lenses. The weaker the objective, the larger the intermediate image and the larger the magnified virtual image produced by the eyepiece. Increasing the power of the eyepiece further increases the magnification of the telescope.

It is desirable that the objective should have a large diameter for two reasons. Firstly, a larger lens gathers more light and produces a brighter image. One of the most striking effects of using a telescope is that stars invisible to the naked eye

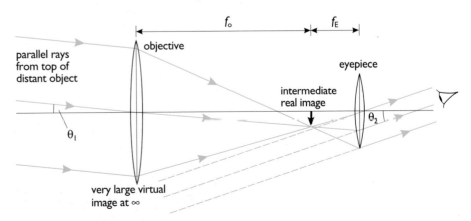

Figure 34.37 Astronomical telescope with the image at infinity

WAVES: TRANSFERRING ENERGY THROUGH SPACE

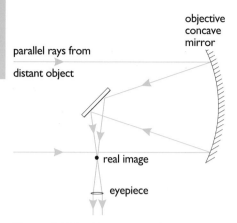

Figure 34.38 Newtonian reflecting telescope

can often be seen through a telescope. Secondly, there is a limit to the ability of a telescope to resolve the detail on, for example, the surface of the Moon or of a planet. This is because the objective lens of the telescope acts as a diffracting aperture. The larger the lens diameter, the smaller the diffraction effects and the better the resolving power of the telescope (← 31.3).

To overcome the problem of chromatic aberration (← 34.3) in the objective lenses of early telescopes, Isaac Newton devised another form of telescope, the **reflecting telescope** (see Figure 34.38). Parallel light from a distant star is brought to a focus by the concave mirror. An observer views the image through an eyepiece, after reflection by a mirror placed on the axis of the concave mirror. Other reflecting telescopes have been devised, such as the Cassegrain. This allows axial viewing, which is often more convenient. Today, large astronomical telescopes are used for photographic imaging and spectral measurements rather than direct viewing.

Question

15a In Figure 34.37, the real image formed at the principal focus of the objective is of height d.
  i Express the small angle θ_1 in terms of d and f_o.
  ii Calculate θ_2 in terms of d and f_E.
  iii Show that the angular magnification of the telescope, when adjusted so that the observed image is at infinity, is equal to f_o/f_E.
b Using Figure 34.37, explain why the eyepiece does not have to be as large in diameter as the objective for all the light falling on the objective to pass through the eyepiece.
c The biggest telescopes in existence are reflecting instruments. What advantages do these have over refracting telescopes, both in their construction and use?
d Telescopes for use on Earth need an 'erecting' lens. Why is this? What sort of lens should this be and where should it be placed?
e Find out how 'opera glasses' work.

MICROSCOPES

Like the telescope, a **compound microscope** is a system consisting of two lenses and able to produce a magnified image. It can produce a much larger image than a single magnifying lens. In principle, it is very similar to the astronomical telescope. A converging lens, the objective, produces an inverted and enlarged image of the object. A second lens, the eyepiece, magnifies the intermediate image (see Figure 34.39).

For maximum magnification, it is usual to position the eyepiece so that the final image is at the near point of the observer's eye. The overall magnification of the instrument is limited by spherical aberration in the high-powered lenses required.

Question

16 Refer to Figure 34.39.
a In a compound microscope, an object of length l_o is placed at the near point of the observer, a distance L from the eye. Calculate the angle θ_1 it subtends.

b The final image is of length l_i and is also at the observer's near point. Calculate the angle θ_2 the image subtends at the eye. Ignore the distance between the eye and the eyepiece.
c Show that the angular magnification of the instrument is the ratio l_i/l_o.
d Use the result in **c** to show that the angular magnification of the compound microscope is equal to the product of the linear magnification produced by the eyepiece and the linear magnification produced by the objective ($\leftarrow$ 34.3).

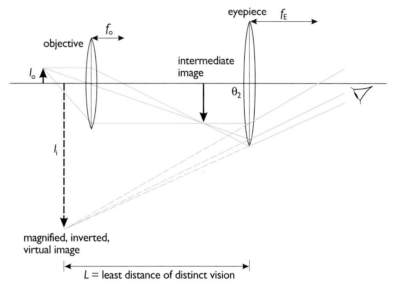

Figure 34.39 Compound microscope with the image at the near point

34.5 HOLOGRAPHY

A lens can recombine light scattered from individual points of an object to produce a set of individual image points a given distance from the lens ($\leftarrow$ 34.3). The lens effectively **decodes** the information contained within the light scattered by the object.

Holography is another increasingly important method of imaging. A laser beam is broadened by a high-power lens in its path. The beam is then split into two coherent beams using a beam splitter, such as a glass plate. One, the reference beam, is allowed to fall directly on a photographic plate which it illuminates uniformly (see Figure 34.40). An object placed in the path of the

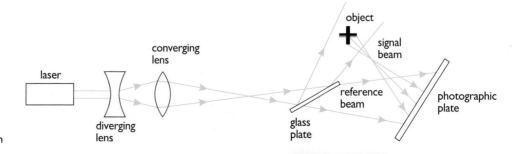

Figure 34.40 Producing a hologram

other beam scatters the laser light. Some of this scattered light also falls on the photographic plate and superposes with the reference beam. When there is no relative movement between the component parts of the system, the two **coherent** beams form a complex and unchanging set of superposition patterns.

Suitable exposure and development of the plate results in a permanent record of the patterns called a **hologram**. Its appearance under ordinary white light bears no resemblance to the original object. However, the hologram contains a particular pattern of superposition for every point on the original object that scattered light on to the photographic plate. When the hologram is broken into many pieces, each piece (if it is not too small), will still contain information about the whole object. This is not true of photographic images obtained using a lens system.

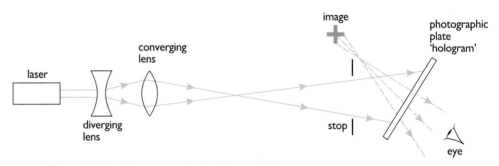

Figure 34.41 Reconstructing the holographic image

When the hologram is viewed under the correct lighting conditions, a three-dimensional image of the original object is reconstructed (see Figure 34.41). The hologram is illuminated with a broad beam of laser light, coming from the same direction as the reference beam. Looking through the hologram, a virtual image is reconstructed from each of the superposition patterns carrying phase and amplitude information about one particular point on the object. The set of patterns on the plate reproduces a complete set of image points corresponding to one particular view of the object. Moving the head slightly to one side, another view of the object is obtained giving the virtual image a three-dimensional appearance.

Figure 34.42 Producing a holographic image

34 OPTICAL SYSTEMS

Holography has generated much interest from research physicists in recent years and has found important applications in engineering and commerce. Holographic techniques have been used: to investigate the vibrational characteristics of the fan blades in jet engines; to examine flaws and cracks in fuel elements in nuclear reactors; for reading bar codes at supermarket check-outs and in the production of credit cards to counter forgery and fraud (see Figure 34.43).

Figure 34.43 Applications of holography

SUMMARY

Rays of light indicate the direction of travel of the energy contained in electromagnetic waves.

Light falling on a plane reflecting surface reflects so that the angle of incidence equals the angle of reflection.

When light passes from one transparent medium into another, it refracts. Snell's law states that:

$$\frac{\sin i}{\sin r} = \text{constant for a given pair of media,}$$

$$= \text{refractive index, } n, \text{ for the pair of media,}$$

where i is the angle of incidence and r the angle of refraction. This ratio is also equal to the ratio of the speed of light in each of the media. When light passes from a vacuum into a transparent medium, the constant, n, in Snell's law is the absolute refractive index of the medium.

The amount of refraction depends on the wavelength as well as on the media involved. The longer the wavelength, the smaller the effect of refraction. Refractive indices are quoted usually for a wavelength of 589 nm.

When light meets a less optically dense medium at an angle greater than the critical angle, total internal reflection occurs. The critical angle c is given by

$$\sin c = \frac{n_1}{n_2},$$

where $n_1 < n_2$. When the less dense medium is a vacuum or air, $n_1 = 1$ and $\sin c = 1/n$, where n is the refractive index through which the light is travelling.

There are two stages in any imaging process. Firstly, an object scatters incident radiation and secondly, the radiation scattered from individual points on the object are recombined to form an image of the original object.

Plane and curved mirrors use reflection of light to create images, while lenses use refraction.

For a thin lens, the object distance u is related to the image distance v and the focal length f of the lens by the relationship:

$$\frac{1}{u} + \frac{1}{v} = \frac{1}{f}.$$

The sign convention adopted is the 'real is positive' convention.

Spherical lenses and mirrors suffer from spherical aberration. Because of the dependence of refractive index on wavelength, lenses also suffer from chromatic aberration.

Holography makes use of superposition of light to create three-dimensional light images.

WAVES: TRANSFERRING ENERGY THROUGH SPACE

SUMMARY QUESTIONS

17 Two wardrobe doors with large plane mirrors are inclined at **i** 30° **ii** 60° **iii** 90° and **iv** 120° to each other. A girl shines a torch at the image of a friend produced by reflection at each mirror and its rays light up her friend's face.
Draw scale diagrams to show how:
a virtual images of the friend are formed by each mirror,
b rays from the torch reach each of the virtual images.
From the angles of incidence and reflection, show that the deviation of the torch beam is twice the angle between the mirrors in each case.

18 You can see a reflected image of yourself when you look in a plane or curved mirror, in the surface of a pool of water, or even in the polished paintwork of a motor car. However, you do not see a reflected image when you look at a sheet of paper, even though the paper may be a good reflector of light. Suggest an explanation for these differences.

19 By drawing accurate ray diagrams, explain how curved mirrors can produce virtual images of an object.
A dentist's mirror has a radius of curvature of 30 mm. How far must it be placed from a tooth to provide a virtual image with a magnification of 5?

20 Discuss the main similarities and differences between a simple camera and the eye. You should consider the processes of focusing, control of light levels, image formation and interpretation. Is a video camera a better model of the human eye?

21 Distinguish between converging and diverging lenses. Draw ray diagrams to show how parallel light wavefronts produce a real or virtual focus in each case.
Show by drawing scale ray diagrams that a converging lens, focal length 50 mm:
a can be used as a camera lens because, for objects 100 mm away or more, it produces real small and inverted images which are between 100 mm and 50 mm from the lens.
b can be used as a magnifying glass because objects less than 50 mm away produce virtual, upright and enlarged images.
Calculate the position and height of images produced in **a** and **b** for 20 mm high objects placed 75 mm and 25 mm respectively from such a lens.

UNIT H

INSIDE THE ATOM

INSIDE THE ATOM

Figure 35.1 Professor J. J. Thomson

Figure 35.2 Electron clouds in orbitals round the nucleus of an atom

Until 1897, it was thought that atoms, of which all matter is composed, could not be divided into smaller constituent parts. Then the English physicist J. J. Thomson (see Figure 35.1) experimented with cathode rays (→ 35.3). He found that within cathode rays there are particles which have a mass and charge, and are smaller than the hydrogen atom. The Irish physicist George Johnstone Stoney called them **electrons**.

Within 20 years, a new picture of the atom had been established, which is similar to that held today. Clouds of electrons surround a tiny, but massive, positive nucleus. The electron clouds exist in particular **orbitals** 10–100 pm across. This is about 10 000 times the diameter of the nucleus. An atom can be excited into definite energy states, each one having a particular arrangement of electron orbitals (see Figure 35.2).

Modern ideas about the atom assume that energy, radiation, matter and charge have a **quantum** nature. When an atom absorbs energy, one or more electrons move to orbitals of greater potential energy, further from the nucleus. Conversely, when an electron moves to a closer orbital of lower potential energy, a **quantum** of radiated energy is emitted. Normally this quantum of energy is radiated as an electromagnetic wave, a **photon**. When an atom is in close proximity to others, such as in a crystal lattice, the energy quantum is often transferred to other atoms as a very high-frequency sound wave, a **phonon**.

Between the innermost electron orbital and the nucleus is apparently empty space, full only of the electric and other fields originating in the nucleus. Just as the energy state of an atom is determined by the levels electrons possess in their orbitals, the nucleus has a definite energy state determined by the energy levels occupied by its constituent **nucleons** – **protons** and **neutrons**. However, their energies are a million or so times larger than those of electrons. A quantum of very high energy, a **γ-ray**, is emitted when a nucleon moves to a lower energy level.

During the twentieth century, a tremendous amount of evidence has been provided, through experiment and theory, for the quantum nature of energy, radiation, matter and charge. The implications this has had for ideas about molecules, atoms, nuclei, energy and particles have revolutionised physics and chemistry.

35 QUANTISATION OF CHARGE: THE ELECTRON

The proton, a positively charged nucleon (→ 40.3), has the same fundamental unit of charge e as the negatively charged electron. No other basic charge seems to exist on its own in nature. Quarks (→ 40.5) have fractional charges of $\pm\frac{1}{3}e$ and $\pm\frac{2}{3}e$, but have yet to be isolated and many physicists believe they never will be. The evidence for, and measurement of, a fundamental unit of charge is of great importance to science.

35.1 CHARGE ON THE ELECTRON

In 1911, the American physicist, Robert Millikan, made the first precision measurement of the basic electronic charge e. He established that the charge Q on any object is made up of an integral number of these basic quantities of charge. (Note that, although Q is the SI symbol for electric charge, e is usually used to denote the charge on a single electron.)

Millikan's experiment is one of the very few pieces of direct evidence for the discrete or quantum nature of charge. In a modern laboratory version of his experiment, an electric field E is produced between two circular horizontal metal plates by applying a voltage between them (see Figure 35.3). The upper plate has

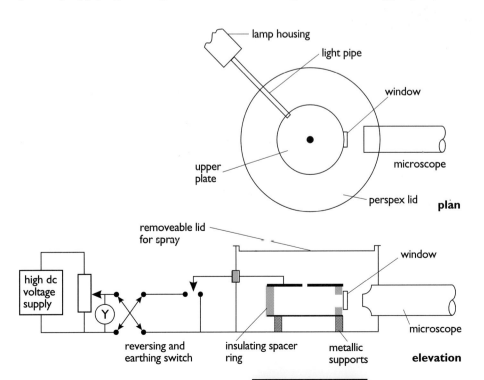

Figure 35.3 Apparatus for Millikan's experiment

a small hole at its centre through which tiny spheres can fall into the space between the plates. Millikan used oil drops from an atomiser spray but today miniature latex spheres are used because they give more reliable results. The space between the plates is illuminated from the side by a light pipe – a collimated beam of light passing along a glass rod. Light scattered from the oil drops makes them appear through a low-power microscope as tiny bright specks against a dark background.

The spheres are charged by friction as they pass into the space between the plates. The charge on any particular sphere can be changed by a burst of radiation from a radioactive source, such as radium 226. This ionises the air around the sphere which, in turn, affects the amount of charge it carries. A sphere carrying charge Q can be stopped from falling under gravity or made to rise by applying and varying the voltage V between the plates. When the sphere is stationary, there is no net force acting on it. So the force F_E due to E acting upwards on the sphere's charge Q is balancing the downward force due to the gravitational field g acting on its mass m:

$$QF_E = mg.$$

Since F_E is equal to V/d (← 16.1), where d is the plate separation, then:

$$Q \cdot \frac{V}{d} = mg$$

$$Q = mg \cdot \frac{d}{V}.$$

So the charge on the sphere is inversely proportional to the balancing voltage V for a given mass of sphere and plate separation.

The mass of the sphere is $\frac{4}{3}\pi r^3 \rho$, where r is the radius of the sphere and ρ its density (see Figure 35.4). Substituting this in the equation above gives:

$$Q = \frac{4}{3}\pi r^3 \rho g \cdot \frac{d}{V}.$$

So by measuring d, V, r and ρ, the experimenter can calculate Q. The experiment is repeated several times with differing amounts of charge on the sphere. Then the experimenter identifies the basic unit of charge common to all the calculated values.

This is a very difficult and tedious experiment to perform. Millikan, using drops of oil, took many years to arrive at a consistent value for the electronic charge. However, there are many computer simulations of what is seen through the microscope and what has to be done to balance the sphere. If you cannot take any real measurements, you could try one of these simulations to appreciate the principle of the experiment.

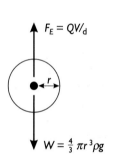

Figure 35.4 Forces on a stationary drop in an **E**-field

Questions

These questions refer to Millikan's experiment described above.

1a When observed through the microscope, the oil drops appear to "fall" upwards when no field is applied. Why?
b What will you observe through the microscope if the plates are not exactly horizontal when the field is applied?

35 QUANTISATION OF CHARGE: THE ELECTRON

2 The following data was obtained for later spheres:

voltage to balance sphere/V	: 337, 234, 279, 337, 425, 840, 564, 1 690
spacing of plates, d/mm	: 6.4 ± 0.1
radius of sphere, r/m (from manufacturer's data)	: $1.03 \pm 0.01 \times 10^{-6}$
density of latex, ρ/kg m^{-3}	: 1 000

a Use the data to find the charge on the sphere for each voltage.
b Show that each charge is an integral multiple of the smallest charge, e. (e)

35.2 CONDUCTION OF ELECTRICITY IN LIQUIDS

Some of the earliest evidence of the particle-like or quantum view of electricity came from Faraday's experiments in 1833, when he formulated the **laws of electrolysis**, that is, the masses of the materials deposited or released at the electrodes of a cell are proportional to the relative atomic masses of the materials (see Figure 35.5). The charge carried by one mole of electrons (or any singly charged ion) is called the **faraday**, F. The number of electrons in one mole is, by definition, the **Avogadro constant**, N_A so from values of F and N_A, you can calculate the charge on an electron (or a proton).

At the end of the nineteenth century, no one had devised a method of finding a very accurate value for N_A. Until relatively recently, the Avogadro constant was calculated from other quantities. Now, however, one of the most accurate and ingenious methods of finding the Avogadro constant does so to one part in a million. An X-ray beam is used to count the number of atoms in a small perfect crystal cube of silicon, so enabling the number of atoms per mole to be found.

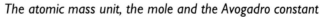

The atomic mass unit, the mole and the Avogadro constant

The Avogadro constant N_A, and the precise value of the atomic mass unit u, are both defined in terms of the nuclide carbon-12, denoted by $^{12}_{6}C$ ($\rightarrow$ 38.1).

A ^{12}C atom has 6 electrons in orbitals surrounding a nucleus containing 6 protons and 6 neutrons. When such particles combine to form an atom, energy is lost. This is called the binding energy, ΔE_B and represents a mass of $\Delta E_B/c^2$ where c is the speed of light ($\rightarrow$ 40.5). So, a ^{12}C atom has a slightly lower mass than that of 6 protons, 6 neutrons and 6 electrons added together.

An atomic mass unit u, is defined as exactly 1/12 of a ^{12}C atom. Its mass, m_u is referred to as the atomic mass constant. Because of the binding energy, the mass of 1 u should be slightly less than that of a proton or neutron, and this turns out to be the case. To three significant figures:

$$m_u = 1.66 \times 10^{-27} \text{ kg}$$

and m_p or $m_n = 1.67 \times 10^{-27}$ kg,

where m_p is the mass of a proton and m_n, the mass of a neutron.

The **Avogadro constant** N_A, is defined as the **number of atoms in 12 g of ^{12}C**. When measured accurately, it is 6.02214×10^{23} mol^{-1}. This leads to the definition of the **mole** as the amount of an element or compound that contains N_A atoms or

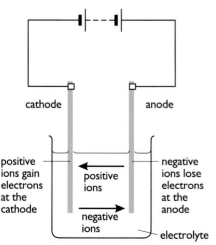

Figure 35.5 Movement of ions in an electrolysis cell

molecules. We can extend the definition to any type of particle:

1 mole = amount of substance that contains N_A entities

$$N_A = 6.02214 \times 10^{23} \, \text{mol}^{-1}.$$

For instance, N_A electrons constitute a mole of electrons of total charge 1 faraday. Clearly, it is essential to specify which type of particle, or entity, is being considered.

Instead of atomic mass, the numerically identical **molar mass** is sometimes quoted. This applies to a mole of substance, rather than an individual particle, atom or molecule. Molar masses are measured in g mol^{-1} or kg mol^{-1}. For example, a mole of $^{12}_{6}$C has a molar mass of exactly 12 g mol^{-1} or 0.012 kg mol^{-1}.

Measuring the Avogadro constant

This is done using perfect crystals of silicon, which can be manufactured up to several centimetres in length and cross-sectional area. The number of atoms in a 0.1 mm length are found by counting the number of atomic planes. With the atomic planes in two crystals aligned, one is moved slowly over the other. The intensity of an X-ray beam passed through both crystals changes as the planes move relative to each other (see Figure 35.6). The X-ray pulses produced are counted by a computer, each pulse corresponding to one layer of atoms.

The crystal length is measured with an error less than the diameter of an atom using optical interference techniques (← 33). From this the volume of a silicon atom, the **atomic volume** v_{Si} is found.

Comparing the masses of silicon atoms to the mass of carbon-12 atoms using a mass spectrometer (→ 40.4) enables the molar mass M_{Si}, of silicon to be established.

Knowing the density of silicon ρ, the molar volume V_{Si} is found. Dividing the molar volume by the atomic volume gives N_A to one part in a million:

$$N_A = \frac{\text{volume of a mole of silicon}}{\text{volume of an atom of silicon}} = \frac{V_{Si}}{v_{Si}}$$

$$= \frac{M_{Si}}{\rho v_{Si}}.$$

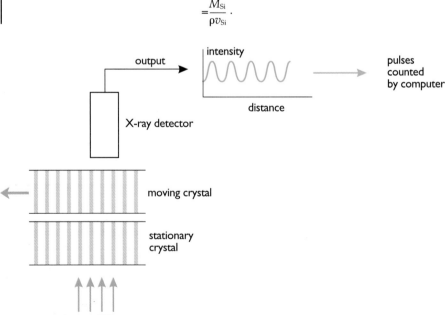

Figure 35.6 Apparatus for counting crystal planes (plane separation magnified about a million times)

35 QUANTISATION OF CHARGE: THE ELECTRON

> **Question**
>
> **3** One **faraday** is the charge possessed by a mole of electrons. What, therefore, is the charge on one electron? (F, N_A)

Faraday's laws enable another important quantity to be calculated – the charge on 1 kg, or the **specific charge**, of an ion. It can be found by measuring the total charge passed during electrolysis and dividing this by the total mass deposited or released at the electrode. In electrolysis of water, for example, 1 kg of hydrogen is released at the electrode when 9.58×10^7 C passes. J.J. Thomson's positive ray analysis experiments of 1910 showed experimentally that the specific charge of the hydrogen ion, the proton, has the same value:

$$e/m_p = 9.58 \times 10^7 \text{ C kg}^{-1}.$$

The specific charge of the hydrogen atom and ion are the same to three significant figures, indicating that the effect of the mass of the electron in a hydrogen atom is small.

35.3 ELECTRICAL DISCHARGES IN GASES: CATHODE RAYS

Air at atmospheric pressure is normally a good insulator. In electric fields greater than 3×10^6 V m^{-1}, air ceases to be an insulator and becomes a good conductor. Such fields cause **ionisation** of the air molecules. This ionisation, or breakdown, of air is accompanied by dramatic sparks, a lightning flash being a natural large-scale example.

Electrical discharges in gases can be created and maintained quite easily at low pressures. Electrodes are placed about 20 cm apart, with a voltage of a few kilovolts between them, in a glass tube. The pressure in the tube is initially about one hundredth that of the atmosphere. Gradual reduction to a pressure of 0.1 Pa, which is about one-millionth that of atmospheric pressure, causes a number of changes in the nature of the discharge. When the pressure is further reduced below 0.01, or 10^{-2} Pa, the air again behaves as a very good insulator.

DISCHARGE TUBE

Figure 35.7a shows a discharge tube in operation. The tube is connected to a vacuum pump and the circular shaped electrodes to a high voltage supply of about 5 kV. When the pressure falls to 10^3 Pa, one or more violet streamers appear between the electrodes. As the pressure is further reduced the colour changes and dark and bright regions appear (see Figure 35.7b). A magnet placed close to the tube has an effect on the appearance of the glow, suggesting that the patterns are related to the movement of charged particles rather than light waves. When the pressure is reduced to less than 0.1 Pa, the tube 'blacks out' and the glass of the tube walls begins to fluoresce. Soda glass glows green and pyrex glass blue. A zinc sulphide coating on the glass glows bright green.

You can explain these observations by considering the collisions of particles in the tube. The electrons and ions accelerated by the electric field between the electrodes collide with air molecules. As the pressure is reduced, there are fewer air molecules and so fewer collisions. At some point the electrons and ions can travel far enough between collisions to gain sufficient kinetic energy to make inelastic collisions with the atoms, increasing their potential energy. The atoms

H INSIDE THE ATOM

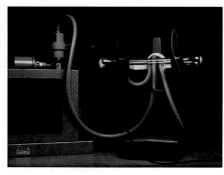

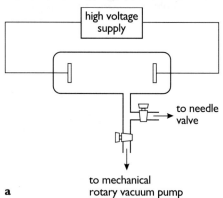

a

Figure 35.7a A gas discharge tube, **b** electrical discharge through air at low pressures

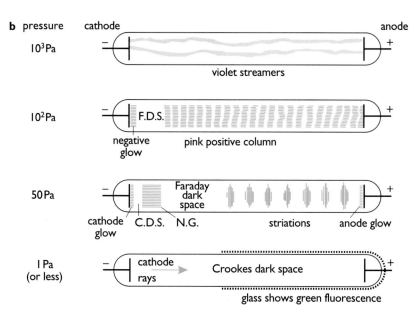

release this extra energy as light (→ 37.1) characteristic of the gases in the tube, and the tube begins to glow.

At the lowest air pressures, the electrons and ions are able to travel the length of tube without any collision. So positive ions striking the cathode have sufficient kinetic energy to cause the cathode to emit secondary electrons, or cathode rays.

Street lighting

Sodium light consists of two very intense emissions with very similar wavelengths: 589.0 nm and 589.6 nm. Sodium lamps used for street lighting have several advantages:

- all the energy is output at wavelengths to which the human eye is sensitive
- the human eye is particularly sensitive to green and yellow light and so can make maximum use of the sodium light available
- only yellow light has to be focused, so chromatic aberration (← 34.3) is reduced.

There are two types of sodium lamps, low pressure and high pressure. Low-pressure lamps have all the above advantages but produce a rather harsh illumination which does not enable people to identify colours. This is because the light is nearly monochromatic. For high roads and motorways this matters little and the efficient, bright low-pressure lamps are used.

High-pressure lamps are a little less efficient but produce a broader spread of wavelengths, giving a softer light. This is due to an effect called **pressure broadening**.

At high pressures, the high concentration of atoms results in multiple collisions, which change slightly the energy of each quantum emitted. So, instead of two narrow spectral lines, a very broad set of wavelengths is emitted, spilling into the red and green parts of the spectrum. Under the nearly white light of high-pressure sodium lamps used in pedestrian areas, people can see shops and their contents in almost their true colours.

SPEED OF ELECTRONS IN AN ELECTRON BEAM

In a discharge tube at low pressures, electrons of charge e are produced by secondary emission from the cathode. They are accelerated uniformly all the way

35 QUANTISATION OF CHARGE: THE ELECTRON

to the anode through the potential difference V between the electrodes. The change in potential energy for an electron of charge e is eV. The electrons have a speed v and a kinetic energy $\frac{1}{2}m_e v^2$ when they reach the anode. The potential energy of the electrons at the cathode has become wholly kinetic energy at the anode. So:

$$\frac{1}{2} m_e v^2 = eV$$

or $$v = \sqrt{\frac{2eV}{m_e}}.$$

Question

4 An electron is accelerated through 5 000 V between a cathode and an anode 20 cm apart in a discharge tube.
a Use the value of the specific charge on an electron e/m_e to calculate the speed of the electron at the anode. (e/m_e)
b How long does it take to travel the 20 cm from cathode to anode?
c How would your answers to **a** and **b** change if the tube was doubled in length?
d How would your answers to **a** and **b** change if the voltage across the tube was doubled?

AN ATOMIC UNIT OF ENERGY

When an electron moves through a potential difference of 1 V, an amount of energy equal to **one electron volt**, or **1 eV**, is transferred to or from the electron. This is a very small amount of energy, but is a convenient unit to use when dealing with particles on an atomic or sub-atomic scale. For example, the ionisation energy of hydrogen is 13.6 eV. This contrasts with the much higher energy, about 25 keV, of an electron as it strikes the screen in a colour television tube. The electron-volt is not an SI unit. The nearest appropriate SI unit is the **attojoule**, **aJ**, which is 10^{-18} J, but it has yet to gain widespread acceptance.

When considering nuclear events which tend to be much more energetic, it is usual to measure energy changes in MeV, millions of electron volts. For example, the kinetic energy of an alpha particle emitted from a nucleus may be of the order of 5 MeV, or 8×10^{-13} J, and is equivalent to the energy an electron would gain when accelerated through 5 MV. Particle accelerators can generate particle energies of the order of GeV, or 10^9 eV.

Questions

5 Calculate the kinetic energy in joules gained by an electron accelerated through a potential difference of 1 volt: 1 electron volt. (e)

6a The minimum energy to separate the electron and proton in a gaseous hydrogen atom is 13.6 eV. Express this energy in joules.
b The ionisation energy of hydrogen is given as 1 312 kJ mol^{-1} in chemical data. Show that this figure is consistent with your answer to **a**.

7 How fast is a 5 MeV alpha particle travelling? (The mass of an alpha particle is 6.6×10^{-27} kg.)

INSIDE THE ATOM

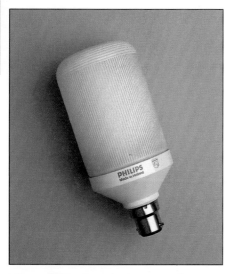

Figure 35.8 A modern compact fluorescent lamp

Fluorescent lamp

The standard fluorescent lighting tube (← 22.2) is an example of a discharge tube. It contains mercury vapour at low pressure. Most of the electromagnetic radiation of the discharge from the excited atoms is beyond the visible region of the spectrum in the ultraviolet (uv). The uv strikes the fluorescent coating of the tube, which re-emits the energy as visible light (→ 37.4). The tube can be made to glow any desired colour through a suitable choice of fluorescent coating material. In many coloured advertising sign tubes, gases rather than a fluorescent coating are used to give the colour. For example, neon gas glows orange-red.

An alternating voltage supply is used in the fluorescent tube, so both electrodes act alternately as cathodes. The cathodes are kept glowing at white heat by the collisions of the positive gas ions. There is no extra electrical heating.

In an ordinary incandescent light bulb, the light source is the glowing metal filament (→ 36.1). Most of the energy supplied to the filament is transferred into making it hotter. A relatively small proportion appears in the radiated light. This is very evident should you get too close to a working light bulb. In the fluorescent tube, the light is produced by the excitation of the gas atoms through collision, not by heating. So the transfer of energy from the supply into light is much more efficient. Another advantage of the fluorescent tube over the light bulb is that it produces a much larger area of light output, giving softer shadows.

The invention of the laser (→ 37.4) has led to renewed interest in gas discharges. Recent advances in understanding the recombination processes has resulted in the design of very compact energy-saving fluorescent lamps (see Figure 35.8).

PROPERTIES OF CATHODE RAYS

When an object is placed between the electrodes inside an evacuated discharge tube, a shadow of the object appears on the glowing glass. The shadow is produced by invisible rays of electrons from the cathode. The name cathode rays is still used for any beam of electrons travelling at high speed in a vacuum as in, for example, the cathode ray tube of an oscilloscope or television (→ 35.6).

The original cathode rays were produced by secondary emission from a cold cathode. Then it was discovered that a heated filament will emit electrons (→ 35.5). Such thermionic emission produces electrons with little spread in energies and is more reliable and controllable. This method is used in all modern tubes, including those in experiments 35.1 – 35.4. The heating filament is inside an **electron gun** (→ 35.5), but the tube itself is still referred to as a discharge tube.

Experiment 35.1a uses a Maltese cross, like the medal in the original demonstrations in the mid-nineteenth century, to show that cathode rays travel in straight lines.

EXPERIMENT 35.1 Investigating cathode rays

Warning: High voltages are used. Ensure that the current-limiting safety resistor is included in the circuit.

a Paths of cathode rays

After setting up a discharge tube as in Figure 35.9a and switching on the heater supply to the gun, you will see a sharp shadow of the cross cast on the screen (see Figure 35.9b). With the high voltage supply switched on, you will

35 QUANTISATION OF CHARGE: THE ELECTRON

Figure 35.9a Arrangement for a typical Maltese cross tube, **b** the image produced, **c** image distortion when anode is disconnected

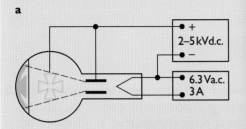
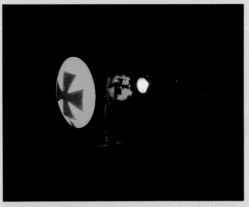
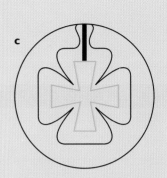

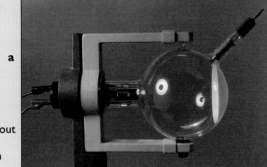
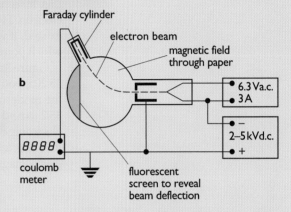

Figure 35.10
a Perrin tube without deflection coils
b circuit for Perrin tube

see that the screen fluoresces except in the region of the shadow. You could also observe the effects of bringing a magnet close to the tube, and of disconnecting the metal cross from the anode of the supply (see Figure 35.9c).

b Sign of charge of rays

This experiment requires a Perrin discharge tube and a magnetic field to deflect the electron beam into the cylinder (see Figure 35.10). You can use a coulombmeter or charged electroscope to detect the sign of the charge in the cylinder when the electrical supply is switched off.

Questions

8 This question refers to experiment 35.1a.
a In which direction does the shadow move when the north pole of a magnet is brought close to the tube in the plane of the metal cross? You will need to think about the direction of the force on a moving charge in a magnetic field (← 17.3). Draw a suitable diagram to illustrate your answer.
   What assumption have you made about the sign of the charge on the electron?
b Explain briefly why distortion occurs, as in Figure 35.9c, when the cross is disconnected from the anode.

H INSIDE THE ATOM

> 9 Experiment 35.1b uses a Perrin tube. Why, in discharge tubes of this type, is it better to earth the anode and have the cathode at a negative potential?

35.4 SPECIFIC CHARGE OF THE ELECTRON

In 1897, during his study of cathode rays, J. J. Thomson was able to show that all cathode rays from whatever source possess a common property – the ratio of their charge to their mass, known as the **specific charge**, e/m_e, is constant. For this, he is regarded as the discoverer of the fundamental particle of electricity, the electron. However he did not prove experimentally that electrons actually existed. It was Millikan who established the constancy of the electronic charge, which pointed directly to the existence of electrons (← 35.1).

Figure 35.11 shows the apparatus Thomson used. By comparing it with Figure 35.20a, you can see that Thomson's apparatus has many of the basic features of the modern cathode ray tube. He used the null-deflection method of experiment 35.2 to find the speed of the cathode rays. This enabled him to find the specific electronic charge, as in experiment 35.4a. He found that the speed of the cathode rays was less than the speed of light. This indicated that the rays were not electromagnetic in nature.

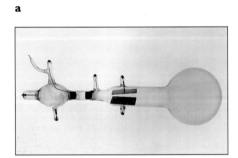
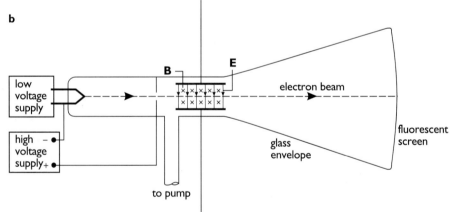

Figure 35.11 a Photograph and **b** diagram of apparatus designed and used by J. J. Thomson to find e/m_e

ELECTRONS MOVING IN ELECTRIC AND MAGNETIC FIELDS

The motion of electrons in both electric and magnetic fields can be investigated using the discharge tubes shown in Figures 35.12 and 35.13.

In the first, the electron gun emits a narrow ribbon of cathode rays from a slit in the anode. The horizontal electron ribbon strikes a vertical mica screen, along its entire length. The screen is coated with a fluorescent material to make the path of the beam visible. When you apply a voltage V between the two horizontal metal plates at the top and bottom of the screen, you create a vertical electric field **E**. The **E**-field takes a constant value in the region between the plates (← 16.1) equal to V/d where d is the separation of the plates. The electron beam is deflected by the **E**-field in a **parabolic path** towards the positively charged plate. The deflection can be varied either by altering the voltage V across the plates or by varying the anode voltage. A reduction in the anode voltage gives the

35 QUANTISATION OF CHARGE: THE ELECTRON

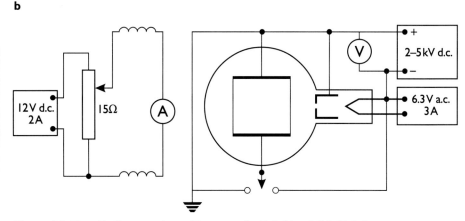

Figure 35.12a Deflection tube and **b** circuits for **B**-field and **E**-field deflections

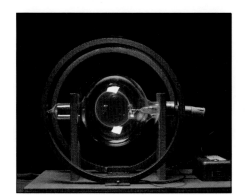
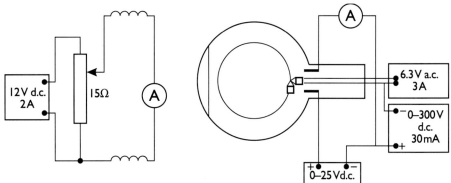

Figure 35.13a Fine-beam tube and **b** circuits for **B**-field and **E**-field deflections

electrons a slower speed. Because they stay in the field between the plates for a longer time, they are deflected by a larger amount.

Question

10 What are the similarities and differences between the motion of electrons between charged plates and the motion of projectiles in the Earth's gravitational field (← 6.3)?

In the tube in Figure 35.13, the electron gun produces a narrow vertical beam. The tube is filled with a gas at low pressure. The gas is ionised by the beam, and glows as the ions and electrons recombine, making the path of the electron beam visible. As in other tubes, the cathode rays stay in a single concentrated beam, showing that all the electrons are moving at the same speed. This happens because the pressure of the gas is such that collisions do not scatter and destroy the electron beam yet there are enough collisions for the glow to be visible. As with the first tube, the speed of the electrons emitted by the gun can be controlled by varying the anode voltage.

In either tube, when you pass a current through the vertical circular **Helmholtz coils** on either side of the tube, you create a uniform horizontal magnetic field **B** which is perpendicular to the electron beam (← 18.3). The resulting force on the electrons turns the beam into the **arc of a circle** within the field region.

INSIDE THE ATOM

Path of an electron in a magnetic field

In Figure 35.14, an electron of mass m_e and charge e travelling at speed v enters a region of **B**-field perpendicular to the plane of its motion. The force F_B on the electron is equal to Bev ($\leftarrow$ 17.4). Because F_B is the only force acting and it acts always at right angles to the motion of the electron, it is a **centripetal force** $m_e v^2/r$ ($\leftarrow$ 7.1). It maintains the electron with a constant speed v in a circular path of radius r. So:

$$Bev = m_e v^2/r$$

$$v = \frac{e \cdot Br}{m_e}.$$

The speed of the electron may be found from its charge to mass ratio e/m_e and the radius of its path in a given **B**-field.

Question

11 This question refers to the deflection tubes described above.
a When the anode voltage is increased how is the path of the beam affected for electric and magnetic deflection?
b What is the effect on the path of **i** an increase in the electric field in electric deflection **ii** a decrease in the magnetic field in magnetic deflection?
c Why will the electron beam be narrow only when all electrons are travelling at the same speed?

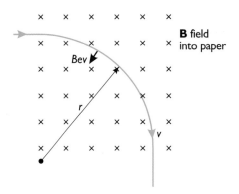

Figure 35.14 Magnetic deflection of an electron beam.

BALANCING PERPENDICULAR ELECTRIC AND MAGNETIC FIELDS

In experiment 35.2, the forces due to the **B**-field and **E**-field are equal but in opposite directions. So an electron beam entering the fields is not deflected.

$$F_B = F_E.$$

$$Bev = Ee,$$

which means that:

$$v = \frac{E}{B}.$$

EXPERIMENT 35.2 Motion of electrons in **E**- and **B**-fields

Using the circuits of Figure 35.12b, you can confirm that an electron beam's trajectory is parabolic for a perpendicular **E**-field and circular for a perpendicular **B**-field.

By arranging for the force due to each to be equal and oppositely directed, so that no net deflection occurs, you can calculate the velocity v of the electron beam from:

$$v = \frac{E}{B}.$$

The **B**-field can be found from the calibration graph or the formula given in experiment 35.3. The **E**-field can be found from the voltage and distance between the plates.

35 QUANTISATION OF CHARGE: THE ELECTRON

Questions

12 Draw a diagram showing the directions of the **E**- and **B**- fields which will enable a beam of electrons to pass undeviated at right angles to both of them.

13 In experiment 35.2, you can arrange that the magnetic force on the beam is equal and opposite to the electric force so that the beam does not deflect from its straight line path. Sketch the apparatus and show the path of an electron beam where:

a $v < \dfrac{E}{B}$.

b $v > \dfrac{E}{B}$.

14 Estimate the strength of the **B**-field required to make electrons of speed 10^7 m s^{-1} travel in a circle of radius 10 cm.

EXPERIMENT 35.3 B-field in the centre of a pair of Helmholtz coils

You can create a **B**-field by connecting a pair of Helmholtz coils to a variable dc supply in series with a 0–5 A ammeter. You can use a calibrated Hall probe or a current balance (← D) to measure the **B**-field in the central region between the coils for several different currents I. Moving the Hall probe in the central region enables you to check how uniform this **B**-field is. (← experiment 18.3)

From a graph of B against I you can check that the **B**-field at the centre of Helmholtz coils of radii r and having N turns is:

$$B = \frac{8}{5\sqrt{5}} \cdot \frac{\mu_o NI}{r}.$$

EXPERIMENT 35.4 Determining the specific charge of the electron

a By using an E-field and B-field together, as in experiment 35.2, you can find the velocity of an electron beam. You can then calculate the value of e/m_e for an electron from:

$$\tfrac{1}{2} m_e v^2 = eV.$$

b Set up the circuit of Figure 35.12b with both plates connected to earth or the fine-beam tube of Figure 35.13b. Using the **B**-field from the pair of Helmholtz coils (see experiment 35.3) you can cause a beam of electrons to travel in a circular path of radius r. By measuring r for different values of the accelerating voltage V and current in the coils, you can calculate e/m_e from:

$$\frac{e}{m_e} = \frac{2V}{B^2 r^2}.$$

INSIDE THE ATOM

Northern and Southern Lights

By placing a bar magnet along the axis of the helix of the beam in a fine-beam tube (see Figure 35.13), the helix can be reversed back on itself. It is this effect, occurring naturally in space around the Earth, which keeps charged particles captive in the Earth's magnetic field. Electrons follow spiral paths around the magnetic field lines approaching the Earth. At the poles, they turn to spiral in reverse direction back out into space (see Figure 35.15a). The path of an electron can be simulated by drawing a straight line across a sheet of transparent paper and then rolling it into a cone.

As they near the poles, some of the electrons collide with nitrogen and oxygen atoms in the upper atmosphere, raising the atoms to excited states ($\rightarrow$ 37.1). Much of the energy is re-emitted as light, producing spectacular and beautiful aurorae 100–200 km above the polar regions (see Figure 35.15b and c). So the Northern Lights (Aurora Borealis) and Southern Lights (Aurora Australis) are really gas discharges ($\leftarrow$ 35.3) on a gigantic scale.

Questions

15 This question concerns experiment 35.4b.
a Show that since:
$$eV = \tfrac{1}{2}m_e v^2$$
and $Bev = m_e v^2 / r$,
then $v = (e/m_e) B r$
and $\dfrac{e}{m_e} = \dfrac{2V}{B^2 r^2}$.

b Estimate the uncertainty in any measurements of **i** V, **ii** B and **iii** r.
c What uncertainty does this give in the calculated value for e/m_e?

16 When a beam of electrons enters a magnetic field region, at right angles to the field direction, it follows a circular path. Show, with the help of a suitable diagram, that when the beam enters the field at any other angle, it follows a helical path along the field direction.

You can demonstrate this using the fine-beam tube of Figure 35.13 by altering the angle of the Helmholtz coils relative to the beam direction.

Electric and magnetic fields are used separately or together to contain, deflect and accelerate beams or pulses of charged particles in machines such as mass spectrometers ($\leftarrow$ D) and high energy accelerators ($\rightarrow$ 40.4). They are also used in research into plasma confinement, controlled fusion reaction ($\rightarrow$ 40.6).

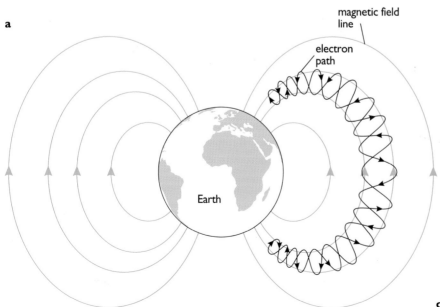

Figure 35.15a Path of charged particles in the Earth's magnetic field, **b** and **c** the Northern Lights

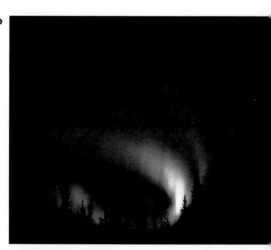

35.5 EMISSION OF ELECTRONS FROM A METAL SURFACE

In a metal, one valence electron from each atom is relatively free to move about within the lattice of positive ions (← 25.2). The ions exert an attractive force on the electrons, keeping them within the metal. These conduction electrons behave rather like a gas contained within a box, the box being the piece of metal. A certain minimum energy is required to remove an electron from the surface, or to escape from the box. This energy is called the **work function** φ of the metal. It is different for each substance but does not depend on how the energy is supplied. Under normal conditions, the kinetic energy of the "free" electrons is much less than the work function so they are unable to escape from the metal's surface.

There are four possible ways of giving electrons energy at least equal to the work function:

  i **thermionic emission**, by heating the metal to high temperatures
 ii **photoelectric emission**, by shining light of appropriate frequency at the surface
iii **secondary emission**, by bombarding the surface with electrons or ions
 iv **field emission**, by applying a very strong electric field, rather greater than 10^9 V m^{-1}, perpendicularly to the surface.

The photoelectric effect is of far-reaching importance and is considered in 36.3. Here, we discuss briefly the other three forms of emission.

THERMIONIC EMISSION

For most metals, significant thermionic emission of electrons only occurs at very high temperatures and many metals melt before reaching such temperatures. Tungsten does not, but its work function is 4.49 eV. This means that a temperature of over 3 000 K is needed for much emission to occur. Coating its surface with caesium reduces the work function to 1.36 eV and with barium oxide, to 1 eV. Such coatings more than halve the temperature of operation needed to produce a certain number of electrons.

Thermionic emission is highly temperature dependent: the higher the temperature, the greater the rate of emission of electrons from the surface. This is similar to the evaporation of molecules from a liquid, where the Boltzmann factor (← 28.4) determines the effect of temperature on the rate at which molecules escape from the liquid surface.

During emission the space above the metal becomes filled with electrons creating a "cloud" of negative charge, often referred to as a **space charge**. Further electrons leaving the metal are both attracted back by the metal surface, which now has a positive charge, and repelled by the negative space charge. A dynamic equilibrium is reached, as in the saturated vapour above a liquid, with as many electrons returning to the surface as escaping from it.

Simple electron gun

Thermionic emission is used to provide the supply of electrons accelerated through an electron gun. In the gun, a cylindrical anode at a positive voltage of about 5 kV is placed in front of an indirectly heated cylindrical cathode to attract electrons away from the cathode (see Figure 35.16). Not all of the electrons hit the anode, and a diverging beam is projected at high speed into the space beyond the anode, as in experiment 35.1a. The beam can be formed into a ribbon or a narrow beam, as in experiments 35.2 and 35.4, by placing a metal plate with a slit or a hole over the end of the anode.

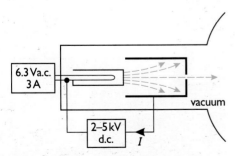

Figure 35.16 A simple electron gun

INSIDE THE ATOM

SECONDARY EMISSION

In the cold cathode discharge tube (← 35.3) the positive ions and electrons are accelerated towards the cathode and anode respectively. With sufficient kinetic energy, the positive ions hit the cathode at high velocity causing electrons to be emitted from the cathode by secondary emission.

Electron multiplier

Radiation cameras or scanners are used increasingly to detect medical disorders. For example, the functioning of the thyroid gland can be investigated by measuring the accumulation in the thyroid of the radionuclide iodine 131. Iodine taken in by the body tends to localise in the thyroid gland. The patient is given, usually orally, a small dose, or **tracer**, of the radionuclide in the form of sodium iodide solution (→ 39.8).

Gamma-ray photons (→ 38.5) emitted by the tracer in the thyroid fall on a crystal within a 'γ-camera', which acts as a **phosphor**. The photons of light it produces fall on a surface with a low work function called the **photocathode**, k, causing electrons to be emitted by the photoelectric effect (→ 36.3).

Each photon can release, at most, one electron, which is not detectable on its own. To amplify the effect, an **electron multiplier** is used (see Figure 35.17a). The emitted electron is accelerated to the first electrode D_1 with enough energy to release several electrons from it by secondary emission. Each secondary electron is accelerated towards the second electrode D_2 to repeat the multiplication process. After 10 or so electrodes, there are sufficient electrons collected at the anode A to give a measurable current. Photons, ions or other particles can be detected separately as individual pulses of current when they arrive at sufficiently spaced time intervals at the front end of a multiplier.

The electron multiplier has many uses. For example, it forms an important part of the detecting system in a mass spectrometer, where it is used to amplify the signal from the ion detector. The photocathode at the entrance window is replaced by a surface that emits electrons when it is struck by the ions to be detected. The **photomultiplier** (see Figure 35.17b), for detecting and amplifying low intensity light, has an electron multiplier as its main component. Photocathodes sensitive to the light being detected are attached to the front, before the first dynode. Electrons emitted from the photocathode are then amplified by the electron multiplier.

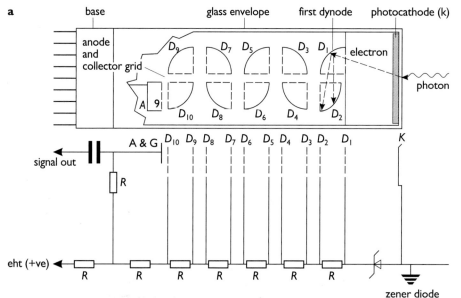
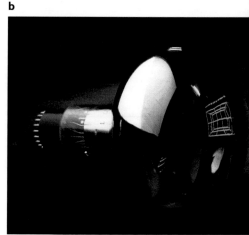

Figure 35.17a Electron multiplier circuit, **b** a photomultiplier tube

35 QUANTISATION OF CHARGE: THE ELECTRON

> **Question**
>
> **17** Suppose each electron incident on a dynode, or secondary emission electrode, can produce three electrons (see Figure 35.17a).
> **a** How many electrons will be emitted at the **i** first electrode, **ii** second electrode, **iii** third electrode?
> **b** There are 10 electrodes in total. How many electrons will be collected at the anode for one detected photon at the photocathode?
> **c** Calculate the current in the detection circuit when 10 photons are detected per second.

FIELD EMISSION

Electrons or ions can be removed from a surface just by increasing the voltage at the surface. Such field emission is useful in an instrument like the **field-ion microscope**, but it can also be a disadvantage. For example, it makes it impossible to increase the charge on a surface beyond a certain limit before it leaks away. Even when the field is below the limit, ionised gas atoms near the surface can be attracted to it, so neutralising some of the surface charge.

The limit is reduced significantly when a metal surface is not smooth. Experiments on the effect of pointed and sharp objects show clearly why this happens. Electric field lines are always perpendicular to a charged conducting surface and the equipotentials around the surface, and so bunch together at a spike (← 16.3). The resultant field at a spike can be sufficient for electrons to leak away from the surface.

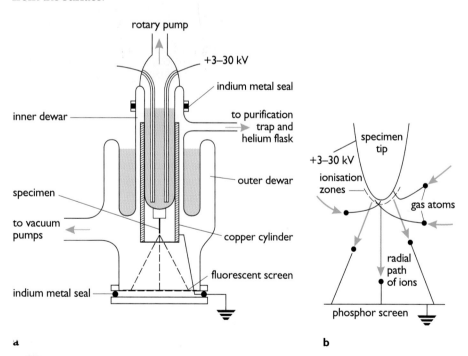

Figure 35.18a Field-ion microscope, **b** detailed view of specimen tip and chamber

> *Field-ion microscope*
>
> The field-ion microscope (see Figure 35.18) makes use of the effect of field emission to achieve magnifications greater than one million times. It allows individual atoms of a metal crystal to be resolved on a fluorescent screen (see Figure 35.19).

H INSIDE THE ATOM

The specimen to be studied is in the form of a needle with a tiny hemispherical tip 10^{-7} m across. This is cooled to 77 K using liquid nitrogen and maintained at a positive potential of about 5 kV. The free electrons in the metal tip are pulled inwards, leaving the surface ions to act as exposed spikes. Gas atoms close to the surface are attracted to the spikes. Each gives up an electron and is then repelled along the field lines from the tip to a fluorescent screen some 10 cm away. The apparatus is at a very low pressure so that few of the ions collide with other atoms on the way. The screen displays a magnified image of the tip of the needle. By increasing the voltage, the surface ions can be pulled from the surface of the needle tip to expose the next layer of atoms. In this way adjacent atomic layers in a specimen can be studied. The technique also ensures a perfectly clean surface clear of contaminating particles and irregularities.

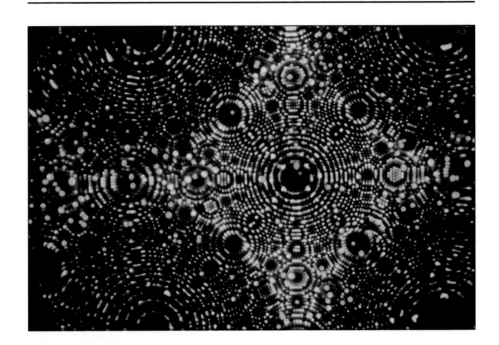

Figure 35.19 Field-evaporated surface

35.6 MASS OF THE ELECTRON

There is no direct method of measuring the mass of an electron. However, from Thomson's experiment (← 35.4), $e/m_e = 1.7585 \times 10^{11}$ C kg^{-1} and from Millikan's experiment (← 35.1), $e = 1.602 \times 10^{-19}$ C. So we can calculate m_e as follows:

$m_e = e/(e/m_e)$

$= 1.602 \times 10^{-19} / 1.7585 \times 10^{11}$ kg

$= 9.110 \times 10^{-31}$ kg.

Question

18 Compare the mass of an electron to the mass of an hydrogen atom. The molar mass of hydrogen is 1.008×10^{-3} kg mol^{-1}. (N_A)

The high speeds of computers and the visual displays of television and cathode ray oscilloscopes are only possible because of the extremely small mass of the electron. Every signal requires electrons to be accelerated and decelerated. Electrical forces can give the very small masses of the electrons very large accelerations, enabling them to change their motion almost instantaneously. However, neither they nor any electrical signals can travel faster than the speed of light. The length of cable along which the electrical signal travels, or the length of the electron beam in space, limits the speed of communication.

Cathode ray tube

This is used for visual picture display in, for example, the television receiver, the computer VDU and the oscilloscope. Figure 35.20 shows a typical layout for an oscilloscope tube. Electrodes inside the electron gun control the brightness and focus the beam on the screen. The cylinder G has a negative voltage with respect to the cathode K. This voltage is varied to allow more or less electrons to reach the first anode A_1, so controlling the brightness of the spot on the screen. The anode A_2 accelerates the electrons. The cylinder between the two anodes, together with the anodes, produces an electric field pattern that converges and focuses the electron beam to a point at the screen. A voltage divider (← 12.7) enables the various voltages required by the anodes and the brightness control grid to be produced from a single high voltage supply. Suitable voltages from other separate sources are applied to each pair of plates, X and Y, to deflect the spot to any point on the screen.

An internal oscillator, producing a sawtooth output as in Figure 35.21, is connected to the X plates. The rate of sweep can be varied to produce the most convenient scale on the x-axis, or time axis, for observing and measuring the shape and frequency of the voltage waveform of the incoming signal connected to the Y-input. The end of the electron beam is moved at a constant rate across the screen from left to right during the positive rise of the sawtooth. The sharp negative fall is the **flyback** signal, pulling the spot rapidly to the left before it repeats its steady motion to the right. The incoming voltage signal is connected through an amplifier to the Y plates so that the most convenient scale can be used for measuring its amplitude. The resultant trace on the screen is a graph of the variation of input voltage against time.

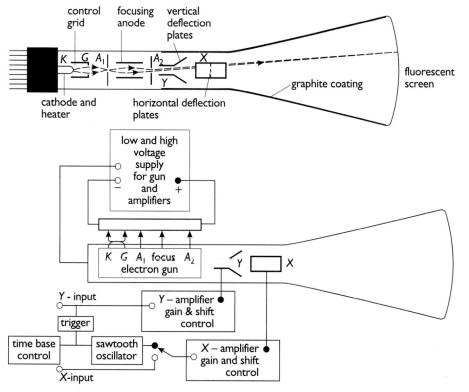

Figure 35.20a A cathode ray tube in an oscilloscope, **b** block diagram showing external controls to the tube

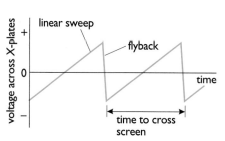

Figure 35.21 Time-base sawtooth voltage waveform

INSIDE THE ATOM

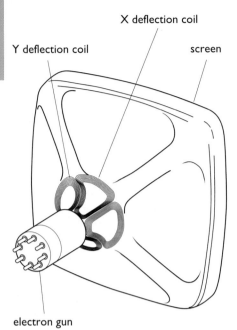

Figure 35.22 Magnetic deflection coils for a television receiver tube

The cathode ray tube in a television does not have the X and Y beam deflecting plates, required for high scanning speeds in oscilloscopes. The beam deflection has to be much greater to cover the larger screen in a television, and this is achieved by having two sets of deflecting coils (see Figure 35.22). The currents in the deflecting coils are synchronised with the input signals to cause the electron beam to sweep across the screen 313 times, while moving steadily from top to bottom, in 1/50 s. The scan is then repeated, but with the lines interlaced between those of the first sweep, to produce a total of 625 lines on the screen. So the image on each point on the surface is renewed every 1/25 s.

Questions

19 The graticule, or grid, on an oscilloscope screen is 10 cm wide. With the time base set to 2 ms cm^{-1}, how many cycles of a 1 kHz alternating waveform will be displayed on the screen?

20 What are the essential differences between a colour television cathode ray tube and a black and white tube?

SUMMARY

Electrons were first observed as cathode rays, produced by secondary emission, in a "cold" discharge tube. Modern instruments use a heated cathode to produce an electron beam by thermionic emission.

The electron carries one quantum of negative charge, $e = -1.60 \times 10^{-19}$ C. This is a fundamental unit of charge, measured, for example, in Millikan's experiment or in electrolysis. Fractional values of e have never been observed.

Electrons can be liberated in a gas by
- heating,
- incident radiation, or
- inelastic collisions.

Electrons can be emitted from a metal surface by
- heating   – thermionic emission,
- incident light   – photoelectric emission,
- incident electrons – secondary emission, or
- a large electric field – field emission.

The work function, the minimum binding energy, is the least energy required to remove an electron from the clean surface of a metal.

The kinetic energy given to an electron accelerated from rest through a voltage V is:

$$\tfrac{1}{2} m_e v^2 = eV.$$

The specific charge of the electron, $e/m_e = 1.76 \times 10^{11}$ C kg^{-1}. It is measured from the path of an electron beam in an electric and/or magnetic field.

The rest mass of the electron, $m_e = 9.11 \times 10^{-31}$ kg, is found indirectly from e and e/m_e.

The force F_E on an electron in an electric field is:

$$F_E = eE$$

in the direction of the field.

35 QUANTISATION OF CHARGE: THE ELECTRON

The force F_B on an electron moving at right angles to a magnetic field is:

$F_B = Bev$

in a direction perpendicular to the plane defined by the direction of the velocity and that of the magnetic field.

SUMMARY QUESTIONS

21 Look back through the experiments in this chapter and note those that
a point directly to the existence of particles of charge,
b provide direct evidence for the negative sign of the charge.

22a Does the amount of bending of a charged beam in an electric or magnetic field reveal anything about either the charge or the mass of the beam?
b What other factor affects the amount of bending?

23 What is meant by the work function of a metal?

24 In experiment 35.2, electrons passing through "opposing" electric and magnetic fields are undeflected when their velocity v equals E/B.
a What would happen to a beam containing electrons of a range of speeds passing through such a field system?
b Why is this arrangement of crossed electric and magnetic fields able to be used as a **velocity selector**?

25a An electron beam is deflected from a straight path. How can you tell whether it has been deflected by an electric or a magnetic field, or both?
b What will happen to an electron beam projected in the direction of
i a uniform electric field?
ii a uniform magnetic field?
c What will happen to an electron beam projected at 45° to
i a uniform electric field?
ii a uniform magnetic field?

26 An electron is accelerated by an electron gun through 2 kV.
a Calculate the value of its kinetic energy when it emerges, in eV and in joules. (m_e, e)
b What is the value of its final speed?
c Which of your answers to **a** and **b** would be different if
i a hydrogen ion of mass 2 000 m_e,
ii an α-particle of mass 8 000 m_e and charge $+2e$,
had been accelerated instead of an electron?

27 The current between anode and cathode in the e/m demonstration tube of Figure 35.12 is 100 μA, when the voltage is 4 kV.
a Calculate the energy reaching the fluorescent screen at the end of the tube per second.
b What is the value of the kinetic energy given to each electron?
c How many electrons hit the end of the tube per second?
d What is the value of the speed of the electrons?

28 In the apparatus of experiment 35.2, a beam of 3 keV electrons moves horizontally across the screen when both an **E**-field and a **B**-field are applied. The upper plate is at 3 kV and the lower plate is earthed. The plates are 5 cm apart.

a Calculate the strength of the **B**-field at the screen caused by the circular coils.

b Draw a diagram showing the relative directions of *E*, *B*, the motion of the electrons and the forces F_E and F_B on the beam caused by the **E**-field and **B**-field respectively.

29 A simple thermionic diode consists of a filament heated white hot and a metal plate placed in front of it in a vacuum. There is an electric current in the circuit only when the plate is positive with respect to the cathode (see Figure 35.23).

a Someone suggests that the light from the filament causes the plate to emit positives charges, which are attracted to the filament, resulting in the current. How would you argue against this hypothesis?

b The curves of Figure 35.23b were plotted using the circuit in **a**. Explain the following features of the curves:

  i there is a current for a negative pd between *A* and *B*
  ii there is a rising region from *B* to *C*,
  iii there is a flat region *DE* or *FG*, depending on the power of the filament supply.

30 The graph in Figure 35.24 shows the time taken for oil drops of different weight to fall 1 mm through air in a Millikan oil drop experiment. Each oil drop is then balanced by applying a suitable voltage between the plates, which are 5.01 mm apart. The results of the timing and balancing experiments are:

time of fall/s	4.39	5.43	6.34	7.50	9.01	11.04	13.83
balancing pd/V	677	813	275	739	278	811	191

Calculate the charge on each drop and show that it agrees with the concept of a basic charge *e*.

31 A portable television set is lined up with the electron gun to centre of screen direction along a north–south line. The set is then rotated horizontally until it is east–west.

a Would you expect the picture to shift significantly, either to one side or up or down?

b Is the picture likely to be distorted near the corners?

c Use the following data to decide how far the central spot on the screen might shift:

The electrons of charge *e* and mass m_e travel about 30 cm between cathode and screen at a speed of about 5×10^7 m s^{-1}. The vertical component of the Earth's **B**-field is 5×10^{-5} T. (*e*, m_e)

You could try this experiment out for yourself when the testcard is being displayed. Use a non-permanent felt tip pen to make a reference mark on the screen before you rotate the set.

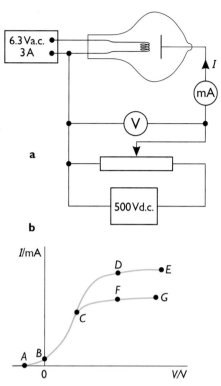

Figure 35.23a Circuit for a thermionic diode, **b** curves plotted using the circuit

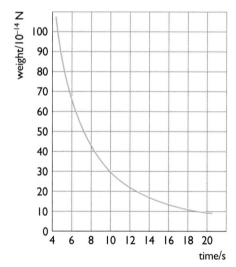

Figure 35.24 Graph of weight against time for oil drops in Millikan's experiment

36 QUANTISATION OF RADIATION

Figure 36.1 The German physicist, Max Planck

Towards the end of the nineteenth century, it was widely believed that electromagnetic radiation, such as light, infrared and radio signals, was wavelike in nature. It exhibited all of the properties of waves, including refraction, diffraction and interference (← 30). But there were a number of observed phenomena, such as the distribution of energy radiated from a hot solid, that could not be explained in terms of the wave model.

In 1900, Max Planck (see Figure 36.1) explained the distribution of energy in radiation from a hot solid by assuming that energy changes occur in quanta of energy. This led to a **quantum model** involving quanta of radiation, or **photons**. In 1905, Albert Einstein extended Planck's ideas to develop a successful model for the photoelectric effect (→ 36.3).

How electromagnetic radiation can have a dual nature, behaving both as wave and particle, is one of the great mysteries in human knowledge. Its theoretical significance is matched only by its implications for technology.

36.1 THERMAL RADIATION

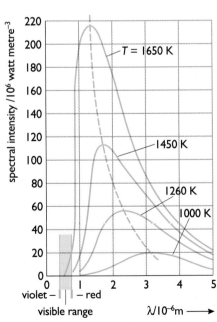

Figure 36.2 Variation of spectral intensity E_λ with wavelength λ

This electromagnetic radiation is emitted by hot bodies and can have a wide range of frequencies. At room temperature though, most thermal radiation is in the infrared region of the electromagnetic spectrum. This is because the vibrational frequency of atoms and molecules at 300 K is about 10^{14} Hz, a frequency in the infrared region of the electromagnetic spectrum (← 32.2).

All solids absorb electromagnetic radiation as well as emitting it. Solids heated to a very high temperature become **incandescent** and emit visible light as well as infrared radiation. The nature of the incandescent light is determined almost entirely by the temperature of the emitter and not by the material of which the solid is made. The proportion of the total radiation emitted by the solid as visible light is very small, as can be seen in Figure 36.2. The amount of radiation emitted or absorbed does, however, depend on the kind of surface. For example, a shiny silver surface is much poorer, both as absorber and emitter, than a dull black surface.

The radiation falling on any surface is absorbed, reflected and transmitted (see Figure 36.3). The relative distribution of the incident energy between these three destinations depends on the wavelength of the radiation as well as the nature of the surface. For example, glass transmits visible light but strongly reflects and absorbs longer-wavelength infrared. Greenhouses make use of this property. Visible light and the near infrared from the Sun penetrate the glass and are absorbed by the plants. But the thermal radiation emitted by the warmer plants is of much longer wavelength which cannot escape from the greenhouse, so

H INSIDE THE ATOM

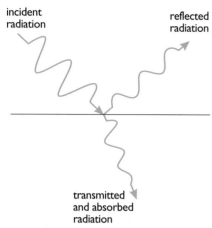

Figure 36.3 Incident energy = reflected + refracted + absorbed energy

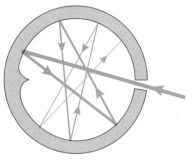

Figure 36.4 Multiple reflections cause the cavity to be a black body absorber

raising its temperature. A similar **greenhouse effect** is caused by the Earth's atmosphere. The atmosphere also has "windows" only allowing radiation of certain wavelengths from space to penetrate to the Earth's surface and absorbing a proportion of the radiation emitted by the Earth's surface (← 32.2).

BLACK BODIES

An object that absorbs all the radiation, at all wavelengths, incident on it is called a **black body**. This ideal body is also a perfect radiator, emitting **black-body radiation**. The concept is analogous to the kinetic model for an ideal gas (← 10). The laws of radiation are invariably stated in terms of a black body. This can be almost realised in practice, just as a real gas can be made to behave nearly ideally under certain conditions.

A black-body absorber can be created by making a small hole in the side of a closed tin can. Even when the inside is shiny, the radiation entering has to make many reflections before any returns to the hole, by which time nearly all of it has been absorbed (see Figure 36.4). So the hole appears black to the observer.

A black-body emitter can be made using an enclosure or cavity at a constant temperature. The radiation escapes through a small hole in the wall. Imagine looking into the embers of a hot coal or log fire through a small hole in the cooler outer material. The outline of the coals deep inside the fire is virtually invisible when they are all at the same temperature. They are then in equilibrium with each other, emitting and absorbing the same range of wavelengths and intensities. The range is characteristic of the temperature. For example, when the temperature of an incandescent solid, such as the tungsten filament of a light bulb, is raised, the filament glows first red-hot, then yellow-hot and finally white hot.

Lummer and Pringsheim measured the intensity of emitted energy with wavelength radiated from a black body at different temperatures. The apparatus of their experiment is shown schematically in Figure 36.5. Experiment 36.1 illustrates the principle of their experiment, but the eye is used as the radiation detector and a grating replaces the infrared spectrometer.

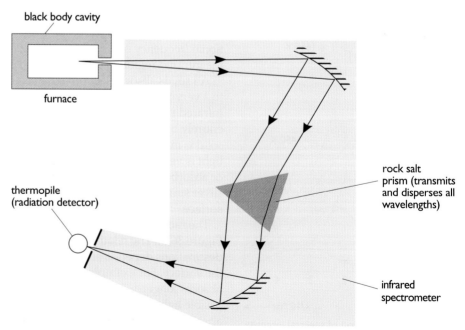

Figure 36.5 Lummer and Pringsheim's apparatus

36 QUANTISATION OF RADIATION

EXPERIMENT 36.1 Colour balance of a lamp filament

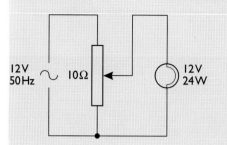

Figure 36.6 Circuit for experiment 36.1

After setting up the circuit of Figure 36.6, increase the lamp voltage from zero to 12 V to observe the radiation. As the filament heats up, you will see a variation in intensity of different parts of the visible spectrum by holding a 300 lines mm^{-1} diffraction grating in front of your eye. Figure 36.7 shows the variation. Note that the filament's temperature is about 2 000 K at full brightness.

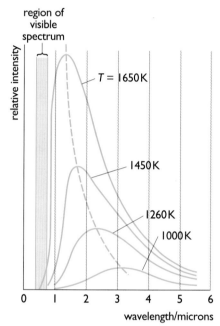

Figure 36.7 Results of Lummer and Pringsheim's experiments: graphs of intensity of radiated energy against wavelength from a black body

In Lummer and Pringsheim's experiment, the black body cavity was maintained at a chosen temperature. The prism formed a spectrum of the radiation from the cavity. The energy contained in each small band of wavelengths was detected and measured by a thermopile, enabling the distribution curves of energy against wavelength to be plotted (see Figure 36.7).

Questions

1 Explain how, in the above experiment, the width of slit in the grating and the dispersion of the prism determine the resolution (← 31.3) of the detector.

2 Colour photographs taken indoors, under normal room lighting conditions, often have a rather yellowish appearance.
a Why is this?
b What two main ways are there of overcoming this problem?

RADIATION LAWS

The experimental distribution curves in Figure 36.7 agree well with two empirical laws. These can also be justified theoretically using thermodynamics. The first is the **Stefan-Boltzmann law**, established in 1879. It states that the total power P emitted by radiation over all wavelengths from a black body is proportional to the fourth power of the thermodynamic temperature T:

$$P = A\sigma T^4,$$

where A is the surface area of the body and σ is the **Stefan-Boltzmann constant**, which has the value 5.67×10^{-8} W m^{-2} K^{-4}.

For a non-black or **grey** body:

$$P = \epsilon A\sigma T^4,$$

where ϵ is the **total emissivity**. This is the proportion of power emitted by the grey body compared to an otherwise identical, black body. Its value lies between 0 and 1.

The second law is known as **Wien's displacement law**. In 1894, Wien showed theoretically that at each temperature T, the peak of the distribution curve occurs at a wavelength λ_{max} such that:

$$\lambda_{max} T = \text{constant}.$$

The value of the constant is 2.9×10^{-3} m K. This law describes how the colour of the visible light emitted by an incandescent body changes with temperature.

H INSIDE THE ATOM

Temperature equilibrium in black bodies

When a black body is placed in an enclosure at a temperature of T_{encl}, it will absorb power P_{absorb} according to the relationship:

$$P_{absorb} = A\sigma T^4_{encl}.$$

However, it will continue to emit power due to its own temperature T. So, the net power output, P_{net} is given by:

$$P_{net} = P - P_{absorb}$$
$$= A\sigma(T^4 - T^4_{encl}).$$

This assumes that the surroundings act as a black body. When the surroundings are hotter than the body, the rate of absorption will be greater than the rate of emission. So the body will rise in temperature until a dynamic equilibrium exists between emission and absorption, at $T = T_{encl}$. The reverse effect will occur when the body is hotter than its surroundings. It will cool to $T = T_{encl}$, assuming the surroundings experience a negligible change in temperature.

However, theory based on the Newtonian model of mechanics cannot explain the detailed shape of the intensity-wavelength graphs. The model predicts the shape of curve labelled as $h = 0$ in Figure 36.8, which is considerably different from the observed distribution.

> **Questions**
>
> 3 Estimate the temperature of the tungsten filament of a 60 W electric light bulb. Take the emissivity to be 0.8, the filament length 0.5 m and the radius 3×10^{-5} m. The filament surface area = $2\pi rl$.
>
> 4 The wavelength at which maximum energy is received from the Sun is 490 nm. Estimate the black body temperature of the luminous surface, or photosphere, of the Sun.

PHOTONS AND QUANTA

In 1900, Max Planck produced a mathematical model resulting in an equation that describes the shape of the observed distributions of wavelength exactly. He suggested that energy is radiated in tiny lumps or quanta called photons, rather than as a continuous wave. Each quantum is associated with radiation of a single frequency. The energy of each quantum is proportional to its frequency f, and:

$$E = hf$$

where h is the Planck constant. Since the velocity of electromagnetic waves c is equal to the product of frequency and wavelength ($\leftarrow$ 29.2):

$$f = c/\lambda$$

and so $E = hc/\lambda$.

Planck calculated the value of h to be 6.63×10^{-34} J s. This value makes a theoretical curve fit exactly with an experimental one (see Figure 36.8). He reached this value by considering the number of short-wavelength, high-energy blue photons observed. As you can see, if the value of h is too small, there have to be more blue photons than observed to make up the total energy. So the curve is shifted to the left relative to the observed curve. Similarly, if the h chosen is too large, there will not be enough energy available for many, if any, blue photons, shifting the curve to the right.

> **Questions**
>
> 5 Show that the SI unit of h is J s.
>
> 6 In arriving at his value of h, why did Planck only need to consider blue photons and not red ones?

36.2 THE PHOTON CONCEPT

Planck's equation $E = hf$ explains why some parts of the electromagnetic spectrum appear to behave more like particles while others behave more like waves.

36 QUANTISATION OF RADIATION

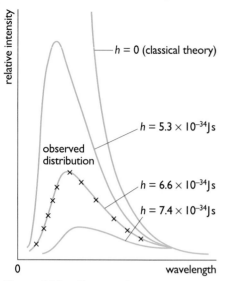

Figure 36.8 Choosing the value of h to fit the theoretical curve to the observed curve

For example, γ-rays are very high-frequency electromagnetic radiation with large photon energies of the order of 1 MeV. γ-ray quanta emitted by atoms must come from energy changes in the nucleus. The rate at which these energetic quanta can be produced from a tiny source is small, so single γ-ray quanta can easily be detected using a Geiger-Muller tube and an electronic counter (→ 38.8).

Radio waves are at the other end of the electromagnetic spectrum (see Figure 36.9). Millions of photons are needed to detect a signal because the energy of each photon is only about 10^{-29} J. So, the wave-like properties of radio waves predominate. The granular nature cannot be observed.

Figure 36.9 relates the quanta emitted in the different regions of the electromagnetic spectrum to what they can do. At the high-energy end, γ-radiation is detected as quanta with a Geiger-Muller tube and counter. At the low energy-end, the quanta are so close together in energy value that radio waves are detected as continuous radiation.

The idea that the energy emitted or absorbed by a system decreases or increases in quanta, or steps, may be extended to include any system. The total energy of any vibrating object, such as a simple pendulum or a mass oscillating on a spring, is quantised. However, the energy steps are far too small to be detected and so any lumpiness is invisible. Quantum effects are only apparent when observing atomic-sized objects, where h is a significant factor in any detectable energy change.

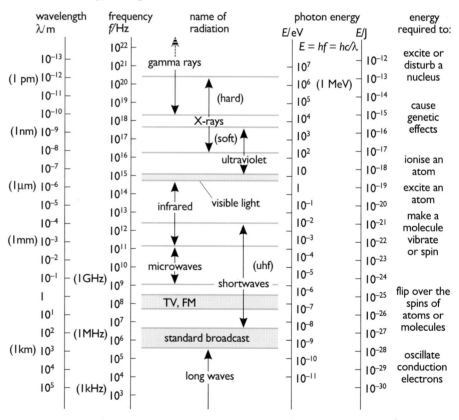

Figure 36.9 Electromagnetic spectrum and photon energies

Questions

7 How much energy is carried by
a a green light photon at $f = 6 \times 10^{14}$ Hz?
b a radio photon at $f = 200$ kHz?
c a cosmic ray photon at $f = 2 \times 10^{22}$ Hz?

INSIDE THE ATOM

8 Use the idea of photons to explain the change in the balance of colours in the lamp spectrum of experiment 36.1 as the lamp increases to full brightness.

9 Estimate:
a the number of photons emitted by a helium-neon laser per second. The output power of the laser is 1 mW and the wavelength of its radiation is 633 nm.
b the number of photons emitted by a radio transmitter per second. The average output power of the transmitter is 1 kW and the wavelength of the broadcast signal is 200 kHz.

10 Show that a photon of energy 9.4×10^{-25} J corresponds to a wavelength of 21 cm.

THE EYE AS A PHOTON DETECTOR

Figure 36.10 shows the visible region of the spectrum, where our eyes can be the detector. Helium-neon (He-Ne) lasers, emit an almost parallel beam of radiation at 633 nm. For a typical 1 mW laser more than 10^{15} photons pass a point in the narrow beam in any second – which is why safety precautions are essential when using a laser.

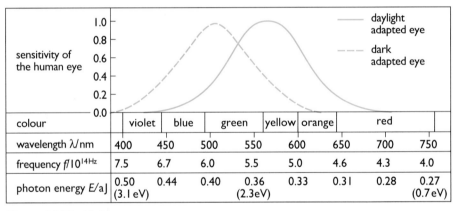

Figure 36.10 Visible spectrum

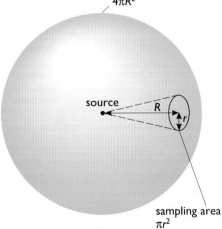

Figure 36.11 Sampling area receives $\pi r^2 / 4\pi R^2$ of the emitted radiation

A tungsten filament lamp has a temperature of about 2 000 K and only emits about 5 % of its total output as visible radiation (see Figures 36.7 and 37.14). In 1 s, a 100 W lamp emits about 2×10^{17} visible photons in all directions. We can consider the photons to be spread out over the surface of a sphere of radius R equal to the distance between source and observer. The observer's eye collects photons over an area πr^2. A fraction $\pi r^2 / 4\pi R^2$ of the emitted photons will be detected (see Figure 36.11). The diameter of the pupil of the eye is about 2 mm, so when looking directly at the bulb from a distance of 1 m, 2×10^{11} photons enter the eye per second. The light is bright enough to leave an after image on the retina, yet it is still 5 000 times weaker than the laser beam.

This analysis uses an **inverse square law** (← 14.2) for the variation of intensity of radiation with distance from the source. The same relationship applies in many other situations, such as the emissions from the Sun's photosphere. Gamma-rays also obey the inverse square law (→ 38.5).

36 QUANTISATION OF RADIATION

Sensitivity of the eye

The eye can just detect light when it absorbs energy at a rate of 10^{-17} W. This corresponds to a rate of 25 photons s^{-1}. When the eye is adapted for maximum sensitivity, it absorbs only 1 in 15 photons incident on it. This means that about 400 photons must enter the eye per second for light to be detected.

At maximum sensitivity, the pupil is wide open, about 6–8 mm in diameter. Then, it is possible for the human eye to detect granularity of the light. The objects in a nearly darkened room look speckly with no sharp outlines. This is because insufficient photons are reaching each point on the retina in the time span required for the brain to detect a single image, that is, about 0.1 s. So no clear picture can be built up. Similarly, as Figure 36.12 shows when the light reaching film emulsion is very weak, the picture remains an apparently random array of spots. This can be fully analysed to good effect ($\rightarrow$ 36.5).

The sensitivity of the eye is very similar to that of the ear ($\leftarrow$ 29.5). The eye absorbing energy at the rate of 10^{-17} W over a pupil area of around 10^{-5} m^2 corresponds to a sound intensity of 10^{-12} W m^2. This is close to the minimum sound intensity that can be detected by the ear.

Questions

11 The eye can just detect light when it absorbs energy at a rate of 10^{-17} W. Show that this corresponds to a rate of 25 photons s^{-1}, given that the average wavelength of light is 500 nm, which corresponds to a frequency of 6×10^{14} Hz.

12 Use the inverse square law to find the minimum visible output power of a light source that can be detected at 2 m by the human eye.

13a Calculate the time between the arrival of successive photons at the eye when, on average, only 400 arrive per second. (c)
b How far apart in distance are they?
c Show that, in question **12**, only one photon is likely to be in flight at any instant.

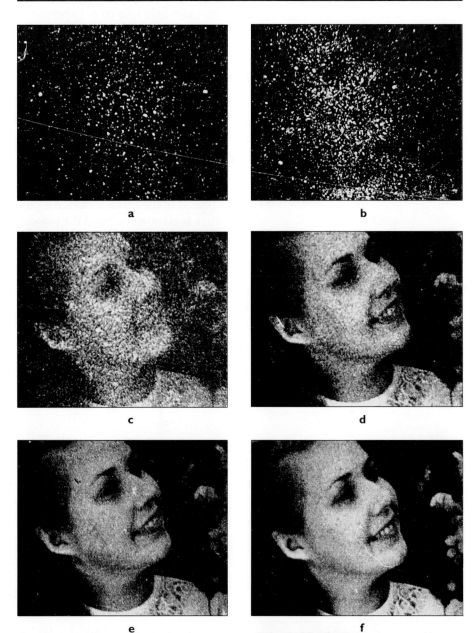

Figure 36.12 Series of photographs showing the quality of the picture obtained from increasing numbers of incident photons

H INSIDE THE ATOM

Individual visible light photons can be detected and counted using a very dim light source and a photomultipler (← 35.5) and counter. The photons reach the cathode of the photomultipler as a stream of single particles. They are detected individually by photoelectric emission and then multiplied to produce a tiny pulse of current, sufficient to be detected electronically.

In 1905, Einstein explained the photoemission of electrons from metals – the **photoelectric effect** – in terms of quanta of energy and so provided major support to the concept of the photon.

36.3 PHOTOELECTRIC EFFECT

The photoelectric effect was discovered by Hertz in 1887. He found that electrons, known as **photoelectrons**, are emitted from a clean metal surface when ultraviolet light falls on it. Experiment 36.2 suggests two simple ways of demonstrating the effect. The first was performed in 1888.

EXPERIMENT 36.2 Demonstrating the photoelectric effect

a A gold-leaf electroscope can be charged by induction (← 16.1) or using an EHT power supply, as in Figure 36.13a. When it is negatively charged, and you shine ultraviolet light from a mercury lamp at a cleaned zinc plate mounted on the cap, the leaf collapses as the charge leaks away. However, a positively charged one does not discharge in this way.

b In Figure 36.13b, the end of a 10 cm length of cleaned magnesium ribbon is placed in the input socket of a picoam meter, and a cylinder of fine copper gauze is clamped around the ribbon but not touching it. When a 12 V dc supply is connected and an ultraviolet lamp is placed close to the gauze, the meter will indicate a small current. You can see how the current changes when the polarity is reversed and what happens as the lamp is moved further away from the ribbon.

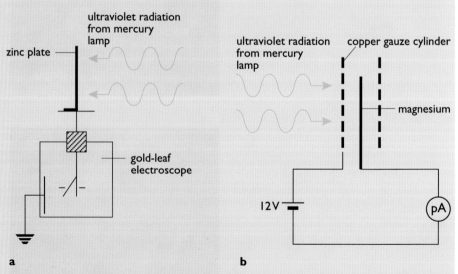

Figure 36.13 Apparatus for demonstrating the photoelectric effect: **a** with a gold-leaf electroscope, **b** with a 'home-made' photo-emissive cell

36 QUANTISATION OF RADIATION

Question

14 How do experiments 36.2a and b indicate the emission of negative charge?

In 1899, Philipp Lenard studied the effect in detail. He found three important results which could not be explained by the wave model for light:

i photoelectrons are emitted with a range of energies up to a maximum value. The value of this maximum energy depends on the particular metal surface illuminated and the frequency of the light
ii there is a minimum frequency below which no electrons are emitted, however intense the light. This **threshold frequency** f_0 varies from metal to metal, too
iii photoelectrons are emitted instantaneously, the intensity of light determining only their number.

Using the wave model, for example, the light's energy would be expected to be distributed uniformly over the metal surface. Calculations show that with a very weak light source it takes many seconds for enough energy to reach an 'electron size' area of the plate and cause emission. Yet emission occurs effectively instantaneously, less than a microsecond after the light is switched on.

PHOTOELECTRIC EQUATION

Einstein explained Lenard's findings by suggesting that the energy within electromagnetic radiation was both generated and absorbed in quanta, known as photons. A photon of the incident radiation could be absorbed by a single electron in the metal surface. The electron would then be emitted with a particular kinetic energy.

An electron needs a certain minimum energy, the **work function** ϕ, to escape from the metal ($\leftarrow$ 35.5). So:

energy of incident photon ≥ work function + its kinetic energy when emitted.

In addition, some energy is absorbed and dissipated in interactions with electric fields and other particles, so:

energy of incident photon = work function ϕ + kinetic energy when emitted + energy lost.

The kinetic energy of the emitted electrons is E_K. The energy absorbed, E_{lost}, will be different for each photoelectron and is given by:

energy of photon, $hf = \phi + E_K + E_{lost}$.

Suppose we consider only those electrons that receive **all** the energy of the incident quanta. These have the maximum possible kinetic energy, $(E_K)_{max}$, and:

$$hf = \phi + (E_K)_{max}$$

or $\quad hf = \phi + \tfrac{1}{2} m_e v_{max}^2$.

This is known as Einstein's **photoelectric equation**. The work function, ϕ is measured in joules but may be quoted in electron volts (or just volts, when it is clear that electrons are being considered).

Einstein's equation can be represented in one-dimensional energy diagrams,

INSIDE THE ATOM

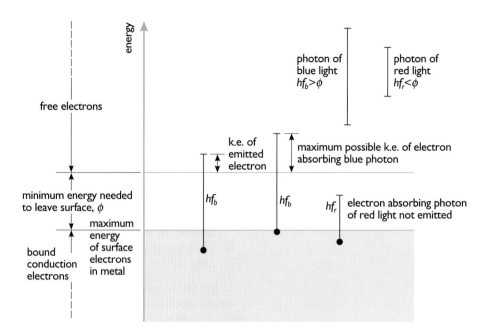

Figure 36.14 One-dimensional energy diagram for the photoelectric effect

such as Figure 36.14. The incident photon energy is shown by a vertical line proportional to its frequency. When an electron in the metal absorbs a photon it jumps vertically an energy equal to that of the photon. When it reaches the free electron energy region, it escapes from the metal. Unless $hf \geq \phi$, no electrons can be emitted.

There is a frequency at which $(E_k)_{max} = 0$ and only the most energetic electrons become free but have zero velocity. This is the threshold frequency, f_0, and at this:

$$hf_0 = \phi$$

$$f_0 = \phi/h.$$

Incorporating this in Einstein's photoelectric equation gives:

$$hf = hf_0 + \tfrac{1}{2}m_e v_{max}^2$$

$$\tfrac{1}{2}m_e v_{max}^2 = h(f - f_0).$$

So, the higher the photon frequency, the greater is the maximum kinetic energy of the photoelectron. For example, some metal surfaces give no photoelectrons with incident red light, very low kinetic energy emission with green light and higher energies with blue light.

Question

15 The work functions of tungsten, sodium and caesium on tungsten are 4.49 eV, 2.28 eV and 1.36 eV, respectively. For each, calculate:
a the threshold frequency,
b the maximum wavelength of the incident radiation that will just cause photoelectron emission.

DETERMINING THE PLANCK CONSTANT

When light of a chosen frequency is incident on the cathode of an isolated photocell, the cathode loses photoelectrons and becomes positively charged with respect to the anode (see Figure 36.15a). The photoelectrons collect at the anode.

The voltage builds up between the electrodes until it reaches a value V_s, the **stopping voltage**, at which even the most energetic photoelectrons just lose all their kinetic energy before reaching the anode. At this value:

$$eV_s = \tfrac{1}{2}m_e v_{max}^2.$$

Under these conditions, Einstein's equation:

$$hf = \phi + \tfrac{1}{2}m_e v_{max}^2$$

becomes $hf = \phi + eV_s$.

Rearranging gives

$$V_s = hf/e - \phi/e.$$

The graph of Figure 36.15b shows the variation of V_s with f. The gradient of the graph is h/e. Knowing e, you can obtain a value of the Planck constant h. The threshold frequency f_0 of the cell is given by the intercept on the x-axis. Using this value and the gradient, you can calculate the work function of the photocathode material hf_0.

Experiment 36.3 allows you to find these values. It also confirms Einstein's photoelectric equation, both by the straight-line graph and by yielding values of h and ϕ that are in agreement with those found by other methods.

EXPERIMENT 36.3 Finding a value for the Planck constant, h

Warning: do not expose the sensitive photocell surface to bright light.

Using the arrangement in Figure 36.15a, illuminate the photocathode of a potassium photocell with light of varying wavelength λ between 630 nm and 380 nm from a diffraction grating. Note the **stopping voltage**, V_s, displayed on a high impedance voltmeter connected across the electrodes of the photocell each time. Plotting a graph of V_s against frequency f, as in Figure 36.15b, enables you to find a value of the Planck constant, and the threshold wavelength and frequency for a potassium surface.

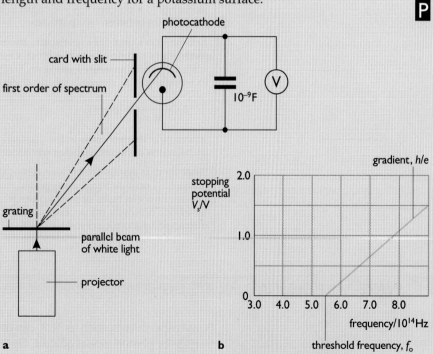

Figure 36.15a Arrangement for experiment 36.3, **b** graph of stopping potential V_s against frequency of incident radiation

H INSIDE THE ATOM

> **Question**
>
> **16a** Suppose experiment 36.3 is repeated with white light of lower intensity. For a given colour of light emitted by the source, what effect will this have on:
>   **i** the energy possessed by each emitted photon?
>   **ii** the number of photons emitted per second?
>   **iii** the maximum kinetic energy of the photoelectrons?
>   **iv** the stopping potential V_s?
> **b** Find a value for the Planck constant h and the work function ϕ from Figure 36.15b.

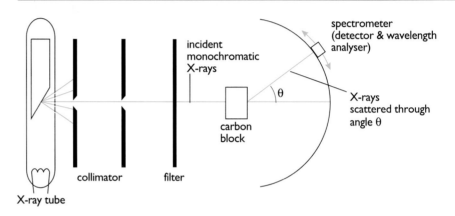

Figure 36.16 Compton's scattering experiment

Compton effect

When high-energy, short-wavelength X-rays or γ-rays, are scattered from loosely bound electrons of light elements, they change their wavelength. In 1923, the American physicist, Compton explained this effect by attributing a **momentum** h/λ as well as energy to a photon of wavelength λ. He considered the scattering to be a mechanical collision between two particles.

In Compton's original experiment, monochromatic X-rays were scattered by the electrons of carbon atoms in a block of graphite. Their longer wavelengths after scattering were measured using an X-ray spectrometer (see Figure 36.16). The change in wavelength $\Delta\lambda$ depends on the angle of scattering θ:

$$\Delta\lambda = \frac{h}{m_e c}(1 - \cos\theta),$$

where m_e is the rest mass of the electron. Experiments and models based on the idea of collisions between particles were found to agree for all angles up to 150°.

Since there is symmetry between energy transferred in light and matter, the French physicist, Louis de Broglie suggested that the same equation $p = h/\lambda$ may be used to explain how electrons of momentum p can have wave-like properties. Electrons, which we normally think of as particles, do exhibit wave-like behaviour when diffracted by atomic crystal lattices (→ 37.6).

36.4 WAVE-PARTICLE NATURE OF LIGHT

Some experiments, such as 36.3, can only be explained theoretically using a particle or photon model of light. Optical and X-ray spectra obtained from excited atoms (→ 37) can also only be explained in terms of the quantisation of

36 QUANTISATION OF RADIATION

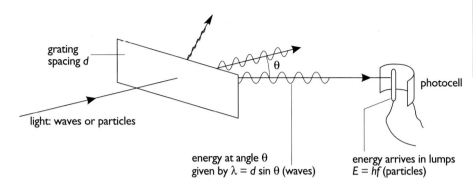

Figure 36.17 Experiment requiring both wave and particle models of light

energy. However, many observations concerning the nature of light require a wave model (← 31). Indeed, experiment 36.3 requires a wave model of light to explain the action of the diffraction grating and a particle model to explain the photoelectric effect (see Figure 36.17).

Light behaving as waves and particles emphasises that these descriptions are both merely **models**. We cannot see light on its journey between a source and a detector and can only observe the effect the light produces at the detector.

In 36.2, Figure 36.12, shows that when very little light is incident on a film, the photons causing the picture interact at random positions rather than in a regular pattern. As more photons arrive, there is a greater chance of them interacting in some places than in others. So the picture is built up.

Similarly, Young's double-slit experiment (← 33.1) can be performed using such a dim light source that only one photon at a time is passing in the slit region of the apparatus. A photomultiplier (← 35.5) is used to count the individual photons arriving at different parts of the interference pattern (see Figure 36.18). Many photons are detected where a bright band of light is expected and only a few where a dark band is expected, as predicted by the wave model. But without the slits, photons arrive randomly, spread uniformly in space as predicted by the particle, or quantum, model.

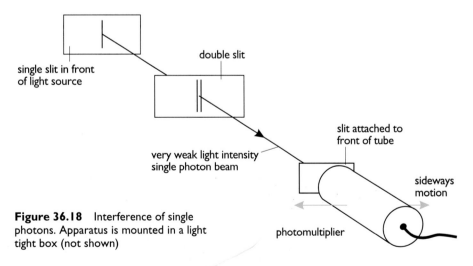

Figure 36.18 Interference of single photons. Apparatus is mounted in a light tight box (not shown)

RECONCILING THE MODELS

The presence of the slits changes the **probability** of where the photons arrive. Each photon is a quantum of energy so the intensity of light on a screen builds up gradually. When only a few have arrived, they give a grainy picture like Figure 36.12a with no recognisable band pattern. When enough photons have arrived, the intensity pattern, or picture, is the same as expected using the wave model.

H INSIDE THE ATOM

The rate at which energy is delivered by a wave motion to a point in space is proportional to the **square of the amplitude** of the wave there. The mechanisms in the wave and particle models for deciding the energy arriving per unit area per second at the detector must give the same result. For this to be so:

probability of a photon arriving at a point in the particle model $\propto$ square of the wave amplitude at that point in the wave model

The probability must be found by observing **large** numbers of photons. So the models can only agree when large numbers of photons are present.

Generally, while light is travelling between a source and detector, it is considered to interact with matter like a wave. At emission and detection, it is considered to act like a particle, especially at higher frequencies. At radio frequencies, it is usually satisfactory to use a wave model throughout ($\leftarrow$ 36.2).

36.5 OPTO-ELECTRONIC DEVICES

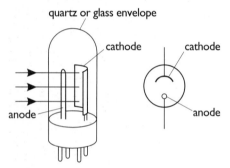

Figure 36.19 Simple photoemissive cell

There are three basic types of photodetector: photoemissive, photoconductive and photovoltaic.

The simple **photoemissive cell**, used in experiment 36.3, relies on the photoelectric effect ($\leftarrow$ 36.3). It is shown in Figure 36.19. The anode is a thin rod of a photoelectrically inert metal. The cathode is coated with a metal that will emit photoelectrons, if radiation of certain wavelengths is incident on it. For example, sodium or potassium emit electrons for visible light and a layer of caesium on oxidised silver emits electrons for infrared light. Most clean metals emit electrons for incident ultraviolet radiation. A problem with the scientific use of the photocell is that the anode can become contaminated with the cathode material and also emit photoelectrons, leading to spurious results.

Uses for photoemissive cells

One commercial application of such cells is in automatic turnstiles in supermarkets. When a light beam is broken, the light reaching the photocell drops to zero, triggering the opening of the doors. The infrared type of cell is part of the circuitry for the optical soundtrack in film projectors. The rate of variation of intensity and the intensity of the light passing through the film are converted into an electrical signal giving the frequency and amplitude of the recorded sound, as shown in Figure 36.20.

A much more sensitive device than the photocell is the **photomultiplier** ($\leftarrow$ 35.5) which has an electron multiplier to amplify the effect of photoelectric emission. All photocells and photomultipliers pass a current even in complete darkness. This limits their sensitivity. The major contribution to the **dark current** comes from thermionic emission ($\leftarrow$ 35.5) of electrons from the cathode. Cooling the tube to liquid nitrogen temperatures (–196 °C or 77 K) greatly reduces this emission and enables the photomultiplier to detect single photons easily.

The **photoconductive** detector is commonly known as the **light dependent resistor**, or **LDR**. It consists of two sets of intersecting metal grids separated by a thin layer of a semiconductor, cadmium sulphide doped with copper. When light falls on the semiconductor, electrons in bound states absorb photons and become free to conduct. The higher the light intensity falling on the cell, the greater the number of electrons released by the incident photons to contribute to

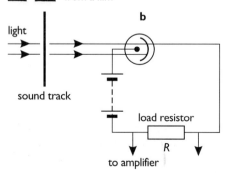

Figure 36.20a Sound track on a film **b** photocell detection circuit

36 QUANTISATION OF RADIATION

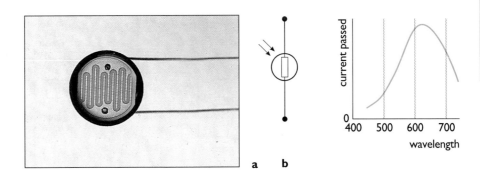

Figure 36.21 The photoconductive cell: **a** ORP 12, **b** circuit symbol, **c** typical spectral response curve

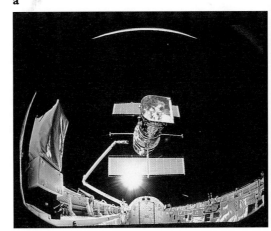

Figure 36.22a Space communications satellite with its solar cell panels deployed, **b** isolated telephone communications centre showing solar cells

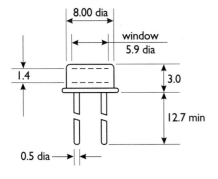

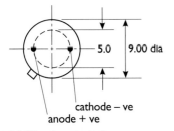

Figure 36.23 A typical silicon photodiode, BPW 21, suitable for use in light monitoring and control

conduction. This causes a greater current for a given voltage across the device: its resistance decreases. The LDR is most sensitive for red and near infrared radiation (see Figure 36.21).

One example of the **photovoltaic** detector is the **solar cell**. Large banks of such cells, arranged in series and parallel, power the electronics of space satellites and isolated telephone communications centres, for example (see Figure 36.22). The solar cell is made from a heavily doped n-type silicon slab with a thin coating of p-type material on its surface ($\leftarrow$ 27.6). Electrons diffuse from the n-type into the p-type and holes move in the opposite direction. This causes a depletion layer, until a sufficient reverse voltage is produced to cause equilibrium. Incident photons that penetrate the p-type coating to the junction liberate electron-hole pairs, one pair per photon. The electrons are pulled into the n-type and the holes into the p-type material by the reverse voltage. When the cell is connected to an external circuit, the electrons flow from the n-type to the p-type through the circuit.

In a **photodiode**, the depletion layer is increased by adding a reverse-biased battery. The current in the circuit depends on the number of photons reaching the depletion layer, as before. The holes and electrons liberated are quickly swept away by the voltage across the depletion layer, causing a current. A typical silicon photodiode is shown in Figure 36.23. This may produce 6 electrons for 10 incident photons, giving a **quantum efficiency** of about 60 %. A **light-emitting diode**, or **LED**, is basically the converse of the photodiode – a pn junction biased

INSIDE THE ATOM

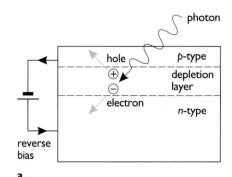
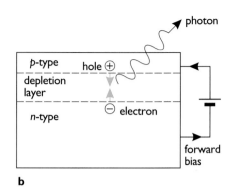

Figure 36.24a Photodiode, **b** LED

in the forward direction. When a hole and an electron meet in the depletion layer, they combine, releasing energy as a visible photon (see Figure 36.24b).

An LED is used in experiment 36.4 to estimate the Planck constant. This experiment assumes that, when the emitted light is very dim each electron in the depletion layer of the LED, which contributes to the current, causes the emission of one photon. The number n of electrons producing photons at the depletion layer per second is I/e, where e is the electronic charge. The power VI delivered in the LED appears in the emitted light. Since n photons are produced per second, each of energy hf:

$$VI = nhf$$

or $$VI = \frac{Ihf}{e}$$

or $$h = \frac{eV}{f}.$$

EXPERIMENT 36.4   Estimating the Planck constant using an LED

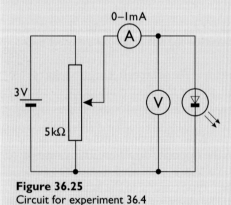

Figure 36.25 Circuit for experiment 36.4

Using the circuit of Figure 36.25 with a red LED mounted at the sealed end of a tube of black paper, adjust the potentiometer until the LED **just** glows and measure the voltage V. By measuring the wavelength λ of the LED's light using a diffraction grating (← 33.3), or estimating it to be 600 nm, you can calculate h from:

$$h = \frac{eV}{f} = \frac{eV\lambda}{c}$$

where c is the velocity of light.

A judgement has to be made as to when the light is just visible. Even so, agreement with the accepted value for h, 6.6×10^{-34} J s, should be remarkably good.

One of the most important opto-electronic devices is the laser (→ 37.4). Opto-electronic systems, such as fibre-optic communication links, use gallium arsenide (GaAs) in infrared LEDs or solid-state lasers and photodiodes as transmitters and receivers. These are rapidly becoming part of our modern communications networks (see Figure 36.26). A single optical fibre can carry 4 000 simultaneous telephone calls. The transmitting and receiving electronic circuits are switched rapidly enough to handle over 500 megabits of information, or pulses, per second. The continuously varying input signal voltage, such as speech, is pulse

36 QUANTISATION OF RADIATION

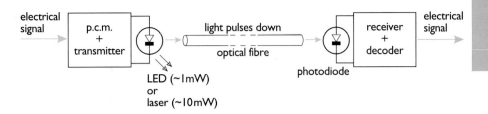

Figure 36.26 Typical optical link

code modulated (← 24), or converted into digital pulses. The pulses are sent down the optical cable and then decoded at the receiver back into a continuously varying signal.

SUMMARY

Planck introduced the concept of the photon to explain the distribution of energy with wavelength radiated from a black body.

The energy of a photon of electromagnetic radiation of frequency f is $E = hf$ where h is the Planck constant, 6.63×10^{-34} J s. Single photons from high-energy sources can be detected and counted as individual events.

Einstein used and extended Planck's idea to explain the photoelectric effect. His photoelectric energy equation is:

(energy of photon) = (work function of metal) + (maximum possible kinetic energy of emitted electron).

Interference and diffraction are phenomena associated with waves. We therefore need both wave and particle models of light to explain observed phenomena.

The wave and particle models only agree when the number of photons arriving at the detector per second is proportional to the square of the wave amplitude at the detector. This will only occur when large numbers of photons are present.

Opto-electronics is a major growth area in electronic communication within computer, telecommunication and sensor systems. GaAs and other Group III-V semiconducting diodes are used as transmitters and receivers.

SUMMARY QUESTIONS

(e, h, c, k and σ)

17 In experiment 36.1 the temperature of the filament of the light bulb was about 2 000 K. Use Wien's displacement law to estimate the wavelength of light at which the maximum energy was emitted. In what region of the spectrum is this?

18a Using Stefan's law and Wien's displacement law, add a sketch curve to Figure 36.7 for a temperature of 6 000 K, the surface temperature of the Sun.
b Estimate the fraction of the total radiation from the Sun that falls within the visible region.
c Explain and justify why the narrow window in the atmosphere at visible wavelengths transmits 83 % of the total energy received at the Earth's surface.

19 A swimmer stands on the edge of a pool. The swimmer's body has a surface area of 2 m², a temperature of 37° C and surface emissivity of 0.6. The temperature of the surroundings is 22° C.

INSIDE THE ATOM

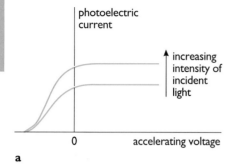

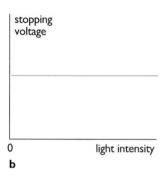

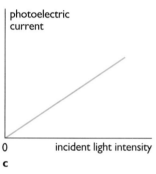

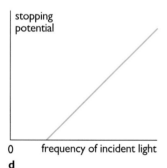

Figure 36.27

a Use the data to estimate the swimmer's net radiative loss of energy per second.
b What other factors affect the body's rate of cooling? Will these be more or less significant than the radiative loss?

20 The power received per square metre perpendicular to the Sun's rays at the top of the Earth's atmosphere is called the **solar constant**. This is the energy incident per second per unit area of the solar cell panels on the satellite in Figure 36.22a.
a The approximate temperature of the Sun's photosphere is 6 000 K. Calculate the power emitted per unit area by the Sun, assuming it to be a blackbody radiator.
b The Sun's radius is 7.0×10^8 m and its mean distance from the Earth is 1.5×10^{11} m. Find the value of the solar constant.
c Suppose the solar cells absorb 10 % of this energy. Estimate the power available per square metre of panel facing the Sun.

21 The graphs in Figure 36.27a–d are plotted from numerical data collected in experiments using a vacuum photocell (see Figure 36.19).
a Describe suitable experiments to produce this data. State clearly which variable quantities are kept constant in each experiment.
b Explain the shape of each graph.

22 A copper surface is illuminated by a mercury lamp, producing intense spectral lines at 253.7 nm, 365.0 nm, 435.8 nm and 546.0 nm. A voltage of 0.24 V is sufficient to stop the emission of photoelectrons from the surface.
a Calculate the threshold wavelength of copper.
b Which of the mercury spectral lines cause photoemission?
c Find the work function of copper.
d Suggest why the copper gauze of the simple photocell of experiment 36.2b is unlikely to emit photoelectrons.

23 The threshold wavelength for a certain metal is 550 nm. The surface is irradiated with blue light of wavelength 450 nm.
a Calculate the maximum kinetic energy of the emitted photoelectrons.
b Calculate the reverse voltage required to stop emission.
c Which of the above quantities will change when:
 i the intensity of the incident light is doubled
 ii the wavelength is doubled?

24 The work function of tungsten is 4.49 eV, or 7.19×10^{-19} J.
a Calculate the maximum wavelength of radiation incident on a tungsten surface to cause photoemission.
b The average energy of a "free" electron in a metal surface is of the order of kT where k is the Boltzmann constant and T is the absolute temperature. Use this information to estimate whether much thermionic emission can be expected from an incandescent tungsten filament at 3 000 K.
c The surface is coated with caesium to reduce the work function to 1.36 V. Is this a suitable material to use for the cathode of an electron gun?

25 Millikan's results from his photoelectric experiments, which were similar to experiment 36.3, with adjustment for the contact voltage between the sodium photocathode and the collecting anode, were:

36 QUANTISATION OF RADIATION

stopping voltage, V/V	0.45	1.02	1.20	1.58	2.12	3.02
frequency of radiation, $f/10^{14}$ Hz	5.49	6.91	7.41	8.23	9.61	11.83

a Plot a suitable graph to find a value for the Planck constant, h.
b Use the graph to find:
 i the threshold frequency of sodium, and
 ii its work function.

26 In 1912, Elster and Geitel observed the almost instantaneous emission of electrons from a metal surface irradiated with light of intensity less than 10^{-10} W m^{-2}.

Using a wave model, the light's energy will be absorbed uniformly over the surface as a steady flow. Estimate how long the experimenters would have had to wait for an atom in the metal surface to absorb sufficient energy to emit an electron, according to this model. Take the work function of the metal surface to be 3×10^{-19} J and the radius of an atom in the surface to be 10^{-10} m.

37 QUANTISATION OF MATTER: ATOMS

If you walk through any shopping centre at night, you will see a multitude of advertising signs. Many contain discharge lamps, emitting a wide variety of colours. The spectra of many gases were investigated during the nine-teenth century to try to find the origin of such light. Early in the twentieth century, it became clear that it is due to energy level changes of electrons in the atoms of the gas concerned.

In 1911, Ernest Rutherford proposed an **atomic model** ($\rightarrow$ 40.2). In it, a tiny positively charged nucleus is surrounded by an appropriate number of negatively charged electrons to create a neutral atom. Almost all of the mass of the atom, in this model, is contained within the nucleus. The electrons are imagined not as particles but as variously shaped clouds of negative charge surrounding the nucleus. These clouds are known as **orbitals**. This atomic model and quantum ideas on energy levels ($\leftarrow$ 36.1) can be developed to explain the spectra of atoms.

The wave and particle models of light can be thought of as two different aspects or views of its wavelength and energy. Such ideas also describe and explain the observed behaviour of the electron in its free state and when bound within the atom.

37.1 EXCITATION AND IONISATION

In the gas discharge tube ($\leftarrow$ 35.3), electrons bombard a gas at low pressure, causing the gas to glow. The electrons are accelerated through only 10–20 V before collisions cause ionisation of the gas. Some of the kinetic energy of the bombarding electron is transferred to an outer bound electron of the gas atom, enabling it to break free of the electric force holding it to the positive nucleus. At a fairly well-defined voltage between the electrodes, called the **ionisation voltage**, the current in the circuit begins to rise sharply due to the increase of free charges in the gas.

The minimum energy to free an electron, the **ionisation energy**, is different for each element. The energy values help to position the elements in the correct order in the periodic table. As Figure 37.1 shows, the inert gases have the highest ionisation energies. This is because they have the most strongly bound electrons. The alkali metals, like lithium and sodium, on the other hand, have a weakly bound single valence electron, giving them a low ionisation energy.

Questions

1 How are ionisation voltage and energy related?

37 QUANTISATION OF MATTER: ATOMS

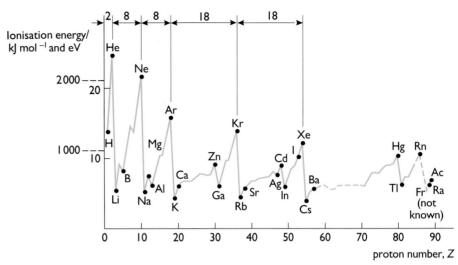

Figure 37.1 Ionisation energies of the elements

2a For a gas molecule at standard temperature and pressure, 273 K and 1 atmosphere, the average kinetic energy is equal to $\frac{3}{2}kT$ ($\leftarrow$ F). What is this energy in eV?
b Using your answer to **a**, estimate at what temperature the gas must be to ionise an oxygen molecule. The ionisation energy of oxygen is 13.6 eV.

Franck-Hertz experiments

The electron-atom collision experiments of Franck and Hertz in 1914 established the existence of energy levels within the atom. Electrons from a heated cathode C were accelerated through a voltage V towards the anode A (see Figure 37.2). The grid G was maintained 0.5 V higher than the anode. The tube contained a small quantity of mercury vapour at a pressure of about 15 mm of mercury, one-fiftieth of normal atmospheric pressure.

At low voltages, the electrons made elastic collisions with the gas atoms, passed through G and were collected at A. The anode current indicated the number arriving at A. As the voltage increased so did the anode current, until the electrons had sufficient energy to make inelastic, or excitation, collisions with the mercury atoms. At this stage, the current dropped because many of the colliding electrons had insufficient energy to reach A and were attracted back to G. A further increase in accelerating voltage caused the current to increase until the electrons had sufficient energy to make second excitation collisions. The voltage between successive current peaks was found to be constant at 4.9 V, as shown in Figure 37.3. This is an increase in energy of 4.9 eV which agrees remarkably well with the quantum energy of the strong mercury spectral line at 253.7 nm in the ultra-violet.

At very low collision energies, the free electrons bounce off the atoms **elastically** ($\leftarrow$ 9.3). The inter-molecular collisions between gas molecules, such as air, at room temperature are also perfectly elastic ($\leftarrow$ 10.3). The random translational kinetic energy of the molecules is just redistributed between them and suffers no net total change. The temperature of the gas must be raised to a temperature approaching 100 000 K before the gas molecules have sufficient kinetic energy to ionise each other.

Some **inelastic** collisions ($\leftarrow$ 9.3) can occur between a free bombarding electron and an atom at energies less than the ionisation energy. Electrons within atoms exist at one of a number of discrete **energy levels** ($\rightarrow$ 37.3). Each atom may absorb some or all of an incoming electron's kinetic energy causing an electron within the atom to be raised to a higher level of potential energy. This process is called **excitation**.

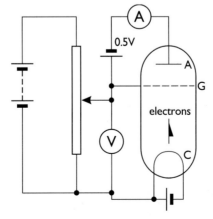

Figure 37.2 Franck and Hertz apparatus

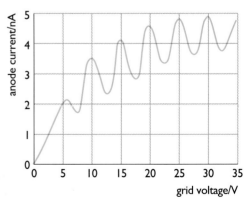

Figure 37.3 Excitation voltages for mercury vapour

H INSIDE THE ATOM

> **Questions**
>
> **3** What would be the effect on Figure 37.3 if the reverse voltage between G and A was **a** reduced to 0.1 V? **b** increased to 1 V?
>
> **4** Show that a photon of energy 4.9 eV has an associated wavelength of 253 nm.

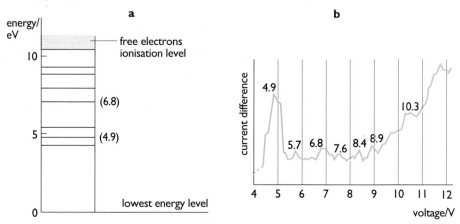

Figure 37.4a Electron energy levels and **b** excitation voltages for mercury vapour (Morris, 1928)

The results of the Franck-Hertz and other experiments on excitation voltages show that the mercury atom can only change its energy by **discrete** amounts with no allowed values in between these. From the excitation voltages, we can build up an energy level diagram for the mercury atom. A similar method can be used for any other gas atoms. Figure 37.4a shows the energy level diagram for mercury constructed from the experimental data in b. These energy levels relate to the line spectrum (→ 37.2) emitted by mercury vapour.

37.2 EMISSION AND ABSORPTION SPECTRA

The electromagnetic radiation from any light source can be split up into various component wavelengths using a **spectrometer** or **spectroscope**. These instruments use a prism or diffraction grating (← 33.3) to disperse the light.

EMISSION SPECTRA

There are three main types of emission spectra, continuous, line and band spectra.

A **continuous** spectrum is produced, for example, by a hot incandescent solid such as the white hot tungsten filament of a lamp (← 36.1). All wavelengths are present and the relative intensity of the radiation depends on the temperature of the solid. At room temperature, most of the energy transitions are small, resulting in mainly infrared photon emission. Temperatures of 2 000 K to 3 000 K are needed to observe a significant amount of visible radiation.

Light radiated from excited, isolated atoms produces a **line** spectrum. It is a series of sharp coloured lines against a dark background. Figure 37.5 shows an example. Some gases, such as air or carbon dioxide, produce more complex **band**

37 QUANTISATION OF MATTER: ATOMS

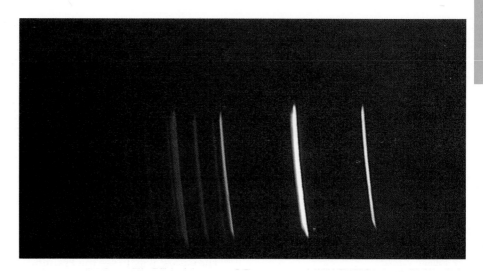

Figure 37.5 Line spectrum from a mercury lamp

spectra. In their molecules, the energy transitions of the electrons in each atom are modulated by the rotation and vibration of the atoms in the molecule. This causes a spread of possible energies in the emitted radiation, changing the lines into bands.

EXPERIMENT 37.1 Emission spectra

Using a diffraction grating of, say, 300 lines per mm you can observe the light from a narrow slit illuminated by discharge lamps containing hydrogen, sodium, mercury and sodium.

In experiment 37.1, some band spectra may be seen from a hydrogen discharge where molecular hydrogen is present. A domestic fluorescent tube will give the same spectrum as the mercury lamp, but not so clearly. In sodium light, the yellow line can be resolved as two lines when observing it in the higher orders through a diffraction grating ($\leftarrow$ 33.3). This is the case for many spectral lines which give the appearance of being single, but their resolution into **doublets** is not always as easy. The double line arises as a result of the **spin** of the electron ($\rightarrow$ 37.3).

The line spectrum produced by each gas is different. This provides a method of finding which elements are present in a light source. A **spectroscopic analysis** of starlight, for example, reveals which elements are present on a distant star. In spectroscopic analysis of any unknown substance, a small quantity is vaporised and its spectrum recorded. The spectrum is then matched against known substances. Traces of chemicals as small as 10^{-13} kg can be detected by this method.

Question

5 Referring to question **2**, how does the spectra of ionised gases, observed from a star, help us to find the star's temperature?

ABSORPTION SPECTRA

When a white light source is placed behind a sodium flame and observed through a diffraction grating, the continuous spectrum is crossed by faint dark

INSIDE THE ATOM

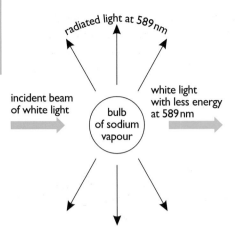

Figure 37.6 Absorption and re-emission by sodium vapour

lines. These are particularly noticeable at the wavelengths where the two bright yellow lines in the sodium spectrum occur, 589.0 nm and 589.6 nm. The energy of each quantum of light at these wavelengths is equal to the difference between two energy levels within the sodium atom. Such quanta therefore easily excite the sodium atoms. The atoms re-radiate the energy at the same wavelengths but in all directions. The intensity in the original direction is reduced, producing dark lines on the bright background (see Figure 37.6).

Absorption spectra are observed whenever light with a continuous spectrum passes through a gas. Fraunhofer first noticed this effect in 1814 when observing the Sun's spectrum. Most of the light received from the Sun is emitted from the **chromosphere**. Certain wavelengths are selectively absorbed by the relatively cool gases surrounding the chromosphere. The wavelengths of these absorption lines enable the gases present in the chromosphere to be identified. Fraunhofer detected hydrogen, sodium and an unknown gas. This became known as helium, after the Greek word for the Sun, *helios*.

The easiest way to see the **Fraunhofer lines** in the Sun's absorption spectrum is to look at sunlight through a diffraction grating.

Warning: never look at the Sun, either directly or indirectly, through any instrument fitted with a telescope or lenses. A safe way to observe the spectrum of sunlight is to reflect the light strongly from a shiny steel pin or needle placed in the path of the sunlight. You can then view the reflected light through a spectroscope, grating or prism.

In experiment 37.2, a band, rather than a line, absorption spectrum is obtained. Light is absorbed by iodine vapour and causes increased molecular vibration of the iodine. The resultant dark bands are equally spaced, suggesting that the vibrational energy levels are also equally spaced.

EXPERIMENT 37.2 Observing absorption spectra

You can view the band spectrum caused by the vibrations of iodine molecules using the arrangement in Figure 37.7a. Instead of using the tube shown, you can place one or two small iodine crystals in a test tube and seal it lightly with a cork. Heating the tube gently over a bunsen flame will produce a purple vapour. Then through the diffraction grating you will see that the red, yellow and green regions of the spectrum are crossed by a series of dark bands, as in Figure 37.7b.

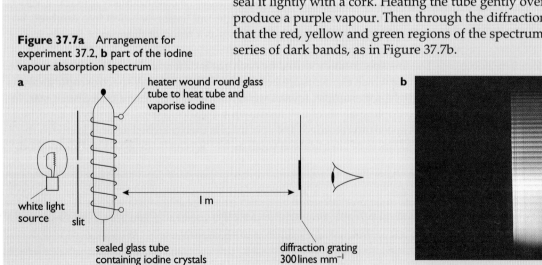

Figure 37.7a Arrangement for experiment 37.2, **b** part of the iodine vapour absorption spectrum

The energy differences in rotational and vibrational spectra usually cause absorption or emission in the infrared region of the spectrum. **Infrared absorp-**

37 QUANTISATION OF MATTER: ATOMS

tion spectroscopy** is one of the most powerful methods of analysis available to the physical chemist. It is used to detect many types of chemical bonding, to detect impurities in substances and to measure the ratio of quantities of substances in mixtures.

37.3 ATOMIC ENERGY STATES OF HYDROGEN

The hydrogen atom can exist in a series of energy states E_1, E_2, E_3 and so on (see Figure 37.8). The lowest level E_1, or **ground state**, represents the minimum energy of the atom: its most stable state. The atom can be excited to a higher energy state E_4 say, by absorbing a quantum of energy ϵ, where

$$\epsilon = E_4 - E_1.$$

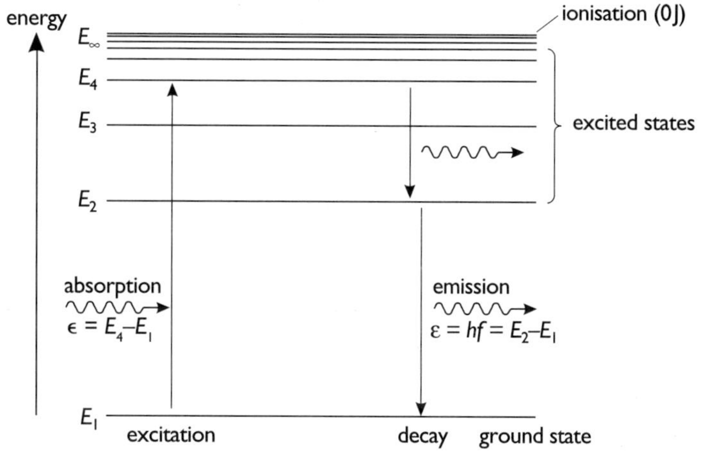

Figure 37.8 Hydrogen atom energy level diagram with some possible transitions

The electron in the atom moves from its lowest energy level 1 to a higher one, in this case level 4.

When the atom reverts in a single **electron transition** to the ground state, it emits this energy ϵ as a photon. The frequency f of the emitted photon is given by:

$$f = \epsilon/h$$

and its wavelength by $\lambda = hc/\epsilon$,

where h is the Planck constant and c the velocity of light ($\leftarrow$ 36.1).
Each frequency in the emission line spectrum of the gas corresponds to a particular electron transition. Normally, almost all the atoms are in the ground state. The most common transitions, causing the brightest spectral lines, are those close to the ground state.

You can draw the energy states of the atom on a vertical scale, as shown in Figure 37.8. The transition between two levels is marked with an arrow, upwards for absorption and downwards for emission. Note that the energy states are closer when the atom is more excited. The further apart the energy states are, the greater the frequency of the photon absorbed or emitted in the transition. The highest energy state E_∞ is zero and occurs when the electron is removed entirely from the atom; to 'infinity'. The ionisation energy E_{ionise} is the energy needed to remove the electron to infinity ($\leftarrow$ 16.3) from orbital 1:

$$E_{\text{ionise}} = E_\infty - E_1$$

$$= -E_1.$$

INSIDE THE ATOM

Bohr's model of single electron atom spectra

In 1913, Neils Bohr produced a model for the atomic energy states of hydrogen to explain the line spectrum obtained from hydrogen. The model was based on a series of circular electron orbits, rather like the Moon revolving around the Earth.

He proposed that the electron can only exist in well-defined circular orbits each, with a sharply defined energy level. No energy is radiated while the electron remains in a particular level, so allowing the electron to maintain its orbit. He also suggested that, when an electron is changing from one orbit to another, a quantum of radiation is emitted or absorbed with an energy equal to the difference between the energies of the two levels, E_m and E_n.

$E_m - E_n = hf.$

Bohr's model is a blend of classical mechanics and the quantum model. It describes a pattern of electron energy levels within hydrogen atoms, which predicts all the spectral lines observed for hydrogen (see Figure 37.9). He also established that, in general, the energy level E_n (in joules) of each orbital is given by:

$$E_n = -21.8 \times 10^{-19} \cdot \frac{1}{n^2}$$

where $n = 1, 2, 3,$ etc.

Many thought that Bohr's idea of electron transitions and photon emission should, in principle, apply to all atoms, whatever their number of electrons. However, he was only able to calculate correctly values for the energy levels of the simplest atoms: those of hydrogen, singly ionised helium and doubly ionised lithium, all of which have a single electron.

The negative sign indicates that energy needs to be supplied to the atom to ionise it. When bound, the electron energy is both kinetic and potential. The potential energy is large and negative so the total electron energy is negative, indicating a bound state: energy must be supplied to remove it from the atom. An electron that is free of the atom has a positive value of energy above E_∞, or zero. This energy is all kinetic.

A hydrogen atom in its first excitation state E_2, can emit a photon spontaneously so that the atom returns to its ground state E_1. It does so within 10^{-7} s, the electron moving from level 2 to level 1. Such short intervals apply to most other transitions involving the emission of a photon. However, when the excitation is higher than E_2, the atom may return to the ground state through intermediate states. In this case, the electron makes a series of jumps between levels, with the emission of a photon at each transition.

The energy, in joules, of each level in the hydrogen atom is given by:

$$E_n = -21.8 \times 10^{-19} \cdot \frac{1}{n^2}$$

where $n = 1, 2, 3,$ etc.

The ground state occurs when the electron is in the first orbital and $n = 1$. So, 21.8×10^{-19} joules needs to be supplied to the atom when the electron is in its lowest level to remove it from the atom. This is the atom's ionisation energy.

Just one-quarter of this energy is required to ionise the atom when the electron is in level 2, one-ninth when in level 3 and so on. The energy levels become closer at higher levels, converging at an energy of 21.8×10^{-19} J above the ground state. This value of the ionisation energy of the hydrogen atom is the same as that measured in collision experiments (← 37.1).

Transitions from higher states to E_1 emit a set of ultraviolet spectral lines, known as the **Lyman series** after their discoverer. The most prominent spectral line is emitted by the transition from E_2 to E_1 at a wavelength of 121 nm. Transitions from higher levels to E_2 emit a set of visible and ultraviolet spectral lines known as the **Balmer series** (see Figure 37.9). The strongest visible line is for the transition from E_3 to E_2 at 656 nm, which is a deep red colour.

Distribution of electrons in atoms

Hydrogen has the simplest spectrum and is the simplest atom, consisting of only two particles – a single nuclear proton and a single orbital electron. Bohr's model successfully predicts the energy levels of hydrogen but is unable to explain the spectra of more complex atoms. In a more sophisticated model, each orbital an electron can occupy is specified by a series of **quantum numbers**. The principal quantum number gives the orbital number, which is all that is required for hydrogen. Three other quantum numbers relate to the orbital angular momentum, orbital orientation, or magnetic effects, and the spin of each electron. All these quantum numbers are needed to give an accurate description of all the possible energy states of a multi-electron atom.

The electron configuration determines the shell structure of multi-electron atoms. This can be defined in terms of quantum numbers using the **Pauli exclusion principle**. It states that no two electrons in any system of electrons can have the same set of quantum numbers. The quantum numbers are written in groups. The first orbital group, n=1, has all the other quantum numbers except spin equal to zero. Only spin up and spin down states exist so there are only two electrons in this first group or **shell**. In the next group, n=2, the orbital angular momentum can be 0 or 1 leading to 8 different sets of available states, in the second shell and so on.

37 QUANTISATION OF MATTER: ATOMS

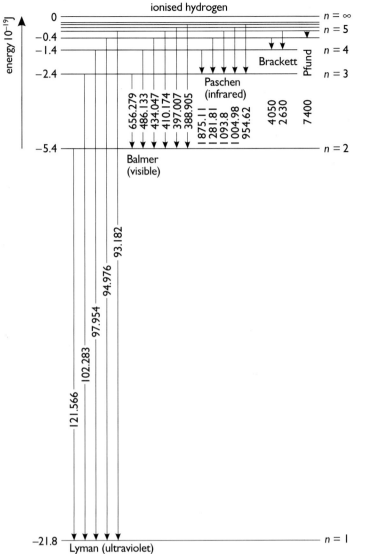

Figure 37.9 Energy levels of the hydrogen atom

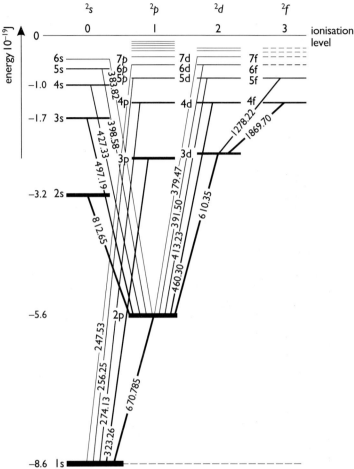

Figure 37.10 Energy levels within the lithium atom

Shells are designated by K for n=1, L for n=2 and so on up to Q for n=7.

The chemical activity of any element can be predicted using its electron shell configuration. Figure 37.1 shows the numbers of electrons in each shell: 2 in the K-shell, 8 in the L-shell, 18 in M and so on. The inert gases, such as helium, have closed shells and so are chemically inactive. Alkali metals, such as lithium, have one electron outside closed shells and are very active. Figure 37.10 is an energy level diagram of lithium showing the spectroscopic notation used and many of the wavelengths observed in its spectrum.

Questions

6a Use the energy values on the vertical scale of Figure 37.9 to check the wavelengths of the Balmer series.
b Explain why the Balmer series contains all the visible spectral lines of hydrogen.

7 What is the relationship between the photon energies of the two lowest frequency emissions of the Lyman series and the lowest frequency emission of the Balmer series (see Figure 37.9)?

INSIDE THE ATOM

RADIO EMISSIONS FROM HYDROGEN IN SPACE

In its ground, or lowest energy, state, the hydrogen atom can exist in two possible arrangements. The proton of the nucleus and its accompanying electron, shown in Figure 37.11, spin about parallel axes. The spinning particles behave like tiny electromagnets, having a magnetic field associated with their rotation. When the two spins are aligned in the same direction, the system has slightly more energy than when the spins are in opposite directions. A spontaneous reversal of one of the spins causes the emission of a radio photon of wavelength 21 cm and energy 9.4×10^{-25} J.

There is a 50 % chance of such an energy transition occurring once in 11 million years. However, there is sufficient hydrogen in interstellar space for us to be able to measure the emission and absorption of this radiation. It enables radioastronomers to learn a great deal about galactic structure and the distribution of matter in space.

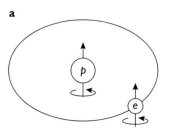
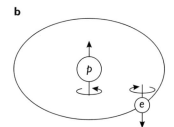

Figure 37.11 Electron spin reverses between **a** and **b** causing emission of a 21 cm-wavelength radio photon

Electron spin

Charged atomic particles, such as electrons or the nuclei of atoms, behave like tiny magnets. This is caused by their charged components spinning around an axis. The **spin** of each electron in an atom can only be aligned parallel or anti-parallel to the spin axis of the nucleus and its magnetic field. So the spin can only have two values.

The effect of spin can be observed in the spectrum of sodium. The bright yellow line can be resolved into two lines, called a **doublet**, at wavelengths 589.0 nm and 589.6 nm (← experiment 37.1). The sodium spectrum is similar to that of hydrogen, being caused by discrete changes in energy of a single valence electron outside closed shells. The photons released in the transition between the first excited and ground states can have two energies differing in value by 3.3×10^{-22} J, or 2×10^{-3} eV, due to the spin of the valence electron. The spin energy contribution is 1/200th part of the transition energy.

Caesium atomic clocks (← 19.1) provide the standard measurement of time. Like sodium, caesium has atoms with a single electron outside closed electron shells. When radio waves at a particular frequency are absorbed by the atom, the spin of this electron changes. The atomic clock consists of a beam of caesium-133 atoms passing a tuned radio transmitter, causing the atoms to resonate in the alternating magnetic field. The tuning is accurate to one part in 10^{11}. This means that it takes 3 000 years for an error of one second to arise between two such clocks.

Questions

8 What is the value of the energy difference between the two spin states in hydrogen gas responsible for the 21 cm wavelength emission?

9 The two spin states of $^{133}_{55}$Cs in its ground state differ in energy by 3.81×10^{-5} eV. What is the value of the frequency of the transition?

37 QUANTISATION OF MATTER: ATOMS

37.4 PHOSPHORS AND LASERS

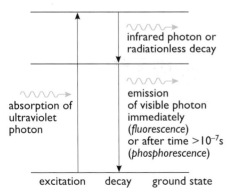

Figure 37.12 Fluorescence and phosphorescence

FLUORESCENCE AND PHOSPHORESCENCE

So far, we have considered only gases, which consist of isolated atoms or molecules. In many solids, an excited atom may pass its extra energy to the lattice vibrations of the solid. This increases the internal energy with no emission. However, there is a category of substances called **phosphors** that, when excited, emit light. Phosphors are used on stamps and dot codes for the automatic optical sorting of letters.

When electrons strike the phosphor coating of a cathode-ray tube screen, light in the form of a line spectrum is emitted. The light is known as **electro-luminescence**. When the invisible ultraviolet light from excited mercury atoms strikes the powder coating of a fluorescent lamp, visible light, called **opto-luminescence**, is emitted.

When the photon is emitted almost immediately, the process is called **fluorescence**. There are materials in which some atoms stay in an excited state for a time much longer than 10^{-7} s. Then the process is called **phosphorescence** (see Figure 37.12). A television screen fluoresces and so can change in 1/100th second. A fluorescent lamp flashes on and off 100 times each second, yet needs an afterglow to smooth out the short dark intervals. A phosphorescent coating provides the afterglow. The effect of the coating on the optical efficiency of a fluorescent lamp is shown in Figure 37.13, contrasted with a domestic filament bulb (see also Figure 36.7).

100 W lamp type	power output/W convection and conduction	radiation in	
		infrared	visible
uncoated fluorescent	35	62	3
coated fluorescent	50	30	20
domestic filament	12	83	5

Figure 37.13 Comparison of power output of some lamps

Sorting and delivery of mail

Phosphors are used by the Post Office to speed sorting and delivery of mail. The phosphors glow when exposed to ultraviolet light and so can be detected optically. This process is independent of the colour background of the envelopes.

Stamps have phosphor bars on them so that automatic letter-facing machines can sense the orientation of a letter and whether it is first- or second-class mail. The correctly orientated letter passes to a member of the Post Office staff, who reads the written postcode and, using a keyboard, marks a set of phosphor dots on the front of the envelope. The pattern of dots identifies the delivery area, each postcode on average relating to about fifteen delivery addresses. Sorting the mail can then be done completely automatically, with optical sensors reading the phosphor dots after exposure to ultraviolet light.

Sometimes, the paper used for envelopes contains a phosphor as a brightening agent. To avoid confusing the sorting machines, a phosphor with a long afterglow is used for the dots. The envelopes are passed over the sensor some minutes after they have been exposed to ultraviolet radiation.

STIMULATED EMISSION

The photon emission of phosphors is spontaneous and random. There is another process by which an excited atom can lose its energy – it can be **stimulated** to

H INSIDE THE ATOM

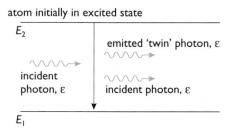

Figure 37.14 Stimulated emission

radiate by a photon of the same energy passing by (see Figure 37.14). The photon produced by stimulated emission is a **twin** of the photon causing the emission. It it identical in frequency, phase, energy and direction, and so is **coherent** (← 33.1).

When such emission happens on a large scale it is called **laser** action. This stands for **l**ight **a**mplification by **s**timulated **e**mission of **r**adiation. The energy of an initial photon is amplified by the addition of other exactly similar photons moving in phase with it as it passes through a substance containing excited atoms. Laser action can be produced in many materials, including solids such as ruby and gases such as helium and neon, and carbon dioxide.

Warning: **Never** look into a laser or allow a reflection to enter your eye. Its intensity is sufficient to damage the retina.

HELIUM-NEON LASER

The He-Ne gas laser, used in school laboratories, has continuous action and produces low intensity radiation of about 1 mW over an area of 2 mm^2. It consists essentially of a discharge tube containing a mixture of about six parts of helium and one part neon gas. The total pressure exerted by the gas is 330 Pa, about 1/300th of normal atmospheric pressure. When a large electric field is applied to the gas, ionisation occurs and the tube glows with the characteristic red colour of neon tubes.

To provide laser action, the ends of the tube have mirrors, usually mounted inside the tube and set parallel and facing each other. The mirrors can be made highly reflecting for a particular wavelength by coating them with a suitable material of given thickness. Typically, one mirror is made 99.9 % reflecting and the other, the output mirror, is partially coated to give only around 99 % light reflectivity. The electrodes at the ends of the tube are usually ring-shaped so that the light between the mirrors is free to reflect back and forth between the mirrors and so be amplified (see Figure 37.15).

Einstein and stimulated emission

In 1916, while studying the absorption and emission of photons from atoms, Einstein discovered the process of stimulated emission. He was awarded the Nobel Prize for this and his work on the photoelectric effect, in 1921. The special property that the emitted photon has the same phase and direction as the stimulating photon was overlooked until the 1950s. The first successful laser was constructed in 1960.

LASER ACTION

When a gas discharge tube is operating normally in thermal equilibrium, absorbing and emitting photons, nearly all of the atoms are in the ground state (← 37.3). Photons passing through the gas are much more likely to meet atoms in the ground state than in a suitably excited state to stimulate emission. So no amplification occurs. When many of the atoms are in the same excited state, a suitable photon entering the gas can cause a chain reaction as atoms decay by stimulated emission. An intense pulse of coherent photons then moves through the gas. So, for laser action, a discharge tube must be designed to produce many atoms in the required excited state rather than in the ground state. This is called a **population inversion**. The photons must also travel along the axis of the system for the pulse intensity to build up rather than being lost through the sides of the tube.

In a helium-neon laser, the population inversion occurs between two excited states of the neon atoms. An electric discharge through the gas excites the helium atoms into **metastable** states. These are 'nearly stable' energy states where the atoms tend to remain excited for a considerable fraction of a second. The helium atoms lose their energy by colliding inelastically with neon atoms, making a radiationless energy transfer. This excites neon atoms causing an electron in each to move to the higher 3s level. As a result, there are many more excited neon atoms with an electron in the 3s level than in the lower 2p level (see Figure 37.16).

By enclosing the gas in a sealed tube with mirrors at both ends, the photons released in this neon atom transition are reflected back and forth causing stimu-

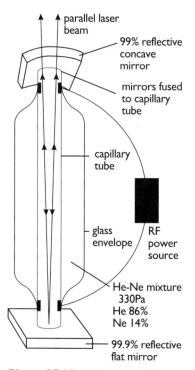

Figure 37.15 Helium-neon laser

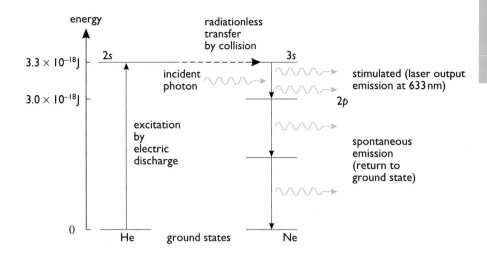

Figure 37.16 Laser action in the He-Ne gas laser

lated emission, building up a coherent beam of light along the tube. A constant supply of excited neon atoms to maintain the beam is provided by collisions with excited helium atoms. A small part of the beam, at a wavelength of 632.8 nm in the red, escapes into the laboratory through the output mirror, which is not a perfect reflector.

LASERS IN OPERATION

There are many laser systems giving a variety of wavelengths and powers in the visible and infrared. The laser is used in many diverse applications from alignment in the building industry and in civil engineering projects, such as road construction and drainage, to measuring distance or even the height of ocean waves. Pulses of laser light, bounced off a reflector left by astronauts on the Moon, monitor its distance from Earth to within a few centimetres.

Continuous lasers, such as the helium-neon laser, tend to be of low power, around 1 mW. Pulsed lasers, such as the solid state ruby laser, can give quite large peak powers, perhaps up to 10^{11} W lasting for as little as 10^{-8} s. The argon laser generates visible spectral lines at steady power levels up to 100 W and also pulsed outputs up to a few tens of kilowatts. Carbon dioxide lasers give a continuous power output of up to 100 kW at 10.6 μm, in the infrared.

In a **semiconductor laser**, as in a light-emitting diode, or LED, n- and p-type layers form a junction from which light is emitted. A laser cavity surrounds the junction. There are various types in use, composite diodes using aluminium gallium arsenide being the most common. They emit high-intensity highly directional infrared radiation of narrow bandwidth, with wavelengths starting around 800 nm. They are of low power in the visible or infrared and can operate continuously or in pulsed mode. Remote control circuits, optical scanning, pollution and fire detection, compact disc and videodisc playing systems and fibre-optic communications are just some of their applications.

Medical uses of lasers

There are a number of medical problems for which lasers have either simplified treatment or have made treatment possible where none was available before.

A laser beam can cut through human tissue. As it does so, the blood vessels are sealed, or cauterised, so that bleeding does not occur. This process makes it possible, for example, to remove a diseased portion of the liver without any haemorrhaging. When used with a fibre-optic endoscope, the laser can vaporise

H INSIDE THE ATOM

and so remove blood clots, stones or tumours. This is a very cost effective procedure and one that causes minimal discomfort for the patient compared with more traditional methods.

Pulsed ruby lasers are used in eye surgery. The retina may become detached from underlying tissue, producing a 'blind' area. A pulse, lasting for less than a millisecond, delivers enough energy into an area of about a square millimetre to 'spot-weld' the retina and restore normal vision.

37.5 X-RAY SPECTRA

These are produced in much the same way as optical spectra (← 37.2). The frequencies of the spectral lines for X-rays are much greater than optical ones, though, because the difference in energies between two states of an atom that emits an X-ray photon is relatively large. We can conclude that X-ray photons are emitted during transitions between energy levels deep in multi-electron atoms close to the nucleus: between the K, L and M electron shells, corresponding to principal shell, or orbital quantum number, n of values 1, 2 and 3.

X-rays are produced when high-speed free electrons are stopped by the surface of a solid. A modern X-ray tube is shown in Figure 37.17a. A cathode is heated by a filament so that thermionic emission occurs. The higher its temperature, the more electrons are produced and the greater the tube current. The resultant electron beam is accelerated through a high voltage of 40–100 kV and focused on a small area of the target, from which X-rays are emitted. The emitted X-rays can be diffracted by the regular layers of atoms of a crystal to produce a spectrum (see Figure 37.17b), just as a diffraction grating can produce a spectrum from a light source (← 32.2 and 33.3).

A considerable amount of internal energy is generated in the target by the electronic bombardment. As a result, only about 1 % of incident energy is converted to X-radiation. To reduce the heating of the target, the anode of most diagnostic tubes is rotated so that the target area is constantly changing and excessive local heating does not occur.

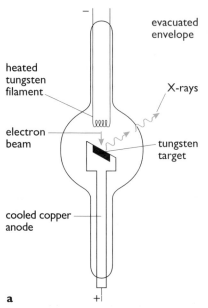
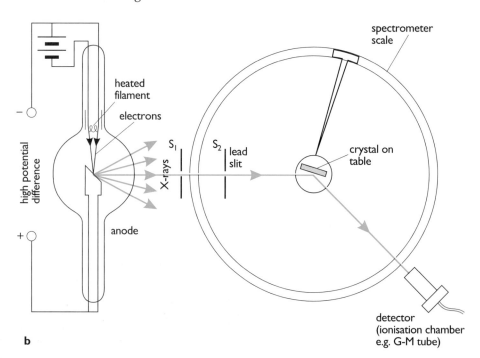

Figure 37.17a A modern X-ray tube, **b** X-ray quanta being emitted from an X-ray tube target

37 QUANTISATION OF MATTER: ATOMS

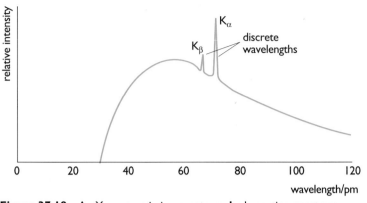
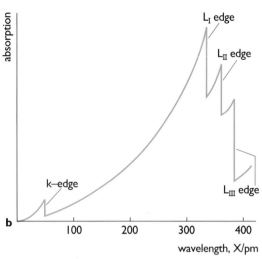

Figure 37.18 An X-ray **a** emission spectrum, **b** absorption spectrum

Continuous X-ray spectra: braking radiation

As electrons strike any solid, they are rapidly decelerated so producing electromagnetic radiation. This is frequently referred to by the German word *bremsstrahlung* which means 'braking radiation'. It forms a continuous spectrum with an upper photon energy limit equal to the maximum electron energy. The radiation arises as a result of the rapid deceleration of the electrons.

Suppose that an incident electron is brought to rest in a single collision and all its kinetic energy is emitted in one X-ray photon. Then:

(kinetic energy of electron) = (maximum energy of X-ray photon)

or $eV = hf_{max}$

where V is the accelerating voltage of the X-ray tube and f_{max} is the maximum frequency possible. The corresponding minimum wavelength is given by

$$\lambda_{min} = \frac{hc}{eV}.$$

A typical X-ray spectrum is shown in Figure 37.18a. It consists of two parts, a continuous spectrum and a line spectrum – the spikes. Each element used in a target produces its own X-ray spikes – a **characteristic spectrum**. The reason for this was explained by Moseley in 1913. According to **Moseley's law**, the frequencies of the X-ray spikes are related to the proton number Z of the target element by:

$$f = a(Z-b)^2$$

where a and b are constants.

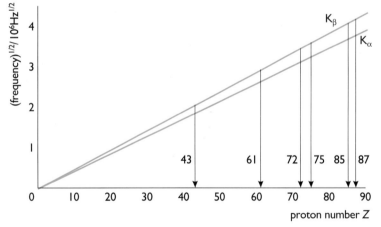

Figure 37.19 Plot of Moseley's law, revealing missing elements

Characteristic X-ray spectra

In 1913, Moseley tried 39 different elements, from aluminium to gold, as targets in his X-ray tube. He found that the frequency of the X-rays was related to the proton number Z of the target element by the relation, known as **Moseley's law**:

$$f = a(Z-b)^2$$

where a and b are constants.

His work helped to determine the proton numbers of elements.

As a result, four pairs of elements, including nickel and cobalt, were reversed in the periodic table. Six gaps in the value of Z appeared, which corresponded to elements that have since been discovered (see Figure 37.19). Moseley's work was one of the major keys in the formation of a consistent picture of atomic structure.

H INSIDE THE ATOM

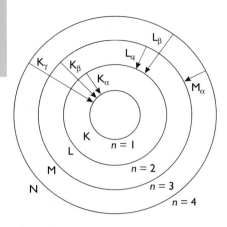

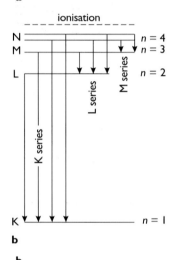

Figure 37.20a 'Shell' structure of an atom, **b** corresponding hydrogen-like X-ray energy level diagram

Moseley used the Bohr model of energy levels (← 37.3) to relate the X-ray spikes with the high-energy photons emitted as a free electron is captured by an atom into its lowest energy levels. A very energetic bombarding electron can knock out an electron close to the nucleus of a target atom. This creates a 'hole' in an inner orbital. When a free electron falls into this, an X-ray is emitted.

The series of X-ray lines give an X-ray energy level diagram similar to that of the hydrogen atom, where $n = 1, 2, 3 \ldots$ correspond to the electron shells K, L, M… of multi-electron atoms. Electron transitions to the K-shell generate a series of X-ray spectral lines K_α, K_β, etc. and similarly for transitions to the L, M… shells (see Figure 37.20).

Questions

10 In what way can X-ray production be thought of as an inverse process to the photoelectric effect?

11 Refer to Figure 37.18a.
a Estimate the voltage across the tube used to generate the X-rays giving this spectrum.
b Calculate the atomic number of the anode of the X-ray tube used. For the K_α line, take $a = 2.5 \times 10^{15}$ Hz and $b = 1$ in Moseley's law.

12 The K_α line for copper has a wavelength of 0.154 nm.
a To what photon frequency does this wavelength correspond?
b What is the value of the energy difference between the K and L energy levels in copper?
c What anode voltage is required to accelerate electrons to a sufficiently high energy to produce the K_α line in a copper target X-ray tube?

13 Optical spectral lines arise in the same way as X-ray spectral lines – electrons within the atom falling from one energy level to a lower one. However, optical wavelengths are some 1 000 times longer than X-ray wavelengths. Why is this?

X-RAY ABSORPTION SPECTRA

As X-rays pass through materials, their intensity gradually reduces. The amount of energy each quantum loses as it interacts with the electrons and electric fields in a material, depends on the nature of the material. This enables medical imaging, **X-ray tomography**, to be carried out (← 32.2).

Analysis by an X-ray spectrometer (← 32.2) shows that when X-rays pass through or are scattered by a material, they give an **absorption spectrum** characteristic of that material.

In particular, a **K-absorption edge** is a feature of such spectra (see Figure 37.18b). This is due to the particular chemical elements contained within the absorbing material. There is a sharp cut-off in the intensity of each spectrum caused by X-ray quanta either having or not having sufficient energy to eject an electron from the K-shell of each atom in the material. Similar, though less detectable, absorption edges occur for X-rays which have just sufficient energy to eject electrons from the L and M shells.

37.6 MATTER WAVES

Electromagnetic radiation behaves as both a stream of particles and as a wave (← 36.4). For example, in Figure 37.17b, the beam of X-rays emerging from the

37 QUANTISATION OF MATTER: ATOMS

a

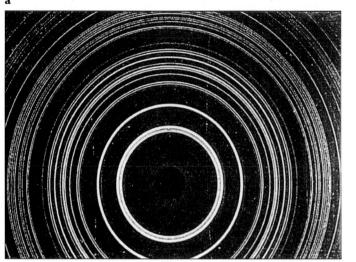

b

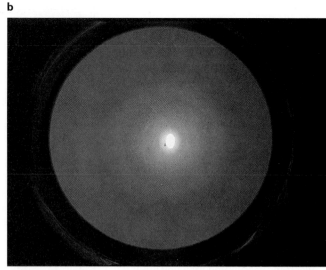

Figure 37.21 Diffraction patterns from aluminium foil produced by **a** X-rays and **b** electrons of the same wavelength

tube consists of a stream of photons. A Geiger tube and counter clearly detect the random arrival of X-ray photons. However, the regular layers of atoms in the crystal diffract the X-ray beam allowing us to calculate, using a wave model, the wavelength associated with the beam.

Similar diffraction occurs when a stream of electrons passes through a thin film of crystalline matter (see Figure 37.21). So electrons also appear to have wave-like properties. You can observe this in experiment 37.3, which is a version of an experiment first performed by Thomson and Reid in 1927.

EXPERIMENT 37.3 Electron diffraction

In the 'electron diffraction' tube shown in Figure 37.22, a narrow beam of electrons is diffracted by a thin film of graphite producing a pattern. By observing the diffraction pattern of two rings on the fluorescent screen and increasing the anode voltage V from 2 500 V to 5 000 V, you should notice that the diameters of the two rings become smaller as the voltage is increased. By measuring the ring diameters D_1 and D_2 for a range of voltages V, you can plot a graph of the diameter of each ring against $1/\sqrt{V}$ and show that

$$mv = \frac{h}{\lambda}.$$

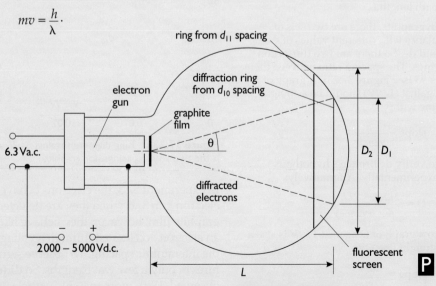

Figure 37.22 Electron diffraction tube

H INSIDE THE ATOM

Question

14 The following results were obtained in an experiment similar to experiment 37.3:

anode voltage V/V	3.0	3.5	4.0	4.5	5.0
smaller ring diam D_1/m	32	29	27	26	24
larger ring diam D_2/m	55	51	48	45	42

a A single electron beam is diffracted by the graphite into two distinct rings. What does this suggest about the nature of the beam and the structure of the graphite?
b For a given wavelength associated with the electron beam, what determines the angle of diffraction θ in Figure 37.22 for a given ring?
c Plot graphs of D_1 and D_2 against $1/\sqrt{V}$.
d Find the ratio of the gradients of each of the two lines. What information does this give you?

Model for electron diffraction

Using de Broglie's hypothesis, electrons of momentum p have an associated wavelength λ given by:

$$\lambda = h/p$$

where h is the Planck constant. Electrons accelerated by a voltage V transfer potential energy eV to kinetic energy:

$$\tfrac{1}{2}m_e v^2 = eV.$$

Momentum $p = m_e v$ and:

$$p^2 = (m_e v)^2 = 2m_e eV$$

so

$$\lambda = \frac{h}{\sqrt{2m_e eV}}.$$

When waves of wavelength λ are diffracted by the crystal lattice, the first maximum occurs when λ is $d.\sin\theta$, where d is the interatomic spacing (← 33.3). Normally d is much less than λ and so θ is quite small and from Figure 37.22:

$$\lambda = d\theta = d\frac{D}{2L}$$

So, you can calculate λ from the ring diameter D around the arc of the tube, the atom spacing d and the electron path length L.

In graphite, there are two sets of planes with different values of d: d_{10} and d_{11}. So there are two diffraction rings. The spacing ratio between them is √3 (see Figure 37.23). A graph of λ against $(1/\sqrt{V})$ should be a straight line with:

$$\text{slope} = \frac{h}{\sqrt{2m_e e}}$$

to verify de Broglie's hypothesis experimentally. Alternatively:

$$D = \frac{2L\lambda}{d} = \frac{2Lh}{\sqrt{2m_e eV}},$$

so a graph of D against $1/\sqrt{V}$ is also a straight line with:

$$\text{slope} = \frac{2Lh}{\sqrt{2m_e e}}.$$

DIFFRACTION OF ELECTRONS

In 1924 Louis de Broglie proposed that the wave-particle duality already found for light should be extended to include all material particles, such as electrons, protons and atoms. He suggested that a particle has a wavelength λ related to its momentum mv so that $\lambda = h/mv$. Experiment 37.3 confirms the physical reality of de Broglie waves. A lower accelerating voltage gives an electron a lower value of momentum. This means a longer wavelength and a larger diffraction angle resulting in a larger ring on the screen.

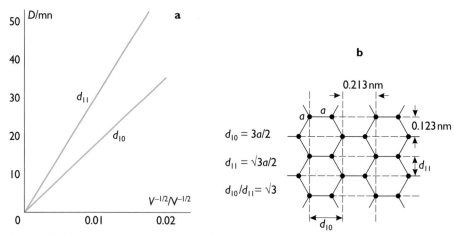

Figure 37.23a Ring diameter against $1/\sqrt{V}$, and **b** the two rows spacings, for graphite

In experiment 37.3, the electrons behave as particles as they are accelerated in the electron gun and when they are detected on the fluorescent screen. Near the graphite film, however, they behave like waves. The wavelength associated with an electron accelerated through a voltage of 5 kV is 1.8×10^{-11} m, about 1/10th of the interatomic spacing. To observe significant diffraction effects requires apertures of only a few wavelengths. So diffraction effects for electrons are easily observed. With more massive particles, the wavelengths are extremely small.

37 QUANTISATION OF MATTER: ATOMS

> **Questions**
>
> **15a** Show that the wavelength associated with a 5 keV electron is 1.8×10^{-11} m.
> **b** Experiment 37.3 shows that the ring diameter is proportional to $1/\sqrt{V}$. Why does this suggest that a particle's momentum, rather than its kinetic energy, is the important factor in determining its wavelength?
>
> **16a** Calculate the momentum neutrons would need to have to give a wavelength of about 10^{-10} m.
> **b** What would be the value of their kinetic energy? (m_n)

ELECTRON MICROSCOPE

The small wavelength of the electron is put to use in the electron microscope. An electron accelerated through 100 kV has a wavelength of about 4 pm. This is about 10^5 times smaller than the shortest visible light wavelength of 400 nm. Electrons can be deflected and focused by electric and magnetic fields (← 35.4). With magnetic lenses, it is possible to construct an electron microscope similar to an optical one, as shown in Figure 37.24.

In theory, the electron microscope has a resolving power (← 31.3), proportional to the wavelength of the electrons, 10^5 times better than its optical counterpart. Magnetic lenses are not as efficient as optical ones. Nevertheless, a gain of 10^3 is still possible, enabling objects only about 1 nm across to be observed.

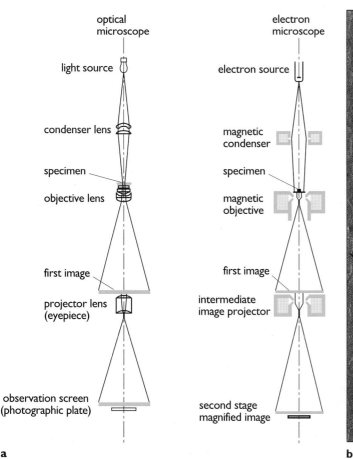
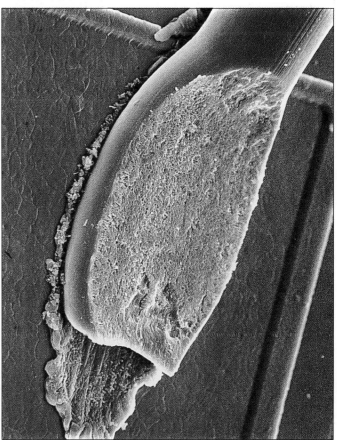

Figure 37.24a Comparison of transmission optical and electron microscopes, **b** output of electron microscope, an integrated circuit connector pin

H INSIDE THE ATOM

Questions

17 Show that when an electron has an associated wavelength of 4 pm, it has an energy of approximately 100 keV.

18 Imagine that a friend of yours reads the following in a biology text book and does not understand it: 'The resolving power of a light microscope is limited by the wavelength of light. Progress in studying cell structures was held up until the development of the electron microscope. Instead of using light as a means of illumination, using electrons with a much shorter wavelength, enabled such details to be resolved.'

Explain this passage to your friend.

19 Compare the wavelength of a proton with that of an electron accelerated through the same voltage. Use the fact that the mass of a proton is approximately 1 840 times that of an electron.

20 A driven golf ball is a moving particle. Show that a suitable de Broglie wavelength for it is about 10^{-34} m.

ELECTRON ORBITALS AND WAVE MECHANICS

In the early picture of the atom, electrons were considered to be orbiting the nucleus, held by electrical attraction. In this model, electrons are always accelerating towards the nucleus. So they should be radiating electromagnetic waves, just as charge oscillating in a radio aerial does. The electrons should radiate away their energy and collapse into the nucleus in less than 10^{-10} s. To explain why they do not, Bohr suggested that a rule of quantum mechanics is that bound electrons do not radiate but can only have certain allowed energies ($\leftarrow$ 37.3). This energy-level model of the atom can be used to explain line spectra. However the model is not sophisticated enough to account for all observed atomic spectra and their characteristics.

A more satisfactory model pictures electrons as clouds of negative charge surrounding the nucleus. The distribution of charge and mass of the electron within the cloud – its density – looks like a three-dimensional standing wave pattern. The mathematical expression for this standing wave pattern, the **wave function**, contains the de Broglie wavelength of the electron and its momentum and energy. It comes from Erwin **Schrödinger's wave equation**, established in 1926 as part of the new model of **quantum mechanics**. This fully accounted for the distribution of electrons around the nucleus of an atom.

The intensity of a wave is proportional to its (amplitude)^2 ($\leftarrow$ 29.2). In much the same way, the **square of the amplitude** of a matter wave at a particular point is proportional to the **probability** of finding the particle within a small region around the point. The patterns that satisfy the conditions are known as **orbitals**.

Schrödinger's equation for matter waves

Assume that a particle of mass m and velocity v has a wave associated with it having displacement ψ. A typical equation for such a wave of period T and wavelength λ is:

$$\psi = A \sin 2\pi (t/T - x/\lambda).$$

To see how this wave is distributed in space, you can differentiate twice with respect to x:

$$\frac{d^2\psi}{dx^2} = -(2\pi/\lambda)^2 \psi.$$

So the wave equation can be written:

$$\frac{d^2\psi}{dx^2} + (2\pi/\lambda)^2 \psi = 0.$$

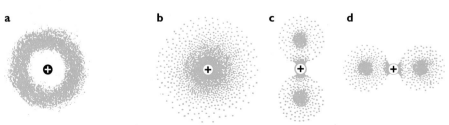

Figure 37.25 Some typical electron orbitals in atomic hydrogen: **a** corresponds to the ground state

For any particle, kinetic energy $E_K = \frac{1}{2}mv^2$ and momentum $p = mv$, so:

$$p^2 = 2mE_K$$
$$= 2m(E-E_P)$$

where E is the total energy of the particle and E_P is its potential energy. From de Broglie's hypothesis, $p = h/\lambda$, which gives:

$$(2\pi/\lambda)^2 = \frac{8\pi^2 m}{h^2}(E-E_P).$$

Substituting this in the wave equation gives:

$$\frac{d^2\psi}{dx^2} + \frac{8\pi^2 m}{h^2}(E-E_P)\psi = 0$$

This is Schrödinger's equation in one-dimension.

The ground-state orbital of hydrogen is spherical. Schrödinger's equation for the amplitude A of a spherical standing wave at radius r can be written in the form:

$$\frac{d^2A}{dr^2} + \frac{8\pi^2 m(E-E_P)}{h^2}.A = 0.$$

The total energy E of the bound electron is constant at -21.8×10^{-19} J $E = E_K + E_P$ and $E_P = -e^2/4\pi\epsilon_0 r$ ($\leftarrow$ 16.3). So A can be found by solving the equation.

The density of the electron charge cloud around the nucleus is proportional to the probability P of finding the electron there:

density $\propto P \propto A^2$.

Similarly:

potential $V \propto A$.

The graph of $P (\propto A^2)$ against r is shown in Figure 37.26. Figure 37.25a gives another representation of the ground-state orbital.

Each orbital has a different energy. Each energy corresponds to a particular momentum with an associated de Broglie wavelength. When this is a sub-multiple of the circumference, a standing wave can exist and the electron is bound to the atom. Figure 37.25 illustrates some electron orbitals in atomic hydrogen.

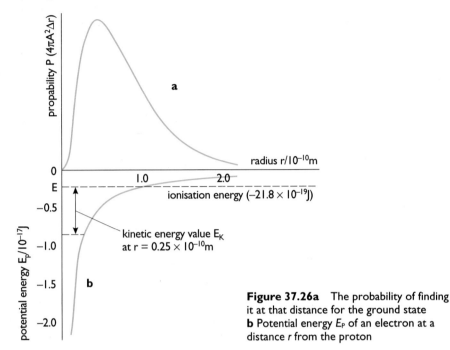

Figure 37.26a The probability of finding it at that distance for the ground state
b Potential energy E_P of an electron at a distance r from the proton

The quantum mechanics model often seems contrary to common sense. When using it, bear in mind that:

- when you want to know where a particle is, use the wave model to calculate where there is a high probability of finding it.
- when you want to know how the particle interacts with matter, use the particle's momentum.

SUMMARY

The ionisation energy of an element is the energy to remove a bound electron to infinity. It is found in collision experiments between electrons and gas atoms.

Evidence for the existence of energy levels in atoms comes from inelastic collision experiments, such as the Franck-Hertz experiment, where excitation energies are measured.

Atoms and molecules absorb and emit electromagnetic quanta at precisely defined wavelengths, producing line spectra. Spectra are explained using the quantum theory of allowed energy changes within an atom, devised by Neils Bohr.

In the Rutherford-Bohr model of the atom, the frequency f and wavelength λ of an absorbed or emitted photon are related to the allowed energy states of the atom by the formula:

$$E_2 - E_1 = hf = \frac{hc}{\lambda}$$

where E_1 and E_2 are the energies of the two states involved in the transition, h is the Planck constant and c is the velocity of light. For the hydrogen atom, the allowed energy states E have values:

$$E_n = -21.8 \times 10^{-19} \frac{1}{n^2} \text{ J}$$

where $n = 1, 2, 3 \ldots$ denotes the electron orbital number.

The concept of spin helps to explain the spectra of multi-electron atoms, the fine details of the hydrogen spectrum and the shell structure of the atom.

Optical spectra are due to movements of electrons in the **outermost**, or valence, shells from higher to lower energy levels in the atom. X-ray spectra are due to movements of electrons between energy levels in shells **close to the nucleus**.

Moseley's analysis of X-ray spectra led to a revised ordering of the elements in the periodic table.

Particles of small mass can be shown to have wave properties. For example, a beam of electrons of momentum p has an associated de Broglie wavelength given by:

$$\lambda = h/p$$

where h is the Planck constant and λ is the wavelength of the particle's associated wave function. Both wave and particle models are needed to describe the behaviour of particles of matter or quanta of radiation.

Schrödinger's wave equation describes how the wave function associated with a particle varies with distance and time. By finding the square of this at a particular position, we can evaluate the probability of finding the particle there. In an atom, three-dimensional standing waves determine the energy level of each electron orbital. Taking all of an atom's electrons together, standing waves determine the state of the whole atom.

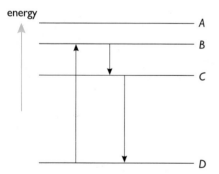

Figure 37.27 Four energy levels of an atom

SUMMARY QUESTIONS

21 Different models of the atom are used to explain different physical situations. Suggest one or more situations where each of the following models is adequate to explain the observed phenomena.
The atom behaves as:
a an elastic or solid sphere connected to its neighbours by springs,
b a sphere where one or more electrons can be removed,
c a tiny nucleus surrounded by a cloud of electrons.

22 Using Figure 37.28a, what is the minimum energy in aJ, that an electron must have to excite a hydrogen atom in its ground state by collision?

23 Figure 37.27 shows four energy levels of an atom and three possible transitions between energy states of the atom.
a Which transitions shown are caused by
 i absorption, and
ii emission, of a photon?
b How many emission transitions are possible between the four levels?
c Which of the transitions shown is related to i the highest frequency photon and ii the longest wavelength photon?
d Which level could be the ground state of the atom?
e Which of the transitions shown could be a possibility for laser action?

37 QUANTISATION OF MATTER: ATOMS

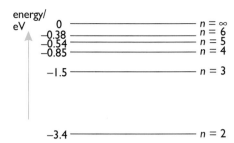
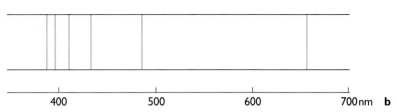

Figure 37.28a Energy levels of the electron in hydrogen, **b** some lines of the hydrogen spectrum

24 Figure 37.28a shows the energy levels of the single electron in a hydrogen atom and Figure 37.28b some of the lines of the hydrogen spectrum.
a Copy Figure 37.28a and insert arrows to show the transitions that the electron makes to produce the spectral lines of Figure 37.28b.
b How can you find hydrogen's ionisation energy from Figure 37.28a?
c One line in the spectrum has a frequency of 2.9×10^{15} Hz. This corresponds to a transition between which two levels?
d Between which levels does the absorption line in the spectrum corresponding to a wavelength of 650 nm arise?

25a Use Figure 37.28a to draw an energy level diagram for singly ionised helium, He^+.
b In what ways do you expect the spectrum of He^+ to be
i similar to and **ii** different from, that of hydrogen, H?
c Why would you expect the effective radius of He^+ in its ground state to be less than that of H in its ground state?
d Would you also expect the effective radius of He in its ground state to be less than that of H in its ground state?

26 Why is the effective radius of the sodium ion Na^+ so much smaller than that of the chlorine ion Cl^- in a NaCl crystal?

27 Figure 37.6 shows one method of observing the absorption spectrum of sodium. The experiment is repeated using a mercury lamp and a bulb of mercury vapour. With a bulb of glass, which absorbs ultraviolet, and a cool vapour, no absorption spectrum is observed. With a bulb of quartz, there are ultraviolet absorption lines. When the bulb is warmed, absorption lines also appear in the visible wavelength range.
   Use the energy level diagram for mercury shown in Figure 37.4 to explain these observations.

28 In a demonstration tube, such as that in Figure 35.9, each electron can be accelerated through 5 000 V. Suppose all of this energy is converted into one photon when the electron is stopped at the screen. Is the photon in the X-ray region of the spectrum?

29 The minimum wavelength of X-rays from an X-ray tube operating at 40 kV is 31 pm. Use this information to find a value for the Planck constant. This is one of the most reliable methods of measuring h. (e, c)

30 Figure 37.21 contrasts the diffraction rings caused by electrons and X-rays of the same wavelength.
a Show that electrons and X-rays have the same wavelength and energy when $\lambda = h/2m_e c$ and find this value.
b Through what voltage would electrons have to be accelerated to achieve this de Broglie wavelength? (h, m_e, c)

38 RADIOACTIVITY

Radioactivity was discovered accidentally in 1896 by the French physicist, Henri Becquerel, while he was investigating fluorescence in uranium salts. He found that radiation from uranium salts caused an image on a photographic plate even when the plate was wrapped in thick black paper or shielded from the salt by a thin metal plate. This radiation is called **radioactivity**. His historic experiment can be repeated quite easily, as described in experiment 38.1.

Becquerel also discovered that the radiation could ionise a gas. This led the New Zealand physicist Ernest Rutherford, working in Canada in 1899, to study the ionising abilities of the radioactive emissions from radium. He also measured the thickness and density of materials needed to absorb the radiation from uranium. He found that there are two different types of radiation. The first is absorbed by a sheet of paper or a few centimetres of air and the second is about 50 times more penetrating. He called these alpha (α) – and beta (β)-rays respectively. A year later, the Frenchman Paul Villard discovered that radium also produces a much more penetrating radiation, which was named gamma (γ)-rays.

Now we know that α–, β– and γ-rays are all emitted from the nuclei of unstable atoms. They are distinguished from each other by their different penetrating powers, ionising effects and their paths through regions of electric and magnetic field (see Figures 38.1 and 38.2). In this chapter, we look at the properties and nature of each radiation in turn, the different methods of detecting them and some practical uses of radioactive materials.

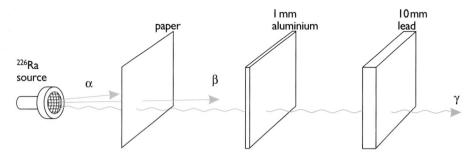

Figure 38.1 Penetration abilities of the radiations from radium

---

EXPERIMENT 38.1   Detecting radioactivity using a photographic emulsion

Placing radium, or some other source, face downwards on a wrapped-up piece of photographic paper for about 30 minutes will cause an image to be produced by the radiation. This can be seen when the paper is developed.

P

38 RADIOACTIVITY

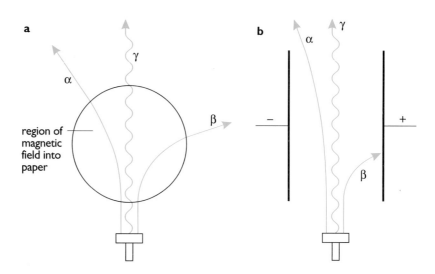

Figure 38.2 Radiations being deflected in **a** a **B**-field, **b** an **E**-field

Question

1 Explain why it takes so much longer to fog the photographic paper in experiment 38.1 when it is wrapped up by considering the penetrating and ionising abilities of the radiations from a radium source. See Figure 38.35 for a summary of radiation properties.

38.1 RADIONUCLIDES AND THEIR CHARACTERISTICS

Radioactive materials are often known as **radionuclides**. They are unstable and emit radioactive particles, such as α-rays and β-rays in the process of changing to a different material, which may itself be stable or unstable. A sample of a radionuclide is often referred to as a **radioactive source**.

The strength or **activity** of a particular source depends both on the quantity and properties of the radionuclide present. Activity is defined as the number of disintegrations or emissions of particles per second. The nature and activity of sealed laboratory radioactive sources are marked on the container and on the source holder. The unit of activity is the **becquerel**, or **Bq**: one disintegration per second.

The original unit of activity is the curie (Ci), which is the activity of one gramme of naturally occurring radium in equilibrium with its daughter nuclides ($\rightarrow$ 39). It is equal to 3.7×10^{10} Bq. It is named after Marie Curie, who gained the Nobel prize in 1911 for isolating the element radium. She died in 1934 as a result of prolonged exposure to radiation, mostly from the radium she had so successfully isolated.

Strictly, the term activity should only be used when one radionuclide is present in a sample and it decays into a stable material. Otherwise the particles coming from the sample will be a mixture of radiations from several different radionuclides. In this case, the term **count rate** is used.

ISOTOPES AND NUCLIDES

These are among the words used to classify atoms. An atom of atomic number Z in the periodic table consists of Z electrons surrounding a tiny massive nucleus containing Z protons and N neutrons. The atomic number Z uniquely identifies

INSIDE THE ATOM

an element. Atoms of an element that have different masses are called **isotopes**. The mass difference occurs because the nuclei have different numbers of neutrons. Since neutrons and protons are the constituents of a nucleus, they are called **nucleons**. The mass of an atom is determined by its **nucleon number**, or **mass number**, A. The isotopes of an element have the same proton number but different nucleon numbers. The word **nuclide** is used when referring to a particular kind of atom with a chosen proton and neutron number. There are only about 100 elements, that is, atoms with different proton numbers, but there are more than 2000 different nuclides when we include all the isotopes of the different elements.

We use the following notation to specify a nuclide:

nucleon number A
proton number Z. $\times$ chemical symbol

Examples of different nuclides are:

$^{1}_{1}H \quad ^{2}_{1}H \quad ^{3}_{1}H \quad ^{4}_{2}He \quad ^{3}_{2}He \quad ^{235}_{92}U \quad ^{238}_{92}U.$

The first three are isotopes of hydrogen: $^{1}_{1}H$, normal, or light; $^{2}_{1}H$, deuterium, or heavy; $^{3}_{1}H$, tritium. The others are two isotopes of helium and two of uranium. A nuclide may also be written in the form *element – A*, such as helium–4 or uranium–238.

Isotopes have different numbers of neutrons. This is the difference between the upper nucleon number and the lower proton number in each nuclide symbol. For example, $^{2}_{1}H$ has 1 neutron in addition to the hydrogen atom's single proton, giving it a nucleon number of 2.

ATOMIC MASS AND ITS UNIT

The mass of an atom (or sub-atomic particle) is usually measured using the **atomic mass unit, u** (← 35.2). This is 1/12th of the mass of an atom of the nuclide $^{12}_{6}C$. So, the atomic mass of an atom of $^{12}_{6}C$ is defined to be precisely 12 u. Frequently, particularly in chemistry, the **relative atomic mass** or atomic weight, is used. This is the number of times heavier a particle is than an atomic mass unit. It is a pure number. Similarly, the mass of a molecule can be expressed in u, but the molecular mass of a chemical compound is normally given as **relative molecular mass**. This is the ratio of the molecular mass of a chemical compound to that of an atomic mass unit, and is also a pure number.

The atomic mass of a particular nuclide, or nuclear species, is slightly different from its nucleon number A. The value of A is the total number of protons and neutrons in the nucleus of one atom and therefore must be an integer. For example, the nuclide $^{238}_{92}U$, has an atomic mass of 238.0507 u. Its nucleon number is 238. Similarly, the calcium isotope $^{40}_{20}Ca$, has an atomic mass of 39.9626 u and a nucleon number of 40.

For all nuclides, other than $^{12}_{6}C$ the atomic mass will not be an exact integer. This is because of the **binding energy** that keeps nuclear particles close together in the nucleus and, to a much smaller extent, holds the orbital electrons. The total binding energy ΔE, results in a reduction of mass, or **mass defect**, Δm which can be calculated from:

$$\Delta E = \Delta m c^2$$

where c is the speed of light (→ 40.5).

The mass of an atomic mass unit in kg is called the **atomic mass constant**, m_u. The mass of a single atom of $^{12}_{6}C$ is found by measurement to be

38 RADIOACTIVITY

1.99265×10^{-26} kg, and so:

atomic mass constant, m_u = mass in kg of 1 atomic mass unit, u

$$= \frac{1.99265 \times 10^{-26} \text{kg}}{12}$$

$$m_u = 1.66054 \times 10^{-27} \text{kg}.$$

Questions

2a Calculate the reciprocal of the atomic mass constant, m_u.
b Compare this with the value of the Avogadro constant, N_A. Why are they so similar? (N_A)

3 Many radioactive sources are still labelled in curie. Find the activity in Bq of sources labelled 1 µCi and 5 µCi.

4 Write down a simple equation relating the number N of neutrons to the mass number A and the atomic number Z for any nuclide.

38.2 RADIATION HAZARDS

Many naturally occurring elements on the Earth emit one or more of the three ionising radiations, as do many that are made artificially in nuclear reactors or by high-energy collisions in particle accelerating machines. The low-level **background radiation** from natural elements is normally considered to be harmless. However, higher levels of exposure can be dangerous. To ensure your safety, and the safety of others, you must follow certain procedures when using laboratory radioactive sources (see Figure 38.3). The following are some basic rules:

- always handle the sources with tongs or a source holder
- always point the source away from yourself and any other person, and never look into the open face of the source
- always return the source immediately after use to its lead-lined box. Never have more than one source out of its container at a time
- never poke any object under the wire mesh. Never bring the source into contact with any chemicals
- always wash your hands carefully when your experiment is finished.

Your body and any biological materials are damaged when they absorb the energy of the radiation. Very high levels of radiation cause burns and radiation sickness. Often damaged tissue can repair itself in the same way as from any other burn. But penetrating radiation can cause molecules to become dissociated or ionised. The molecules may then recombine in a different and dangerous

laboratory source	symbol	emissions	(energy /MeV)	half life/year
americium	$^{241}_{95}\text{Am}$	α	(5.49)	433
		γ	(0.06)	
strontium	$^{90}_{38}\text{Sr}$	β⁻	(0.546)	28.1
cobalt	$^{60}_{27}\text{Co}$	β⁻	(0.31)	5.26
		γ	(1.33)	
		γ	(1.17)	

Figure 38.3 Nature of laboratory sources

form. Important cell molecules, such as enzymes and DNA, carrying genetic codes for future generations, are sensitive to radiation. Extreme exposure to radiation can result in mutations or sterility, or lead to diseases such as anaemia, leukaemia and cancers.

Your everyday exposure to radiation will be very tiny compared with the strengths required to cause damage. The radiation you receive depends on the distance of the source and the time of exposure (see Figure 38.4). For example, a γ-ray source held in tongs for several minutes during the course of an experiment will give a negligible dose compared with the same source held in the hand for a few seconds. Any source of radiation that can be absorbed by the body is especially dangerous. Radioactive powder, dust, liquid or gas can all enter the body easily. The damage they then cause depends partly on the length of time the source is retained by the body. In the case of $^{90}_{38}$Sr, the time can be considerable.

This radionuclide decays by β-emission with a half-life of 28 years. It is a component of radioactive fall-out from atmospheric nuclear bomb tests and its abundance was a major worry in the 1950s before the atmospheric test ban treaty. Strontium has similar chemical properties to those of calcium, which is important in bone development. The radionuclide can find its way into milk through the natural food cycle. It can then replace calcium in the bones, particularly the growing bones of children. The bones become brittle after a number of years as strontium decays into another element, yttrium.

The explosion at Chernobyl in April 1986 released heavy radionuclides into the atmosphere. Many of these have found their way into the human food chain, especially through the meat of animals. Most are short lived but there is a very sinister residue from the disaster – caesium-137 which will still have half its activity left in the year 2016.

The film badge dosimeter (→ 38.6) is the commonest device used for monitoring the radiation received by an individual. The International Commission of Radiological Protection defines maximum permitted levels of exposure (see Figure 38.5). These depend on factors including age, sex and occupation. Students over the age of sixteen are permitted an exposure of an extra half of the normal background radiation in any one year. A single exposure must not exceed one tenth of this.

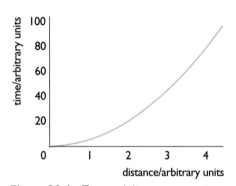

Figure 38.4 Time and distance to receive the same dose from an exposed source

maximum permissible dose rate for general public	5 mSv per year
maximum permissible dose rate for radiation workers	50 mSv per year
natural background dose rate	1.25 mSv per year
dose rate due to industrial, medical, and agricultural use	120 μSv per year
maximum dose rate due to atmospheric bomb testing (1954-61)	12 μSv per year
average dose rate due to nuclear reactors (general population, 1980)	2 μSv per year
threshold for induction of cataract	15 Sv life total
threshold for nausea	1 Sv in a few hours
threshold for fatality	1.5-2 Sv in a few hours
50% fatality within 30 days (from infection)	3 Sv in a few hours
gastro-intestinal death within 3-5 days	10 Sv in a few hours
central nervous system death within hours	≈20 Sv in a few hours

There is some controversy about these figures because no direct evidence.

Reference: American Institute of Physics Handbook

Figure 38.5 Permitted dose rates and the likely effects of various dose rates

ABSORBED DOSE AND DOSE EQUIVALENT

The radiation energy absorbed by the body, called the **absorbed dose**, is measured in **grays, Gy**. One gray is the dose when 1 joule of energy is absorbed in 1 kilogram of tissue.

Equal doses of different radiations do not produce equal biological effects. For the same absorbed dose, α-particles are 20 times more damaging than X-rays. To allow for this, the **dose equivalent**, measured in **sievert Sv**, is used. This equals the absorbed dose multiplied by a biological effectiveness, or **quality**, factor. The factor is unity for X-radiation.

For example, an adult person typically contains 4 000 Bq of potassium-40, which is a naturally occurring β-source. This contributes an annual dose equivalent of approximately 0.2 mSv. The total average annual dose equivalent from background radiation is 1 mSv. So a fifth of this is from potassium-40. Figure 38.6 shows the proportion of radiation from different sources absorbed by the average person.

A further complication is that the same dose equivalent presents a different level of risk to different organs in the body. This must also be taken into account when assessing a person's exposure to radiation. Overall, a whole body dose of 5 Sv is usually considered to be fatal.

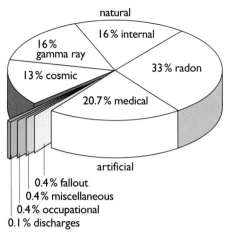

Figure 38.6 Pie-chart showing proportion of radiation from different sources absorbed by average person

Question

5 Given that the quality factor has no units, express the sievert in standard SI units.

NATURAL RADIATION

Figure 38.7 shows the common sources of natural radiation. Cosmic radiation from outer space is easily detected in the laboratory with a Geiger-Muller tube.

Figure 38.7 Sources of natural radiation

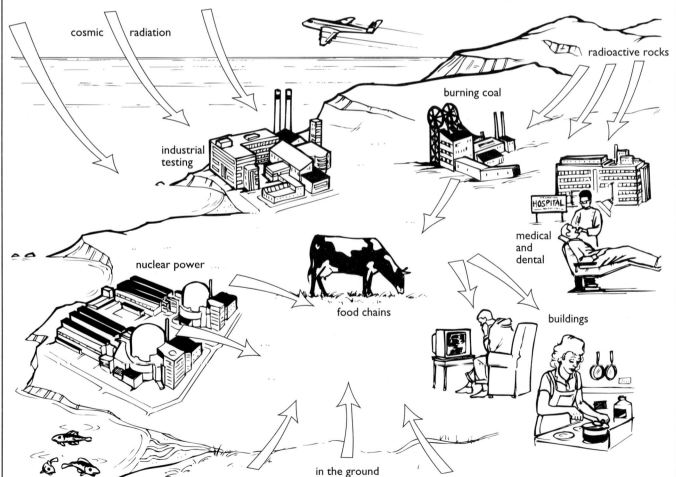

H INSIDE THE ATOM

Other sources include: contamination of apparatus from previous use; naturally occurring elements in the body, such as $^{14}_{6}C$ and $^{40}_{19}K$, which are β-sources; and in nearby rocks, such as $^{238}_{92}U$ and $^{226}_{88}Ra$, which are γ-sources.

The levels of natural radiation vary from place to place and are much higher in areas where the underlying rock is granite. The radioactive gas, radon, seeps out of the ground and from stone house building materials, etc. The highest level ever recorded in Britain was in a building of local stone in Cornwall.

38.3 ALPHA-PARTICLES

Between 1902 and 1910, Rutherford and collaborators counted the number of α-particles emitted from various sources, and measured their charge and energies. They proved that α-particles are helium nuclei – doubly ionised helium atoms, $^{4}He^{++}$. In 1914, Rutherford measured their **specific charge**, Q/m.

Background α-particle radioactivity and radon

A novel method of detecting α-radiation is to use a clear inert polymer material derived from allyl diglycol carbonate. It was developed a few years ago in the USA and is now manufactured under the name of TASTRAK at Bristol University. It is insensitive to γ-rays, X-rays, β-rays and light, and is particularly sensitive to α-particles. So it is ideal for analysing α-particle activity, such as that in the soil from radon gas emanating from underlying rocks.

The radon can contaminate houses and this has become a very important issue. In 1987, the Institute of Physics sponsored a survey of soil activity carried out by schools nationally. The results of this survey are shown in Figure 38.8. Each area of the country suffers from different levels of radon contamination due to variations in the amount and depth of radon producing rocks. Sumps, fans and similar devices may be used to try to reduce radon contamination. There are even building firms that specialise in radon remedial work.

Identifying the α-particle

The nature of the α-particle was found in an elegant and simple experiment performed by Rutherford and Royds in 1909. Radon gas, which emits α-particles, was collected and trapped in a very thin-walled tube A in the apparatus shown in Figure 38.9. The α-particles emitted by the gas atoms passed through the thin wall into the outer originally evacuated chamber B. After several days, the 'α-particle gas' in B was forced into a small space at the top by raising the level of mercury. An induction coil connected between the electrodes C caused a gas discharge. When analysed, this was found to have the spectrum of helium. Further control experiments verified that the helium atoms in tube B must be α-particles.

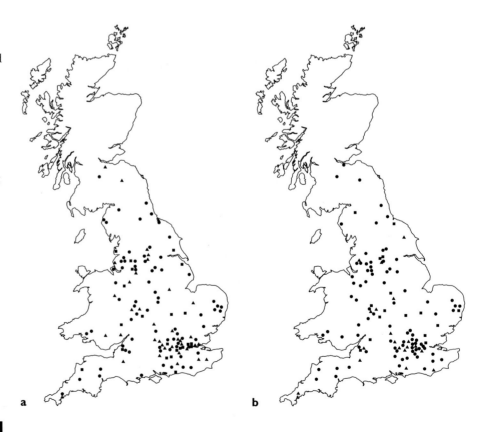

Figure 38.8 Distribution of radon activity in soil in Britain (From IOP survey published in *Physics Education* vol 23 (1988) pp 212–217, Figure 4a and 4b)
a ● <500 ▲ 500 – 1000 ■ >1000 Bq kg^{-1}
b ● <200 ▲ 200 – 400 ■ >400 Bq m^{-3}

38 RADIOACTIVITY

IONISING THE AIR

The α-particles, emitted by a particular radionuclide all have the same kinetic energy. But the value of this kinetic energy differs from one radionuclide to another. The particles travel with initial speeds of 1.4 to 2×10^7 m s^{-1}.

As each α-particle passes through air, it ionises the air molecules. It loses about 35 eV of its energy, on average, as its attractive electric field pulls the outermost electron from an air molecule. The massive positive air ion and free electron so formed are called an **ion pair**. The α-particle is about 7 000 times more massive than an electron. So it does not suffer any appreciable deflection from its straight path, provided it does not approach too closely to the nucleus of an atom.

Eventually, the α-particles are slowed to thermal speeds – the speeds of particles of a gas at room temperature. Then they can no longer cause ionisation. The distance α-particles can travel before losing their ionising ability is known as their **range**. In air at atmospheric pressure, this varies from 30 mm for the least energetic to 70 mm for the most energetic particles.

The number of ionisations produced by an α-particle in air can be measured using an ionisation chamber (see experiment 38.3). All electrical methods of detection of α, β and γ-radiations are based on the fact that they ionise atoms. Figure 38.10 shows quite spectacularly how straight and uniform in length the tracks of α-particles generally are. Their range in air can be measured from such photographs. Alternatively, you can use a spark counter (see Figure 38.11) or an ionisation chamber (see Figure 38.12) as in experiment 38.2.

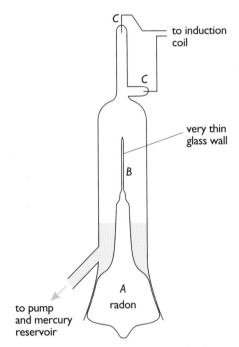

Figure 38.9 Apparatus of Rutherford-Royd's experiment

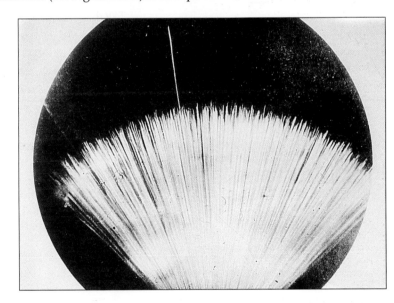

Figure 38.10 Tracks in a cloud chamber from α-particles

Questions

6a On average, how many ion pairs will a 1 MeV α-particle make along its path?
b Suggest why all the α-particles in Figure 38.10 do not have exactly the same range.

7 Use the kinetic model energy formula (← 10.5):

$$\tfrac{1}{2}m_\alpha V^2 = \tfrac{3}{2}kT,$$

where k is the Boltzmann constant, T is the temperature to estimate the final thermal speed of an α-particle. (m_α, k)

H INSIDE THE ATOM

EXPERIMENT 38.2 Finding the range of α-particles

You can use a spark counter, as in Figure 38.11, or an ionisation chamber fitted with a gauze lid, as in Figure 38.12. Moving the source away from the counter or gauze until the sparks cease or the current in the circuit falls to zero gives you the α-particle range. It is a very precise distance because α-particles from a particular nuclide all have the same energy.

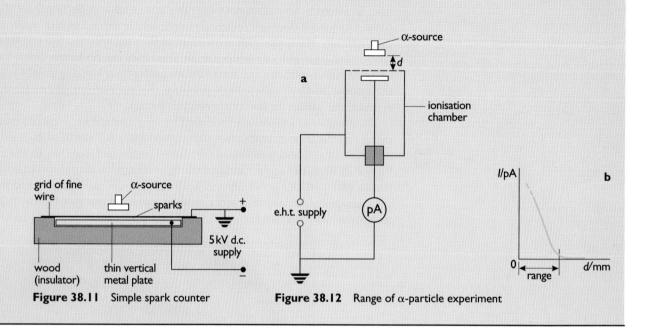

Figure 38.11 Simple spark counter

Figure 38.12 Range of α-particle experiment

Beta-particles and γ-rays do not ionise air molecules as efficiently as α-particles. So demonstrations, such as experiment 38.2, can be used to indicate the presence of α-radiation.

Using apparatus looking very similar to the spark counter, early experimenters found how the ionisation of air varies along an α-particle track (see Figure 38.13a). The mean range of α-particles against initial energy is shown in Figure 38.13b.

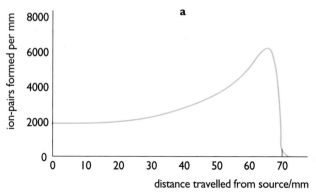
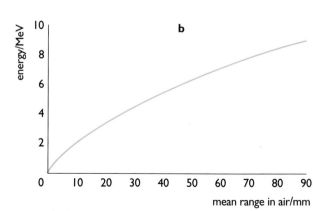

Figure 38.13a Number of ion-pairs per mm formed along the track of an α-particle in air, **b** mean range of α-particles in air at standard temperature and pressure

Questions

8 Suggest why the ionisation per mm in the graph of Figure 38.13a increases towards the end of the α-particle track.

9 How can the number of ion-pairs produced by an α-particle of a given range be found from Figure 38.13a?

10 Use Figure 38.13a and b to estimate the energy required to produce each ion-pair.

11 The thin end-window of a Geiger-Muller (GM) tube slows α-particles. The window is equivalent to about 20 mm of air. Use Figure 38.13b to find the maximum distance from an α-source that a GM tube can be placed to detect 5 MeV particles.

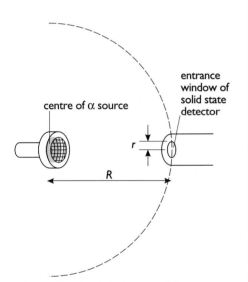

Figure 38.14 Calculating useful activity

Useful activity of a source

The useful activity of a sealed source is rather less than the **actual** activity. Consider a point source emitting α-particles equally in all directions, as in Figure 38.14. Suppose A is the activity, the total number of particles, emitted per second. The surface area of the sphere of radius R centred on the source is $4\pi R^2$ (← 36.2). So the number passing through unit area per second at a distance R is $A/4\pi R^2$.

The activity A_d registered by a perfect detector of window area πr^2 is:

$$A_d = \frac{A}{4\pi R^2} \cdot \pi r^2$$

$$= \frac{A.r^2}{4R^2}.$$

But half of the α-particles are emitted backwards and are absorbed by the source holder. Others are stopped by the side shields of the source holder. The source is a thin coating and not a single point in space at the centre of a sphere, as assumed in the expression above. The detector may also not be 100 % efficient. Taking all these factors into account, possibly only 25 % of the α-particles will actually reach the distance R from the source. A more realistic expression for the useful activity of the source is:

$$A = A_d \cdot \frac{R^2}{r^2}.$$

We can find the number of ionisations per α-particle when we know the useful activity of our source, which we measured in experiment 38.3a.

The constant current I in the ionisation chamber circuit of experiment 38.3b is the only measurement needed to find the number of ion-pairs produced per second in the chamber. Each ion-pair separates in the electric field in the chamber. The electrons are attracted to the central electrode at the source and the positive ions move to the casing. The number N of ion pairs created per second is given by

$$N = I/e,$$

where e is the electronic charge. The number of ion-pairs per α-particle is given by N/A, where A is the **useful activity** of the source. You can also calculate the initial energy of an α-particle, knowing how much energy on average is needed to create an ion-pair, about 35 eV.

H INSIDE THE ATOM

EXPERIMENT 38.3 Measuring activity of an α-source and ion-pair production

a Measuring the activity

Using a solid state detector, as in Figure 38.15, you can count the number of α-particles received over several timed intervals. From the mean number A_d of particles received per second over the sensitive surface area of the detector window, you can calculate the useful activity of the source from:

$$A = A_d \cdot \frac{R^2}{r^2}$$ where R is the distance between source and detector.

b Measuring the number of ion-pairs produced by an α-particle in air

When you place an α-source inside an ionisation chamber, as shown in Figure 38.16, and raise the supply voltage to about 1 000 V, the current I reaches a constant value. The number of ion-pairs produced per second can be calculated from I/e. Dividing this by A from experiment 38.3a gives the number of ion-pairs produced by each α-particle:

$$\text{number of ion-pairs produced} = \frac{I}{Ae}.$$

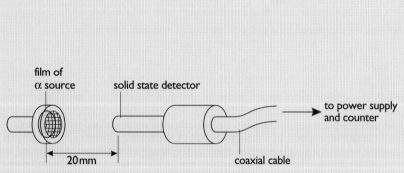

Figure 38.15 Arrangement for experiment 38.3a

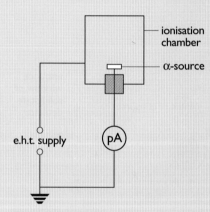

Figure 38.16 Circuit for experiment 38.3b

Questions

12a Why does the current in experiment 38.3b become constant above a particular supply voltage?
b Suggest why the ammeter reading rises and falls after each change of voltage. **Hint:** when a hand is placed close to the ionisation chamber the current rises and then falls back to its former value; when the hand is removed the current falls and then rises again.

13 In experiment 38.3a, a student recorded the following results:
radius of detector = 1.8 mm
distance of detector from source = 20 mm
measured count rate = 300 s^{-1}.
In experiment 38.3b, she obtained:
    ionisation current = 3 × 10^{-10} A.

a What is the useful activity of the source in **i** Bq and **ii** Ci?
b The charge on an ion is 1.6×10^{-19} C. How many ion-pairs are produced per second?
c How many ion-pairs are produced per α-particle?
d Given that, on average, 35 eV is required to produce an ion-pair in air, estimate the energy of each α-particle. What assumptions are you making in arriving at this estimate?

38.4 BETA-PARTICLES

Both α- and β-particles ionise gases and this is the basis of electrical methods to detect them. Beta-particles also lose energy by creating ion-pairs. However their ionising ability is about 100 times less than that of α-particles. As a result, the range of an average β-particle is about 1 metre in air. The range is sufficient to allow you to pass a beam of β-particles between the poles of a magnet and to measure their deflection in the field as they are charged particles, as in experiment 38.4. This is impossible with α-particles as their range in air is too short, only a few centimetres.

EXPERIMENT 38.4 Deflecting β-particles in a magnetic field

By putting a Geiger-Muller (GM) tube in front of a narrow beam of β-particles, you can measure the count rate. If you then place a strong magnet between the source and tube, as shown in Figure 38.17a, you should notice that the count falls to the background value. Moving the GM tube horizontally around the arc of a circle centred on the magnet through about a right angle should enable you to detect a strong signal, again at angle θ (see Figure 38.17b). You could also try the effect turning the magnet over, so reversing the magnetic field.

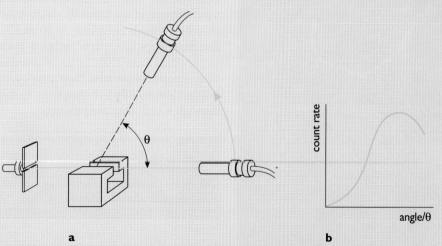

Figure 38.17a Arrangement and **b** plot of deflection of β-particles in a **B**-field

H INSIDE THE ATOM

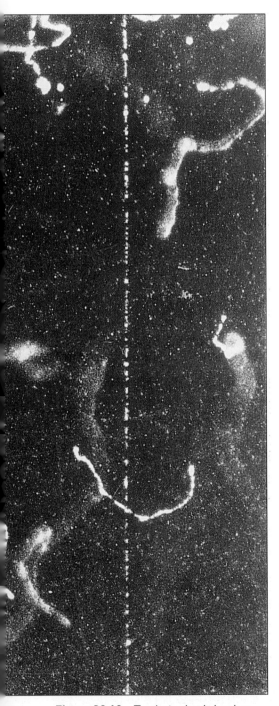

Figure 38.18 Tracks in cloud chamber from β-particles

Questions

14 This question refers to experiment 38.4.
a What happens when the direction of the magnetic field is reversed by turning the magnet over?
b Suggest two reasons why this experiment cannot be performed with an α-particle source.
c How would the experiment have to be modified to show that α-particles are deflected in a magnetic field?

15 Estimate the ratio of the specific charges (Q/m) of an α-particle to a β-particle. (m_α, m_β, e)

The direction and large deflection of the β-particles in experiment 38.4 show that they are negatively charged and have a small mass. The measurement of their specific charge, e/m, by Bucherer in 1908 confirmed that they are fast-moving electrons. α-, β- and γ-radiations are all emitted with energies of similar magnitude. A high-energy β-particle typically has an energy of 3.5 MeV. This gives a particle of the mass of an electron a speed close to that of light, $3 \times 10^8 \text{ m s}^{-1}$. So **relativistic mechanics** (← 9.6) instead of Newtonian mechanics must be used in any calculations on the motion of the particles. In relativistic mechanics, the mass of a particle increases significantly as its speed approaches that of light. This explains the finding of early experimenters that the specific charge of the β-particle decreases as its energy increases. At low speeds, β-particles have the same value of e/m as electrons emitted from a heated cathode and accelerated in an electric field.

COMPARING β- AND α-PARTICLES

The behaviour of the β-particle differs from the α-particle in two important ways. The α-particle brushes aside electrons in the material in which it is travelling. It is only deflected significantly on the rare occasions when it comes close to an atomic nucleus. Since all of the α-particle's collisions cause ionisation, the α-particle loses its energy in a short distance. So the range of the α-particle is well defined and straight. In contrast β-particles, having the same mass as atomic electrons, are easily deflected in encounters with them. So a β-particle may have a path length many times its range. Compare the tracks of the particles in a cloud chamber, shown in Figures 38.10 and 38.18. A piece of paper of the thickness of this page stops all α-particles but β-particles of energy of 1 MeV or more are able to penetrate more than 1 mm of aluminium sheet. They are more penetrating than α-particles because they travel very fast and have a smaller charge, so they interact less readily to lose energy to the atoms or molecules of the material through which they are passing.

Questions

16 Why do slower-moving β-particles have tracks which are less regular and thicker than fast-moving β-particles?

The second major difference between α- and β-radiation is in the kinetic energy of emission. All α-particles emitted from a given radionuclide have the same energy. However, as Becquerel discovered at the beginning of the twentieth

38 RADIOACTIVITY

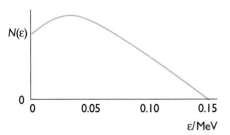

Figure 38.19 β-particle energy spectrum for $^{14}_{6}C$

century, β-particles are emitted with energies ranging from zero to a maximum value.

The β-particle spectrum of $^{14}_{6}C$ is shown in Figure 38.19. $N(\epsilon)$ is the number of β-particles emitted in a small range of energies close to ϵ. The mean energy is 0.05 MeV and the maximum energy, 0.15 MeV. This variation in energy gives rise to the considerable spread in the deflection angle of Figure 38.17b. If all the β-particles from a source were emitted with the same energy, the Geiger-Muller (GM) tube would only detect β-particles over a small range of angles.

RANGE AND ENERGY OF β-PARTICLES

α- and β-particles both radiate energy as X-ray photons when they are slowed by the electric fields of the charged particles in a solid material. The transfer of energy by this process is inversely proportional to the square of the mass of the particle. So the energy loss is some 5×10^7 times greater for β– than α-particles. The various processes causing β-particles to lose energy combine to give, in practice, an almost exponential decrease with thickness in the number of β-particles passing through an absorber (see Figure 38.20).

For β-particles, a better quantity to measure than thickness is **absorber thickness**, or **surface density**, measured in kg m^{-2} (see Figure 38.21).

Absorber thickness = density (kg m^{-3}) × thickness (m).

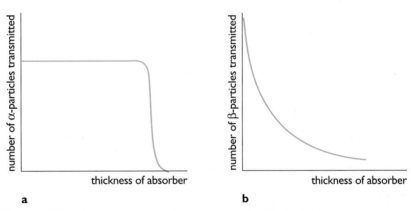

Figure 38.20 Penetration curves for **a** α-particles and **b** β-particles

This has the advantage that the material does not have to be stipulated. The penetration of β-particles is almost the same for any of the light elements. Dividing the **maximum** absorber thickness, or **range**, by the density of the chosen material gives the actual thickness of material required to stop the most energetic β-particles. So, if you know the density of each material, you can use Figure 38.21 to find the absorption effects of a GM tube window material, an aluminium foil absorber and the air between a β-source and its detection, as in experiment 38.5.

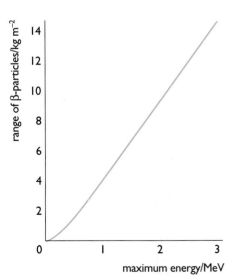

Figure 38.21 Range, or absorbing thickness, of β-particles plotted against maximum energy

Question

17 Calculate the absorber thickness in kg m^{-2} of a material 1 mm thick and density 7200 kg m^{-3}.

H INSIDE THE ATOM

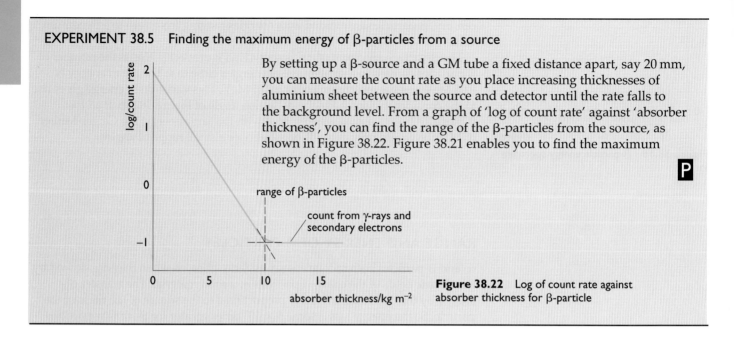

EXPERIMENT 38.5 Finding the maximum energy of β-particles from a source

By setting up a β-source and a GM tube a fixed distance apart, say 20 mm, you can measure the count rate as you place increasing thicknesses of aluminium sheet between the source and detector until the rate falls to the background level. From a graph of 'log of count rate' against 'absorber thickness', you can find the range of the β-particles from the source, as shown in Figure 38.22. Figure 38.21 enables you to find the maximum energy of the β-particles.

Figure 38.22 Log of count rate against absorber thickness for β-particle

38.5 GAMMA-RAYS

A γ-ray is emitted from a nucleus when the nucleons within rearrange themselves and move into a lower energy state. This can happen after an α- or a β-particle has been emitted by a **parent** nucleus (→ 39.1). The nucleons in the daughter nucleus move into a lower energy state, releasing the excess energy as high-energy electromagnetic radiation – a γ-ray.

There are many naturally occurring γ-ray sources. They may also be produced artificially, usually involving bombardment by high-energy particles, such as neutrons and protons, from nuclear reactors and particle accelerators (→ 40.4). In this way, radionuclides of specific activity may be engineered for use in, for example, medicine or scientific research (→ 38.7).

One of the common laboratory α-particle sources, $^{241}_{95}$Am, emits some low-energy γ-radiation. Strictly, it is the daughter element $^{237}_{93}$Np that emits the γ-ray. Similarly a $^{60}_{27}$Co source is used for γ-radiation experiments, and it is usual to talk of $^{60}_{27}$Co γ-rays. However $^{60}_{27}$Co, emits a low energy β-particle and it is $^{60}_{28}$Ni that emits 1.17 MeV and 1.33 MeV γ-rays (see Figure 38.23).

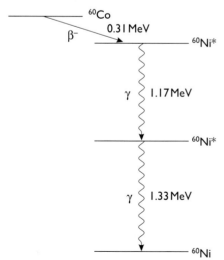

Figure 38.23 Energy levels in decay of $^{60}_{27}$C

Food and γ-radiation

When biological material is exposed to low-level γ-radiation, mutations in the genetic material may occur (← 38.2). Such mutations can help plant breeders to produce crops with new desired characteristics. These include increased resistance to disease and larger yields.

Bacteria are destroyed by γ-radiation. In medicine, materials such as surgical instruments, hypodermic needles and dressings can be sterilised more effectively by γ-rays than by, for example, heat treatment. Similarly low-intensity γ-radiation can be used to sterilise fresh food and so increase its storage life. Governments of a number of countries allow the use of this process and believe it to be neither harmful to consumers nor to affect the taste of food. The process is particularly helpful in hot climates where, without treatment, much of the fresh produce rots before it reaches the consumer.

38 RADIOACTIVITY

Wavelengths of γ-rays are a millionth those of visible light, which are centered around 500 nm, and a thousandth those of X-rays, which are about 100 pm. The energy released in atomic electron transitions as light is of the order of several eV, as X-rays, many keV, and as γ-rays, several MeV.

Light and X-rays emitted by such transitions have precise energies (← 37) and the same is true for γ-rays. β-emission is the only one of the three nuclear radiations with a broad energy spectrum.

> **Question**
>
> **18** Calculate the wavelengths λ associated with the γ-rays from $^{60}_{27}$Co using:
>
> $$E = \frac{hc}{\lambda},$$
>
> where E is their energy, h is the Planck constant and c is the speed of light.
> (h, c)

INTERACTING WITH MATTER

Gamma-rays are not deflected by electric or magnetic fields. Being uncharged, they also cause very little ionisation. But γ-rays can interact with matter in three distinct ways, depending on their energy:

i a low-energy γ-ray can lose all its energy to an atom, creating an ion pair. This is known as photoelectric absorption (← 36.3). In contrast, α- and β-particles, lose their energies in making many thousands of ionisations. The resulting ion pair will be very energetic, having sufficient energy to make further ionisations. It is for this reason that a GM tube can detect γ-rays, although rather inefficiently. A few of the γ-rays entering the tube are absorbed by atoms of its walls. The energetic electrons emitted by the ionised atoms cause ionisations in the gas, so creating a discharge pulse which can be counted (→ 38.8)

ii higher-energy γ-rays interact by the Compton effect (←36.3)

iii very high-energy γ-rays passing near nuclei give rise to electron-positron pairs, which is known as **pair production** (→ 40.3).

Air absorbs little of the energy of γ-rays. So the intensity of γ-rays in air falls off as the inverse square of the distance from the source, in much the same manner as light from a lamp (→ experiment 38.6).

EXPERIMENT 38.6 Inverse square law for γ-rays

By placing a GM tube at several distances R from a γ-ray source, you can find the average count rate N for a time interval of 10 seconds, say, at each distance. You can then plot a graph of distance R against $1/\sqrt{}$ (count rate). For an inverse square law relationship, this will be a straight line. You will need to make corrections for **a** the background count when the count rate is small, and **b** the 'dead time' of the GM-tube (← 38.8) when the count rate is very large.

INSIDE THE ATOM

Questions

19 Experiment 38.6 shows that γ-rays obey an inverse square law. Given that, show that the following equation holds:

$$N_{\text{detected}} = kN_{\text{emitted}} \cdot \frac{r^2}{4R^2}$$

where N_{detected} is the measured count rate, N_{emitted} is the activity of the source, r is the radius of the detector 'window' facing the source, R is the separation of source and detector, and k is a constant related to the efficiency of the counter (← 38.3).

20a The graph in experiment 38.6 is unlikely to pass through the origin. Suggest a reason for this.
b Why is it sensible to plot R against $1/\sqrt{\text{count rate}}$ and not count rate against $1/R^2$?

In solids, such as lead, the absorption of γ-rays follows an exponential law, as shown in experiment 38.7. The absorption is of the form:

$$N = N_0 e^{-\mu x}$$

where N_0 is the initial count rate with no absorber, x is the thickness of the lead and μ is the **linear absorption coefficient**. The thickness at which the count rate, N, falls to half its initial value is known as the **half-value thickness** of lead to the radiation. For a $^{60}_{27}$Co source this thickness is 12.5 mm.

EXPERIMENT 38.7 The absorption of γ-rays by lead

Using the arrangement in Figure 38.24a, you can place increasing thicknesses of lead sheet between a γ-source and a GM tube, keeping the distance between source and detector constant. This allows you to plot a graph of count rate against thickness of lead, and obtain the half-value thickness of lead, as in Figure 38.24b. Compare this with the slope of the log of count rate against thickness.

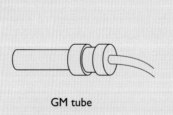

Figure 38.24a Arrangement for experiment 38.7, **b** graph of count rate against thickness of lead

Question

21 In experiment 38.7 why is it preferable:
a to plot 'log of count rate' rather than just 'count rate'?
b to measure the half-value thickness rather than the range, or penetration, of γ-rays through lead?

38.6 FILM BADGE DOSIMETER

The danger from radioactivity depends on the energy, time of exposure and penetration of the radiation (← 38.2). It is easy to protect ourselves from α and β-radiations outside the body as their penetration power is small, although higher-energy β-particles can be dangerous to skin and eyes. Gamma- radiation is much more dangerous. If you pass a GM counter around the outside of the lead box in which the γ-source is kept, you will probably obtain a count rate well above background level.

The most common device used to monitor exposure to radiation is the film badge dosimeter. This looks like a large rectangular plastic brooch. It is worn by anyone who works close to radioactive materials or X-rays. The emulsion of the film is blackened by exposure to β, γ or X-rays. The film is mounted in a special plastic holder containing windows covered by filters of different materials and thicknesses. So the device can distinguish the radiations by their different penetrating powers (see Figure 38.25).

Comparison of the blackening under a plastic filter with that of the open window enables the strength of β-particle energies to be found. Comparison of the blackening under a metal filter with the open window allows the strength of X and γ-ray photons to be measured. (The β-particles are absorbed by the metal sheet.) The film has emulsions of different sensitivities on its two surfaces. The slow-reacting rear surface of the film is used to monitor intense radiations, while the more sensitive front surface monitors weak radiation exposure. So a wide range of exposures can be measured by one film strip.

Figure 38.25 Film badge dosimeter

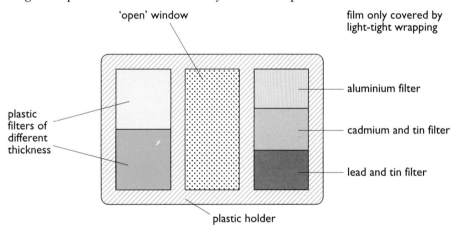

Questions

22 A developed film shows the open area heavily fogged but the two covered areas almost clear. Suggest the nature and strength of the radiation to which the person was exposed.

23 Why is a dosimeter not suitable to detect α-particle radiation?

38.7 APPLICATIONS OF RADIATION

Many radionuclides are made in nuclear reactors (→ 40.6) for use in medicine and industry. The following are some examples of their applications.

MONITORING THICKNESS AND LEVEL

The absorption of β-radiation depends essentially on the thickness of the absorber (← 38.4). So in the manufacture of sheet materials, such as paper, plastic or steel, a constant check can be kept on the thickness of the sheet using a collimated β-source on one side of the sheet and a detector on the other. The detector can be attached to a suitably calibrated meter to indicate the thickness directly. When linked to a control system in the machinery, it can correct automatically any change in thickness.

The absorption property of β-particles can also be used to measure the level of liquids, powders or even pills in opaque containers. A sudden change in count rate occurs when the substance in the container rises above the level of the beam of β-particles. Cracks or cavities in castings or pipes can be detected by moving the source and detector simultaneously along the casting. Any sudden increase in count rate indicates a cavity within the object.

CHECKING THE QUALITY OF COAL

The quality of coal depends on how much stone is mixed with it: the less stone, the higher the coal's value for heating purposes. In coal mining, the quality of coal is monitored as it passes from the coalface along a conveyor belt. The material is kept at a constant thickness. A γ-source is placed above the belt and a detector below. The absorption coefficients for coal and stone are different. So the detector can be calibrated to show the percentage of coal in the material on the belt.

MONITORING THE WEAR OF MOVING PARTS

The wear to moving parts of machinery, such as the moving parts of aircraft engine, may be monitored and tested using radionuclides (see Figure 38.26). Some of the nuclei in the piston rings are made radioactive by bombarding the rings with neutrons for a short time inside a nuclear reactor. The bombarding neutrons can be focused so that only the surface layers down to

Figure 38.26 Using γ-rays to monitor wear on aircraft engine fans

Figure 38.27 Using α-rays in a smoke detector

0.1 mm of the ring become radioactive. This greatly reduces the radiation hazards to workers. The engine, containing the radioactive rings, is run on a test bed. The decrease in activity from the piston rings is proportional to their wear. The lubricating oil now contains radioactive metal atoms. The increase in activity from the oil is also measured. Any radioactivity in the oil must come from the material worn away, providing another check on the rate of wear.

DETECTING SMOKE

One type of smoke detector (see Figure 38.27) uses a low-activity α-source, usually $^{241}_{95}$Am, of activity 3×10^4 Bq. The α-particles ionise the air, causing a small steady current in an otherwise open circuit. When smoke particles enter this 'ionisation chamber', the ionisation current caused by the radioactive source falls. A sufficient change in current triggers an alarm circuit. This detector has the advantage over an optical system that it can detect invisible products of combustion from fast-burning, high-energy fires, such as burning solvents.

MEDICAL USES

Radiotherapy with γ-rays from $^{60}_{27}$Co is used in the treatment of cancer. Sources of activity as great as 10^{13} Bq are used for deep therapy. Such activities are lethal to living tissue, so the exposure has to be carefully calculated and controlled. The γ-rays are focused on to the malignant tissue as shown in Figure 38.28. Elaborate safety precautions are necessary for both patient and attendant medical staff.

Weak γ-rays from such radionuclides as ^{131}I are widely used in medical diagnosis as **tracers** (→ 39.8).

Figure 38.28 Patient undergoing radiotherapy

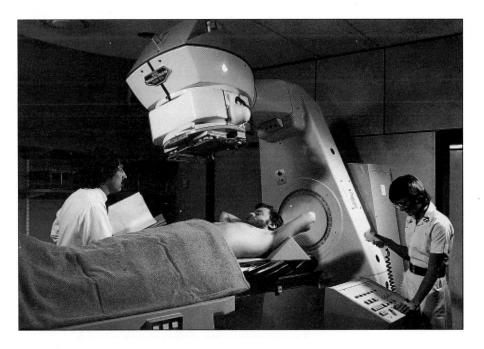

38.8 RADIATION DETECTORS

Many detectors, including the ionisation chamber, the Geiger-Muller tube, the diffusion cloud chamber and the spark counter, respond to the ionisation caused by the passage of nuclear radiations through a gas or vapour. The simple **ionisa-**

INSIDE THE ATOM

tion chamber and spark counter (← 38.3) are frequently used to detect α-particles.

Another α-particle detector, used in experiment 38.3a, is the **solid state detector**. This will also detect β-particles, γ-rays, protons and neutrons. It works in a similar way to a solar cell (← 36.5). Electrons and holes are created by each incident particle reaching the *pn* junction of the diode at the detector entrance window. A voltage applied across the junction then produces a short pulse of current, which can be amplified and counted.

GEIGER-MULLER TUBE

Shown in Figure 38.29, this special ionisation chamber is the most versatile of the detectors of nuclear radiations. It is usually a sealed metal cylinder containing a gas such as argon at about 0.1 of normal atmospheric pressure. The cylinder is the earthed cathode and the central wire, the anode. There is an optimum potential difference of about 450–500 V between the electrodes for the most efficient counting of incident particles. The counting efficiency is approximately 100 % for α– and β-particles, which enter the sensitive region through the thin mica window.

When, for example, a β-particle enters the tube, it ionises the gas along its track. The freed electrons are accelerated towards the wire by the strong electric field close to the wire. This causes a cascade of ions and secondary electrons along the entire length of the wire. The electrons are counted as a single negative pulse of approximately the same size, whatever the energy or path of the incident particle. The entire negative pulse takes less than 1 microsecond. However the positive ions, being many thousand times more massive than the electrons, take several hundred times as long to reach the outer cathode. During this time, called the **dead** or **paralysis time** of the counter, further ionising particles cannot be counted. This means that a Geiger counter cannot count events occurring less than 10^{-4} s apart.

The ions collide with the cathode at speed, causing secondary electrons to be emitted from the surface. These electrons would be accelerated to give further spurious counts were it not for the **quenching vapour**. This is provided by a small amount of bromine added to the gas. The bromine molecules absorb energy from the ions or secondary electrons and dissociate into bromine atoms. The atoms then readily recombine into molecules again for the next pulse.

Because the GM tube is so sensitive, it is affected by cosmic rays, local contamination of apparatus and the other environmental radiations that make up the background radiation (← 38.2). So you should always measure the background count before and after any experiment, and make suitable corrections to the count rate.

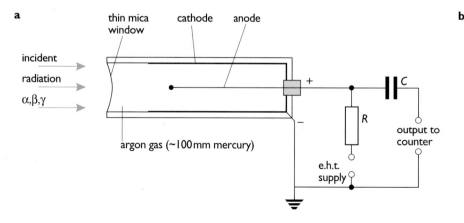

Figure 38.29a Diagram and **b** photograph of a Geiger-Muller tube

38 RADIOACTIVITY

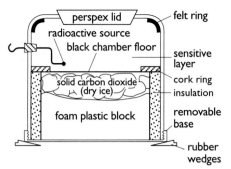

Figure 38.30 Diffusion cloud chamber

DIFFUSION CLOUD CHAMBER

As shown in Figure 38.10 and 38.18, this reveals dramatically the paths taken by any ionising particles. Each track is an indirect effect of the electric forces exerted by the fast-moving charged particle on the atoms of the vapour. The charged particle leaves a trail of ion pairs. The saturated vapour within the chamber condenses around these ions in much the same way as water vapour condenses on the ejected particles from a high-flying aircraft in a clear sky, leaving a visible white vapour trail.

A simple cloud chamber is shown in Figure 38.30. The felt strip inside the top of the chamber is saturated with ethanol. The base of the chamber is in contact with dry ice, which is solid carbon dioxide. The ethanol evaporates continuously, producing a saturated vapour close to the cold floor. When an α– or β-particle passes, the vapour condenses on the ion pairs formed, making the track of the particle visible.

OTHER IONISATION COUNTERS

The bubble chamber (see Figure 38.31) and the multiwire chamber (see Figure 38.32) are used to track high-energy particles. The bubble chamber works rather like a cloud chamber "in reverse". The multiwire chamber is like stacks of rows of GM tubes, with each wire acting as a separate anode. This chamber can record millions of particles per second.

Figure 38.31 Bubble chamber

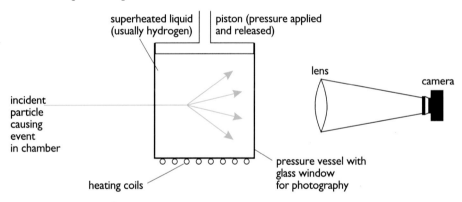

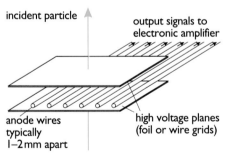

Figure 38.32 Multiwire chamber

> ### Questions
>
> **24a** Why must a GM tube for detecting α-particles have a very thin end window?
> **b** Why does a GM tube for detecting γ-rays not need a window at all?
>
> **25** Rubbing the perspex top of a cloud chamber with a cloth makes the α-particle tracks sharper and clearer. Suggest a reason for this.

SCINTILLATION COUNTER

This detector (see Figure 38.33) uses the fact that all nuclear radiations make certain phosphors fluoresce (← 37.4). Each particle incident on the phosphor causes the emission of a number of optical photons, proportional to the energy of the particle. Zinc sulphide is used as the scintillator for detecting α-particles and doped sodium iodide for γ-rays. The photons from the scintillator are detected by a photomultiplier tube (← 35.5), converted into an electric pulse and counted.

The combination of phosphor, or scintillator, and photomultiplier is called a

H INSIDE THE ATOM

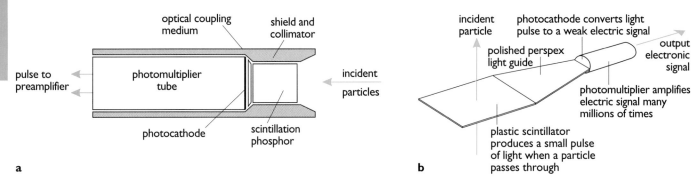

Figure 38.33 Two types of scintillation counter

scintillation counter. Its advantages over the GM tube are that it can:

- determine the relative energies of the incident particles
- detect very low-energy radiation and so can distinguish small energy differences, giving it a high resolution
- detect uncharged particles, such as the γ-ray and neutron, with nearly 100 % efficiency, with the appropriate scintillator material
- count very rapidly, the flashes being as short as 1 nanosecond.

A γ-ray camera consists of a bank of scintillation counters. In high-energy particle physics, large-area low-cost scintillating materials are used to detect particles, as in Figure 38.33b.

SUMMARY

The major properties of the three types of nuclear radiation are listed in Figure 38.34.

The radiations can be distinguished by their penetrating power or deflection in an **E**-field or **B**-field.

The nucleus of an atom with a particular proton and neutron number is known as a **nuclide** and is specified using the notation:

$$^{A}_{Z}\text{X} \quad \begin{array}{l}\text{nucleon number A} \\ \text{proton number Z}\end{array} \quad \text{chemical symbol}$$

where the nucleon number equals the sum of the proton and neutron numbers. There are more than 2000 nuclides, those emitting radiation being called **radionuclides**.

Nuclides of a particular element, that is, those with the same proton number are called **isotopes**.

The activity of a radionuclide, the number of disintegrations per second, is measured in becquerels, Bq.

Radiation can damage all biological matter. Radiation damage depends on time of exposure, distance from source, strength of source and radiation absorption.

Standards of safety are defined in terms of the dose equivalent, measured in sievert (Sv), received over a given period of time. Workers who may be exposed to radiation wear a film badge to measure their exposure.

Most radiation detectors are based on the fact that the radiations have an ionising ability. The GM tube is the most versatile ionisation detector.

The scintillation counter detects the photons emitted when radiation strikes a phosphor. It is the most efficient detector of all three radiations.

property \ radiation	nature of particle	mass/m_u	charge/e	energy	speed/ ms^{-1}	ion-pairs/ mm of air	approx. no. of collisions before absorption	range in air	absorbed by	affects a fluorescent screen	deflection in E- and B-fields
α	helium nucleus	4	+2	constant for given nuclide	~10^7	10^3–10^4	10^5	50–70 mm	paper	strongly	as +ve charge (small)
β⁻	electron	1/1837	−1	0 to max.	up to ~10^8	~10^2	10^5	~1 m	1–5 mm aluminium	yes	as −ve charge (large)
β⁺	positron		+1								as +ve charge (large)
γ	photon	0	0	constant for given nuclide	3×10^8	~0	1	obeys inverse square law	1–10 cm lead	strongly for sodium iodide	none

Figure 38.34 Summary of the nature of α, β and γ-radiation

SUMMARY QUESTIONS

26a The symbol ^A_ZX represents a particular nuclide. Write down the symbol for the helium nuclide – the α-particle.
b Lead has three stable isotopes. How does an isotope differ from a nuclide?
c The percentage abundance of the three isotopes of silicon are ^{28}Si 92 %, ^{29}Si 5 % and ^{30}Si 3 %. Calculate the relative atomic mass of silicon.
d One of the radionuclides used as a tracer in medical diagnosis is $^{131}_{53}$I. How many neutrons are there in this nuclide?

27 A small positively charged sphere is suspended on an insulating thread close to a radioactive source. How will the charge on the sphere change when the source emits: **a** α-particles? **b** β-particles? **c** γ-rays?

28 A radioactive source emits two types of radiation. Outline experiments to check this statement and to identify the radiations.

29 In terms of self absorption, explain why radioactive sources for laboratory work, $^{241}_{95}$Am, $^{90}_{38}$Sr and $^{60}_{27}$Co are prepared as thin coatings on a metal plate. Hint: Think about the self absorption of the radiation by a spherical source.

30 A GM tube and counter are used to measure the count rate from a sample of rock. Suggest some of the problems that arise in determining the total activity of the radioactive material within the lump of rock.

31 The energy necessary to ionise either oxygen or nitrogen is about 14 eV. However, an α– or β-particle loses about 34 eV, **on average**, to create each ion pair. Where do you think the other 20 eV goes?

32 Below an energy of a few MeV, an α-particle is not always a doubly ionised nucleus. It can pick up one or two electrons, so that it exists part of the time as a singly charged ion and part as a neutral atom. These times are short compared with the time it is doubly ionised. What effect will this have on the range of the α-particle?

33 A student suggests that the range of α-particles in air can be studied by placing a solid-state detector a given distance from an α-source in a chamber containing air at low pressure. The pressure is increased until the α-particle count rate falls to zero. The distance between the source and detector can be changed and the experiment repeated.
a Discuss the feasibility of this suggestion.

b Is there any other information about the passage of α-particles through air that could be found from this experiment?

34 Until recently, the unit of activity was the curie. The curie is equal to the total number of α-particles emitted by 1 g of radium in 1 s or 3.4×10^{10} Bq. This is how the unit originated. Rutherford found that the charge lost by 1 g of radium in 1 s was 10.5 nC. What can you conclude about α-particles from this information?

35 $^{90}_{38}$Sr is a common laboratory nuclide. Use Figure 38.21 to estimate the range of a β-particle of energy 8.8×10^{-14} J (550 keV) emitted by ^{90}Sr in:
a air, of density 1.3 kg m^{-3},
b aluminium, of density 2 700 kg m^{-3},
c lead, of density 11 000 kg m^{-3}.

36a A GM tube measures a background count of 25 min^{-1}. A small $^{60}_{27}$Co source emitting γ-rays is placed 25 cm from it. The count rate is 825 min^{-1}. Estimate the count rate you would expect when the source is moved back 25 cm. What assumptions have you made in reaching your answer?
b In another experiment, the γ-source and GM tube are kept a fixed distance apart. Various lead sheets of different thicknesses are placed between the source and the detector. The table of corrected count rate A_d recorded against absorber thickness x is:

A_d s^{-1}	48	40	33	27	23	16	8
x mm	0	3.1	6.2	9.3	12.4	18.6	31.0

Plot a suitable graph to show that the γ-rays follow an exponential absorption law and find the half-value thickness for the $^{60}_{27}$Co γ-rays.

37 Apart from the correction required for background radiation, another correction should be made for the paralysis or dead time of a GM tube. It takes about 300 μs for the tube to recover before it can count the next particle. When two particles arrive within 300 μs of each other, only the first will be detected.
a Suppose the recorded count rate is 300 counts per second. How long is the counter inoperative during each second?
b For what fraction of a second is the counter operative?
c Calculate the true number of particles entering the GM tube.
d Having made these corrections, suggest a design for a simple experiment to measure the efficiency of a GM tube.
e Would you expect to find the same efficiency of detection for β- and γ-radiation?

38a Draw an arrangement of apparatus to check the quality of coal (← 38.7).
b Suggest a suitable γ-ray counter for it.
c Why is it necessary to have the conveyor belt covered with a uniform thickness of mined material?
d Suggest how you might produce a collimated beam of γ-rays.
e What is the fractional change in detected signal when the absorption coefficient changes from 0.40 m^{-1} to 0.39 m^{-1}? (← 38.5)
f Suppose the background count in the mine is 20 min^{-1}. The source strength is 1.0×10^4 Bq in a collimated beam towards the detector. The material on the belt is 0.30 m thick. Is the change in **e** detectable?

39 RADIOACTIVE DECAY

When unstable nuclei disintegrate, they generally emit α-, β- or γ-radiation. Usually γ-radiation is emitted immediately after an α- or β-particle is ejected. Occasionally, it is emitted by itself. Knowing how α-, β- and γ-radiation behave allows you to detect and distinguish between them (← 38).

The changes in a nucleus that produce these emissions are independent of any external factors, such as pressure or temperature. They occur quite randomly and spontaneously. However, the **decay** of a large number of such nuclei in a sample of radioactive material is a highly predictable process. An understanding of the decay process has led to many applications, including the radioactive dating of rocks and other archaeological remains, and the use of radionuclides as power sources and as tracers.

39.1 EQUATIONS OF RADIOACTIVE DECAY

The emission of an α- or β-particle changes an atom into a new element. The nucleus of the atom will increase or decrease its proton number or mass number (← 38.1) depending on the nature of the emission. Orbital electrons are gained or lost, so keeping the atom neutral.

When an atom changes, the **parent** atom is said to disintegrate to form a **daughter** atom of another element. In a **radioactive decay chain** the daughter produces a granddaughter and so on, with the emission of an α- or β-particle at each stage.

In α-**decay**, the parent nuclide loses a helium-4 nucleus. This contains four nucleons, two of which are protons, so A decreases by 4 and Z by 2 (← 38.1). Using the symbols P and D for the parent and daughter elements respectively, α-decay can be expressed in an equation:

$$^{A}_{Z}P \rightarrow ^{A-4}_{Z-2}D + ^{4}_{2}He$$

(see Figure 39.1a). For example, radon-220 decays to polonium-216 emitting an α-particle:

$$^{220}_{86}Rn \rightarrow ^{216}_{84}Po + \alpha.$$

In β-**decay**, a neutron decays to a proton, and a high-energy electron is emitted from the nucleus. So the nucleon number A is unchanged but the proton number Z increases by one. The equation for β-decay is:

$$^{A}_{Z}P \rightarrow ^{A}_{Z+1}D + ^{0}_{-1}e$$

(see Figure 39.1b). Protactinium-234 decays by emission of an electron or β-particle to become uranium-234:

$$^{234}_{91}Pa \rightarrow ^{234}_{92}U + \beta^{-}.$$

INSIDE THE ATOM

Figure 39.1a α-emission
b β-emission **c** β-emission
d γ-emission

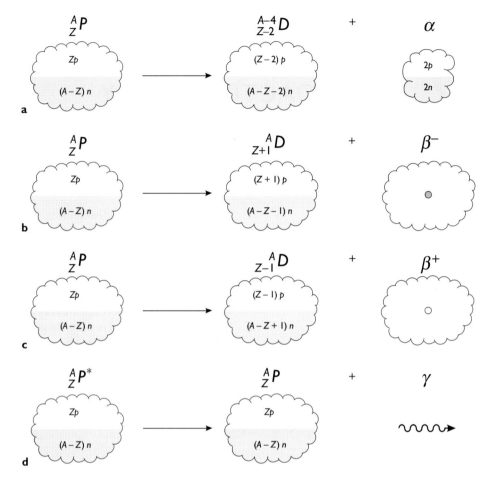

In all β-particle decays another particle, the **neutrino**, is also emitted. This has zero charge and very little, if any, mass. It carries away some of the energy and momentum of the disintegration process.

In γ-emission, there is only a release of energy from the nucleus, so both nucleon and proton numbers are unchanged. The nucleus is initially in an excited state, indicated as P^*.

$$^A_Z P^* \rightarrow ^A_Z P + \gamma$$

(see Figure 39.1d). For example, the radionuclide iodine-131 decays to an excited state of xenon by β–decay. The xenon then emits a γ-ray:

$$^{131}_{53}I \rightarrow ^{131}_{54}Xe^* + \beta^-$$

then $^{131}_{54}Xe^* \rightarrow ^{131}_{54}Xe + \gamma$.

In every decay, the total **number of nucleons** is the **same** on **each side of the equation**. The total charge must also be the same. **Charge is conserved** in the reaction.

Positron emission

There are situations where it is energetically more favourable for a proton to become a neutron (→ 40.5). The particle emitted from the nucleus in this situation is called a **positron** and the decay equation becomes:

$$^A_Z P \rightarrow ^A_{Z-1} D + ^0_{+1} e$$

(see Figure 39.1c). The positron is often denoted by the symbol β⁺, and the symbol β⁻ is used for the more familiar β-particle, the electron.

Questions

1 In a β-decay, a neutrino is also ejected from the nucleus. Explain how this accounts for the continuous spectrum of β-particle energies (← 38.4). Note that kinetic energy is conserved during the emission and that the kinetic energy of the recoiling nucleus is small due to its large mass.

2 The laboratory sources $^{241}_{95}$Am and $^{90}_{38}$Sr decay by α- and β-emission respectively. Using a table of elements, write down the equation of radioactive change in each case.

3 The parent nucleus $^{238}_{92}$U emits three particles in a chain of decays but still ends up as an isotope of uranium. Name the three particles and write down the nucleon and proton numbers of the resulting nuclide.

39.2 LAW OF RADIOACTIVE DECAY

The **rate of decay** of atoms in a given radionuclide at any instant is **proportional to** the **number of undecayed atoms** in the sample at that instant. Each unstable atom has a particular probability that it will decay within a given time. So the more undecayed atoms there are, the more in total are likely to decay in that given time and, hence, the greater the rate at which they decay. For example, a sample containing 1 g of a radionuclide will emit twice as many particles per second as a sample containing only 0.5 g of the same nuclide.

The daughter nuclide produced may be stable and not decay. Then the number of atoms decaying per second is directly proportional to the measured activity of the sample. However, if the daughter also decays, the activity will also include particles emitted from the daughter. In many experiments using naturally occurring radioactive materials, the activity of the sample is due to several radionuclides disintegrating within the sample. Experiments to investigate the decay of a radionuclide, for example experiment 39.2, are designed in such a way that only the particle emissions of that particular nuclide can be detected. This ensures that the count rate measured is due to the nuclide under investigation.

Question

4 Suppose you throw 1 000 dice.
a How many would you expect to turn up 'six'?
b When these are removed, how many would be left?
c How many of the remainder would you expect to turn up 'six' when thrown?
d Repeating the process, sketch a graph showing the number of dice left against the number of the throw.
e If you actually performed the experiment, how might your results differ from your predictions? The experiment can easily be simulated using a computer.
f How is the dice model helpful in understanding radioactive decay?

39.3 HALF-LIFE OF A RADIONUCLIDE

The rate of decay varies from one radionuclide to another. If you start with the same number of atoms of each of two different nuclides, you will find that they decay at different rates. The **half-life** $T_{1/2}$, of a radionuclide is defined as the time taken for half of the atoms of that nuclide present in a sample to disintegrate.

INSIDE THE ATOM

Figure 39.4 shows how the number of undecayed atoms decreases with time. You can see that only one eighth of the atoms remain undecayed after three half-lives. The half-life is also the time for the activity – the number of α- or β-particles emitted per second – to halve.

Half-lives vary enormously in value. For example, both $^{212}_{84}$Po and $^{232}_{90}$Th emit α- particles but the half life of $^{212}_{84}$Po is 3×10^{-7} seconds, whereas that for $^{232}_{90}$Th is 1.4×10^{10} years, some 10^{24} times longer.

Figure 39.2 shows a way of displaying the sequence of changes that take place in radioactive decay. Nucleon number is plotted vertically and proton number horizontally. Horizontal lines correspond to β-decay and diagonal lines to α-decay (→ 39.5).

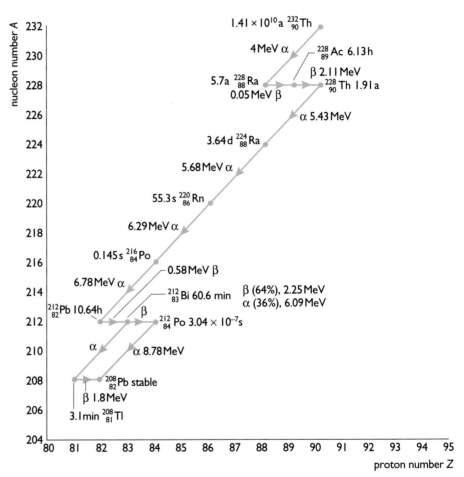

Figure 39.2 Thorium series radioactive decay chain

One easily measured half-life is that of the radionuclide gas $^{220}_{86}$Rn, which decays by the emission of α-particles. The nuclide giving rise to it is $^{228}_{90}$Th, with a half-life of 1.9 years. This decays into the radionuclide $^{224}_{88}$Ra, which decays to $^{220}_{86}$Rn with a half-life of 3.6 days. In experiment 39.1, α-particles emitted from the radon nuclei produce ion pairs in an ionisation chamber. With a potential difference of 100 V across the chamber, the positive ions and electrons move in the electric field to the electrodes, producing an initial current of about 100 pA, or 10^{-10} A, in the circuit. The ionisation current is proportional to the number of α-particles being emitted per second – the activity of the radon. This, in turn, is proportional to the number of undecayed radon atoms.

39 RADIOACTIVE DECAY

Question

5 In a sample, N_o is the initial number of radioactive atoms. Why, after n half-lives, will there be $2^{-n}.N_o$ atoms left undecayed in the sample?

EXPERIMENT 39.1 Measuring the half-life of radon-220

Warning: hazardous radiation – you must be supervised by a suitably qualified person

Fitting a plastic bottle containing thorium hydroxide or carbonate with two tubes and valves provides you with a radon-220 generator. By squeezing the bottle, you can release the gas into an ionisation chamber, as shown in Figure 39.3a. Recording the ionisation current every 15 seconds allows you to produce a graph similar to that in Figure 39.3b.

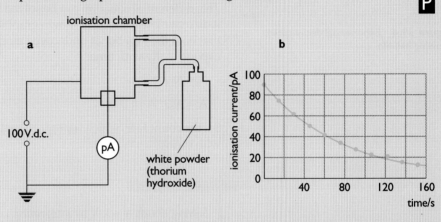

Figure 39.3a Arrangement for experiment 39.1, **b** decay of radon-220

Question

6a From the graph in Figure 39.3b, show that the half-life of radon-220 gas is about 55 seconds.
b Does it matter whether you measure the time for the current to halve from 90 µA or 80 µA or any other value?

39.4 MATHEMATICAL MODEL OF RADIOACTIVE DECAY

The **probability** of an atom decaying in unit time is defined as the **decay constant**, λ. Normal-sized samples of nuclides contain a vast number of atoms and the probability λ equals the **fraction of atoms decaying in unit time**. Because the processes relating to decay are within the nuclei of atoms, λ is independent of all physical factors, such as temperature and pressure.

Suppose at time $t = 0$, N_o atoms of a radionuclide are separated from its parents and after a time t, the number of atoms remaining undecayed is N. A short time Δt later, at $t + \Delta t$, a number of atoms ΔN will have decayed leaving $(N - \Delta N)$ undecayed atoms left.

The change in the number of atoms in time Δt is $-\Delta N$. The negative sign indicates a reduction. So the fraction of atoms decaying in time Δt is $-\Delta N/N$, and the fraction of atoms decaying in unit time is given by:

$$\lambda = \frac{-\Delta N}{N} \cdot \frac{1}{\Delta t}.$$

Alternatively $\Delta N = -N \cdot \lambda \Delta t$

or $\frac{\Delta N}{\Delta t} = -\lambda N.$

This means that the **average rate** at which the atoms in the sample are decaying over an interval Δt is equal to the **product** of the **decay constant** and the **number of undecayed atoms** present.

Taking a smaller and smaller interval Δt, we can write

$$dN/dt = -\lambda N.$$

The quantity dN/dt is the **activity** A of the source.

This differential equation can be solved by an iterative method or by integrating it directly. The solution is shown graphically in Figure 39.4. Analysis indicates that both the number of undecayed atoms and the activity decrease exponentially with time:

$$N = N_0 e^{-\lambda t}$$

$$A = A_0 e^{-\lambda t}.$$

So, the count rate, which is proportional to activity, also decreases exponentially with time.

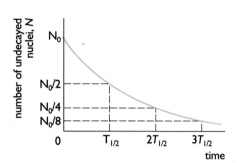

Figure 39.4 Exponential decay curve showing half-life

Solving $dN/dt = -\lambda N$ by integrating

Separating the variables in the differential equation and integrating from $t = 0$ to $t = t$ gives:

$$-\int_{N_0}^{N} \frac{dN}{N} = \int_0^t \lambda dt$$

$$\ln N_0 - \ln N = \lambda t.$$

Rearranging gives: $\ln(N/N_0) = -\lambda t$

$$N/N_0 = e^{-\lambda t}.$$

This is more usually written

$$N = N_0 e^{-\lambda t}.$$

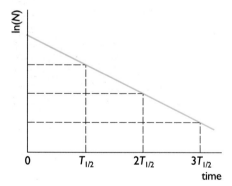

Figure 39.5 Semi-logarithmic plot of exponential decay

The fluctuations in experimental readings sometimes make it difficult to judge where to draw the best smooth decay curve through the points. However, as shown in Figure 39.5, the graph of $\ln N$ against time is a straight line. It is easier to judge the best straight line through plotted points. So a more accurate value for the half-life can be found from such a semi-logarithmic plot ($\rightarrow$ 41.3).

Question

7a Plot a graph of ln (ionisation current) against time for the data shown in Figure 39.3b.

39 RADIOACTIVE DECAY

b Bearing in mind that the ionisation current is proportional to the number of undecayed atoms, how can you obtain the decay constant from this graph?

RELATING DECAY CONSTANT AND HALF-LIFE

Taking logarithms to base e for each term in the equation

$$N = N_o e^{-\lambda t}$$

gives:

$$\ln\left(\frac{N_o}{N}\right) = \lambda t.$$

This is the equation of a straight line, the gradient of which gives the magnitude of the decay constant λ.

The half-life $T_{1/2}$ is related to the decay constant by the equation:

$$\lambda T_{1/2} = \ln 2 = 0.693.$$

This equation shows that radionuclides with the shortest half-lives have the largest decay constants. The long-lived nuclides have very small decay constants, the probability of any chosen atom decaying in each second being very small.

Relation between $T_{1/2}$ and λ

The radioactive decay equation is:

$$N = N_o e^{-\lambda t}$$

By definition, at $t = T_{1/2}$, $N = N_o/2$. So:

$$N_o/2 = N_o e^{-\lambda T \frac{1}{2}}$$
$$1/2 = e^{-\lambda T \frac{1}{2}}$$

giving $\lambda T_{1/2} = \ln 2 = 0.693$.

Questions

8a Radon-222 has a decay constant of 2.1×10^{-6} s^{-1}. What is its half-life in seconds?
b Calculate the decay constant of $^{238}_{92}$U. It has a half-life of 4.5×10^9 years.

9 The half-life of radium-226 is 1 622 years.
a Calculate the decay constant.
b How many atoms are there in 1 g of radium-226 ($\leftarrow$ 35.2). (N_A)
c What is the expected activity, dN/dt, of 1 g of radium-226, assuming minimal absorption, in **i** Bq and **ii** Ci?
d It is a relatively simple matter to find the half-life of a short-lived radionuclide, such as radon-220, as it can be measured directly. Clearly, the half-life of a long-lived radionuclide cannot be measured in the same way. What clues do the above calculations give to a method for measuring long half-lives?

10 You have three new radioactive sources, each of activity 2×10^5 Bq. They are:
  i a strontium −90 source, half-life 28 years,
  ii a radium −226 source, half-life 1 622 years,
  iii a cobalt −60 source, half-life 5.3 years.
a Which source has the largest number of radioactive atoms? Which has the smallest? Explain your answers.
b How many atoms are there in the strontium source? What is the **mass** of this source, in kg?
c Which source would be the most active, assuming that each contained the same number of radioactive atoms?

11 The time constant for the exponential decay of charge in a RC circuit is RC ($\leftarrow$ 13.3). Comparing charge remaining to undecayed nuclei, how are time constant RC and decay constant similar?

INSIDE THE ATOM

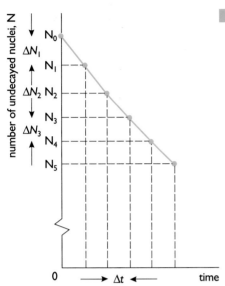

Figure 39.6 A decay curve produced by an iterative method of plotting

Iterative solution of dN/dt = −λN

This equation can be solved by an iterative method, generating the curve of Figure 39.6. Starting at time $t = 0$, the average number ΔN which decay in a small time interval Δt is given by

$$\Delta N = -\lambda N_0 \Delta t$$

where N_0 is the initial number of atoms. After one time interval Δt, the number N_1 of undecayed nuclei is:

$$N_1 = N_0 - \Delta N.$$

Since $\Delta N_1 = -\lambda N_1 \Delta t$, after a second interval Δt, the undecayed number N_2 is:

$$N_2 = N_1 - \Delta N_1$$

and so on.

Suppose you choose a time interval Δt of 5 s and N_0 to be 1 000. The value of λ to make the half-life of Rn 55 s is 1.26×10^{-2} s^{-1}. Putting these figures into the above equation gives:

$$\Delta N = -(1.26 \times 10^{-2}) \times 1\,000 \times 5 = 63$$

and, therefore, $N_1 = 1\,000 - 63 = 937$.

The decrease from N_1 in the next interval Δt is given by:

$$\Delta N_1 = -(1.26 \times 10^{-2}) \times 937 \times 5 = 59$$

and $N_2 = 937 - 59 = 878$.

If you continue the process, you will produce the following table of values:

t s	0	5	10	15	20	25	30	35	40	45	50	55
N	1000	937	878	823	771	722	677	634	594	557	522	489

Note that the number of atoms has fallen to 1/2 sometime between 50-55 seconds. So, the half-life is approximately 50 seconds.

This iterative method emphasises the nature of radioactive decay. A spreadsheet for use with a computer is an ideal way of carrying out such an analysis ($\rightarrow$ 41.3). The effect of changing various parameters can then be seen quickly and easily.

Question

12 This question relates to the box 'Iterative solution'.
a Plot a graph of N against t for the data above. How does the graph compare with that in Figure 39.5b showing the variation in ionisation current with time for radon gas?
b Why does the calculated curve fall below the expected value of $N=500$ after 55 s?
c Would you expect the result to be different if the value of Δt had been taken as 1 s?

39.5 GROWTH AND DECAY

When the parent nuclide disintegrates to a stable daughter nuclide, the sum of the number of parent and daughter atoms in the sample remains constant.

39 RADIOACTIVE DECAY

Figure 39.7 shows the growth of the daughter nuclide N_D against the decay of the parent N_P. Most naturally occurring radionuclides are members of a **decay chain**. So it is usual for the daughter nuclide to decay into a granddaughter. Figure 39.8 shows how the number of parent, daughter and granddaughter atoms in the sample change with time. In this decay chain, the granddaughter is assumed to be stable, for simplicity, and the half-life of the parent is 1.5 times that of the daughter.

As the parent decays into daughter nuclide, the rate of decay is large to begin with. Daughter nuclide grow quickly in number, being produced at a greater rate than they are decaying. As time passes, the rate of decay of the parent falls to a point where it is the same as the rate of decay of the daughter. The number of daughter atoms has reached a maximum. Subsequently, daughter atoms fall in number because they are being produced at a slower and slower rate as the parent decays. Throughout, granddaughter atoms are growing in number until there are N_o of them. Notice that:

$$N_o = N_P + N_D + N_{GD}$$

at all times.

Growth of a stable daughter

Suppose N_o is the initial number of parent atoms and after a time t, N_P and N_D are the number of parent and daughter atoms. Then:

$$N_o = N_P + N_D$$

or $N_D = N_o - N_P$.

The decay of the parent with time is given by:

$$N_P = N_o e^{-\lambda t}$$

where λ is the decay constant ($\leftarrow$ 39.4).

So: $N_D = N_o - N_o e^{-\lambda t}$

$$N_D = N_o (1 - e^{-\lambda t}).$$

Number of daughter atoms

A mathematical model ($\leftarrow$ 39.4) describes the situation as the daughter nuclide decays. The differential equation for the daughter is of the form:

(rate of change of daughter) = (rate of decay of parent) – (rate of decay of daughter)

= (rate of formation of daughter)

or $dN_D/dt = \lambda_P N_P - \lambda_D N_D$

$$\frac{dN_D}{dt} = N_o e^{-\lambda_P t} - \lambda_D N_D. \quad (1)$$

The solution of this equation is:

$$N_D = \frac{\lambda_P N_0}{\lambda_D - \lambda_P} \cdot (e^{-\lambda_P t} - e^{-\lambda_D t}). \quad (2)$$

This solution describes how the number of atoms of a nuclide in a radioactive decay chain changes with time. The number depends on the initial number of parent atoms and the decay constants or half-lives of the parent and the daughter.

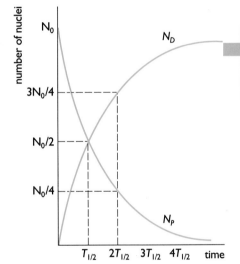

Figure 39.7 Decay of parent N_P and growth of stable daughter nuclide N_D

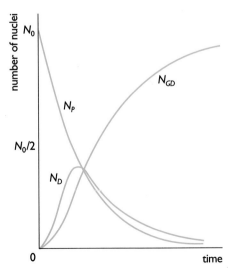

Figure 39.8 Decay of parent to stable granddaughter. Parent half-life is 1.5 times that of daughter

Questions

13 Sketch graphs similar to Figure 39.9 showing the variation of parent, daughter and granddaughter nuclide with time when the half-life of the daughter is
a much shorter and
b much longer than that of the parent.
Assume the granddaughter is stable.

14 The box above gives a differential equation (1) and a solution (2), showing how the number of daughter atoms N_D changes with time.
a In the solution (2), differentiate N_D with respect to time.
b By substitution in the differential equation (1), show that (2) is indeed a solution to it.

H INSIDE THE ATOM

RADIOACTIVE SERIES

When all the naturally occurring radionuclides are plotted on a chart of nucleon number A against proton number Z, they are found to form three distinct chains. Each chain is headed by a different parent nuclide: $^{232}_{90}$Th for the **thorium series** (see Figure 39.2); $^{238}_{92}$U for the **uranium series** (see Figure 39.9), and $^{232}_{92}$U for the **actinium series**. Each of the chains ends in a different stable isotope of lead. In these charts, an α- particle emission is shown by a diagonal arrow and a β-particle emission by a horizontal arrow. Note that the emission of one α-particle and two β- particles, in any order, results in the same proton number Z and so the same element.

The radioactive gas $^{220}_{86}$Rn is one member of the thorium decay series. It is used to demonstrate half-lives and their measurement (← 39.3). In such demonstrations, a plastic bottle, the 'radon generator', contains a white powder, thorium hydroxide. All the nuclides in the decay chain are present in the bottle, but the radon can be isolated from the others because it is a gas.

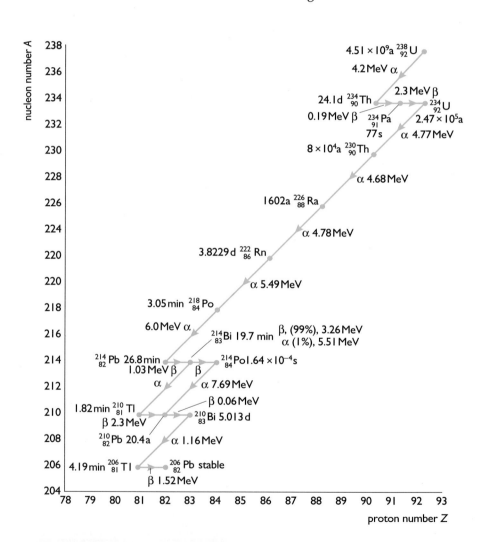

Figure 39.9 Uranium series radioactive decay chain

Questions

15 Explain why all the nuclides in the thorium series decay chain are present in a 'radon generator' bottle.

39 RADIOACTIVE DECAY

16a Describe in words the changes that occur over a very long period of time in a sample of thorium-232, the nuclide at the head of the thorium series (see Figure 39.2).

17 Only three radioactive decay chains occur in nature. These are known as the $4n$, $4n + 2$ and $4n + 3$ series.
a Find out why they are referred to in this way.
b Suggest a reason why a fourth chain, the $4n + 1$ series, does not appear to exist in the Earth's crust.
c Consider the $4n$, $4n + 1$, $4n + 2$ and $4n + 3$ decay series. Why can there be no more than these four decay chains?

Protactinium-234, or $^{234}_{91}Pa$, is a nuclide in the uranium series decay chain. Its half-life can also be measured easily as in experiment 39.2.

Both $^{234}_{91}Pa$ and its parent, thorium-234, decay by β-emission. The grandparent $^{238}_{92}U$ and granddaughter $^{234}_{92}U$ are α-particle emitters. Experiment 39.2 is possible because the β-particles from protactinium are the only particles emitted by the radionuclides in the bottle until much lower down the chain, (see Figure 39.9) with sufficient energy to penetrate the wall of the bottle. The β-particles from the thorium and the α-particles are absorbed in the wall. Also, the measured count rate is only proportional to the surface activity of the sample. Only those β-particles emitted close to the wall of the bottle can escape. The others are absorbed in the liquid.

EXPERIMENT 39.2 Decay and growth of protactinium-234

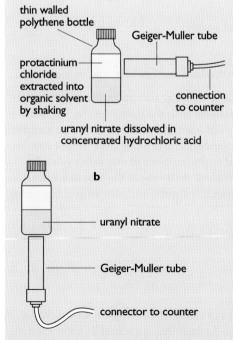

The source can be prepared by dissolving uranyl (VI) nitrate in water and shaking it in a thin-walled plastic bottle with an equal quantity of an organic solvent (see Figure 39.11a). About 95 % of the $^{234}_{91}Pa$ produced by the decay of $^{234}_{90}Th$ present will dissolve in the organic solvent. Leaving the mixture to stand allows the organic solvent containing the $^{234}_{91}Pa$ to float.

Warning: Be careful not to spill the liquid when shaking it. Make sure the cap is tight on the bottle. Do not allow the GM tube to come into contact with the bottle. It could contaminate the tube, giving high background readings in future experiments.

a Decay – With the Geiger Muller (GM) tube almost touching the top half of the bottle, as in Figure 39.10a, you can begin to count the β-particles over a series of, say, 10 s intervals. By finding the average background count over a 10 s interval, you will be able to plot graphs of 'corrected count rate' and 'ln of count rate' against time. These should show that the half life is about 77 s.

b Growth – With the GM tube under the bottle, as in Figure 39.10b, you can count the β-particles produced from the growth of $^{234}_{91}Pa$ as it is formed from $^{234}_{90}Th$. After several minutes, the count rate will become constant. By plotting the growth curve of measured activity against time you can compare its shape with that of the decay curve.

Figure 39.10 Arrangement for measuring **a** decay and **b** growth in experiment 39.2

INSIDE THE ATOM

Questions

18 Do the growth and decay curves obtained in experiment 39.2 look like the parent and daughter curves in Figure 39.7 or 39.8? Explain why with reference to the data on Figure 39.9.

19 The half-lives of $^{234}_{90}$Th and $^{234}_{91}$Pa are 24.1 days and 77 seconds respectively. Show that the decay constant for thorium is much less than that for protactinium.

RADIOACTIVE EQUILIBRIUM

The growth curve of protactinium looks like that of N_D in Figure 39.7. However Figure 39.7 represents the growth of a **stable** nuclide, while the activity of the protactinium is proportional to the number of **unstable** atoms present. The similarity arises because the half-life of protactinium's parent nuclide, thorium-234, is long, so the number of parent atoms N_p is approximately constant over the short time of the experiment. The number of daughter atoms N_D present grows from zero until as many are decaying as are being formed. The count rate then becomes constant.

The condition where the number of daughter atoms present in a sample becomes constant is called **radioactive** or **secular equilibrium**. The daughter half-life is so much shorter than the parent half-life that it only takes a few half-lives before the daughter atoms are decaying at the same rate as they are being formed. When radioactive equilibrium is reached, the rate of change of the number of daughter atoms (dN_D/dt) is zero. From the section 'Number of daughter atoms' this means that:

(rate of decay of parent) = (rate of decay of daughter)

$$\lambda_P N_P = \lambda_D N_D$$

and as the rate of change of the next nuclide in the chain, the granddaughter, is also zero,

$$\lambda_D N_D = \lambda_{GD} N_{GD} \text{ and so on.}$$

Thus $\lambda N = \text{constant}$

for all nuclides in the series except the last which is, of course, stable.

For the three radioactive decay chains still occurring naturally, the first member of each series is very long lived in comparison with the daughter. So radioactive equilibrium exists in nature. All members of a series will be present in a sample of rock or material containing the first member. The relative numbers of atoms of each nuclide will be constant and related to the number of any other nuclide present by the above equation. Each nuclide is decaying at the rate at which it is being formed. Were this not the case, short-lived radionuclides would have long since disappeared from the Earth.

39.6 POWER FROM A RADIOACTIVE SOURCE

Early experimenters noticed that samples of radioactive material were always slightly warmer than their surroundings. At first, this appeared to contradict the laws of thermodynamics. But the experimenters soon found that the energy

39 RADIOACTIVE DECAY

came from the absorption of the radiations by the material itself. The rate at which energy is produced is equal to the kinetic energy of the particles absorbed per second.

We can make use of this energy source. For example, thermocouples embedded in the source will transfer energy and generate electricity. Small quantities of radionuclides are used to generate electricity for cardiac pacemakers, as shown in Figure 39.11.

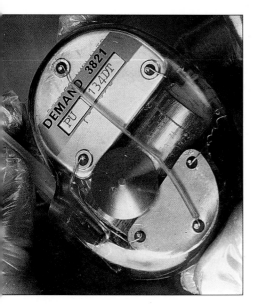

Figure 39.11 Cardiac pacemaker powered by a radionuclide battery

Questions

20 Two sources commonly used in the laboratory are
a strontium-90, which emits β-particles each of energy 0.55 MeV
b americium-241, which emits α-particles, each of energy 5.49 MeV and each accompanied by a γ-quantum of energy 0.06 MeV.
What are the maximum powers obtainable from each source, assuming that they both have activities of 1.85×10^5 Bq, about 5 μCi?

21 A 10 W generator is to be made from a source of cobalt-60. Each nuclear disintegration produces a total energy of about 2.8 MeV, mainly as γ-radiation. The source has a half-life of 5.3 years.
a How many joules of energy are produced per disintegration?
b What minimum number of disintegrations per second is required to produce a 10 W output? Why is this a 'minimum' number?
c How many atoms of cobalt-60 are required to produce this activity?
d What mass of cobalt does this represent?
e Comment on the advantages and disadvantages of using strontium-90 rather than cobalt-60 as a source for this generator. Strontium-90 has a half-life of 28 years, emitting β-particles of energy 0.55 MeV.

Maximum power available from a radioactive source

The maximum power P is that power available when all the particles emitted by a source are absorbed by the source. It is dependent upon the number of disintegrations per second in the sample dN/dt and the kinetic energy E_K of each emission:

$$P = E_K \cdot dN/dt.$$

In most calculations involving radioactivity, you need to know the number N of atoms present in the source because the product λN determines the activity of the source (← 39.4). You can calculate the number N from the source mass m, the molar mass of the atoms M and the Avogadro constant N_A using:

$$N = \frac{m}{M} \cdot N_A$$

(← 35.2). The activity λN is given by:

$$\text{activity} = \frac{\lambda m}{N} \cdot N_A.$$

So the maximum power available is given by:

$$P = \lambda \frac{m}{M} \cdot N_A E_K$$

$$\text{or} \quad = \frac{m N_A E_K \cdot \ln 2}{M \cdot T_{1/2}},$$

where $T_{1/2}$ is the half-life (← 39.4).

Small atomic power sources (see Figure 39.12) are also used, for example, in light buoys, satellites, spacecraft and amplifiers for transoceanic cables. A number of factors have to be considered when designing such a power source.

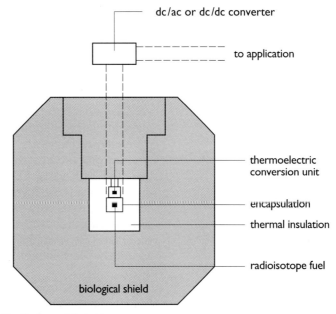

Figure 39.12 Radionuclide battery

INSIDE THE ATOM

The available power will decrease with time, so the maximum and minimum limits of useful output must be known. Suppose the mass or volume of the source is fixed. Then to generate a large power for a short time, a source with a short half-life must be chosen. On the other hand, for steady power generation over a long period, the source should have a long half-life.

The total energy available from two sources with equal mass and energy of particle emission will be the same. When the choice is between α- and β-emitters, the α-particles are more easily absorbed and, on average, release more energy per emission. However, there are many artificially produced, low-mass β-emitters giving a larger number of particles for the same volume.

It will be a disadvantage if the source also emits γ-rays, which are difficult to absorb (← 38.5). Thick shielding will be required to protect people and other living things. The extra shielding will have a large heating capacity, so the temperature difference between the source and its surroundings will be smaller. Also, the power source will be more massive and bulky.

The final consideration is the cost per unit mass of the source and its purity. These together determine the cost of each useful unit of energy produced.

39.7 RADIOACTIVE DATING

The model of radioactive decay can be used to find the ages of archaelogical specimens, rocks and other materials. Consider a sample of rock containing a natural radionuclide, such as an isotope of uranium. Some of the uranium atoms will have decayed through the chain shown in Figure 39.9 to the final stable nuclide, an isotope of lead. The ratio of the number of uranium atoms to the number of stable lead atoms present indicates the age of the rock.

The age of archaeological specimens can be estimated using a similar method called **carbon dating**. Here we are considering a much shorter time scale, a hundred thousand years rather than a thousand million years. Carbon dating was used to date the cloth of the famous Turin Shroud, for example, confirming that it was very much younger than originally believed.

There are three naturally occurring isotopes of carbon, by far the most abundant of which is $^{12}_{6}C$. One, $^{14}_{6}C$, is unstable and decays by β-emission with a half-life of 5 730 years (see Figure 39.13). A tiny quantity of this nuclide exists in the atmosphere due to bombardment of the plentiful $^{14}_{7}N$ nuclei in the outer atmosphere by neutrons from space. The resulting $^{14}_{6}C$ atoms diffuse throughout the Earth's atmosphere as part of carbon dioxide molecules. A very small proportion of all atmospheric carbon dioxide molecules contain $^{14}_{6}C$ atoms, sometimes one, very occasionally two. The distribution of this nuclide throughout the atmosphere reaches equilibrium over a few thousand years when its rate of formation equals its rate of decay.

A tiny proportion of the carbon dioxide absorbed by living organisms is, therefore, radioactive. The ratio of $^{14}C : ^{12}C$ in living organisms is $1:10^{12}$. When these organisms die, their ^{12}C content remains constant, but their radioactive ^{14}C content decreases with time. By comparing the β-particle emission rates of similar living and dead organisms, the time since death can be estimated. The count rates are very small, so the efficiency of the counter must be known accurately. The fluctuation in background count must also be carefully monitored.

The count rate from objects older than some 50 000 years is indistinguishable from the background count. So a different technique has to be used. The actual number of ^{14}C nuclei in a known mass of carbon is counted using a mass

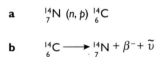

Figure 39.13a Formation and **b** decay equations for $^{14}_{6}C$

a $^{14}_{7}N\ (n,p)\ ^{14}_{6}C$

b $^{14}_{6}C \longrightarrow\ ^{14}_{7}N + \beta^- + \tilde{\nu}$

spectrometer. Even tiny fragments of bone can be dated using this method. However, it is very sensitive to contamination by other nuclides of similar atomic mass, such as ^{14}N, causing large errors.

The date calculated from carbon dating is subject to at least a 10 % uncertainty. For example, the method assumes that the density of ^{14}C in the atmosphere was the same when the organism died as it is now. The burning of fossil fuels, volcanic eruptions, nuclear tests and other widespread pollution of the atmosphere may have upset the balance. Changes in the Sun's activity and the magnetic field of the Earth have also caused slight variations in the ratio of ^{14}C to ^{12}C. Nevertheless, carbon dating is a useful guide, alongside other methods of dating, to the age of an artifact containing plant or animal matter.

> **Questions**
>
> **22** The half-life of carbon-14 is 5 730 years.
> **a** Show that the probable age t, in years, of a dead plant is given by:
>
> $$t = 5\,730 \cdot \frac{\ln(A_t/A_0)}{\ln 2},$$
>
> where A_t is the activity at time t and A_0 is the activity at the time of death.
> **b** Calculate the probable age of an unearthed branch where the $^{14}_{6}$C activity is a fifth of that of a living tree.
>
> **23** The variation of ^{14}C concentration in the atmosphere over the past few thousand years has been found from dating tree rings.
> **a** How do you think this is done?
> **b** How can this help the accuracy of carbon dating?

39.8 RADIOACTIVE TRACERS

Figure 39.14 Radioactive tracers following the movement of waterborne waste, made radioactive, in an estuary, using a waterproof detector

A marker or indicator, called a **tracer**, which can easily be detected is often used to study the motion of fluids. For example, water may have a dye mixed with it to see where and how it flows.

Radioactive tracers have many applications in science research, in industry and in medicine. For example, they are used to study the flow of fluids in situations as different as the build-up of silt in the Thames estuary to the flow of blood through the human body (see Figure 39.14). Figure 39.15 shows how a tracer may be used to measure the flow rate in a pipe or to detect an underground leak. A requirement is that the tracer used should decay into a stable daughter nuclide and have a short half-life.

Any source of radiation that can be absorbed by the body is potentially dangerous. So the amounts of the tracers injected into the body are very small, and the radionuclides chosen have short half-lives and emit low-energy radiation. The radionuclide used must be a γ-ray emitter for the radiation to be detected outside the body. The detector may be a scintillation counter or a γ-ray camera ($\leftarrow$ 38.8). The tracer must also act chemically like the substance it is replacing. One of the most used radionuclides is $^{131}_{53}$I. This isotope of iodine has a half-life of 8 days and emits radiation having an energy of 0.36 MeV. It is used in studies of the function and disease of the thyroid, liver, heart and lungs. It also helps to diagnose general blood disorders and to locate sites of internal bleeding.

In agriculture, the radionuclide phosphorus-32 helps show how plants absorb fertilizer. Phosphorus in the form of phosphates occurs in all fertilizers, playing

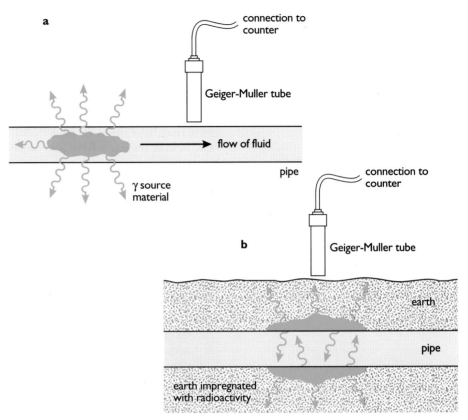

Figure 39.15 Radioactive tracer measurement of **a** flow rate **b** leakage

an important role in the health and growth of the plant. Labelling phosphates with phosphorus-32 can be used to find which phosphate is the most effective and at which stage in the growth of a plant the uptake of phosphorus is greatest. Phosphorus-32 decays by β-emission with a half-life of 14.3 days. Its progress through the plant is indicated by the change in count rate, measured with a Geiger-Muller tube, from its stem and leaves. The knowledge obtained from such investigations allows the farmer to gain maximum benefit from the minimum use of fertilizer.

SUMMARY

Radioactive decay is a spontaneous random process. On a large scale, though, it can be predicted.

Exponential decay is a mathematical model for the experimentally observed decay pattern.

The disintegration rate, the number of atoms decaying per second, is called the activity. It is proportional to the number of undecayed atoms N present and is written symbolically as:

activity = $dN/dt = -\lambda N$

where λ is the decay constant. This is the fraction of atoms that disintegrate per second or the probability that any chosen undecayed atom will disintegrate in the next second.

The half-life $T_{1/2}$ of a radionuclide is the time for half of its undecayed atoms to decay or its activity to halve. Half-life and decay constant are related by:

$T_{1/2} = \ln 2/\lambda$.

39 RADIOACTIVE DECAY

Radioactive equilibrium occurs when parent atoms N_P and daughter atoms N_D exist in the same sample, and daughter nuclei disintegrate at the same rate as they are formed. In this situation:

$\lambda_P N_P = \lambda_D N_D = \lambda N = $ constant.

Radioactive equilibrium exists in the three naturally occurring radioactive series – the thorium, uranium and actinium series.

The maximum power P available from a radionuclide due to its absorption of the radiation is given by:

power = activity × energy of each disintegration.

Rocks, archaeological finds, etc, can be dated by measuring the activity of a radionuclide within the specimen and assuming a knowledge of the radionuclide's original activity.

SUMMARY QUESTIONS

24 Nuclei with the same nucleon number A but different proton numbers Z are called **isobars**. Show that the parent and daughter nuclei in β-decay are isobars. In nuclear physics, isobar means same baryon number while in meteorology, it means same pressure.

25 Look back at Figure 38.3. For each of these parents write down the full symbols of the daughter nuclide to which it decays.

26 The actinium series decays into a stable isotope of lead with nucleon number 207. The parent of the actinium series is $^{235}_{92}U$. In the decay chain there are seven α-emitters and four β-emitters. Does this give the correct proton and nucleon number for the stable nuclide at the end of the chain?

27 When it is not bound inside a nucleus, the neutron acts as a radioactive particle, decaying into a proton and an electron with a half-life of 10.8 min. Calculate the probability that a particular neutron will decay in the next second.

28 Copy Figures 39.4 and 39.5. To each, add curves showing the decay of a radionuclide with a half-life **a** twice and **b** half that of the one drawn.

29a Suggest why it is more difficult to deal with radioactive waste that has a half-life of months or years than waste with a very long or very short half-life. To illustrate your answer, consider the activities of:

1 g of $^{226}_{88}Ra$ with $T_{1/2}$ of 1 620 years and

1 g of $^{90}_{38}Sr$ with $T_{1/2}$ of 28 years.

b What other factor(s) make $^{90}_{38}Sr$ the more dangerous of the two?

30 The school laboratory source $^{241}_{95}Am$ has an activity of 1.85×10^4 Bq when manufactured. Its half-life is 430 years. Calculate:
a the number of atoms in the source (N_A),
b the mass of the radionuclide.

31 In experiment 39.1, a tiny quantity of $^{220}_{86}Rn$ is squeezed from the polythene bottle into the ionisation chamber. It has a half-life of 55 s.
a Calculate the mass of gas entering the chamber for an initial activity of 2 500 Bq.

b The initial current in the ionisation chamber circuit is about 10^{-10} A. Each α-particle produces about 10^5 ion pairs. Show that 5 000 Bq is a reasonable estimate of the initial activity.
c Look at the thorium series decay chain shown in Figure 39.2. If you take into account the decay of the daughter products, is the estimate of the initial activity of the Rn still valid?

32 You are asked to choose a radionuclide, which can be made artificially, to act as a tracer in studying the action of the thyroid gland of a human subject. Suggest properties you would look for when choosing one.

33 Two radionuclides have the following half-lives:

$^{32}_{15}$P has $T_{1/2}$ of 14.3 days, $^{131}_{53}$I has $T_{1/2}$ of 8.07 days.

Both emit β-particles that are absorbed by the body and both are used in the diagnosis and study of liver disease.

To compare them, assume for simplicity of calculation that the phosphorus isotope is actually nuclide ^{32}X with $T_{1/2}$ = 16 days and the iodine isotope is actually nuclide ^{128}Y with $T_{1/2}$ = 8 days.
The same number of atoms X and Y are initially present.
a Which source gives the greater initial activity?
b Which source would have given the greater initial activity with equal **masses** of X and Y present? Assume no absorption.
c Find the ratio of the numbers of atoms of X to the number of Y remaining undecayed **i** after 16 days, **ii** after 32 days. Assume the human body does not remove the radionuclides through its own natural processes.
d Calculate the ratio of the energy liberated into the liver by X to that by Y in **i** 16 days and **ii** 32 days. Each of X's emissions has an energy of 1.71 MeV and each of Y's 0.61 MeV.
e Calculate the ratio of the activity from nuclide X to that from nuclide Y after **i** 16 days, **ii** 32 days.
f By what factor does the activity of nuclide X fall in 1 year?

34 You need to estimate the power available from 1 g of newly prepared $^{226}_{88}$Ra whose half-life is 5.12×10^{10} s. The kinetic energy of each α-particle emitted is 4.78 MeV.
a How many atoms are there in the source?
b What is the activity of the source, in Bq and Ci?
c Marie Curie noticed that her sample of radium warmed up. What is the power available from 1 g?
d How much energy would have been released if all the radium nuclei had disintegrated in 1 second?
e Why is the total energy emitted by the sample over a long period of time likely to be much greater than your answer to **d**?

35 The radionuclide $^{222}_{86}$Rn has a longer half-life than $^{220}_{86}$Rn, whose decay was measured in experiment 39.1. The rate of disintegration of radon-222 is measured for a period of 5 minutes each day and the count rate is corrected for background radiation. The activity on successive days as a percentage of the initial count rate A_0 is:

day number	0	1	2	3	4	5	6	7	8	9	10
count rate as a % of A_0	100	84	70	59	49	41	34	27	24	20	17

39 RADIOACTIVE DECAY

radionuclide	half-life	% by weight in material supplied by manufacturer	main radiations, emitted by one decaying nucleus	energy of radiaton emitted
Tm^{170} (thulium)	128 days	40	β⁻ particle	0.97 MeV
Co^{60} (cobalt)	5.3 years	35	β⁻ particle... and γ-ray...	0.32 MeV about 1.2 MeV
Cm^{244} (curium)	18.1 years	95	α-particle	5.8 MeV

Figure 39.16

a Plot a graph of 'ln of count rate' against time.
b Predict the count rate after 14 days.
c What is the value of the decay constant of $^{222}_{86}$Rn?
d Calculate the half-life of this radionuclide.
e Why is it better to plot the graph of the logarithm of the measured activity rather than the count rate?

36 The three radionuclides in Figure 39.16 are being considered for use as power sources. Using the data provided, write a report on the best choice of nuclide to:
a supply amplifiers to be built into a transatlantic cable,
b supply a space vehicle for a short exploration of the Moon.
Give reasons, including relevant quantitative comparisons, for each recommendation you make.

37 In this question, you are asked to plan an investigation into the absorption of β-radiation by the material emitting it. Figure 39.17 shows a version of experiment 39.2 to measure the half-life of protactinium, which emits β-radiation. A small amount of this element in solution has been extracted from a solution of uranium salt. It has been put in a plastic bottle as quickly as possible after extraction, as the half-life of the element is about 70 seconds. The bottle is placed over a counter, as shown.

The initial count rate seems to be about 5 counts per second. Without any source near the counter, there is still about 1 count per second. The extraction process can be repeated as necessary. The walls of the bottle are about 1 mm thick. When a bottle with walls twice as thick is tried, the initial count rate is markedly reduced. The solution containing the element is some 50 mm deep in the bottle. It can be assumed that the absorption of the β-radiation by the solvent material is unlikely to differ from that of the plastic material of the bottle by a factor of more than 2 or 3.

Write a plan to investigate how much β-radiation emitted by the element in the solution fails to reach the counter because the radiation is absorbed by the solvent. In your plan you should:
a state clearly what you propose to do and why.
b describe the difficulties you would expect to meet and explain how you propose to overcome them.
c where possible, make rough estimates of appropriate quantities, such as count rates to be expected, times over which you would make counts and the dimensions of relevant parts of the apparatus.
d explain how you would obtain useful information from your data.

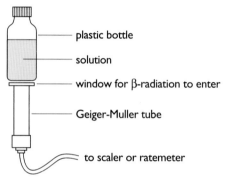

Figure 39.17

40 NUCLEAR MODEL OF THE ATOM

Right in the centre of each atom is its nucleus. It is extremely small, about 10^{-14} m across. Around each nucleus, at about 10^{-10} m, lie a set of electron orbitals. Knowledge of these electrons, how they are arranged and what energies they have, comes mostly from electron collision experiments in gas-filled discharge tubes (← 37). The structure of the nucleus is more elusive, but collisions between particles and nuclei have revealed evidence about it.

An historic experiment using α-particles as probes led Rutherford to his model of the atom. The use of the neutron as a probe explained the forces that hold the nucleus together and the reasons for the release of energy in fission and fusion.

The nomenclature and decay equations for nuclei form an important background to understanding the nucleus (← 38.1 and 39.1). The motion of charged particles in **E**-fields and **B**-fields is also of great importance (← 35.4), as is the electrical potential energy between charged particles (← 16).

40.1 NUCLEAR PROBES: THE KEY TO THE ATOM

It is not possible to look inside an atom with light because the atom's diameter, about 0.1 nm, is much less than the wavelengths of visible light. Other probing methods must be used, together with evidence obtained from particles that can interact or have interacted with the constituents of atoms. Such evidence shows that atoms consist of a central small but massive positive nucleus surrounded by clouds of electrons in a series of orbital shells and a great deal about the constituents of the nucleus, including their size, mass and energy.

Probing the nucleus

The history of the discovery of the constituents of the atom and its structure is a fascinating story. At the turn of the century, the size of the atom was known to be about 100 pm in diameter. Lord Rayleigh in 1899 had found the upper limit to the size of a molecule by measuring the thickness of a layer of oil on a water surface. Similar values for molecular size had already been found using measurements of the Avogadro constant, the molar volume and the density of solids. The problem was how to look inside the atom as it was impossible to make any direct observations.

The α-particles discovered by Becquerel in 1896 were studied by Rutherford. They seemed the obvious tool to probe the atom's structure. They could be aimed at metal foils to see what effect, if any, there would be on the atoms. Early experiments favoured J. J. **Thomson's** 'plum-pudding' **model** of the atom (see Figure 40.1). Then, in 1911, Geiger and Marsden, under the guidance of Rutherford, observed the rebound of a very few α-particles from a thin sheet of gold foil. They concluded that the atom consisted of a tiny nucleus, containing most of the mass of the atom with a diameter some 10^4 times smaller than the atom itself, about 10^{-14} m, surrounded by empty space containing electrons: the **Rutherford-Bohr model** (see Figure 40.2). In

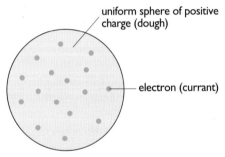

Figure 40.1 Thomson's 'plum pudding' atom (1907)

40 NUCLEAR MODEL OF THE ATOM

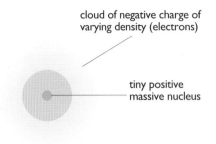

Figure 40.2 Rutherford-Bohr model of the atom (1911–25)

1914, Neils Bohr led the way towards the explanation of why electrons did not collapse into the nucleus, and of some atomic spectra. In the mid 1920s, the idea of wave particle duality and the development of wave mechanics produced a satisfactory model to explain atomic spectra, atomic size and why the nucleus could not contain electrons (← 37). In 1920, Rutherford predicted the existence of a neutral nucleon. Twelve years later, the neutron was conclusively identified by Chadwick, again by firing nuclear particles at stable matter.

The discovery of the neutron gave scientists a new valuable tool in probing nuclei. Unlike a proton or α-particle, the neutron is uncharged and so is not repelled by the positive charge of a nucleus. In 1938, Hahn and Strassmann discovered nuclear fission by bombarding uranium with slow, or thermal, neutrons and releasing considerable energy. In 1942, Fermi established the first nuclear reactor in an old squash court in Chicago. This was the forerunner of modern nuclear power stations.

Today high-energy particles produced in large particle accelerating machines are used to probe deeper into the nucleus. They smash into nuclei and scientists examine the resulting fragments. One of the ultimate goals of nuclear science is the production of a controlled fusion reaction. This will enable us to develop technology on the necessary scale to provide safely, and in an environmentally friendly way, for all our energy needs.

Questions

1 Why would you expect a tiny proportion of α-particles (← 38.3) to be scattered backwards from an atom with a nucleus and electrons in surrounding orbitals – a Rutherford-Bohr atom?

2 Why would you expect no back scattering from a Thomson, or 'plum-pudding' atom, where electrons are embedded in the atom along with the positive charges?

40.2 SCATTERING OF α-PARTICLES

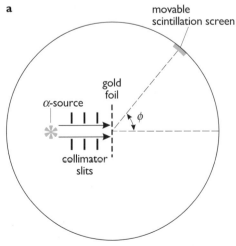

Figure 40.3a Arrangement for Geiger and Marsden's experiments, **b** some of the results

At Rutherford's suggestion, Geiger and Marsden aimed a narrow collimated beam of α-particles at a thin gold foil inside an evacuated chamber. They counted the number of scintillations in a given time interval caused by the scattered particles hitting a zinc sulphide screen placed at different angles to the incident beam. Rutherford used the Coulomb inverse square law of force between electrical point charges (← 16.2) to predict the fraction of the incident α-particles that should be scattered through an angle ϕ. Theory and experiment were found to agree for a wide range of angles. Geiger and Marsden's arrangement and some of their results are shown in Figure 40.3.

ϕ/degree	$1/\sin^4\phi/2$	N/m^2 per s
150	1.15	33.1
120	1.79	51.9
75	7.25	211
45	46.6	1435
30	223	7800
15	3445	132000

H INSIDE THE ATOM

> **Question**
>
> **3** Rutherford's formula for the fraction f of incident α-particles reaching unit area of the screen per second at angle ϕ is given by:
>
> $$f = k/(\sin^4 \phi/2),$$
>
> where k is a constant which depends on the energy of the incident α-particles, the thickness and density of the foil, etc.
>
> Show that the results tabulated in Figure 40.3b are consistent with Rutherford's formula.

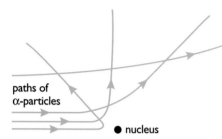

Figure 40.4 Hyperbolic paths of α-particles near a nucleus

A surprising discovery was that a small number of α-particles bounced back from the foil, being deflected through angles greater than 90°. The path of an α-particle in the electric field of the nucleus of a foil atom is a hyperbola (see Figure 40.4). Any α-particles that pass some distance from the nuclei just brush aside the tiny electrons and carry on almost undeviated. As such a small number of the α-particles are scattered significantly, the nuclei must be only a very tiny fraction of the volume of the whole atom. This led Rutherford to predict the model of the atom shown in Figure 40.2.

You can gain an understanding of some of the predictions of Rutherford's model by looking at a gravitational analogue model of the interaction between an α-particle and a nucleus, as in experiment 40.1. The laws of force in both gravitation and electricity are inverse square laws. So the force between point charges or masses is proportional to $1/r^2$ where r is the distance between the charges or masses. The gravitational and electrical potential energies vary as $1/r$ (← 15.3 and 16.3). In the gravitational model, you can represent the α-particle by a small ball, which can roll with little friction on a smooth curved hill. This represents the electrical potential energy of the α-particle and nucleus. The hill is shaped like a cone where the height h above the base is inversely proportional to its distance r from the central vertical axis, as shown in Figure 40.5. When you look down on the ball rolling across the hill, its path will approximate to a hyperbola, like that of an α-particle close to a nucleus (see Figure 40.6).

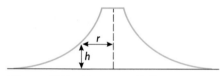

Figure 40.5 $1/r$ potential hill

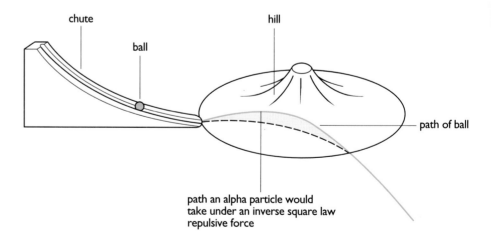

Figure 40.6 Hyperbolic path of an α-particle close to a nucleus

---

EXPERIMENT 40.1   Gravitational analogue of α-particle scattering

By setting up the chute and hill as shown in Figure 40.6 and 40.7a, and releasing the ball from a chosen height on the chute, you can measure the scattering angle ϕ for a particular aiming error p. The value of ϕ depends both on p and on the energy of the ball – how high it is released on the chute. Plotting a graph of p against ϕ should provide you with a similar curve to that in Figure 40.7b.

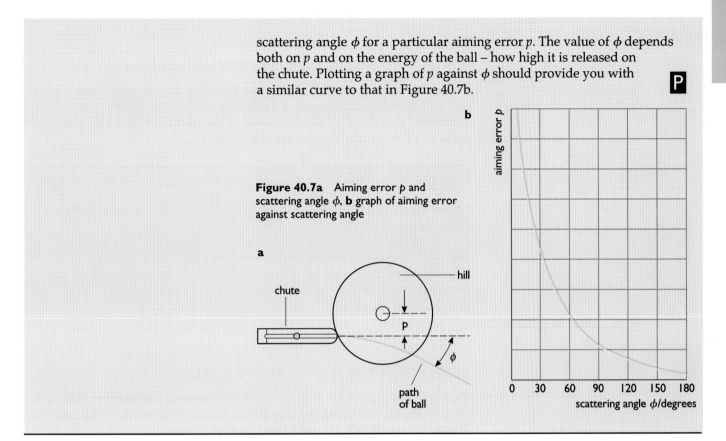

Figure 40.7a Aiming error p and scattering angle ϕ, **b** graph of aiming error against scattering angle

In experiment 40.1, you can measure **the scattering angle ϕ** for a particular **aiming error p** (see Figure 40.7a) as a ball rolls down the chute. However any α-particles scattered through an angle of more than ϕ must travel within an area of collision of πp^2, as shown in Figure 40.8a. So the number N of particles scattered through more than angle ϕ is proportional to p^2. The results of Figure 40.3b obtained from Geiger and Marsden's experiment, have produced the graph of Figure 40.8b. When this is compared to the graph of p^2 against ϕ using data from experiment 40.1, the two curves are found to be the same shape within the

Figure 40.8a α-particles scattered through angle ϕ or more, **b** graph of 'number of α-particles scattered through an angle greater than ϕ' against ϕ

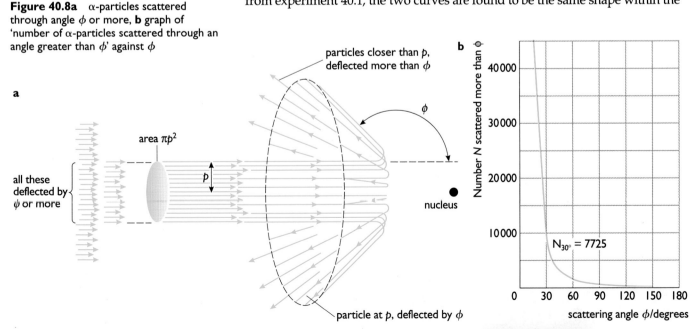

limits of experimental measurement. That for experiment 40.1 has a shape described by:

$$2p \tan \frac{\phi}{2} = \text{distance of closest approach in a head-on collision } (p = 0).$$

This is the equation for a ball subject to a 'repulsive' inverse square law of force as it is deflected from the centre of the hill. By analogy, the scattering in Rutherford's model of the atom is caused by the electric, or Coulomb, force between a tiny positive nucleus and the α-particles.

> **Question**
>
> **4a** In what ways might the gravitational hill model be **i** helpful and **ii** misleading?
> **b** Use Figure 40.7b, or your results from experiment 40.1, to construct a table of values of p^2 against ϕ. Plot these on a scale such that p^2 at 30° is equivalent to the N value at 30° on Figure 40.8b. Compare the two curves. They should be very similar.

ESTIMATING THE SIZE OF THE NUCLEUS

The chute in Figure 40.7a can be adjusted to make $p = 0$. The ball then rolls up the hill and returns back along the same path – it makes a 'head-on' collison and is scattered through 180°. When its kinetic energy is increased by releasing it from higher up the chute, the ball climbs higher before returning. At the point where the ball is stopped momentarily on the hill, all its energy is potential energy and is equal to the incident kinetic energy. This point is at the distance of closest approach of an α-particle to the nucleus (see Figure 40.9). Rutherford used this to calculate an upper limit to the size of the nucleus.

At the distance R of closest approach between an α-particle and a nucleus in a head-on collision, the energy of the system is all electrical potential energy. For

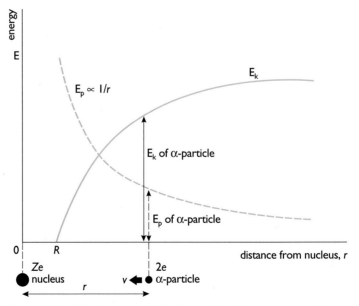

Figure 40.9 Kinetic energy (E_K) of α-particle changes to potential energy (E_P) as α-particle approaches nucleus. Total energy E is constant

40 NUCLEAR MODEL OF THE ATOM

two point charges of magnitude $+2e$ and $+Ze$ a distance R apart, the potential energy E_p is:

$$E_p = \frac{2e \cdot Ze}{4\pi\epsilon_o R}$$

($\leftarrow$ 16.3). In Geiger and Marsden's experiments, the incident α-particles each had a kinetic energy of about 5 MeV and the target nuclei were gold with a proton number Z of 79. Substituting in these values gives a value of R of 4.5×10^{-14} m.

This value of R is the **upper limit** on the 'radius' of the nucleus. The force between two point charges is E_p/R ($\leftarrow$ 16.3). The force of repulsion between the nucleus and α-particle at distance R is 18 N: the same order of magnitude as the weight of this book. However, the mass of an α-particle is only 6.7×10^{-27} kg!

Question

5 Using the equations and data above, show that the force F of repulsion between a nucleus and an α-particle at closest approach is about 18 N.
(e, ϵ_o)

40.3 NUCLEAR TRANSMUTATIONS

THE PROTON

Rutherford continued his studies of the nucleus by firing α-particles at gas atoms, such as hydrogen and nitrogen (see Figure 40.10). Cloud chamber photos show the results of such collisions very clearly ($\leftarrow$ Figure 8.19). The α-particles he used had a maximum range of 70 mm through these gases. By positioning the α-sources at more than 70 mm from the screen, he could ensure that any scintillations on it were caused by protons, which have a longer range in gas.

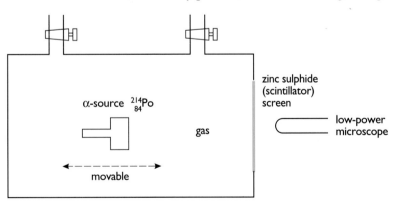

Figure 40.10 Rutherford's apparatus

With hydrogen in the chamber, Rutherford detected a number of proton scintillations. However, he also found some proton scintillations with nitrogen and these protons had greater energy than those in the hydrogen gas experiment. He explained the results by suggesting that, in a few of the collisions, an α-particle was captured by a nitrogen nucleus. The resultant unstable nucleus then emitted a fast proton, leaving behind a stable oxygen nucleus – the first example of **nuclear transmutation**.

You can describe nuclear transmutations using equations similar to those for chemical reactions. That in Rutherford's experiment can be written:

$${}^{4}_{2}\text{He} + {}^{14}_{7}\text{N} \rightarrow {}^{17}_{8}\text{O} + {}^{1}_{1}\text{H}.$$

This is a real equation, so the proton (Z) numbers and nucleon (A) numbers must balance on both sides.

Nuclear transmutations may instead be written in the form:

initial nuclide (in-particle, out-particle) final nuclide.

The transmutation of nitrogen in this form would be:

$${}^{14}_{7}\text{N}\,(\alpha, p)\,{}^{17}_{8}\text{O}.$$

Apart from producing the first transmutation of one element into another, Rutherford's experiment confirmed that the **proton** is a constituent of the nucleus. It also provided evidence for the existence of a short-range, strong attractive force – the **nuclear force**.

Alpha-particles and nitrogen nuclei are both positive and so repel each other. However in a few cases they bind together to form a new nucleus. This indicates that, when an α-particle approaches sufficiently close to a nucleus, a strong nuclear force overcomes the repulsive Coulomb force.

Question

6 Tritium, or ${}^{3}_{1}\text{H}$ is an isotope of hydrogen discovered in 1939. It is formed by bombarding another isotope, deuterium ${}^{2}_{1}\text{H}$ with itself.
a Describe the nuclear transmutation by copying and completing:

$${}^{2}_{1}\text{H}\,(\quad,\quad)\,{}^{3}_{1}\text{H}.$$

b Tritium decays by emission of a β-particle with a half-life of 12.5 years. What is the symbol for the final stable nuclide?

RELEASING ENERGY

In 1932, Cockcroft and Walton accelerated protons using the first particle accelerator. The protons were given sufficient energy, about 750 keV, for a few of them to be able to penetrate the target nuclei of ${}^{7}_{3}\text{Li}$. The result was the first splitting of the atom or **fission** (see Figure 40.11). The entrance of a proton made the nucleus unstable and it split into two α-particles, which moved apart at 180° to each other:

$${}^{1}_{1}\text{H} + {}^{7}_{3}\text{Li} \rightarrow 2\,{}^{4}_{2}\text{He}.$$

The α-particles had more energy than the incident proton. This was the first detection of the release of energy from within the nucleus.

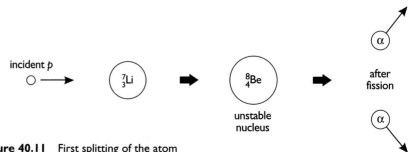

Figure 40.11 First splitting of the atom

40 NUCLEAR MODEL OF THE ATOM

THE NEUTRON

Another example of a nuclear transmutation led to the discovery in 1932 of the neutron. James Chadwick found that very penetrating radiation is produced when α-particles are fired at the element, beryllium. Further work showed that this radiation is due to particles similar in mass to protons but with zero charge, now known as **neutrons**.

In 1930, Chadwick fired α-particles at beryllium (Be) and produced some very penetrating radiation. When he passed this radiation into paraffin wax, a material rich in hydrogen, high-energy protons were knocked out of the wax. He found the energy of the protons by the thickness of aluminium (Al) sheet needed to absorb them (see Figure 40.12).

In 1932, Chadwick studied collisions between this radiation and nitrogen gas in a cloud chamber. He measured the momentum of the recoiling nitrogen nuclei from their tracks in the chamber. The maximum velocities of the nitrogen nuclei and of the protons from the wax, were consistent with his idea that the incident radiation was a neutral particle of approximately the same mass as the proton – the neutron.

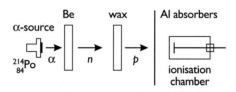

Figure 40.12 Chadwick's apparatus

Chadwick's analysis

After a head-on elastic collision between a mass m with velocity u and a stationary mass M, the final speed v of the target particle M is:

$$v = \frac{2mu}{M + m}$$

(← 9.3). So the ratio of the maximum speeds of the hydrogen nuclei (protons) and nitrogen nuclei in Chadwick's experiments are given by:

$$\frac{v_P}{v_N} = \frac{M_N + m}{M_P + m},$$

where m is the mass of the particle in the unknown radiation. This enables m to be calculated.

Questions

7 The nuclear transmutation that occurs when α-particles are fired at beryllium is:

$$^{9}_{4}\text{Be}\,(\alpha, n)\,^{A}_{Z}X.$$

Identify the nuclide $^{A}_{Z}X$.

8 The maximum values of v_p and v_N measured by Chadwick were:

$v_p = 3.3 \times 10^7 \text{ m s}^{-1}$

$v_N = 4.7 \times 10^6 \text{ m s}^{-1}$.

Show that these are consistent with the mass of the unknown particle m being about equal to that of the proton M_P. Assume $M_N = 14\,M_P$.

ARTIFICIAL RADIOACTIVITY

In 1934, the famous French physicists, the Joliot-Curies, investigated the transmutation:

$$^{27}_{13}\text{Al}\,(\alpha, p)\,^{30}_{14}\text{Si}$$

by directing an α-source at aluminium. They also detected neutrons and so concluded that the following transmutation occurs too:

$$^{27}_{13}\text{Al}\,(\alpha, n)\,^{30}_{15}\text{P}.$$

However, when they removed the α-source, they still detected radiation from the metal, with a half-life of 2.5 minutes. The isotope of phosphorus $^{30}_{15}\text{P}$ is a

INSIDE THE ATOM

radionuclide and does not occur in nature. For their discovery of the first **artificial radionuclide,** the Joliot-Curies received the Nobel prize.

Today, many thousands of radionuclides are made in commercial quantities inside a nuclear reactor core. A suitable material is irradiated by neutrons for a period of a few half-lives of the required radionuclide. For example, $^{32}_{14}Si$ is irradiated by neutrons to produce the radionuclide $^{32}_{15}P$, which is extensively used in agriculture, medicine and metallurgy as a tracer (← 39.8). It decays by β-emission with a half-life of 14.3 days.

THE POSITRON

The radionuclide discovered by the Joliot-Curies emits a **positron** to become an isotope of silicon:

$$^{30}_{15}P \rightarrow {}^{30}_{14}Si + \beta^+.$$

The positron had been discovered in 1932 in America in cloud chamber photographs of collision events caused by incident cosmic rays.

Figure 40.13 shows a modern bubble chamber photograph in which the energy of γ-ray quanta has become electron and positron particles. These are created simultaneously in a process called **pair production**. There is a magnetic field perpendicular to the plane of the picture. This causes the particles to move in opposite helixes, showing they have opposite charges.

Particle-antiparticle annihilation

The positron is the **antiparticle** of the electron. When it comes close to an electron, an 'atom' of positronium is produced. This is similar to an atom of hydrogen but much lighter. In a very small fraction of a second, the two coalesce and annihilate, or destroy, each other. The matter of the two particles becomes radiation energy, producing two γ-ray quanta via a neutral **pion** stage. Figure 40.14 shows a γ-ray initiating a cascade of electron-positron pairs as it hits a lead plate. The electrons and positrons curve away from each other due to a magnetic field into the page. These pairs also produce more γ-rays through annihilation and radiation which hit the next plate and trigger a new cascade.

Figure 40.13 Bubble chamber photograph showing pair production

Figure 40.14 Cascades of electron-positron pairs

Questions

9 Write down the equations for the creation and decay of $^{32}_{15}P$.

10 Using $\Delta E = 2m_e c^2$ (← 9.6) show that the minimum energy of a γ-ray quantum capable of causing pair production is 1.02 MeV. (m_e, c)

40 NUCLEAR MODEL OF THE ATOM

40.4 MASS SPECTROMETERS AND ACCELERATORS

The study of nuclei requires us to be able to measure their charges and masses. The technique for looking inside a nucleus is to hit it with an 'atomic hammer' to break it into 'pieces'. The charges and masses of these particles then have to be analysed. In this section we look at the two instruments used to perform these tasks, the **mass spectrometer** and the **particle accelerator**.

MASS SPECTROMETERS

These are used to compare the masses of different atoms and to measure their atomic masses (← 38.1). In any mass spectrometer, three steps are performed:

 i a beam of charged ions is produced;
 ii the ions are separated, or analysed, according to their masses and velocities;
iii the separated ions are detected.

Modern spectrometers can analyse specimens containing minute traces of elements and can give automatic computer readouts of isotopes of the elements present and of their relative abundances.

> **Question**
>
> 11 This question refers to the box 'First mass spectrometer'.
> a What features of the parabolae in Figure 40.15a indicate that
>   i the magnetic field alternates in direction?
>   ii not all the ions travel at the same speed?
>   iii the two isotopes are not equally abundant?
> b Which parabola is produced by the more massive isotope?

First mass spectrometer

J. J. Thomson created the first mass spectrometer when studying the properties of 'positive rays' in 1913. He used the apparatus shown in Figure 40.15 to deflect positive neon ions. Notice that he used an electric and a magnetic field placed **parallel** to each other. The ions are formed anywhere in the bulb and are then accelerated through the cathode C into a narrow beam. He predicted that ions of the same mass and charge would be deflected to make the shape of a parabola on the screen.

He found two curves one for ^{20}Ne and the other for ^{22}Ne. This was the first successful identification of the isotopes of a light non-radioactive element.

neon gas at low pressure ionised by discharge between A and C

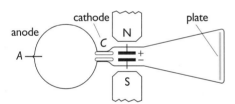

Figure 40.15a Thomson's mass spectrometer

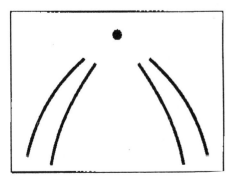

Figure 40.15b

In the modern **single-focusing mass spectrometer** (see Figure 40.16) ions are produced by electron bombardment of a sample in the **ion source** which contains an ionisation chamber called an **ion block** (see Figure 40.17).

A **velocity selector** is sometimes used to ensure that only ions with the same selected velocity emerge from the ion source. The particles are accelerated by a voltage V of several thousand volts. This voltage produces a high-energy particle beam, the velocity of which depends on the mass and charge of the particles. The beam is then deflected in an arc by a **B**–field produced by a large electromagnet. The **B**–field exerts a force perpendicular to the beam at any stage, so providing a centripetal force. The amount each particle in the beam is deflected by the magnetic field depends on its mass (← 35.4). When the radius of a deflected beam is such that it enters the end of the detector, or collector, the signal produced is amplified and displayed on a monitor screen.

By changing either the accelerating voltage or the magnetic field, different beams can be focused on to the end of the detector. Normally the monitor screen is scanned horizontally at the same rate as that at which the accelerating voltage or magnetic field is changed. This is repeated continually, so displaying on the screen a spectrum of the abundance of the various particles in the sample being analysed.

The horizontal position of each peak on the screen is determined by the type of charged particle producing it. The height of each peak is determined by the particle's relative abundance.

INSIDE THE ATOM

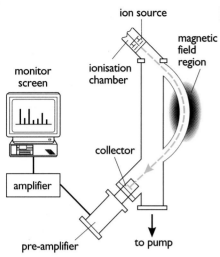

Figure 40.16 Basic features of modern single-focusing mass spectrometer, and path of ion beam

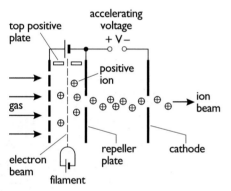

Figure 40.17 Ion source

Separating particles in a mass spectrometer

When particles of mass m and charge Q are accelerated through the voltage V:

$$\text{loss of potential energy} = \text{gain of kinetic energy}$$

$$QV = \tfrac{1}{2}mv^2$$

and so

$$v^2 = \frac{2QV}{m}.$$

The **B**-field bends each particle beam in an arc of radius r as shown in Figure 40.16. The force produced is a centripetal force and:

force due to **B**-field, BQv ($\leftarrow$ 17.4) = centripetal force, $\dfrac{mv^2}{r}$ ($\leftarrow$ 7.1)

$$BQv = \frac{mv^2}{r}.$$

This means that momentum mv equals BQr and

$$v = \frac{BQr}{m}.$$

Substituting this expression for v in equation (1) above gives:

$$\frac{(BQr)^2}{m^2} = \frac{2QV}{m}$$

$$r^2 = \frac{2V}{B^2} \cdot \frac{m}{Q}.$$

So, for particles with a particular value of m/Q, either V or B can be varied until the beam enters the end of the detector. Ion beams having a different mass or charge will require a different value of V or B to focus them on the detector. In practice, a high scanning rate is usually required to prevent flicker on the monitor screen. So the accelerating voltage is changed rather than the **B**-field. The extremely high inductance of the coils in the electromagnet make it difficult to change the current in them quickly.

PARTICLE ACCELERATORS

Until Cockcroft and Walton split the atom using a proton accelerator ($\leftarrow$ 40.3), the only particle with enough energy to bombard the nucleus was the naturally emitted α-particle. Now machines are built to accelerate charged particles to many thousands of MeV. Two such machines are the linear accelerator and the synchrotron. In both, charged particles are accelerated again and again by applying high voltage pulses, each pulse giving the particles further energy.

In a **linear accelerator**, the charged particles move at constant speed through cylinders of increasing length (see Figure 40.18). The cylinders are connected to a high-frequency alternating supply, producing an electric field that accelerates the particles between the cylinders. So the particles move at progressively faster constant speeds in each cylinder. Successive cylinders are longer so that the

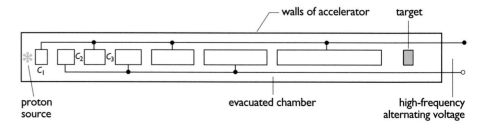

Figure 40.18 Basic components of a linear accelerator

40 NUCLEAR MODEL OF THE ATOM

charged particles spend the same time in each. So they reach the end as the supply polarity reverses.

A linear accelerator would have to be almost impossibly long to produce really high-energy particles. For example, the Stanford University's 3 km linear accelerator can only accelerate electrons to 20 GeV. The **synchrotron** avoids this problem by using magnets to make the charged particles move inside a highly evacuated ring. As the speed of the particles increases, the magnetic field has to be increased to keep them moving in a circle of constant radius. The particles are accelerated once each lap in one small section of the tube by the electric field of a radio-frequency oscillator. The frequency of this electric field also has to be increased as the particles move faster.

The first generation of big proton synchrotrons, in the 1950s at Brookhaven and CERN, accelerated particles to 30 GeV. Elaborate design features and high-precision engineering are required to keep the beam moving in a ring which at Brookhaven, is 256 m in diameter. The magnetic field is supplied by more than 200 small magnets spaced around the tube separated by field free regions. However, this is insignificant beside the latest Super Proton Synchrotron, shown in Figure 40.19, at CERN. It occupies a tunnel some 7 km in diameter and can accelerate protons to 400 GeV. The CERN SPS is now used as the final link in a chain of accelerators which feed the 27 km diameter LEP electron-positron collider.

Synchrotron proton energy

For a proton of charge Q in a circular orbit of radius r:

  centripetal force = force due to the **B**-field

$$\frac{mv^2}{r} = BQv.$$

So the momentum p is given by:

$$p = mv$$
$$= BQr.$$

Protons in a synchrotron reach speeds close to that of light c. So relativistic mechanics must be used ($\leftarrow$ 9.6) in which energy E is given by:

$$E = pc.$$

Therefore, proton energy:

$$E = BQrc.$$

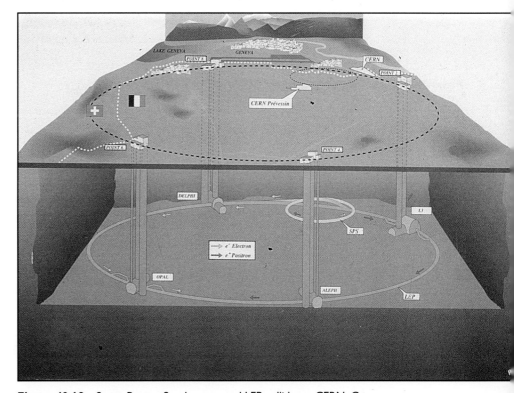

Figure 40.19 Super Proton Synchrotron and LEP collider at CERN, Geneva

Question

12 Show that in the Brookhaven accelerator, a field approaching 1 tesla is required to contain the highest energy protons.

INSIDE THE ATOM

The protons from an accelerator are fired at some target material. In the resulting interaction, various sub-atomic particles, such as π-mesons, K-mesons, neutrinos and hyperons, are formed (→ 40.5). Much higher energy interactions can be achieved by arranging a head-on collision between two beams of particles travelling in **opposite** directions in the machine. For example protons and **antiprotons** which have opposite charge but the same mass can be accelerated at the same time. Once moving at very high speeds, the beams can be brought together, releasing twice the energy available in the fixed target experiments.

40.5 NUCLEAR ENERGY AND FORCE

In the first accelerator experiments, Cockcroft and Walton (← 40.3) found that the two α-particles emitted from the bombardment of a $^{7}_{3}$Li nucleus by a proton had more kinetic energy than the incident proton. When the masses of the individual neutrons, protons and electrons making up a particular atom are added, their total mass is always greater than the measured mass of the atom. For example, the mass of $^{12}_{6}$C is less than the mass, individually, of six protons, six neutrons and six electrons. You can check this using the data in Figure 40.20.

Note also that the mass of a $^{7}_{3}$Li nucleus (the nuclide minus three electrons) and a proton is greater than the mass of two α-particles. The missing mass appears as the extra kinetic energy of the α-particles. The mass difference is called the **mass defect**, or deficit, Δm. It is linked to the extra kinetic energy ΔE by Einstein's equation (← 9.6):

$$\Delta E = \Delta mc^2.$$

particle or nuclide	mass/u
electron	0.0005
proton	1.0073
neutron	1.0087
$^{1}_{1}$H	1.0078
$^{4}_{2}$He	4.0026
α-particle	4.0015
$^{7}_{3}$Li	7.0160
$^{12}_{6}$C	12.0000

Figure 40.20 Masses of some particles and nuclides

BINDING ENERGY

As Einstein's equation shows, when the energy of a body increases so does its mass. The protons and neutrons bound together in a nucleus have less energy than when they are free particles. The energy needed to separate them is their **total binding energy**. This represents the mass defect of the nucleus.

Using the values in Figure 40.20, we can estimate the total nuclear binding energy of $^{7}_{3}$Li as follows:

mass of $3p + 4n + 3e$ = 7.0582 u

mass of $^{7}_{3}$Li = 7.0160 u

so mass defect = 0.0422 u.

From Einstein's equation, this mass defect Δm is linked to the total binding energy ΔE by:

$$\Delta E = \Delta mc^2$$

= 39.3 MeV.

Measurements from mass spectroscopy (← 40.4) give the atomic masses of all nuclides. It is normal practice to consider the **average binding energy per nucleon** $\Delta E/A$ rather than the total binding energy or mass defect. For $^{7}_{3}$Li the binding energy per nucleon is:

$\Delta E/A$ = 39.3/7

= 5.6 MeV.

40 NUCLEAR MODEL OF THE ATOM

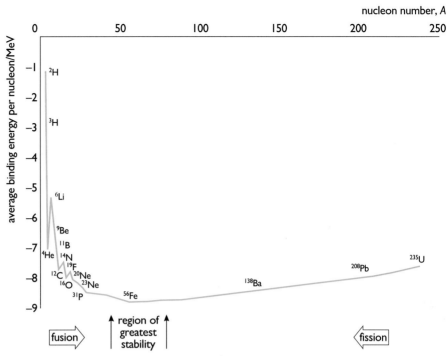

Figure 40.21 Average binding energy per nucleon against nucleon number

Figure 40.21 shows the binding energy per nucleon for nuclides. As you can see, it is generally about 8 MeV, but is less for very light and very heavy nuclides. Notice also the strong binding energy of $_2^4$He. This suggests why an α-particle rather than a proton or neutron is emitted in a natural radioactive decay. The most stable nuclide of all, which has the greatest binding energy per nucleon, is iron $_{26}^{56}$Fe. The region of greatest stability lies betwen nucleon numbers 50 to 80. Other peaks of stability include $_6^{12}$C and $_8^{16}$O. The variations in binding energy are due to the way the nuclear force acts.

Questions

13 Show that the energy of 1 u is 931 MeV = 1.49×10^{10} J.

14 How massive is a 1 MeV electron or β-particle?

15 How does the average binding energy of an atomic electron compare with the average binding energy per nucleon?

In Figure 40.21, the binding energy per nucleon is a negative quantity. Similarly, when two masses are bound by gravity, they have a negative gravitational potential. The zero level of gravitational potential energy is taken when the masses are free of their mutual gravitational attraction, at an infinite distance apart (← 15.3). In the nucleus, zero energy is taken to be the energy equivalent of the mass of the separate nucleons making up the nucleus.

NUCLEAR FORCE

Another useful graph to plot in the study of nuclear stability is neutron number N against proton number Z for stable nuclei, as in Figure 40.22. It shows that the lightest stable nuclides lie close to the line $Z = N$. The heavier nuclides have more

INSIDE THE ATOM

Figure 40.22 Neutron number N against proton number Z for all stable nuclides

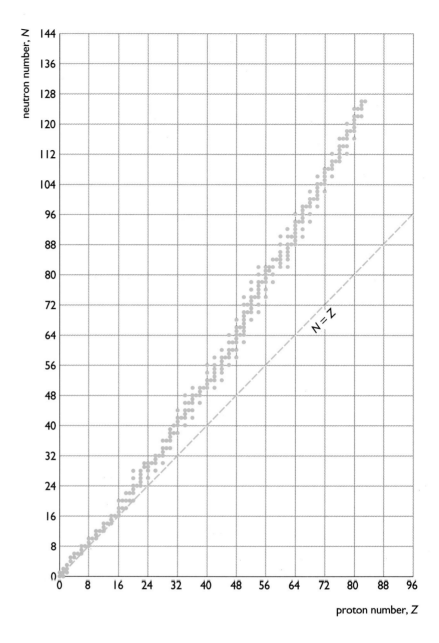

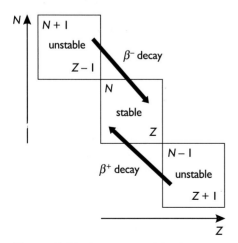

Figure 40.23 Increasing stability by electron or positron emission

neutrons than protons. This is necessary due to the electrical repulsion between protons.

Some idea of the great strength and short range of this force is given by α-particle scattering calculations (← 40.2). An α-particle that approaches a gold nucleus to a distance of 5×10^{-14} m is not captured by the nucleus but repelled by a force of 18 N. A strong nuclear force overcomes the repulsive force between protons and holds the nucleus together. But the repulsive force increases rapidly as the number of protons in the nucleus increases. So stable nuclei are only possible at high nucleon numbers when there are more neutrons than protons.

However, when there are too many neutrons, the nucleus again becomes unstable. The nuclear force seems to favour the binding of particles in pairs – one neutron with one proton and then one pair with another. Evidence for this comes from the great stability of $^4_2$He, $^{12}_6$C and $^{16}_8$O, and the fact that the stable light nuclei lie along the $Z = N$ line. Radionuclides that lie above or below the line of stability decay by electron (β⁻) or positron (β⁺) emission to bring them to a stable nuclide closer to the line (see Figure 40.23).

40 NUCLEAR MODEL OF THE ATOM

Sub-atomic particles

The particles within an atom can be divided into two families – the **leptons**, which do not feel the strong nuclear force, and the **hadrons**, which do. Examples of leptons are electrons, positrons, muons and neutrinos. The **muon** appears to be identical to the electron except that its mass is some 200 times greater and it is unstable.

Hadrons have been discovered in high-energy collision experiments in particle accelerators (see Figure 40.24). The collisions provide sufficient energy to free the hadrons briefly from their bound states in the nucleus. Their rest masses can be calculated from the tracks of the particles in bubble chamber photographs. One set of these particles called **mesons** acts as the proton-neutron 'glue' holding the nucleus together.

By 1964, many hadrons had been observed and a model was developed for them. This suggests that hadrons are made up of particles called **quarks**. All observed hadrons are made up of combinations of three quarks of the same or different types. No free quark has been detected. The present model suggests that it is not possible for a free quark to exist.

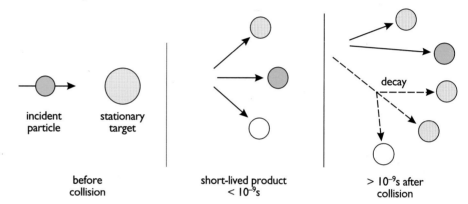

Figure 40.24 Short-lived hadrons are produced when collision energy releases matter

40.6 FISSION, FUSION AND THE NUCLEAR REACTOR

The average binding energy per nucleon curve (see Figure 40.21) indicates that nuclides with low and high nucleon number A are less strongly bound than those in the middle. When a nucleus with a large nucleon number is split into two fragments, the two resulting nuclei are more tightly bound. Their higher binding energies mean that there is a release of energy. This process, **fission**, is used as the source of energy in a nuclear reactor or a nuclear bomb. Similarly, combining two low nucleon number nuclei produces a more strongly bound nucleus and releases energy. This process, **fusion**, occurs inside stars, enabling them to generate and radiate energy for very long periods of time.

FISSION

Heavy nuclides become more unstable with increased neutron number ($\leftarrow$ 40.5). The fission of a nucleus was first achieved in Germany in the late 1930s by causing neutrons to be absorbed by nuclei of $^{235}_{92}U$. The **fission products**, or **fragments**, from such reactions can be various combinations of 34 elements. Three typical reactions are:

$$^{235}_{92}U + ^{1}_{0}n \rightarrow \; ^{236}_{92}U^* + \gamma \rightarrow \; ^{141}_{56}Ba + ^{92}_{36}Kr + 3^{1}_{0}n$$
$$\text{or} \qquad\qquad\qquad\qquad \rightarrow \; ^{90}_{38}Sr + ^{142}_{54}Xe + 4^{1}_{0}n$$
$$\text{or} \qquad\qquad\qquad\qquad \rightarrow \; ^{28}_{12}Mg + ^{205}_{80}Hg + 3^{1}_{0}n.$$

INSIDE THE ATOM

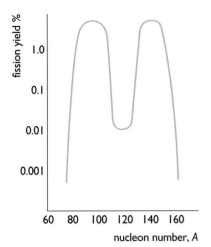

Figure 40.25 Fission yield against nucleon number

The uranium nucleus does not always split up in the same way, so a whole range of elements are fission products. Figure 40.25 shows the distribution of fission fragments against nucleon number A.

Figure 40.26 shows a typical fission reaction for ^{235}U. The uranium nucleus captures a slow **thermal** neutron with a kinetic energy of about 1/40 eV. An excited unstable nucleus of $^{236}_{92}$U* results which breaks into two, usually uneven, nuclei and several neutrons. All the neutrons possess considerable kinetic energy. This energy is transferred to other atoms by collision, increasing the internal energy and, therefore, the temperature of the material very quickly. Once their energy is reduced sufficiently, these **prompt** neutrons can interact with other $^{235}_{92}$U nuclei and continue the fission process. These fission fragments are very rich in neutrons and often shed **delayed** neutrons several seconds after fission has occured.

All the slow neutrons can cause fission in other $^{235}_{92}$U nuclei, giving rise to a **chain reaction** (see Figure 40.27). The disintegration energy released in each fission is about 200 MeV. In the first design of a nuclear bomb, two masses of $^{235}_{92}$U were brought together so that an uncontrolled chain reaction could occur. This released an enormous quantity of energy in a very short time with devastating consequences.

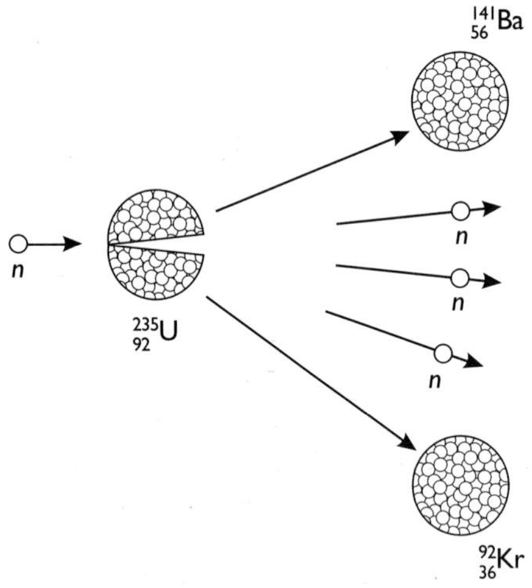

Figure 40.26 Typical fission reaction for uranium-235. The products are unstable

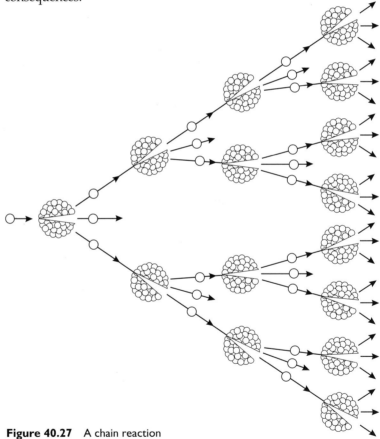

Figure 40.27 A chain reaction

Question

16a How much energy is generated by the fission of 1 kg of $^{235}_{92}$U?
b Comment on this value, comparing it with the 3×10^7 joules of energy transferred during the thermal combustion of 1 kg of coal.

40 NUCLEAR MODEL OF THE ATOM

SUSTAINING A CHAIN REACTION

In nature only 0.7 % of uranium on the Earth's surface is $^{235}_{92}U$. The neutrons from any spontaneous fissions do not cause the other 99.3 % of uranium nuclei $^{238}_{92}U$ to split. This is why there is still natural uranium in the ground. To allow a chain reaction, the percentage of $^{235}_{92}U$ is increased, producing **enriched** uranium.

Whether or not a chain reaction then builds up depends on the average number of neutrons produced in a fission that can induce fission in the next stage or generation of the chain. The **effective multiplication factor** k is the ratio of the number of neutrons in one generation to those in the previous generation of the chain reaction. The factor k depends on the rate of production of neutrons and on their rate of loss. Neutrons may be lost through absorption within the system by various atomic nuclei and through leakage from the system.

When the effective multiplication factor:

- $k < 1$, any spontaneous reactions quickly die away. This occurs in natural uranium where the amount of fissionable ^{235}U is insufficient to generate enough neutrons to maintain the reaction.
- $k = 1$, the reaction is said to become **critical**. When pure fissionable material such as pure ^{235}U, is in spherical form, its mass is described as the **critical** mass. The energy produced by the reactions remains constant and very high temperatures may be produced.
- $k > 1$, catastrophic energy production results in a nuclear explosion or, in the case of a nuclear reactor, a **meltdown**.

NUCLEAR REACTOR

In a nuclear reactor, k must not be allowed to exceed 1 by more than a small fraction for any length of time or the reactor may overheat. However a reactor could never explode like a nuclear bomb because the uranium in the reactor is not sufficiently enriched. In addition, suitable **absorbers**, made of elements such as cadmium and boron, are inserted with the uranium in the reactor to absorb the excess neutrons. These **control rods** can be raised or lowered to start or shut down the reactor (see Figure 40.28). The neutrons are also slowed down by being passed through a material, called a **moderator**, because it moderates their energies. This increases the probability of the neutrons being captured and causing fission. The reactor is also surrounded by a moderator material to reflect escaping neutrons back into the core.

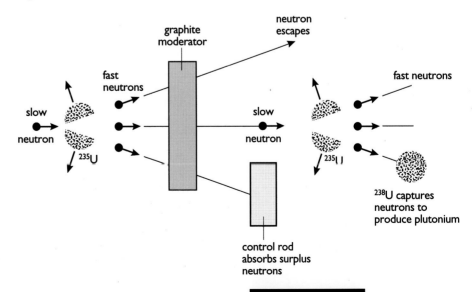

Figure 40.28 Processes occurring in a uranium reactor

INSIDE THE ATOM

Figure 40.29 Enrico Fermi

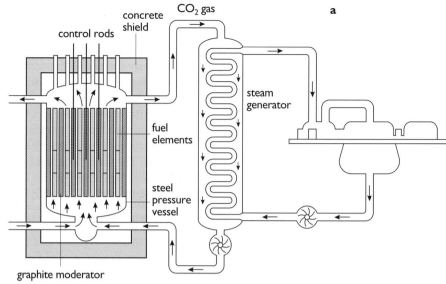

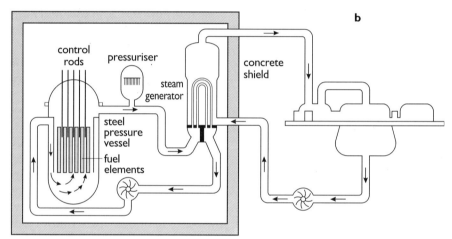

Figure 40.30a Graphite-moderated, Advanced Gas-cooled Reactor (AGR), **b** Pressurised Water-moderated Reactor (PWR)

Nuclear power

The first reactor was constructed in what had been a squash court in Chicago by Enrico Fermi (see Figure 40.29). Natural uranium of mass 6 000 kg was stacked with blocks of pure graphite to act as a moderator. The control rods were made of cadmium. The k factor was 1.0006. When it was allowed to become critical, in 1942, this experimental reactor generated 200 W — the power of two light bulbs! Today there are more than 200 commercial reactors in operation, some with power capacities of more than 1 GW. Small reactors have powered cargo ships, icebreakers and submarines and tiny ones have been used on space missions.

There are two main types of reactor: **gas-cooled**, such as the AGR and MAGNOX types, and **water-cooled**, such as the BWR, PWR and HWR types. In gas-cooled reactors, gas is used to cool and extract internal energy from the high-temperature core (see Figure 40.30a). This is the system used in most contemporary British nuclear power stations. In water-cooled reactors, water is used as the moderator as well as the coolant (see Figure 40.30b).

Most reactors use natural or enriched uranium as fuel and neutrons moderated to thermal energies. In the reactor core, $^{238}_{92}U$, which is a non-fissile material, is converted through the absorption of a slow neutron and two beta decays to ^{239}Pu, which, like ^{235}U, is a fissile material:

$$^{238}_{92}U + ^{1}_{0}n \rightarrow ^{239}_{92}U \ (T_{1/2} = 93.5 \text{ mins}) + \gamma$$
$$\downarrow$$
$$^{239}_{93}Np \ (T_{1/2} = 2.35 \text{ days}) + \beta^-$$
$$\downarrow$$
$$^{239}_{94}Pu \ (T_{1/2} = 24\,000 \text{ yrs}) + \beta^-.$$

Eventually, most of the $^{235}_{92}U$ is used up in a reactor and the fuel rods have to be removed and reprocessed. The $^{239}_{92}U$, $^{238}_{92}U$ and the remaining $^{235}_{92}U$ is removed for reuse. The remaining material, mostly fission fragments, is long-lived radioactive waste and its safe disposal poses a complex environmental problem. The stockpiled $^{239}_{94}Pu$ may also cause problems. It is chemically unstable and has a half-life of 2.4×10^4 years compared to 4.5×10^9 years for the $^{238}_{92}U$ from which it has been manufactured.

40 NUCLEAR MODEL OF THE ATOM

An important type of fast reactor called a **breeder** has $^{239}_{94}\text{Pu}$ as the core fuel surrounded by uranium. As well as generating energy, more plutonium is 'bred' from the $^{238}_{92}\text{U}$ in the uranium.

FUSION

The change in binding energy when two light nuclei fuse together to form a more massive nucleus is much greater than in fission (see Figure 40.21). Unfortunately, fusion is difficult to achieve because of the increasing electrical Coulomb repulsion between nuclei as they approach each other. It is not until a temperature of about 10^7 K is reached that the nuclei have enough kinetic energy to approach close enough to fuse.

Question

17 Two hydrogen nuclei (protons) in a gas can fuse when each of their kinetic energies is just sufficient to provide the potential energy that will exist when they are in contact. At temperature T the average kinetic energy of each nucleus is $\tfrac{3}{2}kT$, where k is the Boltzmann constant ($\leftarrow$ 10.5). Their potential energy E_p is given by:

$$E_p = e^2/4\pi\epsilon_0 r,$$

where ϵ_0 is the permittivity of free space ($\leftarrow$ 16.3). Estimate the temperature required for two protons to be likely to fuse. Assume that r is the nuclear diameter.

stage 1 $p + p \rightarrow {}^2_1\text{H} + \beta^+ + \nu$ $p + p \rightarrow {}^2_1\text{H} + \beta^+ + \nu$

stage 2 ${}^2_1\text{H} + p \rightarrow {}^3_2\text{He} + \gamma$ ${}^2_1\text{H} + p \rightarrow {}^3_2\text{He} + \gamma$

stage 3 ${}^3_2\text{He} + {}^3_2\text{He} \rightarrow {}^4_2\text{He} + p + p$

Figure 40.31 Proton-proton cycle: the formation of helium from hydrogen in the Sun

The Sun's energy is generated by the fusion of hydrogen into helium and then into more massive elements. One possible fusion reaction of hydrogen takes place in three stages with energy being released at each stage (see Figure 40.31).

The probability of the first stage occurring is 10^{-26}, that is, only one collision in 10^{26} will result in the fusion of two protons. If the probability was greater, the fusion of the Sun's material would have been completed long ago and the star would now be dead. There are so many protons in the Sun that deuterium, ${}^2_1\text{H}$, is formed at a rate of 10^{12} kg s^{-1}. The deuterium fuses with another proton to form tritium, ${}^3_1\text{He}$, in a few seconds. But this nucleus may wait on average for up to 10 years before the third stage, the formation of an α-particle. About 26 MeV of energy is released in the complete process. The fusion of 1 kg of hydrogen to helium releases 6×10^{14} J.

A controlled fusion reaction on the Earth using deuterium from the sea is both the dream and research project of many scientists. At the JET fusion project deuterium and tritium are used as fuel to produce fusion energy:

$${}^2_1\text{H} + {}^3_1\text{H} \rightarrow {}^4_2\text{He} + {}^1_0\text{n} + Q.$$

This reaction releases energy Q, per fusion of about 18 MeV, or 2.8×10^{-12} J.

INSIDE THE ATOM

JET fusion project

Matter at 10^7 K is known as a plasma. In a plasma, matter exists as a gas of charged particles, in dynamic equilibrium, which possess collective behaviour and whose motion is governed by electromagnetic forces. The problem is how to contain a plasma once it is created so that it does not expand and cool through interaction with the atoms of the walls of the containing vessel. In a star, gravity holds the plasma together against the radiation pressure trying to blow it apart. On Earth the method of plasma confinement which has been most researched has been the "magnetic bottle", where the plasma is turned back into itself by magnetic fields.

The most successful device to do this is called a Tokamak, invented by the Russian physicist Andrei Sakharov. The hot plasma is trapped by an arrangement of strong magnetic fields inside a doughnut-shaped container or torus. One field is generated by the electric currents in giant water cooled copper rings along the axis of the torus and the other field is around the cross-section of the torus, caused by the current through the plasma itself. The cost of constructing a large Tokamak is such that funding from many nations is required.

The Joint European Torus (JET), built at Culham in Berkshire, is Europe's attempt to try to produce fusion energy from deuterium-tritium fuel. In 1978 the International Atomic Energy Authority initiated the INTOR programme – a pooling of scientific and technical knowledge from the ten or so large Tokamaks in the World – to prove that a controlled thermonuclear fusion reactor is possible. Plasma confinement for very short periods has been achieved at great energy cost.

In early 1992, JET generated more energy from a tritium-deuterium fusion reaction than was supplied. At this critical stage, it became self sustaining. There is still a very long way to go, however, before such methods will be efficient enough to generate electricity.

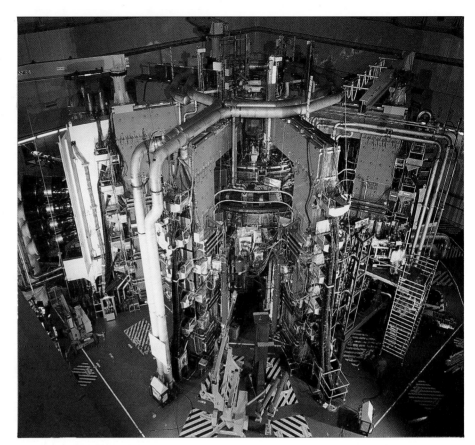

Figure 40.32 JET, the Joint European Torus fusion reactor

Controlled fusion reactors could provide us with all our energy needs almost for ever. However, no machine yet built can raise more than a tiny amount of matter to 10^7 K and contain it in a small space for more than a fraction of a second.

SUMMARY

The atom is about 10^{-10} m in diameter. Its mass is contained in a tiny positive nucleus about 10^{-14} m in diameter. An upper limit to the size of the nucleus can be estimated using Rutherford scattering of α-particles in head-on collisions.

The nucleus consists only of neutrons and protons. The particles are held together by a nuclear force with a range no greater than the nuclear diameter.

The nucleus is probed by firing high-energy nuclear particles at it. When captured, these can cause nuclear transmutations. A nuclear transmutation is sometimes written:

   target nuclide (incoming particle, outgoing particle) final nuclide.

Heavy nuclei, such as $^{235}_{92}U$, may split into two fragments, releasing energy. This is known as fission. Cockcroft and Walton were the first to split the atom, releasing energy in the transmutation $^{7}_{3}Li\,(p, \alpha)\,^{4}_{2}He$.

The nuclei of very light nuclides can combine in high-velocity collisions to form heavier nuclides, releasing energy. This is known as fusion.

The masses of all nuclides are measured very accurately using a mass spectrometer. The atomic mass is usually given in atomic mass units, u.

40 NUCLEAR MODEL OF THE ATOM

1 u = 1.66054 × 10^{-27} kg, which is 1/12th of the mass of a neutral atom of $^{12}_{6}C$.

The difference between the sum of the masses of the individual protons and neutrons of a nuclide and the mass of the nuclide is called the mass defect, Δm. It is related to the total binding energy ΔE of that nuclide by Einstein's equation $\Delta m = \Delta E/c^2$. A mass defect of 1 u corresponds to a binding energy of 931 MeV.

The average binding energy per nucleon of a nuclide of A nucleons is $\Delta E/A$. For most stable nuclides, this is about 8.4 MeV.

In fission and fusion, the binding energy per nucleon of the products of the interaction is greater than the binding energy per nucleon of the initial nuclides.

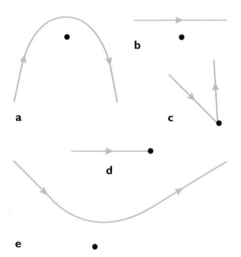

Figure 40.33

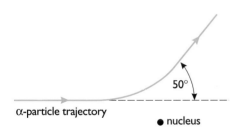

Figure 40.34 α-particle and gold nucleus

SUMMARY QUESTIONS

18 Figures 40.33a to e show the trajectories of a light particle or object close to a massive one. For each:
a Suggest the type of force (e.g. attractive inverse square law) that may exist between the two, and
b give one example of a real situation where it may arise.

19 In α-particle scattering experiments, almost all of the α-particles pass undeviated through the foil or are deflected by only small angles.
a What does this indicate about nuclear size?
b Why, in a head-on collision, is it only possible to find an upper limit to the size of a nucleus?
c Will a value of this upper limit increase or decrease, assuming all other variables remain unchanged, if:
   i the incident particles are protons with the same initial kinetic energy as the α-particles?
   ii the nuclei of the foil have a lower proton number?
   iii the incident α-particles have more kinetic energy?

20 According to Newton's third law of motion, there is an equal and opposite force on a gold nucleus and an approaching α-particle. Describe how the forces on each of the particles varies as the α-particle follows the trajectory shown in Figure 40.34.

21 In the α-particle scattering analogue model shown in Figure 40.6, the height from which the ball is released on the chute is proportional to the distance of closest approach in a head-on collision, when $p = 0$.
a Justify this statement.
b Suggest under what circumstances it will no longer be true.

22a For a given aiming error, will incident protons be deflected more or less than incident α-particles of the same energy?
b Does the deflection increase or decrease when
   i a foil of lower proton number is used?
   ii the α-particle energy is increased?

23 In one of Geiger and Marsden's experiments, the detector was fixed at a particular angle to the foil (see Figure 40.3a). Increasing numbers of thin mica sheets were placed in the incident beam of α-particles to reduce their energy. How would you expect the number of observed scintillations on the screen to change? (Hint: think about the change in deflection of the ball by the hill when it is released from lower heights on the chute.)

INSIDE THE ATOM

24 In a perfectly elastic collision between an α-particle of mass $m = 4$ u and an isolated stationary nucleus of mass $M = 200$ u, the fraction of the incident kinetic energy given to the nucleus is $4mM/(M + m)^2$.
a Calculate this fraction.
b Why will the exchange of kinetic energy in the gold foil experiments be much less than this?

25a A pulse of protons with energy E enters a synchrotron. Their energy increases by a constant factor f on each circuit. Assuming this factor includes any relativistic effects, how many circuits are needed for their energy to double?
b In the CERN synchrotron, depicted in Figure 40.35, the protons are accelerated from 50 MeV to 28 GeV. How many circuits will the protons have to make, if factor f is 1.2?

26 In a linear accelerator, shown in Figure 40.18, the length l of a cylinder is given by the formula:

$$l = v/2f$$

where v is the speed of the particles in the cylinder and f is the frequency of the alternating supply. The particles are accelerated through a voltage V between cylinders.
a What would be the result of doubling
  the voltage of the supply?
  the frequency of the supply?
  the length of the cylinders?
b In each case, what other quantities would have to be altered assuming the accelerator still operates? Would the change lead to an increase or decrease in the energy of the particles incident on the target?

27a In the core of a nuclear reactor, many different fission reactions are possible. Complete the following two possible reaction equations:
  i $^{235}_{92}U + ^{1}_{0}n \rightarrow ^{140}_{54}Xe + ^{A}_{Z}X + ^{1}_{0}n$
  ii $^{235}_{92}U + ^{1}_{0}n \rightarrow ^{139}_{53}I + ^{95}_{39}Y + ^{A}_{Z}X$
b One material in the control rods is boron. When a boron nucleus absorbs a neutron, it decays to lithium. Write a suitable equation for this transmutation. What is the symbol for the emitted particle?
c How many $^{235}_{92}U$ nuclei must undergo fission in 1 second to generate enough energy to light a 100 W light bulb? Take the energy release per fission to be 200 MeV.

28a When an electron and a positron meet, they annihilate producing two γ-rays. Calculate the **minimum** energy of these two rays.
b A β-particle is often emitted with an energy of 1 MeV. Suppose the electron and positron have energies of 1 MeV when they collide. By what factor will the energies of the γ-rays increase over the minimum possible energy? (m_e, c)

29 Estimate the contribution to the kinetic energy of the fission fragments in Figure 40.26 caused by the electrical repulsion between them. Use the Coulomb potential energy formula (← 14).

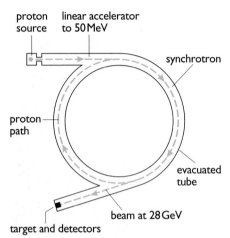

Figure 40.35 The 28 GeV proton synchrotron at CERN

UNIT 1

INFORMATION AND DATA NEEDED IN PHYSICS

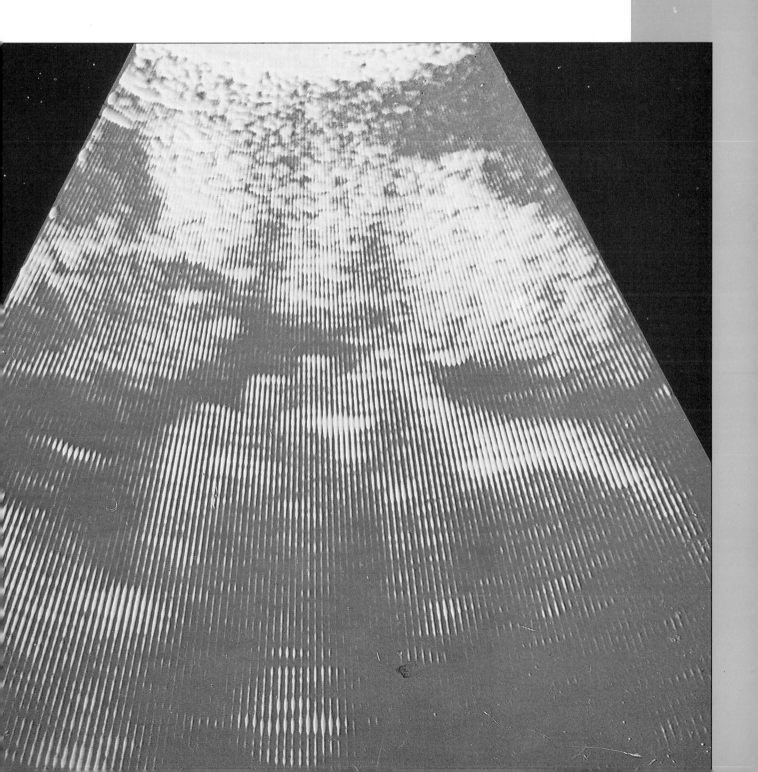

41 INFORMATION AND DATA NEEDED IN PHYSICS

Information needs to be presented in the clearest possible way. It must always be used in such a way that people can easily remember and understand it. Much information is only for reference. It may consist purely of data, or it may contain references and discussion about various aspects of the material.

41.1 TABLES OF CONSTANTS

Approximate values to be used in questions throughout the book:

Fundamental physical constants:		
Speed of light	c	3.0×10^8 m s^{-1}
Permeability of a vacuum	μ_0	$4\pi \times 10^{-7}$ H m^{-1}
Permittivity of a vacuum	ϵ_0	8.9×10^{-12} F m^{-1}
Gravitational constant	G	6.7×10^{-11} N m^2 kg^{-2}
Planck constant	h	6.6×10^{-34} J s
Elementary charge	e	1.6×10^{-19} C
Electron mass	m_e	9.1×10^{-31} kg
Electron specific charge	e/m_e	1.8×10^{11} C kg^{-1}
Proton mass	m_p	1.7×10^{-27} kg
Neutron mass	m_n	1.7×10^{-27} kg
Atomic mass unit	m_u	1.7×10^{-27} kg (931 MeV)
Avogadro constant	L or N_A	6.0×10^{23} mol^{-1}
Faraday constant	F	9.6×10^4 C mol^{-1}
Molar gas constant	R	8.3 J K^{-1} mol^{-1}
Boltzmann constant	k	1.4×10^{-23} J K^{-1}
Stefan constant	σ	5.7×10^{-8} W m^{-2} K^{-4}
Earth constants:		
Mean gravitational field	g	9.8 N kg^{-1} or m s^{-2}
Mean atmospheric pressure	p	1.01×10^5 Pa
Mean distance from Sun	r_E	1.5×10^{11} m
Mean radius	R_E	6.4×10^6 m
Mass	M_E	6.0×10^{24} kg
Escape velocity	V_E	1.1×10^4 m s^{-1}
Solar constant	I_S	1.4×10^3 W m^{-2}

INFORMATION AND DATA NEEDED IN PHYSICS

ACCURATE VALUES

Quantity	Symbol	Value	Unit	Notes
Speed of light in a vacuum	c	299 792 458	$m\,s^{-1}$	definition
Permeability of a vacuum	μ_0	$4\pi \times 10^{-7}$	$H\,m^{-1}$	definition
Permittivity of a vacuum	ϵ_0	$8.854\,187\,817 \times 10^{-12}$	$F\,m^{-1}$	$1/(c^2\mu_0)$
Planck constant	h	$6.626\,075\,5 \times 10^{-34}$	$J\,s$	$\Delta = 40$
Gravitational constant	G	$6.672\,59 \times 10^{-11}$	$N\,m^2\,kg^{-2}$	$\Delta = 85$
Elementary charge	e	$1.602\,177\,33 \times 10^{-19}$	C	$\Delta = 49$
Electron mass	m_e	$9.109\,389\,7 \times 10^{-31}$	kg	$\Delta = 54$
Electron specific charge	e/m_e	$1.758\,819\,62 \times 10^{11}$	$C\,kg^{-1}$	$\Delta = 53$
Proton mass	m_p	$1.672\,623\,1 \times 10^{-27}$	kg	$\Delta = 10$
Neutron mass	m_n	$1.674\,928\,6 \times 10^{-27}$	kg	$\Delta = 10$
Atomic mass unit	m_u	$1.660\,540\,2 \times 10^{-27}$	kg	$\Delta = 10$
Avogadro constant	L or N_A	$6.022\,136\,7 \times 10^{23}$	mol^{-1}	$\Delta = 36$
Faraday constant	F	$9.648\,530\,9 \times 10^{4}$	$C\,mol^{-1}$	$\Delta = 29$
Molar gas constant	R	$8.314\,510$	$J\,K^{-1}\,mol^{-1}$	$\Delta = 70$
Boltzmann constant	k	$1.380\,658 \times 10^{-23}$	$J\,K^{-1}$	$\Delta = 12$
Stefan constant	σ	$5.670\,51 \times 10^{-8}$	$W\,m^{-2}\,K^{-4}$	$\Delta = 19$

Δ = the uncertainty in the last two figures of the quoted value.

41.2 MATHEMATICAL PROCESSES

It is very important to try to understand the mathematics used to describe physical processes. The ideas presented here are an outline to provide you with the basis on which to work. Further reading of books on mathematics is very important and you should do this whenever you get stuck or want to develop your mathematical confidence. Practice is very important and attempting several similar examples in a particular topic in mathematics enables you to reinforce what you have learnt.

CONSTANTS AND VARIABLES

Fundamental physical constants such as the charge on the electron e or the speed of light in a vacuum c are very well known and documented (← 41.1). However, there are many other quantities which, although not fundamentally constant, you will keep constant or assume are constant during an investigation: atmospheric pressure, the mass of a trolley or the voltage of a battery, for example. It is often convenient to name such quantities with a symbol: p, m and v perhaps in these examples. Even though, strictly speaking, such quantities can be changed, they are treated mathematically as constants. Assigning a value to such quantities, at an appropriate stage, is all that is required.

Other quantities are referred to as variables. Those quantities which you control or observe during an experiment are referred to as the **independent**

41 INFORMATION AND DATA NEEDED IN PHYSICS

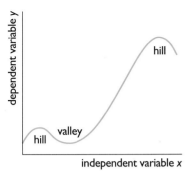

Figure 41.1

variables. Those quantities which change as a result of the effects of the controlled or observed variables are referred to as the **dependent variables**.

Constructing graphs between dependent and independent variables is the normal way to investigate the relationship between variables. The dependent variable is normally plotted on the vertical, or y-axis and the independent variable along the horizontal, or x-axis. This is shown in Figure 41.1 for a height, y, against a horizontal distance, x.

SYMBOLS USED IN MATHEMATICS

A page of mathematics using symbols which are unfamiliar to you can be particularly daunting. However, by isolating each symbol and explaining to yourself what it stands for and what it does, you will begin to gain an insight into what the mathematics is all about. The more you are familiar with mathematical symbols the more maths you will understand and the easier it becomes. Mathematical symbols broadly fall into two types: names and operators.

A **name** is usually a letter used to mean a constant, a variable or a function:

A **constant** has a particular value which does not change when other quantities such as distance, time, temperature or pressure change.

A **variable** has a particular numerical value which changes depending on the values of other quantities.

A **function** is a particular type of variable which consists of a formula or relationship, such as $x^3 + 2y$. It has a value when its constituent variables, in this case x and y, have been defined. A function is frequently written with the names of the variables on which it depends in brackets. So a function **A** which depends upon the variables x and y is written as $\mathbf{A}(x, y)$.

Usually, letters of the Roman alphabet are used for names. Sometimes, however, letters from the Greek alphabet and other special symbols are used. When stating the value of any of them, you need to say what unit it is 'measured in'.

When there is a range of variables of a similar type, each is given the same name but with a **suffix** to indicate this:

a_1, a_2, a_3 and so on up to a_n.

Notice how 'n' is used to represent this suffix in general.

An **operator** indicates what should happen to constants, variables and functions.

Generally, operators tell you what to do with the function to the right. The simplest are the arithmetic operators: **+, −, ×** and **÷**. Another common operator is the left hand bracket **(** which means act on every term in the same way until the right hand bracket **)** is reached. Other operators that you will need in physics include:

Σ – add each term,

$\int$ – integrate,

$\dfrac{\mathrm{d}^n}{\mathrm{d}x^n}$ – differentiate with respect to x, n times.

Sometimes the operator is between two functions such as the ratio symbol **:** or afterwards as with the factorial operator **!** which means multiply the integer n to the left of the symbol by $n-1$, $n-2$, etc. down to 2.

691

INFORMATION AND DATA NEEDED IN PHYSICS

POWERS AND INDICES

An **index** is a number which represents the number of times a quantity, or **base**, is multiplied by itself.

Base	Index	Example:
5	3	5^3 is 5 cubed or 5 raised to the power of 3 $= 5 \times 5 \times 5 = 125$
5	2	5^2 is 5 squared or 5 raised to the power of 2 $= 5 \times 5 = 25$
2	1	2^1 is 2 raised to the power of 1 (no multiplication) $= 2$
2	0	2^0 is 1. Any base raised to the power 0 is 1.

An index can be a fraction:

Base	Index	Example:
2	1/2 or 0.5	$2^{1/2}$ is the square root of $2 = \sqrt{2} = 1.41$ i.e. the number which when squared produces the base, 2
2	1/3 or 0.33	$2^{1/3}$ is the cube root of $2 = \sqrt[3]{2} = 1.26$ i.e. the number which when cubed produces the base, 2

Negative indices also occur:

Base	Index	Example:
10	−1	10^{-1} is 1/10 or 1 tenth $= 0.1$
2	−3	2^{-3} is $1/(2 \times 2 \times 2)$ $= 1/8 = 0.125$
16	−1/4	$16^{-1/4}$ is $1/16^{1/4}$ or the reciprocal of the fourth root of 16 $= \dfrac{1}{\sqrt[4]{16}} = \dfrac{1}{2} = 0.5$

When bases raised to different powers are multiplied you add the indices together. Similarly, when they are divided you subtract them. For example:

$$2^3 \times 2^2 = 2^{(3+2)} = 2^5 = 32$$

$$2^7/2^3 = 2^{(7-3)} = 2^4 = 16.$$

For base 10 this is particularly useful to remember when you are performing calculations using the standard form of expressing numbers and quantities. For example, you can calculate the magnitude of the force F between an electron and a proton in the hydrogen atom when they are 0.1 nm apart as follows:

$$F = \frac{q_1 q_2}{4\pi\epsilon_0 r^2}$$

$$= \frac{1.6 \times 10^{-19} \times 1.6 \times 10^{-19}}{4 \times \pi \times 8.9 \times 10^{-12} \times (0.1 \times 10^{-9})^2}$$

$$= \frac{(1.6 \times 1.6)}{(4 \times \pi \times 8.9)} \times \frac{10^{-19} \times 10^{-19}}{10^{-12} \times (10^{-10})^2}$$

$$= 0.023 \times 10^{(-19-19+12+20)}$$

$$= 0.023 \times 10^{-6}$$

$$= 2.3 \times 10^{-8} \text{ N}.$$

LOGARITHMS

The **logarithm** of a number N is the power to which a base B has to be raised in order to give the number: $N = B^{\log N}$. Now, a number can be expressed as another number to a particular power. For example: $100 = 10^2$; $16 = 2^4$. So, 2 is the logarithm of 100 to base 10 and 4 is the logarithm of 16 to base 2.

The base is usually written as a subscript to the word **log**:

$\log_{10} 100 = 2$ and $\log_2 16 = 4$.

Three examples for a base of 10 are:

$$2 = 10^{0.301} \qquad 10 = 10^1 \qquad 1/2 = 10^{-0.301}.$$

So, $\log_{10} 2 = 0.301$, $\log_{10} 10 = 1$, and $\log_{10}(0.5) = -0.301$.

Three examples for a base of 2 are:

$$2 = 2^1 \qquad 4 = 2^2 \qquad 1024 = 2^{10}.$$

So, $\log_2 2 = 1$, $\log_2 4 = 2$, and $\log_2 1024 = 10$.

The number e has the value 2.718 to 4 significant figures. So when e is used as a base for logarithms:

$$N = e^{\ln N}.$$

Where ln means $\log_e$.

This is often referred to as a natural, or Naperian, logarithm.

Three examples for base e are:

$$2 = e^{0.6931} \qquad e = e^1 \qquad 10 = e^{2.303}.$$

So, $\ln 2 = 0.6931$, $\log_e e = 1$, and $\ln 10 = 2.303$.

You should note that the logarithm of any base is 1.

Because logarithms are indices of a base, they can be added when multiplication occurs or subtracted when division occurs.

Suppose you want to calculate $1.30^3 \times 2.29^2$:

let $N = 1.30$ and $M = 2.29$.

So, for base e: and for base 10:

$\ln N = 0.262$ and $\ln M = 0.829$ $\qquad \log_{10} N = 0.114$ and $\log_{10} M = 0.360$.

So $N^3 \times M^2 = e^{3 \times 0.262} \times e^{2 \times 0.829}$ $\qquad N^3 \times M^2 = 10^{3 \times 0.114} \times 10^{2 \times 0.360}$

$\qquad \qquad \qquad = e^{0.786 + 1.658}$ $\qquad \qquad \qquad \qquad = 10^{0.342 + 0.720}$

$\qquad \qquad \qquad = e^{2.444}$ $\qquad \qquad \qquad \qquad = 10^{1.062}$

$\qquad \qquad \qquad = 11.5$ (using antilog$_e$) $\qquad \qquad = 11.5$ (using antilog$_{10}$).

INFORMATION AND DATA NEEDED IN PHYSICS

Using normal multiplication:

$$N^3 \times M^2 = 1.30^3 \times 2.29^2$$
$$= 2.20 \times 5.24$$
$$= 11.5.$$

Clearly, in this example, any of the three methods can be used. Normally, when using a calculator, the third method is the simplest. However, for very complex calculations, logarithmic methods have some advantages.

EXPONENTS, EXPONENTIAL e AND e^{ax}

Consider the following relationship:

$$y = B^x$$

x is a variable and it is also an index: one which represents the power to which the base, B has to be raised to obtain the value of y. Such a variable is known as an **exponent**.

An **exponential** change in a quantity, therefore, is one where its value depends on an exponent.

The base B can be replaced by any other base, say b or e, because, in general you will see that $B = b^c = e^a$ where a and c are constants.

Usually, in mathematics and in science, the base is taken as e, the **exponential constant**, which to 4 significant figures, has a value of 2.718.

So you can write $y = e^{ax}$ instead of $y = B^x$.

In general an exponential change is written as:

$$y = Ae^{ax} + C \text{ where } A, a \text{ and } C \text{ are constants.}$$

What is so special about e? It is an irrational number rather like π and is defined by the formula:

$$e = (1 + 1/n)^n \text{ as } n \to \text{infinity}$$

so you can write $e^x = (1 + 1/n)^{nx}$ as $n \to$ infinity.

Alternatively, e can be defined by the relationship:

$$\frac{d(e^x)}{dx} = e^x$$

which means that e^x can also be defined by the series:

$$e^x = 1 + x/1! + x^2/2! + x^3/3! + \cdots + x^n/n! \quad \text{as } n \to \text{infinity}$$

where $2! = 1 \times 2$; $3! = 1 \times 2 \times 3$ and so on which means that:

$$e^x = \sum_{n=0}^{n} x^n/n! \quad \text{as } n \to \text{infinity.}$$

So e can also be found from this series by using $x = 1$:

$$e = 1 + 1/1! + 1/2! + 1/3! + \cdots + 1/n! \quad \text{as } n \to \text{infinity.}$$

Question

1 Using a **scientific calculator** satisfy yourself that $e = 2.718$ to 4 significant figures,
a when n is 4824 or more for the formula, and
b when n is 6 or more for the series.

41 INFORMATION AND DATA NEEDED IN PHYSICS

DIFFERENTIATION: RATES OF CHANGE AND SLOPES OF GRAPHS

The gradient, or slope, of a straight line in x and y coordinates can be found by dividing an increase in height Δy by the corresponding increase in horizontal distance Δx:

gradient = $\Delta y / \Delta x$.

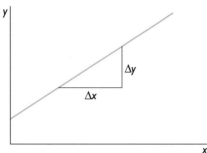

Figure 41.2 **Figure 41.3**

For a curved relationship, you need to take very much smaller increments (see Figures 41.2 and 41.3). For absolute accuracy, these have to be infinitesimal and tend to zero. They are then called **dy** and **dx**. In general, the gradient of any curve, dy/dx is given by:

$$\frac{dy}{dx} = \underset{\Delta x \to 0}{\text{Limit}} \frac{\Delta y}{\Delta x}.$$

This is often called the differential coefficient of y with respect to x. It is the rate of change of y with respect to x. Of course variables other than y or x may be used.

Any function of x can be represented by y. Finding the differential coefficient is called **differentiation** and this is the basis of **differential calculus**.

In general $\dfrac{dy}{dx} = \underset{\Delta x \to 0}{\text{Limit}} \dfrac{(y + \Delta y) - y}{\Delta x}$.

Imagine you have a graph of $y = x^2$.

Then $\Delta y / \Delta x = \dfrac{(x + \Delta x)^2 - x^2}{\Delta x}$

$= \dfrac{2x\Delta x + (\Delta x)^2}{\Delta x}$

$= 2x + \Delta x$.

So, $dy/dx = \underset{\Delta x \to 0}{\text{Limit}} (2x + \Delta x)$

$= 2x + 0$

$= 2x$.

This means that the gradient of the curve $y = x^2$ is always $2x$. So, the curve gets steeper and steeper as y increases at a rate which is twice the value of x as shown in Figure 41.4.

In general, when $y = x^n$,

$dy/dx = nx^{n-1}$.

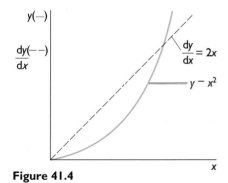

Figure 41.4

INFORMATION AND DATA NEEDED IN PHYSICS

> **Question**
>
> 2 Using the method described in the main text, satisfy yourself that for:
> a x^3 the differential coefficient is $3x^2$
> b x the differential coefficient is 0.

For $\sin x$ and $\cos x$ the relationships are:

$$\frac{dy}{dx} = \frac{d(\sin x)}{dx} = \cos x$$

and

$$\frac{dy}{dx} = \frac{d(\cos x)}{dx} = -\sin x.$$

Note how $(\sin x)$ and $(\cos x)$ replace y in the differential coefficient. This is done to make it clear what function ($\sin x$ or $\cos x$ in this case) is being differentiated.

An apostrophe above a variable's name means that it represents its differential coefficient with respect to an already mentioned variable, a dot indicates that it is with respect to time. So for a:

$$\acute{a} = \frac{da}{dx} \text{ or } \frac{da}{dy} \text{ etc.}$$

$$\dot{a} = \frac{da}{dt}.$$

> **Question**
>
> 3 Find $\frac{d(e^x)}{dx}$, by differentiating each term in the series:
>
> $$e^x = 1 + x/1! + x^2/2! + x^3/3! + \cdots + x^n/n!$$
> $$\text{as } n \to \text{infinity.}$$

The differential coefficient of e^{ax} is ae^{ax}. So you can see that the rate of change of an exponential quantity is proportional to itself. This property of exponential relationships is particularly useful in a number of areas of physics: radioactive growth and decay and the charge and discharge of a capacitor, for example.

It is possible to find the rate of change of the rate of change of a quantity. This is called the **second order differential coefficient**. Acceleration is a good example of this as it is the rate of change of velocity which in turn is the rate of change of distance with respect to time. So acceleration is the rate of change of the rate of change of distance.

It is written like this:

velocity $v = dx/dt$ and acceleration $a = dv/dt$

so acceleration $= \frac{d(v)}{dt} = \frac{d}{dt}\left(\frac{dx}{dt}\right) = \frac{d^2x}{dt^2}$.

$\frac{d^2x}{dt^2}$ is called the **second derivative** or the **second order differential coefficient** of x with respect to time.

41 INFORMATION AND DATA NEEDED IN PHYSICS

By the same reasoning dx/dt is the **first derivative** or **first order differential coefficient** but these terms are not often used.

Two dots above a variable also mean the second order differential coefficient with respect to time:

$$\ddot{x} = \frac{d^2x}{dt^2}.$$

INTEGRATION: SUMMATION, CUMULATIVE EFFECTS, AND AREAS ENCLOSED BY GRAPHS

Whenever you need to know the area under a curve, you can use integration.

For example a graph showing the extension of a spring with a stretching force is a line starting at the origin. Suppose you stretch it 50 cm with a force which as it is applied, increases from zero to 100 N. The graph will be as shown in Figure 41.5a. Imagine this to be made up of a large number of narrow rectangles.

The area under the graph represents the energy stored in the spring.

$$E_p = \text{sum of} \quad (\Delta E_1 + \Delta E_2 + \cdots)$$
$$= \sum_{n=1}^{n} \Delta E_n \quad \text{(this means the sum of } \Delta E_n \text{ as n increases in steps of 1 to the highest value of n considered)}$$
$$= \sum_{n=1}^{n} F_n \Delta x \quad \text{(note that } \Delta x \text{ is the same for every term, no matter what the value of n).}$$

Now imagine that Δx tends to zero and becomes infinitesimally small so that we can call it dx. Then the summation sign Σ is replaced by an **integration** sign, $\int$. Normally, this will be between the lower limit of x_1 and an upper limit x_2:

$$E_p = \int_{x_1}^{x_2} F dx.$$

This is called a **definite integral**, when the lower and upper limits of x are specified.

The graph is shown in Figure 41.5b.

If the limits are not specified, then you will need to write:

$$E_p = \int F dx + C, \text{ where C is a constant.}$$

This is known as an **indefinite integral**. C is usually evaluated after the integration has been performed.

Integration or integral calculus is the opposite process to differentiation. So in general we can write:

$$\text{when } y = x^n, \int y dx = \frac{x^{n+1}}{n+1} + C.$$

Some functions are very difficult to integrate and a list is provided in section 41.3.

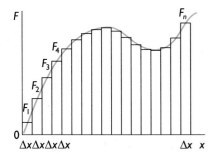

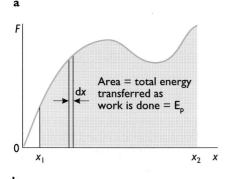

Figure 41.5

DIFFERENTIAL EQUATIONS AND THEIR ANALYTICAL SOLUTION

Sometimes, differential coefficients are shown as such in an equation and in order to find a solution to the equation, you will need to integrate it. This is known as a **differential equation**. Two simple examples met in physics are:

INFORMATION AND DATA NEEDED IN PHYSICS

for radioactive decay: $\dfrac{dN}{dt} = -\lambda N$

which is an example of a first order differential equation;

for simple harmonic oscillations: $\dfrac{d^2x}{dt^2} = -\omega^2 x$

which is an example of a second order differential equation because the differential coefficient is of order 2.

To solve such equations there are a number of techniques which you can use. Some are analytical and others are iterative. The simplest analytical method is to try a simple guess and see whether it satisfies both sides of the equation. For example, in radioactive decay, the graph appears to be an exponential shape. So you might try $N = Ae^{bt}$. In the case of a simple harmonic oscillator, you know that the motion is a continuous oscillation, so perhaps a sine or a cosine function might work such as $x = a\cos(bt)$.

To show that $N = Ae^{bt}$ is a solution to the radioactive decay relationship $dN/dt = -\lambda N$, satisfy yourself that $A = N_0$, the original number of undecayed atoms and that $b = -\lambda$, the decay constant, by doing the following:

Left hand side: $dN/dt = bAe^{bt}$

Right hand side: $-\lambda N = -\lambda Ae^{bt}$.

For these to be equal, $b = -\lambda$.

So $N = Ae^{-bt}$ is a solution.

At $t = 0$, $N = N_0$, so using this solution, $N_0 = Ae^{-\lambda 0}$

$= A$.

The solution is therefore:

$N = N_0 e^{-\lambda t}$.

Question

4 Show that $x = a\cos\omega t$ is a solution to $\dfrac{d^2x}{dt^2} = -\omega^2 x$.

ITERATIVE PROCESSES

Instead of finding an analytical function such as e^x or $\sin\theta$ as a solution to an equation, it is possible to find, by iteration, solutions which are based purely on the starting conditions. Iteration involves calculating what happens in a particular time, distance or other appropriate interval and, using the results from this, calculating what will happen in the following interval. You then repeat this for subsequent intervals, always using the results from the previous interval.

For example, suppose you are told that a car is accelerating from rest at 2 m s^{-2} until it reaches 30 m s^{-1} when it continues to travel at this constant speed for 10 seconds. You can calculate the total distance travelled at a particular time by considering time intervals Δt.

Starting conditions are: speed = 0 m s^{-1}
acceleration = 2 m s^{-2}.

For each iteration you need to choose an appropriate value for Δt, say 1 s, and use the following logical procedure:

while speed < than 30 then speed = previous speed + acceleration × Δt
distance = previous distance + speed × Δt.

So the first three iterations are:

Iteration 1:
(after 1 s)
speed = 0 + 2 × 1 = 2
dist = 0 + 2 × 1 = 2

Iteration 2:
(after 2 s)
speed = 2 + 2 × 1 = 4
dist = 2 + 4 × 1 = 6

Iteration 3:
(after 3 s)
speed = 4 + 2 × 1 = 6
dist = 6 + 6 × 1 = 12

The 14th, 15th and 16th iterations are:

Iteration 14:
(after 14 s)
speed = 28 + 2 × 1 = 30
dist = 30 + 30 × 1 = 60

Iteration 15:
(after 15 s)
speed = 30
dist = 60 + 30 × 1 = 90

Iteration 16:
(after 16 s)
speed = 30
dist = 90 + 30 × 1 = 120

41 INFORMATION AND DATA NEEDED IN PHYSICS

The final iteration is number 24 which is 10 seconds after the speed reaches 30 m s^{-1}.

Iteration 24:
(after 24 s)
speed = 30 = 30 m s^{-1}
distance = 360 + 30 × 1 = 390 m

In this way the variation of distance with time, speed with time or even distance with speed can be logged and plotted on a graph as each iteration is complete. Notice that to be slightly more accurate you could take the **half–interval** times of 0.5, 1.5, 2.5 seconds to calculate the distance. The coarser the interval you take, the more important this is. Conversely, for very small intervals, of say 0.01 s in this case, it is insignificant.

An iteration for capacitor discharge

Possible starting conditions:

$V = 10$ V
$C = 500 \times 10^{-6}$ F
$R = 100 \times 10^3$ Ω
$T = 0$
$\Delta T = 2$ s.

For each iteration:

$Q = C \times V$
$I = V/R$
$\Delta Q = -I \times \Delta T$
$Q = Q + \Delta Q$
$V = Q/C$
$T = T + \Delta T$.

You should try this model for various different values to get a feel for the iterative technique and then try and invent your own for, say, simple harmonic oscillations, radioactive decay, or motion in a circle.

> **Question**
>
> **5** Using the iteration procedure shown, sketch a graph of **a** acceleration **b** speed and **c** distance covered, against time.

Computer programs, spreadsheets and modelling programs, enable quite complicated iterations to be performed at high speed and graphs to be drawn relating any pair of variables. The secret of success in such programming is always to be aware of where discontinuous changes occur and to choose the right interval length. For example, suppose you were to write a program of iterations to simulate the falling of a sky diver. You would need to introduce a term for an upward force when the parachute opens. It is not unknown in such simulations, when program writers have used short interval iterations, for the diver to be catapulted out into space! Alternatively, should you choose too long an interval length, then the diver may well hit the ground before the parachute has a chance to open! Clearly, the real situation must always be kept in mind.

41.3 TABLE OF MATHEMATICAL FORMULAE

CONSTANTS

e = base of natural logarithms ≈ 2.718 28
$\log_{10} e \approx 0.434\,294$ $\log_e 10 \approx 2.302\,59$
$\log_{10} N \approx \log_e N \times 0.434\,3$ $\log_e N \approx \log_{10} N \times 2.302\,6$
1 radian ≈ 57.295 8° ≈ 57° 17' 45" $\pi \approx 3.141\,592\,65$
$\log_{10} \pi \approx 0.497\,15$ $\frac{1}{\pi} \approx 0.318\,31$ $\frac{\pi}{180} \approx 0.017\,45$ $\pi^2 \approx 9.869\,6$

ALGEBRA

$\log_a x = y \Leftrightarrow x = a^y$ $\log_q p = \log_q r \log_r p$
Sum of first n terms of the series $a, a+d, a+2d, \ldots$
$S_n = \frac{1}{2}n[2a + (n-1)d] = n \times$ (average of first and last terms)

$\sum_{r=1}^{n} r = \frac{1}{2}n(n+1)$ $\sum_{r=1}^{n} r^2 = \frac{1}{6}n(n+1)(2n+1)$

$\sum_{r=1}^{n} r^3 = \frac{1}{4}n^2(n+1)^2$ $\sum_{r=0}^{n-1} a^r = \frac{1-a^n}{1-a}$

INFORMATION AND DATA NEEDED IN PHYSICS

If $f(x) = ax^2 + bx + c$, roots α, β of $f(x) = 0$ given by $\dfrac{(-b \pm \sqrt{b^2 - 4ac})}{2a}$. Also $\alpha + \beta = \dfrac{-b}{a}$, $\alpha\beta = \dfrac{c}{a}$

$f(x) > 0$ for all real x and $a > 0$, $c > 0$, $4ac > b^2$

Remainder when polynominal $P(x)$ divided by $(x - a)$ is $P(a)$

Number of combinations of n objects taken r at a time,

$${}_nC_r \text{ or } \binom{n}{r} = \dfrac{n!}{(n-r)!r!}$$ where $n! = n(n-1)(n-2)\ldots 3.\,2.\,1$

Taylor's expansion: $f(a + x) = f(a) + xf'(a) + \dfrac{x^2}{2!}f''(a) + \dfrac{x^3}{3!}f'''(a) + \cdots$

or (Maclaurin's form): $f(x) = f(0) + xf'(0) + \dfrac{x^2}{2!}f''(0) + \dfrac{x^3}{3!}f'''(0) + \cdots$

Expansions (*valid if $|x| < 1$, the rest valid for all x)

$\sin x = \dfrac{x}{1!} - \dfrac{x^3}{3!} + \dfrac{x^5}{5!} - \dfrac{x^7}{7!} + \cdots$

$\cos x = 1 - \dfrac{x^2}{2!} + \dfrac{x^4}{4!} - \dfrac{x^6}{6!} + \cdots$

$e^x = 1 + \dfrac{x}{1!} + \dfrac{x^2}{2!} + \dfrac{x^3}{3!} + \cdots$

*$\log(1 + x) = x - \dfrac{x^2}{2} + \dfrac{x^3}{3} - \dfrac{x^4}{4} + \cdots$

*$(1 + x)^n = 1 + nx + n\dfrac{(n-1)}{2!}x^2 + \dfrac{n(n-1)(n-2)}{3!}x^3 + \cdots + \binom{n}{r}x^r + \cdots$

Newton–Raphson iterative formula for root of $f(x) = 0$: $x_{n+1} = x_n - \dfrac{f(x_n)}{f'(x_n)}$

Distance of centre of mass from the centre (d)

a Hemisphere (radius r) $d = \dfrac{3r}{8}$

b Hemispherical shell $d = \dfrac{r}{2}$

c Sector of circle (angle 2θ) $d = \dfrac{(2r\,\sin\theta)}{3\theta}$

d Arc of circle $d = \dfrac{(r\,\sin\theta)}{\theta}$

e Cone (height h) $d = \dfrac{h}{4}$ (from centre of base)

41 INFORMATION AND DATA NEEDED IN PHYSICS

ANALYSIS

List of derivatives

y	$\dfrac{dy}{dx}$	y	$\dfrac{dy}{dx}$
x^n	nx^{n-1}		
$\sin x$	$\cos x$	$\cos x$	$-\sin x$
$\tan x$	$\sec^2 x$	$\cot x$	$-\csc^2 x$
$\sec x$	$\sec x \tan x$	$\csc x$	$-\csc x \cot x$

List of integrals

$F'(x) = f(x)$	$F(x) = \int f(x)dx$
x^n	$\dfrac{x^{n+1}}{n+1}\quad n \neq -1$
$\dfrac{1}{x}$	$\log|x|$
e^x	e^x
a^x	$\dfrac{a^x}{\log a}$
$\dfrac{1}{\sqrt{a^2 - x^2}}$	$\sin^{-1}\dfrac{x}{a}$

Simpson's rule $\int_a^b f(x)dx \approx \tfrac{1}{3}h(y_0 + 4y_1 + y_2)$ where $h = \tfrac{1}{2}(b - a)$

$(uv)' = u'v + uv'$, $\quad \left(\dfrac{u}{v}\right)' = \dfrac{u'v - uv'}{v^2}$

$\int uv' \, dx = uv - \int u'v \, dx$

Radius of curvature $\rho = \dfrac{\left[1 + \left(\dfrac{dy}{dx}\right)^2\right]^{3/2}}{\dfrac{d^2y}{dx^2}}$

MENSURATION

Area of triangle, (sides a, b, c): $\Delta = \tfrac{1}{2} bc \sin A$
or $\sqrt{s(s-a)(s-b)(s-c)}$ where $2s = a + b + c$
Circle (radius r): Perimeter $= 2\pi r$; Area $= \pi r^2$

Ellipse (axes $2a$, $2b$): Perimeter $\approx 2\pi \sqrt{\dfrac{a^2 + b^2}{2}}$; Area $= \pi ab$

Cylinder (radius r, height h): Area $= 2\pi r(h + r)$, Volume $= \pi r^2 h$
Area of curved surface of cone $= \pi rl$, where $l =$ slant height
Volume of cone or pyramid $= \tfrac{1}{3} Ah$, where $A =$ base area, $h =$ height
Sphere (radius r): Area $4\pi r^2$, Volume $\tfrac{4}{3}\pi r^3$
Area cut off on sphere by parallel planes h apart $= 2\pi rh$

INFORMATION AND DATA NEEDED IN PHYSICS

TRIGONOMETRY

a $\sin(\theta \pm \phi) = \sin\theta \cos\phi \pm \cos\theta \sin\phi$
$\cos(\theta \pm \phi) = \cos\theta \cos\phi \mp \sin\theta \sin\phi$
$\tan(\theta \pm \phi) = \dfrac{\tan\theta \pm \tan\phi}{1 \mp \tan\theta \tan\phi}$
$\sin 2\theta = 2\sin\theta \cos\theta$
$\cos 2\theta = \cos^2\theta - \sin^2\theta = 2\cos^2\theta - 1 = 1 - 2\sin^2\theta$
$\sin 3\theta = 3\sin\theta - 4\sin^3\theta,\ \cos 3\theta = 4\cos^3\theta - 3\cos\theta$
$\sin A + \sin B = 2\sin\tfrac{1}{2}(A+B)\cos\tfrac{1}{2}(A-B)$
$\sin A - \sin B = 2\cos\tfrac{1}{2}(A+B)\sin\tfrac{1}{2}(A-B)$
$\cos A + \cos B = 2\cos\tfrac{1}{2}(A+B)\cos\tfrac{1}{2}(A-B)$
$\cos A - \cos B = -2\sin\tfrac{1}{2}(A+B)\sin\tfrac{1}{2}(A-B)$

b In any triangle: $\dfrac{a}{\sin A} = \dfrac{b}{\sin B} = \dfrac{c}{\sin C} = 2R$ (sine rule)
$a^2 = b^2 + c^2 - 2bc\cos A$ (cosine rule)

41.4 THE GREEK ALPHABET

A	α	Alpha	N	ν	Nu
B	β	Beta	Ξ	ξ	Xi
Γ	γ	Gamma	O	o	Omicron
Δ	δ	Delta	Π	π	Pi
E	ϵ	Epsilon	P	ρ	Rho
Z	ζ	Zeta	Σ	σ	Sigma
H	η	Eta	T	τ	Tau
Θ	θ	Theta	Y	υ	Upsilon
I	ι	Iota	Φ	ϕ	Phi
K	κ	Kappa	X	χ	Chi
Λ	λ	Lambda	Ψ	ψ	Psi
M	μ	Mu	Ω	ω	Omega

41.5 TABLE OF NUCLIDES

The following table lists a selection of nuclides.
Column 1 gives the atomic number, symbol and mass number of the nuclide or particle. The mass numbers of stable nuclides are printed in bold type. An asterisk with the mass number indicates an excited nucleus. **Column 2** gives the abundance, of the isotope in the naturally occurring element and for the unstable isotopes indicates the type of decay by the symbols: α, β^-, β^+, –radiation, p proton emission, n neutron emission, k electron capture, i.t. isomeric transition. **Column 3** gives the atomic masses in mass units. The masses of the nuclei can be obtained from these by subtraction of the masses of the Z electrons of mass 0.000 549 u each. **Column 4** gives for unstable isotopes the maximum energy, **E**, of the emitted particles for several possible disintegrations in the order shown in column 2. The predominant energy is shown in italics. **Column 5** gives the corresponding half-value periods $T_{1/2}$ in s, min, d or y.

41 INFORMATION AND DATA NEEDED IN PHYSICS

1 Nuclide Z Symbol A		2 Decay Type or % age if stable	3 Mass m_u	4 E MeV	5 $T_{1/2}$
−1 e	0	stable	$0.000\,548_9$		
1 p	1	stable	$1.007\,276$		
0 n	1	β^-	$1.008\,665_4$	0.78	10.8 min
1 H	**1**	99.985	$1.007\,825_2$		
D	**2**	0.015	$2.014\,102_2$		
T	3	β^-	$3.016\,049$	0.018	12.3 y
2 He	3	1.4×10^{-4}	$3.016\,030$		
	4	~100	$4.002\,604$		
	5	n	$5.012\,3_0$		$\sim 6 \times 10^{-20}$ s
	6	β^-	$6.018\,9$	3.5	0.82 s
3 Li	5	p	$5.012\,5$		
	6	7.42	$6.015\,12_6$		
	7	92.58	$7.016\,00_5$		
	8	β^-	$8.022\,48_8$	~13	0.84 s
	9	$\beta^- + n$	9.02_7	$\beta \sim 8$	0.17 s
4 Be	7	K	$7.016\,93_1$		53 d
	8	2α	$8.005\,30_8$	0.05	$\sim 3 \times 10^{-16}$ s
	9	100	$9.012\,18_6$		
	10	β^-	$10.013\,54_5$	0.56	2.7×10^6 y
	11	β^-	$11.021\,6_6$	11.5; 9.3	14 s
5 B	8	$\beta^+ + 2\alpha$	$8.024\,61_2$	$\beta\ 14$	0.8 s
	9		$9.013\,33_5$		
	10	19.6	$10.012\,93_9$		
	11	80.4	$11.009\,305$		
	12	$\beta^- (+\alpha)$	$12.014\,35_3$	13.4	0.02 s
	13	β^-	$13.017\,78$		0.04 s
6 C	10	β^+	$10.016\,8_1$	2.1	19 s
	11	β^+	$11.011\,43$	0.96	20.5 min
	12	98.89	12 (Stand.)		
	13	1.11	$13.003\,35_4$		
	14	β^-	$14.003\,242$	0.158	5570 y
	15	β^-	$15.010\,60_0$	9.8; 4.5	2.3 s
7 N	12	$(\beta^+ + 3\alpha)$	$12.018\,7$	16.6; 12.2	0.012 s
	13	β^+	$13.005\,73_9$	1.2	10.1 min
	14	99.63_4	$14.003\,074$		
	15	0.36_6	$15.000\,10_8$		
	16	β^-	$16.006\,0_9$	10.4; 4.3	7.4 s
	17	$\beta^- + n$	$17.008\,4_5$	3.7 (0.9)	4.1 s
8 O	14	β^+	$14.008\,597$	1.8	72 s
	15	β^+	$15.003\,07_2$	1.68	124 s
	16	99.76	$15.994\,915_1$		
	17	0.037 4	$16.999\,13_3$		
	18	0.204	$17.999\,160$		
	19	β^-	$19.003\,58$	4.6; 3.2	29.4 s
9 F	17	β^+	$17.002\,10$	1.75	66 s
	18	$\beta^+(K)$	$18.000\,95$	0.65	110 min
	19	100	$18.998\,40$		
	20	β^-	$19.999\,99$	5.4	11 s
10 Ne	18	β^+	$18.005\,7_2$	3.4	1.3 s
	19	β^+	$19.001\,89$	2.2	18 s
	20	90.92	$19.992\,440$		
	21	0.257	$20.993\,84_9$		
	22	8.82	$21.991\,384$		
	23	β^-	$22.994\,47$	4.4; 3.9	38 s
	24	β^-	$23.993\,6$	2.0; 1.1	3.4 min
11 Na	20	$\beta^+ + \alpha$	20.008_9	$3.5 < E_\beta < 7.3$; $E_\alpha > 2$	0.3 s
	21	β^+	$20.997\,6_4$	2.5	23 s
	22	$\beta^+(K)$	$21.994\,44$	0.54	2.6 y
	23	100	$22.989\,77_3$		
	24	β	$23.990\,97$	1.4	15 h
	*24	i.t., β^-		~6	0.02 s
	25	β^-	24.989_9	3.8; 2.8	60 s
12 Mg	23	β^+	$22.994\,14$	3.1; 2.6	12 s
	24	78.7	$23.985\,04_5$		
	25	10.1	$24.985\,84_0$		
	26	11.2	$25.982\,59_1$		
	27	β^-	$26.984\,35$	1.8; 1.6	9.5 min
	28	β^-	$27.983\,8_8$	0.4	21.3 h
13 Al	24	$\beta^+(+\alpha)$	24.000	$\beta: 8.5; \alpha: 2$	2.1 s
	25	β^+	$24.990\,4_1$	3.2	7.2 s
	26	$\beta^+(K)$	$25.986\,90$	1.2	7×10^5 y
	*26	β^+		3.2	6.5 s
	27	100	$26.981\,53_5$		
	28	β^-	$27.981\,91$	2.9	2.3 min
	29	β^-	$28.980\,44$	2.5; 1.5	6.6 min
14 Si	27	β^+	$26.986\,70$	3.8	4 s
	28	92.21	$27.976\,93$		
	29	4.70	$28.976\,49$		
	30	3.09	$29.973\,76$		
	31	β^-	$30.975\,35$	1.5	157 min
	32	β^-	$31.974\,0$	~0.1	~700 y
15 P	28	β^+	27.992	10.6	0.28 s
	29	β^+	$28.981\,8_2$	3.9	4.3 s
	30	β^+	$29.978\,3_2$	3.2	2.5 min
	31	100	$30.973\,76_3$		
	32	β^-	$31.973\,90_8$	1.7	14.5 d
	33	β^-	$32.971\,73$	0.25	25 d
	34	β^-	33.973_3	5.1; 3.2	12.4 s
16 S	31	β^+	$30.979\,6_0$	4.4	2.6 s
	32	95.0	$31.972\,07_4$		
	33	0.76	$32.971\,46$		
	34	4.22	$33.967\,86$		
	35	β^-	$34.969\,03$	0.167	87 d
	36	0.014	$35.967\,0_9$		
	37	β^-	36.971_0	4.7; 1.6	5.0 min
	38	β^-	37.971_2	3.0; 1.1	2.9 h
17 Cl	32	$\beta^+(+\alpha)$	31.986	9.5; 8.2	0.3 s
	33	β^+	32.977_4	4.5	2.8 s
	34	β^+	$33.973\,7_6$	4.5	1.5 s
	*34	i.t., β^+		2.5; 1.3	32.4 min
	35	75.5	$34.968\,85$		
	36	$\beta^-(K)$	$35.968\,3_1$	0.71	3×10^5 y
	37	24.5	$36.965\,90$		
	38	β^-	$37.968\,0_0$	4.8; 2.8; 1.1	37.3 min
	39	β^-	$38.968\,0$	3.5; 2.2; 1.9	56 min
	40	β^-	39.970	7.5; 3.2	1.4 min
18 Ar	35	β^+	$34.975\,3$	4.96	2 s
	36	0.337	$35.967\,55$		
	37	K	$36.966\,77$		34 d
	38	0.063	$37.965\,72$		
	39	β^-	$38.964\,3_2$	0.56	265 y
	40	99.60	$39.962\,38_4$		
	41	β^-	$40.964\,5_0$	2.5; 1.2	110 min
19 K	37	β^+	$36.973\,4$	5	1.2 s
	38	β^+	$37.969\,1$	2.7	7.7 min
	*38	β^+		5.1	0.95 s
	39	93.10	$38.963\,71$		
	40	$0.011\,8\,\beta^-$, K	$39.964\,01$	1.32	1.3×10^9 y
	41	6.88	$40.961\,83$		
	42	β^-	$41.962\,4$	3.6; 2.2	12.5 h
20 Ca	39	β^+	$38.970\,7$	5.5	0.9 s
	40	96.97	$39.962\,59$		
	42	0.64	$41.958\,63$		
	43	0.14	$42.958\,78$		
	44	2.1	$43.955\,49$		
	45	β^-	$44.956\,19$	0.26	165 d
	46	0.003	$45.953\,6_9$		
	47	β^-	$46.954\,5$	1.94; 0.66	4.7 d
	48	0.18	$47.952\,3_6$		
	49	β^-	$48.955\,6_6$	2.0; 0.9	8.8 min
21 Sc	**45**	100	$44.955\,92$		
	46	β^-	$45.955\,1_7$	0.36	84 d
	*46	i.t.			20 s
	47	β^-	$46.952\,4_0$	0.6; 0.44	3.4 d
	48	β^-	$47.952\,2_1$	0.65	44 h
22 Ti	**46**	7.93	$45.952\,63$		
	47	7.28	$46.951\,7_6$		
	48	73.94	$47.947\,95$		

INFORMATION AND DATA NEEDED IN PHYSICS

1 Nuclide Z Symbol A	2 Decay Type or % age if stable	3 Mass m_u	4 E MeV	5 $T_{1/2}$
49	5.51	48.947 87		
50	5.34	49.944 79		
51	β^-	50.946 6	2.1; 1.5	5.8 min
23 V 48	β^+, K	47.952 2_6	0.70	16 d
50	0.24; K	49.947 17		4×10^{14} y
51	99.76	50.943 98		
52	β^-	51.944 8_0	2.6	3.77 min
24 Cr 50	4.31	49.946 05		
51	K	50.944 79		28 d
52	83.76	51.940 51		
53	9.55	52.940 65		
54	2.38	53.938 88		
55	β^-	54.941$_1$	2.8	3.5 min
25 Mn 54	K	53.940 3_6		280 d
55	100	54.938 05		
56	β^-	55.938 9_1	2.9; 1.0	2.6 h
26 Fe 54	5.82	53.939 6_2		
55	K	54.938 30		2.7 y
56	91.66	55.934 9_3		
57	2.19	56.935 3_9		
58	0.33	57.933 2_7		
59	β^-	58.934 8_7	0.46; 0.27	45 d
27 Co 57	K	56.936 29		
58	K(β^+)	57.935 7_5	0.47	71 d
59	100	58.933 19		
60	β^-	59.933 8_1	0.314	5.29 y
*60	i.t., –		1.5	10.5 min
28 Ni 58	67.9	57.935 3_4		
59	K	58.934 3_4		$\sim 10^5$ y
60	26.2	59.930 7_8		
61	1.2	60.931 0_5		
62	3.7	61.928 3_5		
63	β^-	62.926$_7$	0.067	120 y
64	1.1	63.927 9_6		
65	β^-	64.930 0_4	2.1; 1.0; 0.6	2.6 h
29 Cu 63	69.1	62.929 5_9		
64	β^-, β^+, K	63.929 7_6	β^-0.57 β^+0.66	12.8 h
65	30.9	64.927 7_9		
66	β^-	65.928 8_7	2.6; 1.6	5.1 min
30 Zn 64	48.89	63.929 15		
65	K, β^+	64.929 2_3	0.33	245 d
66	27.81	65.926 0_5		
67	4.11	66.927 1_5		
68	18.57	67.924 8_7		
69	β^-	68.926 7	0.9	55 min
*69	i.t.			14 h
70	0.62	69.925 3_5		
71	β^-	70.928$_0$	2.3	2.2 min
78 Pt 190	0.013 α	189.960$_0$	3.3	10^{12} y
192	0.78 α	191.961 4	~ 2.6	$\sim 10^{15}$ y
193	K	192.963$_3$		<500 y
*193	i.t.			4.4 d
194	32.9	193.962 8		
195	33.8	194.964 8_2		
196	25.3	195.964 9_8		
197	β^-	196.967 3_6	0.67; 0.48; 0.47	19 h
*197	i.t.			83 min
198	7.2	197.967$_5$		
199	β^-	198.970$_7$	1.7 ... 0.8	30 min
79 Au 196	K(β^-)	195.966 5_5	0.3	5.6 d
197	100	196.966 5_5		
198	β^-	197.968 2_4	(1.37); 0.96	2.7 d
199	β^-	198.968 7_5	(0.46); 0.30 0.25	3.15 d
80 Hg 196	0.15	195.965 8_2		

1 Nuclide Z Symbol A	2 Decay Type or % age if stable	3 Mass m_u	4 E MeV	5 $T_{1/2}$
197	K			66 h
198	10.0	197.966 7_7		
199	16.8	198.968 2_6		
200	23.1	199.968 3_4		
201	13.2	200.970 3_2		
202	29.8	201.970 6_2		
203	β^-	202.972 8_5	0.21	47 d
204	6.9	203.973 4_8		
205	β^-	204.976$_2$	1.6; 1.4	5.1 min
81 Tl		201.972$_1$		12.5 d
203	29.5	202.972$_3$		
204	β^-, K	203.973 8_9	0.76	~4 y
205	70.5	204.974 4_6		
205		205.976 0_8	1.6	4.2 min
207	β^-	206.977 4_5	1.44	4.8 min
208	β^-	207.982 0_1	(2.4); 1.8; 1.6; 1.2	3.1 min
210	β^-	209.990 0_0	1.9	1.32 min
82 Pb 203	K	202.973 4		52 h
204	1.5 α	203.973 0_7	2.6	1.4×10^{17} y
206	23.6	205.974 4_6		
207	22.6	206.975 9_0		
208	52.3	207.976 6_4		
209	β^-	208.981 0_9	0.64	3.3 h
210	β^-	209.984 1_8	0.06; 0.018	20 y
211	β^-	210.988 8	1.39; 0.5	36.1 min
212	β^-	211.991 9_0	0.58; 0.34	10.6 h
214	β^-	213.999 8	0.65; 0.59	26.8 min
83 Bi 209	100	208.980 4_2		
210	$\alpha(\beta^-)$	209.984 1_1	α: 4.9	3×10^6 y
*210	$\beta^-(\alpha)$		β^- 1.17	5.0 d
211	$\alpha(\beta^-)$	210.987 2_9	α: 6.6; 6.3	2.15 min
212	β^-, α	211.991 2_7	β: 2.25 α: 6.09; 6.05	60.5 min
214	$\beta^-(\alpha)$	213.998 2_3	β: 3.2; α: 5.5; 5.4	19.7 min
84 Po 209	α	208.982 4_6	4.88	200 y
210	α	209.982 8_7	5.30	138 d
211	α	210.986 6_5	7.44	0.6 s
212	α	211.988 8_6	8.78	3×10^{-7} s
214	α	213.995 1_9	7.68	1.6×10^{-4} s
215	$\alpha(\beta^-)$	214.999 5	α: 7.38	1.8×10^{-3} s
216	$\alpha(\beta^-)$	216.001 9_2	α: 6.78	0.16 s
218	$\alpha(\beta^-)$	218.008 9	α: 6.00	3.05 min
85 At 210	K(α)	209.987$_0$	5.52; 5.4	8.3 h
215	α	214.998 6_6	8.0	$\sim 10^{-4}$ s
216	α	216.002 4_0	7.8	$\sim 3 \times 10^{-4}$ s
218	$\alpha(\beta^-)$	218.008 5_5	α: 6.7	~2 s
86 Rn (Em)				
219	α	219.009 5_2	6.8; 6.5; 6.4	3.92 s
220	α	220.011 4_0	6.28	52 s
222	α	222.017 5	5.48	3.825 d
87 Fr				
223	$\beta^-(\alpha)$	223.019 8	β:1.2; α: 5.3	22 min
88 Ra				
223	α	223.018 5_6	5.87 ... 5.71 ... 5.43	11.7 d
224	α	224.020 2_2	5.68; 5.45	3.64 d
226	α	226.025 3_6	4.78; 4.60	1600 y
228	β^-	228.031 2_3	0.053	6.7 y
89 Ac 227	$\beta^-(\alpha)$	227.027 8_1	β:0.046; α:4.9	22 y
228	β^-	228.031 1_7	2.2 ... 0.5	6.13 h
90 Th				
227	α	227.027 7_7	6.0 ... 5.7	18.2 d
228	α	228.028 7_5	5.42; 5.34	1.91 y
229	α	229.031 6_3	5.0; 4.9; 4.8	7×10^3 y
230	α	230.033 0_8	4.68; 4.62	8.0×10^4 y
231	β^-	231.036 3_5	0.3 ... 0.09	25.6 h
232	100 α	232.038 2_1	4.01; 3.95	1.41×10^{10} y
233	β^-	233.041 4_3	1.23	22 min

41 INFORMATION AND DATA NEEDED IN PHYSICS

1 Nuclide Z Symbol A		2 Decay Type or % age if stable	3 Mass m_u	4 E MeV	5 $T_{1/2}$
91 Pa	234	β^-	234.043 5$_7$	*0.19*; 0.10	24.1 d
	231	α	231.035 9$_4$	5.05 ... 4.67	3.4×10^4 y
	233	β^-	233.040 1$_1$	0.57; *0.26*; 0.15	27.4 d
	234	β^-	234.043 4	1.1; 0.5; 0.3	6.7 h
	*234	β^-(i.t.)		2.3; 1.5; 0.6	1.2 min
92 U	233	α	233.039 5$_0$	4.82	1.6×10^5 y
	234	0.005 6 α	234.040 9$_0$	4.77; 4.71	2.5×10^5 y
	235	0.720 α	235.043 9$_3$	4.56 ...*4.38* .. 4.12	7.1×10^8 y
	236	α	236.045 7$_3$	4.50	2.4×10^7 y
	237	β^-	237.048 5$_8$	0.25	6.8 d
	238	99.27 α	238.050 7$_6$	4.19; 4.14	4.5×10^9 y
	239	β^-	239.054 3$_2$	1.2	23.5 min
93 Np	237	α	237.048 0$_3$	4.9 ... 4.5	2.2×10^6 y
	238	β^-	238.050 9	1.2 ... 0.3	2 d
	239	β^-	239.052 9$_4$	0.72 ... 0.3	2.3 d
94 Pu	239	α	239.052 1$_6$	5.15 ... 4.9	2.4×10^4 y
	240	α	240.053 9$_7$	5.16; 5.12	6.6×10^3 y
	241	$\beta^-(\alpha)$	241.056 7$_1$	β:0.02; α:4.9	13 y
	242	α	242.058 7	4.90	3.8×10^5 y
95 Am	241	α	241.056 6$_9$	5.5 ... 5.3	460 y
	242	β^-, K	242.059 4$_8$	0.6	~100 y
	*242	β^-, K		0.67; 0.63	16 h
	243	α	243.061 3$_8$	5.34 ... 5.17	8×10^3 y
96 Cm	242	α	242.058 8$_0$	6.11; 6.07	163 d
	243	$\alpha(K)$	243.061 3$_8$	6.06 ... 5.63	35 y
	244	α	244.062 9$_1$	5.80; 5.76	18 y
	245	α	245.065 3$_4$	5.45; 5.36	1×10^4 y

1 Nuclide Z Symbol A		2 Decay Type or % age if stable	3 Mass m_u	4 E MeV	5 $T_{1/2}$
97 Bk	248	α, fis. Fis.		5.0	5×10^5 y
	243	$K(\alpha)$	243.062 9$_2$	6.72; *6.55*; 6.20	4.5 h
	245	$K(\alpha)$	245.066 2$_4$	6.37; 6.17; 5.89	5.0 d
	247	α	247.070 1$_8$	5.67; *5.51*; 5.30	10^4 y
	249	$\beta^-(\alpha)$	249.074 8$_4$	β:0.1; α:5.4; 5.0	310 d
	250	β^-	250.078$_5$	1.9; 0.9	3.2 h
98 Cf	246	α	246.068 7$_8$	6.75; 6.71	36 h
	248	α	248.072 3$_5$	6.3	~300 d
	249	α	249.075 7$_0$	6.2; 5.9; 5.8	360 y
	250	α	250.076 5$_5$	6.02; 5.98	10 y
	252	α(fis. Fis.)		6.11; 6.07	2.6 y
	254	fis. Fis.			~60 d
99 Es	251	$K(\alpha)$	251.079 8$_5$	(6.5)	1.5 d
	253	α	253.084 6$_8$	6.63; 6.59 ... 6.18	20 d
	254	$\beta^-(K, \alpha)$	254.088$_1$	β:1.0; α:6.4	38 h
100 Fm	250	α, K	250.079 4$_8$	7.4	30 min
	252	α	252.082 6$_5$	7.0	30 h
	253	K, α		6.9	~5 d
	254	α	254.087 0$_0$	7.2	3 h
	255	α		7.0	21 h
	256	fis. Fis.			3 h
101 Md	255	K, α	255.090$_6$	7.3	0.5 h
102 No	253	α		8.5	~10 min
	254	α		8.8	3 s

41.6 ELECTRICAL AND ELECTRONIC SYMBOLS

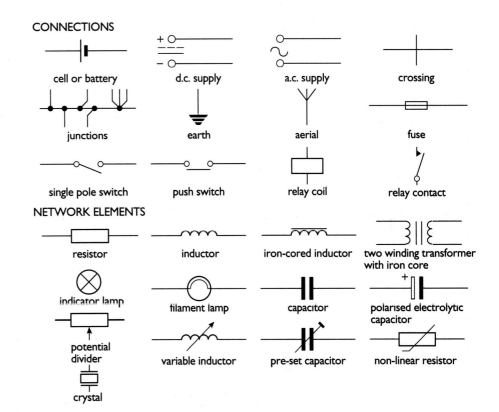

INFORMATION AND DATA NEEDED IN PHYSICS

MEASURING DEVICES

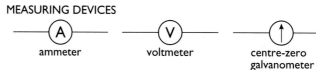

ELECTRONIC DEVICES

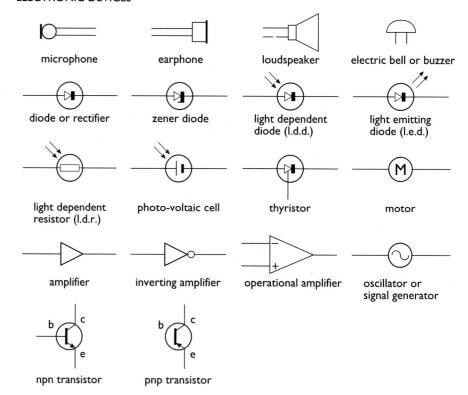

LOGIC SYMBOLS

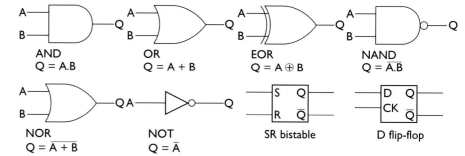

INDEX

absolute refractive index 536, 537
absolute zero of temperature 103, 425, 431
absorbed dose (radiation energy) 624
absorption spectra 283, 601–3
   X-ray 612
acceleration 47–9
   angular 70–71
   centripetal 64
acceleration due to gravity (g) 48, 51, 177, 179
   measuring 52
   variation 179, 186
accommodation (eye) 546
accuracy 11–12
achromatic lens 545
actinium series (radioactive decay chain) 654
activity (radioactive source) 621, 629, 630, 650
Adams, John Couch 180–81
ADC (analogue to digital converter) 367
adiabatic change 121, 424
adsorption 400
aerials 489, 494–5, 520
air resistance 95
air track experiments
   investigating collisions 77–9
   oscillations 253–4
aircraft
   jet propulsion 80–81
   turning 66
algebraic formulae 699–700
alloys 381
α-particles 620, 621, 626–30, 632–3
   detection 640, 641
   detection of background 626
   emission 645
   identifying 626
   ionising power 628, 630
   range 627, 628
   scattering 205, 665–8
   specific charge 626
   use in smoke detector 639

alternating current 298–300
   describing 312–14
   energy and power in 312–13
   generation and transmission 321–4
   and phasors 300
   rectification 135, 334–5
   in RL and RC circuits 302–4
alternators 323
americium-241 623, 634
ammeters 143
Ampère, André 213, 215
ampere (unit) 123
   definition 240–41
amplification factor 329
amplifiers 142
   coupling 320–21
   inverting 340
   non-inverting 342–3
   summing 342
   see also operational amplifier
amplitude 249
   in forced oscillations 277–80
   of sound wave 445
   of wave 436, 438
amplitude modulation 493
analogue computing 347
analogue to digital converter *see* ADC
angle of contact 400–401
   measuring 401
angular acceleration 70–71
angular frequency 259
angular magnification (telescope) 549
angular momentum 71
   conservation 71, 84
angular velocity 64, 65
annealing 380
antinode (wave) 464
antiparticles 672
antiphase 326
antiprotons 676
Archimedes' principle 407
astable multivibrators 362, 365
astronomical telescopes 549–50
asynchronous counter 363

INDEX

atmosphere, absorption of radiation   504
atom   556, 598
   component particles   621–2, 679
   electron distribution in   604–5
   nucleus   664–84
      forces in   173–4, 670, 677–8
      interactions with α-particles   205, 665–8, 669–70
      size   668–9
   transmutation   669–72
atomic clocks   251, 606
atomic mass constant   559, 622–3
atomic mass unit   559, 622, 689, 690
atomic number   621–2
atomic volume   560
aurora   570
Avogadro constant   559–60, 689, 690

B-fields   *see* magnetic fields
back emf   237–8
background radiation   623, 625–6
Balmer series (spectral lines)   604
band-pass filter   329
Barton's pendulums   275, 276–7
base dimensions   8
bat, navigation method   457
BCC   *see* body-centred cubic structure
beat frequency   263, 265
beats   260, 263, 265–7, 279
Becquerel, Henri   620
becquerel (unit)   621
Bernoulli's equation   408–9
β-particles/radiation   620, 621, 631–4
   detection   642
   emission   645–6
   range and energy   633–4
   use in thickness/level monitor   638
bifocal lenses   548
'Big Bang'   174
binary arithmetic circuits   359–60
binding energy   98, 622, 676–7
Biot-Savart law   243
bistable   *see* latch
black bodies   580–81, 582
black holes   395
blood flow, measuring   471
blooming of lenses   523
body-centred cubic (BCC) structure   376
Bohr, Niels, model of atom   604, 616, 664–5
Boltzmann, Ludwig   430
Boltzmann constant   108, 428, 689, 690
Boltzmann factor   428–9, 430
bond energies
   liquids   398
   sodium chloride   396
bond stiffness   394

bonding, in solids   394–5
Boolean algebra   354, 355, 358–9
Boyle, Robert   444
Boyle's law   100–101, 104
Bragg, William   378, 502
Brahe, Tycho   176, 184
braking radiation   *see* bremsstrahlung
braking systems (car)   406
break frequency (filter circuit)   303–4
breeder reactors   683
bremsstrahlung   611
bridge circuits   146–7
Brinell hardness number   386
brittle materials   27, 374
Brown, Robert   100
Brownian motion   100
bubble chamber   641
bubble raft experiments   378–9
bubbles   404
   excess pressure in   402
Bucherer,   632
buffer circuits   343, 355
bulk modulus   24, 447

caesium-137   624
calorimeter   118
cameras   548–9
   γ-ray   642
   pin-hole   542–3
   single lens reflex   539
cancer treatment   502, 639
capacitance   151
   of sphere   204–5
capacitor–resistor circuits, time constant   289–92
capacitors   150–51
   charge growth and decay on   156–62
   charges on plates   208–9
   discharge
      equations for   160
      iteration for   699
   energy storage in   153–6
   reactance   300–301
   resistance   303
   in series and parallel   151–3
   see also reactive circuits
capillarity   403–4
carbon dating   658–9
carbon fibre   384
cardiac pacemaker   657
Carnot cycle   424
Carnot, Sadi   425
cars
   braking system   406
   collisions   79–80
   minimising of drag   95
   petrol ignition system   307

INDEX

shock absorbers 272
suspension system 285
tyres 28–9
Cassegrain telescope 550
catapult field 217
cathode ray tube 142, 575–6
cathode rays 556, 561–5
Cavendish, Henry 178
cavity insulation (structures) 414
cells, internal resistance 130–31
centre of mass 33
centrifugal force 67–8
centrifuge 67
centripetal acceleration 64
centripetal force 65–7, 68
ceramics 381–2
   effect of surface cracks 384
Chadwick, James 671
chain reactions (nuclear) 680–81
chance and probability 426–9
change 5–6
charges
   conservation 123, 124
   on electron 125, 557–8, 689, 690
   flow 123–5, 415–19
      see also electrical conduction
   force between 199–200
   force on, in magnetic fields 222–7
   quantum nature 557–8
   storage 150–53, 206–7, 290
   see also specific charge
Charles's law 100, 103, 104
Chladni's figures 466
chokes 300
chromatic aberration 545
chromosphere (Sun) 602
circular motion 63–71, 258–60
   see also orbital motion
Clausius, Rudolph 426
clay 382
clock pulses 362
clocks 251, 606
cloud chamber 641
coal, quality checking 638
cobalt-60 623, 634
Cockroft, John 670
coefficient of dynamic friction 38
coefficient of static friction 37, 39, 94
coefficient of viscosity 410
coercivity 227
coherent radiation 500, 517–18, 608
   sources 513–14
coherent wave sources 461, 462
collisions 77–82, 82–4
   elastic 90–92
   inelastic 92–3
   of sub-atomic particles 98
colour vision 500
combinational logic circuits 358–60

Comet disasters 388
communication systems 366–8
   optical 540, 594–5
   see also radio broadcasting; television broadcasting
commutator 221–2
compact disc systems 367
comparator circuits 338
compass, magnetic 213
composite materials 383
compression 22
Compton effect 590
computer, use as timer 49
concave (diverging) lens 544, 547
concave mirrors 542
concepts 5
conductance 132
   equivalent 139
conductivity 132
conservation laws 76–162
   angular momentum 71, 84
   charge and energy 123–47
   energy 86, 112–21
   energy and mass 97–8
   momentum 60, 77–84, 86, 90, 98
   nucleons and charge in radioactive decay 646
conservative fields/forces 96
constants 689–90, 690, 691, 699
   unit-dependent 9
control circuits 357
convection 413–15
converging lens *see* convex lens
convex (converging) lens 543, 546–7
convex mirrors 543
Copernicus, Nikolaus 176
copper
   macroscopic properties 380
   microstructure 374
cosmic rays 505, 625
Coulomb, Charles 199
Coulomb's law 199, 204, 390
counters 363–4
couple 41
coupled oscillations 277
covalently bonded solids 395
creep 386
critical angle (total internal reflection) 538
critical damping 273
critical temperature
   gases 103
   superconductivity 136
crop spraying 195
crystal structures 374–9
   determination 502–3
crystals
   defects 377–8
   resonance in 283

INDEX

cup and cone fracture 374
Curie, Marie 621
curie (unit) 621
current 123
   growth/decay in inductor 295–6
   Kirchhoff's law 124
   magnetic fields around 214–16
   transient 150, 289
   unit 123
   in wire 125
   see also alternating current
current balance 218

DAC (digital to analogue converter) 367
damping 249, 263, 271–4
   decrement 272
damping constant 272
Davy, Humphry 373
de Broglie, Louis 590, 614
de Broglie waves 614
de Morgan's theorem 359
debounced switches 362
Debye, P. 395
decay constant (λ) 649, 651
Decca navigation system 521–2
decibel scale 449
definite integrals 697
delta configuration (electrical circuit) 145
demultiplexers 360
density 388
dependent variable 690–91
detergents 404
diamagnetic materials 226
dielectric 207, 496
diesel cycle 425
differential equations 697–8
   solving 348
differentiating circuits
   inductor–resistor 296–7
   resistor–capacitor 291, 293
differentiation 695–7
   list of derivatives 701
diffraction 459–61, 474–8, 513
   electrons 502–3, 613–14
   light 474–8, 480–81, 513
   microwaves 484
   phasor treatment 478–9
   radio waves 484–5
   X-rays 376–7, 502, 503, 612–13
diffraction gratings 528–31
diffusion 109–10, 426–7
digital circuits 352–3
   see also electronic systems, digital
digital to analogue converter *see* DAC
dimensions 8

diode thermometer 138
diodes 135
   current–voltage characteristic 333–4
   as detectors 336–7
   protective action 337
dioptre (unit) 543
dipole–dipole attraction 395
dipole moment 395
disc brakes 406
discharge tube 561–2
   see also fluorescent tube
dislocations (crystal defects) 378–9
dispersion 536
dispersion forces 395
displacement 32
displacement–time graphs 46, 53
dissipative forces 94–7
   and conservation 96
distance–time graphs 6
   see also displacement–time graphs
diver, motion of 68
diverging lens *see* concave lens
domains (magnetic) 226
doping (semiconductors) 416
Doppler effect 469–71
dose equivalent (radiation) 624–5
double glazing 418
doublets (spectral lines) 601, 606
drift velocity (charge carriers) 125
driving force (driver) of oscillations 274
droplets, excess pressure in 402
drum brakes 406
ductile materials 20, 374
dynamic equilibrium 6
   between forces 38
   molecules at liquid surfaces 399–400
dynamic friction 36, 38

E-fields *see* electric fields
ear, human 263, 448
Earth
   escaping from 189
   estimating mass 183
   magnetic field 220
Earth constants 689
earth leakage circuit trip 133
earthing 132–3
echoes 455
eddy currents 321
edge dislocations 378
efficiency 96–7
   of fluorescent tube 317
   of heat engine 425
effusion 109–10
Einstein, Albert
   discovery of stimulated emission 608
   and photoelectric effect 579, 586,

INDEX

587–8, 589
theory of special relativity   97, 176, 676
elastic behaviour   20
elastic collisions   90–92
elastic hysteresis   28
elastic limit   20
elastic potential energy   27–8, 88, 90
electric charge   *see* charges
electric fields   170, 192–210, 206
   motion of charged particle in   209–10
   motion of electron in   566–70
   patterns   194
   strength   195
   uniform   206–8
electric flux   197
electric flux density   198
electric lamp, colour balance   581
electric motors
   dc   221–2, 237–8
   speed regulation   226
electric shocks, prevention   132–3
electrical circuits
   bridge   146–7
   energy transfer in   125–8
   equivalent   144–5
   power in   127–8
electrical conduction
   liquids   417–19
   metals   125, 416
   semiconductors   416–17
   and thermal conduction   418
electrical conductivity   415
electrical oscillations   *see under* oscillations
electrical potential
   around charged conductors   198–9
   around charged sphere   203
   around point charge   201, 202
   zero level   201
electrical potential difference   126, 172
electrical potential energy   125–6, 126–7, 202
   storage in capacitors   153–4
electrical potential gradient   196
electrical potential hill   205
   α-particle/nucleus interaction   666–7
electrical potential well   205
electricity meter   127
electro-luminescence   607
electrolysis   559, 561
electromagnetic damping   273
electromagnetic forces   173
electromagnetic induction   230–33
   in changing fields   236–7
   laws   233–8
electromagnetic radiation   474, 488–511
   coherence   517–18
   generation   489–90

   interference   513–27, 519–20
   laws   581
   polarisation   490, 509–11
   quantum nature   492, 579–95
   spectrum   491–2
   superposition   513–31
   velocity   490, 505–8
   see also light
electromagnets   216
electromotive force   125
   back   237–8
   Kirchhoff's second law   126
electron gun   571
electron microscope   484, 615
electron multiplier   572
electron volt (unit)   563
electronic systems
   analogue   333–49, 367
   digital   352–66
electrons   556
   charge   125, 557–8, 689, 690
   diffraction   502–3, 613–14
   distribution in atoms   604–5
   emission from metal surface   571–4
   energy levels   599, 604–5
      metastable   608
   mass   561, 574–5, 689, 690
   motion in electric fields   210, 566–9
   motion in magnetic fields   223, 566–70
   orbitals   556, 598, 616
   –positron pairs   672
   specific charge   566–70, 569, 689, 690
   speed in electron beam   562–3
   spin   601, 604, 605
   wave-like properties   590
   see also β-particles; cathode rays
emf   *see* electromotive force
emission spectra   600–601, 603–4
   X-ray   610–12
   see also stimulated emission
emissivity, grey body   581
empirical laws   4
enable signal   357, 360
endoscope   539–40
energy
   and fields   171–2
   internal   109, 116–17
   of moving mass in changing field   187
   of oscillating spring   250
   quantisation   428
   of rolling cylinder   69, 70
   of rotating object   70
   of simple harmonic oscillator   261–3
   storage in capacitors   153–6
   transfer   114–19
      across walls   414–15
      by thermal conduction and convection   411–15
      describing   421

INDEX

    in electrical circuits 125–8, 312–13
    in mechanical systems 86–98
  unit 87, 563
energy flux 170
energy flux density 170–71
energy product 214
energy resonance 277
enthalpy change 116–17
entropy 430
equation of state (ideal gas) 104
equations
  differential 697–8
    solving 348
    nuclear 670
equations of motion
  constant acceleration 50, 51
  simple harmonic oscillator 254
equations of radioactive decay 645–6
equilibrium
  of coplanar forces 41
  forces in 32–4
  radioactive 656
  rotational 40–41
  translational 41
  see also dynamic equilibrium
equilibrium position (oscillation) 249
equilibrium separation (particles) 390, 392, 394
  in liquids 399
equivalent capacitance 151–2
equivalent circuits 144–5
equivalent conductance 139
equivalent resistance 139–40
error bars 11
escape velocity 189
excitation (atom) 598–600, 603
exclusive OR gate 359–60
experiments 4
  designing 7
  uncertainties in 9–12
exponential changes 158, 160, 694, 696
exponential constant (e) 694
exponential integration 347
exponents 694
extensometer 26
eye, human 263, 481–2, 500, 545–8
  as photon detector 584–5
  sensitivity 585

face-centred cubic (FCC) structure 376
falling objects 51–2, 167–8
farad (unit) 151
Faraday, Michael 230, 559
faraday (unit) 559
Faraday constant 689, 690
Faraday's law 234
fatigue 374, 387–8

static (polyethene) 386
FCC *see* face-centred cubic structure
feedback 339–40
  see also positive feedback circuits
feedback factor 339
Fermi, Enrico 665, 682
ferrites 214
ferromagnetic materials 226
field emission (electrons) 571, 573
field ion microscope 376, 573–4
fields 166, 169–71
  conservative and non-conservative 171
  energy and 171–2
  variation with distance 170–71
  see also electric fields; gravitational fields; magnetic fields
film badge dosimeter 624, 637
filters 303–4, 329
Fizeau, A.H.L. 505
flash gun 155–6
flaw detection, ultrasonic device 455
Fleming's left-hand rule 217
Fleming's right-hand rule 235
flexible materials 24
flotation 407
fluid flow 407–11
  streamlined 407–9
  tracer studies 659
  viscous 409–11
fluorescence 607
fluorescent tube 316–17, 501, 564, 607
flux 170–71
flux density 170–71
focal length
  lens 543, 544
  mirror 542
food, radiation sterilisation 634
forced oscillations *see under* oscillations
forces 166–7
  acting at distance 167–8
  adding 34–5
  between particles 390–92
  centrifugal 67–8
  centripetal 65–7, 68
  coplanar, equilibrium 41
  dissipative 94–7
  in equilibrium 32–4
  fundamental 172–4
  and motion 45–61
  paired 60
  resolving 35–6
  turning 39–41
  units 8, 57
  see also impulse
formulae 4
  mathematical 699–702
Foucault, J.B.L. 505
Fourier analysis 253

fracture, mode of 374
Franck-Hertz experiments 93, 599
Fraunhofer lines 602
free surface energy (liquids) 393, 399
frequency 249
   angular 259
   independent of amplitude 254
   of sound waves, related to pitch 448
   of wave 438
frequency modulation 493
friction 36–9, 94, 116
   and centripetal forces 65
full adder 360
full scale deflection (fsd) 143
functions 691
fundamental note 450, 467, 468

Galileo Galilei 55
γ-rays 503, 556, 583, 620, 621, 634–5
   emission 646
   interaction with matter 635–6, 637
   scattering 590
   uses 638–9
gas constant
   per mole *see* universal gas constant
   per particle *see* Boltzmann constant
gas discharge tubes 572, 598
gas turbine engine 80–81
gases
   electrical discharge in 561–5
      liquefaction 103
      thermal conduction in 413
   see also diffusion; effusion; 'ideal' gases
Gauss's law 197–8, 200, 207–8
Geiger–Marsden experiment 205, 664, 665–6, 667, 669
Geiger–Muller tube 640
generators 235–6, 323–4
geostationary orbits (satellites) 182–3, 499
germanium, n- and p-type 225
glass-reinforced plastic 383
glass temperature 385, 387
glasses 382–4
   effect of surface cracks 382–3, 384
   resistance variation with temperature 139
   stress-strain graph 25, 27
gluons 174
gold-leaf electroscope 586
Graham's law of effusion 110
graphite, electron diffraction pattern 613
graphs 11, 691
grating equation 530
gravitational fields 96, 169–70, 176–89, 206

conservative nature 171
   strength 179
gravitational forces 172–3, 177–8, 181–2
gravitational potential 185–9
   zero 187
gravitational potential difference 171, 186, 188
gravitational potential energy 88, 89, 187
gravitational potential gradient 186
gravitational potential well 188–9
gravity 63–4
   see also falling objects
gray (unit) 624
Greek alphabet 702
greenhouse 579
'greenhouse effect' 580
grey body, emissivity 581
Griffith cracks 383–4
ground state 603, 604

hadrons 684
Hahn, Otto 665
half adder 359, 360
half-life (radionuclides) 647–8, 650, 651–2, 702–5
   analogy in *RC* circuits (time constant) 158, 161
half-wave dipole 494
Hall effect 224–6, 416
Hall probe 234
hardness 386
harmonics *see* overtones
HCP *see* hexagonal close pack structure
   hearing 448–50
   human 444
heat capacities *see* molar heat capacities; specific heat capacities
heat engines 422–4
   efficiency 425
heat pumps 423
heat treatment (metals) 381
helium 602
Helmholtz coil system 244, 567
   B-field in centre 569
Helmholtz resonator 283
henry (unit) 238, 295
Herschel, William 180
Hertz, Heinrich 489, 586
hertz (unit) 249, 438
hexagonal close pack (HCP) structure 375
hip joint, artificial 26
holography 551–2
Hooke, Robert 20
Hooke's law 5, 20
hover-train 168

INDEX

human body
    infrared emission   498
    resonances   275
Huygens, Christian   458, 477, 536
Huygens' principle   476, 477
hydraulic braking system   406
hydrogen atom
    electron orbitals   616
    emission of radio photon   606
    energy states   603–6
    potential energy   205
hydrogen bonding   119
hydrometer   407
hypermetropia   546–7
hypotheses   4
hysteresis   346
    elastic   28
    magnetic   28, 226–7

ideal fluids   408
ideal gas temperature scale   103
'ideal' gases   100–10
    compression   113–14
    equation of state   104
    internal energy   109
    properties   100–104
    and real gases   109
imaging systems   540–41
    thermal   498
    underwater   445
    see also X-ray tomography
impedance   302–3
    and power factor   316
impedance matching   344
impulse   58, 60, 86
incandescence   579
indefinite integrals   697
independent variable   690–91
index notation   10, 692–3
induced current, direction   235
induced emf/voltage   231, 233–4
    and magnetic flux   232–3
    polarity   234–5
inductance   238
    measuring   305–7
    and rate of change of current   295
induction heating   321
inductors   293
    current growth/decay in   295–6
    reactance   300–301
    resistance   303
    see also reactive circuits
inelastic collisions   92–3
inertia   56–7
    rotational   68–9
infrared absorption spectroscopy   602–3
infrared radiation   495, 497–9, 504

infrared spectroscopy   504
ink-jet printers   196
instantaneous acceleration   48
instantaneous velocity   46, 47
insulators (electrical)   138–9
integrated circuits   333, 352, 354
integrating circuits   354
    inductor–resistor   296–7
    resistor–capacitor   291, 292, 293, 342
integration   697
    list of integrals   701
integrator   342, 347
intensity
    sound   448–9
    wave   438
interference (waves)   461–4, 513–27, 519–20
internal energy   109, 116–17
internal resistance   130–31
inverse square laws   170–71
    electrical forces   170–71, 199–200, 391
    gravitational fields   170, 177
    magnetic forces   214
    radiation intensity   584, 635
inverting amplifiers   340
investigations   13–16
iodine
    absorption spectrum   602
    ^{131}I decay   646
    ^{131}I tracer studies   572, 659
ionic solids   394–5
    see also sodium chloride
ionisation chamber   627, 628, 639–40, 648
ionisation energy   598, 603–4
ionisation voltage   598
ionising radiation   620
    detectors   639–42
    hazards   623–6
    monitoring   *see* film badge dosimeter
    sources of natural   625–6
ionosphere   504
ions   417
    migration in liquid   419
    specific charge   561
irreversible processes   426
isobaric change   113
isochronous motion   251
isothermal change   113, 120–21, 424
isotopes   622
iterative processes   698–9

JET fusion project   683–4
jet propulsion   80–81
Jodrell Bank radio telescope   485
Joliot-Curie, Jean and Irène   671–2
Joule, James Prescott   114, 115

INDEX

joule (unit) 87
joulemeter 127

Keesom, W.H. 395
Kelvin, Lord *see* Thomson, William
Kelvin (thermodynamic) temperature scale 103, 426
kelvin (unit) 103
Kepler, Johannes 176
   laws of planetary motion 184
kinetic energy 88–9
   in elastic collisions 90–92
   in inelastic collisions 92–3
kinetic model (ideal gas) 100, 104–9
   and effusion 110
   and pressure 105–7
   and temperature 107–9
Kirchhoff, Gustav 123
Kirchhoff's laws
   first law 124
   second law 126
klystron 496
Kundt, August 466

lasers 500, 517–18, 594, 608–10
   infrared emission of semiconductor diode 499
   use in holography 551–2
   wavelength of emitted light 480
latch (bistable) 357, 361
latent heats 118–19
   of vaporisation 118–19, 393, 398
lattice energy 396
lattice structures *see* crystal structures
laws (physics) 4, 5
   of conservation *see* conservation laws
   of electrolysis 559, 561
   of electromagnetic induction 233–8
   of radiation 581
   of radioactive decay 647
   of thermodynamics
      first law 120–21, 422
      second law 426, 430
      third law 431
      zeroth law 421
   see also inverse square laws
lead, absorption of γ-rays 636
Lee's method (thermal conductivity measurement) 412
Lenard, Philipp 587
lenses 542–5
   blooming 523
   defects 545
Lenz's law 235
leptons 679

lever 39
Leverrier, Urbain 180–81
light
   diffraction
      at circular aperture 480–81
      at slit 474–8
      Young's experiment 513
   interference 513–17
      in air wedge 524
      at multiple slit 525–7
      in thin films 523
   polarisation 509–11
   reflection 536
      phase change on 519–20
      see also mirrors; total internal
   refraction 536–7
   resolution of sources 481–4
   velocity (c) 251, 505–6, 507, 669, 670
      and refractive index 537
   wave nature 474–6
   wave–particle model 5, 590–92
   wavelength 479–80
      measuring 530–31
      and Young's fringes 516–17
light-beam galvanometer 273–4, 278
light-dependent resistor 135, 592–3
light-emitting diode 135, 335–6, 593–4
light gate 46, 49
lightning 197
lightning conductors 202–3
line spectra 500
linear accelerator 674–5
linear magnification (lens) 544
liquids 398
   electrical conduction 559–61
   electrical resistance 417
   thermal conduction in 413
Lissajous' figures 263, 264–5, 446
lithium atom, energy levels 605
Lloyd's mirror 520
load cells 147
lodestone 213, 214
logarithms 673–4
logic gates 353–7, 359–60
   see also combinational logic circuits; sequential logic
logic symbols 686
London, H. 395
long sight 546–7
longitudinal pulse 435–6
loudness 449
loudspeaker 252
Lummer, 580–81
Lyman series (spectral lines) 604

Madelung constant 396
Magnadur magnets 214

INDEX

magnetic compass 213
magnetic conductivity 242
  see also permeability
magnetic fields 213
  around currents 214–16, 239
  around flat coil 243
  around straight conductor 239, 240
  electron motion in 223, 566–70
  force on current-carrying conductor 216–22
  force on moving charge 222–7
  inside coil 243–4
  inside long solenoid 239, 241–2
  strength see magnetic flux density
magnetic flux 232–3
  around currents 239–44
magnetic flux density 219, 232
magnetic forces 168
magnetic hysteresis 28, 226–7
magnetic levitation 168
magnetic pole 213
magnetic susceptibility 226
magnetron 496
magnification
  angular 549
  linear 544
magnifying glass 544
mail sorting, automatic 607
mains electricity 132–3
manometer 405
mass 59
  and relativity theory 97
  units 8
mass defect 98, 622, 676
mass number 622
mass spectrometry 572, 673–4
mass-spring systems
  coupled oscillations 277
  energy 261–2
  oscillations 249–51, 254–7, 267
    forced 278, 280, 281
  resonances 282
master-slave data latch 363
matching 92
matching transformer 320
mathematical models 4, 4–5
matter waves 612–17
Maxwell, James Clerk 173, 489, 506
Maxwell-Boltzmann distribution 105, 426, 428
mean speed (particles) 106
mean square speed (particles) 106
Melde's experiment 465
memory blocks 358
meniscus 400–401
mensuration, formulae 701
mercury, emission spectrum 601
mercury vapour discharge tube 501, 564

mesons 679
messenger particles 174
metals 374–81
  bonding in 395
  electrical conduction in 125, 129, 416
  electron emission from surface 571–4
  fatigue 374, 387–8
  resistivity and temperature 136
  thermal conduction in 413
  whiskers 381
  Young modulus, measuring 441
metre, definition 506
Michelson, A.A. 505
microprocessors 352
microscopes
  electron 484, 615
  optical 484, 550–51
microwave oven 496–7
microwaves 495–7
  diffraction 484
  phase change on reflection 519
  polarisation 509
  superposition 519
Millikan, Robert 557, 566
mirrors 541–2
mode of fracture 374
models 5
  see also mathematical models
modems 368
modulation 263, 266
  pulse code 366, 368
molar gas constant see universal gas constant
molar heat capacities 112–13, 114, 115, 116, 117–18
molar mass 560
mole (amount of substance) 559–60
molecular solids 395
molecules, resonance in 283
moment (of force) 39–40
moment of inertia 69
momentum 57–8, 98
  angular 71, 84
  conservation of 60, 77–84, 86, 90, 98
  linear 77
monostable circuits 365–6
Moon
  estimating mass 184
  motion around Earth 177
Moseley's law 611–12
motion 45–61
  under gravity 51
  see also equations of motion; Newton's laws of motion
moving coil meter 221, 273
  protection 337
multiplexers 360
multiwire chamber 641
mumetal 227

INDEX

muons 679
musical instruments 283, 467–9
musical sounds 448
  quality 450
mutation 634
mutual inductance 238, 306
  measuring 307
myopia 547

national grid 322
natural frequency 250
nautilus 81
navigation systems
  Decca 521–2
  ultrasonic 457–8
negative feedback 339
negative feedback circuits 340–45
Neptune (planet), discovery 180–81
neptunium-257 634
neutrino 646
neutrons 556, 665, 671
  mass 689, 690
  in nuclear fission reactions 679–81
Newton, Isaac 53
  law of gravitation 176, 177, 179
  laws of motion 71, 79
    first law 55–6
    second law 57
    third law 60
  reflecting telescope 550
newton (unit) 57
Newtonian liquids 410
nickel-60 634
node, nodal point (wave) 282, 464
noise 449
nominal tensile stress 23
non-conservative fields 96
normal force 32
Northern Lights *see* aurora
nuclear fission 670, 679–83
nuclear forces 670
  strong 173–4, 677–8
  weak 174
nuclear fusion 679, 683–4
nuclear power 682
nuclear reactors 681–3
nucleon number 622
nucleons 556, 622
nuclides 622, 702–5
  table 702–5

Oersted, Hans 213, 214–15
Ohm, Georg Simon 129
ohm (unit) 128
ohmic conductors 129

Ohm's law 129
Onnes, Kamerlingh 136
operational amplifier (op-amp) 142, 333, 337–40
operators 691
optical fibres 539–40, 594–5
optical systems 535–53
opto-electronic devices 592–5
opto-luminescence 607
orbital motion 181–3
orbitals *see* electrons, orbitals
order of accuracy 11–12
oscillations 248
  and circular motion 258
  coupled 277
  electrical 289–307
    natural, in LC circuit 327
    resonant frequency 305
  forced 252, 271, 274–7
    amplitude and phase in 277–80
    analysis 281
  mechanical 249–67
  see also damping; simple harmonic motion
oscilloscope 142, 575
Otto cycle 425
over-damping 273
overtones 450, 467, 468–9

paint spraying 195
pair production 672
paired forces 60
parallel resonance circuits 326–7
paramagnetic materials 226
particle accelerators 674–6
pascal (unit) 24
Pascal's vases 405
Pauli exclusion principle 604
pearly nautilus 81
Peltier effect 423
pendulum 249
period (oscillation) 249, 438
  of mass-spring system 250–51
permanent magnets 214
permeability 240
  relative 242
  in vacuum/free space 240, 240–41, 506, 689, 690
permittivity 197, 200, 207, 506, 689, 690
  of free space 200
  relative 200
'perpetual motion' 86
phase (oscillations) 260
  in forced oscillations 277–80
phase constant 260
phasors 34, 260–61
  and ac 300

INDEX

and alternating currents 300
and diffraction 478–9
and impedance 302–3
and interference patterns 526–7
and transformer action 319
phonons 556
phosphorescence 607
phosphors 607, 641
phosphorus, ^{32}P tracer studies 659–60
photo multiplier 572
photoconductive detector 592–3
  see also light-dependent resistor
photodiode 593, 594
photoelastic stress analysis 510
photoelectric emission (electrons) 571, 586–7
photoemissive cell 592
photomultiplier 592
photons 492, 556, 579, 582–3, 587, 591
  detection by eye 584–5
  momentum 590
phototransistor 46
photovoltaic detector 593
piezoelectric effect 251, 445
piezoelectric transducers 445
pin-hole camera 542–3
pitch (sound) 448
Planck, Max 579, 582
Planck constant 582, 689, 690
  determining 588–9, 594
planets
  discovery of outermost 180–81
  motion 181, 184
plastic behaviour 20
Pluto (planet), discovery 181
plutonium 682–3
Poiseuille's equation 410–11
polar orbits (satellites) 499
polarising filter 509
polaroid 509, 510, 511
polyethene, static fatigue (creep) 386
polymers 384–7
population inversion 608
positive feedback 339, 345
positive feedback circuits 345–9, 361, 362
positrons 646, 672
potential difference
  electrical 126, 172
  gravitational 171, 186, 188
potential divider 140–42
potential energy 88
  between two particles 391–2
  hills 205, 666–7
  wells 392–3, 398
  electrical 205
  gravitational 188–9
see also elastic potential energy; electrical potential energy; gravitational potential energy
potentiometer 141–2, 142–3
  as voltmeter 143
power 93–4
  in ac circuits 312–13
  in electrical circuits 127–8
  from radioactive sources 656–8
  of lens 543
  nuclear 682
  in reactive and resistive circuits 314–18
  smoothing of supplies 156
  units 93, 127
  wattless 323
  wave 438
power factor 316
powers 692–3
presbyopia 548
pressure
  in fluids 405–6
  in gases 101–4
  and kinetic model 105–7
pressure equation 106–7
Pringsheim, 580–81
problem solving 16
projectiles 54–5
protactinium, ^{234}Pa decay 645, 655
proton-proton collisions 84
protons 556, 669–70
  force between 173–4
  mass 689, 690
  specific charge 561
pulsatance 259
  see also angular frequency
pulse code modulation 366, 368
pulse generators 362–6
pumped storage systems 322

Q-factor 280–81
  in musical instrument design 283
  in tuned circuits 329
quantisation 556
  of charge 557–76
  of energy 428
  of matter 598–617
  of radiation 579–95
quantum detectors 497–8
quantum mechanics 616–17
quantum numbers 604
quarks 174, 557, 679

radar 484, 541
radial B-fields 221
radiation sterilisation 634
radio aerials 495

INDEX

radio broadcasting 493
radio receiver 320, 344
radio telescope 485
    stellar interferometer 526–7
radio waves 490, 492, 504, 583
    diffraction 484–5
    emission by hydrogen in space 606
    polarisation 509
    superposition 490, 524
radio window 504
radioactive dating 658–9
radioactive decay 158, 174, 645–60, 698
    law 647
    mathematical model 649–52
radioactive decay chains 653–6
radioactive equilibrium 656
radioactive series 654–6
radioactive sources 621
    power from 656–8
    useful activity 629
radioactivity 620–42
    artificial 671–2
    detection, using photographic paper 620
radiography 502
radionuclides 621, 678
    applications 637–9
    artificial 671–2
    half-life 647–8, 650, 651–2, 702–5
    table 702–5
    see also radioactive decay; tracer studies
radiotherapy 639
radon 626
    ^{220}Rn decay 645, 648, 654
raindrops 402
ramp generator 342
Ramsay, W. 100
Rankine, William M. 426
Rankine cycle 425
rate of change 6
Rayleigh, Lord 664
Rayleigh criterion (resolving sources) 483
rays (light) 535
reactance 300–301
    and power factor 316
reaction (force) 32
reactive circuits, power in 314–15
rectifier circuits 334–5, 336
reed switch 243
reflection 455–8, 536
    phase change on 519–20
    see also mirrors; total internal reflection
refraction 458–9, 536–7
refractive index 536–7
refrigerator 424
relative atomic mass 622
relative molar mass 622

relative permeability 242
relative permittivity 200
relative uncertainty 11–12
relative velocity 46–7
relativistic mechanics 632
    see also special relativity theory
relaxation oscillator 347
reluctance 242
remanence 227
residual current circuit breaker 133
resistance 128–35, 134
    equivalent 139–40
    internal 130–31
    measuring 143
    variation with temperature 136–9
resistance thermometers 136
    bridge circuit for 147
resistive circuits, power in 313, 314
resistivity 131
    variation with temperature 136
resistors 133–4
    light-dependent 135, 592–3
    in series and parallel 139–40
resolution of sources 481–4
resonance 271, 279
    of air in tube 466
    in atoms and molecules 283
    band-width of curve 281
    in Barton's pendulums 277
    in child's swing 274–5
    in continuous media 281–2
    in crystals 283
    in electrical circuits 324–9
    in human body 275
    in machines 284–5
    in musical instruments 283
    in structures 284
resultant 35
retina 500
ripple tank experiments
    interference patterns 462–3
    wave diffraction 460–61
    wave reflection 456–7
    wave refraction 458–9
ripple through counter
    see asynchronous counter
ripple voltage 335
rocket propulsion 81–2
rocks, radioactive dating 658
Römer, O.C. 505
Röntgen, Wilhelm 502
root mean square speed (particle) 106, 107
root mean square voltage (ac circuit) 313–14
rotating objects 68–71
rotational equilibrium 40–41
rotational inertia 68–9
Royds, T. 626, 627

INDEX

rubber 385–6
  elastic hysteresis 28
Rutherford, Ernest
  atomic model 598, 666
  and Geiger-Marsden experiment 664–5
  identification of α-particles 626, 627
  nuclear transmutation 669
  study of ionising radiation 620

Sakharov, Andrei 684
Salter's duck 443
satellites 63–4, 168
  motion 181–3
  television broadcasting by 497
  transmitting dish 484
  weather 499
saturated vapour pressure 400
scalar quantities 32
Schmitt trigger circuit 346, 352
Schrödinger, Erwin 616
scintillation counter 641–2
search coil 234
Searle's method (thermal conductivity measurement) 412
second, definition 251
secondary emission (electrons) 571, 572
secular equilibrium *see* radioactive equilibrium
Seebeck effect 422–3
seismic waves 437
self-inductance 238, 294–5
  measuring 305, 306
semiconductor lasers 609
semiconductors 137
  electrical conduction 416–17
  Hall effect 224–5, 416
  Seebeck and Peltier effects 422–3
  thermal runaway 138
sequential logic 361–2
series-resonance circuit 325, 327, 328
servo mechanisms 345
shear modulus 24
shear stress 24
shock absorbers 272
short sight 547
SI units 7–8
siemens (unit) 132
sievert (unit) 625
significant figures 10–11
silica 382
silicon, doping 417
simple harmonic motion 258–61, 698
  integration 348
simple harmonic oscillations 253–61
  energy 261–3
sinusoidal motion 253

skiing 38
slip (in crystal structures) 377–8
smoke detector 639
Snell's law 536, 538
soaps 404
sodium
  absorption spectrum 601–2
  emission spectrum 601, 606
sodium chloride
  resonance in crystals 283
  structure 395–6
sodium lamps 562
solar cell 593
solar constant 689
solar-synchronous orbits (satellites) 499
solid state radiation detector 640
solids
  macroscopic properties 373–4
  thermal conduction in 413
sonar 458
sound
  in gases 444–7
  and hearing 448–50
  levels 449
  speed of 264
    in gases 445–7, 447
    in metal rod 440–41
sound waves 437
  diffraction 459–60
  Doppler effect 469–70
  interference 462, 463
  refraction 459
  stationary 466
  wavelength 447
Southern Lights *see* aurora
space charge 571
space diagrams 33–4
Space Shuttle 82
spark counter 627, 628, 640
special relativity theory 97, 98, 632
specific charge
  α-particles 626
  electron 566–70, 569, 689, 690
  ions 561
  proton 561
specific heat capacities 112, 115, 116, 117
specific latent heats 118
specific stiffness 373
spectra
  electromagnetic 491–2
  see also absorption spectra; emission spectra
spectrometer (spectroscope) 600
speed 45
spherical aberration
  lens 545
  mirror 542
spring constant 20

INDEX

springs 5, 19–21
   energy stored in 697
   reflection of pulses in 455–6
   in series and in parallel 20–21
   speed of transverse pulse in 442
   stationary wave on 464
   superposition of pulses in 454
   tranverse and longitudinal pulses 435–6
   see also mass-spring systems
square law detectors 263
SR bistables 361, 362
standing (stationary) wave 282, 464–9
star configuration (electrical circuit) 145
static friction 36–7, 94
statistical model
  diffusion 426–8
  and energy 428–9
  and second law of thermodynamics 429–30
stator 323
steel 26, 380–81
Stefan-Boltzmann constant 581, 689, 690
Stefan-Boltzmann law 581
stellar interferometer 526–7
stiff materials 24
stiffness
  specific 373
  of spring systems 5, 20–21
    see also spring constant
stimulated emission 607–10
Stoney, George Johnstone 556
stopping voltage (photoelectrons) 589
strain 23
  see also stress-strain graphs
strain gauge 147
strakes 284
Strassmann, 665
streamlined flow 407–9
street lighting 562
stress 22–5
stress analysis 510
stress-strain graphs 25
  polymers 386
stringed instruments 467
strings, waves in 437
strontium-90 623, 624
structures
  in equilibrium 31–42
  oscillations 274
  resonance in 284
summing amplifier 342
Sun
  estimating mass 183–4
  infrared emission 498
  nuclear fusion in 683
  spectrum 602
  ultraviolet emission 500

superconductivity 136, 168
superposition principle 453–4, 464
  radio waves 490, 524
surface free energy, measuring 403
surface tension 399
  measuring 401
surface wetting 400–401
susceptibility *see* magnetic susceptibility
swing 274–5
switches
  debounced 362
  reed 243
symbols
  electrical and electronic 705–6
  mathematical 691
synchrotron 675
systems approach (electronics) 333

Tacoma Narrows bridge 248, 284
TASTRAK (α-particle detector) 626
Taylor, G.I. 378
telescopes 549–50
  see also radio telescopes
teletext systems 368
television broadcasting 493
  by satellite 497
  signal reception 520–21
television receiver tube 576
temperature 421
  absolute zero 103, 425, 431
  and electrical resistance 136–9
  ideal gas scale 103
  Kelvin (thermodynamic) scale 426
  and kinetic model 107–9
  and speed of sound 447
tensile stress and strain 22, 23–4
tension 22
terminal velocity 51–2
tesla (unit) 220
thermal conduction
  and electrical conduction 418
  mechanisms 412–13
thermal conductivity 411–15, 412–13
  measuring 412
thermal imaging camera 498
thermal infrared detectors 497
thermal insulation 414
thermal radiation 579–82
thermal resistance 418
thermal runaway (semiconductors) 138
thermionic emission (electrons) 564, 571
thermistor 134–5, 137, 497
thermocouples 423
thermodynamics 421–31
  see also laws, of thermodynamics

INDEX

thermoelectric effect 422–3
thermometers
  constant volume gas 102
  diode 138
  resistance 136
    bridge circuit for 147
  thermistors 137
thermopile 497
thin film interference 523–4
Thomson, J.J.
  atomic model 664
  and cathode rays 556, 566
  and mass spectrometer 637
  and specific charge on proton 561
Thomson, William (Lord Kelvin) 426
thorium series (radioactive decay chain) 648, 654
threshold of audibility 449
threshold of feeling (pain) 449
thrombosis, detection 470
ticker-tape timer 49
timbre 450
time, measuring 46, 251
time constant
  in LR circuits 296
  in RC circuits 158, 161–2, 289–92
torque 39–40
  and angular acceleration 70–71
  provided by couple 41
torsional stress 23
total internal reflection 538–9
tough materials 382
tracer studies 572, 639, 659–60
transfer characteristic
  circuit 341
  logic gate 356
transformers 227, 238, 318–20
  action of, and phasors 319
  applications in electronics 320–21
  see also variable voltage transformers
transients 150, 289
translational equilibrium 41
transmission cables 322–3
transport equations 411, 415
transverse waves 435–6
travelling waves 436–8
triangle of forces 33
tribology 37
trigonometry, formulae 702
triple point 103
trolley experiments
  measuring acceleration 49
  model of wave transmission 439–40
truth tables 353, 354
tuned circuits 329
tungsten lamps, infrared emission 498–9
turning forces 39–41

U value 414
ultimate tensile stress 24, 387
ultrasonic waves 437, 445, 453
  flaw detection using 455
  navigation and ranging using 457
  use to measure blood flow 470
ultraviolet radiation 500–501
uncertainties 9–12
underwater imaging 445
unified model of forces 174
unit cube 376
unit-dependent constants 9
units 7–8
  absorbed dose of radiation energy 624
  activity of radioactive source 621
  atomic mass 559, 622, 689, 690
  capacitance 151
  conductance 132
  conductivity 132
  dose equivalent of radiation 624–5
  electric current 123, 240–41
  emf 125
  energy 87, 563
  force 8, 57
  frequency 249, 438
  inductance 238
  lens power 543
  magnetic flux 233
  magnetic flux density 220, 233
  power 93, 127
  resistance 128
  resistivity 131
  self-inductance 295
  sound intensity 449
  spring constant 20
  stress 24
  temperature 103
  thermal conductivity 411
  torque 40
  viscosity 411
  Young modulus 24
universal constant of gravitation (G) 177–8, 689, 690
universal gas constant 104, 689, 690
uranium
  nuclear fuel 681–2
  ^{235}U fission 680
uranium series (radioactive decay chain) 654, 655
Uranus (planet), discovery 180

van der Waals forces 395
vapour pressure 400, 429
variable voltage transformers ('variacs') 141
variables 690–91

INDEX

'variacs' *see* variable voltage transformers
vector diagrams 33–4
vector polygon 33–4
vector quantities 31–3
vectors, resolving 35–6
velocity 32, 45–7
   angular 64, 65
   instantaneous 46, 47
   relative 46–7
   terminal 51–2
Villard, Paul 620
virtual earth 340
viscosity 411
viscous flow 409–11
visible spectrum 500
vision 481–2, 500
   defects 546–8
volt (unit) 125
voltage 126
   root mean square (a.c. circuit) 313–14
   variable 140–43
voltage–current relationships 133–5
   see also Ohm's law
voltage divider 140–42
voltage-follower circuit 321, 343
voltmeters 143
von Guericke, Otto 444
vortex shedding 284, 469

Walton, Ernest 670
water
   potential energy well for molecules 393, 398
   thermal properties 119
   triple point 103
water waves 437
   diffraction 460–61, 475–6
   as energy source 443
   interference 462–3
   reflection 456–7
   refraction 458–9
   speed 442–3
'waterproof' fabrics 401
watt (unit) 93, 127
wattless power 323
wave equation 616–17
wave function 616
wave profile 436
wavefront 535
waveguide 496
wavelength 438
   light 479–80
   sound waves 447
waves 435–50
   diffraction 459–61, 474–8, 513
   see also X-ray crystallography

   energy 443
   interference 461–4, 513–27, 519–20
   reflection 455–8
      phase change on 519–20
   refraction 458–9
   sources 252
   speed 438, 439–40, 442–3
   stationary 282, 464–9
   superposition 453–4, 464, 513–31
   transverse 435–6
   travelling 436–8
   see also sound waves; ultrasonic waves; *and under* light
wear monitor 638–9
weather satellites 499
weber (unit) 233
weight 32–3, 33, 59
wetting of surfaces 400–401
Wheatstone, Charles 146
whiskers, metal 381
Wien's displacement law 581
wind instruments 468–9
work 86, 87, 89, 421–2
   done in compression of gases 113–14
   energy transfer by 115–16
   external and internal 116
work function (metals) 571, 587
work hardening 379–80, 380

X-ray crystallography 502–3
X-ray spectra
   absorption 612
   emission 610–12
X-ray tomography 612
X-rays 502
   diffraction 376–7, 502, 503, 612–13
   medical uses 502, 612
   scattering 590

yagi array 494
yield stress 24
Young modulus 24
   and bond stiffness 394
   metals 441
   polymers 386
Young's experiment 513
Young's interference fringes 514–15, 591
   using laser light 518
   and wavelength of light 516–17

zeroth law of thermodynamics 421